Fundamental Interactions
Cargèse 1981

NATO ADVANCED STUDY INSTITUTES SERIES

A series of edited volumes comprising multifaceted studies of contemporary scientific issues by some of the best scientific minds in the world, assembled in cooperation with NATO Scientific Affairs Division.

Series B: Physics

Recent Volumes in this Series

Volume 76 – Metal Hydrides
edited by Gust Bambakidis

Volume 77 – Nonlinear Phenomena at Phase Transitions and Instabilities
edited by T. Riste

Volume 78 – Excitations in Disordered Systems
edited by M. F. Thorpe

Volume 79 – Artificial Particle Beams in Space Plasma Studies
edtied by Bjørn Grandal

Volume 80 – Quantum Electrodynamics of Strong Fields
edited by W. Greiner

Volume 81 – Electron Correlations in Solids, Molecules, and Atoms
edited by Jozef T. Devreese and Fons Brosens

Volume 82 – Structural Elements in Particle Physics and Statistical Mechanics
edited by J. Honerkamp, K. Pohlmeyer, and H. Römer

Volume 83 – Physics of Ion-Ion and Electron-Ion Collisions
edited by F. Brouillard

Volume 84 – Physical Processes in Laser-Materials Interactions
edited by M. Bertolotti

Volume 85 – Fundamental Interactions: *Cargèse 1981*
edited by Maurice Lévy, Jean-Louis Basdevant, David Speiser, Jacques Weyers, Maurice Jacob, and Raymond Gastmans

This series is published by an international board of publishers in conjunction with NATO Scientific Affairs Division

A Life Sciences B Physics	Plenum Publishing Corporation London and New York
C Mathematical and Physical Sciences	D. Reidel Publishing Company Dordrecht, The Netherlands and Hingham, Massachusetts, USA
D Behavioral and Social Sciences E Applied Sciences	Martinus Nijhoff Publishers The Hague, The Netherlands

Fundamental Interactions

Cargèse 1981

Edited by

Maurice Lévy and Jean-Louis Basdevant

Laboratory of Theoretical Physics and High Energies
Université Pierre et Marie Curie
Paris, France

David Speiser and Jacques Weyers

Institute of Theoretical Physics
Université Catholique de Louvain
Louvain-la-Neuve, Belgium

Maurice Jacob

Theory Division
C.E.R.N.
Genève, Switzerland

and

Raymond Gastmans

Institute of Theoretical Physics
Katholieke Universiteit Leuven
Leuven, Belgium

PLENUM PRESS • NEW YORK AND LONDON
Published in cooperation with NATO Scientific Affairs Division

Library of Congress Cataloging in Publication Data

Cargèse Summer Institute on Fundamental Interactions (1981)
Fundamental interactions, Cargèse, 1981.

(NATO advanced study institutes series. Series B, Physics; v. 85)
"Published in cooperation with NATO Scientific Affairs Division."
Includes bibliographical references and index.
1. Nuclear reactions—Congresses. 2. Particles (Nuclear physics)—Congresses. 3. Quantum field theory—Congresses. I. Lévy, Maurice, 1922- . II. North Atlantic Treaty Organization. Division of Scientific Affairs. III. Title. IV. Series.
QC793.9.C37 1981 539.7'54 82-10164
ISBN 0-306-41116-4

Proceedings of the 1981 Cargèse Summer Institute on Fundamental Interactions, held July 14-31, 1981, in Cargèse, Corsica, France

A Division of Plenum Publishing Corporation
233 Spring Street, New York, N.Y. 10013

Printed in the United States of America

PREFACE

The 1981 Cargèse Summer Institute on Fundamental Interactions was organized by the Université Pierre et Marie Curie, Paris (M. LEVY and J.-L. BASDEVANT), CERN (M. JACOB), the Université Catholique de Louvain (D. SPEISER and J. WEYERS), and the Katholieke Universiteit te Leuven (R. GASTMANS), which, like in 1975, 1977 and 1979, had joined their efforts and worked in common. It was the 22nd Summer Institute held at Cargèse and the 6th one organized by the two institutes of theoretical physics at Leuven and Louvain-la-Neuve.

This time, while the last school was dominated by the impressive advances which were made in the field of perturbative quantum chromodynamics and its applications to high energy phenomena involving strongly interacting particles, the 1981 school clearly reflected a period of transition, where the new insights gained by experiment and theory are digested and put in order. Place of pride among the experiments belonged this time to DESY. On the theoretical side the reader will find a more thorough interpretation and understanding of the experiments as well as approaches to new theories. Finally several talks were devoted to experiments of the future.

We owe many thanks to all those who have made this Summer Institute possible!

Thanks are due to the Scientific Committee of NATO and its President for a generous grant and especially to the head of the Advanced Study Institute Program, Dr. R. Chabbal and his collaborators for their constant help and encouragements.

We also thank the "Direction des Recherches, Etudes et Techiques du Ministere de la Défense" and the "Institut National de Physique Nucléaire" et de Physique des Particules" (France) as well as the National Science Foundation (USA) for their financial assistance.

Special thanks are due to the Université de Nice for having put at our disposal the facilities of the Institut de Cargèse.

We wish to thank Miss M.-F. HANSELER, Mrs M. McClean, M D. Bernia and all others from Paris, Leuven, Louvain-la-Neuve and especially Cargèse for their collaboration.

We thank Mrs V. van CAILLIE and Miss A. DENIS for typing the manuscript, and Mrs F. de VILLE, Miss D. GUSBIN, MM M. DENIS and D. VERSTEGEN for their help in correcting the proofs and Miss M. CARTER (Plenum Press) for her collaboration.

Finally, the financial contributions from the Brouwerijen Artois, the Generale Bankmaatschappij and the Kredietbank are gratefully acknowledged. They helped to give this Summer Institute a broader international audience.

Mostly however, we would like to thank all lecturers and participants who came from over 20 countries: the willingness of the former to answer all questions and the keen interest of the latter provided the stimulus which made (we hope) this institute a success.

M. LEVY D. SPEISER

J.-L. BASDEVANT R. GASTMANS

M. JACOB J. WEYERS

Le présent document a été établi en exécution du contrat N° 81-1130 passé par la Direction des Recherches Etudes et Techniques, Direction Scientifique, Section Soutien à la Recherche, Délégation Générale pour l'Armement.

CONTENTS

FUNCTIONAL METHODS IN QUANTUM FIELD THEORY

B. de WIT

NIKHEF - H, Amsterdam

I. INTRODUCTION

In 1933 Dirac pointed out that it would be desirable to have a formulation of quantum mechanics in close correspondence to the Lagrangian method in classical mechanics, rather than to the more conventional Hamiltonian framework[1]. The Lagrangian method is based upon an action, defined as the time integral of the Lagrangian, and the principle of least action expresses the equations of motion in terms of a variational principle. The action is a relativistic invariant, and therefore the obvious advantage of this approach is that relativistic invariance is manifest at all stages. Feynman, in his pioneering work, fully developed this line of thought, and applied his methods to a large variety of problems[2,3]. His work led to the notion of an integral over all paths, which is an integration in the space of functionals. Such integrations had actually been studied in the mathematical literature (for a review, see (4)).

In the last decade path integrals have become an important tool for the study of quantum field theory for essentially the same reason that was given by Dirac. The time-ordering in the standard operator approach had led to cumbersome problems for theories with derivative interactions, and in particular for nonabelian gauge theories, which formed a severe obstacle for further progress. It is straightforward to deal with such problems in the covariant path integral approach, and functional methods have indeed been an essential ingredient in the study of gauge theories of all kinds[5]. Since the path integral respects all symmetries of the action it leads to a method of quantization in which the consequences of these symmetries can be derived in an elegant and straightforward way. This is of particular importance for nonlinear symmetries where the transformation rules are nonlinear in the fields, and for local gauge symmetries where the transformation parameters depend on space-time.

These lectures are intended as an introduction to functional methods in field theory, in which we will outline the main concepts and principles that underly the path integral quantization. We will often do this by discussing specific examples or by pointing out certain applications. We have not attempted to give a full survey of functional techniques and their applications (for useful references, see (6-17)), nor have we strived for mathematical rigour (a well-defined theory of functional integrals does not exist). One obvious omission is that we do not discuss the quantization of gauge fields, but on the other hand the most important elements that play a role in this subject are introduced.

The organization of these lectures is as follows. Section II introduces the path integral starting from Feynman's original postulates, where we devote some attention to the definition of functional integrals. The relevance of the path integral for quantum mechanics and statistical equilibrium physics is pointed out. In Section III we derive the representation of a path integral in terms of Feynman diagrams. Because of its manifest covariance the path integral is an effective tool for the derivation of the Feynman rules. As an application we discuss the effective potential both for zero and finite temperature. This section and the remaining ones deal to a large extent with the diagrammatic aspects of the path integral. As far as the perturbative aspects are concerned the path integral can be viewed as a convenient tool for summarizing complex diagrammatic manipulations. This point of view may not be correct when studying the nonperturbative aspects, a subject that has been left aside in these lectures. Section IV considers the application of functional techniques to fermions on the basis of anticommuting numbers. Section V reviews various aspects of unphysical fields (ghosts), that have no direct connection to the physical excitations of the system in question, and that are frequently employed in the functional approach.

II. PATH INTEGRALS

1. The classical action

In classic mechanics [18] one describes a system in terms of a set of generalized coordinates, which we shall denote generically by $q(t)$. Every function (or a set of functions) $q(t)$ specifies a path of the system. Usually one considers paths with certain boundary conditions, such as

$$q(t_1) = q_1,$$
$$q(t_2) = q_2. \qquad (1)$$

Of course, there are many possible paths that interpolate between the endpoints q_1 and q_2, but in general one expects that only one particular path is consistent with the dynamics of the system in question. That path is the classical trajectory, which is determined by the equations of motion for the system. In principle the choice of the boundary conditions is such that the solution of these equations becomes uniquely determined.

It is possible to derive the equations of motion, and thus the classical trajectory, from a variational principle. To this end we define the action for a path q(t) as

$$S[q(t)] = \int_{t_1}^{t_2} L(q(t),\dot{q}(t))dt, \qquad (2)$$

where $L(q(t),\dot{q}(t))$ is the Lagrangian for the system expressed in terms of generalized coordinates and velocities. Hence the action maps functions q(t) into numbers. Such objects are called functionals, rather than functions. In principle the Lagrangian may explicitly depend on time, but in the applications discussed in these lectures this will not be the case.

Variations in the path q(t) cause a change in the action S[q(t)]. To examine this we write the variation in q(t) as

$$q(t) \to q(t) + \delta q(t)$$

The corresponding variation of the velocity is then

$$\delta\dot{q}(t) = \frac{d}{dt}\,\delta q(t).$$

According to (2) the resulting change in the action is

$$\delta S[q(t)] = S[q(t) + \delta q(t)] - S[q(t)] = \int_{t_1}^{t_2} \left(\frac{\partial L}{\partial \dot{q}}\delta\dot{q} + \frac{\partial L}{\partial q}\delta q\right)dt$$

$$= p(t)\delta q(t)\Big|_{t_1}^{t_2} - \int_{t_1}^{t_2} \delta q(t)\left(\dot{p}(t) - \frac{\partial L}{\partial q}\right) dt, \qquad (3)$$

where we have performed a partial integration. We have also introduced the conjugate momenta p(t), defined by

$$p(t) = \frac{\partial L}{\partial \dot{q}} \tag{4}$$

We now consider the action for paths that satisfy the boundary condition (1). The principle of least action then states that the classical trajectory is that path for which the action is stationary. Inspection of (3) shows that this requirement implies the following equation along the classical trajectory $q_o(t)$

$$\frac{d}{dt} p(t) - \frac{\partial L}{\partial q} = 0 \tag{5}$$

These equations are the classical equations of motion, which are also called Euler-Lagrange equations. When we insert the solution of (1) and (5) into the action we find the stationary value S_{cl}, which is a function of t_1, t_2, q_1 and q_2. This function is known as "Hamilton' s principal function".

Problem 1. The Lagrangian for a free particle is $L(q,\dot{q}) = 1/2\, m\dot{q}^2$. Prove that the classical action takes the form

$$S_{cl} = \frac{m}{2(t_2 - t_1)} (q_2 - q_1)^2 \tag{6}$$

Problem 2. The Lagrangian for the harmonic oscillator is $L(q,\dot{q}) = 1/2\, m\dot{q}^2 - 1/2\, m\omega^2 q^2$. Prove that the classical action is given by

$$S_{cl} = \frac{m\omega}{2\sin\omega(t_2 - t_1)} [(q_1^2 + q_2^2) \cos\omega(t_2 - t_1) - 2q_1 q_2] \tag{7}$$

The Hamiltonian is defined by means of a Legendre transformation

$$H(p,q) = p\dot{q} - L(q,\dot{q})$$

Using (4) and (5) one directly derives Hamilton's equations

$$\frac{\partial H}{\partial p} = \dot{q} \quad ; \quad \frac{\partial H}{\partial q} = - \dot{p} \tag{8}$$

If H has no explicit dependence on time these equations ensure that the hamiltonian is a conserved quantity

$$d/dt \;\; H(p(t),q(t)) = 0,$$

where p(t) and q(t) are solutions of (8). The value of the Hamiltonian is the energy of the corresponding solution.

Problem 3. Use (3) to show that the energy can also be found by calculating the derivative of the classical action with respect to the time of one of the end points (but keeping the endpoints fixed).

The above results can be extended straightforwardly to systems with an infinite number of degrees of freedom. For instance, one may consider a one-dimensional string, described by a displacement field $\phi(x,t)$. Hence $\phi(x,t)$ represents the displacement of the string from equilibrium at position x and at time t. The action for a string takes the form

$$S[\phi(x,t)] = \int_{t_1}^{t_2} dt \int_0^L dx \left[\frac{1}{2}\rho\left(\frac{\partial\phi}{\partial t}\right)^2 - \frac{1}{2} T \left(\frac{\partial\phi}{\partial x}\right)^2 \right] \tag{9}$$

with ρ the mass per unit length, and T the tension. The x integration in (9) extends over the whole string. The boundary conditions are now given in terms of functions of x

$$\phi(x,t_1) = \phi_1(x),$$

$$\phi(x,t_2) = \phi_2(x).$$

The derivation of the equations of motion proceeds as before, and we find the field equation

$$\rho \frac{\partial^2\phi}{\partial t^2} = T \frac{\partial^2\phi}{\partial x^2}$$

Eq. (9) is a simple example of a classical field theory, which describes an infinite number of degrees of freedom. It is easy to find similar field theories that are relativistically invariant. Three important Lagrangians, which describe free massive particles of spin 0, 1/2 and 1, are

$$L = -\frac{1}{2}(\partial_\mu\phi)^2 - \frac{1}{2} m^2\phi^2 \tag{10}$$

$$L = -\bar{\psi}(\gamma^\mu\partial_\mu + m)\psi \tag{11}$$

$$L = -\frac{1}{4}(\partial_\mu V_\nu - \partial_\nu V_\mu)^2 - \frac{1}{2} m^2 V_\mu^2 \tag{12}$$

where ϕ, ψ, and V_μ are a space-time dependent scalar, spinor, and vector field, respectively. The field equations corresponding to (10)-(12) are

$$(\partial^2 - m^2)\phi = 0, \qquad \text{(Klein-Gordon equation)} \quad (13)$$

$$(\not\partial + m)\psi = 0 \qquad \text{(Dirac equation)} \quad (14)$$

$$\partial_\nu(\partial_\mu V_\nu - \partial_\nu V_\mu) + m^2 V_\mu = 0 \qquad \text{(Proca equation)} \quad (15)$$

In the first part of these lectures we consider a rather general variety of theories, but later we shall mostly concentrate on relativistic field theories, such as those defined by (10)-(15).

For classical mechanics the principle of least action results in the determination of a unique path followed by a system between predetermined endpoints. In quantum mechanics the endpoints can be considered as fixed (for instance by making measurements on the system before and after its time-evolution), but the notion of a unique path followed by the system at intermediate times is lost. This leads to a formulation of the quantum theory in which one considers a system as following all possible paths between the endpoints, instead of one unique path. Intuitively one expects that certain paths are more likely to occur than others, and in particular in the classical limit one should find that the system must again follow the original classical path. A more precise formulation of these ideas will now be given below.

2. The Feynman Postulates

The first suggestion that the formulation of classical mechanics in terms of an action principle should have its analogue in quantum mechanics was made by Dirac [1]. As he pointed out one of the advantages of such an approach was that relativistic invariance would be manifest. This in contrast with the Hamiltonian theory, where a particular time direction is selected from the beginning. The idea was reintroduced and further developed by Feynman, who summarized his formulation in terms of two postulates [2].

Postulate I. The probability (amplitude) for a system to go from state A to state B equals the sum of the probabilities (probability amplitudes) of going from state A to state B extending over all possible paths that connect A and B.

To elucidate the expression "sum over paths" we will first discuss the simple example of the one-dimensional discrete random walk. Because this is a classical system we use probabilities rather than probability amplitudes. The system consists of one particle, which after every time interval Δ will move one step with fixed length ε, either to the left or to the right, each with equal probability 1/2. Given that at $t = 0$, the particle was at the origin, we want to calculate $p_k(n)$, the probability that at time $t = k\Delta$ its position is at $n\varepsilon$. We may solve this by using postulate I. Consider all possible paths of k steps, beginning at the origin and ending at $n\varepsilon$. Clearly the probability for each path is $(1/2)^k$. The total number of possible paths can be derived from the observation that every path must have precisely n steps more to the right than to the left. The number of steps to the right thus equals $(k + n)/2$, and to the left $(k - n)/2$. Each path can be fully specified by the times on which the steps to the right are taken: thus the number of possible paths is

$$\begin{pmatrix} k \\ \frac{k+n}{2} \end{pmatrix}$$

Therefore we obtain as the solution

$$p_k(n) = \left(\frac{1}{2}\right)^k \begin{pmatrix} k \\ \frac{k+n}{2} \end{pmatrix}$$

Of course solving the problem by use of the difference equation

$$p_{k+1}(n) = \frac{1}{2} p_k(n+1) + \frac{1}{2} p_k(n-1) \tag{16}$$

would lead to the same result. This example illustrates the general significance of postulate I as a probability-theoretical principle.

The second postulate specifies the probability amplitude that must be assigned to each path in a quantum theory.

<u>Postulate II</u>. The probability amplitude for every path has the same magnitude, and a phase given by the classical action for that path measured in units of $\hbar$.

The quantum mechanical aspects of this postulate are apparent from the fact that we have defined a complex probability amplitude rather than a probability, and from the introduction of $\hbar$, Planck's constant divided by 2π.

According to Feynman's postulates the probability amplitude for a system to go from state A to state B is given by summing $\exp(i\hbar^{-1}S[q(t)])$ over all possible paths between A and B. This leads to the definition of the transition amplitude in terms of a <u>path integral</u>. Hence the transition amplitude to go from q_1 at

$t = t_1$ to q_2 at $t = t_2$ is given by

$$W(q_2,t_2;q_1,t_1) = \int [dq] \exp\left(\frac{i}{\hbar} S[q(t)]\right) \tag{17}$$

where the integrand is defined by (2). The integral $\int[dq]$ indicates that we sum over all possible paths that satisfy the boundary condition. The definition of the path integral implies the following properties

$$\int dq_2 W(q_3,t_3;q_2,t_2)W(q_2,t_2;q_1,t_1) = W(q_3,t_3;q_1,t_1) \tag{18}$$

with $t_3 \geqslant t_2 \geqslant t_1$, and

$$W(q_2,t;q_1,t) = \delta(q_2 - q_1) \tag{19}$$

The summation procedure in (17) may be intuitively clear for discrete paths, like in the example of the random walk, but for the continuous case a more precise mathematical definition is required. Usually such a <u>functional integral</u> is defined by means of a limiting procedure in the following way [3), 4)]. Let $F[q(t)]$ be an arbitrary functional, i.e. a mapping of the functions $q(t)$ onto numbers. We now write the path integral as

$$W = \frac{1}{A_N} \int_{-\infty}^{\infty} dq(t_1) \ldots \int_{-\infty}^{\infty} dq(t_{N-1})\, F(q(t_1),\ldots,q(t_{N-1})) \tag{20}$$

Notice the equivalence with the random walk; the N-1 integrations represent the sum over all paths, whereas the function F corresponds to the probability for each path. The position variable q has been taken continuous, but time is still discrete: $t = k\Delta$ where Δ is a small time interval. In this limiting procedure the functional $F[q(t)]$ has been replaced by the (N-1)-parameter function $F(q(t_1), \ldots, q(t_{N-1}))$, where we ignore the dependence on the endpoints of the path. If we now take the limit $\Delta \to 0$, keeping the total time interval of the path $T = N\Delta$ constant, it is not obvious that (20) will exist. Therefore we have allowed a normalization factor $(A_N)^{-1}$ into the integral which may become infinite as N approaches infinity, in such a way that (20) remains finite. If this procedure works then multiplication with the constant A_N leads to a well-defined functional measure.

One of the most important examples of a functional measure is the so-called <u>Wiener measure</u>. It is defined by the normalization condition

$$\int dq_2\ W(q_2,t_2;q_1,t_1) = 1, \tag{21}$$

where W is the path integral of the functional

$$F[q(t)] = \exp(-\frac{1}{2}\int_{t_1}^{t_2} (\dot{q}(t))^2 dt) \tag{22}$$

This functional is the Euclidian analogue of the quantum-mechanical path integral (17) for a free particle, which plays a role in the theory of diffusion and Brownian motion. Application of the limiting procedure (20) shows that the Wiener measure corresponds to

$$A_N = (2\pi\frac{t_1 - t_2}{N})^{N/2} \tag{23}$$

The Wiener measure can be used to define the functional integral for a larger class of functionals than just (22). For instance one may include certain well-behaved "potentials" $V(q)$ into $F[q(t)]$

$$F[q(t)] = \exp\ [\int_{t_1}^{t_2} (-\frac{1}{2}(\dot{q}(t))^2 - V(q(t))\ dt]$$

Of course the condition (21) will no longer be satisfied in the presence of interactions, so that an additional normalization is needed to make contact with (18)-(19).

Most of the discussion on path integrals is within the context of the Wiener measure. Nevertheless in field-theoretic applications one considers the integral over all sorts of functionals, including those for which the Wiener measure is not appropriate. For instance one may have functionals which do not depend on velocities, but only on the coordinates, such as

$$F[q(t)] = \exp\ (-\frac{1}{2}\int_{t_1}^{t_2} dt\ (q(t))^2) \tag{24}$$

Functionals such as (24) put the whole discussion on path integrals in a different perspective. First of all, the Euler-Lagrange equation for this case is

$$q(t) = 0,$$

so that imposing boundary conditions such as (1) has no classical significance. Secondly, the functional integral over (24) will have important contributions from highly irregular paths with arbitrary large velocities. This in contrast to the case of (22), where the presence of the kinetic term will suppress such paths. Consequently there is no correlation between the endpoint values in the path integral over (24). Therefore this integral can be expressed as the product of functions of each endpoint value separately. Imposing the normalization condition (21) then leads to obvious contradictions. In any case to give a meaning to integrals over functionals of type (24) an entirely different measure is required.

In spite of these subtleties a large variety of functionals is used in field theory, without paying too much attention to boundary conditions and normalization conditions. The reason is that boundary conditions are usually imposed at $t = \pm\infty$, so that normalization is only a matter of a (finite or infinite) constant. However, such normalization constants do not enter in the calculation of Green's functions and S-matrix elements. Various manipulations on functional integrals are therefore freely performed, such as redefinitions of integration variables and partial integrations. Furthermore, one often expresses functional determinants in terms of Gaussian integrals over real or complex fields

$$(\det A(x,y))^{-1/2} = \int [d\phi] \exp(-\frac{1}{2\hbar} \int dxdy \phi(x) A(x,y) \phi(y))$$

$$(\det A(x,n))^{-1} = \int [d\phi] [d\phi^*] \exp(-\frac{1}{\hbar} \int dxdy \phi^*(x) A(x,y) \phi(y)) \tag{25}$$

where $A(x,y)$ is a generalized matrix in the space of functions. The choice of the functional measure is implicit in (25), and depends on the nature of $A(x,y)$. If $A(x,y) = \partial^2\delta(x - y)$, then one must choose the Wiener measure; the right-hand side of (25) will then be the path integral for a real or complex Klein-Gordon field. On the other hand, if $A(x,y) = \lambda\delta(x - y)$ with λ some constant, then one needs a measure which is appropriate for functionals of type (24). Also the space of functions used in (25) is left unspecified. For example, one may have functions that are periodic over a certain interval, or functions that vanish at the boundary of a certain domain. Another formal result which is often used is the expression for a determinant in terms of the exponential of a generalized trace

$$(\det A(x,y))^{-n} = \exp(-n \int dxdy\ \delta(x - y) \ln A(x,y) \tag{26}$$

However, there are situations where the question of functional measure and normalization is important. In nonrelativistic quantum mechanics one assumes boundary conditions at finite times, t_1 and t_2.

The measure then depends on the time difference, as we have seen in (23). However, in that context one rarely encounters functionals that cannot be treated within the framework of the Wiener measure. The situation is somewhat different for field theories at finite temperature, which we shall discuss in subsection 6. In that case the measure depends on the temperature, whereas one applies similar manipulations as in zero-temperature field theory. Therefore the question of normalization requires special attention, at least when one calculates the partition functions [19].

The introduction of new fields in (25), which are not necessarily related to dynamic degrees of freedom, is characteristic for the use of functional methods in field theory. In this way formal manipulations may lead to formulations of the theory in question for which the validity of certain physical principles is no longer manifest. Properties such as causality or unitary must then be proven a posteriori.

Problem 4. Show that the Euler-Lagrange equations (15) for the Proca lagrangian (12) are equivalent to

$$(\partial^2 - m^2)V_\mu = 0,$$

$$\partial_\mu V_\mu = 0.$$

According to the second equation only transversal polarizations occur as dynamic degrees of freedom, which is appropriate for massive spin - 1 particles. The path integral requires a functional integration over all polarizations. Try to argue why the integration over three of them requires a Wiener measure. This part can then be considered as a generalization of what one does for Klein-Gordon fields. The integration of the fourth polarization is of the type that one uses for functionals such as (24).

3. Quantum Mechanics

It is possible to justify Feynman's postulates within the framework of conventional quantum mechanics [20]. Let $|q,t>$ be the state vector which at time t is an eigenstate of the position operator Q in the Schrödinger picture. Then W is given by

$$W(q_2,t_2;q_1,t_1) = < q_2,t_2|q_1,t_1 >$$

Using the evolution operator $U(t,t') = \exp(i\hbar^{-1}H(t' - t))$, with H the hamiltonian, we can express W as a matrix element taken between eigenstates $|q>$ of the position operator Q at some fixed time (Heisenberg picture)

$$W(q_2,t_2;q_1,t_1) = \langle q_2|U(t_2,t_1)|q_1\rangle \tag{27}$$

It is sufficient to evaluate W for infinitesimally small time intervals $t_2 - t_1$. For finite intervals W can then be constructed by means of a limiting procedure. Namely, we divide up the time interval into N sections

$$W(q_2,t_2;q_o,t_o) = \int dq_{N-1} \cdots \int dq_1 \tag{28}$$

$$W(q_N,t_N;q_{N-1},t_{N-1})W(q_{N-1},t_{N-1};q_{N-2},t_{N-2})\cdots$$

$$W(q_1,t_1;q_o,t_o) \tag{28}$$

with $t_N \geqslant t_{N-1} \geqslant \cdots \geqslant t_o$. Subsequently we take the limit $N \to \infty$, keeping $t_N - t_o$ fixed. Before calculating W for a small time interval we introduce the eigenstates $|p\rangle$ of the momentum operator, again in the Heisenberg picture. We can then write the matrix elements of the hamiltonian H(P,Q) as

$$\langle p|H(P,Q)|q\rangle = \frac{1}{\sqrt{2\pi\hbar}}\, e^{-\frac{i}{\hbar}pq}\, H(p,q) \tag{29}$$

where H(p,q) is the hamiltonian with the operators P and Q replaced by their respective eigenvalues p and q. We have used the following result:

$$\langle p|q\rangle = \frac{1}{\sqrt{2\pi\hbar}}\, e^{-\frac{i}{\hbar}pq} \tag{30}$$

Note that there is an essential ambiguity in equations such as (29) related to the fact that we have no definite prescription for ordering the operators P and Q. This aspect is not crucial here; for a discussion of the ordering and the limiting procedure, see for instance refs 13, 16.

Using (27)-(30) we write

$$W(q_2,t+\Delta;q_1t) = \int dp\, \langle q_2|p\rangle \langle p|\exp(-\frac{i}{\hbar}H(P,Q)\Delta)|q_1\rangle$$

$$= \int\frac{dp}{2\pi\hbar}\exp\{\frac{i}{\hbar}(p(q_2-q_1) - H(p,q_1)\Delta)\} \tag{31}$$

where we have made use of the completeness of the states $|p\rangle$.

For small values of $\hbar$ the integral may be approximated by choosing a value for p such that the exponent becomes stationary (this corresponds to taking the classical limit, which we shall discuss in more detail in the next subsection). Hence we write

$$W(q_2,t+\Delta;q_1,t) = \frac{1}{2\pi\hbar} \exp \frac{i\Delta}{\hbar} \{p \frac{q_2 - q_1}{\Delta} - H(p,q_1) + O(\hbar)\} \tag{32}$$

where

$$\frac{\partial H}{\partial p} = \frac{q_2 - q_1}{\Delta}$$

Identifying $(q_2 - q_1)\Delta^{-1}$ as the velocity $\dot{q}$, the exponent is then proportional to the action for a path with endpoints q_1 and q_2 at times t and $t + \Delta$, respectively. This shows the correspondence with (17).

Problem 5. Show that (32) is exact if the hamiltonian is of the form [8), 9), 20)]

$$H(P,Q) = \frac{1}{2m} P^2 + V(Q),$$

by performing the Gaussian integral over p.

Problem 6. Find the corrections to (17) when the hamiltonian is of the form [21)]

$$H(P,Q) = \frac{1}{2m} P^2 f(Q) + V(Q).$$

It is obvious from the representation (27) that W satisfies the Schrödinger equation. Together with (18)-(19) this implies that W is in fact the evolution operator of the Schrödinger equation. If the state of the system at time $t = t_o$ is determined by a wave function $\psi_o(q)$, then the wave equation at later time t is given by

$$\psi(q,t) = \int W(q,t;q',t_o)\psi_o(q')dq'.$$

A useful representation of the evolution operator is obtained by inserting a complete set of eigenstates $|\phi_n>$ of the hamiltonian into (27)

$$W(q_2,t_2;q_1,t_1) = \sum_n \phi_n(q_2)\phi_n^*(q_1)e^{-\frac{i}{\hbar}E_n(t_2 - t_1)}, (t_2 \geq t_1) \tag{33}$$

where E_n is the eigenvalue of the hamiltonian for the wave function ϕ_n, and the sum extends over all eigenfunctions. Notice that this representation manifestly satisfies (18)-(19).

4. The Classical Limit

It is an obvious requirement that the Feynman postulates yield the classical results in the limit of vanishing $\hbar$. Therefore we consider the path integral representation of W as given by (17). If we let $\hbar$ become very small $\exp(i\hbar^{-1}S[q(t)])$ becomes a rapidly oscillating function of the action value, so that the contributions to the path integral of the different trajectories will tend to average out. Of course, this negative interference will be less effective, and hence the contribution to W will be relatively larger, there where the action depends less strongly on the paths. Thus when we take the limit $\hbar \to 0$, the asymptotically dominant contribution is given by that trajectory q(t), for which S[q(t)] is stationary under arbitrary variations; this is exactly the trajectory given by the classical principle of least action.

To obtain the quantum corrections to the classical result we may expand the functional S[q(t)] in a Volterra series about a classical trajectory $q_o(t)$:

$$S[q(t)]= S[q_o(t)]$$

$$+ \frac{1}{2} \int dt_1\, dt_2 \left.\frac{\delta^2 S[q(t)]}{\delta q(t_1)\delta q(t_2)}\right|_{q=q_o} (q(t_1)-q_o(t_1))(q(t_2)-q_o(t_2))$$

$$+ \frac{1}{6} \int dt_1\, dt_2\, dt_3 \left.\frac{\delta^3 S[q(t)]}{\delta q(t_1)\delta q(t_2)\delta q(t_3)}\right|_{q=q_o} \{(q(t_1)-q_o(t_1))$$

$$(q(t_2)-q_o(t_2))(q(t_3)-q_o(t_3))\} + \ldots$$

where we have used that $\delta S[q(t)]$ is zero for the classical trajectory $q_o(t)$, and the functional derivatives of the action are evaluated at $q = q_o$. Now we can write W as in (34).

From this expression we see that for small $\hbar$ only those paths q(t) will contribute to W, for which $(q(t)-q_o(t))$, the deviation from the classical trajectory, is of the order of $(\hbar)^{1/2}$. Thus in the limit $\hbar \to 0$ we may approximate W by the probability amplitude for the classical trajectory.

$$W = \exp\left(\frac{i}{\hbar} S[q_o(t)] \right) \int [dq]$$

$$\times \exp\{\frac{i}{2\hbar} \int dt_1 dt_2 \left.\frac{\delta^2 S[q(t)]}{\delta q(t_1)\delta q(t_2)}\right|_{q=q_o} [\,(q(t_1)-q_o(t_1))$$

$$(q(t_2)-q_o(t_2))] + \ldots\} \qquad (34)$$

It is possible to obtain the first-order quantum correction to the classical action in closed form. Namely, we replace the integration variables $q(t)$ in (34) by $q(t) + q_o(t)$. Then we use (25) and (26) to write

$$W = \exp\{\frac{i}{\hbar} (S[q_o(t)] + \frac{1}{2} i\hbar \int dt_1 dt_2 \delta(t_1-t_2) \ln(\left.\frac{-i\delta^2 S}{\delta q(t_1)\delta q(t_2)}\right|_{q=q_o}$$

$$+ O(\hbar^2) \qquad (35)$$

Problem 7. If the action is at most quadratic in $q(t)$ then (35) is exact. Use this fact to show that the dependence on the end-point values q_1 and q_2 is entirely contained in the contribution from the classical action, so that

$$W = f(t_1,t_2) \exp(\frac{i}{\hbar} S_{cl}),$$

where f is an unknown function of t_1 and t_2. This function can be determined by requiring W to satisfy the Schrödinger equation, by making use of (18)-(19), or by imposing normalization conditions such as (21). In this way show that the path integral for a free particle is

$$W = \left(\frac{m}{2\pi i\hbar(t_2 - t_1)}\right)^{1/2} \exp\left(\frac{im}{2\hbar(t_2 - t_1)} (q_2 - q_1)^2\right) \qquad (36)$$

by using (6). Likewise prove from (7) that the path integral for a harmonic oscillator is given by

$$W = \left(\frac{m\omega}{2\pi i\hbar \sin\omega(t_2-t_1)}\right)^{1/2} \exp\left(\frac{im\omega}{2\hbar \sin\omega(t_2-t_1)}((q_1^2+q_2^2)\cos\omega(t_2-t_1)\right.$$

$$\left. - 2q_1q_2)\right) \qquad (37)$$

Note the appearance of factors i in these results. How to deal with those is the subject of the next subsection.

5. The Euclidean Postulate

Previously we have defined functional integrals by means of a limiting procedure. However, in doing so we have implicitly assumed that all the integrals in the finite-dimensional case were well defined. In a quantum theory, however, the path integral for W is in general not convergent in the finite-dimensional case because of the occurrence of the factor i in the exponent. Even in the simple example where the action is only quadratic in q, we encounter integrals of the form

$$\int_{-\infty}^{\infty} dq\, e^{-iq^2}$$

which are ill-defined because of the factor i. However, if we replace i by a general complex parameter z, we may describe integrals of this type as the result of an analytic continuation from the region of the complex z-plane where the integral is defined, i.e. where Re(z) > 0. More precisely, we would perform all our calculations for $z = i + \varepsilon$, and only in the final result consider the limit $\varepsilon \downarrow 0$.

The Euclidean postulate now prescribes which analytic continuation one must choose. We will discuss this for a relativistic field theory. According to this postulate the theory in Minkowski space with real times t is to be considered as an analytic continuation from the theory calculated in Euclidean space with imaginary times $t \equiv -i\tau$, where τ is a real parameter. As an example take the action given by

$$S[\phi] = \int dx\, (-\tfrac{1}{2}(\partial_\mu\phi)^2 - \tfrac{1}{2}m^2\phi^2),$$

with $x = (\mathbf{x}, it)$ a vector in Minkowski space. Following the Euclidean postulate we set $\tau = it$ and work with the Euclidean coordinates $x_E \equiv (\mathbf{x},\tau)$, where x_E is a vector in the Euclidean space. The expression for $i\hbar^{-1}S$ becomes

$$\frac{i}{\hbar} S[\phi] = -\frac{1}{2\hbar}\int dx_E ((\partial_\mu{}^E\phi)^2 + m^2\phi^2) \qquad (38)$$

where we have used $dt = -id\tau$, and $(\partial_\mu\phi)^2 \equiv (\nabla\phi)^2 - (\partial_t\phi)^2 = (\nabla\phi)^2 + (\partial_\tau\phi)^2 \equiv (\partial_\mu{}^E\phi)^2$. Now the expression on the right hand side of (38) is indeed negative definite, so that the path integral will exist.

In Fourier transformations an analytic continuation of t by rotating over $-\pi/2$ in the complex plane requires to make a simultaneous continuation of the energy variable k_0 by rotating over $\pi/2$ in the complex k_0 plane. Therefore, we must have

$$\begin{aligned} dx &= -\, idx_E, \\ dk &= idk_E, \\ \delta(x) &= i\delta(x_E), \\ \delta(k) &= -\, i\delta(k_E). \end{aligned} \tag{39}$$

Subsequently we absorb a minus sign into the definition of the energy component of the Euclidean momentum, to make all expressions manifestly covariant under real rotations. This does not affect (39), and we have

$$\begin{aligned} (\vec{x}, it) &= x_E = (\vec{x}, \tau) \\ (\vec{k}, ik_0) &= k_E = (\vec{k}, \omega) \end{aligned} \tag{40}$$

In order to perform the analytic continuation as prescribed by the Euclidean postulate the momentum integrals that we encounter should allow rotations over $\pi/2$ in the complex energy plane. This implies that the function over which one integrates should be free of singularities in the right-upper and left-lower quadrant of this plane. In this way the Euclidean postulate tells us how to deal with the poles caused by $(k^2 + m^2)^{-1}$ terms in the propagators. Namely, we replace $k^2 + m^2$ by $k^2 + m^2 - i\varepsilon$, where the limit $\varepsilon \downarrow 0$ is to be taken at the end of the calculation. In that case we have poles at $k_0 = \pm(\sqrt{k^2 + m^2} - i\varepsilon)$, so that the analytic continuation can be performed. Thus the sign of the $i\varepsilon$-term is directly prescribed by the Euclidean postulate. One can show that this sign has important physical consequences for the theory, related to causality and unitarity. We wish to add that the Euclidean postulate is not always a sufficient prescription for dealing with propagator poles. For instance the prescription will not work for vanishing mass, where one encounters infrared divergencies.

6. Field Theory at Finite Temperature

From the previous discussion it should be clear that Euclidean path integrals are better defined in a mathematical sense. Therefore the more rigorous results are usually obtained for Euclidean field theory [22]. Independent of this one may wonder what the physical relevance is of the Euclidean field theory. Because the phase factor for each path is now replaced by a real number, the most apparent applications are in the theory of random processes. Indeed the solution of the random walk discussed in section II.2 is of that type. When we take the appropriate continuum limit the probability distribution p(x,t) is a solution of the diffusion equation, which is the analogue of the difference equation (16)

$$\frac{\partial p(x,t)}{\partial t} = \frac{\partial^2 p(x,t)}{\partial x^2}$$

Not surprisingly, this is precisely the Schrödinger equation for a free particle at imaginary time.

Another application of the path integral is found in quantum statistical mechanics [23), 3)]. The thermodynamic properties of a quantum system in equilibrium at some temperature $T = k/\beta$ (k is Boltzmann's constant) follow from the partition function Z_β

$$Z_\beta = \sum_n \rho_n e^{-\beta E_n}. \tag{41}$$

The quantity ρ_n specifies the number of states with energy E_n. From (41) the Helmholtz free energy is obtained

$$F(\beta) = \beta^{-1} \ln Z_\beta ,$$

from which all thermodynamic quantities can be evaluated in the standard way.

The quantum-mechanical partition function can be expressed in terms of the density matrix

$$d_\beta(q',q) = \sum_n \phi_n(q')\phi_n^*(q) e^{-\beta E_n}, \tag{42}$$

where the $\phi_n(q)$ denote the normalized eigenfunctions of the hamiltonian for the system with eigenvalue E_n. The relation between Z_β and d_β is

$$Z_\beta = \int dq\, d_\beta(q,q). \tag{43}$$

The density matrix has a strong resemblance to the path integral represented according to (33). Both expressions are identical if we replace $t_2 - t_1$ by $-i\hbar\beta$. Since we are dealing with systems without an explicit dependence on time, we can thus express the density matrix in terms of a path integral over imaginary time $\tau = it$

$$d_\beta(q,q') = \int [dq] \exp(\frac{1}{\hbar} \int_0^{\hbar\beta} L(q(\tau),\dot{q}(\tau)) \tag{44}$$

where the integral extends over paths (i.e. functions of the imaginary time variable τ) with boundary conditions

$$q(0) = q,$$
$$q(\beta\hbar) = q'.$$

The partition function can therefore be written as a path integral over all periodic paths, i.e. paths with

$$q(0) = q(\beta\hbar)$$

Hence we have

$$Z_\beta = \int_{\text{periodic}} [dq] \exp (\frac{1}{\hbar} \int_0^{\hbar\beta} L(q(\tau),\dot{q}(\tau))$$

with

$$\dot{q}(\tau) = \frac{d}{d\tau} q(\tau).$$

For simple systems where the (Euclidean) lagrangian can be written as

$$L(q(\tau),\dot{q}(\tau)) = - \frac{1}{2} m\dot{q}^2 - V(q), \tag{45}$$

we can evaluate the path integral for high temperatures $\beta < \hbar^{-1}$. In that case the paths can never deviate too much from their boundary values, since this will induce large values for the "kinetic" term in (45). Hence we can approximate the integral over the potential by

$$\int_0^{\hbar\beta} d\tau\, V(q(\tau)) = \hbar\beta V(\frac{q + q'}{2})$$

This leads to the following expression for the density matrix

$$d_\beta(q,q') = e^{-\beta V(\frac{q+q'}{2})} \int [dq] \exp(-\frac{m}{2\hbar} \int_0^{\hbar\beta} (\dot{q}(\tau))^2 d\tau) \tag{46}$$

The integral in (46) has already been evaluated in (36) for real time. Using that result we find straightforwardly

$$Z_\beta = \left(\frac{m}{2\pi\hbar^2\beta}\right)^{1/2} \int dq\; e^{-\beta V(q)},$$

which agrees with Boltzmann's result for classical statistical mechanics.

It is in principle straightforward to extend the finite temperature description to relativistic field theories. As we shall discuss later, the partition function is given by the path integral for the corresponding Euclidean action, with periodic or antiperiodic boundary conditions

$$\phi(\vec{x},\hbar\beta) = \pm\, \phi(\vec{x},0) \tag{47}$$

Problem 8. Prove that the density matrix for an harmonic oscillator is (cf. (37))

$$d_\beta(q_1,q_2) = \left(\frac{m\omega}{2\pi\hbar\sinh\omega\hbar\beta}\right)^{1/2}\exp\left\{\frac{-m\omega}{2\hbar\sinh\omega\hbar\beta}\left((q_1^2+q_2^2)\cosh\omega\hbar\beta-2q_1q_2\right)\right\} \tag{48}$$

Show that the corresponding partition function is equal to

$$Z_\beta = (2\sinh\beta\hbar\omega/2)^{-1} \tag{49}$$

and compare this to the standard sum

$$Z_\beta = \sum_{n=0}^{\infty} e^{-\beta\hbar\omega(n+\frac{1}{2})}$$

Evaluate the partition function also for antiperiodic boundary conditions, and notice that the result can be written as

$$Z_\beta = \left[\sum_{n=0}^{1} e^{-\beta\hbar\omega(n-\frac{1}{2})}\right]^{-1} \tag{49}$$

III. FEYNMAN DIAGRAMS

1. A Graphical Representation of the Path Integral

It is obvious that one cannot explicitly calculate the path integral for any given action. However, it is generally possible to express the integral as an infinite series of so-called Feynman diagrams. To show this we consider a general action which is not only quadratic in the fields, but also contains terms of arbitrary order in the fields. After expanding about some classical solution, we divide this action into a part quadratic in the fields and the remaining terms

$$S[\phi,\phi^*] = S_0[\phi,\phi^*] + S_I[\phi,\phi^*] ,$$

where S_0 is the part of S quadratic in the fields. We shall discuss the case of independent fields ϕ and ϕ^*, which are related by complex conjugation. However, the precise relation between ϕ and ϕ^* is not essential, since we do not insist on a real action. We can now write the path integral as

$$W = N \int [d\phi][d\phi^*] \ \exp(S_0[\phi,\phi^*]) \ \exp S_I[\phi,\phi^*]).$$

Note that we have allowed for an arbitrary normalization constant N. Since we are only interested in the generic structure of the path integral we have suppressed the inverse of Planck's constant, and a possible factor i. Furthermore we shall be using a schematic notation, where S_0 will be written as

$$\begin{aligned} S_0[\phi,\phi^*] &= (\phi^*,D\phi) \\ &= \sum_{ij} \phi_i^* D_{ij} \phi_j , \end{aligned}$$

with the labels i and j running over all different space-time points and possible additional field indices. In the continuum limit this can be written as

$$(\phi^*,D\phi) = \int dxdy \ \phi^*(x)D(x,y)\phi(y).$$

Ignoring S_I for the moment, we consider the expression

$$\int [d\phi][d\phi^*] \ \exp((\phi^*,D\phi) + (J^*,\phi) + (\phi^*,J)), \qquad (50)$$

where we have added

$$(J^*,\phi) = \int dxJ^*(x)\phi(x),$$

$$(\phi^*,J) = \int dx\phi^*(x)J(x),$$

with J(x) a complex external source. The expression (50) can be calculated by shifting the integration variables to $\phi \to \phi - D^{-1}J$ and $\phi^* \to \phi^* - D^{-1}J^*$, and performing the Gaussian integral. The results of this integral will no longer depend on J, but only on the determinant of D. Therefore modulo a normalization factor, which we shall later absorb into the overall normalization of W, (50) can be written as

$$\int[d\phi][d\phi^*] \exp((\phi^*,D\phi) + (J^*,\phi) + (\phi^*,J)) = \exp(-(J^*,D^{-1}J)) \tag{51}$$

This result enables us to find the following simple density

$$\int[d\phi][d\phi^*]\phi_{i_1} \cdots \phi_{i_m}\phi_{j_1}^* \cdots \phi_{j_n}^* \exp(S_o[\phi,\phi^*])$$

$$= \int[d\phi][d\phi^*] \frac{\delta}{\delta J_{i_1}^*} \cdots \frac{\delta}{\delta J_{i_m}^*} \cdots \frac{\delta}{\delta J_{j_1}} \cdots \frac{\delta}{\delta J_{j_n}}$$

$$\exp((\phi^*,D\phi) + (J^*,\phi) + (\phi^*,J))\Big|_{J=0}$$

$$= \frac{\delta}{\delta J_{i_1}^*} \cdots \frac{\delta}{\delta J_{i_m}^*} \cdots \frac{\delta}{\delta J_{j_1}} \cdots \frac{\delta}{\delta J_{j_n}} \exp(-(J^*,D^{-1}J))\Big|_{J=0}. \tag{52}$$

To compute the full path integral we expand $\exp(S_I[\phi,\phi^*])$ in the path integral as a power series in $S_I[\phi,\phi^*]$. Using (52) then leads to the following representation of the path integral

$$W = \int[d\phi][d\phi^*] \exp(S[\phi,\phi^*]) = \exp(S_I[\frac{\delta}{\delta J^*}, \frac{\delta}{\delta J}]\exp(-(J^*,D^{-1}J))\Big|_{J=0} \tag{53}$$

This representation can be used to generate a graphical expansion of the path integral. Notice that the previous contruction only requires that the matrix D is invertable. Since the actual value of the Gaussian integral that led to (51) is not relevant, any further restriction on D such as to ensure the convergence of the integral, are immaterial.

Fig. 1

Expression (53) defines an expansion in terms of Feynman diagrams. To show this we consider the power series expansion of both the exponents. Because we shall put J = 0 at the end, only those combinations of terms will give a nonvanishing contribution to the path integral, in which the number of factors J equals the number of factors $\delta/\delta J$. We now represent each term in S_I by a vertex where each field $\phi(\phi^*)$ present in this term corresponds to an open line with incoming (outgoing) orientation arrow (see Fig. 1). Since a field corresponds to taking a derivative with respect to J, each such line must be connected to the operator $-D^{-1}$ with one additional J. Therefore $-D^{-1}$ corresponds to a line, called the propagator (see Fig. 2).

The propagator connects ϕ to ϕ^* lines, and therefore carries an orientation as well. Notice that the left index of matrix D^{-1} corresponds to the endpoint of the propagator line with incoming arrow. The orientational arrows of these lines should be distinguished from the momentum assignments of Feynman diagrams, which are often indicated by arrows as well. Because we let the source J vanish eventually, we must again apply a differentiation with respect to the source at the other endpoint of the propagator. In other words, at both endpoints the propagator must be connected to a vertex. Hence, only those structures are nonvanishing in the limit J = 0 that are closed in the sense that all propagators are connected to vertices, and all lines emerging from a vertex to propagators. Therefore we obtain diagrams such as given in Fig. 3.

Notice that the presence of an orientation poses restrictions on the kind of diagrams that one can construct. Such restrictions are not present for real fields, where one must connect propagators and vertices in all possible ways.

Problem 9. Repeat the diagrammatic expansion of the path

Fig. 2

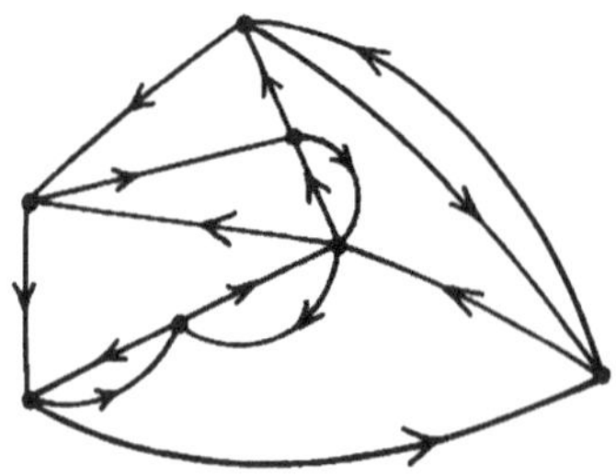

Fig. 3

integral for real fields. In that case one writes

$$S_o[\phi] = \frac{1}{2}(\phi, D\phi),$$

and adds a source term (J,ϕ) in (50).

It is often convenient to refrain from taking $J = 0$ in (53). In that case W is a functional of the source J. Differentiation of the logarithm of $W[J,J^*]$ with respect to J and J^*, and then putting $J = 0$ defines the connected Green's functions. For instance, the one- and two-point functions are

$$< \phi_i > = \frac{\delta}{\delta J_i^*} \ln W[J,J^*] \Big|_{J=0}$$

$$= \frac{1}{W[0]} \int [d\phi][d\phi^*] \, \phi_i e^{S[\phi,\phi^*]},$$

$$< \phi_i \phi_j^* > = \frac{\delta}{\delta J_i^*} \frac{\delta}{\delta J_j} \ln W[J,J^*] \Big|_{J=0}$$

$$= \frac{1}{W[0]} \int [d\phi][d\phi^*] \phi_i \phi_j^* e^{S[\phi,\phi^*]} - < \phi_i^* >< \phi_j > \tag{54}$$

It is straightforward to include the nonvanishing source into the diagrammatic evaluation, by introducing a vertex linear in the fields. This observation allows one to write $W[J,J^*]$ in terms of diagrams. When expanding $W[J,J^*]$ as a Volterra series in J the path integral leads to all Green's functions.

Problem 10. Show that the Green's functions $< \phi_i >$ and

$< \phi_i \phi^*_j >$ correspond to the sum of all connected diagrams with one or two external lines, respectively. Realize that W[0] corresponds to the sum of all "vacuum" diagrams, i.e. diagrams with no external lines. Note that the two-point function is the propagator in the presence of interactions.

2. Propagators

This section is devoted to a brief discussion of the propagator functions that occur for relativistic field theories of real scalar fields. Hence the Lagrangian is

$$L = -\frac{1}{2}(\partial_\mu \phi)^2 - \frac{1}{2} m^2 \phi^2. \tag{55}$$

In the Minkowski case we perform a Fourier transformation

$$\phi(x) = \int dk\, e^{ikx}\, \phi(k). \tag{56}$$

The action corresponding to (55) then takes the form

$$S[\phi] = -\frac{1}{2}(2\pi)^d \int dk\, \phi(-k)(k^2 + m^2)\phi(k),$$

where we have assumed d space-time dimensions. This expression, which should be compared to (53), defines the propagator

$$\Delta(x,x') = \Delta(x - x') = \frac{-i}{(2\pi)^d} \int dk \frac{e^{ik(x - x')}}{k^2 + m^2 - i\varepsilon}$$

We have included the $i\varepsilon$ in the denominator as prescribed by the Euclidean postulate. It is customary to ignore Planck's constant in the definition of the propagators. Obviously a propagator carries a factor $\hbar$, whereas vertices have $\hbar^{-1}$.

Problem 11. Show that the number of propagators (P) and vertices (V), is related to the number of closed loops (L), i.e. the number of unrestricted momentum integrations, according to

$$P - V = L - 1,$$

for any connected diagram. Argue that an expansion in $\hbar$ corresponds to a diagrammatic expansion in terms of the number of closed loops (for finite temperature $\hbar\beta$ is to be kept fixed).

It is possible to further evaluate the positive and negative frequency parts of the propagator. Therefore we write

$$\Delta(x) = \frac{-i}{(2\pi)^d} \int \frac{d\vec{k}}{2\omega(\vec{k})} \int dk_o e^{-ik_o x_o} \left(\frac{1}{k_o + \omega(\vec{k}) - i\varepsilon} - \frac{1}{k_o - \omega(\vec{k}) + i\varepsilon} \right) \tag{57}$$

where we have used the definition $\omega(\vec{k}) = \sqrt{\vec{k}^2 + m^2}$; We now integrate over k_o by means of a contour integration. For positive time we can close the contour in the lower half plane, whereas for negative time we must choose the upper half plane (see Fig. 4). Therefore the first term in (57) only contributes for negative time, and the second one for positive time. Application of Cauchy's theorem yields

$$\Delta(x) = \theta(x_o) \frac{1}{(2\pi)^{d-1}} \int \frac{d\vec{k}}{2\omega(\vec{k})} e^{i\vec{k}.\vec{x} - i\omega(\vec{k})x_o}$$

$$+ \quad \theta(-x_o) \frac{1}{(2\pi)^{d-1}} \int \frac{d\vec{k}}{2\omega(\vec{k})} e^{i\vec{k}.\vec{x} + i\omega(\vec{k})x_o}$$

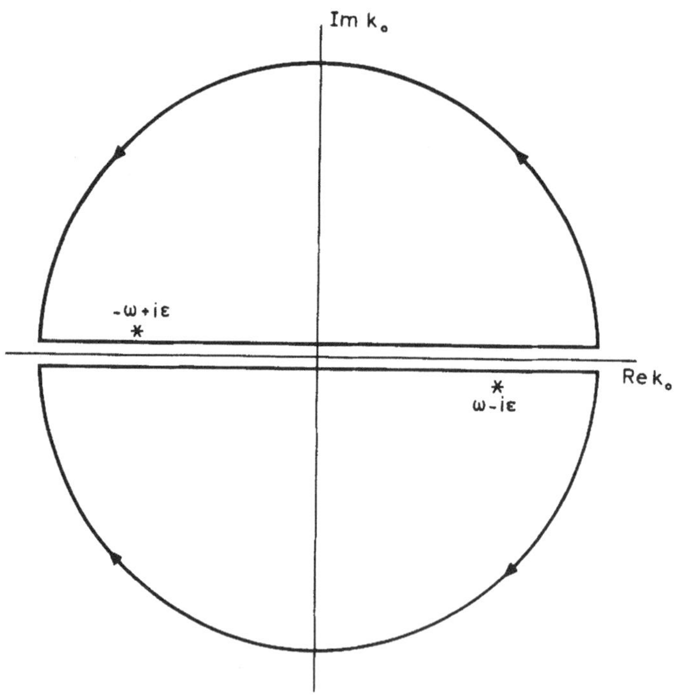

Fig. 4

One recognizes the crucial role of the $i\varepsilon$ terms in this calculation.

The evaluation of the Euclidean propagators at finite temperature proceeds in a similar fashion. In this case we are interested in periodic or antiperiodic fields (see (47)). Therefore we expand the fields in a Fourier series

$$\phi(\vec{x},\tau) = \sum_n \int d\vec{k}\, e^{i\vec{k}.\vec{x} + i\omega_n\tau} \phi_n(\vec{k}), \tag{58}$$

where the summation runs over even (odd) integers n for periodic (antiperiodic) boundary conditions with

$$\omega_n = \frac{\pi}{\hbar\beta} n.$$

The action at finite temperature can be written as

$$S_\beta[\phi] = -\frac{1}{2}\frac{\hbar\beta}{2\pi}(2\pi)^d \sum_n \int d\vec{k}\, \phi_{-n}(-\vec{k})(\omega_n^2 + \vec{k}^2 + m^2)\,\phi_n(\vec{k}),$$

where we have used

$$\int_o^{\hbar\beta} e^{i(\omega_m - \omega_n)\tau} d\tau = \hbar\beta\delta_{mn} \tag{59}$$

for both n and m even or odd. The counterpart of (59) is

$$\sum_n e^{i\omega_n(\tau - \tau')} = \hbar\beta\delta(\tau - \tau').$$

The propagator is now defined by

$$\Delta_\beta(\vec{x},\tau) = \frac{2\pi}{\hbar\beta}\frac{1}{(2\pi)^d} \sum_n \int d\vec{k}\, \frac{e^{i\vec{k}.\vec{x} + i\omega_n\tau}}{\omega_n^2 + \vec{k}^2 + m^2} \tag{60}$$

Apart from being symmetric in x and τ the propagator satisfies a periodicity condition

$$\Delta_\beta(\vec{x},\tau + \beta\hbar) = \pm\,\Delta_\beta(\vec{x},\tau).$$

Therefore it is sufficient to consider (60) for values of τ in the interval $0 < \tau < \hbar\beta$. To illustrate how one deals with frequency sums [24] we further evaluate (60). We first write

$$\Delta_\beta(\vec{x},\tau) = \frac{2\pi}{\hbar\beta}\frac{1}{(2\pi)^d}\sum_n \int \frac{d\vec{k}}{2\omega(\vec{k})}\, e^{i\vec{k}.\vec{x} + i\omega_n\tau}\left(\frac{1}{i\omega_n+\omega(\vec{k})} - \frac{1}{i\omega_n-\omega(\vec{k})}\right) \tag{61}$$

Subsequently we express the frequency sum in terms of a contour integral

$$\frac{2\pi}{\hbar\beta}\sum_n \frac{e^{i\omega_n\tau}}{i\omega_n-\omega} = i\oint_C dz\, \frac{e^{z\tau}}{1 \mp e^{\hbar\beta z}}\,\frac{1}{z-\omega} \tag{62}$$

where the minus (plus) sign holds for the sum over even (odd) values of n. The contour is given in Fig 5a, where the dots (crosses) denote the poles of the integrand along the imaginary axis for the periodic (antiperiodic) sum. It is now easy to verify that the

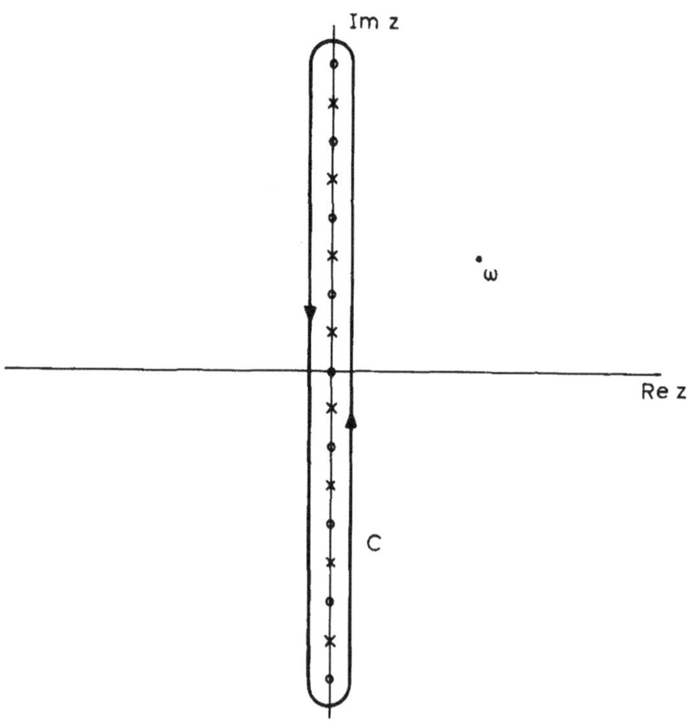

Fig. 5a

contour C may be deformed as indicated in Fig. 5b into two closed contour C_1 and C_2. The contributions along the arcs vanish asymptotically for $0 < \tau < \hbar\beta$. This proves that we can write (62) as

$$\frac{2\pi}{\hbar\beta} \sum_n \frac{e^{i\omega_n\tau}}{i\omega_n - \omega} = \frac{2\pi\, e^{\omega\tau}}{1 \mp e^{\hbar\beta\omega}} \qquad (0 < \tau < \hbar\beta) \tag{63}$$

Notice that a similar result can be derived if the left-hand side of (63) is multiplied by an arbitrary polynomial of ω_n. This observation is useful if one wishes to repeat this calculation for fields with spin. Using (63) we may write (61) as

$$\Delta_\beta(\vec{x},\tau) = \frac{1}{(2\pi)^{d-1}} \int \frac{d\vec{k}}{2\omega(\vec{k})}\, e^{i\vec{k}.\vec{x}}\, \frac{e^{\omega(\vec{k})(\hbar\beta - \tau)} \pm e^{\omega(\vec{k})\tau}}{e^{\omega(\vec{k})\hbar\beta} \mp 1} \tag{64}$$

3. Feynman Rules

Making use of the same manipulations as in the previous subsection, it is straightforward to derive the Feynman rules in momentum

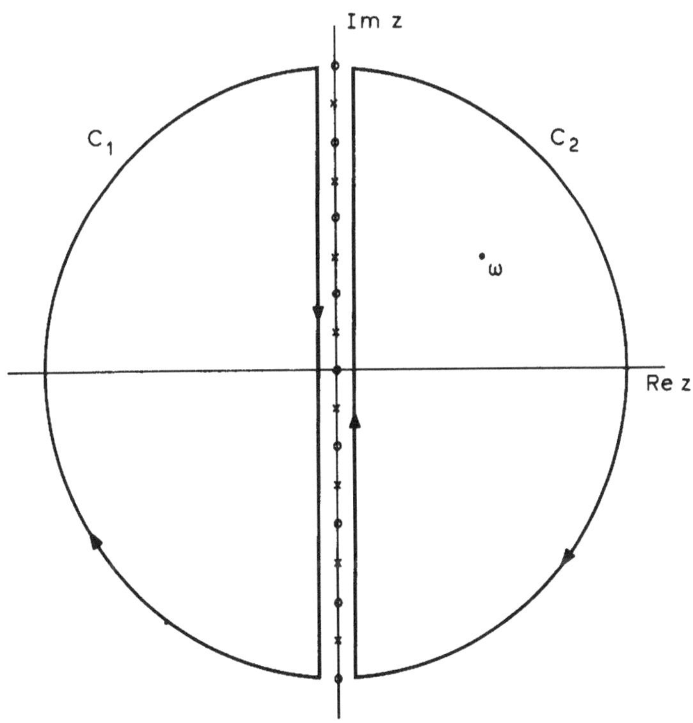

Fig. 5b

space for Minkowskian or for periodic Euclidean field theory. After performing the appropriate Fourier transforms (56) or (58) we find that the vertices correspond to

$$\text{vertex:}\ (2\pi)^d\ \varepsilon\ \delta(\sum_i q_i) \tag{65}$$

where the q_i denote the (incoming) momenta corresponding to each line of the vertex. The factor $(2\pi)^d\ \varepsilon$ is the result of the Fourier transform, where ε is given by

$\varepsilon = i$, for Minkowski field theory

$\varepsilon = \hbar\beta/2\pi$, for Euclidean field theory with (anti)periodicity interval $\hbar\beta$.

The momentum conservation at each vertex is expressed by the δ-function in (65). In the Euclidean case the δ-function over the imaginary energy variable is replaced by a Kronecker symbol, which requires the sum of the discrete energy values to vanish.

It is possible to have interactions that contain space-time derivatives of fields. These derivatives lead to corresponding momentum factors iq_μ, where q_μ represents the (incoming) momentum of a given line.

We have already defined the propagators in subsection 2. We write for each propagator

$$\text{propagator:}\ (2\pi)^{-d}\ \varepsilon^{-1}\ \Delta(q),$$

where $\Delta(q)$ is directly related to the negative inverse of the matrix that characterizes the quadratic terms in the action. Hence, for Klein-Gordon fields we have

$$\Delta(q) = \frac{1}{q^2 + m^2}\ .$$

Finally we have to integrate over all momenta that are involved, keeping track of the δ-functions at each vertex. The above rules can readily be generalized to fields with spin, or with internal degrees of freedom. In that case the momentum space propagators are simply matrices, and the endpoint of each line carries spinorial or internal indices.

Problem 12. Derive the Feynman rules for massive spin - 1 fields from the Proca Lagrangian (12). Show that the propagator is now a d x d matrix given by

$$\Delta_{\mu\nu}(q) = \frac{\delta_{\mu\nu} + q_\mu q_\nu m^{-2}}{q^2 + m^2}$$

Verify that its longitudinal part has no pole at $q^2 = -m^2$:

$$q_\mu \Delta_{\mu\nu}(q) = m^{-2} q_\nu .$$

What remains is to establish the combinatorial weight of each given diagram. This weight originates from the number of possible differentiations with respect to the source J in (53) that lead to the same diagram. Diagrammatically it simply corresponds to the number of ways in which one can form the diagram by connecting the various vertices by means of propagator lines. However, vertices that occur in an identical way in the diagram require a factor 1/n! where n is the number of such vertices.

Problem 13. Consider a field theory with selfinteraction for real and complex scalar fields

$$L_I(\text{real}) = -\lambda\phi^4$$

$$L_I(\text{complex}) = -\lambda|\phi|^4 .$$

Show that the combinatorial factors of the selfenergy diagram given in Fig. 6 are equal to 12 and 4 respectively.

4. Diagrammatic Evaluation of Gaussian Integrals

In section II.2 we have argued that one may formally express the determinant of a generalized matrix A(x,y) in two different ways. One was in terms of a Gaussian integral, and the other made use of the fact that the logarithm of the determinant of a matrix equals the trace of its logarithm (cf. (25) and (26)). Combining the two results leads to a formal identity

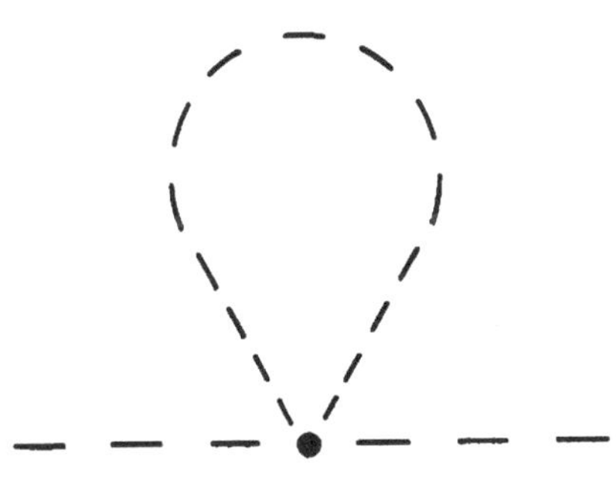

Fig. 6

$$\int[d\phi]\ \exp(-1/2\hbar \int dxdy\phi(x)A(x,y)\phi(y))$$

$$= \exp(-\ 1/2 \int dxdy\delta(x - y)\ln A(x,y) \qquad (66)$$

We shall now rederive such identities by means of Feynman diagrams. We distinguish two cases

1. $A(x,y) = (1 + f(x))\delta(x - y),$ (67)

2. $A(x,y) = (\partial^2 + f(x))\delta(x - y),$ (68)

where f(x) is some arbitrary function of space and time. The quantity A(x,y) is to be inserted in a Gaussian integral over real or complex scalar fields $\phi(x)$. Hence we consider a field theory of fields ϕ coupled to an external source f(x), defined in terms of the action

$$S = \int dxdy\ \frac{1}{2}\ \phi(x)A(x,y)\phi(y), \qquad \text{for real fields,}$$

$$S = \int dxdy\ \phi^*(x)A(x,y)\phi(y), \qquad \text{for complex fields.}$$

The propagators follow directly from the first term in A(x,y), and we find

1. $\Delta(k) = -\ 1$ for (67)

2. $\Delta(k) = 1/k^2$ for (68) (69)

The vertex that describes the coupling to f(x) is the same in both cases.

We now consider all Feynman diagrams without external ϕ lines. We start by evaluating a connected graph with n vertices, and all ϕ lines connected into a closed loop (see Fig. 7)). For complex

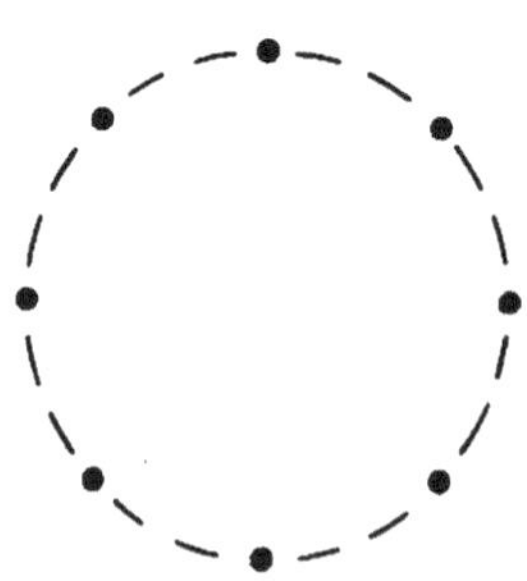

Fig. 7. One-loop diagram without external ϕ lines and eight $f\phi^2$ vertices; in the text we consider such diagrams with an arbitrary number of vertices.

scalar fields this diagram occurs with weight (n-1)!, since there are (n-1)! ways of connecting the ϕ lines such that one obtains a connected diagram. For real fields the combinatorial factor is (n-1)!/2 because each vertex has an extra factor 1/2. This factor tends to cancel, since at each vertex one has the possibility of connecting the ϕ lines in two different ways. But the cancellation is incomplete and one factor 1/2 remains. Hence the result for complex fields is simply twice that for real fields, which is consistent with the observation that a complex field can be viewed as two real fields. Therefore we shall confine ourselves to real fields henceforth.

The diagram of Fig. 7 with n vertices then corresponds to

$$\text{diagram} = \frac{1}{2}(n-1)! \int dk\, \Delta(k)\Delta(k+q_1) \ldots \Delta(k+q_1+q_2+\ldots+q_{n-1})$$
$$\times\, f(q_1)f(q_2)\ldots f(q_n)\ \delta(q_1+q_2+\ldots+q_n)$$

We now discuss the case (67). Inserting the value of the propagator we have

$$\text{diagram} = \frac{1}{2}(n-1)!\ (2\pi)^d(-)^n f(q_1)f(q_2)\ldots f(q_n)$$
$$\times\, \delta(0)\delta(q_1+q_2+\ldots+q_n). \qquad (70)$$

This result can be viewed as the result of an effective Lagrangian , which contains an interaction of the form $(f(x))^n$. Having such an interaction would lead to a tree diagram corresponding to

$$n!\,\hbar^{-1}(2\pi)^d\ \varepsilon f(q_1)f(q_2)\ldots f(q_n)\ \delta(q_1+q_2+\ldots+q_n)$$

where the factor n! originates from the n! ways of connecting the sources f(x) to the vertex. This observation shows that the sum of all connected diagrams of the type depicted in Fig. 7 will arise from the following effective Lagrangian

$$L_{eff} = \frac{\hbar}{2\varepsilon} \sum_{n=1}^{\infty} \frac{1}{n} (-f(x))^n\ \delta(0) = -\frac{\hbar}{2\varepsilon}\ \delta(0)\ \ln(1 + f(x)) \qquad (71)$$

It is illustrative to compare this result to the exponent of the right-hand side of (66). Using the fact that the logarithm of A(x,y) can be written as

$$\ln A(x,y) = \delta(x - y)\ \ln(1 + f(x)) , \qquad (72)$$

where it is crucial to realize that A(x,y) is an operator in function space, and that $\delta(x - y)$ corresponds to the identity matrix. Inserting (72) into (66) it is obvious that this formal result is consistent with the result of the diagrammatic evaluation leading to (71).

Problem 14. Show that (71) also correctly reproduces the disconnected diagrams.

We now turn to the second case (68), where we again try to reconcile (70) with some effective interactions of the source f(x). In this case the effective interaction will also contain space-time derivatives of f(x), which makes that the diagrammatic result is no longer expressible in closed form. Therefore we confine ourselves to calculating the terms that do not lead to derivatives. The coefficient of this term is found directly by taking f(x) equal to a constant, which means that we select its Fourier component with $q = 0$. The diagram of Fig. 7 with n vertices corresponds to

$$\text{diagram} = \frac{1}{2}(n-1)! \int dk(\Delta(k)f(0))^n\delta(q_1 + q_2 + \ldots + q_n).$$

Following the same arguments that led to (71) and inserting the appropriate form of $\Delta(k)$ given in (69), we find that the effective interaction is given by

$$L_{eff} = -\frac{\hbar}{2\varepsilon}\left(\frac{1}{2\pi}\right)^d \int dk \ln\left(1 - \frac{f(x)}{k^2 - i\varepsilon}\right) + \text{terms containing } \partial_\rho f(x).$$

Adding an f-independent constant to this result leads to

$$L_{eff} = -\frac{\hbar}{2\varepsilon}\left(\frac{1}{2\pi}\right)^d \int dk \ln\left(\frac{k^2 - i\varepsilon - f(x)}{\mu^2}\right)$$
$$+ \text{terms containing } \partial_\rho f(x), \tag{73}$$

where μ is some arbitrary dimensionful parameter. Notice that although this modification corresponds to some irrelevant constant in Minkowski space, it leads to temperature-dependent corrections in the periodic Euclidean space. We shall see later why the modification in (73) is appropriate in this case. Again the result (73) may be compared to the formal result based on (66).

Problem 15. It is straightforward to generalize the above results to fields with spin or internal indices. In that case the diagrams require a trace over the appropriate indices. For instance, for a Proca field coupled to a scalar source the analogue of (73) is

$$L_{eff} = -\frac{\hbar}{2\varepsilon}\left(\frac{1}{2\pi}\right)^d \int dk \, \mathrm{tr} \ln\left(\frac{(k^2 + m^2 - i\varepsilon - f(x))\,\delta_{\mu\nu} - k_\mu k_\nu}{\mu^2}\right)$$

Show that this can be written as

$$L_{eff} = -\frac{\hbar}{2\varepsilon}\left(\frac{1}{2\pi}\right)^d \int dk\left\{3\ln\left(\frac{k^2 + m^2 - i\varepsilon - f(x)}{\mu^2}\right) + \ln\left(\frac{m^2 - f(x)}{\mu^2}\right)\right\} \tag{74}$$

by using ln det = tr ln. Give a qualitative explanation for the two terms in (74). Compare this to the result of problem 4.

5. The Effective Potential

The Gaussian integral that we have discussed above can be used to calculate the one-loop quantum corrections for a self-interacting field theory. For instance, take the Lagrangian

$$L = -\frac{1}{2}(\partial_\mu \phi)^2 - V(\phi)$$

If we choose the source $f(x)$ equal to

$$f = -V''(\phi), \tag{75}$$

where $V''(\phi)$ is the second derivative of the potential with respect to ϕ, then the Gaussian integral with kernel

$$A(x,y) = (\partial^2 + f(x))\,\delta(x - y)$$

generates precisely all one-loop diagrams for this theory. The correctness of this statement can easily be verified by considering diagrams of the type depicted in Fig. 7, but now with the source given by (75). Actually, we have already encountered this result when considering the corrections to the path integral in (35). In that case the field ϕ had to be a solution of the classical field equations, but since the diagrams that we have been considering depend only on the second functional derivative, the result (35) can be used for general field configurations. In this way we obtain from the analogue of (35)

$$- \frac{1}{2}\hbar \int dxdy\ \delta(x - y)\ \ln\left(\frac{-\delta^2 S[\phi]}{\delta\phi(x)\delta\phi(y)}\right)$$

$$= - \frac{1}{2}\ \hbar \int dxdy\ \delta(x - y)\ \ln\ \{[-\partial^2 + V''(\phi(x))]\ \delta(x - y)\} \tag{76}$$

for Euclidean field theory. Modulo an irrelevant constant this expression leads directly to (73) and (75)

In the previous subsection our aim was to write the result of the Gaussian integral in the form of an effective interaction expressed in terms of f(x) and its space-time derivatives. When we substitute (76), and add this result to the original action $S[\phi]$, we obtain the so-called effective action in the one-loop approximation. The tree diagrams constructed from the effective action thus contain all one-loop diagrams. The notion of an effective action can be generalized to all orders in perturbation theory. When expanded about $\phi = 0$ the expansion coefficients of the effective action correspond precisely to the one-particle irreducible diagrams (also called proper vertices) of the theory. This aspect was manifest in the previous subsection. It is not our aim to give a complete discussion of the effective action here, and we refer to the literature for further details [25), 6)]. Instead we concentrate on that part of the effective action which only depends on the fields and not on their space-time derivatives. For obvious reasons this part is called the effective potential, and it is this potential which is relevant for studying translationally invariant field configurations, which usually characterize the ground state of a system [8, 26), 27), 28)].

We can now use the result of the previous subsection to find the effective action in the one-loop approximation. Taking into account that the potential occurs with a minus sign in the action, (73) and (75) lead straightforwardly to the following result

$$V^{eff}(\phi) = V(\phi) + \frac{\hbar}{2\varepsilon}\left(\frac{1}{2\pi}\right)^d \int dk \ln\left(\frac{k^2 + V''(\phi) - i\varepsilon}{\mu^2}\right) \tag{77}$$

Problem 16. Prove that the one-loop contribution to the effective potential for a complex field with Lagrangian

$$L = - |\partial_\mu \phi|^2 - V(|\phi|),$$

is given by the sum of two one-loop terms for a real field (cf. (77)), one with

$$V'' = \delta^2 V(|\phi|)/\delta\phi\delta\phi^* + |\delta^2 V(|\phi|)/(\delta\phi)^2|$$

and the other one with

$$V'' = \delta^2 V(|\phi|)/\delta\phi\delta\phi^* - |\delta^2 V(|\phi|)/(\delta\phi)^2|.$$

We shall first discuss the effective potential at zero temperature in four space-time dimensions. Obviously the momentum integral in (77) diverges in that case. To separate these infinities it is convenient to re-expand the logarithm as follows

$$V^{eff}(\phi) = V(\phi) - \frac{\hbar}{2i}\left(\frac{1}{2\pi}\right)^d \int dk \sum_{n=3}^{\infty} \frac{1}{n}\left(\frac{-V''(\phi) + m^2}{k^2 + m^2 - i\varepsilon}\right)^n$$

$$+ \frac{\hbar}{2i}\left(\frac{1}{2\pi}\right)^d \int dk \left\{\ln\left(\frac{k^2+m^2-i\varepsilon}{\mu^2}\right) + \frac{V''(\phi)-m^2}{k^2+m^2-i\varepsilon} - \frac{1}{2}\left(\frac{V''(\phi)-m^2}{k^2+m^2-i\varepsilon}\right)^2\right\} \tag{78}$$

We have expanded in powers of $V''(\phi)-m^2$ to avoid spurious infrared divergences. The parameter m^2 may be chosen such that it cancels exactly the contribution of a mass term in $V''(\phi)$. In that case $V''(\phi)-m^2$ will not contain a constant term.

In four space-time dimensions the last three terms in (78) represent the ultraviolet divergent contribution to the effective potential. If the theory in question is renormalizable, one must be able to absorb these infinities into the definition of the parameters that occur in the original potential. This is for instance possible when the potential is a polynomial in the fields containing no higher powers than ϕ^4. There is a divergent constant in (78), which represents the zero-point energy. Usually constants in the potential are discarded, but as we shall see later, it is important to consider their temperature dependence in calculations of the partition function.

We now concentrate on the calculation of the finite terms in (78). The necessary momentum integrals are given by [29)]

$$\int dk (k^2 + m^2 - i\varepsilon)^{-n} = i\pi^{d/2} \frac{\Gamma(n - d/2)}{\Gamma(n)} m^{-2n+d}$$

$$= i\pi^2 \frac{m^4}{(n-1)(n-2)} \left(\frac{1}{m^2}\right)^n$$

for $n > 2$, $d = 4$ (79)

Substituting this result into (78) we find

$$(V^{eff})_{finite} = -\frac{\hbar}{2i}\left(\frac{1}{2\pi}\right)^d \int dk \sum_{n=3}^{\infty} \frac{1}{n}\left(\frac{-V''(\phi)+m^2}{k^2+m^2-i\varepsilon}\right)^n$$

$$= -\frac{\hbar m^4}{32\pi^2}\sum_{n=3}^{\infty}\frac{1}{n(n-1)(n-2)}\left(\frac{-V''(\phi)+m^2}{m^2}\right)^n$$

After writing $(n(n-1)(n-2))^{-1}$ as $(2n)^{-1} + (2(n-2))^{-1} - (n-1)^{-1}$ we resum the series. The result takes the following form

$$(V^{eff})_{finite} = \frac{\hbar}{64\pi^2}\{(V''(\phi))^2 \ln\left(\frac{V''(\phi)}{m^2}\right) - \frac{3}{2}(V''(\phi) - \frac{2}{3}m^2)^2 + \frac{1}{6}m^4\} \tag{80}$$

Notice that the effective potential becomes complex when $V''(\phi)$ is negative. This is a reflection of the fact that the theory has tachyons for those field values; these field configurations are therefore unstable. We shall come back to this in the next section. In any case it should be clear that it may be misleading to base discussions of possible spontaneously broken realizations of symmetries on the original potential of the theory. Quantum corrections such as those given by (80) may entirely change the qualitative features of the potential [30].

6. The Effective Potential at Finite Temperature

We now evaluate the effective potential (77) for finite values of the temperature [31], [32]. Hence we consider

$$V_\beta^{eff}(\phi) = V(\phi) + \frac{1}{2}\hbar\,\frac{2\pi}{\hbar\beta}\left(\frac{1}{2\pi}\right)^d \int d\vec{k} \sum_n \ln\left(\frac{\omega_n^2+\omega^2(\vec{k})}{\mu^2}\right) \tag{81}$$

where $\omega(\vec{k})$ is defined by

$$\omega(\vec{k}) = \sqrt{\vec{k}^2 + V''(\phi)}$$

To calculate the sum over discrete energies ω_n, we note that

$$\sum_n \ln\left(\frac{\omega_n^2+\omega^2(\vec{k})}{\mu^2}\right) + 2\sum_n \ln\beta - 2\ln\left(e^{\frac{1}{2}\beta\hbar\omega(\vec{k})} \mp e^{-\frac{1}{2}\beta\hbar\omega(\vec{k})}\right) \tag{82}$$

must be a constant independent of the potential and of the

temperature. To prove this we differentiate with respect to $\omega(\vec{k})$ and use the result (64) of the periodicity sum for the finite temperature propagator (60). The independence of β can then be deduced from the fact that (82) is actually a function of $\beta\omega(\vec{k})$, rather than of β and $\omega(\vec{k})$ separately.

The previous observation allows us to express (81) as

$$V_\beta^{eff}(\phi) = V(\phi) + \frac{1}{\beta}\left(\frac{1}{2\pi}\right)^{d-1} \int d\vec{k}\, \{ \ln(e^{\frac{1}{2}\beta\hbar\,\omega(\vec{k})} \mp e^{-\frac{1}{2}\beta\hbar\,\omega(\vec{k})}) - \sum_n \ln\beta \} \tag{83}$$

where we have dropped the irrelevant constant. To clarify this result, we first assume that $V(\phi)$ contains only a mass term, such that $V''(\phi) = m^2$. Reinserting (83) into the action multiplied by $\hbar^{-1}$ we find for the one-loop corrections

$$\frac{V}{(2\pi)^{d-1}} \int d\vec{k}\, \{ \ln(e^{\frac{1}{2}\beta\hbar\omega(\vec{k})} \mp e^{-\frac{1}{2}\beta\hbar\omega(\vec{k})}) - \sum_n \ln\beta \} \tag{84}$$

where V is the spatial volume arising from the integral over Euclidean space

$$\int d\vec{x}d\tau = \hbar\beta V$$

We now recall that (84) represents the result of the Gaussian path integral, and it should therefore correspond to the logarithm of the partition function for a field theory without interaction. Indeed the first term in (84) can be compared to the partition function for a system of harmonic oscillators (cf. (48) or (49)), but the second term has no counterpart. It is important to realize that this term is independent of the theory in question, and therefore we may absorb it into the measure of the functional integral. In fact this term is precisely of the kind that is prescribed by the Wiener measure (cf. (23)), so that the necessity for making such a uniform subtraction should not come as a surprise. We repeat that this question of normalization is only important in calculations of the partition function, and does not enter in the determination of Green's functions [19].

The above observations lead to the following result for the effective potential

$$V_\beta^{eff}(\phi) = V(\phi) + \frac{\hbar}{(2\pi)^{d-1}} \int d\vec{k}\, \{\frac{1}{2}\,\omega(\vec{k}) + \frac{1}{\hbar\beta} \ln(1 \mp e^{-\beta\hbar\omega(\vec{k})}) \tag{85}$$

The first term of the integrand represents the result for zero temperature, which has been considered in the previous subsection. It corresponds precisely to the zero-point energy of an infinite number of harmonic oscillators. The finite temperature contributions are given by the second term in (85). In contrast to the zero-temperature result, these terms do not suffer from ultraviolet divergencies, [31), 32)]. This means that the zero-temperature renormalizations suffice to render the theory finite at arbitrary values of the temperature.

For many applications one is interested in the effective potential at high temperature. In order to evaluate the most relevant terms, we perform the integral over angular variables in the last term of (85), and rescale the remaining integration variable. The result is

$$V_\beta^{eff}(\phi) = V(\phi) + \left(\frac{1}{2\pi}\right)^{d-1} \int d\vec{k}\, \frac{1}{2}\hbar \sqrt{\vec{k}^2 + V''(\phi)}$$

$$+ \hbar \left(\frac{1}{\hbar\beta}\right)^d \frac{\Omega^{d-1}}{(2\pi)^{d-1}} \int_0^\infty d\kappa\, \kappa^{d-2} \ln\left(1 \mp e^{-\sqrt{\kappa^2+\beta^2\hbar^2 V''(\phi)}}\right) \qquad (86)$$

with Ω^{d-1} the surface area of the unit sphere in d-1 dimensions

$$\Omega^{d-1} = \frac{2\pi^{(d-1)/2}}{\Gamma((d-1)/2)}$$

The last integral is then expanded in terms of the inverse temperature

$$\int_0^\infty d\kappa\, \kappa^{d-2} \ln\left(1 \mp e^{-\sqrt{\kappa^2 + \beta^2\hbar^2 V''(\phi)}}\right)$$

$$= \int_0^\infty d\kappa\, \kappa^{d-2} \ln(1 \mp e^{-\kappa}) - \frac{1}{2}\beta^2\hbar^2(d-3)V''(\phi) \int_0^\infty d\kappa\, \kappa^{d-4} \ln(1 \mp e^{-\kappa})$$

$$+ \ldots \} \qquad (87)$$

where we have performed a partial integration in the second term. In four space-time dimensions we thus obtain the result

$$V_\beta^{eff}(\phi) = V_{\beta=\infty}^{eff}(\phi)$$
$$\mp \hbar \{\frac{15 \pm 1}{16} \frac{\pi^2}{90} (\hbar\beta)^{-4} - \frac{3 \pm 1}{96} V''(\phi)(\hbar\beta)^{-2} + \dots\} \tag{88}$$

where we have used

$$\int_0^\infty d\kappa\, \kappa^m \ln(1 \mp e^{-\kappa}) = - \sum_{n=1}^\infty \frac{1}{n} (\pm)^n \int_0^\infty d\kappa\, \kappa^m e^{-n\kappa}$$
$$= \mp\, 2^{-(m+2)}\Gamma(m+1)\, [(2^{m+2} - 1) \pm 1]\, \zeta(m+2)$$

for $m \geqslant 0$; $\zeta(m)$ is the Riemann zeta function, defined by

$$\zeta(m) = \sum_{n=1}^\infty (1/n)^m$$

which is related to the Bernouilli numbers for even values of m. For our purpose it suffices to know $\zeta(2) = \pi^2/6$ and $\zeta(4) = \pi^4/90$.

The effective potential plays an important role in the study of symmetry restoration at high temperature [33]. To illustrate this briefly we consider a theory of a complex scalar field, defined by

$$L = - |\partial_\mu\phi|^2 + m^2 |\phi|^2 - \lambda|\phi|^4 \tag{89}$$

The mass term in (89) has the wrong sign, so that the potential has a minimum at

$$|\phi| = V(\beta = \infty) = \sqrt{\frac{m^2}{2\lambda}}$$

In this case the theory is said to be in the ordered phase (in analogy with ferromagnetic spin systems). In the ordered phase the field (the analogue of the magnetization) takes a macroscopic value due to the phenomenon of Bose condensation [35]. In this realization the symmetry under phase transformations

$$\phi \to \phi' = e^{i\Lambda}\phi \tag{90}$$

is no longer manifest: the theory is realized in a spontaneously broken way. Therefore the two degrees of freedom contained in ϕ

have unequal masses. Excitations corresponding to the radius $\rho = |\phi|$ have a mass $2m^2$, whereas the excitations related to the phase θ of ϕ remain massless. The latter correspond to the Goldstone bosons [26], whose presence is implied by the spontaneous breakdown of the symmetry (90).

We have already pointed out that the quantum corrections to the potential may change the qualitative features of the potential. The same phenomenon may occur at finite temperature, as we will show by considering the high-temperature contributions to the effective potential. Straightforward application of (88) for a complex field (see problem 16) leads to

$$V_\beta^{eff}(\phi) = m^2(\beta)|\phi|^2 + \lambda|\phi|^4 \tag{91}$$

where we have ignored field-independent terms as well as the one-loop zero-temperature contribution (80). Furthermore we have introduced a temperature-dependent mass, defined by

$$m^2(\beta) = -m^2 + \frac{\lambda}{3\hbar\beta^2}$$

We have imposed periodic boundary conditions as is appropriate for bosons; this will be discussed in section IV.3. The implications of (91) are now clear. At temperatures higher than a critical temperature, given by

$$\beta_c^{-1} = \sqrt{\frac{3\hbar m^2}{\lambda}}$$

the mass term of the effective potential changes sign, and the minimum is achieved at

$$|\phi| = v(\beta) = 0, \qquad \beta^{-1} > \beta_c^{-1}$$

In that case the symmetry (90) is manifest. It is straightforward to find the minimum of the potential and the mass spectrum at arbitrary temperature. The results, which are shown in Fig. 8, demonstrate the existence of two phases. Below the critical temperature the theory is in the spontaneously broken phase. At $\beta = \beta_c$ there is a phase transition, and symmetry restoration takes place. Because the masses and $v(\beta)$ are continuous at β_c, this is a second order transition. Above the critical temperature the theory is in the symmetric phase.

The above discussion was based on an approximation of the full one-loop result. However, it is not quite possible to repeat the analysis on the basis of (86), because the effective potential becomes complex for small field values due to the fact that $V''(\phi)$ becomes negative for $\phi < \sqrt{m^2/2\lambda}$ (cf. problem 16). This phenomenon tends to make the determination of the critical temperature and of the nature of the phase transition unreliable. On the other hand it is important to realize that the value of ϕ where the effective potential turns complex is at this stage entirely determined by the classical potential. Higher-order corrections will certainly affect this result. Therefore it is important to include certain higher-

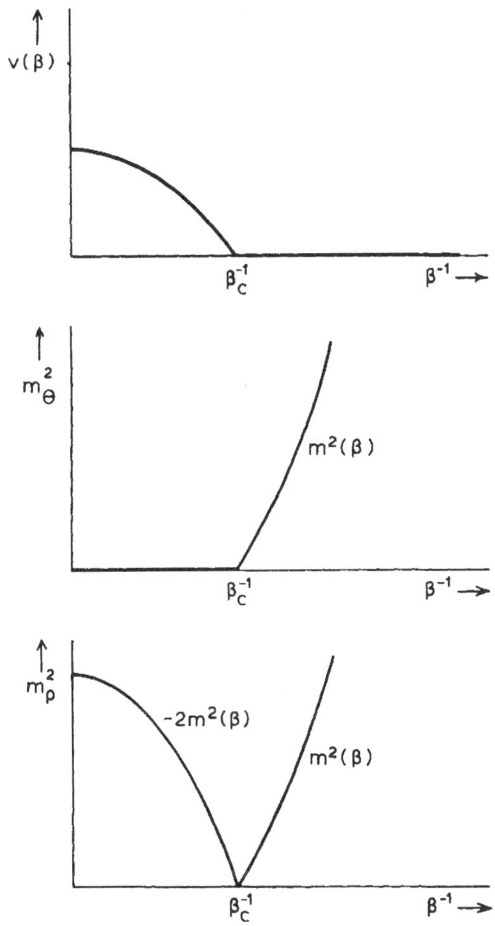

Fig. 8. Field value at the minimum of the potential and the mass of the excitations described by the field ϕ as a function of the temperature; these graphs are based on an analysis of (91).

order contributions at an earlier stage in a self-consistent way, in order to study the phase transition. We refer to the literature for further details [31) - 34), 36)].

IV FERMIONS

1. The Grassmann Algebra

The use of functional methods for fermions is based on anticommuting c-numbers [37)]. These can be discussed within the context of Grassmann algebras. We start by considering anticommuting c-numbers θ_i, which satisfy

$$\theta_i\theta_j = -\theta_j\theta_i .$$

On the basis of n such anticommuting objects one defines an n-dimensional Grassmann algebra with n generators $\theta_1, \ldots, \theta_n$. Each element $P(\theta)$ of the algebra can be decomposed in the following way

$$P(\theta) = p^{(o)} + \theta_i p_i^{(1)} + \theta_i\theta_j p_{ij}^{(2)} + \ldots + \theta_n\theta_{n-1}\ldots\ \theta_1 p^{(n)}, \tag{92}$$

where summation over repeated indices is implied. The total number of terms in (92) is at most 2^n. A monomial $\theta_{i_1}\ldots\ \theta_{i_p}$ is called a monomial of degree p. Monomials of odd (even) degree are anticommuting (commuting) objects. The square of a monomial vanishes unless its degree is zero, in which case we have an ordinary c-number. Monomials of degree higher than the dimension of the Grassmann algebra vanish identically. It is not difficult to define differentiation and integration on the Grassmann algebra. As we shall see below, the application of functional methods to fermions does not require any essential modifications.

Differentiation on the Grassmann algebra. We distinguish two different derivatives on the Grassmann algebra, which are called right and left derivatives. The derivative of a general element of the algebra is obtained by differentiating its monomials and resumming the result. To calculate the right (left) derivative with respect to θ_i we must, in every monomial, permute θ_i to the right (left) and then drop it. If ε^i is the sign of the permutation needed to bring θ_i to the right (left), and $\varepsilon^i = 0$ when θ_i does not occur in the monomial, then the right (left) derivative of a monomial can be written in the following way

$$\frac{\partial}{\partial\theta_{i_k}}(\theta_{i_1}\cdots\theta_{i_{k-1}}\theta_{i_k}\theta_{i_{k+1}}\cdots\theta_{i_m})$$

$$=\varepsilon^{i_k}(\theta_{i_1}\cdots\theta_{i_{k-1}}\theta_{i_{k+1}}\cdots\theta_{i_m}) \tag{93}$$

Unless we specify otherwise, we shall always use left derivatives.

Although the expression $P(\theta)$ for an element of the Grassmann algebra is not a function of θ in the mathematical sense, most of the properties of left and right derivatives of $P(\theta)$ are equivalent to the properties of derivatives of ordinary functions. For example if $\delta\theta_i$ denotes an additional "infinitesimal" anticommuting c-number we can symbolically write for (93)

$$\frac{\partial}{\partial\theta}P(\theta) = \lim_{\delta\theta_i\to 0}\frac{1}{\delta\theta_i}\{P(\theta_1,\ \ldots,\ \theta_i+\delta\theta_i,\ldots\ \theta_n)$$

$$-\ P(\theta_1,\ \ldots,\ \theta_i,\ \ldots\ \theta_n)\}$$

The chain rule holds in the same form as for commuting numbers

$$\frac{\partial}{\partial\theta_i}P(\alpha(\theta)) = \frac{\partial}{\partial\theta_i}\alpha_j(\theta)\frac{\partial}{\partial\alpha_j}P(\alpha),$$

where P is an element of a Grassmann algebra on the basis of the α's, and α is an element of a Grassmann algebra with the θ's as its generators. Also Leibniz'rule has its analogue. For instance, for left derivatives we have

$$\frac{\partial}{\partial\theta}(f(\theta)g(\theta)) = (\frac{\partial}{\partial\theta}f(\theta))g(\theta) \pm f(\theta)(\frac{\partial}{\partial\theta}g(\theta)),$$

where the plus (minus) sign is valid for an even (odd) $f(\theta)$.

Integration over the Grassmann algebra. We will now construct the analogue of the one-dimensional integral

$$\int_{-\infty}^{\infty} dx\ f(x)$$

for the case of anticommuting c-numbers. We shall write this integral as

$\int d\theta \; p(\theta)$.

As a starting point for the construction we require that the following property of the integral over commuting parameters

$$\int_{-\infty}^{\infty} dx \; f(x) = \int_{-\infty}^{\infty} dx \; f(x + a) \tag{94}$$

valid for any finite a, holds for an integral over anticommuting parameters as well. Also we assume that the standard rules for summing integrals remain valid. We now construct the integral. In the one-dimensional case, a general element of the Grassmann algebra is decomposed as follows

$$P(\theta) = a + \theta b, \tag{95}$$

with a and b arbitrary, and θ anticommuting. The analogue expression (94) for anticommuting numbers is given by

$$\int d\theta \; P(\theta) = \int d\theta \; P(\theta + \alpha), \tag{96}$$

for any anticommuting α. If we now substitute (95) in (96), we find that (96) requires

$$\int d\theta \; \alpha b \equiv 0,$$

for any α and b. Hence we must define

$$\int d\theta \; (\text{any element not depending on } \theta) = 0. \tag{97}$$

The remaining integral from (96) is normalized by defining

$$\int d\theta \; \theta \equiv 1 \tag{98}$$

Multiple integrals can be understood as iterated integrals, leading to the anticommuting symbols $d\theta_i$. Integration over anticommuting numbers turns out to be equivalent to taking the left derivative

$$\int d\theta \; P(\theta) = \frac{\partial}{\partial\theta} P(\theta),$$

Also the rules of partial integration apply for integrals over anticommuting parameters. Analogous to

$$\int_{-\infty}^{\infty} dx \; \frac{\partial}{\partial x} f(x) = 0,$$

one can easily show that

$$\int d\theta \frac{\partial}{\partial\theta} P(\theta) = 0.$$

2. Gaussian Integrals and the Superdeterminant

In section III generalized Gaussian integrals played an important role. We now discuss their evaluation for anticommuting variables. The generalization of a Gaussian integral over complex commuting variables

$$\int \frac{dz_i d\bar{z}_i}{2\pi} \exp(-(\bar{z},Az)) = (\det A)^{-1} \tag{99}$$

is most easy to find. Using the integration rule for anticommuting quantities one derives straightforwardly

$$\int d\theta_i d\bar{\theta}_i \exp(\bar{\theta},A\theta) = \det A \tag{100}$$

Notice that the definition of (99) requires A to be a positive-definite hermitean matrix in order that the integral converges, whereas (100) is valid for arbitrary A.

It is also possible to define Gaussian integrals over real variables. We give the result without further derivation

$$\int \frac{dx_i}{\sqrt{2\pi}} \exp(-\frac{1}{2}(x,Ax)) = (\det A)^{-1/2} \tag{101}$$

$$\int d\theta_i \exp(\frac{1}{2}(\theta,A\theta)) = \pm (\det A)^{1/2} \tag{102}$$

In this case the matrix A should be symmetric in (101) and antisymmetric in (102). In addition A must be positive definite in the first case to ensure convergence of the integral. The quantity $(\det A)^{1/2}$, with A antisymmetric, is called a Pfaffian in the mathematical literature. One can show that the Pfaffian is a monomial in each of its variables, a property which is obvious from (102).

Gaussian integrals can be used to define the determinant of a matrix acting in superspace [38]. Vectors in this space are decomposed in terms of commuting and anticommuting variables, x and θ, respectively. Linear transformations in superspace can be written as matrices acting on these coordinates

$$\begin{pmatrix} x \\ \theta \end{pmatrix} \to \begin{pmatrix} x' \\ \theta' \end{pmatrix} = \begin{pmatrix} A & D \\ C & B \end{pmatrix} \begin{pmatrix} x \\ \theta \end{pmatrix} \tag{103}$$

where the submatrices A and B are commuting, and C and D are anti-commuting. One can now construct the superdeterminant of a super-space matrix M

$$M = \begin{pmatrix} A & D \\ C & B \end{pmatrix}$$

by calculating a generalized Gaussian integral over commuting and anticommuting variables

$$\frac{1}{\det M} = \int \frac{dxd\bar{x}}{2\pi} d\bar{\theta}d\theta \exp(-(\bar{x},Ax) - (\bar{x},D\theta) - (\bar{\theta},Cx) - (\bar{\theta},B\theta)) \tag{104}$$

A formal evaluation of this integral by making a shift in integration variables

$$\theta \to \theta - B^{-1}Cx$$

$$\bar{\theta} \to \bar{\theta} - \bar{x}DB^{-1}$$

and using (99), (100), leads to the following expression of the superdeterminant

$$\det M = \frac{\det(A - DB^{-1}C)}{\det B} = \frac{\det A}{\det(B - CA^{-1}D)} \tag{105}$$

The second form can be obtained by performing a shift in the integration variables x and $\bar{x}$. This result for the superdeterminant becomes plausible when we write M as a product of two matrices

$$M = \begin{pmatrix} A & 0 \\ C & 1 \end{pmatrix} \begin{pmatrix} 1 & A^{-1}D \\ 0 & B - CA^{-1}D \end{pmatrix} \tag{106}$$

or alternatively

$$M = \begin{pmatrix} 1 & D \\ 0 & B \end{pmatrix} \begin{pmatrix} A - DB^{-1}C & 0 \\ B^{-1}C & 1 \end{pmatrix} \tag{107}$$

Problem 17. Show that also the determinant of M can be expressed as a Gaussian integral by changing the commuting and anticommuting character of the integration variables in (104).

Precisely as for the conventional determinant one has a product rule for the superdeterminant

$$\det M_1M_2 = \det M_1 \det M_2 \tag{108}$$

To show this we first introduce the notion of the supertrace [39)]

$$\operatorname{tr} M = \operatorname{tr} A - \operatorname{tr} B$$

where the trace operation on the right-hand side is the conventional one applied to the submatrices A and B. Owing to the minus sign in front of the submatrix B, which acts entirely in the anticommuting sector, the supertrace satisfies the familiar cyclicity property

$$\operatorname{tr} M_1M_2 = \operatorname{tr} M_2M_1 \tag{109}$$

This allows one to straightforwardly derive

$$\operatorname{tr} \ln M_1M_2 = \operatorname{tr} \ln M_1 + \operatorname{tr} \ln M_2 \tag{110}$$

Problem 18. Prove the cyclicity property (109). Use the Campbell-Baker-Hausdorff formula for ordinary matrices

$$(\exp A)(\exp B) = \exp(A + B + \text{repeated commutators of A and B}),$$

to show the validity of (110). The exponent and the logarithm of a matrix in superspace are defined by a series expansion, precisely as for ordinary matrices.

We now write M as the product of two triangular matrices according to (106) or (107), and use (110) to prove

$$\ln \det M = \operatorname{tr} \ln M$$

This then suffices to establish the product rule (108) by means of (110).

An important aspect of determinants is that they occur in the definition of the Jacobian of a transformation. A similar situation exists for superdeterminants. Indeed for a transformation in superspace

$$(x,\theta) \to (x(\hat{x},\hat{\theta}),\theta(\hat{x},\hat{\theta}))$$

we may write

$$\int dx d\theta \; P(x,\theta) = \int d\hat{x} d\hat{\theta} J(x(\hat{x},\hat{\theta}),\theta(\hat{x},\hat{\theta}))P(x(\hat{x},\hat{\theta}),\theta(\hat{x},\hat{\theta}))$$

where

$$J(x(\hat{x},\theta),\theta(\hat{x},\hat{\theta})) = \pm \det \begin{pmatrix} \frac{\partial x}{\partial \hat{x}} & \frac{\partial x}{\partial \hat{\theta}} \\ \frac{\partial \theta}{\partial \hat{x}} & \frac{\partial \theta}{\partial \hat{\theta}} \end{pmatrix} \tag{111}$$

In (111) the fermionic derivatives are right derivatives. The correctness of (111) is entirely obvious for linear transformations. For instance, we may redefine the integration variables (x,θ) of the Gaussian integral (104) according to the inverse of (108), which leads directly to (111). But (111) holds for general nonsingular transformation as well.

Problem 19. An intermediate step in the proof of (111) is to show that (111) holds for an integral over a two-dimensional Grassmann algebra. Parametrize the transformation $\theta_i(\hat{\theta}_1,\hat{\theta}_2)$ according to (92), and demonstrate the validity of (111) by explicit calculation. Show also the validity of (111) for an integral over one commuting and one anticommuting parameter.

3. Applications

Using the results for the Gaussian integrals over anticommuting variables it is straightforward to derive the Feynman rules for fermions. The derivation is entirely analogous to the one for commuting fields presented in section III, except that one encounters minus signs because of the anticommutativity of the quantities involved. One characteristic minus sign occurs for closed loops of anticommuting fields. If this is taken into account in the diagrammatic evaluation of Gaussian integrals of section III.4 the result is consistent with (100) and (102).

Problem 20. Derive the Feynman rules for fermions, and show that the propagator for the Dirac Lagrangian $L = -\bar{\psi}(\not{\partial} + m)\psi$ is given by

$$\Delta(k) = \frac{1}{i\not{k} + m}$$

For the same reason the contributions of fermions to the effective potential have a relative minus sign. This is particularly relevant in the calculation of the partition function for fermions. To consider this in somewhat more details, let us first consider the analogue of (85) for a Majorana (= real) fermion [40] coupled to a scalar source

$$L = -\frac{1}{2}\,\bar{\psi}(x)(\not{\partial} + m)\psi(x) + \frac{1}{2}\,f(x)\bar{\psi}(x)\psi(x)$$

which is given by (cf. problem 15)

$$L_{eff} = \frac{\hbar}{2\varepsilon}\left(\frac{1}{2\pi}\right)^{d} \int dk\ \mathrm{tr}\ \ln\left(\frac{i\not{k} + m - i\varepsilon - f(x)}{\mu}\right)$$

$$+ \text{ terms containing } \partial_{\rho}f(x) \qquad (112)$$

The trace extends over the $2^{d/2}$-dimensional spinor space. Eq. (112) can be written as

$$L_{eff} = \frac{\hbar}{2\varepsilon}\, 2^{d/2-1}\left(\frac{1}{2\pi}\right)^{d} \int dk\ \ln\left(\frac{k^{2} + (m - i\varepsilon - f(x))^{2}}{\mu^{2}}\right)$$

$$+ \text{ terms containing } \partial_{\rho}f(x) \qquad (113)$$

For a Dirac (= complex) fermion the result is twice that of (113). The partition function for Majorana fermions is now straightforward to derive (cf. (84)), and we compare the results to that of a real boson field

$$\ln Z_{\beta}(\text{Majorana fermion}) = \frac{V2^{d/2-1}}{(2\pi)^{d-1}} \int d\vec{k}\ \ln(e^{\frac{1}{2}\beta\hbar\omega(\vec{k})}\quad e^{-\frac{1}{2}\beta\hbar\omega(\vec{k})})$$

$$\ln Z_{\beta}(\text{real scalar}) = -\frac{V}{(2\pi)^{d-1}} \int d\vec{k}\ \ln(e^{\frac{1}{2}\beta\hbar\omega(\vec{k})}\quad e^{-\frac{1}{2}\beta\hbar\omega(\vec{k})}) \qquad (114)$$

where V is the (d-1)-dimensional spatial volume. Both expressions can now be compared to the standard partition function of a single bosonic and fermionic harmonic oscillator with appropriate ground-state energies $\pm\hbar\omega/2$

$$Z_\beta = \sum_{n=0}^{\infty} e^{-\beta\hbar\omega(n+\frac{1}{2})} \quad \text{(bosons)} \ , \quad Z_\beta = \sum_{n=0}^{\infty} e^{-\beta\hbar\omega(n-\frac{1}{2})} \quad \text{(fermions)} \tag{115}$$

The results (114) and (115) are indeed compatible if we assign periodic boundary conditions to bosons and antiperiodic ones for fermions. As we have already mentioned before, various manipulations in the functional integral make that not all fields will correspond to physical excitations of the system. For those fields consistency often requires to choose precisely the opposite boundary conditions. In that case it is possible that fermionic and bosonic degrees of freedom cancel in the partition function, as follows from (114). This will be further discussed in the next section.

V. GHOSTS

1. Ghost Fields and Fictitious Particles

We have already emphasized that formal manipulations in the functional integral have the tendency to bring the theory in a form in which its physical relevance is no longer manifest. This is particularly the case when new fields are introduced that are not related to the dynamic degrees of freedom of the original version of the theory. They are only used because of certain technical advantages, and at the end one may wish to verify that the physics has not been affected in this way. Depending on the context such fields are called ghosts, auxiliary fields, or compensating fields, and if they seem to imply new dynamic degrees of freedom, those are called fictitious particles. However, one often does not make a distinction between different kinds of unphysical fields, in which case one uses the generic name ghosts.

Let us discuss three typical examples where ghosts may occur

a. We consider a conventional field theory of scalar self-interacting fields

$$L = -\frac{1}{2}(\partial_\mu\phi)^2 - \frac{1}{2}m^2\phi^2 - \lambda\phi^4 . \tag{116}$$

It is now possible to remove the ϕ^4 interaction by introducing another field. Namely, we choose the following Lagrangian

$$L = -\frac{1}{2}(\partial_\mu\phi)^2 - \frac{1}{2}m^2\phi^2 + \frac{1}{2}\mu^2\sigma^2 + g\sigma\phi^2 \tag{117}$$

where $g^2 = 2\lambda\mu^2$. In the path integral σ can be eliminated by performing the integration over this field, upon which (117) reduces to (116). Since the field σ has no kinetic terms it does not lead to dynamic degrees of freedom.

b. The Lagrangian for the Proca field

$$L = -\frac{1}{4}(\partial_\mu V_\nu - \partial_\nu V_\mu)^2 - \frac{1}{2}m^2V_\mu^2 \tag{118}$$

is equivalent to

$$L = -\frac{1}{4}(\partial_\mu V_\nu - \partial_\nu V_\mu)^2 - \frac{1}{2}m^2V_\mu^2 - \frac{1}{2}(\partial_\mu\phi)^2 - m\phi\partial_\mu V_\mu \tag{119}$$

In this case the ghost field ϕ does seem to lead to new particles. However, the two Lagrangians (116) and (117) are still equivalent, because the second one is invariant under the following local gauge transformations

$$\delta V_\mu(x) = \partial_\mu \Lambda(x),$$

$$\delta\phi(x) = m\Lambda(x)$$

This transformation can be used to eliminate ϕ from (119), as can be seen from a redefinition of V_μ

$$V_\mu \to V_\mu + \frac{1}{m}\partial_\mu\phi \tag{120}$$

Upon this substitution (119) reduces to the original Lagrangian (118). The formulation (119) is due to Stueckelberg [41)].

c. Suppose we wish to make a field redefinition by expressing the field ϕ in terms of another field $\hat{\phi}$.

$$\phi = \phi(\hat{\phi})$$

Making this substitution we must also rewrite the functional measure in the path integral in terms of the new fields. Hence we have

$$[d\phi] = [d\hat{\phi}]\, J[\phi(\hat{\phi})]$$

with J the Jacobian corresponding to the parameter change $\phi(\hat{\phi})$. As we have argued in section IV, the Jacobian is the superdeterminant of $\partial\phi/\partial\hat{\phi}$, which can be expressed as a Gaussian integral over new fields. If we have commuting fields ϕ and anticommuting fields ψ then the Jacobian for the transformation

$$\phi \to \phi(\hat{\phi},\hat{\psi})$$

$$\psi \to \psi(\hat{\phi},\hat{\psi})$$

can be written as

$$J(\phi(\hat{\phi},\hat{\psi}),\psi(\hat{\phi},\hat{\psi})) = \det \begin{pmatrix} \frac{\partial\phi(x)}{\partial\hat{\phi}(y)} & \frac{\partial\phi(x)}{\partial\hat{\psi}(y)} \\ \frac{\partial\psi(x)}{\partial\hat{\phi}(y)} & \frac{\partial\psi(x)}{\partial\hat{\psi}(y)} \end{pmatrix}$$

$$\propto \int[d\alpha][d\alpha^*][d\beta][d\beta^*] \exp\{\int dxdy\,[\alpha^*(x)\frac{\partial\psi(x)}{\partial\hat{\psi}(y)}\alpha(y)$$

$$+ \alpha^*(x)\frac{\partial\psi(x)}{\partial\hat{\phi}(y)}\beta(y) + \beta^*(x)\frac{\partial\phi(x)}{\partial\hat{\psi}(y)}\alpha(y) + \beta^*(x)\frac{\partial\phi(x)}{\partial\hat{\psi}(y)}\beta(y)]\} \tag{121}$$

where $\alpha^*(x)$ and $\alpha(x)$ are commuting and $\beta^*(x)$ and $\beta(x)$ are anticommuting ghost fields (cf. problem 17). Hence the effect of this parameter change is the addition of new terms to the Lagrangian, quadratic in certain ghost fields. Note that for periodic Euclidean field theory the ghosts must have the same periodicity properties as the corresponding fields.

In each of these examples the ghost fields serve a different purpose. In the first example the field σ, which is sometimes called an auxiliary field, decreases the amount of nonlinearity in the theory in the sense that a quartic interaction is replaced by a cubic one. (In the case that a symmetry is present they may also be required for group-theoretic reasons. See, for instance [42]). The ghost in the second example is called a compensating field, since in (120) the second term compensates for the effect of the gauge transformation of V_μ. Its presence enables one to discuss the theory within a framework with a higher degree of symmetry. The ghosts in the third example often occur in the context of quantization of gauge theories, where one may distinguish so-called Faddeev-Popov ghosts [43] and Nielsen-Kallosh ghosts [44]. The necessity for such ghost fields was first noted by Feynman [45]; some subsequent work is listed in [46]. The important aspect of ghost fields in this context is that they facilitate the use of Feynman diagrams for an effective Lagrangian that contains the logarithm of a determinant.

2. Ghost Fields in Feynman Diagrams

In order to get acquainted with the presence of ghosts in Feynman diagrams we present a few calculations. In principle the evaluation in terms of Feynman diagrams is equivalent to the more formal manipulations that one performs in the functional integral. Feynman diagrams are not always well defined because some of the corresponding momentum integrals may be divergent, but there are

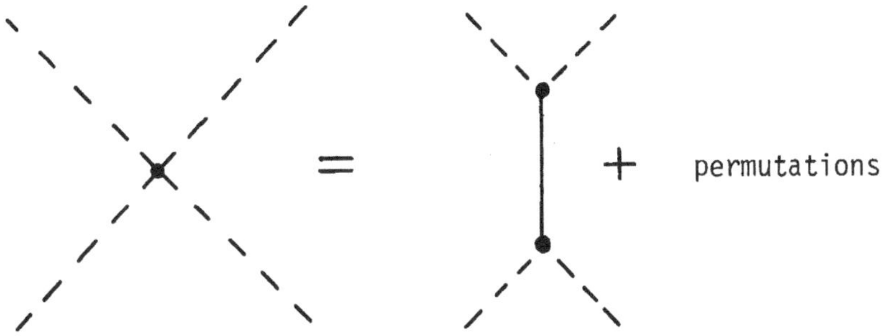

Fig. 9

standard methods of regularization and renormalization to deal with this. In contrast with this a direct analogue of these methods does not exist for the path integral. But we shall ignore such complications here, and confine ourselves to a direct determination of certain diagrams.

It is sometimes easy to see that diagrams based on different Lagrangians lead to the same result. For example, the ϕ^4 interaction in (116) is generated by a sum of diagrams based on (117) in which one has an auxiliary field propagator. Since this propagator is constant it is straightforward to verify the equivalence. Schematically we have Fig. 9, where the solid line denotes the σ propagator.

Problem 21. Verify the diagrammatic identity of Fig. 9 in the one-loop approximation (see Fig. 10).

Let us now consider a more complicated example, where we have the Lagrangian of a free fermion and a free boson field

$$L = -\bar{\psi}(\not\partial + m)\psi - \frac{1}{2}(\partial_\mu\phi)^2 - \frac{1}{2}\mu^2\phi^2 \qquad (122)$$

We perform the following transformation

$$\psi = (1 + g_1\phi)\chi \ , \quad \bar{\psi} = (1 + g_2\phi)\bar{\chi} \qquad (123)$$

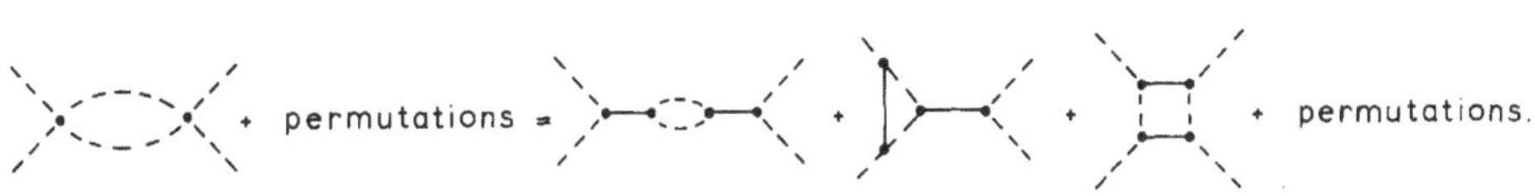

Fig. 10

The spinor field ψ is a Dirac (complex) field. Because ψ and $\bar{\psi}$ occur as independent quantities in expression (122), we may perform independent transformations on ψ and $\bar{\psi}$. Moreover, ψ and $\bar{\psi}$ are spinors, so that the transformation (123) must be carried out for each component of ψ and $\bar{\psi}$ separately.

We now introduce independent commuting ghost fields α and $\bar{\alpha}$, with conjugate fields α^* and $\bar{\alpha}^*$ and an anticommuting ghost field β with conjugate field β^*. The ghosts are used to write the Jacobian (121) corresponding to (123). In this expression α, $\bar{\alpha}$ and β are related to the physical fields ψ, $\bar{\psi}$ and ϕ, respectively. Both α and $\bar{\alpha}$ are spinor fields with the same number of components as ψ. The effective Lagrangian then becomes

$$L_{eff} = - \bar{\chi}(\not{\partial} + m)\chi(1 + g_1\phi)(1 + g_2\phi) - g_1(1 + g_2\phi)\bar{\chi}\gamma^\mu\chi\partial_\mu\phi$$

$$- \frac{1}{2}(\partial_\mu\phi)^2 - \frac{1}{2}\mu^2\phi^2 + \alpha^*\alpha(1 + g_1\phi) + \bar{\alpha}^*\bar{\alpha}(1 + g_2\phi)$$

$$+ g_1(\alpha^*\chi)\beta + g_2(\bar{\chi}\bar{\alpha}^*)\beta + \beta^*\beta \tag{124}$$

The last three terms in (124) can be omitted because of the orientation carried by the ghost field lines. We only have a transition of α and $\bar{\alpha}$ into β, but no transition of β back into α or $\bar{\alpha}$. Therefore these diagrams will never yield a diagram with a closed loop. This is directly related to the fact that the off-diagonal elements of a triangular matrix do not contribute to its determinant. Of course, one may wish to consider diagrams with external ghost lines. In that case the off-diagonal elements are relevant. Diagrams with external ghost lines occur for instance in the Ward identities for gauge field theories [5)-11), 46)].

We will now calculate the one-point function $\langle\phi\rangle$ and the two-point function $\langle\phi\phi\rangle$ in the one-loop approximation. According to the Lagrangian (122) $\langle\phi\rangle$ is zero, so it must vanish also for the effective Lagrangian (124). We find contributions from the following three diagrams (Fig. 11):

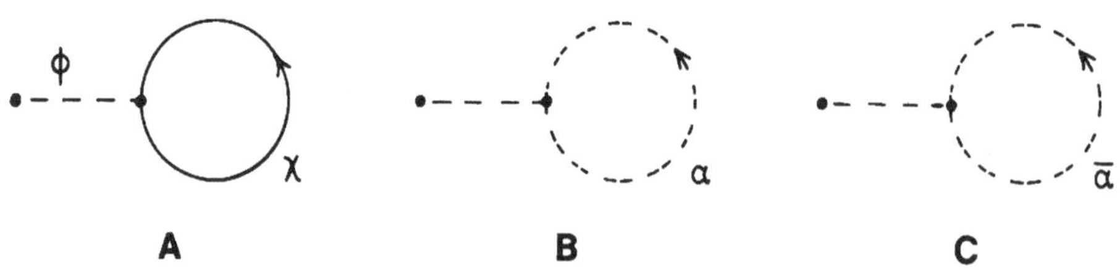

Fig. 11.

The vertices in diagrams (A), (B) and (C) originate from the terms $-(g_1+g_2)\bar{\chi}(\not\partial+m)\chi\phi$, $g_1\alpha^*\phi\alpha$ and $g_2\bar{\alpha}^*\phi\bar{\alpha}$ in L_{eff}, respectively. The external line in all three diagrams yields a common factor. In diagram (A) the fermion loop requires a minus sign. Summing over the spinor components of the fields χ, α and $\bar{\alpha}$ in each diagram, we find straightforwardly

$$\langle\phi\rangle \propto \text{tr}\,\{(-1)^2(g_1 + g_2)(i\not q + m)\frac{1}{i\not q + m} + g_1(-1) + g_2(-1)\} = 0$$

Notice that it was essential for this result that the spinor fields α and $\bar{\alpha}$ are commuting spinor fields with the same number of components as the Dirac field ψ.

According to (122) the two-point function $\langle\phi\phi\rangle$ is that of a free Klein-Gordon field. After the transformation (123) there are higher-order contributions to $\langle\phi\phi\rangle$, which must therefore cancel. We will now demonstrate this in the one-loop approximation. In that order we have the following contributions (see Fig. 12).

All the diagrams have equal numbers of propagators and vertices, which give rise to a common factor. The overall momentum Q is taken from the left to the right. The vertices in diagram (12A) originate from the terms $-(g_1+g_2)\bar{\chi}(\not\partial+m)\chi\phi - g_1\bar{\chi}\gamma^\mu\chi\partial_\mu\phi$ in the effective Lagrangian (124). This yields the following contribution:

$$- \int dpdq\ \delta(p - q - Q)\ \text{tr}\ \{[-(g_1 + g_2)(i\not q + m) - ig_1\not Q\,]\frac{1}{i\not q + m}$$

$$[-(g_1 + g_2)(i\not p + m) + ig_1\not Q\,]\frac{1}{i\not p + m}\}$$

$$= - \int dpdq\ \delta(p - q - Q)\ \text{tr}\ \{(g_1 + g_2)^2$$

$$- g_1(g_1+g_2)i\not Q(\frac{1}{i\not p+m} - \frac{1}{i\not q+m}) + g^2\not Q\,\frac{1}{i\not q+m}\,\not Q\,\frac{1}{i\not p+m}\,\}$$

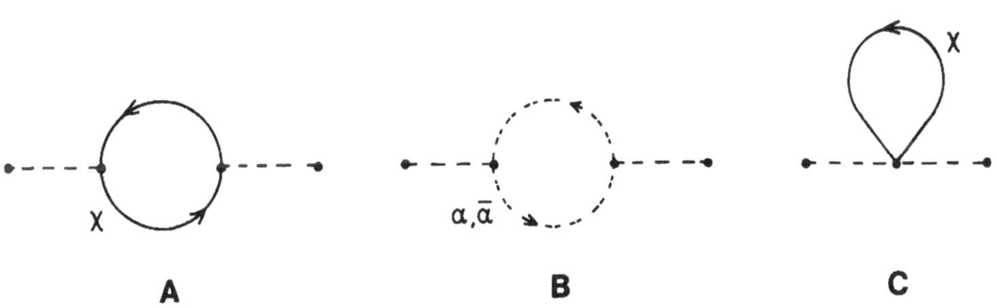

Fig. 12

The second and third terms cancel after elimination of the δ-function and relabeling of the integration variables. Substituting

$$i\not{\phi} = (i\not{p} + m) - (i\not{q} + m)$$

for the second $\not{\phi}$ in the last term, this term vanishes too by the same arguments, so that we are left with the expression

$$- (g_1 + g_2)^2 \int dp \, \mathrm{Tr}\, \{1\} \tag{125}$$

for diagram (12A). The vertices in diagram (12B) originate from the terms $g_1\alpha^*\phi\alpha + g_2\bar{\alpha}^*\phi\bar{\alpha}$ in L_{eff}. They contribute

$$\int dp \, \mathrm{tr}\, \{g_1^2(-1)^2 + g_2^2(-1)^2\} = (g_1^2 + g_2^2) \int dp \, \mathrm{tr}\, \{1\} \tag{126}$$

The vertex in diagram (12C) corresponds to the terms $- g_1g_2\bar{\chi}(\not{\partial}+m)\chi\phi^2 - g_1g_2\bar{\chi}\gamma^\mu\chi(\partial_\mu\phi)\phi$ in L_{eff}. The second term does not contribute, since the two possible diagrams corresponding to this term have opposite signs. The first term leads to the expression

$$- \int dq \, \mathrm{tr}\, \{- 2g_1g_2(i\not{q} + m) \frac{1}{i\not{q} + m}\} = 2g_1g_2 \int dq \, \mathrm{tr}\, \{1\} \tag{127}$$

where the factor 2 arises from combinatorics. Adding the contributions (125)-(127) then indeed gives zero. Notice that in obtaining this result it was essential, that the ghost fields α and $\bar{\alpha}$ are independent commuting spinor fields with the same number of components as the Dirac field ψ. Furthermore, for finite temperature it is crucial that α, $\bar{\alpha}$ and χ satisfy the same periodicity property.

Problem 22. Calculate the one-loop corrections to the fermion propagator, and explain the result. Furthermore prove from the previous results that the two-loop vacuum diagrams (i.e. the diagrams with two loops without external lines) vanish.

We have already mentioned that diagrams such as the ones we have been considering, do not always lead to convergent momentum integrals. To make the above manipulations well-defined one needs a regularization scheme, which should allow us to go through the same manipulations. Actually, this is a rather weak requirement, since we have not done much more than a relabeling of integration variables. One suitable regularization method is dimensional regularization [29] which is obviously applicable since none of the cancellations found above depend on the dimension of space-time.

3. Ghost Counting

From the above examples it seems that there are hardly any

restrictions on the number of ghosts that may be present in a field theory. On the other hand, one would at least expect a restriction on the number of propagating ghosts, i.e. those ghosts that seem to lead to dynamic degrees of freedom. One way to see this is to consider the theory at finite temperature, and to determine the partition function. The number of modes that occur there should be equal to the number of physical modes that the theory is supposed to describe. Hence in the presence of propagating ghosts cancellations should take place between the fictitious degrees of freedom. Such cancellations can only take place if we have both commuting and anticommuting fields present that are subject to the same periodicity requirement (this point was emphasized in [19]).

Let us first outline how this argument works when one quantizes a free Maxwell field in d dimensions. Since gauge-field quantization is not discussed in these lectures we will only give a qualitative description of what happens. The Maxwell Lagrangian

$$L = -\frac{1}{4}(\partial_\mu A_\nu - \partial_\nu A_\mu)^2 \tag{128}$$

is invariant under local gauge transformations

$$A_\mu \rightarrow A_\mu + \partial_\mu \xi$$

It is not possible to quantize this theory by straightforward use of the methods described in these lectures. One must first choose a gauge condition, such as

$$\partial_\mu A_\mu(x) = \lambda(x) \tag{129}$$

To ensure that the path integral remains independent of $\lambda(x)$ it is necessary to introduce complex anticommuting ghost fields, the so-called Faddeev-Popov ghosts [43), 45), 46)], in the Lagrangian. Hence we add to (128)

$$L_{ghost} = -\frac{1}{2}\partial_\mu c^* \partial_\mu c \tag{130}$$

The general prescription is that one needs a complex ghost for every real gauge transformation parameter ξ. Apart from the fact that they are complex and have opposite commutativity properties, the ghosts have exactly the same properties as the transformation parameters. In particular they should satisfy the same periodicity requirement as the parameters, whose periodicity coincides with that of the corresponding gauge field.

The addition of the ghosts has made the path integral independent of $\lambda(x)$. Therefore we may integrate over $\lambda(x)$ with a

Gaussian weight factor

$$\int [d\lambda] \exp (- \frac{1}{2\hbar} \int dx\, \lambda^2(x)) \tag{131}$$

Hence we add a term $- 1/2\, \lambda^2(x)$ to the gauge field Lagrangian. Using the fact that λ is already constrained by (129) the effective Lagrangian (in the so-called Feynman gauge) becomes

$$L_{eff} = - \frac{1}{2} \partial_\mu A_\nu \partial_\mu A_\nu \tag{132}$$

Let us now count the degrees of freedom as they occur in the partition function. The Lagrangian (132) describes d degrees of freedom, whereas (130) implies 2 degrees of freedom. We now recall that the ghosts have the same periodicity as A_μ, but opposite commutativity properties. Therefore their contribution cancels two of the modes that originate from A_μ, so that the number of modes equals d - 2. This is precisely the number of polarizations that one expects for a massless spin - 1 particle in d dimensions.

This counting argument can be repeated for massless bosons of arbitrary spin. Gauge fields of spin s are described by a symmetric s-rank tensor field $\phi_{\mu_1 \ldots \mu_s}$, which is subject to a constraint[47)-49)]

$$\phi_{\rho\rho\sigma\sigma\mu_5 \ldots \mu_s} = 0$$

The fields transform under gauge transformations

$$\phi_{\mu_1 \ldots \mu_s} \to \phi_{\mu_1 \ldots \mu_s} + \partial_{(\mu_1} \xi_{\mu_2 \ldots \mu_s)}$$

where the brackets denote a symmetrized sum over s terms. Consistency requires that $\xi_{\mu_2 \ldots \mu_s}$ is symmetric and subject to

$$\xi_{\rho\rho\mu_4 \ldots \mu_s} = 0$$

Let us now count the number of degrees of freedom contained in ϕ and ξ (and thus in the complex ghosts which have twice the number of degrees of freedom of ξ). A symmetric n-rank tensor in d dimensions has

$$\begin{pmatrix} n + d - 1 \\ d - 1 \end{pmatrix}$$

independent components. Hence ϕ and ξ contain

$$n_\phi = \begin{pmatrix} s + d - 1 \\ d - 1 \end{pmatrix} - \begin{pmatrix} s + d - 5 \\ d - 1 \end{pmatrix}$$

$$n_\xi = \begin{pmatrix} s + d - 2 \\ d - 1 \end{pmatrix} - \begin{pmatrix} s + d - 4 \\ d - 1 \end{pmatrix} \qquad (133)$$

The number of physical modes is thus $n_\phi - 2n_\xi$, which is equal to

$$n_\phi - 2n_\xi = \begin{pmatrix} s + d - 3 \\ d - 3 \end{pmatrix} - \begin{pmatrix} s + d - 5 \\ d - 3 \end{pmatrix} \qquad (134)$$

This is the number of components of an s-rank symmetric traceless tensor in d - 2 dimensions, which is the number of polarizations appropriate for a spin-s massless particle in d dimensions.

The previous discussion showed the necessity for the ghost fields without going through a detailed treatment of the quantization. It is important to recall why in standard discussions of quantum-electrodynamics these ghosts never played a role. The reason is that the underlying gauge symmetry is abelian, so that the ghosts are non-interacting (at least in linear gauges). Therefore ghosts can be ignored in calculations of quantum corrections. However, the argument presented above show that this is only allowed at zero temperature. At finite temperature the ghosts will still be important to cancel fictitious degrees of freedom.

We now wish to show that interesting complications occur in the case of fermionic gauge fields. Massless fermions of spin s are described by a rank-s tensor spinor $\psi_{\mu_1 \dots \mu_s}$, symmetric in indices $\mu_1, \dots, \mu_s$.

We will assume that ψ is a Majorana spinor. The fields are subject to the constraint[48)-51)]

$$\gamma_\rho \psi_{\rho\sigma\sigma\mu_4 \dots \mu_s} = 0$$

and transform under gauge transformations according to

$$\psi_{\mu_1 \dots \mu_s} \to \psi_{\mu_1 \dots \mu_s} + \partial_{(\mu_1} \varepsilon_{\mu_2 \dots \mu_s)}$$

The transformation parameters $\varepsilon_{\mu_2 \dots \mu_s}$ are symmetric rank-(s-1) tensor spinors, that satisfy the condition

$$\gamma_\rho \varepsilon_{\rho\mu_3 \ldots \mu_s} = 0$$

A Lorentz-covariant gauge condition is now specified by

$$\gamma_\rho \psi_{\rho\mu_2 \ldots \mu_s} - \frac{1}{2s} \gamma_{(\mu_2} \psi_{\mu_3 \ldots \mu_s)\rho\rho} = \lambda_{\mu_2 \ldots \mu_s}$$

where λ is now a tensor spinor, subject to (135).

When we quantize according to the same procedure as described above, we must add complex <u>commuting</u> ghosts $c_{\mu_2 \ldots \mu_s}$, which satisfy (135). The numbers of components of ψ and ε (and thus of λ and c) are now easily determined. We find

$$n_\psi = 2^{d/2} \left\{ \begin{pmatrix} s+d-1 \\ d-1 \end{pmatrix} - \begin{pmatrix} s+d-4 \\ d-1 \end{pmatrix} \right\}$$

$$n_\varepsilon = 2^{d/2} \left\{ \begin{pmatrix} s+d-2 \\ d-1 \end{pmatrix} - \begin{pmatrix} s+d-3 \\ d-1 \end{pmatrix} \right\} \qquad (137)$$

However, the numbers of polarizations appropriate for a massless fermion of spin s is not equal to $n_\psi - 2n_\varepsilon$, as one would naively expect. This is a signal that more ghosts are present. It is not difficult to see what their origin is. The gauge condition (136) is now free of derivatives.

Therefore, in order to ensure that all components of ψ occur uniformly in the effective action, one prefers to integrate the path integral with a Gaussian integral weight factor that contains a derivative

$$\int [d\lambda] \exp\left(-\frac{1}{2}\int dx \bar{\lambda}_{\mu_2 \ldots \mu_s}(x) \not{\partial} \lambda_{\mu_2 \ldots \mu_s}(x)\right) \qquad (138)$$

rather than with the analogue of (131). However, (138) introduces new dynamic degrees of freedom in the theory. Again those may not be important at zero temperature, where $(\det \not{\partial})^{1/2}$ is a field-independent constant, but at finite temperature the situation is different. To compensate for (138), which is nothing more than $(\det \not{\partial})^{1/2}$, we introduce a real anticommuting and a complex commuting ghost, $a_{\mu_2 \ldots \mu_s}$ and $b_{\mu_2 \ldots \mu_s}$, respectively. We then add to the Lagrangian

$$L_{\text{extra ghost}} = -\frac{1}{2}\bar{a}_{\mu_2\ldots\mu_3}\not{\partial}a_{\mu_2\ldots\mu_s} - \bar{b}_{\mu_2\ldots\mu_s}\not{\partial}b_{\mu_2\ldots\mu_s}, \tag{139}$$

which lead to Gaussian integrals in the path integral. The ghost $a_{\mu_2\ldots\mu_s}$ thus yields $(\det \not{\partial})^{1/2}$, whereas the ghost $b_{\mu_2\ldots\mu_s}$ gives $1/(\det\not{\partial})$, for every independent component. Together they cancel exactly the $(\det \not{\partial})^{1/2}$ generated by (138). Notice that we need two ghosts here, because the Dirac operator $\not{\partial}$ is antisymmetric.

All the ghosts have the same periodicity as the original gauge field ψ, and we can now count the number of degrees of freedom. We recall that anticommuting components have a negative weight in the counting. Hence the number of physical fermionic components n follows from

$$\begin{aligned} -n &= -n_\psi - n_a + n_b + n_c \\ &= -n_\psi - n_\varepsilon + 2n_\varepsilon + 2n_\varepsilon \\ &= -n_\psi + 3n_\varepsilon \end{aligned}$$

Direct calculation shows that n can be written as

$$n = 2^{d/2}\left\{\binom{s+d-3}{d-3} - \binom{s+d-4}{d-3}\right\} \tag{140}$$

This is precisely the number of components appropriate for a massless particle with half-integer spin. Namely,

$$\binom{s+d-3}{d-3} - \binom{s+d-4}{d-3}$$

is the number of independent components of a symmetric s-rank traceless tensor in d-2 dimensions; the factor $2^{d/2}$ finds its origin in the number of spinor components in d dimensions.

Hence we have seen how a simple counting argument reveals the need for additional ghosts. These ghosts are called Nielsen-Kallosh ghosts [44]. We should emphasize that it depends on the particular type of gauge-fixing whether all these ghosts are present or not. However, the above discussion is far from complete. There are examples known where one has a more complex ghost structure. A notorious example is the antisymmetric tensor gauge field, where the Faddeev-Popov ghost Lagrangian exhibits its own gauge invariance. In order to quantize the system one must now also introduce ghosts for ghosts, and there are cases where an infinite sequence of ghosts seems required. However, this is far outside the scope of these lectures, and we refer to the literature for a more explicit presentation of these phenomena [52].

REFERENCES

1. P.A.M. Dirac, Phys. Z. der Sowjetunion 3, reprinted in "Quantum Electrodynamics", ed. J. Schwinger (Dover 1958); "The Principles of Quantum Mechanics" (The Clarendon Press, 1958).
2. R.P. Feynman, Rev. Mod. Phys. 20, 267 (1948).
3. R.P. Feynman and A.R. Hibbs, "Quantum Mechanics and Path Integrals"(Mc Graw-Hill 1965).
4. J.M. Gel'fand and A.M. Yaglom, J. Math. Phys. 1, 48 (1960).
5. For a review, see M. Veltman, proc. Int. Symp. on Electron and Photon Interactions at High Energies, Bonn (1973) (North-Holland).
6. E.S.Abers and B.W. Lee, Phys. Rep. 9, 1 (1973).
7. M. Veltman, lectures at the Basko Polje Summer School, (1974).
8. S. Coleman, proc. of the 1975 International School of Subnuclear Physics "Ettore Majorana" (Acad. Press, 1975).
9. L.D. Faddeev, in Methods in Field Theory, eds. R. Balian and J. Zinn-Justin (North-Holland, 1976).
10. B.W. Lee, in Methods in Field Theory, eds. R. Balian and J. Zinn-Justin (North-Holland, 1976).
11. J.C. Taylor, Gauge Theories of Weak Interactions (Cambridge Univ. Press, 1976).
12. V.N. Popov, Functional Integrals in Quantum Field Theory and Statistical Physics (in Russian), (Atomizdat, 1976); CERN preprint TH.2424, 1977.
13. C. De Witt-Morette, A. Maheswari and B. Nelson, Phys. Rep. 50, 255 (1979).
14. L.D. Faddeev and A.A. Slavnov, Gauge fields: Introduction to Quantum Theory (Benjamin Cummings, 1980).
15. C. Itzykson and J.B. Zuber, Quantum Field Theory (Mc Graw-Hill, 1980).
16. M.S. Marinov, Phys. Rep. 60, 1 (1980).
17. P. Ramond, Field Theory; a modern primer (Benjamin Cummings, 1981).

18. See for instance, H. Goldstein, Classical Mechanics (Addison-Wesley, 1950).

19. C.W. Bernard, Phys. Rev. D9, 3312 (1974).

20. R.P. Feynman, Phys. Rev. 84, 108, (1951)

21. The effective action for the velocity-dependent potential was first obtained by T.D. Lee and C.N. Yang, Phys. Rev. 128, 885 (1962).

22. J. Schwinger, Proc. Nat. Acad. Sci. 44, 956 (1958);
K. Symanzik, proc. Int. School "Enrico Fermi", ed. R. Jost (Acad. Press, 1969);
K. Osterwalder and R. Schrader, Comm. Math. Phys. 31, 83 (1973).

23. R.P. Feynman, Phys. Rev. 91, 1291 (1953); Statistical Mechanics: A Set of Lectures (Benjamin, 1972).

24. See for instance, A. Fetter and J. Walecka, Quantum Theory of Many-Particle Systems (Mc Graw-Hill, 1965).

25. G. Jona-Lasinio, Nuovo Cim. 34, 1790 (1964).

26. J. Goldstone, A. Salam and S. Weinberg, Phys. Rev. 127, 965 (1962).

27. S. Coleman and E. Weinberg, Phys. Rev. D7, 1888 (1973).

28. R. Jackiw, Phys. Rev. D9, 1686 (1974).

29. G. 't Hooft and M. Veltman, Nucl. Phys. B44, 189 (1972);
C.G. Bollini and J.J. Giambiagi, Phys. Lett. 40B, 566 (1972);
J.F. Ashmore, Lett. Nuovo Cim. 4, 289 (1972).
G.M. Cicuta and E. Montaldi, Lett. Nuovo Cim. 4, 329 (1972).

30. S. Weinberg, Phys. Rev. Lett. 36, 294 (1976);
A.D. Linde, Zh. Eksp. Teor. Fiz. Pis. Red. 23, 73 (1976) (JEPT Lett. 23, 64 (1976)).

31. L. Dolan and R. Jackiw, Phys. Rev. D9, 3320 (1974).

32. S. Weinberg, Phys. Rev. D9, 3357 (1974).

33. D. Kirzhnits and A. Linde, Phys. Lett. 42B, 471 (1972).

34. A.D. Linde, Rep. Prog. Phys. 42, 389, (1979).

35. G. 't Hooft, this volume.

36. J. Iliopoulos and N. Papanicolaou, Nucl. Phys. B111, 209 (1976).

37. F.A. Berezin, The Method of Second Quantization (Acad. Press, (1966).

38. B. de Wit, Phys. Rev. D12, 1628 (1975).

39. R. Arnowitt, P. Nath and B. Zumino, Phys. Lett. 56B, 81 (1975).

40. A. De Rújula, this volume.

41. E.C.G. Stueckelberg, Helv. Phys. Acta 11, 225 (1938).

42. P. van Nieuwenhuizen, Phys. Rep. 68, 189 (1981).

43. L.D. Faddeev and V.N. Popov, Phys. Lett. 25B, 29 (1976).

44. N.K. Nielsen, Nucl. Phys. B140, 499 (1978)
R.E. Kallosh, Nucl. Phys. B141, 141 (1978).

45. R.P. Feynman, In Relativistic Theories of Gravitation, (Pergamon Press 1964); Acta Phys. Polonica 24, 697 (1963).

46. B.S. De Witt. Phvs. Rev. Lett. 12, 742 (1964); Phys. Rev. 162, 1195, 1239 (1967).
M. Veltman, Nucl. Phys. B21, 288 (1970);
E.S. Fradkin and J.V. Tyutin, Phys. Rev. D2, 2841 (1970);
G.'t Hooft, Nucl. Phys. B33, 173 (1971);
G. 't Hooft and M. Veltman, Nucl. Phys. B50, 318 (1972);
B.W. Lee and J. Zinn-Justin, Phys. Rev. D7, 1049 (1973).

47. C. Fronsdal, Phys. Rev. D18, 3624 (1978).

48. T.L. Curtright, Phys. Lett. 85B, 219 (1979).

49. B. de Wit and D.Z. Freedman, Phys. Rev. D21, 358 (1980).

50. J. Fang and C. Fronsdal, Phys. Rev. D18, 3630 (1978).

51. C. Aragone and S. Deser, Phys. Rev. D21, 352 (1980).

52. M.A. Namazie and D. Storey, Nucl. Phys. B157, 170 (1979);
P.K. Townsend, Phys. Lett. 88B, 97 (1979);
W. Siegel, Phys. Lett. 93B, 170 (1980).

INTRODUCTION TO ELECTRO-WEAK INTERACTIONS

Gino Segrè

Department of Physics
University of Pennsylvania, Philadelphia
Pennsylvania, 19104, U.S.A.

I. INTRODUCTION

The 1970-1980 decade will probably be remembered in physics histories as the decade in which gauge theories emerged. In the first years of this period, the QCD theory of strong interactions[1.1] and the $SU_2 \times U_1$ theory of electroweak interactions [1.2,1.3,1.4] were extensively studied, and by the end of the decade more speculative gauge theories such as those for grand unification and for dynamical symmetry breaking came under scrutiny.[1.5]

Both QCD and $SU_2 \times U_1$ are not truly confirmed; on the one hand a fully satisfactory picture of quark confinement still eludes us and on the other hand the $W^{\pm}$ and Z have not yet been observed experimentally. Their fates are also interwoven, e.g. if the theory of strong interactions was a Yukawa type theory with colored elementary scalars, violations of order α to C,P,T, could be caused by gauge loop corrections even if the Lagrangian was invariant.[1.6] Our point of view for these lectures, however, will be that both theories are correct, though not necessarily complete.

Since you will be hearing detailed lectures by Steinberger on the experimental status of $SU_2 \times U_1$, by Susskind on dynamical symmetry breaking, by Glashow on flavour dynamics and by de Rujula on neutrino masses, I will concentrate on providing the necessary introductory structure without going into these specialized topics in detail.

Let me begin by setting the notation we will use, following Langacker's review[1.5]

μ, ν, σ, τ : Lorentz indices

i,j,k, = 1...N : label of group generators

α,β,γ : color indices

r,s,t : flavor indices

m,n,p,q : horizontal indices

Repeated indices in all spaces are summed over and the metric and Dirac matrix notation are those of Bjorken and Drell[1.7] (except for the sign of γ_5). Operators are denoted by capital letters T^i, Q, Y, etc.. their eigenvalues by t^i, q, y, and their matrix representations by L^i, L_Q, L_Y. Gauge boson masses are denoted by M, Higgs masses by μ and quark masses by e.g. m_u, m_c, m_s, m_b..., though we will use M^u, M^d respectively as symbols for the diagonal charge 2/3 and charge -1/3 quark mass matrices, i.e.

$$M^u = \begin{pmatrix} m_u & & 0 \\ & m_c & \\ 0 & & m_t \end{pmatrix} \qquad M^d = \begin{pmatrix} m_d & & 0 \\ & m_s & \\ 0 & & m_b \end{pmatrix} \tag{1.1}$$

and similarly for leptons. Yukawa couplings will be denoted by Γ and the following acronyms will be employed :

SSB = spontaneous symmetry breaking
VEV = vacuum expectation value
IRREP = irreducible representation.

Finally, anti-particles will be denoted by a c superscript (e.g. u^c is the anti u quark) qnd x is always four dimensional.

II. GAUGE INVARIANCE

Non-Abelian Local Symmetries

Our starting point is the transformation properties of fields under rotations through $\vec{\beta}(x)$ in an internal space. The local symmetries based on the invariance under such transformations for non-abelian groups dictates the usual generalization of global symmetries that

$$\psi_a(x) \to U_\psi(\beta(x))_{ab}\,\psi_b(x) = (e^{-i\vec{\beta}(x)\cdot\vec{L}_\psi})_{ab}\psi_b(x) \tag{2.1a}$$

$$\phi_a(x) \rightarrow U_\phi(\beta(x))_{ab}\phi_b(x) = (e^{-i\vec{\beta}(x).\vec{L}_\phi})_{ab}\phi_b(x) \tag{2.1b}$$

where L^i_ψ, L^i_ϕ are the appropriate representation matrices for the ψ and ϕ fields, e.g. Pauli matrices for internal spin equal to one half. The problem for local gauge symmetries is that $\partial_\mu\phi_a$ does not transform under the gauge group G as ϕ_a does because of $\partial_\mu\beta^a(x)$ terms so $(\partial_\mu\phi)^\dagger_a(\partial^\mu\phi)_a$ is not gauge invariant. This requires, just as in ordinary electromagnetism, the introduction of gauge fields, complicated here by the non-abelian nature of the symmetry.(2.1)

The solution is to replace the usual derivative

$$\partial_\mu\phi_a(x) \rightarrow \partial_\mu(U_{ab}(x)\phi_b(x)) = \partial_\mu U(x)_{ab}\phi_b(x) + U_{ab}(x)\partial_\mu\phi_b(x) \tag{2.2}$$

in the Lagrangian by the gauge covariant derivative

$$\begin{aligned}(D_\mu\phi)_a &= (D_\mu)_{ab}\phi_b = (\partial_\mu - ig\vec{A}_\mu.\vec{L})_{ab}\phi_b \\ &= (\partial_\mu\delta_{ab} - igA^i_\mu L^i_{ab})\phi_b\end{aligned} \tag{2.3}$$

where the gauge fields A^i_μ transform to cancel the $\partial_\mu U_{ab}$ piece in (2.2)

$$A^i_\mu(x)L^i \rightarrow A'^i_\mu(x)L^i = UA^i_\mu(x)L^iU^{-1} - i/g(\partial_\mu U)U^{-1} \tag{2.4}$$

so that

$$\begin{aligned}D_\mu\phi &\rightarrow U\partial_\mu\phi + \partial_\mu U\phi - igUA^i_\mu L^iU^{-1}U\phi - ig(-i/g\ \partial_\mu UU^{-1})U\phi \\ &= U\partial_\mu\phi - igUA^i_\mu L^i\phi = UD_\mu\phi\end{aligned} \tag{2.5}$$

The infinitesimal transformation properties of the A^i_μ, which are non-trivial even for global transformations are obtained by letting $\vec{\beta}(x) \rightarrow 0$ in (2.4), and keeping only terms of zeroth and first order in $\vec{\beta}(x)$. We find

$$A^i_\mu \rightarrow A^i_\mu + C_{ijk}\beta^jA^k_\mu - 1/g\ \partial_\mu\beta^i \tag{2.6}$$

with the totally antisymmetric constant C_{ijk} given by the jk element of the matrix L_i in the adjoint representation.

$$-iC_{ijk} = (L^i)_{jk} \tag{2.7}$$

These are the structure functions of the algebra.[2.2] The gauge invariant kinetic term for the A^i fields is easily found

$$\mathcal{L}_{KIN} = -1/4\ F^i_{\mu\nu}F^{i\mu\nu} \tag{2.8}$$

where

$$F^i_{\mu\nu} = \partial_\mu A^i_\nu - \partial_\nu A^i_\mu + gC_{ijk}A^j_\mu A^k_\nu \tag{2.9}$$

the last term being due of course to the non Abelian nature of the symmetry.

$$[T^i, T^j] = iC_{ijk}T^k \tag{2.10}$$

The representation matrices of the group satisfy the same algebra of course

$$[L^i, L^j] = iC_{ijk}L^k \tag{2.11}$$

(remember T^i are operators and L^i are n x n matrices for a representation of dimension n). The L can always be defined so that

$$\mathrm{Tr}(L^iL^j) = T(L)\delta^{ij} \tag{2.12}$$

and then (2.8) can be rewritten as

$$\mathcal{L}_{KIN} = \frac{-1}{4T(L)}\ \mathrm{Tr}(F^i_{\mu\nu}L^iF^{j\mu\nu}L^j) \tag{2.13}$$

$\mathcal{L}_{KIN}$ in (2.9) or (2.13) obviously also includes cubic and quartic gauge field self-coupling terms.

The most general gauge invariant Lagrangian involving spin zero, spin 1/2 and gauge fields is of the form

$$\mathcal{L} = -1/4\ F^i_{\mu\nu}F^{i\mu\nu} + \bar{\psi}(i\not{D} - m)\psi + 1/2(D^\mu\phi)(D_\mu\phi) + \bar{\psi}\Gamma\phi\psi - V(\phi) \tag{2.14}$$

where m and Γ are matrices and $V(\phi)$ is a polynomial. They are chosen so that $\mathcal{L}$ is invariant under G (invariance under global transformations is sufficient for these terms since they do not involve derivatives). Finally $V(\phi)$ must not have terms in ϕ of power higher than four by the requirement of renormalizability. Note that scalar bosons can have masses and so can fermions unless the gauge group is chiral and forbids terms like $m\bar{\psi}_L\psi_R$. The gauge bosons do however have to be massless since a mass term

$$M^2_{ij} A^i_\mu A^{j\mu} \tag{2.15}$$

destroys local gauge invariance, which is necessary for renormalizability.

III. SPONTANEOUS SYMMETRY BREAKING

a) Introduction

This is the ingenious means by which gauge bosons acquire a mass(3.1) without destroying the renormalizability of the theory. (3.2) Even if the notions involved do not survive, there is a general feeling that the electroweak theory will behave in several ways as if the Higgs mechanism was responsible for SSB.

Let $V(\phi)$ be such that the invariance under G is broken by some of the fields ϕ_a acquiring non-zero VEV. The simplest case in which this occurs is for the function of a single scalar field

$$V(\phi) = \mu^2/2\ \phi^2 + \lambda/4\ \phi^4 \tag{3.1}$$

For $\mu^2 < 0$ the minimum of the potential is not when the VEV of ϕ equals zero (3.3), but rather $\langle 0|\phi|0\rangle = \pm(-\mu^2/\lambda)^{1/2}$. In general let ϕ be an n component real field with a general VEV ($|0\rangle$ is the physical vacuum state)

$$v \equiv \langle 0|\phi|0\rangle = \begin{pmatrix} v_1 \\ v_n \end{pmatrix} \tag{3.2}$$

such as might be the case for an O_n (orthogonal group in n dimensions) generalization of $V(\phi)$ in (3.1) to

$$V(\phi) = \mu^2/2\ \phi_a\phi_a + \lambda/4(\phi_a\phi_a)^2 \qquad a = 1\ldots n \tag{3.3}$$

The condition that this be an extremum of $V(\phi)$ is

$$\left.\frac{\partial V}{\partial \phi_a}\right|_{\phi=v} = 0 \tag{3.4}$$

Rewriting the Lagrangian in terms of shifted fields $\hat{\phi}_a(x) = \phi_a(x)-v_a$ which have no VEV we find mass terms appearing in $\mathcal{L}$ as

$$\mathcal{L}_{MASS} = -1/2\ \mu^2_{ab}\hat{\phi}_a\hat{\phi}_b - 1/2\ M^2_{ij}A^i_\mu A^{j\mu} - \bar{\psi}M\psi \tag{3.5}$$

with μ^2_{ab}, M^2_{ij} and M given by

$$\mu^2_{ab} = \left.\frac{\partial^2 V(\hat{\phi})}{\partial\hat{\phi}_a\partial\hat{\phi}_b}\right|_{\hat{\phi}=0} = \left.\frac{\partial^2 V(\phi)}{\partial\phi_a\partial\phi_b}\right|_{\phi=v} \tag{3.6a}$$

$$M^2_{ij} = g^2 v^T L^i L^j v = g^2\langle v|L^i L^j|v\rangle = g^2\langle L^i v|L^j v\rangle \tag{3.6b}$$

$$M = m - \Gamma_a v_a \tag{3.6c}$$

where the M^2_{ij} term comes entirely from the $1/2\ D^\mu\phi D_\mu\phi$ term in (2.14) with ϕ replaced by its VEV. The change in the fermion mass term is obvious. We will not discuss in these lectures the question of Feynman diagram calculations in these theories, nor will we address the issue of renormalizability.

b) Goldstone Bosons

$V(\phi)$ is invariant under G, i.e. $UVU^{-1} = V$. For an infinitesimal transformation we find

$$0 = \delta V(\phi) = \frac{\partial V}{\partial\phi_a}\delta\phi_a = -i\frac{\partial V}{\partial\phi_a}\vec{\beta}\cdot\vec{L}_{ab}\phi_b \tag{3.7}$$

$$\Rightarrow 0 = \frac{\partial V}{\partial\phi_a}L^i_{ab}\phi_b$$

Taking the derivative of (3.7) with respect to ϕ_c and evaluating it at $\phi = v$, we find

$$\left.\frac{\partial^2 V}{\partial\phi_a\partial\phi_c}L^i_{ab}\phi_b\right|_{\phi=v} + \left.\frac{\partial V}{\partial\phi_a}L^i_{ab}\delta_{bc}\right|_{\phi=v} = 0 \tag{3.8}$$

Using (3.4) and (3.6a),

$$\mu^2_{ab}(L^i v)_b = 0 \tag{3.9}$$

Let S be the M dimensional subgroup (of the N-dimensional invariance group G) which is left unbroken by SSB. For convenience choose its generators as the last M of the N generators of G. We then have

$$(L^i v) \neq 0 \qquad i = 1 \ldots N\text{-}M$$

$$(L^i v) = 0 \qquad i = N\text{-}M+1 \ldots N \tag{3.10}$$

μ^2_{ab} is an n x n symmetric matrix with n eigenvalues. (3.9) and (3.10) imply that for each of the linearly independent N-M eigen-vectors $(L^i v) \neq 0$, μ^2_{ab} will have a vanishing eigenvalue. These zero mass meson states are the so-called Nambu-Goldstone(3.4) bosons. In general, we also expect n - (N-M) non-zero mass eigenstates of μ^2, which correspond to ordinary massive scalar mesons. Note that $V(\vec{\Phi})$ involves no derivatives so everything we have done so far holds as well for global symmetries ($\vec{\beta}$ independent of x) as well as local symmetries. Of course, it may be that N=M in which case the n eigen-values of μ^2 are all non-zero, i.e. there are no Nambu-Goldstone bosons.

c) Higgs Mechanism

Massless scalar particles, the zero mass eigenstates of μ^2 are not seen in nature. Happily, the gauged symmetry provides a means for reconciling this experimental fact with the prediction of the Goldstone theorem(3.4) for SSB. To see how this occurs let us re-express the n quantum fields as N-M Nambu-Goldstone fields ξ^i, $i = 1 \ldots N\text{-}M$, which are massless and p = n - (N-M) ordinary Higgs fields(3.1) which we call $\eta \ldots, \eta_p$. We may then rewrite $\Phi(x)$ as

$$\Phi(x) = e^{i\sum_{i=1}^{N-M}\xi^i L^i}(v + \eta) = e^{i\sum_{i=1}^{N-M}\xi^i L^i}\begin{pmatrix} 0 \\ \vdots \\ v_1 + \eta_1 \\ \vdots \\ v_p + \eta_p \end{pmatrix} \tag{3.11}$$

Let us explain this a little more. Remembering (3.10) we have brought the VEV by a set of group rotations to the form where the non-zero entries are the last p, $v_1 \ldots v_p$. The accompanying p quantum fields $\hat{\Phi}_1 \ldots \hat{\Phi}_p$ have been called $\eta_1 \ldots \eta_p$; finally the first N-M fields have been parametrized as exponentials. As a trivial example, let G be SU_2 or O_3 with ϕ the triplet representation; by a rotation the VEV of ϕ can be chosen along the third direction so

$$\langle 0 | \phi | 0 \rangle = \begin{pmatrix} 0 \\ 0 \\ v \end{pmatrix} \tag{3.12}$$

Now parametrize ϕ as (the exponential looks neater with a $1/v$ in this case)

$$\phi = e^{i/v\,(\xi^1 L^1 + \xi^2 L^2)} \begin{pmatrix} 0 \\ 0 \\ v+\eta \end{pmatrix} = \begin{pmatrix} \xi^1 \\ \xi^2 \\ v+\eta \end{pmatrix} + \begin{matrix}\text{higher} \\ \text{order} \\ \text{terms}\end{matrix} \qquad (3.13)$$

so $\xi_{1,2}$ would be the Goldstone bosons. We now perform a local gauge transformation on (3.11) with U equal to

$$U(\xi) = e^{-i \sum_{i=1}^{N-M} \xi^i L^i} \qquad (3.14)$$

under which

$$\phi \to \phi' = U\phi = v + \eta \qquad (3.15a)$$

$$A^i_\mu \to A^i_\mu - \frac{1}{g}\partial_\mu \xi^i + 0(\xi^2) \qquad (3.15b)$$

i.e. ϕ loses N-M of its components. The gauge boson mass matrix is now calculated from $1/2 \; D_\mu\phi \, D^\mu\phi$, the kinetic term in (2.14)

$$\frac{1}{2} D_\mu\phi D^\mu\phi = \frac{1}{2} (v + \eta)^T (\partial_\mu + igA^i_\mu L^i)(\partial^\mu - igA^{j\mu} L^j)(v + \eta)$$

$$= \frac{1}{2} v^T L^i L^j v \; A^i_\mu A^{j\mu} + \ldots\ldots$$

$$= \left(\frac{M^2}{2}\right)^{ij} A^i_\mu A^{j\mu} + \ldots\ldots \qquad (3.16)$$

Since (3.10) tells us that $L^i v) \neq 0$ for the first (N-M) components, M^2 is an (N-M) x (N-M) matrix with in general non-zero eigenvalues To sum it up, N-M Higgs fields have disappeared, and N-M gauge fields have acquired masses, or alternatively speaking, longitudinal components.

d) Pseudo-Goldstone Bosons

This is an important topic in many recent applications of SSB. In particles the existence of so-called pseudo-Goldstone bosons(3.5) is predicted by many models of dynamical symmetry breaking(3.6). To see what they are and how they arise in a conventional model, let us return to our examination of the potential $V(\phi)$ and consider the case in which the invariance of V, $\bar{G}$, is larger than the gauged invariance group G of the Lagrangian.

We may impose directly a larger global symmetry on V, but it may also be that imposing discrete symmetries on $\mathcal{L}$ is sufficient to give V(ϕ) additional global symmetries. Now when SSB takes place, we may break some of the global symmetries of V which are contained in $\bar{G}$, but not in G. In zeroth order we would classify the associated particles as real Goldstone bosons since they are not "eaten" by the gauge bosons of G. In higher order, however, since $\bar{G}$ is not a symmetry of $\mathcal{L}$, they acquired a mass and hence are called pseudo-Goldstone bosons.

These additional symmetries make the choice of V due to SSB ambiguous. Given a group element g in G with representation matrix D(g) D(g)v is equivalent to v, but D(g)v is not if g lies in G, but not in G. The question of which is the true minimum of the potential has been solved by Weinberg(3.5) who shows that the condition is that the one loop potential be minimized at the physical value of v.

The mass matrix of the pseudo-Goldstone bosons has been calculated by Weinberg(3.5) and is expressed in terms of derivatives with respect to v of the one loop potential evaluated at v.

It would be inappropriate in introductory lectures such as these to go into this formalism, but we would like to conclude this section with an example (which we shall return to again later in VI and VII) of a V(ϕ) with a pseudo-Goldstone boson. Let G be $SU_2 \times U_1$ and let V be a function of two isodoublets ϕ_1 and ϕ_2. In addition we demand that $\mathcal{L}$ be invariant under the global symmetry X (α is not a function of x)

$$\phi_1 \to e^{ix_1\alpha}\phi_1$$

$$\phi_2 \to e^{ix_2\alpha}\phi_2 \qquad (3.17)$$

Let us spell out the global symmetry question in a little more detail. Associated with the invariance (3.17) are two linearly independent conserved currents

$$J_{x_{1,2}} = x_1\phi_1^\dagger\partial^\mu\phi_1 + x_2\phi_2^\dagger\partial^\mu\phi_2 \qquad (3.18)$$

With associated charges $Q(x_1, x_2)$. Both symmetries are broken spontaneously if ϕ_1 and ϕ_2 acquire non zero VEV's and so, by the Goldstone theorem, we would expect two Goldstone bosons. Q(1/2,1/2) coincides however with the gauged U_1 so one Goldstone boson will be eaten by the Higgs mechanism.

The most general Higgs potential consistent with (3.17) is (3.7)

$$V(\phi_1,\phi_2) = \lambda_1(\phi_1^\dagger\phi_1 - v_1^2)^2 + \lambda_2(\phi_2^\dagger\phi_2 - v_2^2)^2 + \lambda_3((\phi_1^\dagger\phi_1 - v_1^2) + (\phi_2^\dagger\phi_2 - v_2^2))^2 + \lambda_4((\phi_1^\dagger\phi_1)(\phi_2^\dagger\phi_2) - (\phi_1^\dagger\phi_2)(\phi_2^\dagger\phi_1)) \tag{3.19}$$

Choosing all of λ_i positive and $\phi_{1,2} = (\phi_{1,2}^+,\phi_{1,2}^0)$ we readily see the minimum of $V(\phi_1,\phi_2)$

$$<0|\phi_1|0> = \frac{1}{\sqrt{2}}\begin{pmatrix}0\\ v_1\end{pmatrix} \qquad <0|\phi_2|0> = \frac{1}{\sqrt{2}}\begin{pmatrix}0\\ v_2\end{pmatrix} \tag{3.20}$$

The global symmetry has forbidden terms like $(\phi_1^\dagger\cdot\phi_2)(\phi_1^\dagger\cdot\phi_2)$ from appearing in V. We can now parametrize the Higgs fields as

$$\phi_1 = \begin{pmatrix}\phi_1^+\\ v_1+\eta_1+i\xi_1\end{pmatrix} \qquad \phi_2 = \begin{pmatrix}\phi_2^+\\ v_2+\eta_2+i\xi_2\end{pmatrix} \tag{3.21}$$

Let us focus on the neutral particles; their contribution to V is seen by using (3.21) in (3.19)

$$V_{neutral} = \lambda_1(2\eta_1 v_1 + \eta_1^2 - \xi_1^2)^2 + \lambda_2(2\eta_2 v_2 + \eta_2^2 - \xi_2^2)^2 + \lambda_3((2\eta_1 v_1 + \eta_1^2 - \xi_1^2) + (2\eta_2 v_2 + \eta_2^2 - \xi_2^2))^2 \tag{3.22}$$

so the terms in V which are quadratic in the fields, i.e. the neutral meson mass matrix entries, are

$$4\eta_1^2 v_1^2(\lambda_1+\lambda_3) + 4\eta_2^2 v_2^2(\lambda_2+\lambda_3) + 8\eta_1\eta_2 v_1 v_2 \lambda_3 \tag{3.23}$$

ξ_1 and ξ_2 are massless! A term such as $\lambda_5|\phi_1^\dagger\phi_2 - v_1 v_2|^2$, forbidden by X, does not shift the minimum of V, but gives a mass term to the ξ's of

$$\lambda_5(v_1\xi_2 - v_2\xi_1)^2 \tag{3.24}$$

so labelling

$$G_0 = \frac{v_1\xi_1 + v_2\xi_2}{\sqrt{v_1^2 + v_2^2}} \qquad P_0 = \frac{v_1\xi_2 - v_2\xi_1}{\sqrt{v_1^2 + v_2^2}} \qquad (3.25)$$

We have G_0 is the true Goldstone boson, i.e. remains massless, while P_0 is massless only because of the X global symmetry. It will acquire a mass if X is not a symmetry of $\mathcal{L}$ as a whole.

The analysis is readily extended to the charged Higgs mesons. We find

$$H^{\pm} = \frac{v_1\phi_2^{\pm} - v_2\phi_1^{\pm}}{\sqrt{v_1^2 + v_2^2}} \qquad (3.26)$$

acquires a mass, proportional to λ_4, while the orthogonal combination remains massless. When the symmetry is gauged the massless charged boson will be eaten by $W^{\pm}$.

IV. THE STANDARD MODEL

a) Couplings

The so-called standard model is based on the gauged group $SU_2 \times U_1$ where SU_2 and U_1 are respectively referred to as weak isospin T^i and weak hypercharge Y. The transformation properties of the fields and the SSB are chosen such that electric charge Q equals $T^3 + Y$ and the symmetry associated with $T^3 + Y$ is unbroken. This fact ensures the masslessness of the photon, by the analysis of the previous section. The fact that $SU_2 \times U_1$ is not a simple group, but a product of simple groups, means that the gauge bosons of SU_2 and U_1 couple with strengths g,g', where g'/g is unconstrained. (In that sense $SU_2 \times U_1$ is not a truly "unified" theory of electromagnetic and weak interaction.) Calling A_μ^i and B_μ the gauge bosons of SU_2 and U_1 respectively, the pure bosonic part of $\mathcal{L}$ is

$$\mathcal{L}_{\text{bosonic}} = -\frac{1}{4} F^i_{\mu\nu} F^{i\mu\nu} - \frac{1}{4} B_{\mu\nu} B^{\mu\nu} + (D_\mu\phi)^\dagger D^\mu\phi - V(\phi) \qquad (4.1)$$

with ϕ being an isospin doublet

$$\phi = \begin{pmatrix} \phi^+ \\ \phi^0 \end{pmatrix} \qquad (4.2)$$

We see that the indicated charge assignment of the fields requires that $Y_\phi = 1/2$ (we could e.g. equally well have picked $Y_\phi = 3/2$ so that the fields would have been ϕ^{++} and ϕ^{+} with $Q = 1,2$, but then a VEV for either of the fields would have violated charge conservation.) It will shortly turn out to be useful to employ not only $\phi^\dagger$, but

$$\tilde{\phi} = i\tau^2\phi^* = \begin{pmatrix} \bar{\phi}^0 \\ -\phi^- \end{pmatrix} \tag{4.3}$$

which of course has isospin one half and Y equal to minus one half.

With these assignments, remembering that the j k elements of the representation matrices for isospin in the spin one and spin one half representation are ε^{ijk} and $(\tau^i/2)_{jk}$, we have with

$$F^i_{\mu\nu} = \partial_\mu A^i_\nu - \partial_\nu A^i_\mu + g\varepsilon^{ijk} A^j_\mu A^k_\nu$$

$$B_{\mu\nu} = \partial_\mu B_\nu - \partial_\nu B_\mu$$

$$D_\mu \phi = (\partial_\mu - ig \frac{\tau^i A^i_\mu}{2} - i \frac{g' B_\mu}{2})\phi$$

$$V(\phi) = 1/2\, \mu^2 (\phi^\dagger \phi) + \lambda/4\, (\phi^\dagger \phi)^2 \tag{4.4}$$

In order to discuss the fermions we must distinguish left and right handed fermions which we label respectively as ψ_L and ψ_R.

$$\psi_{L,R} = (\frac{1 \pm \gamma_5}{2})\, \psi \tag{4.5}$$

Some of the fermions are quarks which we will label as $p_{mL,R}$ and $n_{mL,R}$ reserving the various u,d,c,s, etc. for the mass eigenstates. We will drop the color indices at this time, but keep the family index m, where m = 1...F. Left handed quarks and leptons are assigned to SU_2 doublets with hypercharges of respectively 1/6 and 1/2, designated by

$$q_{mL} = \begin{pmatrix} p_m \\ n_m \end{pmatrix}_L \qquad \ell_{mL} = \begin{pmatrix} \nu_m \\ e_m \end{pmatrix}_L \tag{4.6}$$

It is of course clear that two of the four vector bosons will be charged and two neutral. The charged ones will be identified with the $W^\pm$. The evidence available points to purely left handed couplings of the $W^{\pm}$, i.e. charged currents of the form

$$J^{\pm}_{\mu} = \bar{\psi}_L \gamma_\mu \tau^{\pm} \psi_L \tag{4.7}$$

with ψ_L being in weak isospin doublets. The absence of charged right handed currents leads to the assigning of p_{mR}, n_{nR}, e_{mR} to isospin singlets so Y coincides with Q, i.e. is equal to 2/3, -1/3 and -1 respectively. For the moment we also exclude ν_R fields. The kinetic part of the fermion Lagrangian

$$\mathcal{L}_{\text{fermion kin.}} = \sum_{m=1}^{F} \{\bar{q}_{mL} i \not{\partial} q_{mL} + \bar{\ell}_{mL} i\not{\partial}\ell_{mL} + \bar{p}_{mR} i\not{\partial} p_{mR}$$

$$+ \bar{n}_{mR} i\not{\partial} n_{mR} + \bar{e}_{mR} i\not{\partial} e_{mR} \} \tag{4.8}$$

is then uniquely specified with, e.g.

$$D_\mu q_{nL} = (\partial_\mu - ig \frac{\tau^i A^i_\mu}{2} - i \frac{g' B_\mu}{6}) q_{nL} \tag{4.9a}$$

$$D_\mu p_{mR} = (\partial_\mu - ig' \frac{2}{3} B_\mu) p_{mR} \tag{4.9b}$$

With this classification of particles, no bare mass terms are allowed for fermions. To spell it out $m\bar{\psi}_L\psi_L$, $m\bar{\psi}_R\psi_R$ vanish by chirality and $m\bar{\psi}_L\psi_R$ is not allowed by weak isospin invariance since the left handed particles are in SU_2 doublets and the right handed particles in SU_2 singlets. Fermion masses arise, after SSB, from the only allowed term in the Lagrangian we have not discussed so far, the Yukawa coupling

$$\mathcal{L}_{\text{Yukawa}} = \sum_{m,n=1}^{F} \{\bar{q}_{mL} \Gamma^d_{mn} n_{nR} \phi + \bar{\ell}_{mL} \Gamma^\ell_{mn} e_{nR} \phi + \bar{q}_{mL} \Gamma^u_{mn} p_{nR} \tilde{\phi} + \text{h.c.}\} \tag{4.10}$$

where $\Gamma^{d,u,\ell}$ are arbitrary F x F matrices. The joint requirements of invariance and renormalizability forbid the appearance of other terms. The fermion eigenstates are found by replacing ϕ by its VEV in (4.10) and then diagonalizing the mass matrix which results. The mixings which this procedure leads to will be discussed in a later chapter.

b) Gauge Boson Masses

The complex scalar field ϕ may be decomposed into its real and imaginary parts $\phi = \phi_1 + i\phi_2$, where ϕ_1 and ϕ_2 are real isospin doublets

$$\phi^\dagger\phi = \phi_1\phi_1 + \phi_2\phi_2 \tag{4.11}$$

so we see that the potential $V(\phi)$ in (4.4) has in fact an O(4) invariance under rotation of the four Higgs fields. By a global transformation then any VEV for ϕ may be rotated to lie along a specified direction, e.g.

$$<0|\phi_2|0> = \begin{pmatrix}0\\0\end{pmatrix} \qquad <0|\phi_1|0> = \begin{pmatrix}0\\v_o\end{pmatrix} \tag{4.12}$$

with v_o real because ϕ_1 is Hermitian. V(v) is then of the form

$$V(v_o) = \mu^2/2\ v_o^2 + \lambda/4\ v_o^4 \tag{4.13}$$

so that for $\mu^2<0$, the minimum of the potential is at $v_o = (-\mu^2/\lambda)^{1/2}$ (of course for $\mu^2>0$, the minimum is $v_o = 0$). We see, following (3.10) that

$$T^i v \neq 0 \qquad Yv \neq 0 \qquad (T^3 + Y)v = 0 \tag{4.14}$$

Here v denotes the column vector $<\phi>_o$ and v_o is just a number defined in (4.12). $(T^3 + Y)v = 0$ means that Q is left unbroken and the photon, the associated gauge boson, will remain massless. The Higgs field can be re-parametrized as in Section IIIc

$$\phi = e^{\sum_{i=1}^{3} L^i\xi^i} \begin{pmatrix} 0 \\ \dfrac{v_o + \eta}{\sqrt{2}} \end{pmatrix} \tag{4.15}$$

where $L^{1,2} = T^{1,2}$, $L^3 = T^3 - Y$; after a gauge transformation all we have left is one real physical field, η. The other three have been "eaten" by the three gauge bosons coupling to $T^{1,2}$ and $T^3 - Y$. The mass of η is read off from the Higgs potential

$$\begin{aligned} V(v_o + \eta) &= \frac{\mu^2}{2}(v_o + \eta)^2 + \frac{\lambda}{4}(v_o + \eta)^4 \\ &= -\frac{\mu^2}{4\lambda} - \mu^2\eta^2 + \lambda v_0\eta^3 + \frac{\lambda}{4}\eta^4 \end{aligned} \tag{4.16}$$

So η has a mass $-2\mu^2 > 0$. As in (3.6b), the gauge boson mass matrix M^2_{ij}, (modified by presence of two rather than one coupling constant since $SU_2 \times U_1$ is semi-simple) is read off from

$$D_\mu\phi^\dagger D^\mu\phi\Big|_{\phi=v} = v^T \{g \frac{\tau^i A^i_\mu}{2} + \frac{g'}{2} B_\mu\} \{ g \frac{\tau^j A^{j\mu}}{2} + \frac{g'}{2} B^\mu \} v$$

$$= \frac{1}{2} (0, v_o) (g \frac{\tau^i A^i_\mu}{2} + \frac{g'}{2} B_\mu)^2 \begin{pmatrix} 0 \\ v_o \end{pmatrix}$$

$$= \frac{v_o^2}{2} (0,1) \{ \frac{g^2}{2} \tau^- \tau^+ A^+_\mu A^{-\mu} + (- \frac{g}{2} A^3_\mu + \frac{g'}{2} B_\mu)^2 \} \begin{matrix} 0 \\ 1 \end{matrix} \tag{4.17}$$

where $A^\pm_\mu = \frac{1}{\sqrt{2}} (A^1_\mu \mp iA^2_\mu)$ is usually labelled $W^\pm_\mu$. We then have as gauge boson mass terms

$$M_W^2 W^+_\mu W^{-\mu} + \frac{M_Z^2}{2} Z_\mu Z^\mu \tag{4.18}$$

with Z_μ defined by

$$Z_\mu = - \frac{gA^3_\mu + g'B_\mu}{\sqrt{g^2 + g'^2}} = - \cos \theta_W A^3_\mu + \sin \theta_W B_\mu$$

$$\tan \theta_W = g'/g \tag{4.19}$$

The orthogonal combination is the photon

$$A_\mu = \frac{g'A^3_\mu + gB_\mu}{\sqrt{g^2 + g'^2}} = \sin \theta_W A^3_\mu + \cos \theta_W B_\mu \tag{4.20}$$

which of course is massless, associated with the unbroken $U_1^{e.m.}$ subgroup generated by $T^3 + Y$. M_W and M_Z are given by

$$M_W^2 = \frac{g^2 v_o^2}{4} \qquad M_Z^2 = (g^2 + g'^2) \frac{v_o^2}{4} \tag{4.21}$$

In particular note the identity

$$M_W^2 = M_Z^2 \cos^2 \theta_W \tag{4.22}$$

In terms of $W_\mu^\pm$, Z_μ and A_μ, the gauge couplings are

$$\frac{g}{2\sqrt{2}} (J_W^\mu W_\mu^- + J_W^{\mu\dagger} W_\mu^+) + g\sin\theta_W J_{e.m}^\mu A_\mu - \frac{g}{2\cos\theta_W} J_Z^\mu Z_\mu \tag{4.23}$$

where

$$J_W^\mu = \sum_{m=1}^{F} \{\bar{e}_m\gamma^\mu(1+\gamma_5)\nu_m - \bar{n}_m\gamma^\mu(1+\gamma_5)p_m\} \tag{4.24a}$$

$$J_{e.m}^\mu = \sum_{m=1}^{F} \{2/3\ \bar{p}_m\gamma^\mu p_m - 1/3\ \bar{n}_m\gamma^\mu n_m - \bar{e}_m\gamma^\mu e_m\} \tag{4.24b}$$

$$J_Z^\mu = \sum_{m=1}^{F} \{\bar{\psi}_m\gamma^\mu T^3(1+\gamma_5)\psi_m\} - 2\sin^2\theta_W J_{e.m}^\mu \tag{4.24c}$$

This gives us an effective low momentum four Fermi charged current interaction

$$\mathcal{L}_{eff.} = \frac{G_F}{2\sqrt{2}}(J_W^\mu J_{W\mu}^\dagger + J_W^{\mu\dagger} J_{W\mu}) \tag{4.25}$$

with

$$\frac{G_F}{\sqrt{2}} = \frac{g^2}{8\, M_W^2} = \frac{1}{2\, v_o^2} \tag{4.26}$$

and fixes

$$e = g \sin\theta_W \tag{4.27}$$

In terms of the arbitrary mixing angle θ_W we then have

$$M_W = \frac{gv_0}{2} = \sqrt{\frac{\pi\alpha}{\sqrt{2}G_F}}\ \frac{1}{\sin\theta_W} = \frac{37.3 \text{ Gev}}{\sin\theta_W}$$

$$M_Z = \frac{M_W}{\cos\theta_W} = \frac{74.6 \text{ Gev}}{\sin 2\theta_W} \qquad (4.28)$$

with higher order corrections modifying these predictions so that the numerators are 38.53 Gev and 77.06 Gev.[(4.1),(4.2)] The value of $\sin^2\theta_W$ is determined by measurements of the neutral current[(4.3)] For recent reviews of non-leptonic weak interactions see e.g. refs. (4.4), (4.3).

V. ANOMALIES

We will be brief on this subject. The diagrams in which three gauge bosons couple to a fermion triangle can lead to divergences which must be cancelled. Alternatively one could say that the Compton scattering amplitude of fig.1 will violate unitarity bounds at high energy. These are called anomalies, because, when computing the divergence of an axial current, these terms do not appear by formal manipulation of the equations of motion.[(5.1)] They are particularly important because the renormalizability of the theory then places the important restriction on the particle spectrum that it be such that the anomalies cancel.[(5.2)]

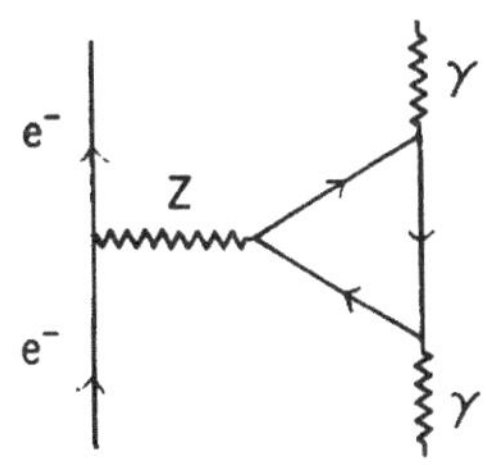

Fig. 1. Compton scattering amplitude with possible anomaly.

To see this let us label left and right handed fermions as $\psi_{L,R}$ and the respective matrices which couple them to gauge boson A^i_μ as $\Lambda^i_{L,R}$, where the Λ's are hermitian matrices

$$\mathcal{L}_{int.} = gA^i_\mu \{ \bar{\psi}_L\gamma^\mu\Lambda^i_L\psi_L + \bar{\psi}_R\gamma^\mu\Lambda^i_R\psi_R\} \tag{5.1}$$

Since $\Lambda^i_{L,R}$ are representation matrices of the gauge group G, they satisfy

$$[\Lambda^i_L,\Lambda^j_L] = ic_{ijk}\Lambda^k_L \qquad [\Lambda^i_R,\Lambda^j_R] = ic_{ijk}\Lambda^k_k \tag{5.2}$$

with, in general, $\Lambda_L \neq \Lambda_R$ since left and right handed fermions transform differently.

The anomaly, which appears in the axial vector-vector-vector coupling is easily shown to be proportional to $D^{ijk}_L - D^{ijk}_R = \Delta^{ijk}$, where

$$D^{ijk}_{L,R} = \mathrm{tr}\ \{\{\Lambda^i_{L,R},\Lambda^j_{L,R}\}\Lambda^k_{L,R}\} \tag{5.3}$$

and $D^{ijk}_{L,R}$ are of course totally symmetric matrices in the i,j,k indices. Δ can of course be zero because D_L and D_R cancel or because they vanish separately. The former are known as vector like theories, SU_3 of color being an obvious example. Theories with only real representations, i.e. in which $\Lambda_{L,R}$ and their complex conjugates are related by a unitary transformation

$$\Lambda^i = -U^\dagger\Lambda^{i*}U \tag{5.4}$$

are automatically anomaly free. This is seen by

$$\begin{aligned} D^{ijk} &= \mathrm{tr}\ \{\{\Lambda^i,\Lambda^j\},\Lambda^k\} = \mathrm{tr}\ \{\{\Lambda^i,\Lambda^j\},\Lambda^k\}^\dagger \\ &= \mathrm{tr}\ \{\{\Lambda^{i*},\Lambda^{j*}\}\ ,\Lambda^{k*}\} \\ &= -\ \mathrm{tr}\ \{\{\Lambda^i,\Lambda^j\}\ ,\Lambda^k\} = -\ D^{ijk} \\ &\Rightarrow D^{ijk} = 0 \end{aligned} \tag{5.5}$$

The only simple groups with anomalies are SU_n with $n \geqslant 3$ (only SU_n, SO_{4n+2} and E_6 have complex representations and for the latter two D^{ijk} vanishes anyway). For SU_2 the anomalies vanish (for e.g. the τ^i matrices, $U = i\tau_2$), so the only anomalies in $SU_2 \times U_1$

turn out to be those caused by the presence of the U_1. Labelling by Y the U_1 generator we must, as an example, have

$$tr\ (Y_L)^3 - tr\ (Y_R)^3 = 0 \tag{5.6}$$

A certain number of identities now prove useful; first of all we know that $Q = T_3 + Y$ and second that $Q_L = Q_R$. The absence of anomalies for $^3SU_2$ means that

$$tr\ (T_L^3)^3 = tr\ (T_R^3)^3 = 0 \tag{5.7}$$

Rewriting $Y_{L,R}$ in terms of $Q_{L,R} = Q$ in (5.6) we find that anomaly cancellation requires

$$tr\ (T_L^2 Q) - tr\ (T_R^2 Q) = 0 \tag{5.8}$$

For the standard model we place left handed particles in isospin doublets and right handed particles in isospin singlets ($T_R^3 = 0$) so the lepton doublet contribution to (5.8) is 1/4 x -1, while the quark contribution is 1/4 (2/3 - 1/3) x 3 so they cancel. Note of course the essential factor of three that comes in because we must count all three colors of quarks. This is of course one more example of the connection between SU_3^c and $SU_2 \times U_1$.

The anomalies need not cancel if they are associated with global rather than gauged symmetries since the anomalous current does not then have a gauge boson coupled to it. A well known example of this is the hadronic U_1 axial vector current for which there is the Adler-Bell-Jackiw anomaly.

$$\partial^\mu j_{\mu 5} = \frac{N\, g_c^2}{16\, \pi^2}\, \varepsilon^{\mu\nu\lambda\sigma} F^\alpha_{\mu\nu} F^\alpha_{\lambda\sigma} \tag{5.9}$$

Here N is the number of quark flavors and $F^\alpha_{\mu\nu}$ is the SU_3^c gauge field tensor, and g_c is the strong coupling constant.

VI. THE FERMI MASS MATRIX

a) Mixing Angles and Phases

We assume the conventional model in which left-handed quarks are in isospin doublets $\psi_{mL} = \begin{pmatrix} p_m \\ n_m \end{pmatrix}_L$ and right-handed quarks are in isospin singlets, p_{nR}, n_{nR}, with m, n=1...F the family index. Assuming that there are K Higgs isodoublets which couple to fermions, we have as Yukawa interaction

$$\mathcal{L}_{\text{Yukawa, quarks}} = \sum_{m,n=1}^{F} \{ \bar{\psi}_{mL}\Gamma_{a,mn} n_{nR}\phi_a + \bar{\psi}_{mL}\Gamma'_{a,mn} p_{nR}\tilde{\phi}_a \} \tag{6.1}$$

with a = 1...K and $\tilde{\phi}_a = i\tau_2\phi_a^\dagger$. Γ and Γ' are arbitrary F x F matrices. The physical quark fields, namely the mass eigenstates, are labelled by u_m, d_m. They are related to the fields appearing in (6/1) by unitary transformations

$$d_L = U^\dagger n_L \qquad\qquad d_R = V^\dagger n_R$$

$$u_L = U'^\dagger p_L \qquad\qquad u_R = V'^\dagger p_R \tag{6.2}$$

chosen so that when the ϕ_a fields are replaced by their VEV's we get the diagonal mass amtrices

$$U^\dagger \Gamma_a v_a V = M_d \tag{6.3a}$$

$$U'^\dagger \Gamma_a v_a V' = M_u \tag{6.3b}$$

$M_{u,d}$ are the real diagonal matrices defined in (1.1). U and U' can most easily be determined by multiplying (6.3a,b) by their adjoints

$$U^\dagger (\Gamma.v)(\Gamma.v)^\dagger U = M_d^2 \tag{6.4}$$

$$U'^\dagger (\Gamma'.v)(\Gamma'.v)U' = M_u^2 \tag{6.5}$$

so given $(\Gamma.v)$ and the physical masses, we can determine U, U', V,V' up to some arbitrary phases[6.1]. These latter can best be seen by the fact that we can transform from e.g., u_{mL}, u_{mR} to

$$\hat{u}_L = P_u u_L \qquad\qquad \hat{u}_R = P_u u_R$$

$$P_u = \begin{pmatrix} e^{i\varphi_1} & & 0 \\ & \ddots & \\ 0 & & e^{i\varphi_F} \end{pmatrix} \tag{6.6}$$

without altering the mass terms.

As far as the gauge couplings are concerned, $J_{e.m}$ and J_Z, defined in (4.24b,c) are unchanged by the unitary transformations. This is obvious for $J_{e.m}$, but for J_Z the extra assumption has be made that all left handed fields with the same charge have also the same values of T^3 and Y_L.[6.2] The charged current is however affected by the unitary transformations

$$\sum_{m=1}^{F} \bar{n}_m \gamma^\mu (1+\gamma_5) p_m = \sum_{m,n=1}^{F} \bar{d}_m A_{mn} \gamma^\mu (1+\gamma_5) u_n$$

$$A_{mn} = (U^\dagger U')_{mn} \tag{6.7}$$

where A is a unitary F x F matrix. The matrix A can therefore be expressed in terms of F^2 real parameters; of these 2F-1 can be absorbed into the definitions of the fields (F by P_u, F by P_d and one however is an overall phase which does not alter the Lagrangian). We therefore have A described by $F^2-(2F-1)$ free parameters. The orthogonal group in F dimension, O_F, has F(F-1)/2 parameters, which can be parametrized as real angles of rotations so overall A is described by F(F-1)/2 rotations and $F^2-(2F-1)-F(F-1)/2 = (F-1)(F-2)/2$ phase angles. For two families we have one rotation angle, known as the Cabibbo angle, and no phase

$$A = \begin{pmatrix} \cos\theta_c & -\sin\theta_c \\ \sin\theta_c & \cos\theta_c \end{pmatrix} \tag{6.8}$$

while in three dimensions we have three angles $\theta_{1,2,3}$ whose cosines and sines we will call $c_{1,2,3}$, $s_{1,2,3}$ and one phase δ.

$$A = \begin{pmatrix} c_1 & s_1c_3 & s_1s_3 \\ -s_1c_2 & c_1c_2c_3 + s_2s_3e^{i\delta} & c_1c_2s_3 - s_2c_3e^{i\delta} \\ -s_1s_2 & c_1s_2c_3 - c_2s_3e^{i\delta} & c_1s_2s_3 + c_2c_3e^{i\delta} \end{pmatrix} \tag{6.9}$$

A for three families is often referred to as the Kobayashi-Maskawa or KM matrix.[6.3] θ_c is believed to be 13.7°, but the accurate determination of $\theta_{1,2,3}$ and δ will obviously be an experimental and theoretical challenge for the coming years. It will require accurate measurement of b quark decay and eventually, of the t quark (assuming it is found)decays.(6.4,6.5) The phase δ in this model is assumed to be the cause of CP violation in K meson decays, inducing a $K^{\circ} - \bar{K}^{\circ}$ transition by the box diagram shown in Fig. 2. It also leads to a small electric dipole moment of the neutron by the diagram of Fig. 3.(6.6)

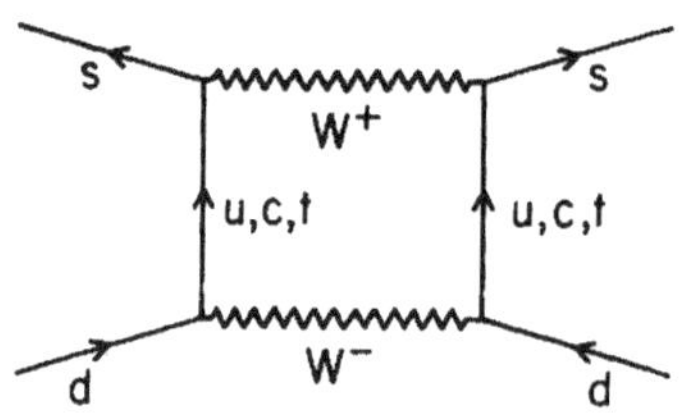

Fig. 2. Effective $\Delta S = 2$ transition.

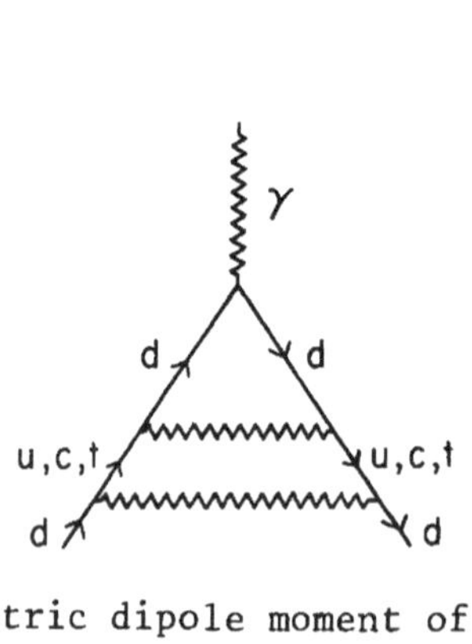

Fig. 3. Electric dipole moment of the neutron.

The beauty of the model is that only small CP violations occur in the K meson system despite the fact that CP was not assumed to be a symmetry of the system, i.e. the Γ matrices were allowed to be arbitrary complex matrices. For further discussion, see the later section on CP.

The matrix A becomes more complicated of course if more families exist.

b) Symmetries of Yukawa Couplings

A subject of considerable speculation has been the calculation on plausible theoretical grounds of the various angles and phases that enter the unitary matrix A of the charged current in (6.7). These attempts, which by now have an almost twenty years history starting soon after the introduction of the Cabibbo angle in 1963, have provided interesting ideas but have not met with convincing success. Typically they involve imposing of so-called horizontal or family symmetries on the quarks

$$n_L \to K_L n_L \qquad n_R \to K_R n_R \qquad \phi_a \to \mathcal{D}_{ab}\phi_b \tag{6.10}$$

where the symmetries may be either discrete or continuous. The charged 2/3 quarks may have a different set of symmetries

$$p_L \to K_L p_L \qquad p_R \to K_R' p_R \qquad \phi_a \to \mathcal{D}'_{ab}\phi_b \tag{6.11}$$

where $K_L' = K_L$ since n_L, p_L are in a gauge doublet. Invariance of $\mathcal{L}_{\text{Yukawa}}$ requires

$$K_L^\dagger \Gamma_a K_R = \mathcal{D}^*_{ab}\Gamma_b \tag{6.12}$$

$$K_L^\dagger \Gamma'_a K'_R = \mathcal{D}'^*_{ab}\Gamma'_b \tag{6.13}$$

So far, so good. The point, however, is that the unitary transformations U, U', V, V' which took us to the physical quark basis, do not necessarily diagonalize the matrices Γ_a, Γ'_a e.g.

$$U^\dagger \Gamma_a V \neq \text{diagonal matrix} \tag{6.14}$$

The diagonalize $\Gamma_a v_a$, but not each individual Γ_a; the latter is what is required if we are to have no flavor changing Higgs mediated neutral currents. In general then we may expect Higgs fields coupling to e.g. $\bar{d}_R s_L$, $\bar{u}_R c_L$, $\bar{u}_L c_R$, $\bar{d}_L s_R$ etc. The limits on such couplings are quite severe; to get a qualitative feeling for their magnitude, the 2W exchange box diagram of Fig. 2 gives and effective strangeness two interaction of (neglecting t quark contributions)

$$\mathcal{L}^{\text{eff.}}_{\Delta S=2}\Big|_{WW} \sim -(\bar{d}_L\gamma^\mu s_R)(\bar{d}_L\gamma_\mu s_R)\ \frac{G_F\ \alpha}{\sqrt{2}\ 4\pi}\ \frac{m_c^2\cos^2\theta_c\sin^2\theta_c}{M_W^2\sin^2\theta_W} \qquad (6.15)$$

in rough agreement with experimental values for the K° - $\bar{K}$° transition mass. A ΔS = 2 Higgs mediated reaction would go like

$$\mathcal{L}^{\text{eff.}}_{\Delta S=2}\Big|_{\text{Higgs}} \sim (\bar{d}_L s_R)(\bar{d}_L s_R)\ \frac{f^2}{M_H^2} \qquad (6.16)$$

with f the effective coupling and M_H the Higgs mass. As one sees the limits are quite severe, typically that $M_H \geqslant 1$ Tev.

There are two cases where the problem of flavor changing neutral currents does not arise. They are i) there exists only a single Higgs doublet coupling to fermions so $\Gamma_a v_a \to \Gamma v$ and ii) there are two Higgs doublets but a discrete symmetry is imposed so that one couples only to n_R and the other only to p_R; then $\Gamma_a v_a \to \Gamma v$, $\Gamma'_a v^*_a \to \Gamma' v'^*$.(6.2) In both cases it is obvious that the transformations which diagonalize the mass matrix diagonalize the Higgs couplings as well since the VEV of the Higgs field is an overall multiplicative constant. Unfortunately in both these cases, there is no possible restriction to be placed on the couplings by discrete symmetries and hence no calculable mixing angles.

This problem has been studied further by Gatto, Morchio and Strocchi(6.7) and by Segrè and Weldon(6.1). They have shown essentially that, except for the two cases mentioned above, imposing NFC (natural flavor conservation) on the Higgs couplings essentially restricts all mixing angles to be trivial (0 or π/2). We remind you here that NFC means that the Higgs meson couplings can be written in diagonal form in the physical basis regardless of the VEV's of the Higgs fields. Rather than going into a lengthy discussion of this matter, we will give a simple, but illustrative example. Consider a two family case, i.e. only four quarks, and let the K symmetry be the permutation symmetry S_3.(6.8) Now place ψ_L, n_R,p_R and two Higgs doublets $\phi_{1,2}$ all in two dimensional representations of S_3. By an additional discrete symmetry let $\phi_{1,2}$ couple only to n_R and then introduce two Higgs isodoublet S_3 singlets χ_o,χ_o' which couple only to p_R. The most general $SU_2 \times U_1 \times S_3$ Yukawa coupling turns out to be

$$\mathcal{L}_{\text{Yukawa}} = f\,[\,(\bar{\psi}_1 n_{2R} + \bar{\psi}_2 n_{1R})\phi_1 + (\bar{\psi}_1 n_{1R} - \bar{\psi}_2 n_{2R})\phi_2]$$
$$+ g(\bar{\psi}_1 p_{1R} + \bar{\psi}_2 p_{2R})\tilde{\chi}_0 + h(\bar{\psi}_1 p_{2R} - \bar{\psi}_2 p_{1R})\tilde{\chi}'_0 \qquad (6.17)$$

The Yukawa interaction can be written in matrix form, with obvious notation, as

$$\Gamma_a \phi_a + \Gamma'_a \tilde{\chi}_a = \begin{pmatrix} 0 & f \\ f & 0 \end{pmatrix} \phi_1 + \begin{pmatrix} f & 0 \\ 0 & -f \end{pmatrix} \phi_2$$
$$+ \begin{pmatrix} g & 0 \\ 0 & g \end{pmatrix} \tilde{\chi}_0 + \begin{pmatrix} 0 & h \\ -h & 0 \end{pmatrix} \tilde{\chi}'_0 \qquad (6.18)$$

Amusingly enough, with

$$U = \frac{1}{\sqrt{2}} \begin{pmatrix} 1 & 1 \\ -i & i \end{pmatrix} \qquad V = \frac{1}{\sqrt{2}} \begin{pmatrix} -i & i \\ 1 & 1 \end{pmatrix}$$

$$U' = V' = \frac{1}{\sqrt{2}} \begin{pmatrix} 1 & 1 \\ i & -i \end{pmatrix} \qquad (6.19)$$

we diagonalize the Yukawa coupling

$$U^\dagger \Gamma_a \phi_a V + U'^\dagger \Gamma'_a \tilde{\chi}'_a V' = \begin{pmatrix} f & 0 \\ 0 & f \end{pmatrix} \Phi_1 + \begin{pmatrix} -if & 0 \\ 0 & if \end{pmatrix} \Phi_2$$
$$+ \begin{pmatrix} g & 0 \\ 0 & g \end{pmatrix} \tilde{\chi}_0 + \begin{pmatrix} ih & 0 \\ 0 & -ih \end{pmatrix} \tilde{\chi}'_0 \qquad (6.20)$$

so we have NFC, but the charged current coupling is trivial.

$$A = U'^\dagger U = \begin{pmatrix} 1 & 0 \\ 0 & 1 \end{pmatrix} \qquad (6.21)$$

If, on the other hand, we go to a variation of this model where there are no χ fields and p_{1R} and p_{2R} are both in S_3 singlets, the interaction is

$$\mathcal{L}_{\text{Yukawa}} = f\,\{(\bar{\psi}_1 n_{2R} + \bar{\psi}_2 n_{1R})\phi_1 + (\bar{\psi}_1 n_{1R} - \bar{\psi}_2 n_{2R})\phi_2\}$$

$$+ (\bar{\psi}_1\tilde{\phi}_1 + \bar{\psi}_2\tilde{\phi}_2)(g p_{1R} + g' p_{2R}) \qquad (6.22)$$

U and V are the same matrices as in (6.19), but

$$U' = \frac{1}{\sqrt{|v_1|^2 + |v_2|^2}} \begin{pmatrix} v_2 & v_1^* \\ -v_1 & v_2^* \end{pmatrix} \qquad (6.23)$$

so we have charm changing neutral currents, i.e. the neutral components of ϕ_1 and ϕ_2 couple to currents to the form $\bar{c}_L u_R$ and $\bar{c}_R u_L$. The matrix A is most conveniently written defining $\rho_{1,2} = |v_1 \pm iv_2|/\sqrt{2}$ and absorbing phases into the quark fields as

$$A = \frac{1}{(\rho_1^2 + \rho_2^2)^{\frac{1}{2}}} \begin{pmatrix} \rho_1 & -\rho_2 \\ \rho_2 & \rho_1 \end{pmatrix} \qquad (6.24)$$

so the Cabibbo angle has $\tan\theta_c = \rho_2/\rho_1$. Looking back at the ϕ_a part of the mass matrix in (6.20) (this is the same for both models) and replacing $\phi_{1,2} \to v_{1,2}$

$$U^\dagger \Gamma . vV = \begin{pmatrix} f & 0 \\ 0 & f \end{pmatrix} v_1 + \begin{pmatrix} -if & 0 \\ 0 & if \end{pmatrix} v_2 \qquad (6.25)$$

so, after absorbing phases

$$\frac{m_d}{m_s} = \frac{f|v_1 - iv_2|}{f|v_1 + iv_2|} = \frac{\rho_2}{\rho_1} = \tan\theta_c$$

so here we have an example with a calculable angle. The diagonalization of the charge 2/3 quark sector is trivial since only one linear combination of p_{1R} and p_{2R} enters into $\mathcal{L}_{\text{Yukawa}}$. We define

$$c_R = \frac{g p_{1R} + g' p_{2R}}{\sqrt{|g|^2 + |g'|^2}} \qquad c_L = \frac{v_1 p_{1L} + v_2 p_{2L}}{\sqrt{|v_1|^2 + |v_2|^2}} \qquad (6.26)$$

and we see that the c quark has mass $m_c = \{(|g|^2 + |g'|^2)(|v_1|^2 + |v_2|^2)\}^{\frac{1}{2}}$, while the u quark is massless. As we already observed there is here no NFC for Higgs mediated neutral currents since the couplings in

(6.22) connect c quarks to u quarks. This is of course in accordance with our general result.

c) Leptons

We have left out the leptons so far. In the past one usually made the assumption that neutrinos were left-handed and massless; one did not introduce ν_R fields and hence the analogue of (6.1) was just

$$\mathcal{L}_{\text{Yukawa, leptons}} = \sum_{m,n=1}^{F} \bar{\chi}_{mL} \tilde{\Gamma}_{a,mn} \ell_{nR} \phi_a \tag{6.27}$$

where χ is the lepton doublet and ℓ_{nR} are the right handed lepton fields. Just as before, one finds the mass eigenstates e^-, μ^-, τ^- ... by diagonalizing (6.27) with $\phi_a \to v_a$. In place of the unitary matrix A in (6.7), we have the unit matrix; to see this, call ν^o_L and $\tilde{A}$ the leptonic analogues of u_L and A. Now define $\nu_{mL} = (A\nu^o_L)_m$; these are of course also mass eigenstates, since all neutrinos are massless, and now the charged current is diagonal. The neutral current is also diagonal so in weak interaction processes N_e, N_μ and N_τ are conserved separately as well as lepton number in gauge couplings, but not generally in Higgs meson couplings.[6.9]

If neutrinos are ordinary Dirac fields, with ν_R as well as ν_L present, the situation becomes exactly analogous to what we have already discussed for quarks and we need not repeat the arguments, except to emphasize that there is no a priori reason that the leptonic and quark mixing matrices A and $\tilde{A}$ be related in any way. Massive neutrinos lead to the extremely interesting phenomenon of neutrino oscillation. The mass eigenstates do not coincide with the weak interaction eigenstates, e.g. for two doublets, labelling $\nu_{1,2}$ the mass eigenstates, the weak interaction doublets are $(\nu_e, e^-)_L$, $(\nu_\mu, \mu^-)_L$ with

$$\mu_e = \nu_1 \cos\gamma + \nu_2 \sin\gamma \qquad \nu_\mu = -\nu_1 \sin\gamma + \nu_2 \cos\gamma \tag{6.28}$$

We produce for instance in π decay a ν_μ but this may then oscillate into ν_e

$$\nu_\mu(t) = -\nu_1 \sin\gamma \, e^{-iE_1 t} + \nu_2 \cos\gamma \, e^{-iE_2 t} \tag{6.29}$$

since $\nu_{1,2}$ are the energy eigenstates.

Finally for neutrinos we have the possibility of fermion number and lepton number violating Majorana mass terms of the form

$$m\nu_L^T C \nu_L \tag{6.30}$$

with C being the Dirac charge conjugation matrix equal to $i\gamma_2\gamma_0$ and ν_L a two-component spinor.

Charge conservation says such terms are only possible for neutral fermions ($\psi\psi$ has twice the charge of the ψ field, whereas $\bar{\psi}\psi$ is of course neutral). A term like (6.30) is invariant under Lorentz transformations where

$$\psi \to e^{i\omega_{\mu\nu}\sigma^{\mu\nu}}\psi \tag{6.31}$$

since $C^{-1}\sigma_{\mu\nu}^{T}C = -\sigma_{\mu\nu}$. It is not invariant under the global symmetry (α is constant)

$$\psi \to e^{i\alpha}\psi \tag{6.32}$$

whose generator is fermion number, but there is nothing sacred about fermion number.

If ψ is a two component field, e.g. $\psi = \nu_L$, eq. (6.30) simplifies

$$C(1+\gamma_5) = \begin{pmatrix} \varepsilon & 0 \\ 0 & 0 \end{pmatrix} \qquad \varepsilon^{\sigma\tau} = \begin{pmatrix} 0 & 1 \\ -1 & 0 \end{pmatrix}$$

$$\Rightarrow m\psi^{T}C\psi \to m\varepsilon^{\sigma\tau}\nu_{L\sigma}\nu_{L\tau} \tag{6.33}$$

where $\sigma,\tau = 1,2$ are Lorentz indices. This is of course just the statement that two (1/2, 0) fields must be combined antisymmetrically to form a Lorentz scalar.[(6.10)]

The Majorana mass term is not allowed in our minimal model because of the simultaneous requirements of $SU_2 \times U_1$ invariance and renormalizability. ψ_2 is in SU_2 doublet with Y = 1/2 so (6.30) is clearly not $SU_2 \times U_1$ invariant. On the other hand

$$\varepsilon^{\sigma\tau}(\phi^0\nu_{L\sigma} - \phi^+\ell^-_{L\sigma})(\phi^0\nu_{L\tau} - \phi^+\ell^-_{L\tau}) \tag{6.34}$$

is $SU_2 \times U_1$ invariant, and with ϕ^0 having a non-zero VEV even leads to a Majorana mass term, but it is not renormalizable.

VII. CP

a) K^o-$\bar{K}^o$ System Phenomenology

Consider an arbitrary state in the K^o, $\bar{K}^o$ basis

$$|\psi(t)\rangle = u_1(t)|K^o\rangle + u_2(t)|\bar{K}^o\rangle \tag{7.1}$$

It satisfies the Schrodinger equation

$$i\frac{\partial}{\partial t}|\psi(t)\rangle = H|\psi(t)\rangle \Rightarrow i\dot{u}_i = X_{ij}u_j \tag{7.2}$$

with

$$X_{ij} = M_{ij} - i\frac{\Gamma_{ij}}{2} \tag{7.3}$$

where M and Γ are Hermitian matrices known respectively as the mass and decay matrix (this is the general form[7.1]). By conservation of probability

$$\frac{d}{dt}\langle\psi|\psi\rangle = \langle\dot{\psi}|\psi\rangle + \langle\psi|\dot{\psi}\rangle = -\langle\psi|\Gamma|\psi\rangle$$

$$= -2\pi\sum_f \rho(f)|\langle f|H|\psi(t)\rangle|^2 \tag{7.4}$$

where $\rho(f)$ is the density of final states per unit energy and f is a particular final state such as $\pi^+e^-\nu$ or $\pi^+\pi^-$. $|K^o\rangle$ and $|\bar{K}^o\rangle$ are the eignestates of $H_{strong} + H_{e.m}$ with strangeness of +1 and -1 respectively. They are not however the eigenstates of H, these are labelled as $|K_{S,L}\rangle$, where we take the state at time t = 0 unless the specific time dependence is written in

$$X|K_{S,L}\rangle = (M_{S,L} - i\Gamma_{S,L}/2)|K_{S,L}\rangle \tag{7.5a}$$

$$|K_{S,L}(t)\rangle = e^{-i(M_{S,L} - i\Gamma_{S,L}/2)t}|K_{S,L}(0)\rangle \tag{7.5b}$$

The 2x2 X matrix is conveniently expressed by the usual Pauli matrices as

$$iX = \Gamma/2 + iM = D + i(E_1\sigma_1 + E_2\sigma_2 + E_3\sigma_3) \tag{7.6}$$

where D, $E_{1,2,3}$ are of course complex since M and Γ are. The states

$K_{S,L}$ are related to $|K^o\rangle$, $|\bar{K}^o\rangle$ by

$$|K_S\rangle = \frac{1}{\sqrt{|r|^2 + |s|^2}} \{r|K^o\rangle + s|\bar{K}^o\rangle\} \tag{7.7a}$$

$$|K_L\rangle = \frac{1}{\sqrt{|p|^2 + |q|^2}} \{p|K^o\rangle + q|\bar{K}^o\rangle\} \tag{7.7b}$$

It is convenient to parametrize $E_{1,2,3}$ in terms of three complex quantities, E, θ, ϕ where $E_3 = E\cos\theta$, $E_1 \pm iE_2 = E\sin\theta\, e^{\pm i\phi}$. We then solve (7.5a) to find s/r and q/p. The standard parametrization, designed to emphasize the small deviations of K_S and K_L from being CP eigenstates involves rather quantities $\varepsilon_{1,2}$ where

$$\frac{1-\varepsilon_1}{1+\varepsilon_1} = \tan\theta\, e^{i\phi} \qquad \frac{1-\varepsilon_2}{1+\varepsilon_2} = \cot\theta/2\, e^{i\phi} \tag{7.8}$$

so that

$$|K_S\rangle = \frac{1}{\sqrt{2(1+|\varepsilon_1|^2)}} \{(1+\varepsilon_1)|K^o\rangle + (1-\varepsilon_1)|\bar{K}^o\rangle\} \tag{7.9a}$$

$$|K_L\rangle = \frac{1}{\sqrt{2(1+|\varepsilon_2|^2)}} \{(1+\varepsilon_2)|K^o\rangle - (1-\varepsilon_2)|\bar{K}^o\rangle\} \tag{7.9b}$$

and $\varepsilon = (\varepsilon_1+\varepsilon_2)/2$, $\Delta = (\varepsilon_1-\varepsilon_2)/2$. Since we know experimentally that violations of CP are small.

$$|\varepsilon|\,,\ |\Delta| \ll 1 \Rightarrow |\phi|\,,\ |\theta-\pi/2| \ll 1 \tag{7.10}$$

which in turn implies

$$|E_2/E_1| \ll 1 \qquad |E_3/E_1| \ll 1 \tag{7.11}$$

Neglecting quadratic terms in ε and Δ

$$\varepsilon \approx -i\frac{\phi}{2} \approx -\frac{i}{2}\frac{E_2}{E} \qquad \Delta \approx -\frac{1}{2}(\theta-\frac{\pi}{2}) \approx \frac{E_3}{E} \tag{7.12}$$

The eigenvalues of iX are $D \pm iE$ since $E.\sigma$ has eigenvalues $\pm E$.

Therefore, examining (7.5a) and (7.6)

$$2iE = \frac{\Gamma_S - \Gamma_L}{2} + i(M_S - M_L)$$

$$2iE_3 = i(X_{11} - X_{22})$$

$$2E_2 = i(X_{12} - X_{21}) \qquad (7.13)$$

Since Γ and M are Hermitian $\Gamma^*_{12} = \Gamma_{21}, M^*_{12} = M_{21}$ so

$$\varepsilon = \frac{\Gamma_{12} - \Gamma^*_{12} + i(M_{12} - M^*_{12})}{\Gamma_S - \Gamma_L + 2i(M_S - M_L)} \qquad \Delta = \frac{\Gamma_{11} - \Gamma_{22} + i(M_{11} - M_{22})}{\Gamma_S - \Gamma_L + 2i(M_S - M_L)} \qquad (7.14)$$

If we know impose CPT invariance

$$\langle K^o | (CPT)^{-1} H (CPT) | K^o \rangle = \langle \bar{K}^o | H | \bar{K}^o \rangle \qquad (7.15)$$

we find that $X_{11} = X_{22}$ so Δ equals zero, $\varepsilon_1 = \varepsilon_2$ in (7.9b) and

$$\langle K_S | K_L \rangle = 2 \text{ Re } \varepsilon \qquad (7.16)$$

Now let us turn to the two pion decay modes of the neutral kaons. Since the orbital angular momentum of the two pions is zero, Bose statistics dictates that the isospin wave function be symmetric, i.e. I = 0 or 2. If we label with an o superscript the two pion stationary states (i.e. without final state interactions) we have, by isospin decomposition,

$$\langle \pi^+ \pi^- | = \frac{1}{\sqrt{3}} \{ e^{i\delta_2} \langle 2\pi(I=2)|^o + \sqrt{2} e^{i\delta_o} \langle 2\pi(I=0)|^o \}$$

$$\langle \pi^0 \pi^0 | = \frac{1}{\sqrt{3}} \{ \sqrt{2} e^{i\delta_2} \langle 2\pi(I=2)|^o - e^{i\delta_o} \langle 2\pi(I=0)|^o \} \qquad (7.17)$$

We know define the amplitudes ($H_W = H_{weak\ int}$)

$$A_o = \langle 2\pi(I=0)|^o H_W | K^o \rangle \qquad A_2 = \langle 2\pi(I=2)|^o H_W | K^o \rangle \qquad (7.18)$$

The corresponding amplitudes for $|\bar{K}^o\rangle$ decays are A^*_o and A^*_2 by the

CPT invariance of H_W. We can define A_o to be real; to see this remember that $|K^o\rangle$ and $|\bar{K}^o\rangle$ were the eigenstates of $H_{strong} + H_{e.m}$ with strangeness S = +1 and S = −1. Now the unitary transformation $U = e^{iS\kappa}$ commutes with $H_{strong} + H_{e.m}$, but changes $|K^o\rangle$ and $|\bar{K}^o\rangle$ by

$$|K^o\rangle \to U|K^o\rangle = e^{i\kappa}|K^o\rangle \qquad |\bar{K}^o\rangle \to e^{-i\kappa}|\bar{K}^o\rangle \tag{7.19}$$

so the relative phase of $|K^o\rangle$ and $|\bar{K}^o\rangle$ is not measureable; we choose it so that A_o is real. We then have, using (7.9a,b) with $\varepsilon_1 = \varepsilon_2 = \varepsilon$ and neglecting ε^2

$$\langle\pi^+\pi^-|H_W|K_S\rangle = \sqrt{2/3}\,\{ReA_2e^{i\delta_2} + i\varepsilon ImA_2e^{i\delta_2} + \sqrt{2}A_oe^{i\delta_o}\}$$

$$\langle\pi^0\pi^0|H_W|K_S\rangle = 2/\sqrt{3}\{ReA_2e^{i\delta_2} + i\varepsilon ImA_2e^{i\delta_2} - A_o/\sqrt{2}e^{i\delta_o}\}$$

$$\langle\pi^+\pi^-|H_W|K_L\rangle = \sqrt{2/3}\,\{iImA_2e^{i\delta_2} + \varepsilon ReA_2e^{i\delta_2} + \varepsilon\sqrt{2}A_oe^{i\delta_o}\}$$

$$\langle\pi^0\pi^0|H_W|K_L\rangle = 2/\sqrt{3}\,\{iImA_2e^{i\delta_2} + \varepsilon ReA_2e^{i\delta_2} - A_o/\sqrt{2}e^{i\delta_o}\} \tag{7.20}$$

Defining

$$\eta_{+-} = \frac{\langle\pi^+\pi^-|H_W|K_L\rangle}{\langle\pi^+\pi^-|H_W|K_S\rangle} \qquad \eta_{oo} = \frac{\langle\pi^0\pi^0|H_W|K_L\rangle}{\langle\pi^0\pi^0|H_W|K_S\rangle}$$

$$\varepsilon' = \frac{1}{\sqrt{2}}\left(\frac{ImA_2}{A_o}\right)e^{i(\pi/2 + \delta_2 - \delta_o)} \tag{7.21}$$

and remembering that the $\Delta I = 1/2$ rule in weak interactions suggests that $A_2/A_o < 1/20$, we can express the measureable quantities η_{+-}, η_{oo}.

$$\eta_{+-} \approx \varepsilon + \varepsilon' \qquad \eta_{oo} = \varepsilon - 2\varepsilon' \tag{7.22}$$

we have neglected terms of order ε^2 and $\varepsilon\, A_2/A_o$ in obtaining these relations. To summarize, $K_L \to 2\pi$, assuming CPT invariance, is characterized by the two complex numbers ε and ε'. The imaginary part of A_2 is both CP violating and $\Delta I = 1/2$ violating (it is the amplitude for $(|K^o\rangle - |\bar{K}^o\rangle)$ to go to $|2\pi(I=2)\rangle$); we therefore expect ε' to be smaller than ε by at least a factor of 20 (the $\Delta I = \frac{1}{2}$ rule). Experimentally[7.2]

$$|\eta_{+-}| = (2.274 \pm .022) \times 10^{-3}$$

$$|\eta_{+-}| = (2.33 \pm .08) \times 10^{-3}$$

$$\text{Arg}\,(\eta_{+-}) = \phi_{+-} = (44.6 \pm 1.2)^\circ$$

$$\text{Arg}\,(\eta_{oo}) = \phi_{oo} = (54 \pm 5)^\circ \qquad (7.24)$$

We will not discuss the fascinating experiments which determine the above numbers, but rather several models of CP violation in the $K_L \to 2\pi$ mode. It is clear that this violation can result from either a $\Delta S = 2$ mixing term in the X matrix or direct CP violation in the $K \to 2\pi$ decay. These two possibilities are said to arise from respectively superweak and milliweak (or microweak) interactions. For the sake of brevity we will limit our discussion of the effects of CP violation to the $\pi\pi$ decay modes of the neutral kaons. We should at least mention however that other tests of CP (or T) violation in the kaon system are the measurement of

$$\delta = \frac{\Gamma(K_L \to \pi^- \ell^+ \nu) - \Gamma(K_L \to \pi^+ \ell^- \bar{\nu})}{\Gamma(K_L \to \pi^- \ell^+ \nu) - \Gamma(K_L \to \pi^+ \ell^- \bar{\nu})}$$

and of the transverse polarization of the muons in $K^o \to \pi^- \mu^+ \nu$, $K^+ \to \pi^o \mu^+ \nu$ decays.

b) Models of CP Violation

i) Superweak Model This is a sense the simplest model of CP violation[7.3]. It postulates a new type of interaction which gives an effective CP violating $\Delta S = 2$ term in the Hamiltonian, with a characteristic magnitude (in units where the nucleon mass is set equal to unity) of $\sim 10^{-8} G_F$. Phenomenologically this corresponds to a small, but non-zero Im M_{12} and Im $\Gamma_{12} = 0$. The state $(|K^o\rangle - |\bar{K}^o\rangle)$ is not connected by H_W to $|2\pi(I=2)\rangle$ so $\varepsilon' = 0$. The predictions of the superweak model, using (7.14) and (7.22), are (remember that $\Gamma_S \gg \Gamma_L$)

$$|\eta_{+-}| = |\eta_{oo}|$$

$$\phi_{+-} = \phi_{oo} = \text{Arg}\varepsilon = \tan^{-1}\left(\frac{2(M_S - M_L)}{\Gamma_S}\right) = 43.8 \pm 0.2^\circ \qquad (7.24)$$

so, comparing (7.23) and (7.24) we see that the superweak model is entirely compatible with present data. The smallness of the superweak interaction ensures that its effect will only be seen in the neutral kaon system where the second order amplitude $A(K_L \to K_S \to 2\pi)$ is amplified by the K_S-K_L mass difference, i.e.

$$A \sim \frac{(10^{-8}) \times (G_F)^2}{M_L - M_S} \sim 10^{-3} G_F$$

ii) Kobayashi-Maskawa We already mentioned, cf. (6.15) the effective $\Delta S = 2$ interaction due to the W^+-W^- exchange box diagram (fig. 2.). Using the matrix (6.9) which specifies the gauge coupling couplings we saw that the diagram of (fig. 2) gave a reasonable value for $\Delta M = M_S - M_L$ using only four quarks. If we include the (t,b) doublet, the result for ΔM is not changed substantially, and in addition we now see that the gauge couplings are complex because of the phase δ. Note that we have not assumed CP invariance for the Lagrangian; it is the structure of the theory and Hermiticity which has allowed us to define away the CP violating phases until the (t,b) doublet was introduced. Evaluating the diagram of (fig.2), using the couplings of (6.9) we can already evaluate $\varepsilon_m = \text{Im}\, M_{12}/\Delta M$. There are obviously various kinematic limits depending on the value of the t quark mass, but $\varepsilon_m \sim 10^{-3}$ is obtained for $|s_2 s_3 \sin\delta| \sim 0(10^{-3})$. This is the basis of the common statement that the K.-M. model reduces to a superweak model.[7.4] In fact that there are significant differences which we now discuss.

From (7.14) we have

$$\text{Arg}\,\varepsilon = \tan^{-1}\left(\frac{2\Delta M}{\Gamma_S - \Gamma_L}\right) + \phi_D \qquad \phi_D = \tan^{-1}\left(\frac{\text{Im}\Gamma_{12}}{2\text{Im}\, M_{12}}\right) \tag{7.25}$$

In superweak theories, as already stated, $\varepsilon' = 0$ and $\phi_D = 0$. In milliweak theories ϕ_D comes from the CP violating phase difference between the amplitude for $K^o \to 2\pi(I=0)$ and the other decay modes the $K^o, \bar{K}^o$, while ε' comes from the CP violating phase ξ between the $K^o \to 2\pi(I=0)$ and $K^o \to 2\pi(I=2)$ amplitudes. Remembering that our convention is that A_o, the amplitude for $K^o \to 2\pi(I=0)$, is real, we find $|\phi_D| \sim 0(10^{-4})$ and hence is negligible. Recent estimates of ε' suggest however that this parameter may be accessible to experiment in the near future.

Let us sketch an estimate of ε' in a milliweak theory. We now assume there is CP violation in the $K^o \to 2\pi$ decay amplitude as well as in the mass matrix; if A_2 is complex and A_o real, we can evaluate ε' directly from eq. (7.21). If the CP violation in $K^o \to 2\pi$ is exclusively in the $I = 1/2$ channel, we still can get a non-zero ε' even though A_2 would appear to be real. The reason is that we have adopted a convention where A_o is real, so if the CP violation introduces a phase ξ in A_o, we rotate it away by a unitary transformation (see eq. (7.19)

$$|K^o> \to e^{-i\xi}|K^o> \qquad\qquad |\bar{K}^o> \to e^{i\xi}|\bar{K}^o>$$

so Im M_{12} is shifted by

$$\begin{aligned}\text{Im } M_{12} &= \text{Im} <K^o|H_W|\bar{K}^o> \to \text{Im } e^{2i\xi}<K^o|H_W|\bar{K}^o> \\ &\approx \text{Im} <K^o|H_W|\bar{K}^o> + 2\xi\text{Re} <K^o|H_W|\bar{K}^o>\end{aligned} \tag{7.26}$$

and A_2 acquires a phase $-\xi$ so, using (7.21) with $A_2^{(o)}$ equal to the (real) amplitude for $K^o \to 2\pi(I=2)$

$$\begin{aligned}\varepsilon' &= \frac{1}{\sqrt{2}}\text{ Im } \left(\frac{e^{-i\xi}A_2^{(o)}}{A_o}\right) e^{i(\pi/2 + \delta_2 - \delta_o)} \\ &\approx -\frac{\xi}{\sqrt{2}}\, e^{i(\pi/2 + \delta_2 - \delta_o)} \left(\frac{A_2^{(o)}}{A_2}\right)\end{aligned} \tag{7.27}$$

Since $(M_S - M_L) = \Delta M = 2\text{Re } M_{12}$ and $\Gamma_S \approx -2\Delta M$, we can rewrite ε, from (7.14) and (7.26) with Im $\Gamma_{12} = 0$, as

$$\varepsilon \approx \frac{i\{2i\text{ImM}_{12} + 4i\xi\text{ReM}_{12}\}}{-4\text{ReM}_{12} + 4i\text{ReM}_{12}} = \frac{\varepsilon_m + 2\xi}{2(1-i)} \tag{7.28}$$

where $\varepsilon_m = \text{Im } M_{12}/\text{Re } M_{12}$. With $A_2^{(o)}/A_o \sim \frac{1}{20}$, this leads to

$$\frac{\varepsilon'}{\varepsilon} \approx \frac{1}{20}\left(\frac{-\xi}{\sqrt{2}}\right) e^{i(\pi/2 + \delta_2 - \delta_o)} \frac{2(1-i)}{\varepsilon_m + 2\xi} \tag{7.29}$$

Since $\delta_2 - \delta_o = -53 \pm 6°$, $e^{i(\pi/2 + \delta_2 - \delta_o)} \left(\frac{1-i}{\sqrt{2}}\right) \approx 1$

which implies,

$$\frac{\varepsilon'}{\varepsilon} \approx \frac{-i}{20} \frac{2\xi}{\varepsilon_m + 2\xi} \tag{7.30}$$

the relation which follows assuming the CP violating $K^o \to 2\pi$ amplitude is pure $\Delta I = 1/2$. This is the case when e.g. the CP violating decay amplitude is due to a so-called penguin diagram (fig. 4).

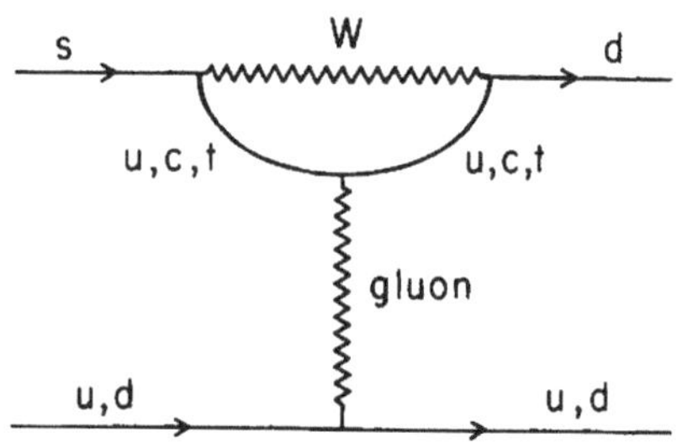

Fig. 4. "Penguin" Diagram for Kaon Decay.

Calculating ξ, the authors of ref. (7.5) find typically that ε'/ε is positive and of order 10^{-2} (the value is subject to the indetermination of the strong coupling constant and may be smaller(7.5) if a "low" value of the Λ parameter, i.e. $\Lambda^2 \sim 0.01$ Gev^2 is used to define g_c. Experimentally(7.6)

$$\frac{\varepsilon'}{\varepsilon} = -\ 0.003 \pm 0.015 \tag{7.31}$$

so we are at the threshold of being able to distinguish superweak models from K.-M. type milliweak models.

iii) Higgs Models If the model contains only a single weak isospin Higgs doublet, there are no effects on CP violation. The one surviving neutral field has real couplings, proportional to the mass matrix.

We emphasized earlier that the K.-M. scenario we discussed in the previous sections follows without imposing CP symmetry on the Lagrangian. As we shall see later in this chapter there is a strong

CP violation, so many authors have been let to consider the consequences of imposing CP as a symmetry of the Lagrangian, thereby banishing this violation. One consequence is that the Yukawa couplings are real as well as the gauge couplings; if only one Higgs doublet is present, there is then no CP violation at all, contrary to experiment. If two or more Higgs doublets are present, we can consider the attractive notion that CP is violated spontaneously, i.e. by complex VEV's of the scalar fields.[7.7] In general, the Higgs bosons in such models must be superheavy (masses $\geqslant$ 1 Tev) because of the flavor changing Higgs mediated neutral currents.[7.8] Weinberg[7.9] has proposed a model with three Higgs doublets, $\phi_{1,2,3}$ with a discrete symmetry such that ϕ_1 couples only to n_{nR} quarks, ϕ_2 only to p_{nR} and ϕ_3 to neither. For historical accuracy we point out that his models did not impose CP as a symmetry of $\mathcal{L}$; it was designed for a four quark model where the K.M. matrix is real. Changing to a three Higgs doublet model where CP is a symmetry, we have

$$\mathcal{L}_{\text{Yukawa, quarks}} = \sum_{m,n=1}^{F} \{\bar{\psi}_{mL}\Gamma_{1,mn}n_{nR}\phi_1 + \bar{\psi}_{mL}\Gamma_{2,mn}p_{nR}\tilde{\phi}_2\} \tag{7.32}$$

with

$$\langle 0|\phi_i|0\rangle = v_i\, e^{i\omega_i} \qquad i = 1,2,3 \tag{7.33}$$

The form of equation (7.32) ensures NFC. The CP invariance of $\mathcal{L}$ forces $\Gamma_{1,2}$ to be real; the phases $\omega_{1,2}$ may be absorbed into the definition of n_{nR} and p_{nR} by

$$n_{nR} \to e^{i\omega_1} n_{nR} \qquad\qquad p_{nR} \to e^{i\omega_2} p_{nR} \tag{7.34}$$

so the mass matrix is real and hence may be diagonalized by orthogonal transformations, rather than the unitary ones we needed in Chapter 6. The K.-M. matrix coupling quarks to gauge bosons, e.g. (6.9), is therefore real[7.10], and hence CP conserving. For the charged Higgs bosons, the argument is more subtle, one linear combination of $\phi_1^\pm$, $\phi_2^\pm$, $\phi_3^\pm$ is eaten by the $W^\pm$, but two linear combinations, $H_{1,2}^\pm$ remain as physical particles.[7.11] They couple to the quarks as

$$\sum_{m,n=1}^{F} \frac{g}{\sqrt{2}M_W} \{ \alpha_i \bar{u}_{mR}(M_u K)_{mn} d_{nR} + \beta_i \bar{u}_{mL}(M_d K)_{mn} d_{nR} \} H_i^+ + \text{h.c.} \tag{7.35}$$

where α_i, β_i with i = 1,2 are complex numbers, $M_{u,d}$ are the diagonal mass matrices and K is real mixing matrix.

Using this model kaon decays have been reanalyzed (7.12). The real part of M_{12} is still dominated by the box diagram of Fig. 2, but Im M_{12} arises from a box diagram where one W and one H are exchanged and the dominant "penguin" type diagram involves a H^- loop instead of a W^- loop (see Fig. 4). The experimental limit on $|\varepsilon'/\varepsilon|$, eq. (7.31), implies, using eq. (7.30)

$$|\xi/\varepsilon_m| = +\,0.03\,{}^{+\,0.25}_{-\,0.13} \tag{7.36}$$

whereas the calculations of ref. (7.12) typically give a value for $|\xi/\varepsilon_m|$ which is an order of magnitude larger. Subsequent refinements by Donoghue et al (7.13) only serve to reinforce the argument of ref. (7.12)

We tentatively conclude therefore that a spontaneous origin of CP violation is unlikely.

c) Neutron Electric Dipole Moment

The experimental limit on the electric dipole moment of the neutron is (7.14)

$$(D/e)_{\text{neutron}} \leqslant (0.4 \pm 1.5) \times 10^{-24} \text{ cm} \tag{7.37}$$

Theoretical calculations using the K.-M. model of section ii) give (6.6)

$$(D/e)_{\text{neutron}} \approx \frac{G_F \alpha \sin\delta\; s_1^2 s_2 c_2 s_3\; (m_t^2 - m_c^2)(m_b^2 - m_s^2) m_d}{\pi^3 M_W^4} \tag{7.38}$$

which is expected to be of order 10^{-30} cm : (the above assumes $m_t \ll M_W$, but the estimate does not change substantially even for $m_t > M_W$). Higgs mediated models of CP violation (see previous section for references) contribute at the one loop level and typically give values of $(D/e)_{\text{neutron}}$ of $10^{-25\pm2}$ cm.

d) Strong CP Violation and the Axion

We will discuss in this subsection another kind of CP violation, due now to the strong interactions. Aside from the intrinsic interest of the subject, our purpose is to use this as an example of pseudo-Goldstone bosons in general, and the axion in particular.

The problem is the following : the most general SU_3^c invariant, renormalizable strong interaction Lagrangian for gluons and quarks preserves C,P and T except for a term of the form

$$\mathcal{L}_\theta = \frac{\theta g_c^2}{64\pi^2} \varepsilon_{\mu\nu\sigma\tau} F_\alpha^{\mu\nu} F_\alpha^{\sigma\tau} \tag{7.39}$$

which violates P and T (θ is a free parameter). Naively we would be tempted to ignore this term since it is of the form $\theta g_c^2/32\pi^2\ \partial_\mu C^\mu$

$$\partial^\mu C_\mu = \frac{1}{3} \partial^\mu \varepsilon_{\mu\nu\lambda\sigma} (6A_\alpha^\nu \partial^\lambda A_\alpha^\sigma + 2g_c f_{\alpha\beta\gamma} A_\alpha^\nu A_\beta^\lambda A_\gamma^\sigma) \tag{7.40}$$

and we are accustomed to discarding total divergences in a Lagrangian density. In this case, however, non-pertubative instanton effects give surface terms which cannot be neglected. This turns out to be a blessing since it solves the famous U_1 problem, namely in the limit of massless u,d quarks, the strong interaction Lagrangian is invariant under (we will limit ourselves to chiral $SU_2 \times SU_2$)

$$\psi_r \rightarrow e^{-i\alpha\gamma_5} \psi_r \qquad r = 1,2 \tag{7.41}$$

with an associated, presumably conserved current,

$$j_{\mu 5} = \sum_{r=1,2} \bar{\psi}_r \gamma_\mu \gamma_5 \psi_r \tag{7.42}$$

This symmetry is broken spontaneously so, by Goldstone's theorem, there is an isoscalar pseudoscalar meson associated with the breaking. The likely candidate, the η, is too heavy; the problem is at first glance solved by the anomaly discussed in section V. formula (5.9), which says j_5^μ is not conserved. Here the plot thickens since we can define a gauge variant, but conserved current $j^\mu_5 - 1/16\pi^2 C^\mu$. How this problem is solved by the use of instantons is a fascinating story, but outside the scope of these lectures.[7.15] Our problem is how to solve the strong T violation problem implied by the presence of (7.19)

A first step is to realize that because of the anomaly equation (5.9), we can rotate away $\mathcal{L}_\theta$. To see this simply, consider a field $\phi(x)$ undergoing an infinitesimal gauge transformation through $\Lambda(x)$.[7.16]

$$\phi(x) \rightarrow \phi(x) + \Lambda(x) f(\phi(x)) \tag{7.43}$$

We then have

$$\frac{\delta\mathcal{L}}{\delta\Lambda} = \frac{\delta\mathcal{L}}{\delta\phi} + \frac{\delta\mathcal{L}}{\delta(\partial_\mu\phi)}\partial_\mu f \tag{7.44a}$$

$$\frac{\delta\mathcal{L}}{\delta(\partial_\mu\Lambda)} = \frac{\delta\mathcal{L}}{\delta(\partial_\mu\phi)} f \tag{7.44b}$$

taking the divergence of (7.44b) and using the Euler-Lagrange equation

$$\partial_\mu\frac{\delta\mathcal{L}}{\delta(\partial_\mu\Lambda)} = \partial_\mu\frac{\delta\mathcal{L}}{\delta(\partial_\mu\phi)} f + \frac{\delta\mathcal{L}}{\delta(\partial_\mu\phi)}\partial_\mu f$$

$$= \frac{\delta\mathcal{L}}{\delta\phi} f + \frac{\delta\mathcal{L}}{\delta(\partial_\mu\phi)}\partial_\mu f = \frac{\delta\mathcal{L}}{\delta\Lambda} \tag{7.45}$$

Defining the current associated with the gauge transformation as

$$J^\mu \equiv - \frac{\delta\mathcal{L}}{\delta(\partial_\mu\Lambda)} = - \frac{\delta\mathcal{L}}{\delta(\partial_\mu\phi)} f \tag{7.46}$$

we then have the desired relation

$$\partial_\mu J^\mu = - \frac{\delta\mathcal{L}}{\delta\Lambda} \tag{7.47}$$

This means that for the case of interest to us, an axial U_1, rotation of the Lagrangian is proportional to the divergence of the U_1 axial current. By the anomaly condition this is in turn proportional to $F_{\mu\nu\alpha}F^{\mu\nu}_{\alpha}$ so we can see that by an U_1 axial rotation we can remove the θ term, equation (7.39), from the Lagrangian. This is not a solution to the strong CP problem because the u,d quarks are not really massless and hence the U_1 axial rotation alters the quark mass matrix : for $q \to e^{-i\alpha\gamma_5}q$

$$\mathcal{L}_{\text{mass, quarks}} = -\bar{q}_L M q_R \to - \bar{q}_L M q_R e^{2i\alpha} \tag{7.48}$$

so there is a phase change in M of $2N\alpha$ for N flavors. We see however that the quantity

$$\bar{\theta} = \theta + \text{Arg det } M \tag{7.49}$$

is left invariant under the axial rotation so that if the mass matrix is made real, the coefficient of the strong CP violating term is $\bar{\theta}$. The bound (7.37) on $(D/e)_{\text{neutron}}$ implies[7.17] that $\bar{\theta} \leqslant 10^{-9}$.

This is the strong CP problem, namely why is $\bar{\theta}$ so small. We could ad hoc set to zero $\bar{\theta}$ in $\mathcal{L}$, but renormalization effects due to CP violation which must be present in the theory ($K_L \to 2\pi$ after all) would generate a $\bar{\theta}$ back again(7.18).

Several ways of dealing with this problem have been suggested, but none has been generally accepted. We will talk about one type of solution in which the quark fields have a global chiral U_1 symmetry left over after the quark mass matrix has been brought to the proper real diagonal γ_5 free form. There are two possible alternatives; the first is that the symmetry is unbroken in which case we may rotate away the offending term (7.39). An unbroken symmetry means however that we have a strictly massless quark (a phase rotation of e.g. u_R leaves $\mathcal{L}$ invariant even after SSB). The limits(7.19)

$$\frac{m_u}{m_d} = 0.47 \pm 0.11 \qquad \frac{m_d}{m_s} = 0.042 \pm 0.007 \tag{7.50}$$

suggest this hypothesis is untenable. Alternatively we may have a global symmetry such as the X symmetry defined in (3.17) extended to quarks and gauge bosons. This is easily done; a by product is that ϕ_1 couple only to charge 2/3 quarks and ϕ_2 to charge -1/3 quarks which is desirable since it assures, we saw it earlier, natural flavor conservation for the neutral Higgs meson couplings. Peccei and Quinn(7.20) have argued plausibly that in such a theory has $\bar{\theta} = \mathrm{Arg}(e^{i\theta}M)$ equal zero as a symmetry relationship. This would then give a so-called natural solution (i.e. parameters require no adjustment) to the strong CP problem. Before explaining how this comes about, let us lay a little groundwork in a fairly pedestrian manner.

We begin by reviewing the breakdown of chiral symmetry due to QCD. To simplify matters, let us restrict ourselves to two quark flavors, u and d; neglecting the weak Yukawa couplings, we see that the theory has a $U_2 \times U_2$ symmetry. Vector U_1 is unbroken, axial U_1 has an anomaly and the remaining $SU_2 \times SU_2$ is broken dynamically down to vector SU_2; this comes about because the strong QCD forces lead to the formation of fermion condensates (non vanishing VEV's of $\bar{\psi}\psi$) which we label v_u and v_d for respectively the u and d fields. The spontaneous breaking of axial SU_2 means there are three pseudoscalar Goldstone bosons. These, the pions, are in reality only pseudo-Goldstone bosons because of the quark mass term in $\mathcal{L}$ (due to Higgs SSB)

$$-\,(m_u\bar{u}u + m_d\bar{d}d) \tag{7.51}$$

which explicitly breaks axial SU_2.

The pion mass is calculated in a rather straightforward way by current algebra[(7.21)] : let A^a_μ, a = 1,2,3 be the three axial vector currents with divergence D^a and

$$Q^a_5(0) = \int A^a_o(\vec{x},o)d\vec{x} \tag{7.52}$$

The pion decay constant f_π is defined by

$$<0|A^a_\mu(o)|\pi^b(k)> = if_\pi k_\mu \delta_{ab} \tag{7.53}$$

The Ward identity gives

$$ik^\mu \int d^4x\, e^{-ikx} < 0|T\,\{A^a_\mu(x), \partial^\nu A^b_\nu(0)\}|\,0>$$

$$= \int d^4x\, e^{-ikx} < 0|T\{\partial^\mu A^a_\mu(x), \partial^\nu A^b_\nu(0)\}|\,0>$$
$$+ < 0|[Q^a_5(0), \partial^\nu A^b_\nu(0)]\,|0> \tag{7.54}$$

As $k_\mu \to 0$, the l.h.s. of (7.54)vanishes, the first term on the r.h.s. is assumed to be dominated by the pion pole and we get then the formula for the mass^2,μ^2_π, of the dynamically generated pseudo-Goldstone bosons (the pions)

$$f^2_\pi \mu^2_\pi = -i< 0|[Q^a_5(0),\ D^a(0)]\,|0> \tag{7.55}$$

The divergence $D^a(0)$ of the current is found by evaluating the commutator of $Q^a_5(0)$ with the Hamiltonian. Calling H' the part of the total Hamiltonian which does not commute with $Q^a_5(0)$ (in our case H' is given by the negative of (7.51)) we find

$$f^2_\pi \mu^2_\pi = m_u v_u + m_d v_d \tag{7.56}$$

This is a particular example of a general formula for the pseudo-Goldstone boson mass matrix (for charges Q_a and couplings F_a defined as in eq. (7.53)

$$\mu^2_{ab} = \frac{-1}{F_a F_b} < 0|[[H',Q_a\,],Q_b]\,|0> \tag{7.57}$$

With this formalism introduced, let us now turn back to the Peccei-Quinn solution of the strong CP problem. The X symmetry of (3.17) is extended to quarks in such a way that ϕ_1 couples only to

n_R and ϕ_2 only to p_R by having quarks transform under X as

$$n_R \rightarrow e^{-iX_1\alpha} n_R$$

$$p_R \rightarrow e^{-iX_2\alpha} p_R$$

$$\psi_L \rightarrow \psi_L \tag{7.58}$$

As we saw in IIId, there are two neutral Goldstone bosons in the Higgs potential, G_o and P_o, but G_o is eaten by the gauge bosons. Let us focus our attention therefore on the P_o content of the model. Neglecting the charged Higgs bosons and $\eta_{1,2}$ (which are massive) we may express the relationship between $\phi_{1,2}$ and $\xi_{1,2}$ as

$$\phi_{1,2} = \frac{v_{1,2}}{\sqrt{2}} \begin{pmatrix} 0 \\ 1 \end{pmatrix} e^{i\xi_{1,2}/v_{1,2}} \tag{7.59}$$

(see equations (3.13) and (3.21)). Using eq. (3.25) and defining $\lambda = \tan^{-1}(v_2/v_1)$ we see that the P_o content of $\phi_{1,2}$ is

$$\phi_1 = \frac{v_1}{\sqrt{2}} \begin{pmatrix} 0 \\ 1 \end{pmatrix} e^{-iP_o(\sin\lambda)/v_1} \qquad \phi_2 = \frac{v_2}{\sqrt{2}} \begin{pmatrix} 0 \\ 1 \end{pmatrix} e^{iP_o(\cos\lambda)/v_2} \tag{7.60}$$

Let us repeat, we are aiming to calculate the P_o mass, generated by the explicit breaking of the X symmetry(7.22). The terms in $\mathcal{L}$ where P_o is present using (7.60) and keeping for simplicity only u and d quarks so p = u, n = d are therefore

$$\mathcal{L}(P_o) = \frac{1}{2}\partial_\mu P_o \partial^\mu P_o - \{m_n \bar{n}_L n_R e^{-iP_o(\sin\lambda)/v_1} + m_p \bar{p}_L p_R e^{-iP_o(\cos\lambda)/v_2} + \text{h.c.}\} \tag{7.61}$$

remembering that ϕ_1 couples to n_R and ϕ_2^* to p_R. We see now that $\mathcal{L}$ is invariant under a combined axial rotation of n and p and a translation of P_o

$$n \to e^{-i\alpha v_2\gamma_5/2v_1} n \qquad p \to e^{-i\alpha v_1\gamma_5/2v_2} p$$

$$P_o \to P_o + \alpha\sqrt{v_1^2 + v_2^2} \tag{7.62}$$

Labelling $\sqrt{v_1^2 + v_2^2}$ by f and $y = v_1/v_2$, we see from (7.46), taking $\delta\mathcal{L}/\delta(\partial_\mu \alpha)$, that the associated conserved current is

$$J_\mu = f\partial_\mu P_o + \frac{1}{2y}\bar{n}\gamma_\mu\gamma_5 n + \frac{y}{2}\bar{p}\gamma_\mu\gamma_5 p \tag{7.63}$$

The argument takes a twist now; so far we have been saying that P_o is a pseudo-Goldstone boson, whose mass vanishes unless $m_{p,n} \neq 0$. Peccei and Quinn(7.20), who developed this line of reasoning, pointed out however that the current in equation (7.63) was anomalous and therefore, just as the SU_2 singlet η meson is massive in an $SU_2 \times SU_2$ model, one might expect P_o to be massive. Weinberg and Wilczek(7.23) observed however that P_o only coupled very weakly to instantons (in quarks) and therefore its mass should be small. This precipitated searches for the P_o, usually called the axion. These have not met with success so far.(7.24)

A simple way to see why the axion is so light is to utilize the fact that the anomaly contributes equally to the divergences of $\bar{n}\gamma_\mu\gamma_5 n$ and $\bar{p}\gamma_\mu\gamma_5 p$ (it is after all a U_1 anomaly) to construct the anomaly free current,

$$\tilde{J}_\mu = J_\mu - (\frac{1}{2y} + \frac{y}{2}) \frac{1}{1+Z} \{Z\bar{n}\gamma_\mu\gamma_5 n + \bar{p}\gamma_\mu\gamma_5 p\} \tag{7.64}$$

with Z an arbitrary constant. A second anomaly free axial vector current of a similar form is the axial isospin current

$$J^3_{5\mu} = \frac{1}{2}(\bar{p}\gamma_\mu\gamma_5 p - \bar{n}\gamma_\mu\gamma_5 n) \tag{7.65}$$

Both J_μ and $J^3_{5\mu}$ are conserved as $m_p, m_n \to 0$. Each is associated with a spontaneously broken symmetry and therefore, in the $m_p, m_n \to 0$ limit, there are two massless pseudoscalar Goldstone bosons, respectively the axion and the neutral pion.

Formula (7.57) is then employed to determine the mass of the pion and the axion. The analogue for the axion of f_π, defined in eq. (7.53) is found from the matrix element of J_μ in (7.63) between P_o and the vacuum.

$$F_{axion} = f = \sqrt{v_1^2 + v_2^2} = (\sqrt{\sqrt{2}\, G_F})^{-1} \tag{7.66}$$

Z is fixed so that the boson mass matrix is diagonal, i.e. in eq. (7.57) we have $\mu^2_{AX,\pi^0} = 0$.

$$< 0 | [[H', Q^3_5], \tilde{Q}] | 0 > = 0 \tag{7.67}$$

We immediately see that the masses squared of the axion and the pion are in the ratio f^2/f^2_π or $f^2_\pi G_F\sqrt{2} \sim O(10^{-7})$. A detailed calculation[7.22] gives,

$$\frac{\mu^2_{axion}}{\mu^2_\pi} = \sqrt{2}\, G_F f^2_\pi \,(y + \frac{1}{y}) \,\frac{Z}{(1+Z)^2} \tag{7.68}$$

with $Z = m_p v_p / m_n v_n$ and $v_{p,n}$ given respectively by the p and n quark condensates. The axion to pion decay amplitudes similarly scale as f_π/f.

Interest in the axion, experimental evidence[7.24] to the contrary, has been recently revived by the suggestion that the model include a scalar singlet ϕ_o as well as the two doublets ϕ_1 and ϕ_2.[7.25] The scalar's VEV, v_s does not contribute to gauge boson or fermion masses so we may have $v_s \gg v_{1,2}$. Letting ϕ_o transform non-trivially under X means that the axion is now a linear combination of ϕ_o, ϕ_1 and ϕ_2; for $v_s \gg v_{1,2}$, we have that the axion mass and coupling to matter are both reduced by $\sqrt{v_1^2 + v_2^2}/v_s \ll 1$, so that it could have easily escaped detection. This idea has been further pursued[7.26] by embedding the model in a grand unified theory where v_s may naturally be presumed to be the grand unification scale.

Finally (perhaps this should have come first rather than last), let us present the argument of why θ defined in eq. (7.49) is zero in the Peccei-Quinn model.[7.27] Let the quark Yukawa coupling be

$$\mathcal{L}_M = - m_n \bar{n}_L n_R \phi_1 / v_1 - m_p \bar{p}_L p_R \phi^*_2 / v_2 \tag{7.69}$$

with the phases of the quark fields chosen so that m_n/v_1, m_p/v_2 are real and positive, and $< \phi_{1,2} > = v_{1,2}\, e^{i\omega_{1,2}}$ as in (7.33) so that

$$\text{Arg det } M = \omega_1 - \omega_2 \tag{7.70}$$

We would like to find the minimum of the effective potential, a problem we already mentioned in IIId. This corresponds to finding the true vacuum of the theory from among the vacua connected by unitary transformations U(g), representing the global symmetry group G. Dashen[7.28] showed this was done by minimizing

$$V(g) = < 0 | U(g)^{-1}\, H'\, U(g) | 0 > \tag{7.71}$$

where H' is the $\bar{G}$ symmetry breaking part of the Hamiltonian. In our case H' is given to lowest order by $-\mathcal{L}_M$ with $\phi_{1,2}$ replaced by their VEV's.

To proceed further, we must calculate the fermion condensates. We assume QCD does not break CP spontaneously, so for $\theta = 0$

$$< 0|\bar{u}_L u_R|0 > = ve^{i\sigma} \qquad < 0|\bar{d}_L d_R|0 > = ve^{-i\sigma} \tag{7.72}$$

i.e. the u and d quarks have VEV's which are equal in magnitude (SU_2 symmetry) and can be brought to a real form by the anomaly free axial transformation generated by Q_5^3. This degree of freedom is associated with the pion. For $\theta \neq 0$, we perform an axial U_1 transformation so that the $\theta\tilde{F}.F$ term is absent and proceed as before; the change of course is a θ phase in (7.72) so we now have

$$< 0|\bar{u}_L u_R|0 > = ve^{i(\theta/2 + \sigma)} \qquad < 0|\bar{d}_L d_R|0 > = ve^{i(\theta/2 - \sigma)} \tag{7.73}$$

and, from equation (7.71)

$$V(\theta,\sigma,\omega_1,\omega_2) = 2v \{m_n \cos(\theta/2 - \sigma + \omega_1)$$

$$+ m_p \cos(\theta/2 + \sigma - \omega_2)\} \tag{7.74}$$

We have chosen $v, m_n, m_p > 0$ so the minimum of V is for

$$\cos(\theta/2 - \sigma + \omega_1) = \cos(\theta/2 + \sigma - \omega_2) = -1 \tag{7.75}$$

which implies the desired condition

$$\bar{\theta} = \theta + \omega_1 - \omega_2 = \theta + \text{Arg det } M = 0 \tag{7.76}$$

We see that the theory "naturally" takes on the value $\bar{\theta} = 0$, i.e. this is the true vacuum which minimizes the effective Hamiltonian. Recapitulating, we have four free angular parameters : θ, ω_1, ω_2 and σ. The minimization of the effective Hamiltonian fixes $\bar{\theta}$ equal to zero; the remaining three variables correspond to one true Goldstone boson eaten however by the Z gauge boson, and two pseudo Goldstone bosons, the pion and the axion. The three are associated respectively with $\omega_1 + \omega_2 - \sigma$, and $\omega_1 - \omega_2$.

This concludes our discussion of CP symmetry violation; there are several topics we have not discussed, such as dynamically generated CP symmetry violation(7.29), or the connection between CP symmetry and the baryon asymmetry of the universe (7.30), but we have to stop somewhere.

VIII. THINGS TO LOOK FOR

There are many things we anticipate discovering about electroweak interactions during the next few years. Let us recapitulate and mention a few ones:

a) W, Z Bosons

We anticipate not only discovering them, but as mentioned in Section IV we predict their masses to be related by $M_W/M_Z = \cos\theta_W$. This prediction was based on the assumption that only isodoublet Higgs had VEV's. The general formula for Higgs fields with VEV, v_n and T, T^3 equal to t, t_n^3 is

$$\rho \equiv \frac{M_W^2}{M_Z^2 \cos^2\theta_W} = \frac{\sum_n \{t_n(t_n+1) - (t_n^3)^2\} v_n^2}{2\sum_n (t_n^3)^2 v_n^2} \tag{8.1}$$

Of course if $t_n = 1/2$ for all n, ρ is identically one. Experimentally(4.3)

$$\rho = 0.992 \pm 0.017(\pm 0.011) \tag{8.2}$$

An additional cause of a deviation of ρ from one is the radiative corrections to the masses of the W and Z due to heavy particles (e.g. a $\bar{t}b$ loop would contribute to W while $\bar{t}t$ and $\bar{b}b$ loops contribute to the Z propagator). Their effect is(8.1)

$$\rho = 1 + \frac{G_F}{8\pi^2}\{ m_1^2 + m_2^2 - \frac{2m_1^2 m_2^2}{m_2^2 - m_1^2} \ln (\frac{m_2^2}{m_1^2}) - 3M^2\tan^2\theta_W \ln(\frac{\mu^2}{M^2}) \} \tag{8.3}$$

where m_1, m_2 are the masses of a heavy fermion doublet and M,μ are the masses of the gauge boson and Higgs boson.

b) More Gauge Bosons

Several groups have continued to explore the possibility of the gauge group being larger than $SU_2 \times U_1$, such as $SU_2 \times U_1 \times U_1$ or $SU_2 \times SU_2 \times U_1$.(8.2) In particular the idea of a left-right symmetry being restored at high energy with one SU_2 coupling to left-handed fermions and the other to right handed fermions has continued to interest many authors.

c) New Fermions

We will not discuss technifermions etc. The t quark remains to be discovered; Buras[6.5] has recently set an upper limit on the t quark mass, but this may be overly conservative. It is possible of course that yet more massive exist; several authors have speculated however that the theory as we know may be inconsistent, if e.g. quarks with masses greater than 200-300 Gev exist.[8.3]

d) Higgs Bosons

We have already discussed these somewhat, with a mass range that goes from ev in the case of the axion to Tev (rough estimate of allowable lower bound on neutral flavor changing Higgs boson mass). We emphasize here that if more than one Higgs doublet exists, there necessarily have to be charged Higgs bosons present since the W "eat" only one linear combination. The arguments are not compelling for the existence of several multiplets, but we remind the reader that calculations of the baryon asymmetry of the universe fail by at least 10^8 - 10^{10} in the minimal model (3 families and one Higgs doublet.)[7.30]

e) Rare Decay Modes

Any kind of anomalous decays such as double beta decay or $\mu \to e\gamma$ or $K^+ \to \pi + \nu\bar{\nu}$ would provide invaluable information about the structure of the electromagnetic and weak interacions. Beyond all this are perhaps surprises we have not even contemplated yet.[8.5]

REFERENCES

Section 1

1. See e.g. W. Marciano and H. Pagels, Phys. Reports 36 (1978), 137.
2. E.S. Abers and B. W. Lee, Phys. Reports 9 (1973), 1.
3. M.A. Beg and A. Sirlin, Ann. Rev. Nucl. Science 24 (1974), 379.
4. J.C. Taylor, Gauge Theories of Weak Interactions (Cambridge University Press, 1976).
5. For a recent overall review of gauge theories see P. Langacker, "Grand Unified Theories and Proton Decay", Univ. of Penn preprint 1980, to be published in Phys. Reports.

6. S. Weinberg, Phys. Rev. D8 (1973), 605,
Phys. Rev. D8 (1973), 4482.

7. J.D. Bjorken and S. D. Drell, Relativistic Quantum Fields (McGraw Hill, N.Y. 1965).

Section 2

1. C. N. Yang and R Mills, Phys. Rev. 96 (1954) 91.
R. Utiyama, Phys. Rev. 101 (1956), 1597.
M. Gell-Mann and S. L. Glashow, Ann. Phys. (N.Y.) 15 (1961), 437.

2. See e.g., R. Gilmore, Lie Groups, Lie Algebras and Some of their Applications (Wiley, N.Y. 1974) for a relevant group theory text.

Section 3

1. P. W. Higgs, Phys. Rev. Lett. 12 (1964), 132,
Phys. Rev. Lett. 13 (1964), 508,
Phys. Rev. 145 (1966), 1156;
F. Englert and R. Brout, Phys. Rev. Lett. 13 (1964), 321;
G. S. Guralnik, C. R. Hagen and T. W. B. Kibble, Phys. Rev. Lett. 13 (1964), 585.

2. G. 't Hooft, Nucl. Phys. B33 (1971), 173, ibid. (1971), 1967;
G. 't Hooft and M. Veltman, Nucl. Phys. B50 (1972), 318;
B. W. Lee and J. Zinn-Justin, Phys. Rev. D5, 3121, 3137, 3155.

3. For a review see, B. W. Lee, Chiral Dynamics, (Gordon and Breach, N.Y. 1972).

4. J. Goldstone, Nuovo Cimento 19 (1961), 15;
Y. Nambu, Phys. Rev. Lett. 4 (1960), 380;
Y. Nambu, and G. Jona-Lasinio, Phys. Rev. 122 (1961), 345.

5. The standard reference is S. Weinberg, Phys. Rev. Lett. 29 (1972), 1658 and Phys. Rev. D7 (1973), 2887.

6. For a recent example see e.g. S. Raby, S. Dimopoulos and L. Susskind, Nucl. Phys. B169 (1980), 373.

7. For a lengthier discussion of a similar potential see H. E. Haber, G. L. Kane and T. Sterling, Nucl. Phys. B161(1979), 493.

Section 4

References (1.2), (1.3), and (1.4) give the details of this chapter. We have followed closely reference (1.5) here.

1. W. J. Marciano, Phys. Rev. D20 (1979), 274;
M. Veltman, Phys. Lett. 91B (1980), 95.

2. D. A. Ross and J. C. Taylor, Nucl. Phys. B51 (1973), 125; A. Sirlin , Phys. Rev. D22 (1980), 971.

3. For a review of neutral current data see J.E. Kim, P. Langacker, M. Levine and H.H. Williams, Rev. Mod. Phys. 53 (1981), 211.

4. H. Fritzsch and P. Minkowski, Flavordynamics of Quarks and Leptons, Munich Preprint 1980, to be published in Phys. Reports.

5. M. K. Gaillard, Proceedings of 1979 Symposium on Lepton and Photon Interactions (T. W. Kirk, H. D. Abarbanel, eds. Fermilab) 375.

Section 5

1. S. Adler, Phys. Rev. 177 (1969), 2426; J. S. Bell and R. Jackiw, Nuovo Cimento 51 (1969), 47.

2. For a discussion of this point in $SU_2 \times U_1$ see J. C. Bouchiat, J. Iliopoulos and P. Meyer, Phys. Lett. 38B, (1972), 519.

Section 6

1. We are following the notation of G. Segrè and H. A. Weldon, Ann. Phys. (N.Y.) 124, (1980), 37.

2. S. L. Glashow and S. Weinberg, Phys. Rev. D15 (1973) 1958; E. Paschos, Phys. Rev. D15 (1977), 1966.

3. M. Kobayashi and M. Maskawa, Prog. Theor. Phys. 49 (1973), 652.

4. For a review see J. J. Sakurai, 8th Topical Conference in Particle Physics, ed. V. Z. Peterson and S. Pakvasa (Hawaii, 1980), 375.

5. For a recent paper see e.g., A. Buras, Phys. Rev. Lett. 46 (1981), 1354.

6. For a classic paper on the subject of box diagrams see J. Ellis, M. K. Gaillard and D. V. Nanopoulos, Nucl. Phys. B109 (1976) 213; see also ref. 4.5 and H. Harari, Phys. Reports 42C (1978) 235.

7. R. Gatto, G. Morchio and F. Strocchi, Phys. Lett. 74B (1979) 265, ibid 83B (1979), 348; R. Gatto, Proceedings of Jerusalem Einstein Centennial Symposium (Y.Ne'eman ed. Addison Wesley, publ. 1981) 185.

8. The two examples that follow are from reference (6.1).

9. J. D. Bjorken and S. Weinberg, Phys. Rev. Lett. 38 (1977), 622.

10. For an elementary discussion, see E. Witten, First Workshop on Grand Unification (Math. Sci. Press 1980) and lectures by

De Rujula at this school.

Section 7

1. Recent overall views of CP violation are contained in V. L. Fitch, Rev. Mod. Phys. 53 (1981), 367, J. W. Cronin, Rev. Mod. Phys. 53 (1981), 373. A review of the K^o-$\bar{K}^o$ phenomenology is K. Kleinknecht, Ann. Rev. Nucl. Sci. (1976), 1. For a more thorough discussion, see e.g. T. D. Lee and C. S. Wu, Ann. Rev. Nucl. Sci. (1966) 511.
2. Particle Properties Data Booklet 1980, Rev. Mod. Phys. 52, 2 (1980).
3. L. Wolfenstein, Phys. Rev. Lett. 13 (1964) 562. For a more recent variation see e.g. R. N. Mohapatra, J. C. Pati and L. Wolfenstein, Phys. Rev. D11 (1975), 3319.
4. For the classic papers on this subject see J. Ellis et al., Ref. (6.6). S. Pakvasa and H Sugawara, Phys. Rev. D14 (1976) 305. L. Maiani, Phys. Lett. 62B (1976) 183.
5. F. J. Gilman and M. Wise, Phys. Rev. D20 (1979) 2392; B. Guberina and R. D. Peccei, Nucl. Phys. B163 (1980), 289.
6. R. Bernstein et al., Fermilab Esperiment No. E 617 (1979).
7. T. D. Lee, Phys. Rev. D8 (1973), 1226; Phys. Reports 96 (1979) 143.
8. For an early example see P. Sikivie, Phys. Lett. 65B (1976), 141. For a more recent example see e.g. V. Goffin, G. Segrè and H. A. Weldon, Phys. Rev. D21 (1980), 1410.
9. S. Weinberg, Phys. Rev. Lett. 37 (1976), 67.
10. For a recent discussion of this point see G. Branco, Phys. Rev. Lett. 44 (1980), 504.
11. For a discussion of this formalism and in particular of possible searches for CP violation in heavy quark decay, see C. Albright, J. Smith and S. H. Tye,, Phys. Rev. D21 (1980), 711; K. Shizuya and S. H. Tye, Phys. Rev. D23 (1981), 1613.
12. N. Deshpande, Phys. Rev. D23 (1981), 2654; A. Sanda, Phys. Rev. D23 (1981) 2647.
13. G. F. Donoghue, J. S. Hagelin and B. R. Holstein, "Higgs Exchange Models of CP violation and $K \to 2\pi$", U. Mass preprint (1981).
14. N. F. Ramsey, W. Dress, P. Miller, P. Perrin and J. Pendlebury, Phys. Rev. D15 (1977), 9. I. S. Altarev et al., Phys. Lett. 102B (1981), 13.
15. For a discussion of this section see e.g. S. Coleman, "The Uses

of Instantons", Lecturer at 1977 Ettore Majorana School of Physics, Erice.

16. See e.g. S. L. Adler and R. F. Dashen "Current Algebra and its Application" (W. A. Benjamin, publishers, 1967).

17. V. Baluni, Phys. Rev. D19 (1979), 2227;
R. Crewther, P. Di Vecchia, G. Veneziano and E. Witten, Phys. Lett. 88B (1979) 123.

18. Attempts to circumvent this difficulty include :
H. Georgi, Had. Jour. 1 (1978),155;
M. A. B. Beg and H. S. Tsai, Phys. Rev. Lett. 41 (1978), 278;
N. Mohapatra and G. N. Senjanovic, Phys. Lett. 79B (1978), 283;
G. Segrè and H. A. Weldon, Phys. Rev. Lett. 42 (1979); 141;
S. Barr and P. Langacker, Phys. Rev. Lett. 42 (1979), 1654.

19. P. Langacker and H. Pagels, Phys. Rev. D19 (1979), 2070.

20. R. Peccei and H. Quinn, Phys. Rev. Lett. 38 (1977) 1440;
Phys. Rev. D16 (1977), 1791;

21. For a review see H. Pagels, Phys. Reports 16, 5 (1975).

22. We are following the procedure of W. A. Bardeen and S. H. Tye, Phys. Lett. 74B (1978), 229. Our forms look somewhat different since they have chosen ϕ_1 and ϕ_2 to have hypercharge 1/2 and - 1/2, whereas, for pedagogical reasons, both our ϕ_1 and ϕ_2 have Y = 1/2.

23. S. Weinberg, Phys. Rev. Lett. 40, (1978), 223;
F. Wilczek, Phys. Rev. Lett. 40 (1978), 279.

24. T. Donnelly, S. J. Freedman, R. S. Lytel, R. D. Peccei and M. Schwartz, Phys. Rev. D18 (1978), 1607. Some suggestive positive evidence has however been reported by H. Faissner, Proceedings of Moriond Conference (1981).

25. M. Dine, W. Fischler and M. Srednicki, "A Simple Solution to the Strong CP Problem with a Harmless Axion", Princetón preprint, 1981. For earlier somewhat similar work see J. E. Kim, Phys. Rev. Lett. 43 (1979), 43; M. A. Shifman, A. I. Vainshtein and V.I. Zakharov, Nucl. Phys. B166 (1980), 493.

26. M. Wise, H. Georgi and S. Glashow "SU_5 and the Invisible Axion" Harvard preprint HUTP 81/A019.
H. Georgi, L. Hall and M.Wise "Grand Unified Models with an Automatic Peccei-Quinn Symmetry", Harvard preprint HUTP 81/A031.

27. We are following here the argument as presented in chapter III of J. S. Preskill, "Unified Gauge Theories without Elementary Scalar Fields", Ph. D. thesis, Harvard, 1980. Part of this thesis is an elaboration of recent work on CP by E. Eichten, K. Lane and J. Preskill, Phys. Rev. Lett. 45 (1980), 225.

28. R. Dashen, Phys. Rev. D3 (1971), 1879.

29. See ref. 27; for a recent attempt of a different nature, see A. Masiero, R.N. Mohapatra and R.D. Peccei, "Fermion-Fermion Condensates and CP-Violation", Max-Planck preprint MPI-PAE 25/81 (1981).

30. For an overall view of particle physics and astrophysics see, e.g. A.D. Dolgen and Ya. B. Zeldovich, Rev. Mod. Phys. 53 (1981),1.

Section 8

1. See e.g. M. Veltman, Proceedings of the 1979 International Symposium on Lepton and Photon Interactions (T. W. Kirk, H. D. Abarbanel eds, Fermilab) p. 523.

2. For recent examples see e.g. V. Barger, E. Ma and K. Whisnaut Phys. Rev. Lett. 46 (1981), 1501; E. H. de Groot, G. J. Gounaris and D. Schildknecht, Phys. Lett. 90B (1980), 427; T. G. Rizzo and G. Senjanovic, Brookhaven Lab Report 29397 (1981); R. N. Mohapatra and G. Senjanovic, Phys. Rev. D23 (1981), 105.

3. L. Maiani, G. Parisi and R. Petronzio, Nucl. Phys. B136 (1978), 115; D. Politzer and S. Wolfram, Phys. Lett. 82B (1979), 242; 83B (1979), 421(E).

4. See ref. 7.30.

5. For a recent review of electroweak interactions in which several interesting suggestions are made see J. D. Bjorken "Electro Weak Interactions" Fermilab preprint, Conf. 80/86 (1980).

THE WEAK INTERACTIONS IN THE CONFINING PHASE

E. Farhi

6-412 Physics Department
MIT
Cambridge Ma 02139, U.S.A.

In this talk I will report on work done in collaboration with L.F. Abbott. The basic idea[1] is to describe the weak interactions as arising from the structure of composite left-handed quarks and leptons. The right-handed quarks and leptons are point like at the scale of the weak interactions $G_F^{-1/2} \simeq 250$ GeV. Our model can produce the same effective four Fermi interaction Lagrangian as the standard model. However for Q^2 bigger than zero the predictions of our model differ from those of the standard model.

The scale of physics in the conventional spontaneously broken weak interaction model [2] is determined by the vacuum expectation value of the scalar field. In our approach there is no spontaneous symmetry breakdown. The scale of the weak interactions is the energy where the confining gauge group, $SU(2)_L$, gets strong which we assume to be $G_F^{-1/2} \simeq 250$ GeV. Thus the weak interaction scale is determined by the dynamics of the underlying gauge group just like the scale of hadronic physics is the energy at which the QCD coupling constant gets big, Λ_{QCD}.

We start with the same renormalizable Lagrangian as the standard model [2]. The gauge group is $SU(2)_L \times U(1)$ and there are N left-handed fermion doublets ψ^a (a = 1, N) with hypercharges y_a ($y = 1/6$ for quarks and $- 1/2$ for leptons). There are 2N right-handed fields U_R^a and D_R^a with hypercharges $y_a + 1/2$ and $y_a - 1/2$. U and D can stand for quarks or leptons and I have included the right-handed neutrinos but they are not essential. There is also a fundamental scalar ϕ which is a $SU(2)_L$ doublet with hypercharge $-1/2$.

At the scale $G_F^{-1/2}$ we assume confinement occurs. There is no spontaneous symmetry breakdown. All physical particles are singlets under the gauge group $SU(2)_L$. For the moment I would like to focus our attention on certain gauge singlet composite states. For each fundamental fermion ψ^a we can make two gauge singlet fermions by combining with the scalar ϕ. These are

$$U_L^a = \phi^* \cdot \psi^a \quad ; \qquad D_L^a = \phi \varepsilon \psi^a \tag{1}$$

which correspond to the two ways to make a singlet out of two SU(2) doublets. These states have hypercharges $y_a + 1/2$ and $y_a - 1/2$ which are the sums of the hypercharges of the constituents. Now the hypercharge is unbroken and the hypercharge gauge boson is massless. It couples to the fermions U_L^a, U_R^a, D_L^a, D_R^a with strength proportional to $y_a \pm 1/2$ which are the electric charges. Thus the hypercharge is electric charge and the U(1) coupling constant has the value e where $e^2/4\pi = 1/137$. These fermions in the confined picture correspond to the usual fermions in the broken picture [3].

There are also spin one states in the confined picture which have charges $\pm$ 1 and 0 like the conventional $W^\pm$ and Z. These are the gauge singlet combinations

$$\phi^* \varepsilon D_\mu \phi$$

$$\phi^* D_\mu \phi$$

$$\phi \varepsilon D_\mu \phi \tag{2}$$

There is a spin zero bound state $\phi^*\phi$ which has no electric charge and looks like the neutral Higgs scalar of the conventional model. Thus there is a correspondance [3] between the states in the broken picture and some subset of the physical states in the confining picture. This is the sense of complementarity discussed in ref. 3.

I will now show that this confining version of the weak interactions is phenomenologically successful [1]. There are two parts to this : the spectrum and the interactions. We have postulated that the observed left-handed quarks and leptons are bound states at the scale $G_F^{-1/2} \simeq 250$ GeV. However they are observed to have masses which are near zero on this scale. How can this be ?

To answer this it is necessary to identify the global symmetries of the Lagrangian we are using. At the weak interaction scale we can neglect electromagnetism and color interactions. Also, for the moment, turn off the Yukawa couplings. There are N left-handed fermion doublets ψ^a so there is an SU(N) symmetry. (The associated U(1) has an anomaly from $SU(2)_L$). There is also an SU(2N) symmetry associated with the right-handed particles but this is not important now since the right-handed fields do not feel the strong force. In addition the scalar sector has an SU(2) global symmetry. To see this write the scalar field as a 2 x 2 matrix

$$\phi = \begin{pmatrix} \varphi_1 & -\varphi_2^* \\ \varphi_2 & \varphi_1^* \end{pmatrix} \tag{3}$$

Now the Lagrangian is invariant under the transformation

$$\phi \to U_L \phi U_R^\dagger \tag{4}$$

where U_L and U_R are independent SU(2) matrices. The transformations associated with U_L are gauged and those given by the action of U_R are the SU(2) global. Thus the strong interaction $SU(2)_L$ theory has the global symmetry group SU(2) x SU(N).

Let us return now to the question of fermion masses. There are two types of masses to consider: Dirac masses connecting right-handed to left-handed fields and Majorana masses which would involve only left-handed fields. In the absence of Yukawa couplings the appearance of mass terms would violate the global symmetries discussed above. Therefore you could say that masses are explicitely prohibited by symmetry considerations. However we know from QCD that strong interactions can spontaneously break chiral symmetries. Do the strong $SU(2)_L$ interactions break these chiral symmetries ?

For the case of Dirac masses connecting left-handed to right-handed fields, this seems impossible because the right-handed fields do not feel the strong force. It would be like imagining that QCD induced a condensate involving the proton and the electron. For a mass term involving only the left-handed fields, it seems possible. However we have shown that the assumption of no chiral symmetry breakdown is consistent with 't Hooft's anomaly conditions and is therefore quite plausible.

't Hooft has argued [4)] that for a strong interaction force to produce massless composite fermions certain algebraic conditions must be satisfied. These state that any anomalous triangle diagram with three global symmetry generators at the vertices should be computable using the fields in the Lagrangian or the physical massless bound states. The result must be the same if the global symmetry is not broken. In our case the only non trivial triangle diagram involves three SU(N) currents. This can be computed using the fundamental fermions ψ^a which come in two's because they are doublets under $SU(2)_L$. Or we can use the bound states $\phi^*\psi^a$ and $\phi\varepsilon\psi^a$ which again form two

$\underline{N}$'s of SU(N). The results agree and it is consistent to view our bound state fermions as massless.

In the presence of Yukawa couplings these chiral symmetries are explicitely broken. Yukawa couplings are characterized by dimensionless couplings λ and we expect these to give rise to fermion masses $\sim \lambda G_F^{-1/2}$. Since the λ parameters in the standard Lagrangian are $G_F^{1/2}\, m_{particle}$ (a very small number) we get masses which are of the right magnitude. Unfortunately we shed no light on the origin or size of the Yukawa couplings which appear in the Lagrangian.

I have shown that the observed fermions have masses which are small compared to the weak interaction scale. I will now show that the interactions amongst these particles are consistent with experimental observation. At low energy all weak interaction data are consistent with the following effective four Fermi interaction:

$$\frac{4G_F}{\sqrt{2}} \{J^+_{\mu L} J^-_{\mu L} + (J^3_{\mu L} - \sin^2\theta J^\mu_{em})^2\} \tag{5}$$

This effective four Fermi interaction has only been tested at $Q^2=0$ in the sense that no propagator effects have ever been positively measured.

First I will show that by looking only at the interactions of our composite bound state fermions with heavy vector mesons, the interaction 5 follows easily. Then I will consider other possible interactions and argue that they are small.

To begin, consider the interaction of the bound state fermions with the composite spin one particles of equation (2). Call these particles $W^\pm_\mu$ and W^o_μ. Under the global SU(2) symmetry they form a triplet and therefore they have equal masses up to electromagnetic corrections and corrections involving the Yukawa couplings. (These latter corrections are proportional to mass differences within fermion multiplets). The two fermions U^a_L and D^a_L of equation (1) form a (2,N) under the global group SU(2) x SU(N). (Remember that all of these particles are $SU(2)_L$ singlets). The operator of lowest dimension which characterizes the interaction of these particles must be of the form:

$$\frac{g}{2} W^\chi_\mu \sum_a (\bar{U}^a_L \bar{D}^a_L)\gamma^\mu \tau^\chi \begin{pmatrix} U^a_L \\ D^a_L \end{pmatrix} \tag{6}$$

if it is to respect the SU(2) x SU(N) symmetry.

Now the exchange of the two W particles gives rise to an

effective four Fermi interaction of the form:

$$\frac{4g^2}{8M^2}\left(J^+_{\mu L}J^-_{\mu L} + J^3_{\mu L}J^3_{\mu L}\right) \tag{7}$$

where

$$J^X_{\mu L} = \sum_a \left(\bar{U}^a_L \bar{D}^a_L\right)\gamma^\mu \tau^X \begin{bmatrix} U^a_L \\ D^a_L \end{bmatrix} \tag{8}$$

and M is the mass of the W particle. You will notice that (7) is just (5) when $\sin^2\theta = 0$ if you identify

$$\frac{G_F}{\sqrt{2}} = \frac{g^2}{8M^2} \tag{9}$$

Now g is an effective coupling constant determined by the underlying strong interaction dynamics. Its value cannot be calculated by present methods. However it is probably a number of order 1 just like the effective g-N-N coupling of hadrons. This determines $M \sim 125$ GeV although this is very rough and it could be much larger. (The work of Bjorken [5], however, indicates that the mass of the W^+ cannot be larger than 167 GeV). The exact value is not important for low energy physics but would be precisely measured at LEP or in $p\bar{p}$ colliders.

There is, of course, more to our effective interaction if we no longer ignore electromagnetism. The bound state left-handed fermions will have electromagnetic form factors due to their structure. For example at $Q^2 \neq 0$, photons will interact with neutrinos if they are bound states. Let us make a simple model for these form factors based on analogies with vector meson dominance [6]. Suppose the photon interacts with the left-handed fermions by first becoming a strongly interacting W^o. Characterize the effective interaction between the photon A_μ and the W^o by a single dimensionless number, X, near $Q^2 = 0$, so that their mixing may be written as

$$XQ^2 A_\mu W^{\mu o} \tag{10}$$

Then the two diagrams in fig. 1. will give the full neutral current of equation (5) if X is related to the measured $\sin^2\theta$.

In this theory the parameter X is determined by the underlying dynamics and it is also not calculable using known techniques. It can be estimated by analogies with the similar $g\gamma$ mixing of hadronic physics and this estimate gives a value of X about one tenth of the measured value. It is important to understand how to estimate X precisely since the W^o is very different than the g. Perhaps X could be computed on the lattice.

At this point you may object and say that there must be an infinite variety of interactions induced by the strong $SU(2)_L$ which compete with the ones I have described. Suppose we go to the limit where SU(N) x SU(2) global is an exact symmetry, that is, all fermion masses are equal and we ignore electromagnetism. Any effective four Fermi interaction at low energies must respect this global symmetry. The most general four Fermi interactions we can write down, consistent with the symmetries and remembering that the bound states are left-handed, are:

$$\sum_X J^X_{\mu L} J^X_{\mu L} \quad \text{and} \quad J^o_{\mu L} J^o_{\mu L} \qquad (11)$$

where $J^X_{\mu L}$ is given by equation (8) and $\tau^o = 1$. Now the first term is exactly the one we want: the usual charged weak interaction and part of the neutral current (see equation (7)). The second term is not observed experimentally and must have a coefficient which is less than about 1/20 times the coefficient of the charged current interaction.

The second term can be generated by $SU(2)_{global}$ isoscalar exchange just as the first term comes from isovector exchange. If we try to construct the isoscalar partner of the $W^{\pm}W^o$ made out of ϕ's we find that it vanishes by bose statistics. The interaction can be generated in higher orders or through other states but it will be different than the first term. Perhaps this is why it is somewhat suppressed relative to the first term.

The interactions of equation (11) do not permit any flavor changing neutral currents. However by introducing mass differences between the fermions we break the global symmetries and flavor changing

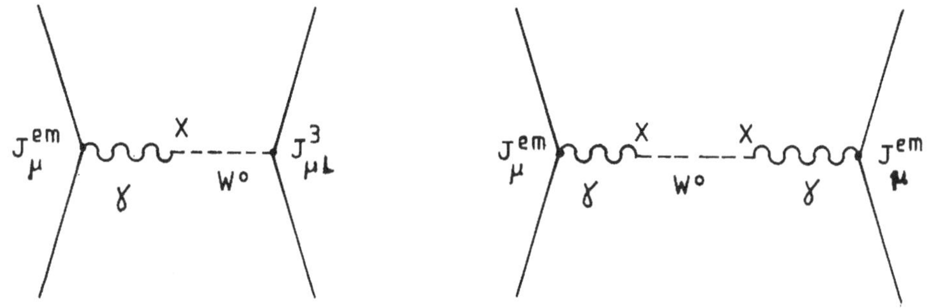

Fig. 1

neutral currents are induced, which vanish when all fermion masses are equal. This model has the GIM[7)] mechanism built in at the start. In fact if you consider any diagram which gives rise to neutral flavor changing currents you find that it is highly convergent and its value is determined by loop momenta of the order of the quark masses. At these low momenta the bound state nature of quarks, leptons and vector bosons is not evident and we are guaranteed to have the same numerical results as the standard model. Thus our calculation of the $K_L K_S$ mass difference agrees with that of Gaillard and Lee [8)].

You might also worry that strong $SU(2)_L$ instantons [9)] might induce baryon violations or strong CP problems which are not present in the standard model. However we have shown that the baryon violations are orders of magnitude below the experimental limits and the strong $SU(2)_L$ CP violating phase parameter can be rotated away.

Thus by making very few dynamical assumptions we have been able to construct a model of the weak interactions without spontaneous symmetry breakdown. The theory is based on a renormalizable field theory and it obeys 't Hooft's consistancy conditions. At energies near 100 GeV the strong nature of the $SU(2)_L$ force should be apparent. The physics in high energy colliding beams will be wild and exciting.

REFERENCES

1. L.F. Abbott and E. Farhi, Phys. Lett. 101B, 69, (1981) and Nucl. Phys. B189, 547, (1981)

2. S. Weinberg, Phys. Rev. Lett. 19 (1967), 1264
A. Salam in Elementary Particle Theory, Nobel Symp. N° 8, ed. N. Svartholm (Almquist and Wiksell, Stockholm, 1968)

3. T. Banks and E. Rabinovici, Nucl. Phys. B160; 349, (1979)
E. Fradkin and S.H. Shenker, Phys. Rev. D19, 3682, (1979)
G. 't Hooft, Cargèse Summer Institute (1979)
S. Dimopoulos, S. Raby, L. Susskind, Nucl. Phys. B173, 208,(1980)

4. G. 't Hooft in reference 3

5. J.D. Bjorken, Phys. Rev. D19, 335, (1979)

6. See J.D. Bjorken in reference 5 as well as P.Q. Hung and J.J. Sakurai, Nucl. Phys. B143, 81, (1978)

7. S.L. Glashow, J. Iliopoulos and L. Maiani, Phys. Rev. D2, 1285, (1970)

8. M.K. Gaillard and B.W. Lee, Phys. Rev. D10; 897 (1974)

9. G. 't Hooft, Phys. Rev. Lett. 37, 8, (1976)

HEAVY QUARK SYSTEMS

A. Martin

CERN, Geneva

ABSTRACT

Properties of systems containing heavy quarks, i.e. beautiful, charmed, and even strange quarks are studied. First we discuss quarkonium in the flavor-independent potential approach, then mass predictions for objects with visible beauty, and finally compare forces between quarks inside mesons and baryons and give as an illustration a calculation of the Ω^- mass.

A detailed version of this talk can be found, under the same title, in the Proceedings of the EPS International Conference on High Energy Physics, Lisbon, July 1981, J. Dias De Deus and J. Soffer Editors.

ELECTRON-POSITRON INTERACTIONS AT HIGH ENERGIES

G. Wolf

DESY
Notkestrasse 85
2000 Hamburg 52

I. INTRODUCTION

Electron-positron annihilation is a fascinating part of particle physics. Unlike in hadronic collisions the final states produced in e^+e^- annihilation possess well defined quantum numbers - those of the mediating current. Furthermore, at high energies, the current couples directly to the fundamental constituents of matter. In fact, all reactions observed in the nonresonant region follow from the same underlying process, pair creation of pointlike charged spin 1/2 fermions (either charged leptons or quarks). For production by the electromagnetic current - which is by far the dominant contributor at present energies - the cross sections for these processes can be calculated. As a result, e^+e^- annihilation is ideally suited to search for new charged leptons, new quarks or any other type of particle that is charged and pointlike. It is also a good place to search for new currents. If they couple to e^+e^- they lead to a breakdown of QED. First evidence for the presence of the weak neutral current has been seen just recently at PETRA. The angular distribution for muon pair production when combining the data from all PETRA experiments shows a distinct forward-backward asymmetry.

High energy e^+e^- annihilation has also become a testing ground for theories of strong interactions. The acceleration of quarks in $q\bar{q}$ formation, $e^+e^- \to q\bar{q}$, becomes so large that quarks have a fair chance to radiate energetic noncollinear field quanta that can be detected. The three-jet events observed at PETRA have given strong support to OCD where such events follow from bremsstrahlung of gluons. The rate of three-jet production provides a direct measurement of the strong coupling strength α_s.

Electron-positron interactions offer also the opportunity to study photon photon scattering. Experiments have just started to explore this field and many of the data obtained are still very preliminary. Nevertheless, the indications are that a wealth of new physics is waiting for discovery.

These lectures concentrate on electron-positron interaction data obtained above the upsilon family, $W \gtrsim 12$ GeV. They also includes material presented only after the school at the Bonn Conference. No attempt is made to cover all experimental data, rather some topics are selected which appear to be of particular interest. More information on e^+e^- interactions at high energies can be found in recent review articles (see e.g. Wiik & Wolf 1979, Wolf 1980, Söding & Wolf 1981, Renard 1981), and in rapporteur reports from the Lisbon and Bonn conferences (Berger 1981, Duinker 1981, Holder 1981, Marshall 1981, Branson 1981, Braunschweig 1981, Bürger 1981, Felst 1981, Fournier 1981, Hollebeek 1981, Litke1981, Wedemeyer 1981).

II. PETRA AND PEP

1. PETRA

Most of the experimental data available at present were obtained at the DESY storage ring PETRA (Fig. 1) which was commissioned in autumn 1978. PETRA has four short and four long straight sections with a total circumference of 2.3 km. Two of the long straight sections are used for the accelerating structures. The maximum total c.m. energy W attained is 36.8 GeV. It is planned to increase in 1982 the maximum energy to 41 GeV by doubling the RF-power and to 46 GeV in 1983 by doubling the number of cavities. The maximum luminosity at W = 30 GeV in 1979 and 1980 was around $5.10^{30}cm^{-2}s^{-1}$ which lead to an integrated luminosity of 100 - 150 nb^{-1} per day. At the end of 1980, by installing additional quadrupoles in the interaction region (minibeta scheme), the luminosity was increased by a factor of three to $1.7.10^{31}$ $cm^{-2}s^{-1}$ and 400 - 600 nb^{-1}/day. This gives roughly 100 hadronic annihilation events per day and experiment.

Five large detectors were constructed and took data, CELLO, JADE, MARK J, PLUTO and TASSO. CELLO and PLUTO share the same interaction pit in a push-pull fashion. PLUTO was taking data in 1979, CELLO in 1980 and in the first half of 1981. All detectors have almost complete solid angle coverage for charged particles and photons, and, except for MARK J, use a solenoidal magnet filled with cylindrical drift and proportional chambers for charged particle tracking. The MARK J detector is a calorimeter.

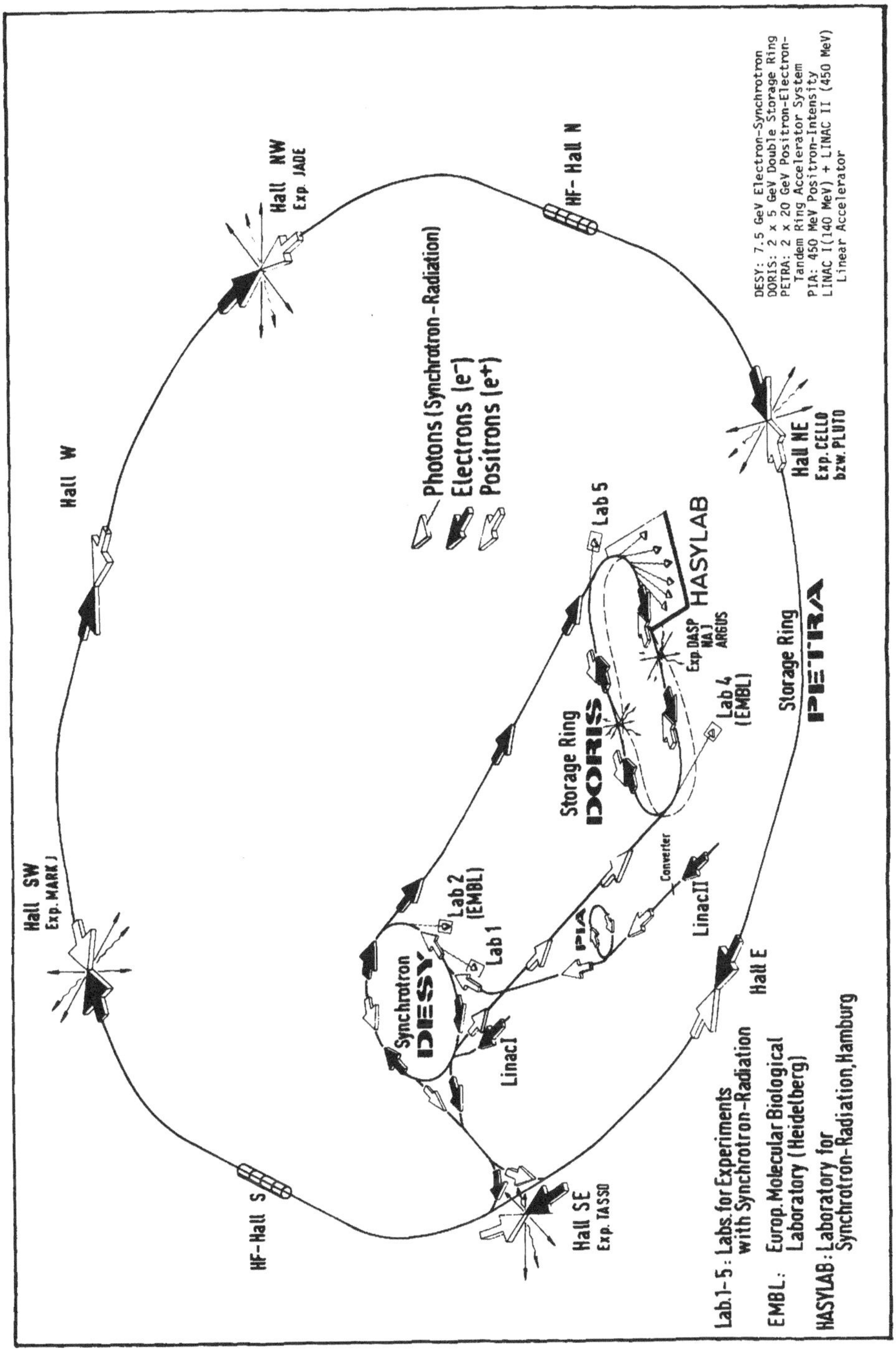

Fig. 1. Layout of the DESY accelerator complex

2. PEP

The SLAC storage ring PEP (see Fig. 2) has been put into operation in the summer of 1980. It has a total circumference of 2.2 km and a maximum design c.m. energy of W = 36 GeV. There are six interaction regions and seven approved experiments: MARKII, MAC, Free Ouark Search, Monopole Search, DELCO, HRS, and TPC. First preliminary results from the first four experiments were reported at the Bonn Conference (Hollebeek 1981, Litke 1981).

III THE STRUCTURE OF LEPTONS

Muon pair production,

$$e^+e^- \to \mu^+\mu^- \qquad (1)$$

is the simplest of all QED reactions and the prototype of almost all e^+e^- annihilation processes at high energies. It proceeds via time-like photon exchange (Fig. 3).

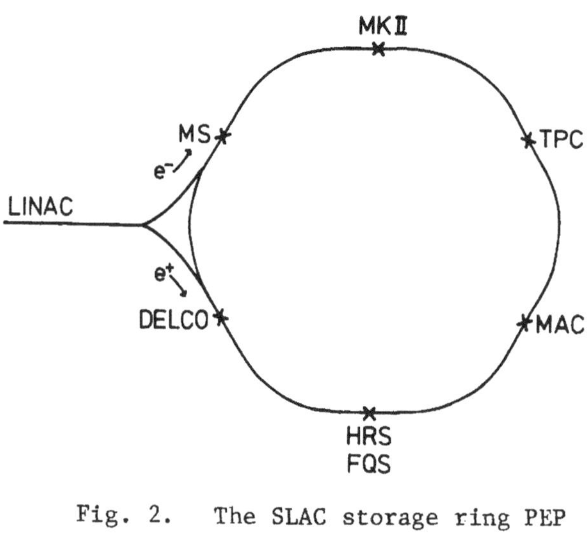

Fig. 2. The SLAC storage ring PEP

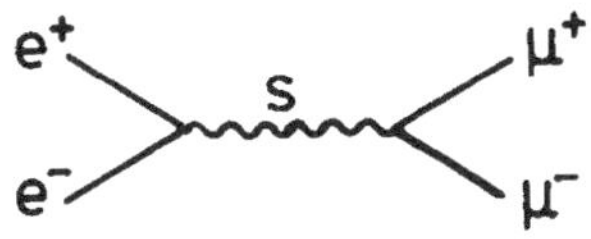

Fig. 3. Muon pair production

The differential cross section reads

$$\frac{d\sigma}{d\Omega} = \frac{\alpha^2}{4s} \beta_\mu \{(1 + \cos^2\theta) + (1 - \beta_\mu^2) \sin^2\theta \} \qquad (2)$$

where $s = W^2$, $\beta_\mu = p_\mu/E_\mu$. For $p_\mu \simeq E_\mu$

$$\frac{d\sigma}{d\Omega} = \frac{\alpha^2}{4s} (1 + \cos^2\theta)$$

The integrated cross section is given by

$$\sigma_{\mu\mu} = \frac{4\pi}{3} \frac{\alpha^2}{s} = \frac{87.6 \text{ nb}}{s} \qquad (s \text{ in GeV}^2) \qquad (3)$$

At W = 30 GeV the muon pair cross section is of the order of 0.1 nb. An integrated luminosity of 500 nb^{-1}/day will yield a maximum detectable event rate of 50 per day.

If electron or muon are extended objects a deviation from the QED prediction will be observed. This deviation can only depend on the mass squared s of the virtual photon and can be parametrized by a cut-off parameter Λ,

$$\sigma = \sigma_{QED} \left(1 \mp \frac{s}{s - \Lambda_\pm^2}\right)^2 \qquad (4)$$

Clearly, the higher the energy, the deeper one can probe for a structure of electron and muon. The parametrization (4) includes also corrections to the photon propagator or the exchange of heavy photon-like objects.

Other second order QED processes which were studied to test QED are:

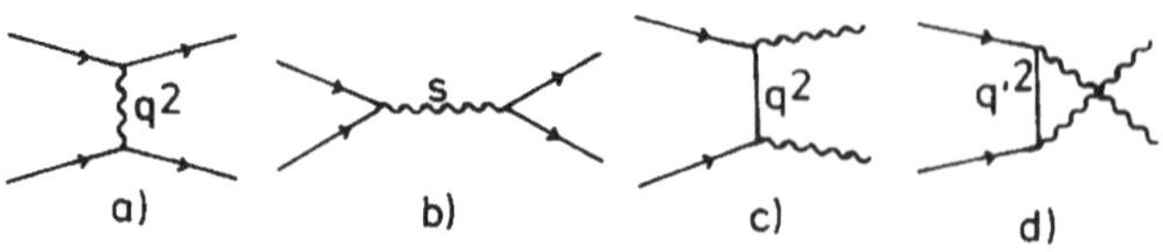

Fig. 4. The space-like (a) and time-like (b) diagrams for Bhabha scattering and the diagrams for two-photon annihilation (c, d)

Bhabha scattering (Fig. 4a,b): $e^+e^- \rightarrow e^+e^-$

Two-photon annihilation (Fig. 4.c,d): $e^+e^- \rightarrow \gamma\gamma$

τ-pair production: $e^+e^- \rightarrow \tau^+\tau^-$

Besides the electron structure, Bhabha scattering tests the photon propagator in the space-like and time-like region. Two-photon annihilation is sensitive to contributions from a seagull term (Fig. 5a) and from excited electrons (Fig. 5b). All four processes have the same s-dependence, $d\sigma/d\Omega \sim s^{-1}$. Fig. 6 summarizes their differential cross sections as predicted by QED. For a recent review see Dittmann & Hepp (1981).

The experimental data have to be radiatively corrected before a comparison with QED can be made:

$$\frac{d\sigma}{d\Omega}^{corr} = \frac{d\sigma}{d\Omega}^{meas} (1 - \delta) \tag{5}$$

The corrections account for photon emission, for electromagnetic

Fig. 5. Seagull term (a) and excited electron exchange (b) for $e^+e^- \rightarrow \gamma\gamma$

modifications of the vertices and for vacuum polarization due to leptons and hadrons (see e.g. Berends and Kleiss 1981).

Possible contributions from weak neutral currents will be discussed in the next chapter.

Fig. 7 displays the total cross section data on μ pair production. They are seen to agree rather well with the QED prediction (solid line). Agreement with QED is also observed for Bhabha scattering (Fig. 8), for two-photon annihilation (Fig. 9) and for τ pair (Fig. 10a).

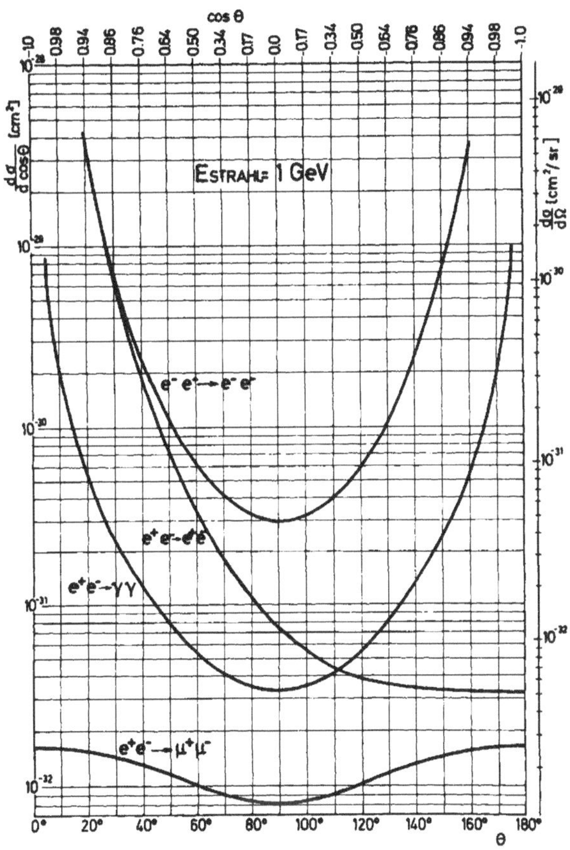

Fig. 6. Differential cross sections for $e^-e^- \to e^-e^-$, $e^+e^- \to e^+e^-$, $e^+e^- \to \gamma\gamma$ and $e^+e^- \to \mu^+\mu^-$ for a beam energy of 1 GeV.

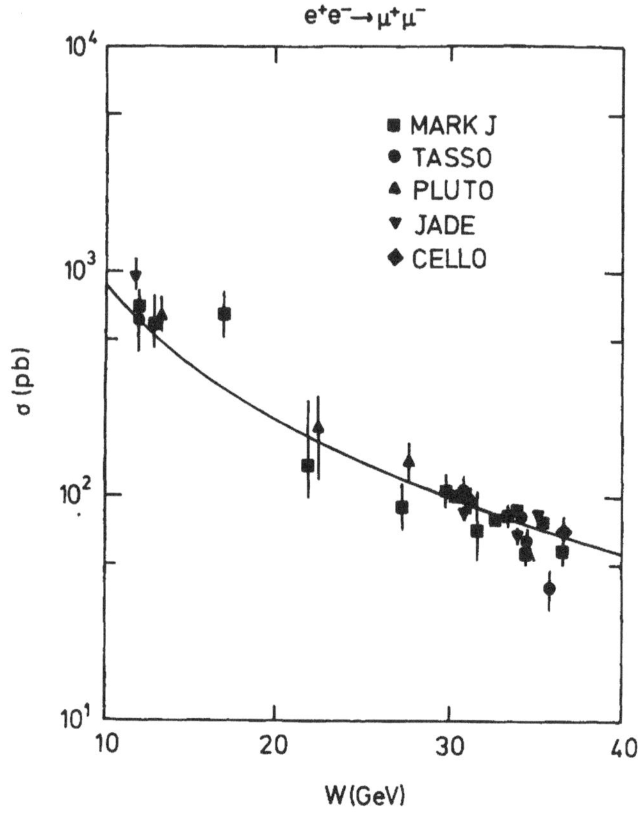

Fig. 7. The total cross section for $e^+e^- \to \mu^+\mu^-$. The solid curve shows the QED prediction $\sigma_{\mu\mu} = 4\pi\alpha^2/3s$. From Branson (1981).

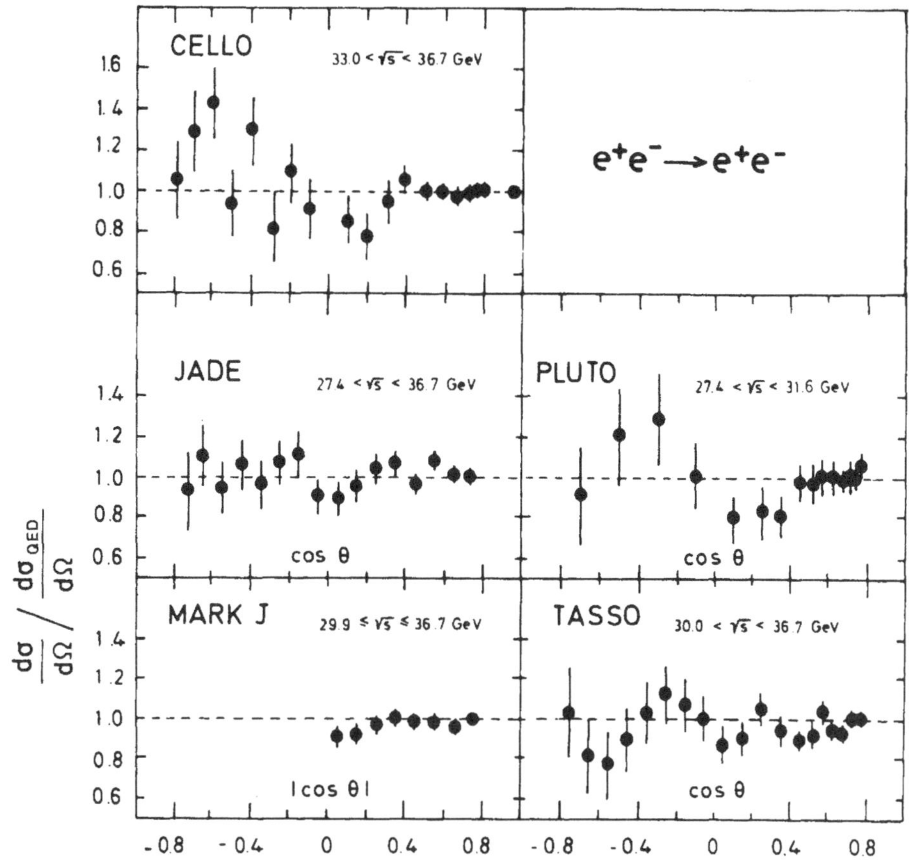

Fig. 8. Differential cross section for the reaction $e^+e^- \rightarrow e^+e^-$ divided by the QED prediction. The data are corrected for radiative effects and hadronic vacuum polarization. From Dittmann & Hepp (1981)

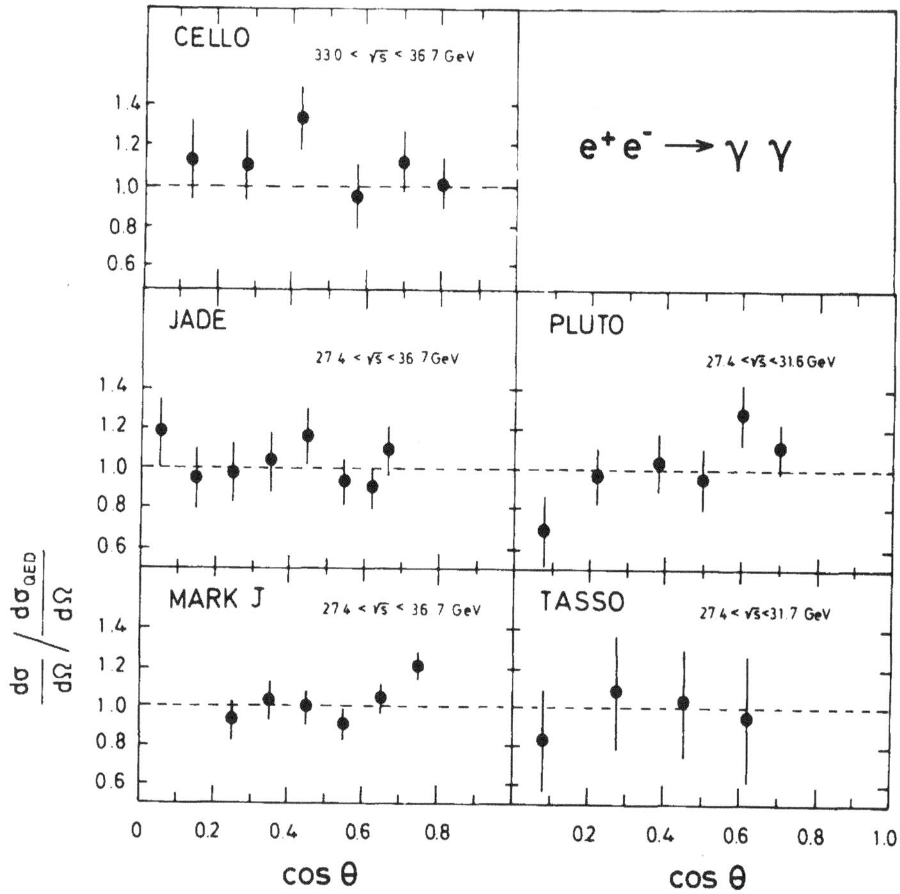

Fig. 9. Differential cross section for the reaction $e^+e^- \rightarrow \gamma\gamma$ divided by the OED prediction. The data are corrected for radiative effects. From Dittmann & Hepp (1981)

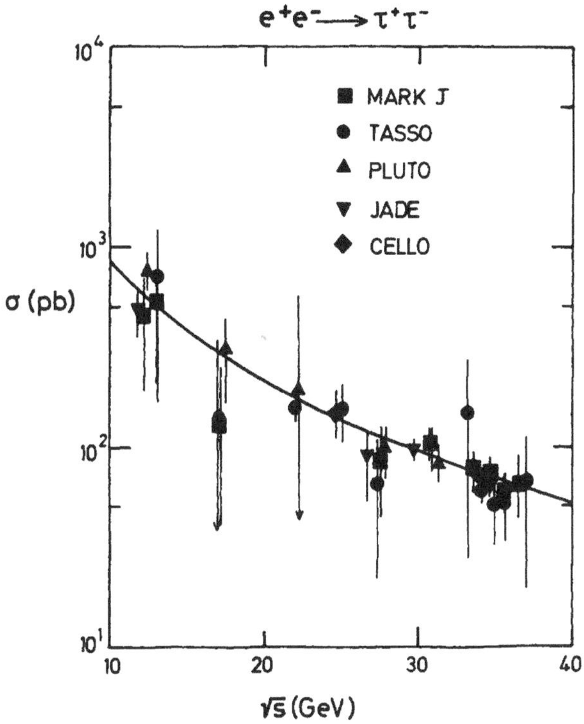

Fig. 10a. The total cross section for $e^+e^- \rightarrow \tau^+\tau^-$. The solid curve shows the QED prediction. From Branson (1981)

Table 1 summarizes the 95% confidence lower limits on the cut-off parameter Λ. The lower limits on Λ are found to be in the range from 100 to 200 GeV. Bearing in mind that we made the implicit assumption that whatever modifies the photon propagator or the lepton-lepton photon vertex has the coupling strength e the results of table 1 can be rephrased by saying that QED has been tested down to distances of 1 - 2.10^{-16} cm and/or that e,μ and τ are pointlike down to this distance.

Table 1. QED cut-off parameters: 95% confidence lower limits in GeV. From Branson (1981) and Hollebeek (1981)

Experiment	$e^+e^- \to e^+e^-$ Λ_+	Λ_-	$\mu^+\mu^-$ Λ_+	Λ_-	$\tau^+\tau^-$ Λ_+	Λ_-	$\gamma\gamma$ Λ_+	Λ_-
CELLO	83	155			139	120	43	48
JADE	112	106	142	126	111	93	47	44
MARK J	128	161	194	153	126	116	55	38
PLUTO	80	234	107	101	79	63	46	-
TASSO	140	296	127	136	104	189	34	42
MARKII							50	41

The study of τ pair production provided first measurements on the τ lifetime. Fig. 10b shows a typical $\tau^+\tau^-$ event with one τ decaying into a μ and the other one into three charged particles. Using the latter tracks one can triangulate the decay vertex and obtain the τ lifetime T from the distance between beam position and decay vertex. The expected flight path at W = 30 GeV is typically 700μ. The TASSO group (Brandelik et al. 1980a. and Woodworth (1981) 1981) found

$$< T_\tau > = - 0.25 \pm 3.5 \; 10^{-13} \text{ sec}$$

and a 95% C.L. upper limit of 5.7 10^{-13} sec. A more precise value was obtained by the MARKII group (Hollebeek 1981). Fig. 10c shows their τ decay length distribution. There is an excess of τ's with positive decay lengths. From this distribution a τ lifetime of

$$< T_\tau > = 4.9 \pm 1.8 \; 10^{-13} \text{ sec}$$

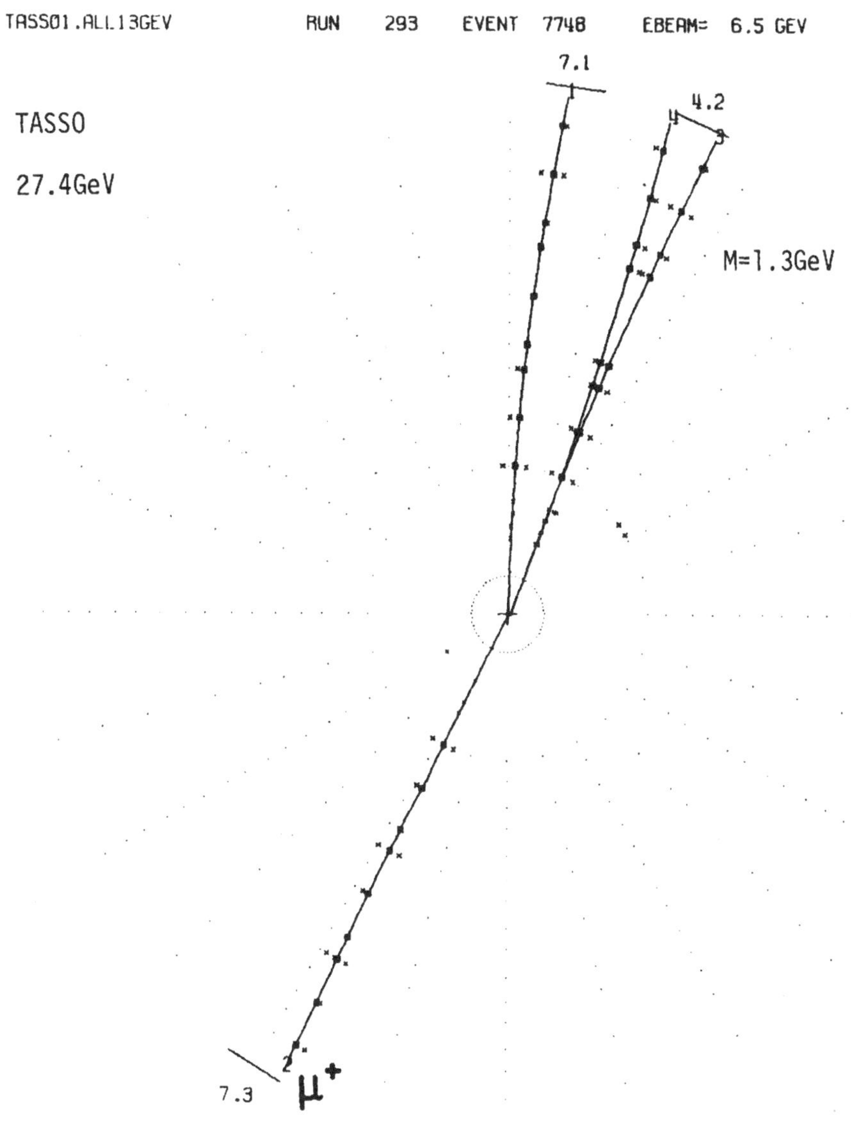

Fig. 10b. Candidate for τ pair production, as observed by TASSO at 13 GeV.

$e^+e^- \to \tau^+ \quad \tau^-$

$\quad\quad \hookrightarrow \mu^+\nu\nu \quad \hookrightarrow \nu + 3$ charged

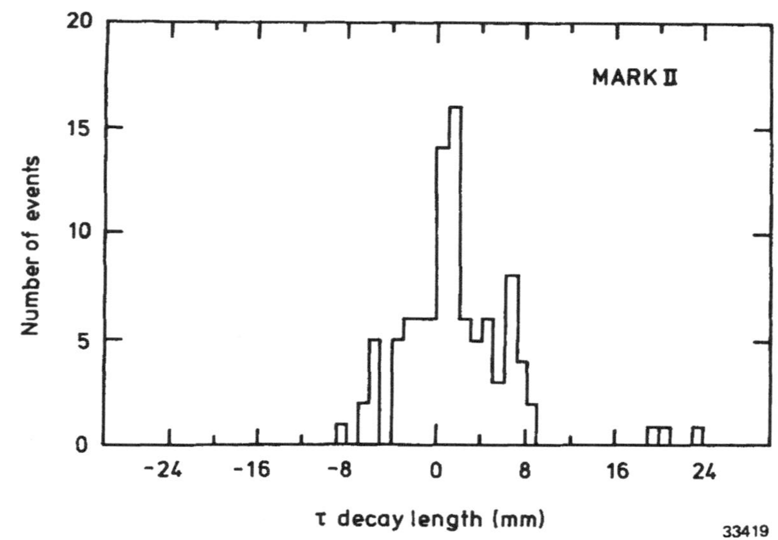

Fig. 10c. The decay length distribution for τ's measured by MARK II. From Hollebeek (1981)

is deduced. Based on the assumption that the $\tau - \nu_\tau$ coupling to the weak charged current is of the same strength as for $e - \nu_e$ theory predicts $< T_\tau > = 2.8\ 10^{-13}$ sec, which is in agreement with the data.

IV. WEAK NEUTRAL CURRENT CONTRIBUTIONS TO LEPTON PAIR PRODUCTION

The standard theory predicts neutral weak current contributions via Z^o exchange to lepton pair production, $e^+e^- \to \ell^+\ell^-$ (see e.g. Fig. 11).

The matrix element for the diagrams in Fig. 11 is given by (see e.g. Hepp and Dittmann 1981)

$$M = -\frac{4\pi\alpha}{s}\,\bar{\ell}\gamma_\mu \ell \bar{e}\gamma^\mu e - \left(\frac{G_F/(2\sqrt{2})}{\frac{s}{m_Z^2} - 1 + i\frac{\Gamma}{m_Z}}\right) \bar{\ell}\gamma_\mu(v + a\gamma_5)\ell\bar{e}\gamma^\mu(v + a\gamma_5)e \tag{6}$$

where $v \equiv 2g_V$, $a \equiv 2g_A$, g_V, g_A are the vector and axial vector couplings, $G_F = 1.02\ 10^{-5}/m_p^2\ GeV^{-2}$, m_Z, Γ_Z are the Z^o mass and width. In writing eq.(6) lepton universality, i.e. g_V and g_A to be the same for e and μ, was assumed.

It is convenient to define

$$D \equiv \frac{G_F}{8\sqrt{2}\pi\alpha}\ \frac{s}{s/m_Z^2 - 1} \equiv g\cdot \frac{s}{s/m_Z^2 - 1} \tag{7}$$

where $\Gamma_Z \ll m_Z$ was assumed and $g = 4.5\ 10^{-5}\ GeV^{-2}$. At PETRA/PEP energies where $s \ll m_Z^2$

$$D \simeq -\,g\cdot s \tag{8}$$

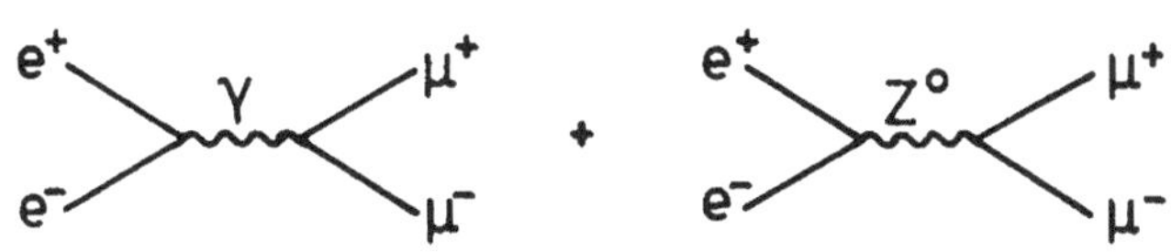

Fig. 11. The electromagnetic and weak contributions to μ pair production

For $W = 35$ GeV, $D \simeq -0.065$. The differential and total cross sections for $\gamma + Z^o$ exchange read

$$\frac{d\sigma}{d\Omega} = \frac{\alpha^2}{4s} [(1 + \cos^2\theta) \{ 1 + 2v^2 D + (v^2 + a^2)^2 D^2 \} + 4 \cos\theta \{ a^2 D + 2v^2 a^2 D^2 \}] \tag{9}$$

$$\sigma(\gamma + Z^o) = \sigma_{\mu\mu}(\gamma) [1 + 2v^2 D + (v^2 + a^2)^2 D^2] \tag{10}$$

The Z^o contribution leads to a deviation from the total cross section predicted by QED and to a forward-backward asymmetry due to the presence of a $\cos\theta$ term.

In the standard model (Glashow 1961, Salam 1962, Weinberg 1967) the size of these effects can be calculated. For all leptons:

$$g_A = -0.5 \qquad g_V = -0.5 (1 - 4\sin^2\theta_W) \tag{11}$$

Taking the current value for θ_W from the neutrino nucleon data, $\sin^2\theta_W = 0.228$, one finds $g_V \simeq 0$. The weak contribution to $\sigma_{\mu\mu}$ expected at present energies is therefore very small,

$$\sigma_{\mu\mu}(\gamma + Z^o) \simeq \sigma_{\mu\mu}(1 + a^4 D^2) \simeq 1.003\, \sigma_{\mu\mu}$$

and cannot be detected with the available statistics.

The forward-backward asymmetry is given by

$$A = \frac{d\sigma/d\Omega(\theta < 90°) - d\sigma/d\Omega(\theta > 90°)}{d\sigma/d\Omega(\theta < 90°) + d\sigma/d\Omega(\theta > 90°)}$$

$$= \frac{3}{2} \frac{a^2 D + 2v^2 a^2 D^2}{1 + 2v^2 D + (v^2 + a^2)^2 D^2} \tag{12}$$

which in the standard model reduces to

$$A \simeq 1.5\, D$$

At $W = 35$ GeV the expected asymmetry is sizeable, $A = -0.10$, and should be observable with a few thousand muon pair events.

In extracting A from the data one has to correct $d\sigma/d\Omega$ for radiative effects which also produce a forward-backward asymmetry. Fig. 12 depicts some of the QED diagrams of order α^2 to α^4 which have to be included in the calculation (see e.g. Berends, Gaemers and Gastmans 1973).

The occurrence of a forward-backward asymmetry can qualitatively be understood by noting that the muon pairs produced from some of the diagrams have negative C parity - in particular from the lowest order diagram (a), but also from (b), (d) - (f) - while other diagrams such as (g), (h) and the box diagrams (i), (j) produce $C = + \mu$ pairs. The interference between the $C = -$ and $C = +$ diagrams will produce a forward-backward asymmetry. Its size depends on the experimental conditions; e.g. different cuts in the acollinearity angle ζ between the muons or in the muon energies E_μ alter the relative contributions of (e) - (h) and therefore the amount of interference. Fig. 13 shows as an example the radiative corrections δ, defined as

$$\frac{d\sigma^{corr}}{d\Omega} = \frac{d\sigma^{meas}}{d\Omega} \quad (1 - \delta)$$

as a function of $\cos\theta$ for the choice $\zeta < 20^o$ and $E_\mu > 0.5\ E_{beam}$. For a typical acceptance of $|\cos\theta| < 0.8$ the forward-backward asymmetry due to radiative effects is + 1.5%, i.e. it is small and has

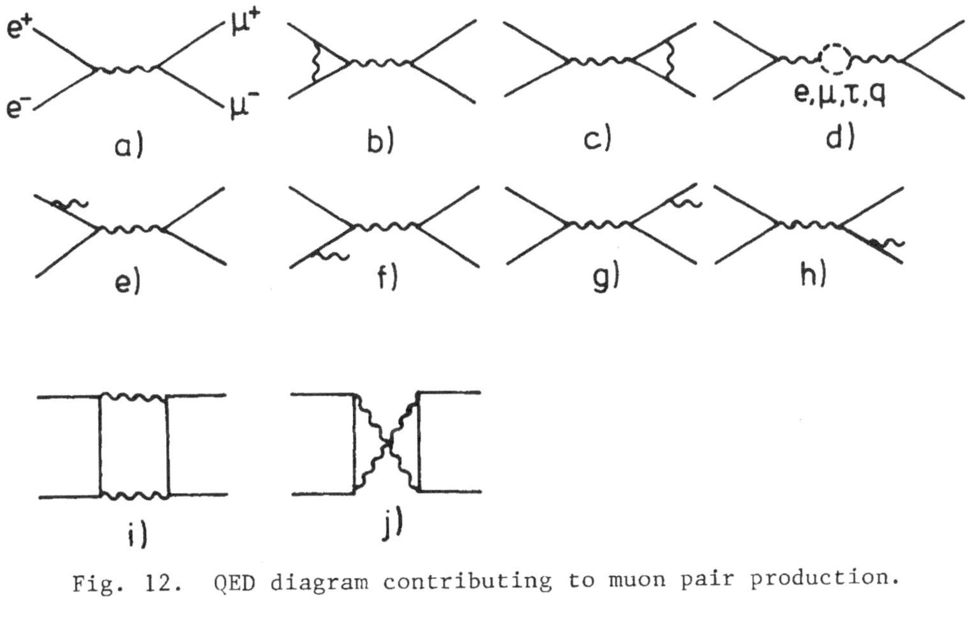

Fig. 12. QED diagram contributing to muon pair production.

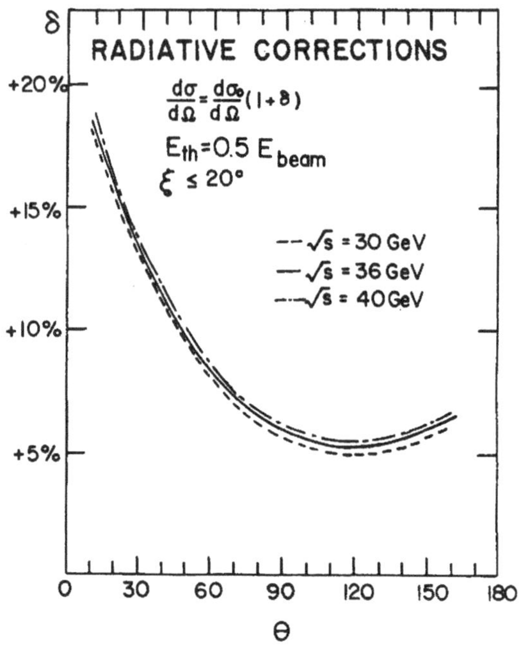

Fig. 13. The radiative corrections δ (see text) for μ pair production as a function of production angle θ for an acollinearity angle $\xi < 20^o$ and μ energy $E_\mu > 0.5\ E_{beam}$ for c.m. energies of 30, 36 and 40 GeV.

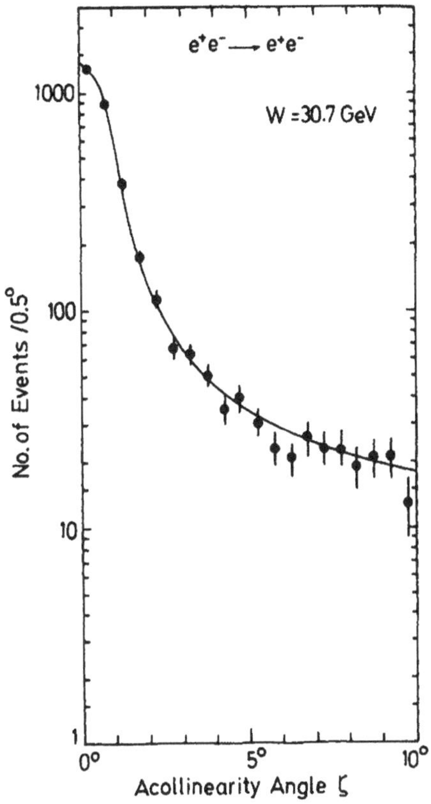

Fig. 14. Distribution of the acollinearity angle ζ in the Bhabha scattering at $W = 30.7$ GeV as measured by TASSO

the opposite sign from that expected for the weak contribution. As a test of the radiative corrections Fig. 14 presents the acollinearity angle distribution for Bhabha scattering. The agreement between data and theory is impressive.

Fig. 15 shows the radiatively corrected differential cross section for muon pair production as measured at an energy of W = 33.5 GeV by the JADE group (Bartel et al. 1981a). The data indicate an angular asymmetry: less μ^+ than μ^- are emitted in the direction of the incoming e^+.

Table 2 summarizes the measured and expected asymmetry values for the five PETRA experiments. The total number of muon pairs that enter Table 2 is about 2700. The average over all PETRA experiments yields

$$\bar{A}_{\mu\mu} = -7.7 \pm 2.4\ \%$$

at a mean $\bar{s}$ = 1100 GeV^2. This is the first evidence for a weak neutral current contribution in e^+e^- annihilation.

Table 2. The muon pair asymmetry as measured at PETRA. For TASSO $A_{\mu\mu}$ has been extrapolated to $|\cos\theta| < 1$. For the other experiments $A_{\mu\mu}$ integrated over the detector acceptance, typically $|\cos\theta| < 0.8$, is given. The theoretically expected asymmetry values were calculated from the standard model. From Branson (1981)

Experiment	s (GeV^2)	$A_{\mu\mu}^{meas}$ (%)	$A_{\mu\mu}^{theor}$ (%)
CELLO	900 - 1340	-1.3^{+8}_{-10}	- 5.8
JADE	900 - 1340	- 11 ± 4	- 7.8
MARK J	900 - 1340	- 3 ± 4	- 7.1
PLUTO	900	+ 7 ± 10	- 5.8
TASSO	900 - 1340	-11.3 ± 5.0	- 8.7

The asymmetries measured at PEP (Hollebeek 1981) for W = 29 GeV are -4.0 ± 3.5 % (MARK II) and (-0.9 ± 5.2 ± 1.5 %) (MAC).

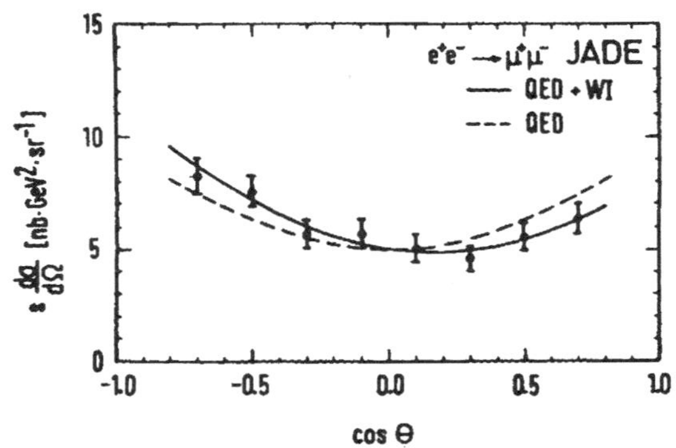

Fig. 15. Differential cross section for μ pair production of an average energy of 33.5 GeV. The curves show the QED prediction (dashed) and the QED plus weak neutral current contributions (solid). From Bartel et al. (1981a).

The observed size of A agrees with the standard model prediction, $A_{theor} = -7.8\%$. It leads to the following value for the axial vector coupling:

$$|g_A| = 0.50 \begin{matrix} +0.07 \\ -0.09 \end{matrix}$$

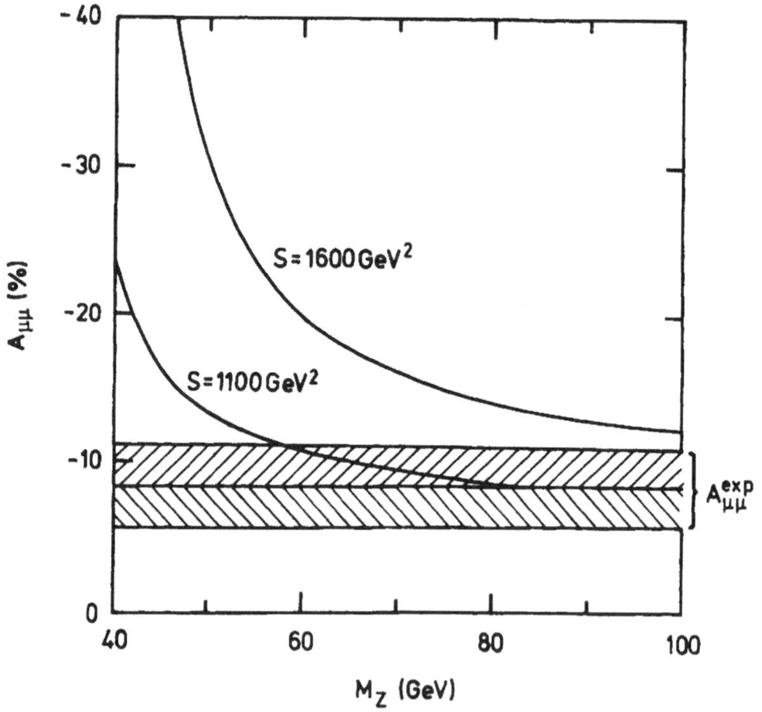

Fig. 16. The asymmetry $A_{\mu\mu}$ measured by the PETRA experiments at $s = 1100\ GeV^2$ (dashed band). The curves show the expected asymmetry as a function of the Z^o mass for $s = 1100\ GeV^2$ and $s = 1600\ GeV^2$.

Fig. 16 compares the measured $A_{\mu\mu}$ value with the asymmetry predicted by the standard model for $s = 1100\ GeV^2$ and $s = 1600\ GeV^2$ as a function of the Z^o mass. The present data limit m_Z only from below, m $m_Z > 49$ GeV with 95% C.L. Much higher statistics and preferably higher energies are required to distinguish between a four fermion interaction ($m_Z = \infty$) and a theory with a finite Z^o mass. The standard model with $\sin^2\theta_W = 0.228$ predicts $m_Z = 89$ GeV.

As discussed above the size of the cross sections for lepton pair production is sensitive to the vector coupling (g_V) and therefore to $\sin^2\theta_W$. The fact that no deviations from QED were observed provides rather tight limits on $\sin^2\theta_W$. The average over all PETRA experiments yields

$$\sin^2\theta_W = 0.24 \pm 0.07$$

at $\bar{s} = 1100\ GeV^2$. The nontrivial result is that $\sin^2\theta_W$ appears to be the same at this mass squared as measured in νN scattering at Q^2 of 50 - 100 GeV^2.

The data can also be used to put limits on weak models with more than one Z^o and/or W doublet. More information can be found in (Branson 1981).

V. SEARCH FOR NEW PARTICLES

As mentioned above an electron-positron storage ring is an ideal place to search for new particles. Any charged and pointlike particle will be produced with a sufficiently large cross section. For spin 1/2 fermions the pair production cross section is given by the expression for μ pair production (eq (2)). For charged pointlike scalars the cross section has a different rise near threshold and its asymptotic value is one quarter of $\sigma_{\mu\mu}$ since the spin is zero:

$$\frac{d\sigma}{d\Omega}(e^+e^- \to H^+H^-) = \frac{\alpha^2}{8s}\beta^3 \sin^2\theta$$

and

$$\sigma(e^+e^- \to H^+H^-) = \frac{\beta^3}{4}\sigma_{\mu\mu}$$

The PETRA experiments conducted intensive searches for a great variety of new particles. So far the findings have been negative. Some of these results put stringent limits on new symmetry schemes. The discussion of these results will follow the review by Bürger (1981).

1. Excited Leptons

The existence of excited leptons (e^*, μ^*, τ^*) will give the ground state leptons a structure. As discussed in section III the lepton pair data rule out a structure down to 10^{-16} cm. The presence of an excited electron could be seen directly in two-photon annihilation, $e^+e^- \rightarrow \gamma\gamma$, where it adds a contribution described by diagram 5b. The deviation of the cross section from the QED prediction can be written as

$$\frac{d\sigma}{d\Omega}(e^+e^- \rightarrow \gamma\gamma) = \frac{d\sigma}{d\Omega}_{QED} \left(1 + \frac{s}{2\Lambda^4} \sin^2\theta\right) \quad (14)$$

where $\Lambda = m_e*.\sqrt{\alpha/\alpha^*}$, m_e* is the mass of the excited electron, α^* its coupling to electron and photon. As shown by Fig. 9, there is no evidence in the data for any deviation from QED. Assuming $\alpha^* = \alpha$ the experimental lower limit on m_e* is 50 GeV (see Table 3).

Table 3. Lower limit (95% C.L.) on the mass m_e* of a possible excited electron, assuming $\alpha^* = \alpha$. From Bürger (1981).

Experiment	W(GeV)	m_e*(GeV)
CELLO	35	> 43
JADE	35	47
MARK J	35	58
PLUTO	31	46
TASSO	35	34
MARKII	29	50

Next we consider excited muons which could be produced either pairwise,

$$e^+e^- \rightarrow \mu^{*+}\mu^{*-} \quad (15)$$

or together with a μ,

$$e^+e^- \rightarrow \mu^*\mu \quad (16)$$

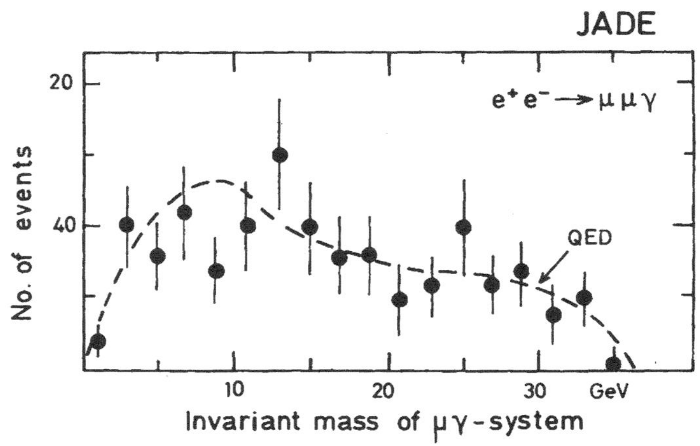

Fig. 17. The reaction $e^+e^- \to \mu^+\mu^-\gamma$ at W = 22 - 37 GeV as measured by JADE. Events were selected with $E_\gamma > 1$ GeV, $M_{\mu\mu} > 1$ GeV, angle between γ and μ larger than 15° and angle sum > 350° From Bürger (1981).

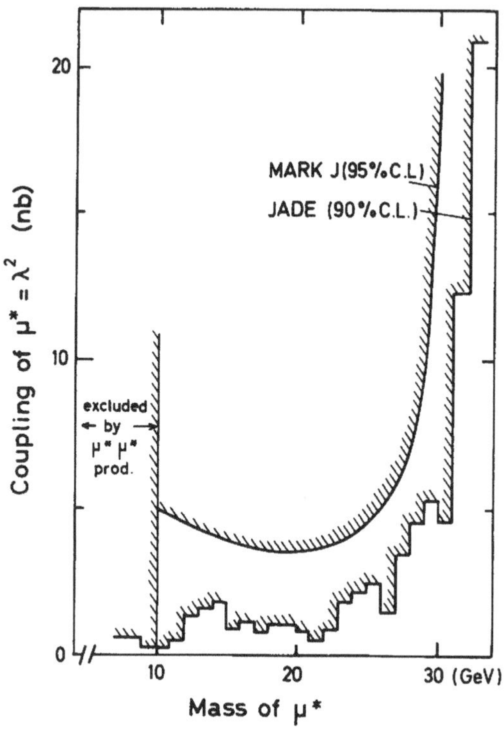

Fig. 18. Upper limit on the coupling constant λ^2 versus the μ^* mass. From Bürger (1981)

For spin 1/2 μ^* the cross section for (16) can be written as

$$\frac{d\sigma}{d\Omega}(e^+e^- \to \mu^*\mu) = \lambda^2\alpha^2 \frac{(s + m^2_{\mu^*})^2}{s^3} \{(s + m^2_{\mu^*}) - (s - m^2_{\mu^*})\cos^2\theta\} \tag{17}$$

where the $\mu^*\mu\gamma$ vertex factor is $2\lambda e\sigma_{\mu\nu}$. The MARK J group searched for the reaction (15) in $\mu^+\mu^-\gamma\gamma$ final states. A μ^* with a mass $m_{\mu^*} < 10$ GeV is ruled out by the data. MARK J and JADE studied events of the type $\mu^+\mu^-\gamma$ to look for reaction (16). The $\mu\gamma$ mass distribution (Fig. 17) as measured by JADE agrees well with QED and shows no evidence for a narrow μ^*. The data are sensitive to μ^* masses up to 30 GeV provided the coupling λ is sufficiently large. If m_{μ^*} is arbitrarily fixed to $m_{\mu^*} = 16$ GeV the data limit the cross section to $\sigma_{\mu^*\mu} < 0.0075.\sigma_{\mu\mu}$ with 90% C.L. (see Fig. 18).

2. Search for a Heavy Sequential Lepton

We know presently of three generations of leptons, (e,ν_e), (μ,ν_μ) and (τ,ν_τ). Whether there should be a fourth one, (L,ν_L), or even more is an open question. The cross section of pair production of L,

$$e^+e^- \to L^+L^-$$

is given by eq. (4).

The possible decay modes of L (see Fig. 19) are either leptonic $L^- \to \nu_L\bar{\nu}_e\ell^-$ or semileptonic, $L^- \to \nu_L\bar{q}q'$ and can be computed in the standard weak theory following the same lines used in the case of the τ (see e.g. Bjorken & Llewellyn-Smith 1973). Neglecting mass effects and possible enhancements one expects all leptonic and semileptonic channels to have the same branching ratio (note: the $\bar{q}q'$ channels count three times because of the colour factor), hence for a lepton of mass 15 GeV, $B(L^- \to \nu_L e^-\bar{\nu}_e) = 1/9$. The possible decay channels lead to characteristic event signatures which readily identify L^+L^- production.

Table 4 summarizes the various event topologies studied by different experiments. No evidence was found and the existence of a heavy sequential lepton is excluded for lepton masses below 18 GeV.

3. Neutral Heavy Leptons

The existence of heavy neutrinos was discussed in connection with the atomic experiments on parity violation (see e.g. Harari 1977). A possible candidate is the E^o which has the same lepton number as the electron. It can be produced by W exchange (see

Table 4. Mass limits on a new heavy sequential lepton. From Bürger (1981)

Experiment	95 % lower limit on M_L (GeV)	event signature
MARK J Barber et al. (1980a)	16	single muon recoiling against many hadrons
TASSO Brandelik et al. (1981a)	15.5	single charged part. recoiling against many hadrons
PLUTO Berger et al. (1981a)	14.5	Single muon recoiling against many hadrons
JADE	18.1	two acollinear jets

Fig. 20a) via

$$e^+e^- \to \bar{E}^o \nu \tag{18}$$

and will decay in the standard way (Fig. 20b), i.e. leptonically ($E^o \to e^-e^+\nu_e$ with a branching ratio of 11%) and semileptonically ($E^o \to e^-\bar{q}q'$). The cross section for reaction (18) assuming a V-A coupling is genen by Bletzacker & Nieh (1977) (see also Bjorken & Llewellyn-Smith 1973).

$$\frac{d\sigma}{d\Omega} = \frac{G_F^2}{32\pi^2} \frac{(1 - M^2/s)^2}{(1 - q^2/M_W^2)^2} \{2s(1 + \cos\theta) - (s - M^2) \sin^2\theta\} \tag{19}$$

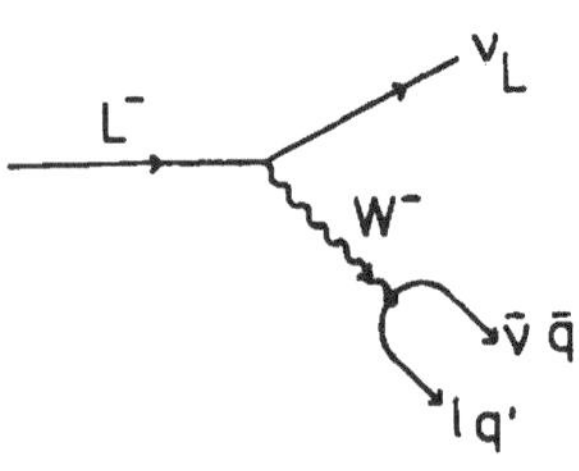

Fig. 19. Diagram for L decay into leptonic and semileptonic final states

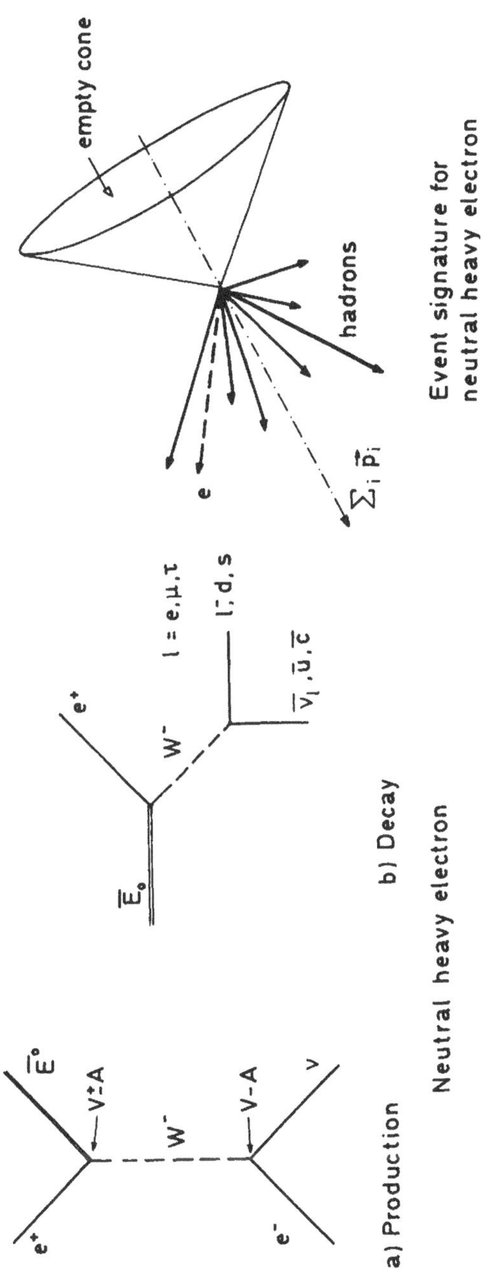
e+
e-
$\bar{E}^{0}$
V±A
V-A
W-
ν
a) Production
$\bar{E}_{0}$
e+
W-
l = e, μ, τ
l-, d, s
$\bar{\nu}_l$, ū, c̄
b) Decay
Neutral heavy electron
empty cone
hadrons
e
$\sum_i \vec{p}_i$
Event signature for
neutral heavy electron

Fig. 20

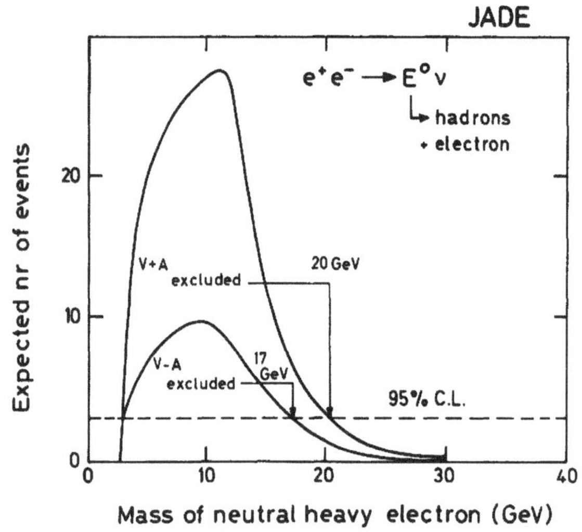

Fig. 21 The reaction $e^+e^- \to \bar{E}^0\nu$: comparison of expected number of events as a function of E^0 mass with observed upper limit. From Bürger (1981).

where M is the mass of E^o and q^2 the square of the momentum transfer between e^+ and $\bar{E}^o$. The predicted cross section is surprisingly large, for W = 35 GeV, $\sigma(\bar{E}^o\nu) \simeq 3$ pb which is roughly 4 % of $\sigma_{\mu\mu}$.

For E^o masses not too close to the c.m. energy W, reaction (18) leads to events with particles in only one hemisphere (Fig. 20c). The JADE group has searched for such events and found none. Fig. 21 compares their upper limit with the number of events predicted for V - A and V + A coupling. Assuming V - A (V + A) coupling E^o masses below 17 GeV (20 GeV) can be excluded.

4. Supersymmetric Scalar Leptons

In supersymmetry the fundamental constituents of matter: leptons, quarks, quanta of fields with spin J have supersymmetric partners with spin J ± 1/2 (see e.g. Gol'fand & Likhtman 1971, Volkov & Akulov 1973, Wess & Zumino 1974, Fayet & Ferrara 1977, Farrar 1978).

In particular, there should exist scalar electrons (S_e) and muons (S_μ) as supersymmetric partners of e and μ. They can be produced in e^+e^- interactions by photon exchange with a cross section given by eq. (13); scalar electrons can also be produced by the exchange of a space-like photino or goldstino (Fig. 22a) (see Farrar and Fayet 1980). The photino is the supersymmetric partner of the photon and the goldstino is associated with the breaking of the symmetry. Both have J = 1/2. The S_e and S_μ are expected to be short lived and to decay by photino emission, $S_e \to \ell$ + photino (Fig. 22b). The photino behaves like a neutrino in its interaction with matter.

Production of scalar leptons (S_e and S_μ) was searched for by several experiments analysing events with two acoplanar leptons:

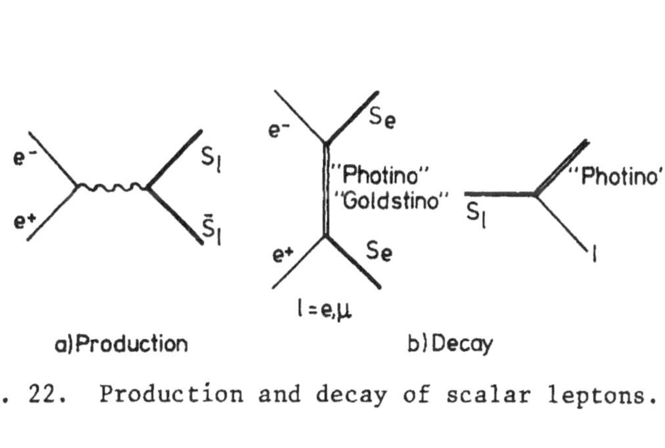

Fig. 22. Production and decay of scalar leptons.

$e^+e^- \to \ell^+\ell^-$ + missing momentum

Background from events of the type $e^+e^- \to e^+e^-\gamma, e^+e^- \to \tau^+\tau^- \to$ $\to e^+e^- + \nu$'s, $e^+e^- \to e^+e^-e^+e^-$ (and similarly for μ's) required special attention. No evidence was found for scalar leptons. As Table 5 shows scalar electrons (muons) are excluded for masses between 2 (5.5) GeV and 16(15) GeV .

Table 5. Mass limit on scalar leptons.

Experiment	Mass range excluded by experiment (GeV) S_e	S_μ
JADE (Cords 1980)	>16	
MARK J (Barber et al. 1980)		3 - 15
PLUTO (Splitzer 1980)	>13	
CELLO	2 - 16.6	5.5 - 15

5. Free Quarks

Using the dE/dx information in the central drift chamber the JADE group (Bartel et al. 1980) searched for free quarks produced in paris, and together with ordinary hadrons:

$$e^+e^- \to q\bar{q} \tag{20}$$

$$e^+e^- \to q\bar{q} + \text{hadrons} \tag{21}$$

at energies between 27 and 37 GeV. The ionization loss dE/dx for quarks with charge Q = 2/3 (1/3) will be 1/4 (1/9) of that for a charge Q = 1 particle provided the velocity is the same. The detection efficiency for Q = 2/3 (1/3) quarks lies between 0.15 - 0.36 (0.11 - 0.27) depending on the quark mass. Fig. 23 shows the mean energy loss dE/dx for charged particles measured as a function of the apparent momentum P/Q. Basically all particles lie in one of the regions expected for e, π, k, p, d or t. The p, d and t points result mainly from beam gas scattering. In the region where Q = 2/3

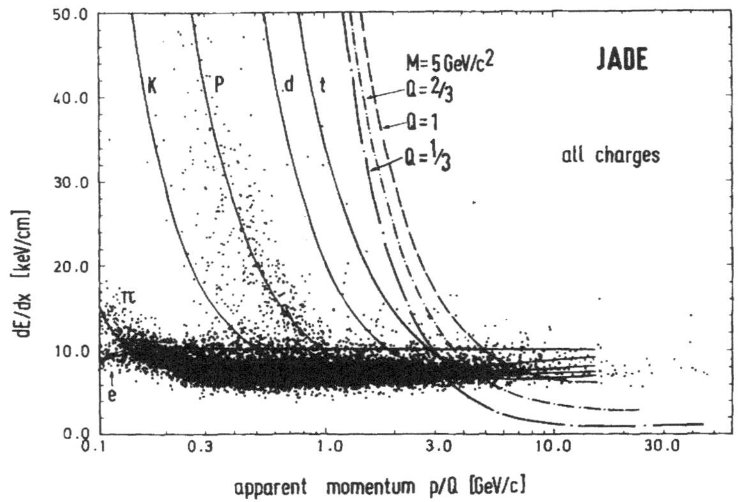

Fig. 23. Mean Energy loss dE/dx as a function of the apparent momentum P/Q where Q is the particle charge. From JADE.

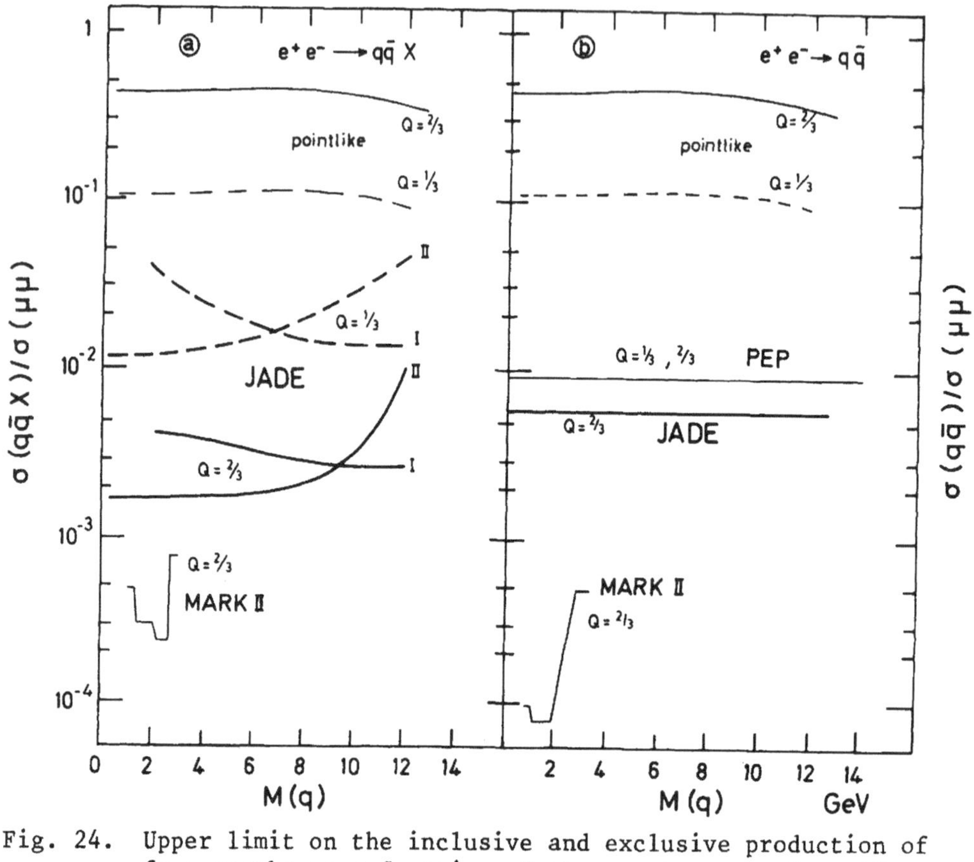

Fig. 24. Upper limit on the inclusive and exclusive production of free quarks as a function of the quark mass. From Bürger (1981)

or 1/3 particles can clearly be separated from the known particles no events are found. Fig. 24 shows the upper limits for free quark production in (20) and (21) as a function of the quark mass. There are two different curves for inclusive quark production (21) based on different assumptions of the quark momentum distribution. Also shown in Fig. 24 are the upper limits obtained by MARKII (Weiss et al. 1981) at lower energies and by the free quark search experiment at PEP (Litke 1981). The upper limit are roughly 10^{-2} - 10^{-4} below the cross section for muon pair production.

In summary, as the title of the section: "Search for new particles" suggested, no evidence has been found yet for excited electrons or muons, for new heavy leptons, for neutral heavy leptons, for supersymmetric scalar leptons, nor for free quarks. The upper limits obtained exclude for most of these particles mass values below $\sim$ 16 - 20 GeV.

VI. JET FORMATION IN e^+e^- ANNIHILATION

1. The Quark Parton Model

In the quark parton model e^+e^- annihilation into hadrons,

$$e^+e^- \rightarrow \text{hadrons}$$

proceeds in two steps (see Fig. 25). First, e^+ and e^- annihilate into a photon which then materializes into quark (q) and antiquark ($\bar{q}$). In the second step q and $\bar{q}$ fragment into hadrons. The first step is almost identical to the production of a pair of muons.

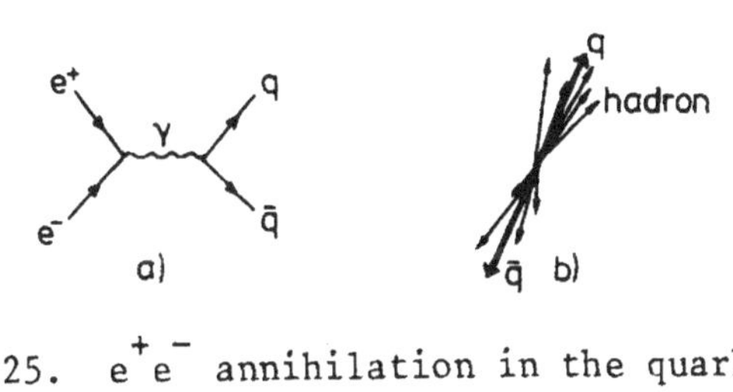

Fig. 25. e^+e^- annihilation in the quark model

The quark model description of e^+e^- annihilation into hadrons has been extensively tested and has been found in surprising agreement with experiment. The evidence for the model rests on the following observations:

a. The total cross section. The total cross section is readily calculated. The cross section for producing a free $q\bar{q}$ pair is the same as for producing a $\mu^+\mu^-$ pair except that the quark charge e_q replaces the muon charge of unity. Assuming that the produced $q\bar{q}$ pairs turn into hadrons with unit probability one finds

$$R \equiv \sigma_{tot}/\sigma_{\mu\mu} = 3 \sum_{q=u,d,s,\ldots} e_q^2 \tag{22}$$

where the factor of three accounts for the fact that quarks come in three colours. At energies above W = 10 GeV where all five known quarks: u,d,s,c,b can be pair produced the predicted value of R is

$$R = 3\left\{ \left(\frac{2}{3}\right)^2 + \left(\frac{1}{3}\right)^2 + \left(\frac{1}{3}\right)^2 + \left(\frac{2}{2}\right)^2 + \left(\frac{1}{3}\right)^2 \right\} = \frac{11}{3}$$

Fig. 26 shows R up to the highest energy measured at PETRA, W = 36.6 GeV (Felst 1981). The error bars indicate the statistical uncertainties. The systematic errors are typically of the order of 7 - 10%. The outstanding features of R are spikes due to the excitation of vector states, ρ, ..., J/ψ, ..., Υ... and the fact that in between these families of vector states and above R is rather constant. The data above 1.5 GeV indicate two steps. Between 1.5 and 3.8 GeV R is approximately 2 - 2.5. Near charm threshold (at 4 GeV) R rises sharply to reach a new level around 4 which persists up to 36.6 GeV and which is in striking agreement with the quark model prediction of 11/3 (see curve in Fig. 26).

The high energy data on R can be used to test the pointlikeness of quarks (Söding & Wolf 1981). In terms of the quark electric and magnetic formfactors G_E, G_M the ratio R is given by

$$R = 3 \sum_q e_q^2 \left\{\frac{2m_q^2}{s} |G_E(s)|^2 + |G_M(s)|^2\right\} \tag{23}$$

where m_q is the quark mass and $m_q^2 \ll s$ has been assumed. If $G_E \sim G_M$ and all quarks have the same form factor we find:

$$R/R_o = |G_M(s)|^2 \tag{24}$$

with $R_o = 3 \sum_q e_q^2$. With the ansatz

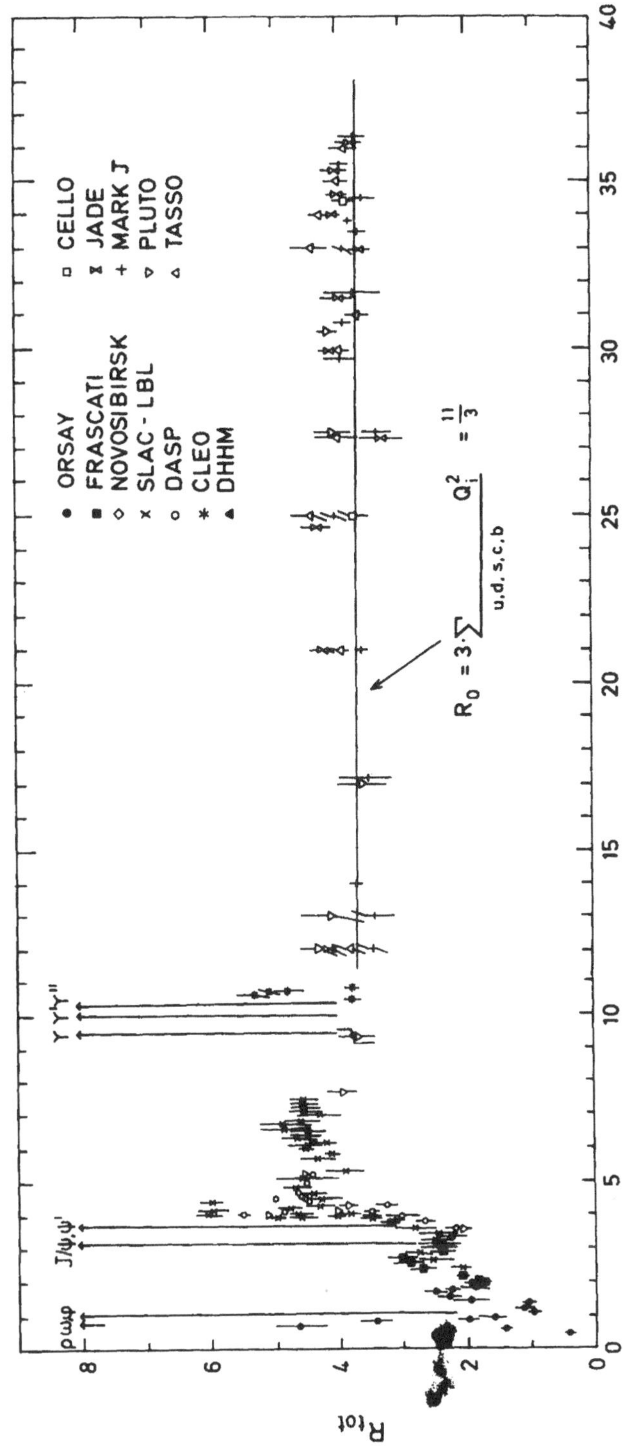

Fig. 26. The ratio R of the total cross section for e^+e^- annihilation into hadrons to the μ pair cross section $\sigma_{\mu\mu} = 4\pi\alpha^2/3s$. From Felst (1981)

$$G_M(s) = (1 - s/M^2)^{-1} \tag{25a}$$

treating R_o as a free parameter and including the systematic uncertainties of σ_{tot} the data yield for $W \gtrsim 17$ GeV, $M > 124$ GeV with 95 % C.L. If quarks are composites of 3 subquarks G_M may have a dipole behaviour:

$$G_M(s) = (1 - s/M_D^2)^{-2} \tag{25b}$$

For this case the fit gives $M_D > 176$ GeV with 95 % C.L. Converting the M parameters into a length we conclude that quarks behave point-like down to distances of $\sim 1.10^{-16}$ cm.

An analysis in terms of the cut-off parameter Λ yields $\Lambda_+ =$ 190 GeV, $\Lambda_- = 285$ GeV (Branson 1981).

b. The occurrence of jets. The quark-parton model predicts the occurrence of jets: at high energies the hadrons will be collimated around the direction of the primary $q\bar{q}$ pair. The can be understood in the following way. Consider a hadron emitted by a quark with longitudinal and transverse momentum components $p_{||}$ and p_T relative to the quark direction of flight (Fig. 27). If the average p_T is energy independent and the average particle multiplicity $\langle n \rangle$ grows only slowly with energy then for high energies

$$\langle p_{||} \rangle \simeq \langle p \rangle \simeq \frac{W}{\langle n \rangle} \tag{26}$$

and

$$\frac{\langle p_T \rangle}{\langle p_{||} \rangle} \sim W^{-1} \tag{27}$$

The produced hadrons will be emitted more and more closely to the

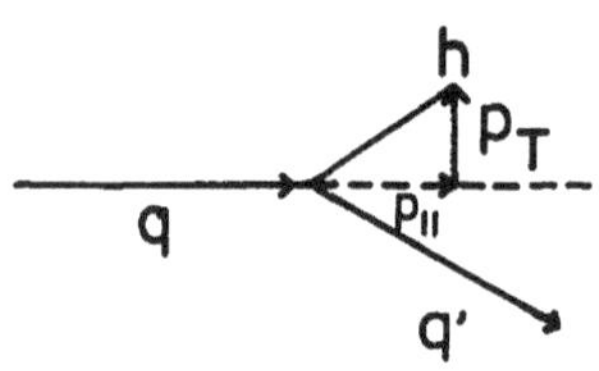

Fig. 27. Hadron emission by a quark

primary quark direction as the energy increases. The mean half opening angle of the jet cones is given by

$$\langle\delta\rangle \simeq \frac{\langle p_T\rangle}{\langle p_{\parallel}\rangle}$$

thus the jet cones shrink roughly $\sim W^{-1}$. (Actually, in a realistic calculation of the quark model using the fragmentation functions of Field and Feynman (1978) one finds $\langle\delta\rangle \sim W^{-1/2}$).

The presence of a two-jet structure is commonly tested in terms of sphericity S (Bjorken and Brodsky 1970) and thrust T (Brandt et al. 1964, and Fahri 1977):

$$S = 3/2 \ \mathrm{Min}\ (\Sigma p_{Ti}^2)/(\Sigma p_i^2) \qquad 0 \leqslant S \leqslant 1$$

$$T = \mathrm{Max}\ (\Sigma |p_{\parallel i}|)/(\Sigma p_i) \qquad 0.5 \leqslant T \leqslant 1 \tag{28}$$

Sphericity measures approximately the square of the jet cone half opening angle,

$$S \simeq 3/2 \ \langle\delta^2\rangle$$

Extreme jettiness ($\delta = 0$) leads to $S = 0$ and $T = 1$ while for spherical events $S \to 1$ and $T \to 0.5$.

The two-jet structure of hadronic events produced by e^+e^- annihilation was first observed in 1975 by the SLAC-LBL group (Hanson et al. 1975). Fig. 28 shows the energy dependence of the average sphericity as measured at DORIS and PETRA. At low energies $W \lesssim 4$ GeV, the observed $\langle S\rangle$ values are close to those predicted by phase space, $\langle S\rangle \simeq 0.4$. Above 5 GeV $\langle S\rangle$ decreases rapidly with increasing W, i.e. the particles become more and more collimated in clear distinction to a phase space behaviour. A power law, $\langle S\rangle = 0.8\ W^{-1/2}$ describes the data well. The jet cone half opening angle as inferred from $\langle S\rangle$ shrinks from $\sim 31^{\circ}$ at $W = 4$ GeV to $\sim 17^{\circ}$ near 36 GeV.

Fig. 29 shows a typical two-jet event at W = 35 GeV.

c. _The jet angular distribution_. The angular distribution of the jet axis relative to the beam axis is of the form (Schwitters et al. 1975)

$$W(\cos\theta_{jet}) \sim 1 + \cos^2\theta_{jet}$$

in agreement with what is expected for production of a pair of spin 1/2 particles. For comparison spin 0 particles would lead to

$$W(\cos^2\theta_{jet}) \sim 1 - \cos^2\theta_{jet}.$$

Fig. 30 shows a measurement of $W(\cos\theta_{jet})$ for two-jet events at high energies (W = 30 - 36 GeV). The agreement with the $1 + \cos^2\theta_{jet}$ form (solid curve) is good.

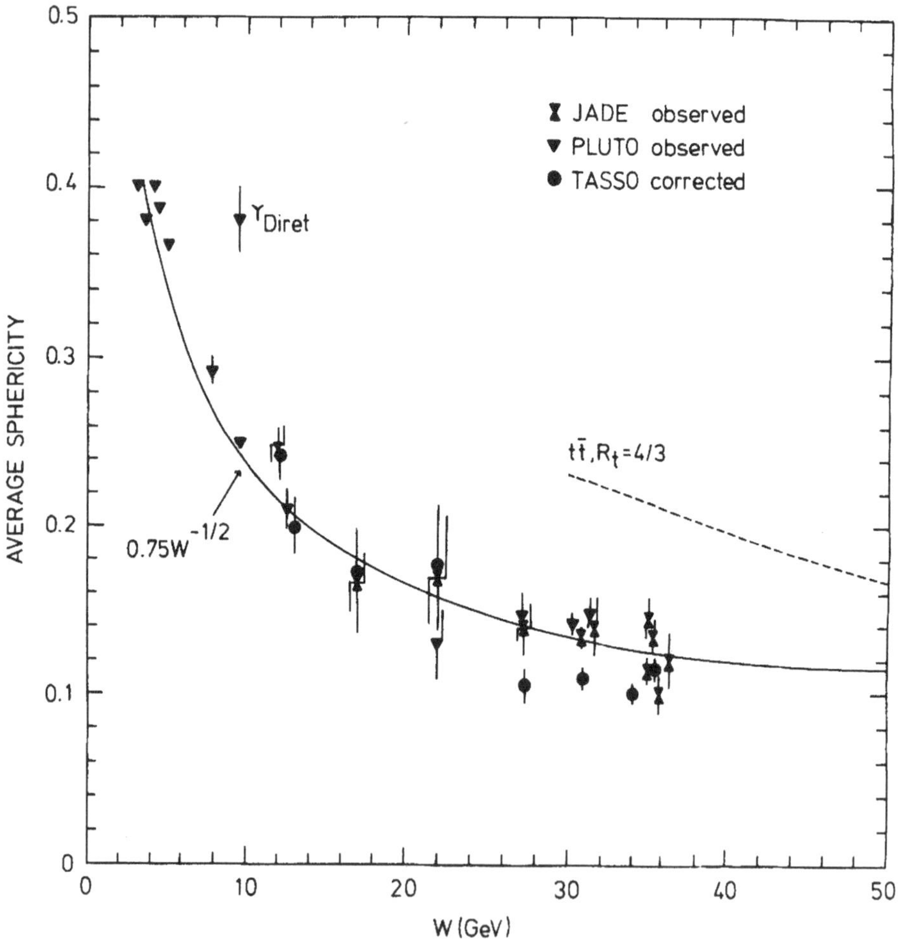

Fig. 28. The average sphericity as measured by the JADE, PLUTO and TASSO groups (See Cords 1980)

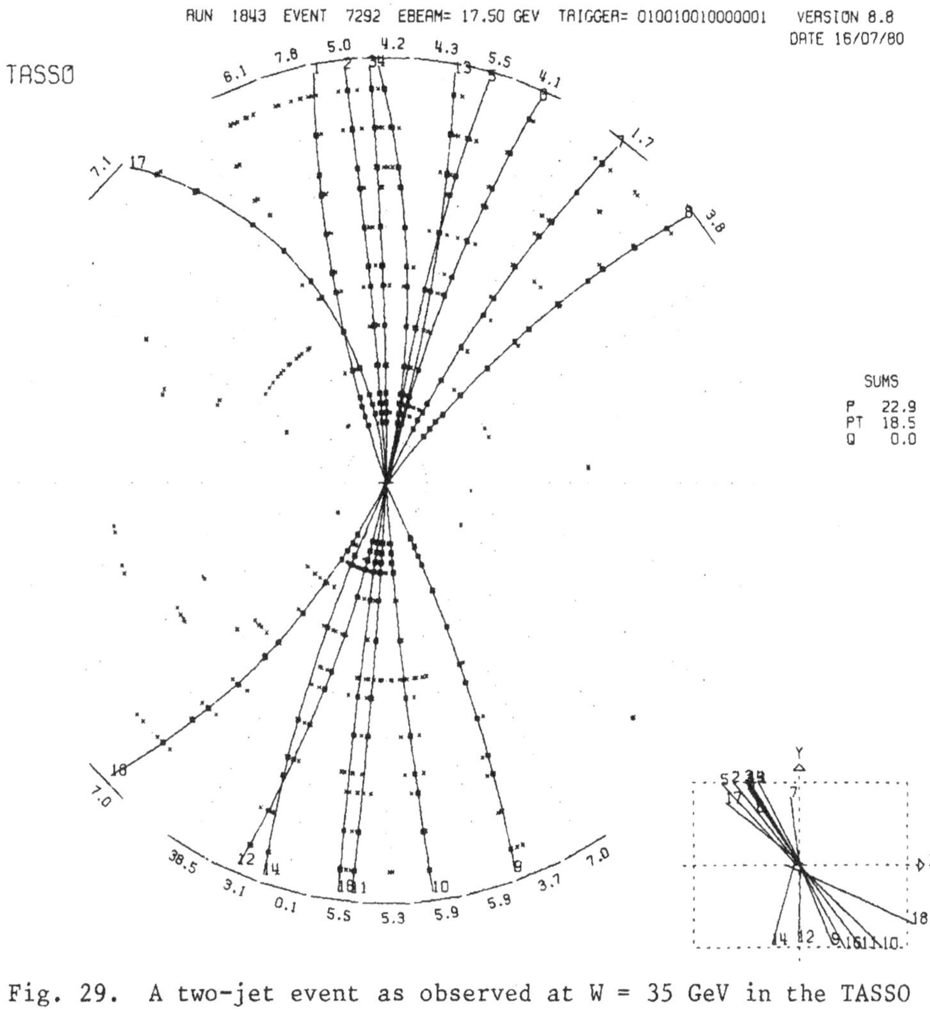

Fig. 29. A two-jet event as observed at W = 35 GeV in the TASSO detector.

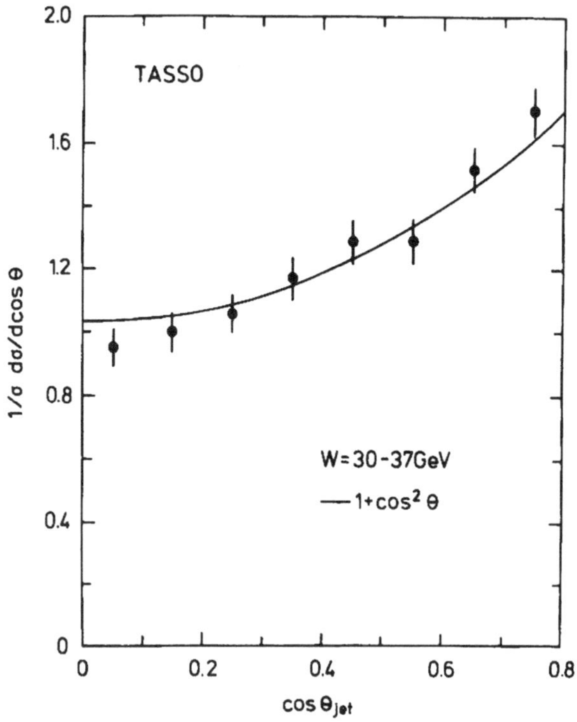

Fig. 30. Angular distribution of the jet axis for two-jet events as measured by TASSO at W = 30 - 37 GeV.

d. Long range charge correlations. The back-to-back produced quarks have opposite charge. Since in the standard picture of quark fragmentation the primary quark is found predominantly in one of the fast particles, one expects a correlation between the charges of leading particles in opposite jets. Besides this long range correlation also a short range correlation is expected since neighbouring particles in the jet cascade should have preferrentially opposite charges. First evidence for long range correlations was presented by the TASSO group (Brandelik et al. 1981b). The effect was established in two ways, from a study of the weighted jet charges and from charge correlations measured as a function of rapidity along the jet axis. The first method goes back to a suggestion by Field and Feynman (1978) who proposed to use the momentum weighted jet charge

$$q_{jet}(\gamma) = \sum_{i=1}^{n_{jet}} e_i x_i^{\gamma} \tag{29}$$

as a measure of the charge of the parton that initiated the jet. In eq. (29) the charge of the i'th particle e_i in the jet is weighted with a power γ of its fractional momentum $x_i = p_i/p_{beam}$; n_{jet} is the number of particles in the jet. According to current ideas on fragmentation a fast particle has a larger probability to contain the primary parton as a constituent than a slow one. Therefore the weighted charge $q_{jet}(\gamma)$ for positive exponents γ is expected to be more strongly correlated to the primary parton charge than the unweighted jet charge $Q_{jet} = q_{jet}(0)$. It is convenient to take out the strong γ dependence of q_{jet} by introducing the quantity q'_{jet}:

$$q_{jet}(\gamma) \rightarrow q'_{jet}(\gamma) = \frac{\Sigma x_i^{\gamma}}{\Sigma x_i^{2\gamma}} q_{jet}(\gamma) \tag{30}$$

Consider now the average product of weighted charges

$$P'(\gamma) = -\langle q_1'(\gamma) q_2'(\gamma) \rangle \tag{31}$$

of the two jets in an event where < > denotes the average over the event sample. Fig. 31a shows $P'(\gamma)$ for all two-jet events at W = 27.4 - 36.7 GeV. The data show a small but positive value for $P'(\gamma)$. However, this could be a trivial consequence of charge conservation: if there is a net positive charge in one jet there has to be a net negative charge of the same magnitude in the opposite jet. The contribution of charge conservation to $P'(\gamma)$ was determined by randomly redistributing in each jet the observed particle

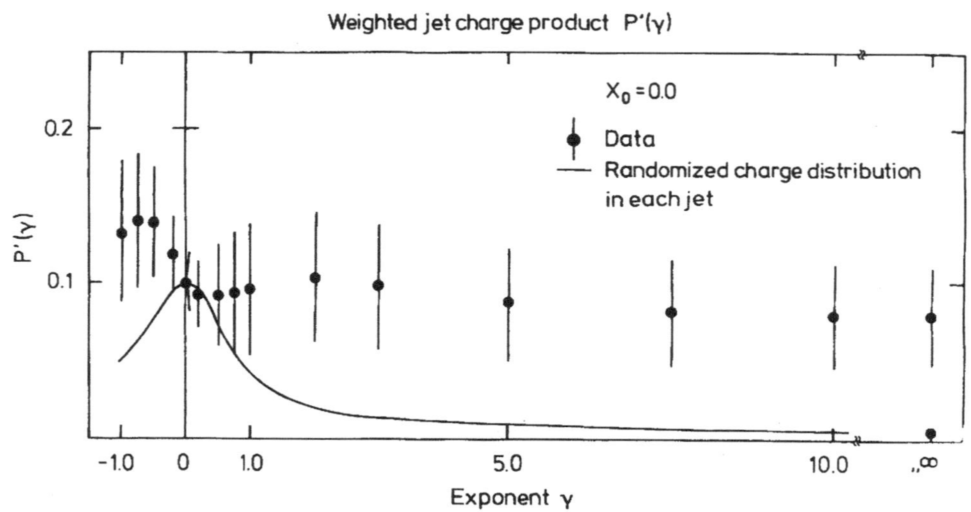

Fig. 31a. The momentum-weighted jet charge product $P'(\gamma)$ as a function of the exponent γ for events having in each jet at least one particle with fractional momentum $x > x_0$. The solid line shows the expectation for a charge distribution randomized in each jet. From TASSO (Brandelik et al. 1981b)

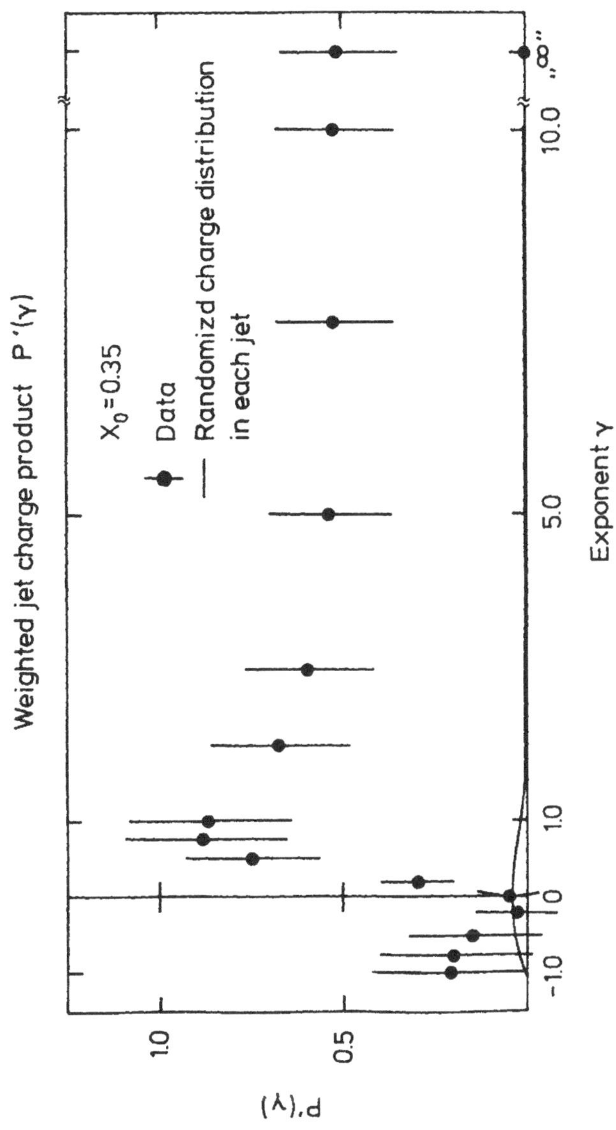

Fig. 31b. Same as Fig. 31a for $X_o > 0.35$

charges amongst the particles of the jet. This yields $P'^{rand}(\gamma)$ which is shown by the solid curve. For $\gamma \leqslant 1$ it is seen to be of almost the same size as $P'(\gamma)$, i.e. most of the value of $P'(\gamma)$ is a result of charge conservation. This is no longer so if one requires each jet to have at least one particle with fractional momentum $x > x_0 = 0.35$ (see Fig. 31b). In this case $P'(\gamma)$ is clearly larger that P'^{rand} which shows that the fastest particles in opposite jets know about their relative charge.

We shall discuss the study of charge correlations with respect to the rapidity y in a rather superficial manner. For the mathematical details the reader is referred to the paper of Brandelik et al. (1981). The analysis is illustrated in Fig. 32, where y is defined in terms of the particle energy and longitudinal momentum $p_{\parallel}$ relative to the jet axis,

$$y = 0.5 \ \ell n \ [(E + p_{\parallel})/(E - p_{\parallel})]$$

We consider now a reference particle with charge $e_i(y')$ at position y' and ask where is its charge compensated. The answer is given by the charge compensation probability which, roughly speaking, is defined by

$$C(y,y') = \frac{N^{+-}(y,y') + N^{-+}(y,y') - N^{++}(y,y') + N^{--}(y,y')}{N(y')} \tag{32}$$

where $N^{+-}(y,y')$ is the two-particle density for having a positive particle in the interval $y, y + \Delta y$ and a negative particle in $y', y' + \Delta y$; similar definitions hold for N^{-+}, N^{++}, and N^{--}. $N(y')$ measures the particle density at y', $y' + \Delta y'$.

Consider an idealized two-jet event (Fig. 32c) with the fastest particles in jet 1 and jet 2 being positive and negative, respectively, and exactly one positive and one negative particle in all other y intervals. If the fastest particle in jet 1 (positron y' in Fig. 32d) is chosen as the reference particle, C is zero everywhere except for the interval containing the fastest particle in the opposite jet. This is called a long range correlation. If, on the other hand, the reference particle is in the central region (y' near zero, see Fig. 32e) we have a short range correlation: C is non zero at $y = y'$ and zero everywhere else.

Fig. 33 shows the properly defined charge compensation probability $\tilde{\Phi}_r(y,y')$ as a function of y for various positions y' of the reference particle, indicated by the shaded area. Also shown is $\tilde{\Phi}_r^{rand}$ which is obtained when the charges are distributed randomly amongst the charged particles of an event (open circles). In contradistinction to $\tilde{\Phi}_r^{rand}$, $\tilde{\Phi}_r$ shows a clear peak centered near

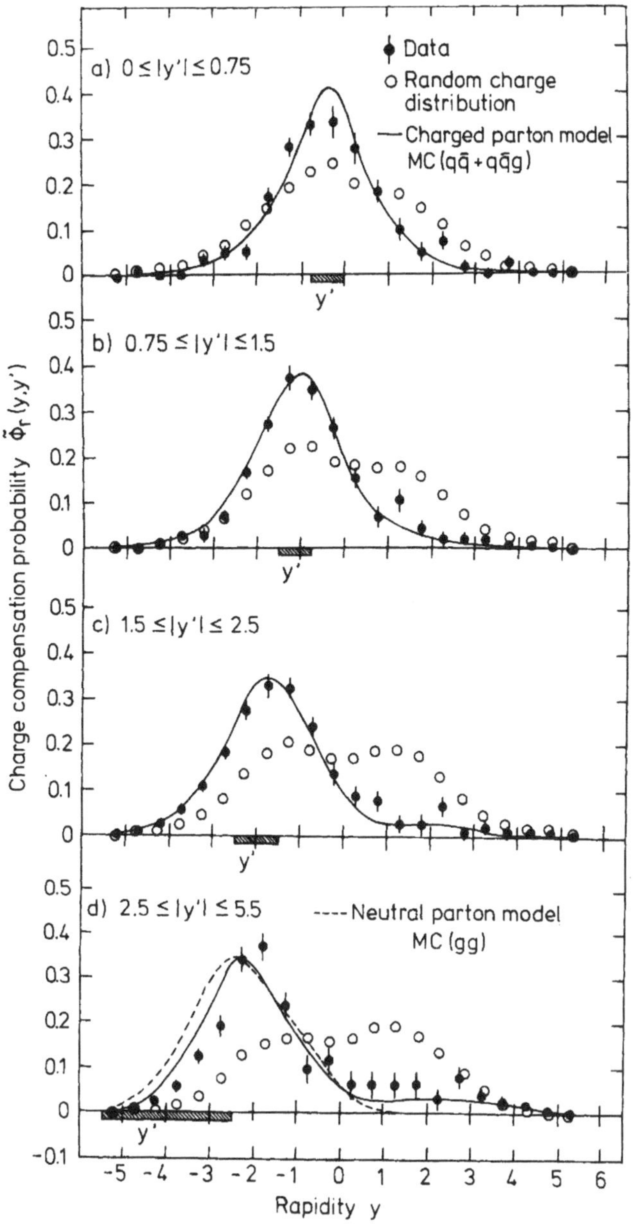

Fig. 33. Charge compensation probability $\tilde{\Phi}_r(y,y')$ as a function of rapidity y for a particle produced at y', as measured by TASSO (Brandelik et al. 1981b). The open circles show the expectation for a charge distribution randomized over the whole event, $\tilde{\Phi}_r^{rand}(y,y')$. The solid line shows the prediction of model MC($q\bar{q}$ + $q\bar{q}g$), the dashed line the prediction of a neutral parton model MC(gg).

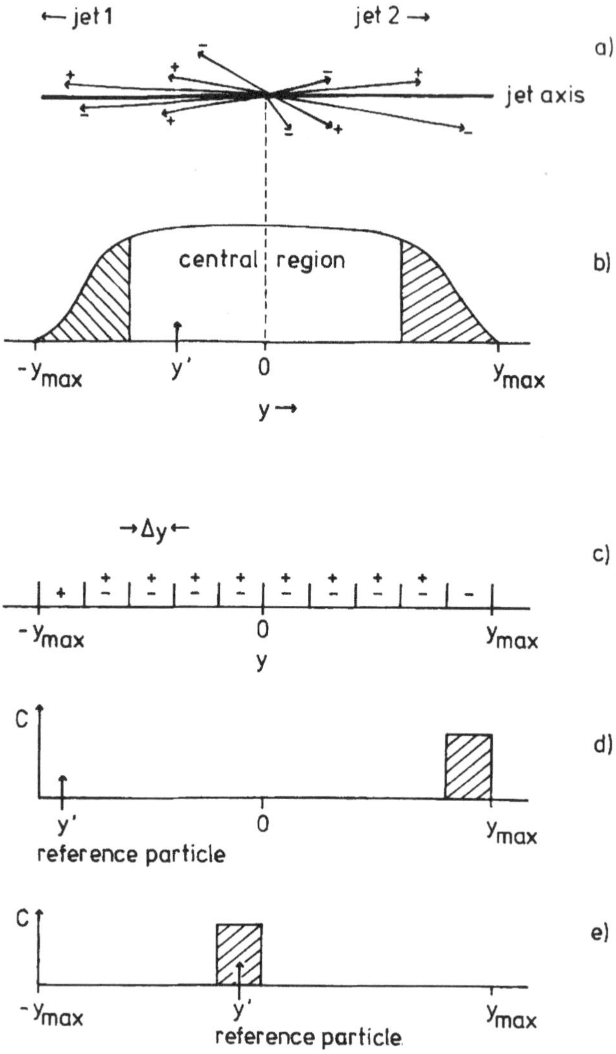

Fig. 32. Illustration of the charge correlation studies
a) particle vectors in a two-jet event
b) rapidity distribution; y' indicates the position of the reference particle
c) idealized charge distribution along the rapidity axis. For this case (d) shows the value of the charge compensation C if the reference particle is at y' in the fragmentation region of jet 1 and (d) in the central region

the position y' of the reference particle which moves with y' as y' is moved from the central region to the fragmentation region of one jet. This is evidence for the presence of short range charge correlations also observed in hadron-hadron collisions (Drijard et al. 1979, 1980).

When moving out to higher values of $|y'|$ $\tilde{\Phi}_r$ exhibits, besides the peak at $y \simeq y'$, an increasing skewness, showing finally for $-2.5 > y' > -5.5$ a long tail in the opposite jet, $y > 0$. This demonstrates the presence of long range charge correlations. The probability that the charge of a particle at $y' < -2.5$ is compensated in the opposite jet at $y > 1$ is found to be (15.4 ± 2.6) %. The observed long range correlation proves that the primary partons which produce the particle jets are charged, in agreement with the assumption that these partons are quarks.

The strong evidence for the quark parton model make e^+e^- annihilations into hadrons a good place to test strong interaction theories, in particular, Quantum Chromodynamics (QCD). Before comparing QCD with the data we remind the reader of some of the ingredients of QCD.

2. Elements of QCD

QCD (Nambu 1966, Fritzsch & Gell-Mann 1972, Fritzsch et al. 1973, Weinberg 1973, Gross & Wilczek 1973) describes the interaction between quarks. The force is provided by the exchange of gluons (see Fig. 34a) in much the same way proton and electron are bound in the hydrogen atom by photon exchange (Fig. 34b).

The properties of gluons are:

- mass $m = 0$
- spin, parity $J^P = 1^-$
- gluons are colour octet states. There are 8 different gluons. They interact with the colour charges of the quarks (see Fig. 35).

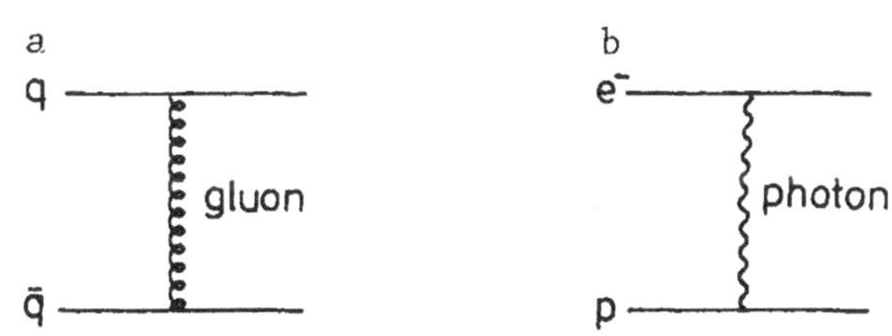

Fig. 34a. $q\bar{q}$ system bound by gluon exchange
b. Electron and proton bound in the hydrogen atom by photon exchange.

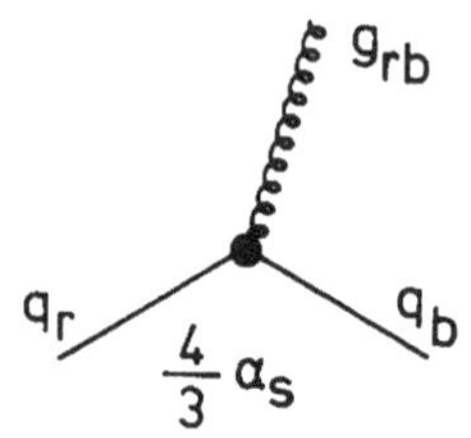

Fig. 35. Quark-gluon interaction

- The gluon quark coupling strength α_s is independent of the quark type (or flavour), i.e. α_s is the same for u, d, s, ...

- In contrast to photons, which are electromagnetically neutral, gluons carry colour charge. As a result gluons can interact with themselves which leads to the presence of three- and four gluon vertices (Fig. 36) in the theory.

- The coupling strength α_s is not constant (running coupling constant). In lowest order

$$\alpha_s = \frac{12\pi}{(33 - 2N_f)\ln Q^2/\Lambda^2} \tag{33}$$

where N_f is the number of different quark types, Q^2 the characteristic momentum transfer and Λ is the QCD scale parameter. The Q^2 behaviour of α_s is sketched in Fig. 37. The coupling strength decreases with increasing Q^2 consistent with asymptotic freedom. That is to say, for small distances (large Q^2) between quarks α_s becomes small and the quarks behave as free. For low Q^2 or large distances ($Q^2 \to \Lambda^2$) α_s will become of order one and confinement may

Fig. 36. Three- and four-gluon vertices

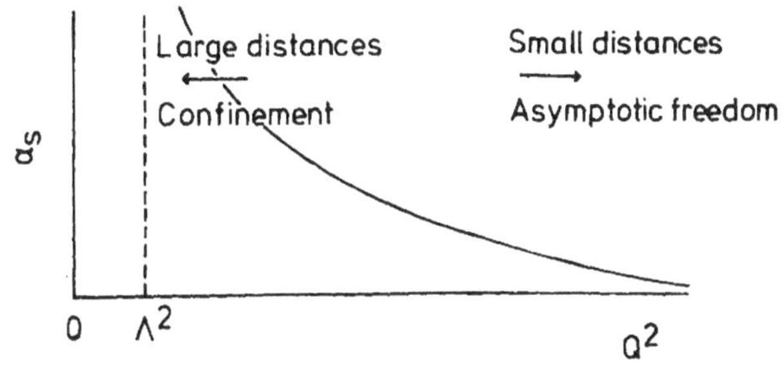

Fig. 37. Qualitative behaviour of the strong coupling strength α_s

arise: quarks never become free; in this regime perturbation theory becomes invalid.

The most stringent tests of QCD are provided by large Q^2 processes where perturbative methods can be used. Examples are deep inelastic lepton nucleon scattering and e^+e^- annihilation at high energies. We shall concentrate on the latter field with the emphasis on R and jet production.

3. QCD Modifications to R

The predictions of the quark parton model are modified by the emission of gluons (see Fig. 38). For R the QCD corrected expression is:

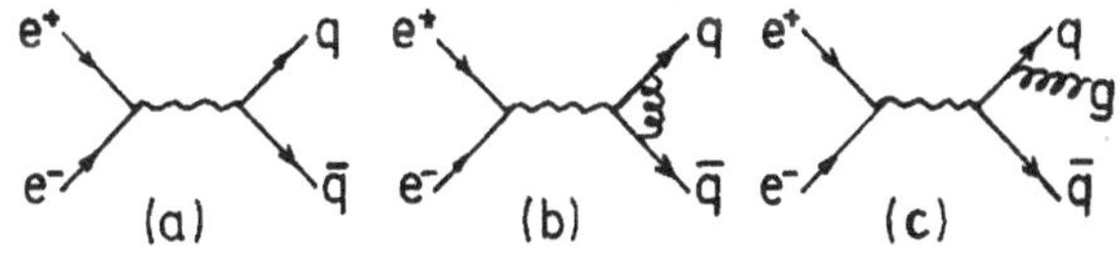

Fig. 38. Quark parton diagram and lowest order QCD diagrams for $e^+e^- \rightarrow q\bar{q}$, $q\bar{q}g$

$$R = 3 \sum_q e_q^2 \{1 + \frac{\alpha_s}{\pi} + C_2(\frac{\alpha_s}{\pi})^2 + \ldots\} \tag{34}$$

The corrections have been computed up to second order (Dine & Sapirstein 1979, Chetyrkin et al. 1979, Celmaster & Gonsalves 1979). In the $\overline{MS}$ scheme

$$C_2 = 1.99 - 0.12 \, N_f$$

where N_f is the number of quark flavours.

After a careful examination of the theoretical problems at low energy (W < 2 GeV) Eidelman et al. (1979) deduced from low energy σ_{tot} data a value of $\Lambda \approx 0.1$ GeV. QCD fits to the intermediate energy region (4 to 7 GeV) have been performed by Barnett et al. (1980). We compare the QCD expression to the data measured above 20 GeV. This energy region is well above heavy quark thresholds furthermore, higher twist effects should be negligible. The average over all PETRA experiments gives

$$\overline{R} = 3.91 \pm 0.04 \text{ (stat.)} \pm 0.3 \text{ (syst.)}$$

for $\overline{W}$ = 33 GeV. After corrections for weak neutral current contributions (which amount to $\Delta\overline{R}_{weak} = 0.03$ for $\sin^2\theta_W = 0.228$) one obtains $\alpha_s = 0.16 \pm 0.03$ (stat.) $\pm$ 0.26 (syst.). In Fig. 39 the QCD prediction for Λ = 0.3 GeV is compared to the data (note that only statistical errors are shown). The agreement is good. Unfortunately, the large systematic uncertainty of R prevents a precise determination of α_s. Still, the data show that the second order contribution is smaller that the first order one, i.e.

$$C_2(\frac{\alpha_s}{\pi})^2 < \frac{\alpha_s}{\pi}$$

suggesting that the perturbative calculation is reliable.

For a detailed discussion of possible weak contributions to R see Bartel et al. (1981b), Barber et al. (1981a) and Branson (1981).

4. Gluon Bremsstrahlung

At high e^+e^- energies the acceleration of the quarks in $q\bar{q}$ formation becomes so large that the quarks have a fair chance to radiate a hard gluon. This offers the possibility to observe gluons directly (Polyakov 1975, Ellis et al. 1976, De Grand et al. 1977, Hoyer et al. 1979, Kramer et al. 1978, 1979, 1980).

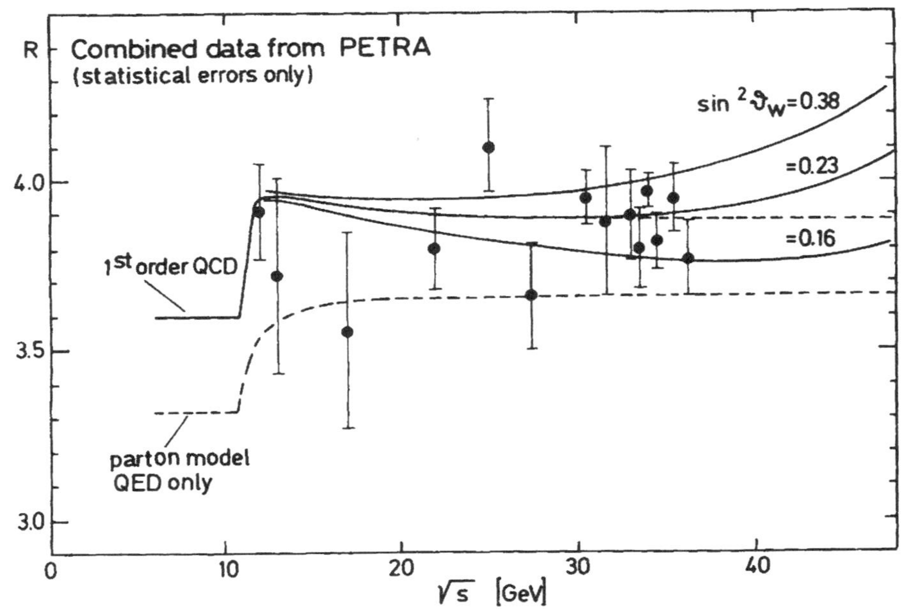

Fig. 39. The ratio R for the high energy region. The data points are averages over the PETRA experiments. The dashed curve indicates the quark parton prediction corrected for quark mass effects. The solide curves show the QCD prediction with Λ = 0.3 GeV for different values of $\sin^2\theta_W$. From Felst (1981).

Denoting by x_1, x_2 the fractional energies of the quarks, $x_i = 2E_i/W$, the cross section for gluon emission (see diagram in Fig. 40),

$$e^+e^- \to q\bar{q}g \tag{35}$$

is given by (Ellis et al. 1976, De Grand et al. 1977):

$$\frac{d\sigma(q\bar{q}g)}{dx_1 dx_2} = \frac{2\alpha_s}{3\pi} \sigma_o \frac{x_1^2 + x_2^2}{(1 - x_1)(1 - x_2)} \tag{36}$$

where $\sigma_o = 3\sigma_{\mu\mu}\Sigma e_q^2$. The energy and angular distribution of the radiated gluon is equal to that of bremsstrahlung photons. For small gluon energies K and angles θ (see Fig. 41)

$$\frac{d\sigma}{dKd\theta} \simeq \frac{\alpha_s}{K\sin\theta} \tag{37}$$

Since gluons are coloured we cannot observe them as free particles. Like quarks they materialize into a jet of hadrons. The emission of hard noncollinear gluons changes the event structure from back-to-back jets to a more complicated one. The most conspicuous effects predicted are (Ellis et al. 1976, De Grand et al. 1977)

a. Broadening of the transverse momentum (p_T) distribution of hadrons relative to a common jet axis with increasing c.m. energy W. The reason for this is that the average transverse momentum of the gluon, K_T, grows with W like

$$< K_T > \sim \alpha_s W \tag{38}$$

b. Presence of planar events since q, $\bar{q}$ and g lie in a plane.

c. Occurrence of events with a three-jet structure.

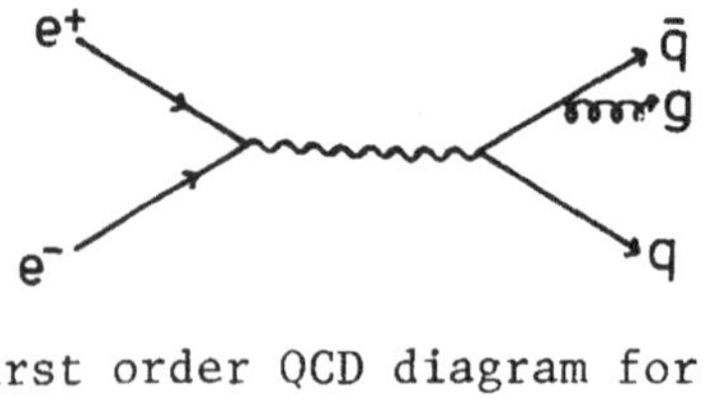

Fig. 40. First order QCD diagram for $e^+e^- \to q\bar{q}g$

Fig. 41. Illustration of the process $e^+e^- \rightarrow q\bar{q}g$

All three effects have been observed almost instantaneously when PETRA reached energies around 30 GeV in summer 1979 (Brandelik et al. 1979, Barber et al. 1979, Berger et al. 1979, Bartel et al. 1980b). Roughly 10 % of the hadronic events observed at W = 30 - 36 GeV are non two-jet events and half of these have a clear three jet structure. Two examples of three jet events are shown in Fig. 42.

In discussing the experimental evidence, we begin by showing in Fig. 43 the transverse momentum distribution $1/\sigma_{tot} d\sigma/dp_T^2$ of charged particles evaluated with respect to the jet axis for c.m. energies between 12 and 36 GeV (Brandelik et al. 1979, 1981c). The broadening of the p^2 distribution with rising energy is clearly borne out by the data. The 12 GeV distribution is well fitted by the $q\bar{q}$ model using a gaussian p_T^2 distribution with σ_q = 0.3 GeV/c. To fit the high energy data with the same model σ_q had to be increased to 0.45 GeV/c.

One of the several methods devised to study event shapes and test for planar events starts from the momentum tensor (Bjorken & Brodsky 1970). The method is similar to the characterization of an extended body by its ellipsoid of inertia. The tensor is formed from the N measured particle momenta:

$$M_{\alpha\beta} = \sum_{j=1}^{N} p_{j\alpha} p_{j\beta} \qquad \alpha,\beta = x,y,z \tag{39}$$

Let $\hat{n}_1$, $\hat{n}_2$ and $\hat{n}_3$ be the eigenvectors of this tensor associated with the eigenvalues Λ_1, Λ_2 and Λ_3 which have been ordered such that $\Lambda_1 < \Lambda_2 < \Lambda_3$. The Λ_i give the sum of the squares of the momentum components with respect to the $\hat{n}_i$ axis:

$$\Lambda_i = \sum_j (p_j \hat{n}_i)^2 \tag{40}$$

The principal axis is the $\hat{n}_3$ direction which is identical to the jet axis. The event plane is spanned by $\hat{n}_2$ and $\hat{n}_3$; $\hat{n}_1$ defines the direction along which the sum of the momentum components is minimal (see Fig. 44).

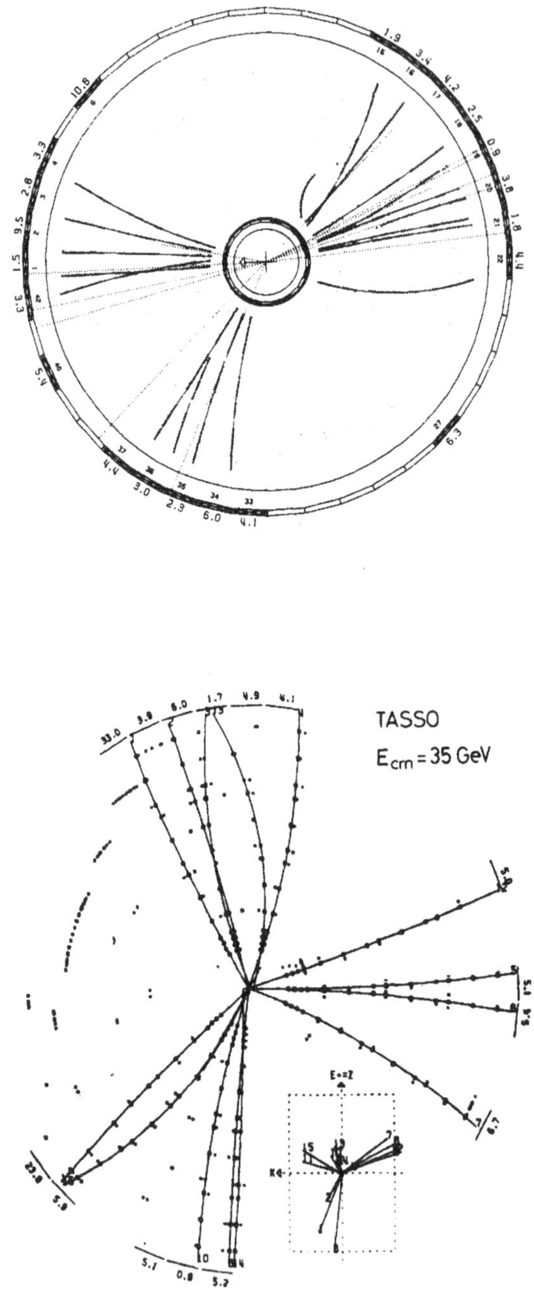

Fig. 42. Examples of the three-jet events as observed in the TASSO and JADE detectors.

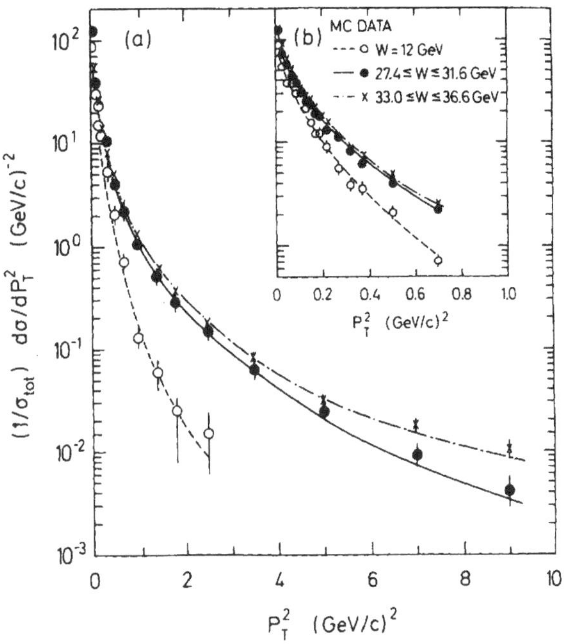

Fig. 43. Distribution of the square of the transverse momentum of charged particles with respect to the jet axis at 12, 27.4 - 31.6 and 35 - 36.6 GeV as measured by the TASSO group (Brandelik et al. 1979a, Brandelik et al. 1981b). The curves show the QCD prediction for $\alpha_s = 0.17$

Fig. 44. The momentum tensor ellipsoid and a 3 jet configuration

The Λ_i measure the flatness (Λ_1), width (Λ_2) and length (Λ_3) of an event. In Fig. 45 the distribution of

$$\langle p^2_{out} \rangle = \frac{1}{N} \Sigma (\vec{p}_j \cdot \hat{n}_1)^2 = \frac{\Lambda_1}{N} \qquad (41)$$

(= square of momentum component normal to the event plane) is compared with that of

$$\langle p^2_{Tin} \rangle = \frac{1}{N} \Sigma (\vec{p}_j \cdot \hat{n}_2)^2 = \frac{\Lambda_2}{N} \qquad (42)$$

(= square of momentum component in the event plane and perpendicular to the jet axis). The data show little increase in $\langle p^2_{out} \rangle$ from low (W = 12 GeV) to high energy (W = 30 GeV to 36 GeV). There are basically no events with $\langle p^2_{Tout} \rangle > 0.4$ (GeV/c)^2. The distribution of $\langle p^2_{Tin} \rangle$, however, becomes much wider at high energies with a long tail of events with high $\langle p^2_{Tin} \rangle$. The tail extends well beyond 1.2 (GeV/c)^2, i.e. for a fraction of the events $\langle p^2_{Tin} \rangle \gg \langle p^2_{Tout} \rangle$. This demonstrates that the data contain a certain number of planar events (Brandelik et al. 1979, Berger et al. 1979, Bartel et al. 1980b). The same conclusion is reached when the energy flow around the jet axis is studied (Barber et al. 1979).

In order to test whether for planar events the particle momenta are distributed uniformly in the plane or collimated into three jets the TASSO group (Brandelik et al. 1979, 1980a) analysed all planar noncollinear events at W = 30 GeV as three jet events and determined the three jet axes. Fig. 46 shows the distribution of p^2_T of the charged particles where the p_T of each particle was calculated with respect to the jet axis it had been associated with. The average transverse momentum is found to be about 0.3 GeV/c. The distribution is compared with the p^2_T distribution of particles from events at W = 12 GeV analysed as two jets. The p^2_T behaviour is found to be the same in both cases, i.e., the particles from planar events at high energies are collimated as closely around three coplanar axes as particles from lower energy events are collimated around a

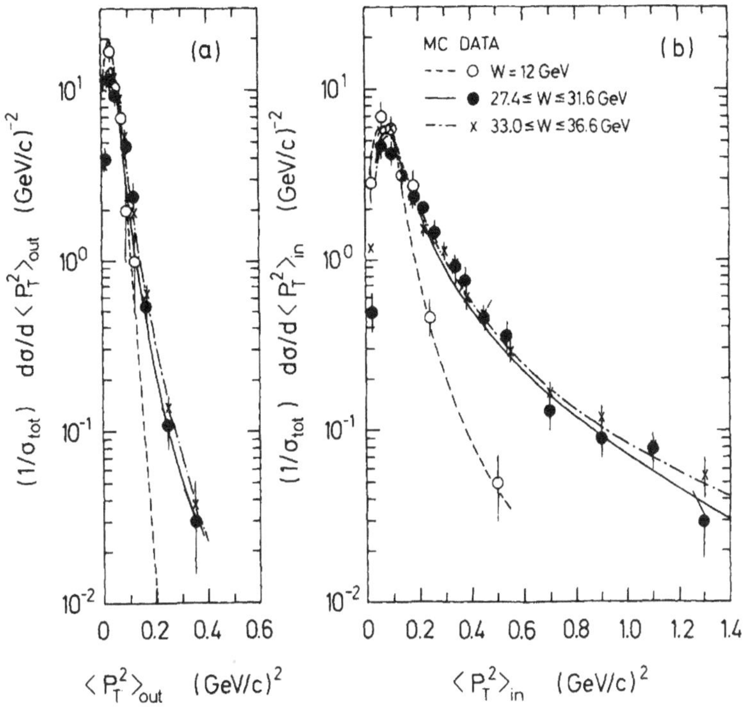

Fig. 45. Distribution of the mean transverse momentum squared per event for charged particles, normal to ($\langle p^2_{Tout}\rangle$) and in the ($\langle p^2_{Tin}\rangle$) event plane as measured by the TASSO group (Brandelik et al. 1979a, 1981b) at c.m. energies of 12, 27.4 - 31.6 and 33 - 36.6 GeV. The curves show the predictions for $\alpha_s = 0.17$

single jet axis.

The three jet structure of planar events can also be seen in the following way (Marshall 1981). The event is first analysed as a two-jet event. After summing the transverse momenta of particles in each jet, $\sum_{jet\alpha} p_{T_i}$, the narrow jet which has the smaller Σp_{T_i} is found.

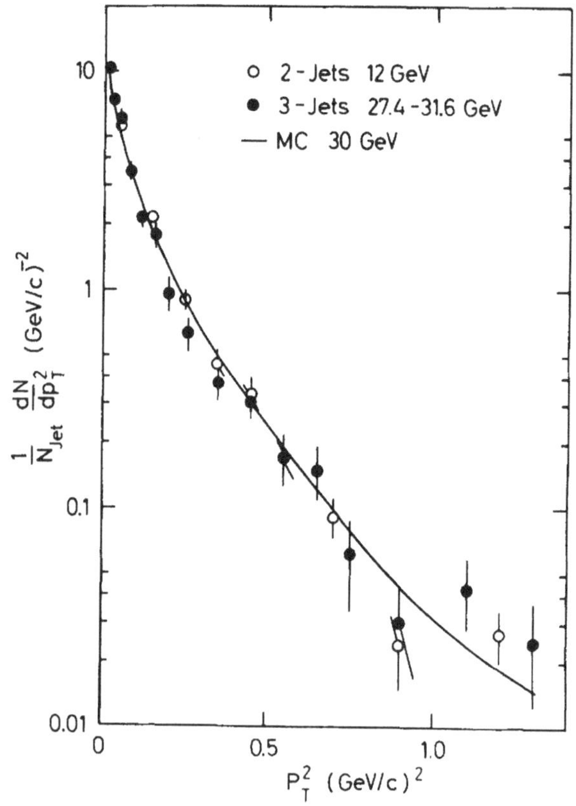

Fig. 46. Distribution of the square of the transverse momentum of charged hadrons from planar events with respect to the three axes found by the generalized sphericity method at W = 30 GeV (•). It is compared to the p_T^2 distribution relative to the sphericity axis for all events at W = 12 GeV analyzed as two jet events (o). The curve shows the QCD prediction for $\alpha_s = 0.17$. The analysis was made by the TASSO group (Brandelik et al. 1980a).

Then, as illustrated in Fig. 47, the angular distribution of particles in the event plane is found with respect to the direction of the narrow jet. θ is the angle between particle i projected into the event plane and the direction of the narrow jet. For a planar event with three-jet structure (Fig. 47a) no or only few particles will be found in the direction opposite to the narrow jet and the particle yield will show a dip at θ = 180°. On the other hand, for a two-jet event (Fig. 47b) the angular distribution will show no dip at θ = 180°. Fig. 48 shows the momentum weighted angular distribution for planar events. There is a narrow peak at θ = 0° which represents the particles in the narrow jet. In addition there is a broad distribution around θ = 180° which has a distinct dip at θ = 180° giving evidence to the three-jet nature of the events. The two bumps result from the two jets opposite the narrow jet.

5. Quantitative Comparison of Jet Production with QCD

The observation of three-jet events is one of the great successes of QCD. It does, of course, not rule out other mechanisms as possible sources of these events. Therefore quantitative tests are called for. The quantitative comparison with QCD faces the problem that QCD predicts the parton distributions (e.g. for q, $\bar{q}$ and g in

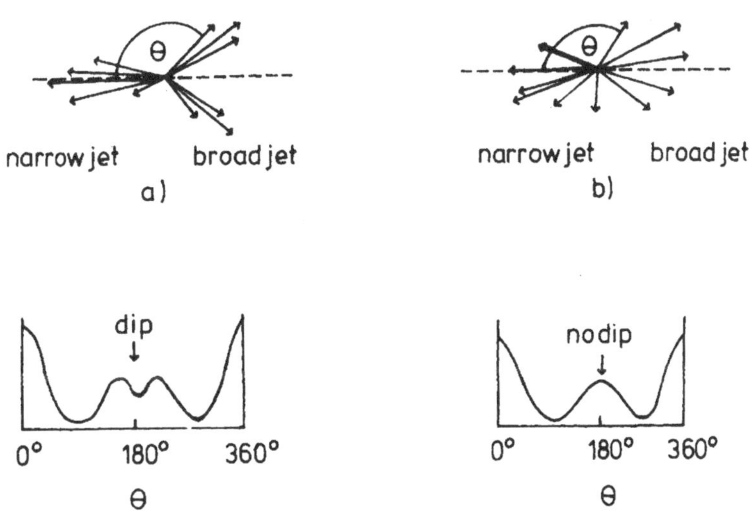

Fig. 47. Particle angular distribution in the event plane expected for three-jet production (a) and two-jet production (b).

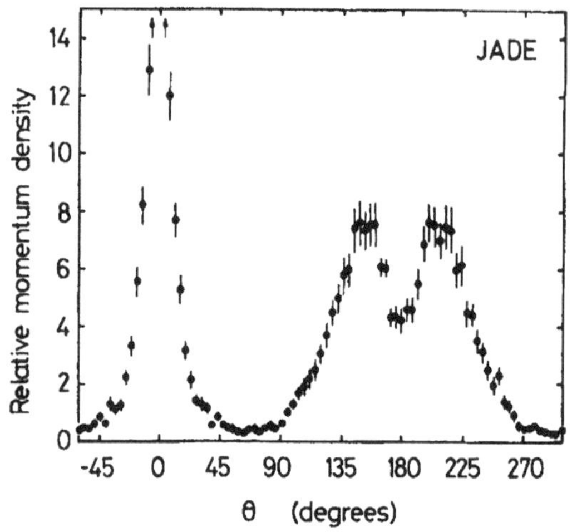

Fig. 48. The momentum weighted angular distribution of particles from planar events measured in the event plane $\theta = 0°$ is the direction of the narrow jet. From Marshall (1981)

$e^+e^- \to q\bar{q}g$) but what can be observed experimentally are the particle jets. Since the hadronization process is not calculable in perturbative QCD one has to combine the QCD calculations with a model that describes quark and gluon fragmentation into hadrons. Most of the analyses have been done using the model of Field and Feynman (1978) where quark fragmentation into hadrons is treated as an iterative process: (see Fig. 49) a quark q fragments into a hadron h plus another quark q'. Then q' fragments into h' plus q" and so on until the available energy/momentum is used up. The (three) fragmentation parameters such as the average transverse momentum of the produced hadrons have to be determined from the data itself.

In the quantitative comparison of data with the QCD prediction, two elaborate Monte Carlo programs have played a major role, that of Hoyer et al. (1979) and an extension by Ali et al. (1980). The first one considers only contributions up to first order in α_s (first 3

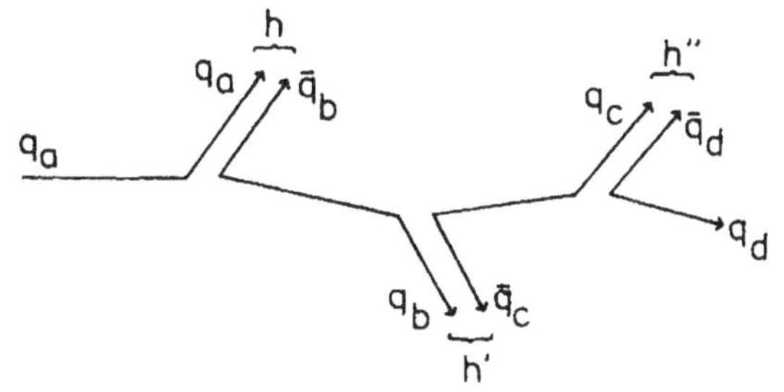

Fig. 49. Quark fragmentation into mesons in the Field-Feynman model

diagrams in Fig. 50), the second one includes those second order terms that lead to four-jet events (diagrams at bottom of Fig. 50).

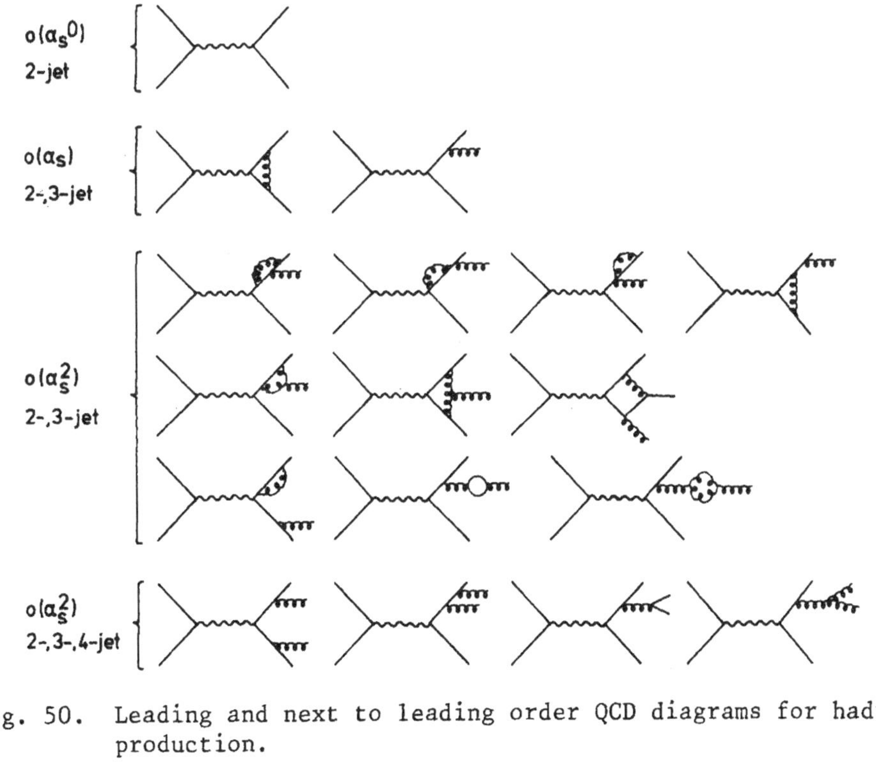

Fig. 50. Leading and next to leading order QCD diagrams for hadron production.

Both programs use the Field-Feynman scheme to describe quark fragmentation. In the analysis of the TASSO group (Brandelik et al. 1980a) the strategy followed was to determine first the fragmentation parameters and the strong coupling constant α_s, using as little of the measured information as possible (to be precise , the distributions at W = 30 GeV for $\langle p^2_{Tout}\rangle$, for the charge particle momenta and multiplicity) and then to use the remaining information for a quantitative check which then no longer involves any free parameter*. The fit yielded $\alpha_s = 0.17 \pm 0.02$ (stat;) $\pm$ 0.03 (syst.) at W = 30 GeV. The QCD predictions obtained in this way for p_T^2, $\langle p^2_{Tout}\rangle$, $\langle p^2_{Tin}\rangle$ and for the 3-jet p_T^2 are shown by the curves in Figs. 43, 45 and 46. The agreement is excellent. In particular the strong c.m. energy dependence of the data for p_T^2 and $\langle p^2_{Tin}\rangle$ is well reproduced by QCD.

Similar conclusions have been reached by the other PETRA groups. Some of these results are presented in the Figs. 51 - 54.

6. The Gluon Spin

The angular correlation between q, $\bar{q}$ and gluon (see Fig.55a) depends on the spin of the gluon. For vector gluons the first order cross section in terms of the scaled parton energies $x_i = 2E_i/W$ is given in eq. (36). From this the distribution of the angles θ_i between the partons θ_i can be calculated with the following relation:

$$x_i = \frac{2\sin\theta_i}{\sin\theta_1 + \sin\theta_2 + \sin\theta_3} \qquad (43)$$

In order to see that the angular correlation between the jet axes is indeed sensitive to the spin of the gluon we also consider a theory with scalar gluons which lead to (Ellis et al 1976):

$$\frac{d\sigma}{dx_1 dx_2} = \frac{\tilde{\alpha}_s}{3\pi}\,\sigma_o\,\frac{x_3^2}{(1-x_1)(1-x_2)} \qquad (44)$$

where $x_3 = 2E_3/W$ for the gluon and $\tilde{\alpha}_s$ is the gluon quark coupling for the scalar case. The vector expression (eq. 36) has both collinear and infrared divergencies, whereas for the scalar case there is only a collinear divergence which leads to a somewhat weaker dependence on x_1. Unfortunately, the difference is largest when x_1 (or x_2) is near 1 where one approaches the two jet configuration, and where the quark and gluon jets become difficult to separate.

---

* A more thorough description of the treatment of hadronization can be found in Söding and Wolf (1981).

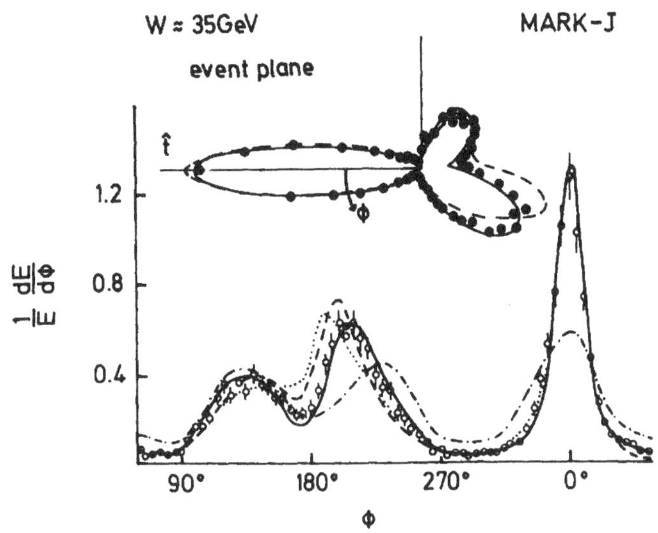

Fig. 51. Energy flow observed for planar events as a function of the azimuthal angle in the event plane relative to the thrust axis. From the MARK J group (Barber et al. 1981b). The curves show fits by i) a QCD calculation (full curve) with $\alpha_s = 0.18$; ii) a two-jet $q\bar{q}$ model with a Gaussian (dashed) or exponential (dotted) p_T distribution; iii) a pure phase space distribution (dashed-dotted curve).

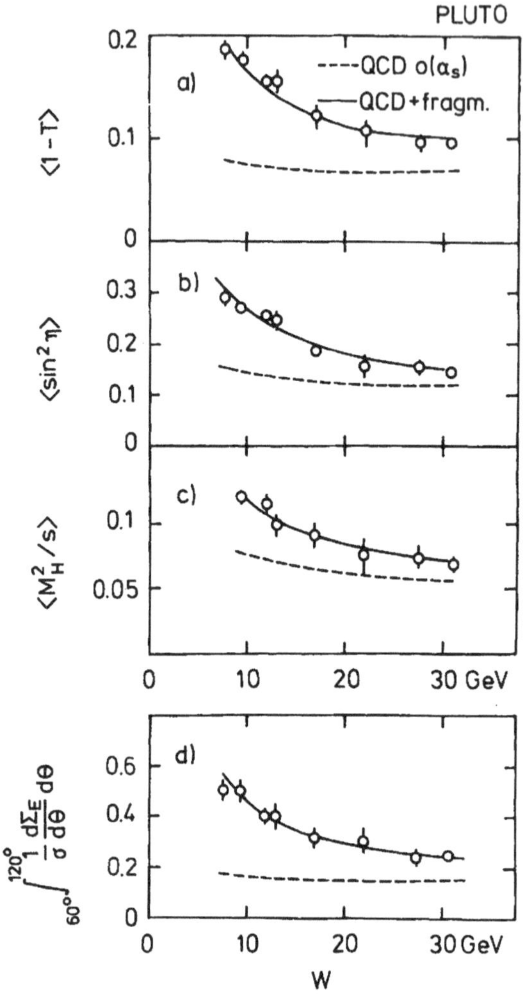

Fig. 52. W-dependence of average thrust, energy-weighted jet broadness, squared invariant mass of the "heavy" jet, and of the integral over the wide-angle part of the energy-energy correlation function in the reaction $e^+e^- \to$ hadrons measured by the PLUTO group (C. Berger et al . 1981b). The curves show the results of fits by a two-term formula $a/\ln(W^2/\Lambda^2) + b/W$; the first term describes the logarithmic W-dependence of the $O(\alpha_s)$ QCD contributions (on the parton level), the second term approximates hadronization effects. From the size of the perturbative QCD terms one extracts $\alpha_s = 0.18 \pm 0.02$.

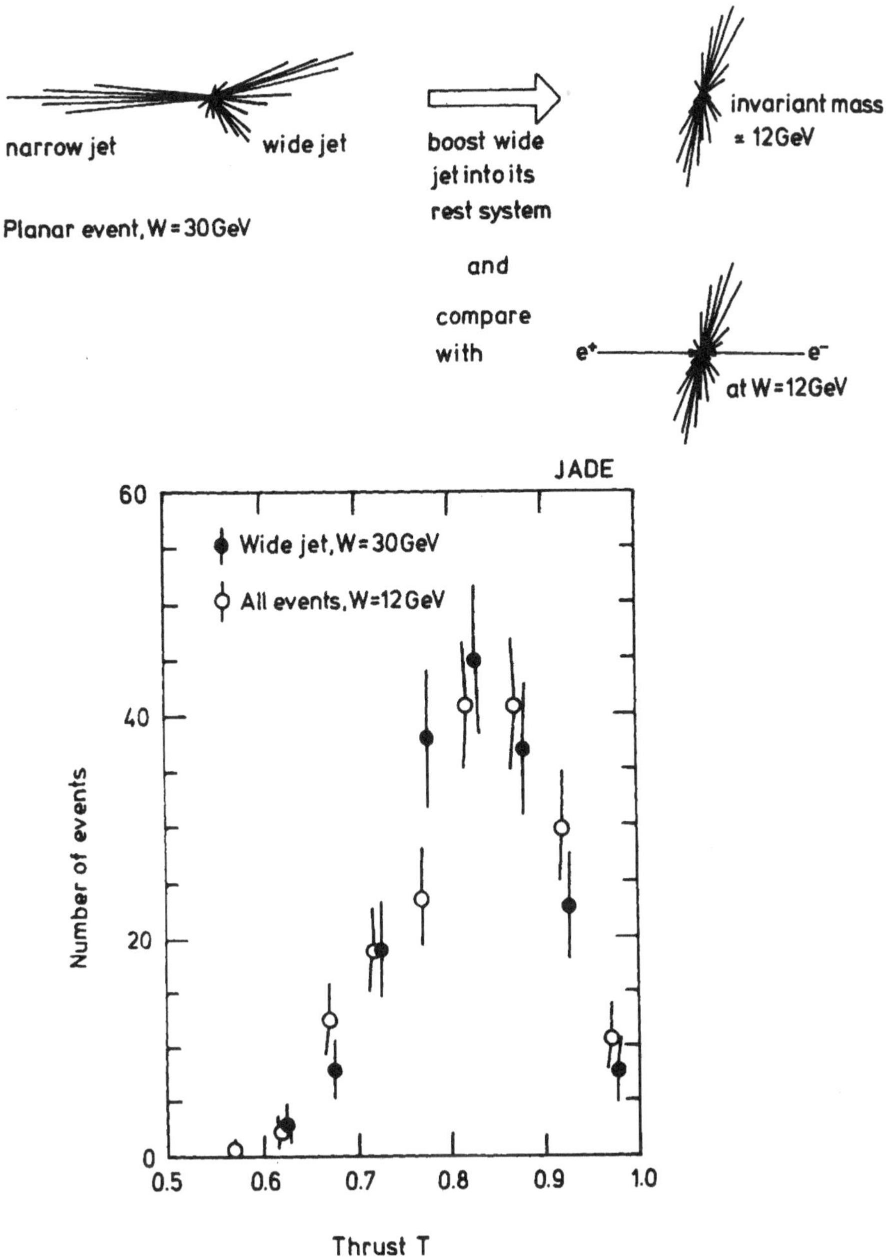

Fig. 53. On the top, a procedure of studying the structure of the wide jet in planar events from the reaction $e^+e^- \to$ hadrons is shown. The lower part shows the comparison of event shapes measured by thrust (T = 1/2 for spherical, T = 1 for collinear states), i) for the wide jet (transformed to its rest system) of planar events at 30 GeV, and ii) for complete events at 12 GeV (nearly all being 2-jet events). From the JADE group (Bartel et al. 1980b)

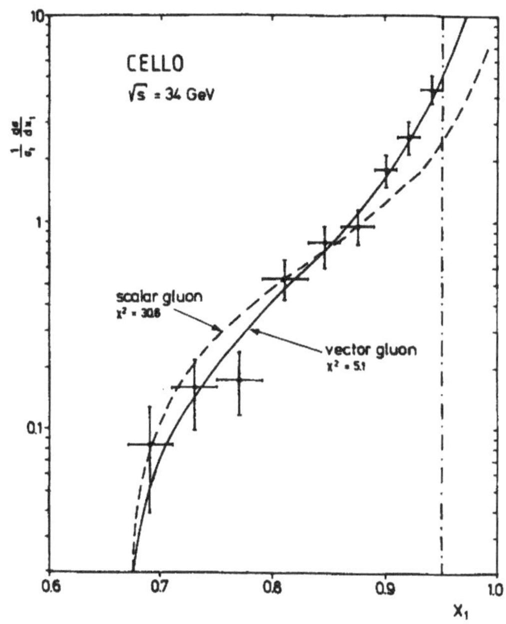

Fig. 54. Parton thrust distribution for three jet events as measured at W = 34 GeV by the CELLO group (Behrend et al. 1981). The curves show the QCD prediction for $\alpha_s = 0.1$ (solid line) and the prediction for scalar gluons (dashed line)

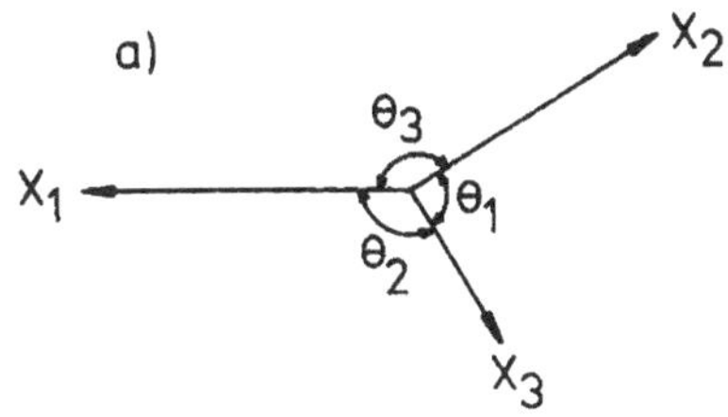

Fig. 55a. Momenta and angles of a $q\bar{q}g$ final state in the c.m. frame

In the analysis of the TASSO group (Brandelik et al. 1980c) the angles θ_i were reconstructed from the jet axes determined from the charged particle vectors. From the angles the parton energies were determined and ordered such that $x_1 \geqslant x_2 \geqslant x_3$. Using Monte Carlo generated events it was ascertained that the x_i obtained by this approach agree closely with those of the parent partons. The resolution in x_1 is 6 % and systematic biases are very small (of the order 1 %).

In order to avoid the problematic area near $x_1 = 1$ only events with $x_1 \leqslant 0.9$ were considered. A quantity particularly sensitive to the question of vector or scalar gluons is the angle $\tilde{\theta}$ illustrated in Fig. 55b (Ellis & Karliner 1979). The $q\bar{q}g$ system is Lorentz boosted into the c.m. frame of partons 2 and 3 (one of which by virtue of the ordering of the x_i is more likely to be the gluon) and $\tilde{\theta}$ measures the angle between the 2, 3 axis and the parton 1. Assuming massless partons

$$\cos\tilde{\theta} = \frac{x_2 - x_3}{x_1} = \frac{\sin\theta_2 - \sin\theta_3}{\sin\theta_1} \qquad (45)$$

The measured distribution of $\cos\tilde{\theta}$ is shown in Fig. 56. The comparison with the vector gluon hypothesis (QCD) involves no free parameter since α_s has already been determined. The agreement between the data and QCD is perfect. The prediction for scalar gluons fails badly; the observed deviations correspond to 5 s.d. The scalar gluon hypothesis is inconsistent with the data (see also Berger et al. 1980b). This conclusion is strenghtened by the fact that the $\tilde{\theta}$ distribution predicted for scalar gluons at the parton level is hardly altered when fragmentation is included (see Fig. 56).

7. Alternative Explanations of the Three-Jet Events ?

After the observation of three-jet events several models were

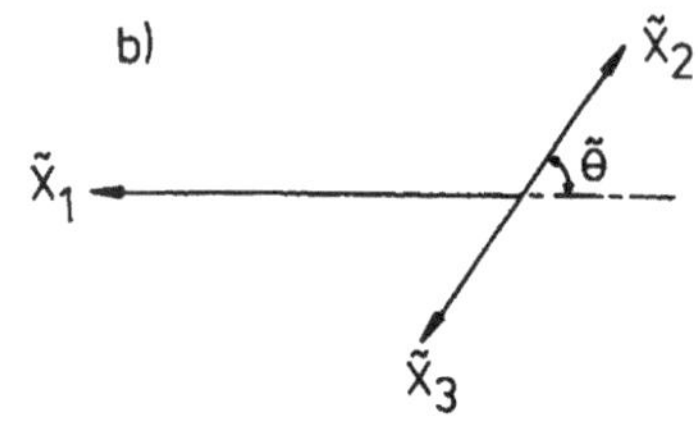

Fig. 55b. The $q\bar{q}g$ final state transformed to the rest frame of particles 2 and 3

proposed which maintain that the three jet events are not necessarily due to gluon bremsstrahlung but can be explained by a different mechanism. A detailed comparison of these models with experiment was recently presented by Söding (1981) using the TASSO data. In this comparison the most sensitive distributions were found to be the angular correlations between the jet axes, or equivalently, the distributions of the fractional energies x_i.

The jet Dalitz plot determined from the TASSO experiment for an average c.m. energy W = 33 GeV with 4945 events is shown in Fig. 57a. Collinear two-jet configurations (for which $x_1 = 1$) lie along the base line of the plot, symmetric three-star configurations at the top corner. The QCD prediction for $\alpha_s = 0.17$ predicts a very similar event distribution (Fig. 57b). The projection onto parton thrust x_1, normalized to σ_{tot} is shown in Fig. 58. The QCD prediction agrees well with the data (solid curve).

We turn now to the comparison of the data with the alternative models. In the constituent interchange model (CIM) (T.A. De Grand et al. 1977) a hard (large p_T) meson or resonance is emitted instead of a gluon. This is a higher twist contribution. Extending an earlier analysis by the PLUTO group (Berger et al. 1980b) the TASSO data have been compared (Söding 1981) with the CIM making the conservative assumption that a whole spectrum of mesons can be emitted. The quark meson coupling strength has been adjusted such that the model fits the p_T^2 distribution at W = 12 GeV (Fig. 43). To describe the data at W = 33 GeV the coupling constant has to be increased by a factor of four (which means a factor of 16 in the cross section). Moreover the model fails to describe the parton thrust (x_1) distribution as shown by the dashed-dotted curve in Fig. 58. As a result no more than 5 % of the three-jet events can be attributed to higher twist effects.

A somewhat similar mechanism is suggested by the massive quark model (Preparata 1981). This model gives a good description of the p_T^2 distribution at W = 12 and 30 GeV. Whether the model can also account for the properties of the planar events such as the parton

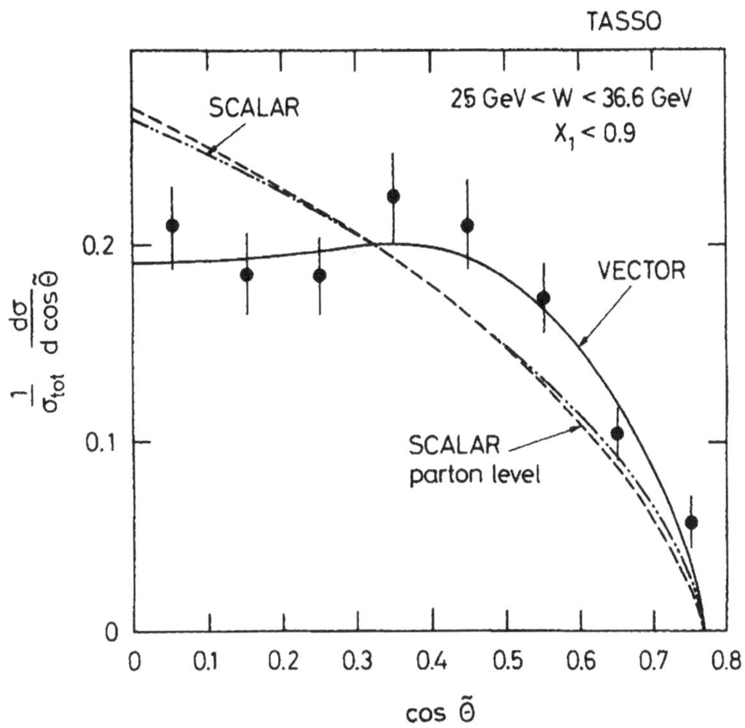

Fig. 56. Observed distribution of the Ellis-Karliner variable $\cos\tilde{\theta}$ for events in the three-jet region defined by parton thrust $x_1 < 0.9$ for the reaction $e^+e^- \to$ hadrons at <W> = 33 GeV (from the TASSO experiment Brandelik et al. 1980c). The full curve shows the QCD prediction to $O(\alpha_s)$ with $\alpha_s = 0.17$ (vector gluons), the dashed-dotted curve the prediction for scalar gluons. The dashed curve shows the scalar gluon prediction on the parton level, i.e. before correction for hadronization, radiative effects, and detector imperfections.

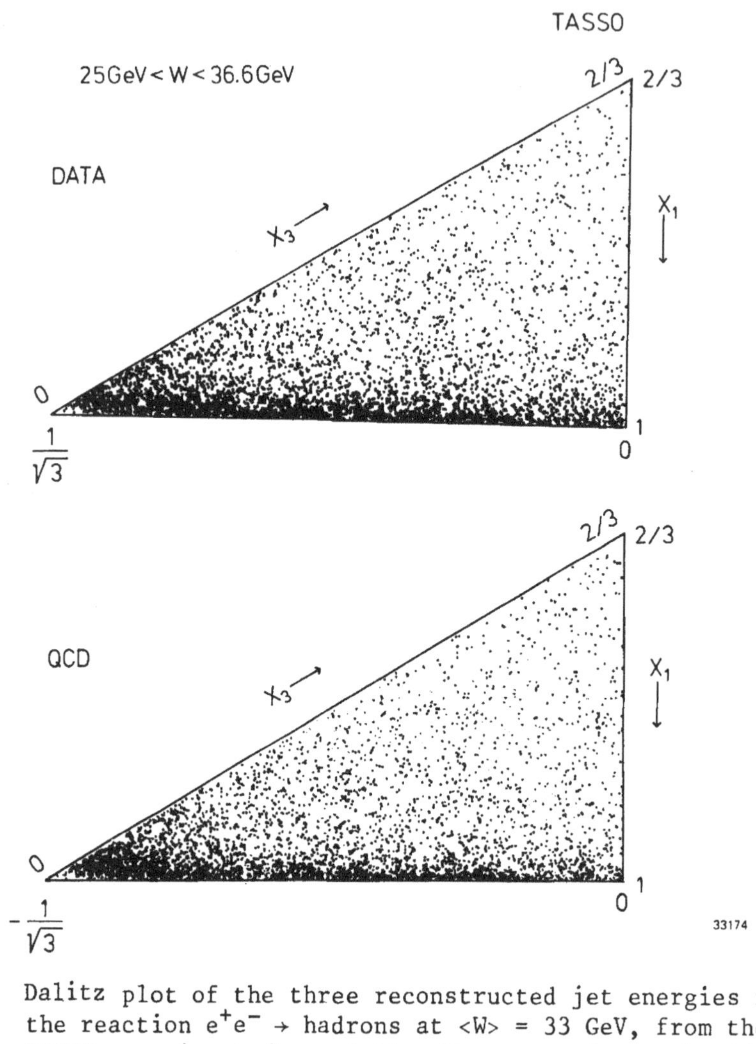

Fig. 57. Dalitz plot of the three reconstructed jet energies for the reaction $e^+e^- \rightarrow$ hadrons at <W> = 33 GeV, from the TASSO experiment (top plot). For comparison, the QCD prediction with $\alpha_s = 0.17$ is shown using Monte Carlo simulations with the same number of events (bottom). From Söding (1981)

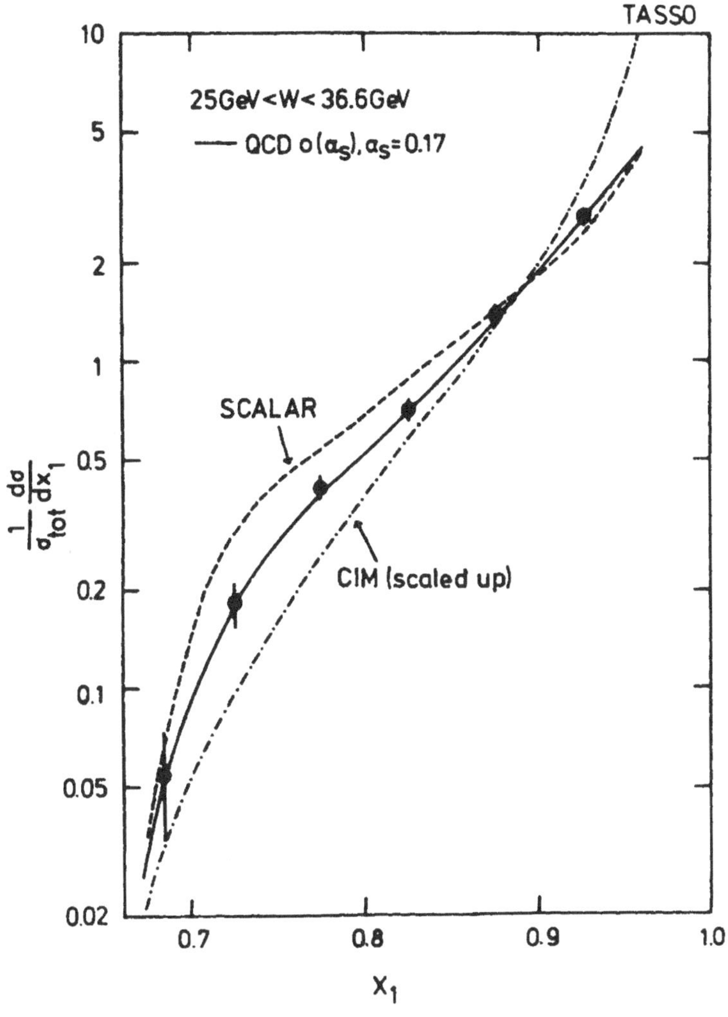

Fig. 58. Distribution of the maximum reconstructed jet energy x_1 (= parton thrust for three-jet events) in the reaction $e^+e^- \to$ hadrons at $\langle W \rangle$ = 33 GeV, from the TASSO experiment. The full curve shows the QCD prediction to $O(\alpha_s)$ with α_s = 0.17, the dashed curve the prediction for scalar gluons (with the coupling α_s = 1.6 optimized to furnish the best overall description of a number of event variables). The dashed-dotted curve is the prediction from the constituent interchange model with a scaled-up coupling constant such as to fit the p_T^2 distribution at the same energy. From Söding (1981)

thrust (x_1) distribution or the distribution of p_T^2 with respect to the three jet axes has not yet been tested.

The emission of a large p_T meson from a quark in the process $e^+e^- \to q\bar{q} \to q\bar{q}M$ may also involve a contribution with the same W dependence as the leading process $e^+e^- \to q\bar{q}$ but with an intrinsic p_T^{-4} tail of the hadron distribution. For comparison, gluon bremsstrahlung (which is point-like) has a p_T^{-2} behaviour. On the other hand the most common assumption for hadronization is a gaussian p_T distribution as used e.g. in the Field-Feynman model (1978). It turns out that while it is possible to adjust the p_T^{-4} tail so as to reasonably fit the TASSO p_T^2 distribution (Fig. 43) e.g. at W = 12 GeV the strong W dependence of the p_T^2 distribution is not reproduced. Furthermore, the model predicts too few three-jet events with large angles between them (i.e. close to the symmetric three-star configution).

The large tail of the p_T^2 distribution at high W could also be due to the fragmentation of the heavy quarks c and b which might impart a much larger intrinsic p_T^2 to their fragmentation products than light quarks do. In other words, the intrinsic transverse momentum q_T distribution could have two terms:

$$\frac{dp}{dq_T^2} \sim Ae^{-q_T^2/2\sigma_q^2} + Be^{-q_T^2/2\sigma_Q^2}$$

where the first is due to the light quarks (u,d,s), the second is due to c and b quarks, and $\sigma_Q \gg \sigma_q$. In order to reproduce the observed p_T^2 distribution σ_Q has to be rather large (σ_Q = 0.8 GeV/c compared to σ_q = 0.3 GeV/c). This has the consequence that planar events from this model show a much weaker collimation around the three jet axes than the data shown in Fig. 46.

In conclusion none of the models which have been proposed as alternatives to QCD can describe all features of three jet production. Whether the massive quark model is more successful remains to be seen.

8. The Value of α_s and the Next to Leading Order Corrections

The number of three-jet events is directly related to the value of the quark gluon coupling strength α_s. The α_s values obtained at energies of 30 GeV are summarized in Table 6. Some of the results were determined in leading order (labelled α_s), some were obtained using the Ali et al. program which includes some of the next-to-leading order diagrams (labelled some α_s^2).

Table 6. Determination of α_s from three-jet production near W = 30 GeV. First error is statistical, second systematic. From Braunschweig (1981).

Experiment	Order in α_s	α_s
CELLO	α_s	0.15 ± 0.015 ± 0.03
JADE	α_s	0.18 ± 0.03 ± 0.03
MARK J	some α_s^2	0.17 ± 0.02
PLUTO	α_s	0.16 ± 0.02
TASSO	some α_s^2	0.17 ± 0.02 ± 0.03

All four experiments are seen to agree on the value of α_s, yielding an average of α_s = 0.17 ± 0.01 with a systematic uncertainty of 0.03.

A still open question is the effect of the complete second order corrections. An evaluation of all second order terms (which are displayed in Fig. 50) was recently made by three groups, R.K. Ellis et al. (1980), Fabricius et al. (1980) and Vermaseren et al. (1980). Although working all in the $\overline{\text{MS}}$ renormalization scheme they arrived at apparently conflicting results which led to different conclusions. Fabricius et al. found that the $O(\alpha_s^2)$ corrections are comparatively small, while the two other groups concluded that the second order corrections to three-jet production are roughly 30 - 40 % of the first order contribution (Kunszt 1981, Ali 1981). All three groups found that the energy partition between the three jets and therefore the angular correlation between the jet axes is almost the same in $O(\alpha_s^2)$ as in $O(\alpha_s)$ which implies amongst other things that the conclusion on the gluon spin is unaffected by the second order corrections (see e.g. Kunszt 1981).

The reason for the dicrepancy between the three theoretical analyses appears now to be understood (Buras 1981). It is due to different definitions of what is called a three-jet event and of thrust, the quantity directly compared with experiment. Fabricius et al. work with a Sterman-Weinberg (1977) type definition; a three-jet event is defined as an event which has all but a fraction $\varepsilon/2$ of its total energy distributed within three separated cones of full opening angle δ (see Fig. 59). Values used for ε, δ in the calculation of Fabricius et al. were $\varepsilon = 0.2$, $\delta = 45°$. The use of this type of jet definition is suggested by physics: if for a fixed W

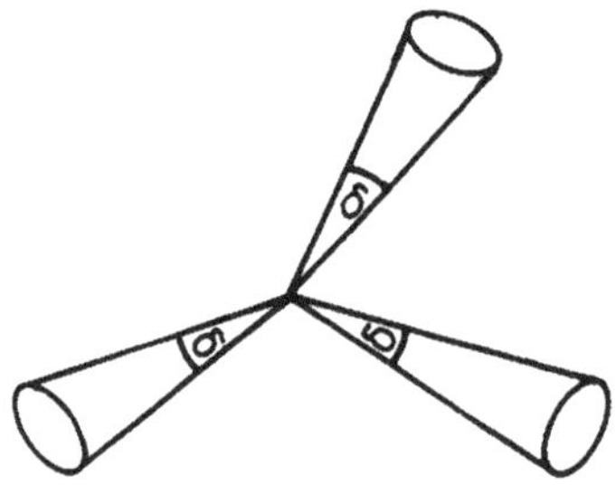

Fig. 59. Definition for a three-jet event used by Fabricius et al.

the angular separation of, say a quark and a gluon is below some minimum angle (defined by the hadronization process) it is impossible to separate the fragments from the two partons, and, consequently to observe two separate partons.

Now consider the diagram for two gluon emission, $e^+e^- \rightarrow q\bar{q}gg$, in Fig. 60.

The effective mass squared of the parton system 1, 2 is given by the product of the four momentum vectors, $M_{12}^2 = 2p_1p_2$. Because of the cut-off in δ partons 1 and 2 are defined to be in one jet if $y_{12} \equiv 2p_1p_2/W^2 \geqslant 10^{-2}$. In the calculations of Ellis et al. and Vermaseren et al. the diagram shown in Fig. 60 is calculated on the parton level and is considered to represent a four jet event down to extremely small y_{12} values ($y_{12} \simeq 10^{-5}$). The differences between the α_s values extracted from the data at W = 30 GeV using one or the other approach are presently of the order of 20 %: Fabricius et al. find α_s = 0.17 while Ali (1981) and Kunszt (1981) based on the work of Ellis et al. obtain α_s = 0.12 - 0.13. These results should be considered as preliminary.

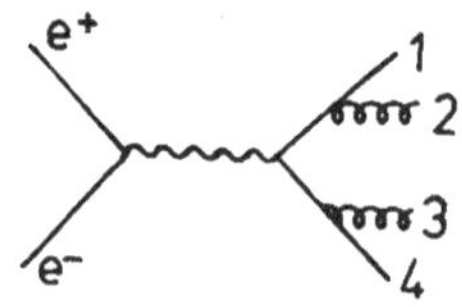

Fig. 60. QCD diagram for $e^+e^- \rightarrow q\bar{q}gg$

The comparison with OCD presented above dealt primarily with processes where a single hard gluon is emitted. Work has also started to search for effects of multiple, soft gluon emission which could affect, e.g. the two particle correlations. The conclusions from these tests are strongly model dependent since the effects are intimately related to the question of hadronization. Reference to these studies can be found e.g. in the review of Criegee and Knies (1981).

9. Summary of the QCD Tests

The observation of three-jet events in high energy e^+e^- annihilation into hadrons may be considered a major triumph of the theory. OCD had predicted such events to occur as a result of hard gluon bremsstrahlung. Close examination of the details of three-jet production at PETRA shows consistency with QCD and confirms the vector nature of the gluon. The rate of three-jet events yields in leading order of QCD $\alpha_s = 0.17 \pm 0.02$.

The observed topology of the direct decays of the Υ, $e^+e^- \to \Upsilon$ (9.4 GeV) $\to$ hadrons (Brandt 1979, Berger et al. 1980a, Niczyporuk et al. 1981) which has not been discussed in this review - exclude colourless gluons (Walsh & Zerwas 1980). In conclusion, the experimental results show a strong preference for a theory with coloured vector gluons - just what QCD is. Two important characteristics of QCD have not yet been verified by experiment, the existence of the three-gluon vertex, and asymptotic freedom, i.e. the decrease towards zero of α_s with increasing Q^2. The first is related to the existence of bound gluon states or glue balls. The first glue balls might have been observed recently in radiative decays of the J/ψ (Scharre 1981). The second point will require higher Q^2 than presently available or a study of "cleaner" processes as $\gamma\gamma$ scattering at large Q^2.

VII. QUARK AND GLUON FRAGMENTATION

As seen above the hadronic final states in high energy e^+e^- annihilation are a result of quark and gluon induced jet formation. By analyzing the final states one can hope to piece the jet fragments together and reconstruct the properties of the primordial parton, such as charge, flavour, etc. This section summarizes briefly what is known on the final states at high energies.

1. Energy Carried by Neutrals

The energy fraction ρ_γ carried by photons (either from π^0, η ... decay or from direct production), and ρ_N by neutrals obtained by subtracting the charged particles energies,

$$\rho_\gamma = \Sigma E_{\gamma j}/W$$

$$\rho_N = 1 - \Sigma E_{CHI}/W \qquad (46)$$

is shown in Fig. 61. ρ_N includes the energy carried by K^o and Λ particles. The fraction of energy carried by photons and by neutrals appears to be energy independent. Roughly speaking, one third of the energy goes into neutrals. The difference $\rho_N - \rho_\gamma$ provides an upper limit on the fraction of energy ρ_ν taken up by neutrinos. The JADE group found $\rho_\nu < 10$ % (95 % C.L.) which disfavours the free quark model of Pati and Salam (1978) where $\rho_\nu = 18 - 28$ % was predicted.

2. Charged Particle Multiplicity

In Fig. 62 the average charged particle multiplicity* $<n_{CH}>$ is plotted as a function of c.m. energy. Above $\sim$ 7 GeV $<n_{CH}>$ is seen to rise (logarithmically) faster than at lower energies. Phase space predicts $<n_{CH}> \sim s^{1/4}$ which fits the data rather well (see curve B); it fails, however, to reproduce the dispersion of the multiplicity distribution,

$$D = \sqrt{<n_{CH}^2> - <n_{CH}>^2}$$

which was found to increase with energy like $<n_{CH}>$.

The simplest form of scaling of charged particle production leads to $<n_{CH}> = a + b \ln s$. This is certainly at variance with the data if the full energy range is considered. The observed rise of $<n_{CH}>$ cannot be attributed to the onset of $b\bar{b}$ production which is found to yield an increase by $\sim$ 0.2 units. In QCD an increase of $<n_{CH}>$ over the scaling curve is predicted due to the additional contribution from gluon fragmentation. The exact form of the resulting energy dependence is not yet clear. If the result for infinitely heavy quarks is taken for guidance, one expects $<n_{CH}>$ to grow like (Ellis 1980, Furmanski et al. 1979)

$$n_{CH} = n_o + a \exp(b\sqrt{\ell n(s/\Lambda^2)}).$$

Fits of this form with $\Lambda = 0.3$ GeV yield $n_o = 2.0 \pm 0.2$, $a = 0.027 \pm 0.01$, $b = 1.9 \pm 0.2$ and reproduce the trend of the data (see curve A in Fig. 62). One may therefore be tempted to attribute the rapid

---

* $<n_{CH}>$ includes the $\pi^\pm$ coming from $K_s^o \to \pi^+\pi^-$ decay. This contribution amounts to 0.4 units at 7.4 GeV, 0.6 at 12 GeV and 1 unit at 30 GeV.

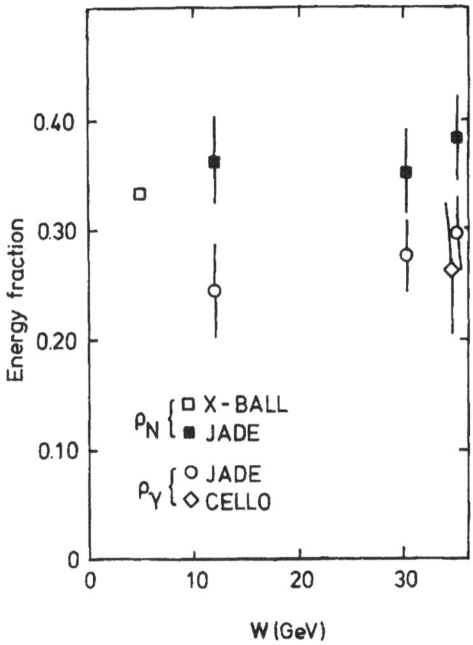

Fig. 61. The energy fractions carried by photons (ρ_γ) and by neutrals (ρ_N)

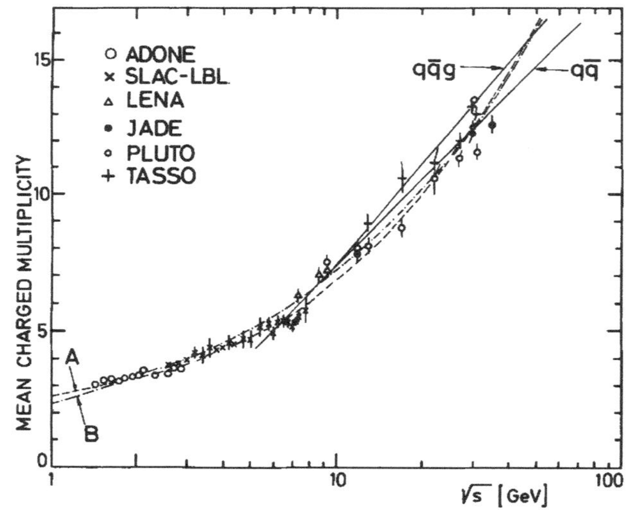

Fig. 62. Mean charged multiplicity versus center-of-mass energy. The straight line labelled $q\bar{q}$ is a typical prediction for $q\bar{q}$ production with subsequent fragmentation of quarks into mesons. The line labelled $q\bar{q}g$ includes gluon bremsstrahlung. The curve labelled (A) is the best fit of the form $\langle N_{ch} \rangle = N_o + a.\exp(b\sqrt{\ln(s/\Lambda^2)})$ described in the text. Curve (B) shows the dependence $\langle N_{ch} \rangle = 2.3.s^{1/4}$(s in GeV^2).

rise to hard gluon effects. The solid lines in Fig. 62 compare the $<n_{CH}>$ data with the $q\bar{q}$ model, $e^+e^- \to q\bar{q} \to$ hadrons using the Field-Feynman fragmentation functions but without hard gluon contributions. The model accounts well for the rise seen above $\sim$ 5 GeV*. The inclusion of hard gluon emission raises the prediction by a negligible amount below 10 GeV; at 35 GeV it adds 0.8 units. We conclude therefore that the rapid rise of $<n_{CH}>$ is mostly due to the growing phase space: particle masses matter less as the energy increases.

In Fig. 63 the e^+e^- multiplicity distributions are compared in a KNO plot (Koba et al. 1972) with pp and $\bar{p}p$ data. Plotted is $P_{CH} \cdot <n_{CH}>$ where P_{CH} is the probability for observing a final state with n_{CH} charged particles versus $n_{CH}/<n_{CH}>$. The e^+e^- data obey KNO scaling between 9.4 and 35 GeV. They agree well with the $\bar{p}p$ data but disagree with the pp data which have a larger dispersion.

3. Inclusive Particle Spectra without Particle Identification

The differential cross section for producing a particle h with momentum and energy P, E and angle θ relative to the beam axis can be expressed in terms of two structure functions $\bar{W}_1$ and $\bar{W}_2$ which are closely related to W_1 and W_2 measured in inelastic lepton hadron scattering (Drell and coworkers 1969)

$$\frac{d^2\sigma}{dx d\Omega} = \frac{\alpha^2}{s} \beta x \{m\bar{W}_1 + \frac{1}{4}\beta^2 x \nu\bar{W}_2 \sin^2\theta\} \qquad (47)$$

where m is the mass of h, $\beta = P/E$, $x = E/E_{beam} = 2E/\sqrt{s}$ and ν is the energy of the virtual photon as seen in the h rest system, $\nu = (E/m)\sqrt{s}$.

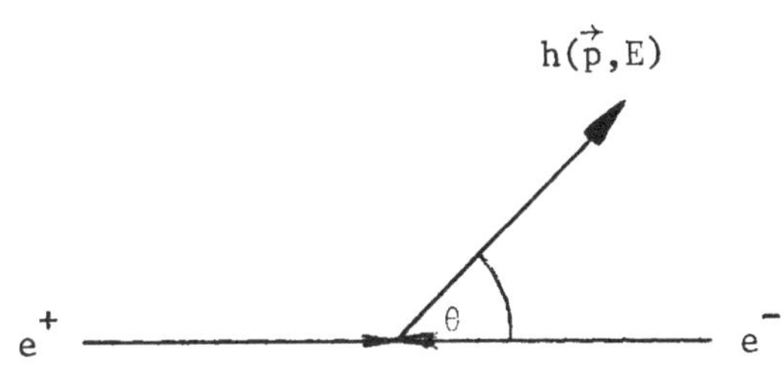

Diagram for inclusive particle production

* At lower energies the model is presumably less reliable because of the approximations made for the fragmentation.

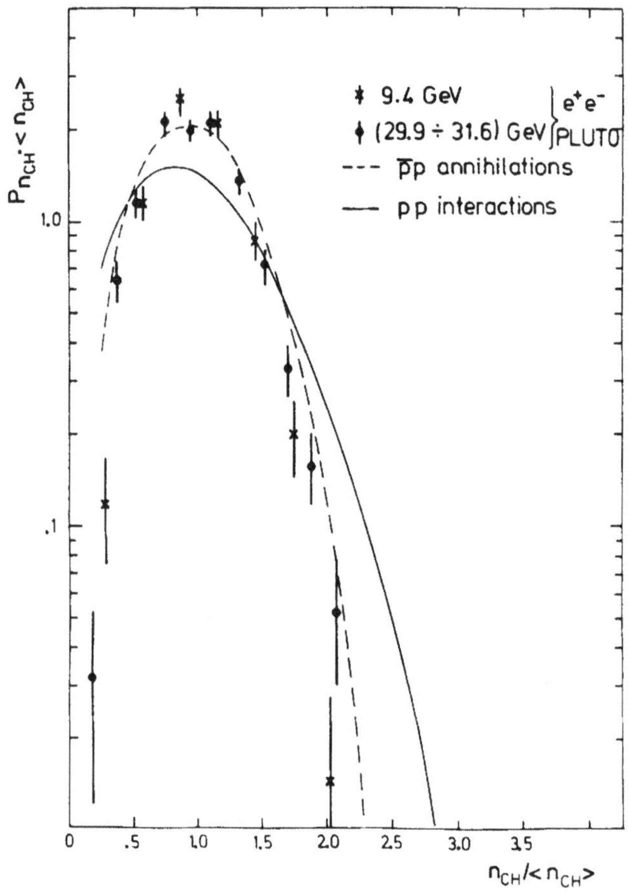

Fig. 63a. KNO plot for charged particle multiplicities measured in e^+e^- annihilation. The dashed and solid curves show the results for $\bar{p}p$ annihilation and pp scattering. $P_{n_{CH}}$ is the probability of observing the charged multiplicity n_{CH}.

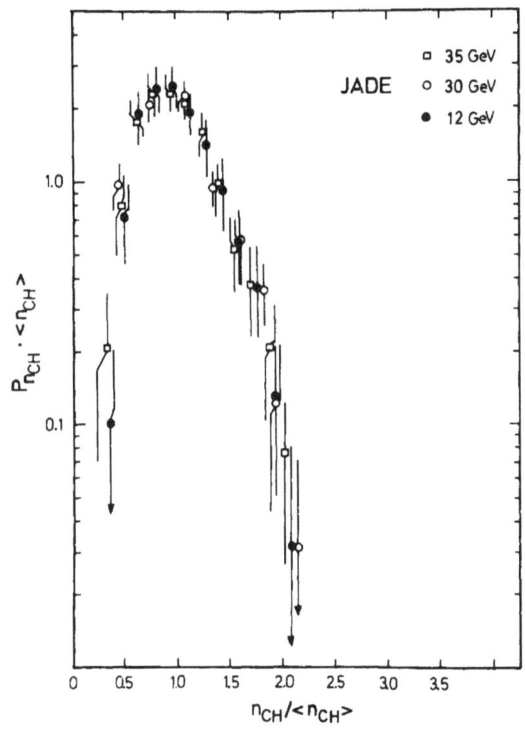

Fig. 63b. Same as Fig. 63a.

At particle energies large enough that particle masses can be neglected, x can be replaced by the normalized momentum $x = P/E_{beam}$ and the scaled cross section reads

$$s d\sigma/dx \simeq 4\pi\alpha^2 \, x \, \{m\bar{W}_1 + \frac{1}{6} \, x \, \nu\bar{W}_2\} \qquad (48)$$

The structure functions in general depend on two variables, e.g. x and s. If scale invariance holds $\bar{W}_1$ and $\nu\bar{W}_2$ are functions of x alone and $s d\sigma/dx$ is energy independent.

Scaling behaviour is e.g. expected in the quark parton model: at energies large enough that particle masses can be neglected, the number of hadrons h produced by a quark q with fractional energy x, $D_q^h(x)$, is independent of s. This leads to

$$\frac{d\sigma}{dx}\,(e^+e^- \rightarrow q\bar{q} \rightarrow h) = \sigma_{q\bar{q}} \cdot 2D_q^h(x) = \frac{8\pi\alpha^2}{s}\, e_q^2 D_q^h(x) \qquad (49)$$

Fig. 64 displays scaled cross sections not separated according to particle type. In this case the particle energy is unknown and x is defined as $x = 2p/W$. The data in Fig. 64 were measured at c.m. energies between 5 and 35 GeV. At low x values ($x < 0.15$) the particle yield shows a dramatic rise when the c.m. energy is increased from 5 to 35 GeV. This rise is related to the multiplicity growth seen above. At $x > 0.2$ only little energy dependence is observed; most of the variation with energy occurs between W = 5 and 12 GeV, with the low W data being higher. A similar behaviour is seen by MARK II comparing data taken at SPEAR and PEP (Hollebeek 1981). This type of scale breaking is expected in QCD where at high W gluon emission depletes the particle yield at large x and increases it at small x. However, scale breaking effects are better studied with the cross sections for specific particle types (π, k, p, ...) rather than with the unseparated cross sections which at lower energies are strongly affected by mass effects. For instance, proton production passes its phase space maximum only at a momentum of 1 GeV/c (see below) which corresponds to an x_p value of 0.4 at W = 5 GeV.

Fig. 65 shows the rapidity distributions normalized to the total cross section as measured by TASSO between 13 and 31.6 GeV. The thrust axis was taken as the jet axis since in this way the corrections for the y-axis relative to the true axis were found to be smaller at large y than if the sphericity axis had been chosen. To compute y, the particles were assumed to be pions. Also plotted in Fig. 65 are the data from SLAC-LBL measured at low energies (4.8 and 7.4 GeV). They were determined with respect to the sphericity axis. One observes a plateau near y = 0 that broadens with increasing c.m. energy. The height of the plateau is not constant but

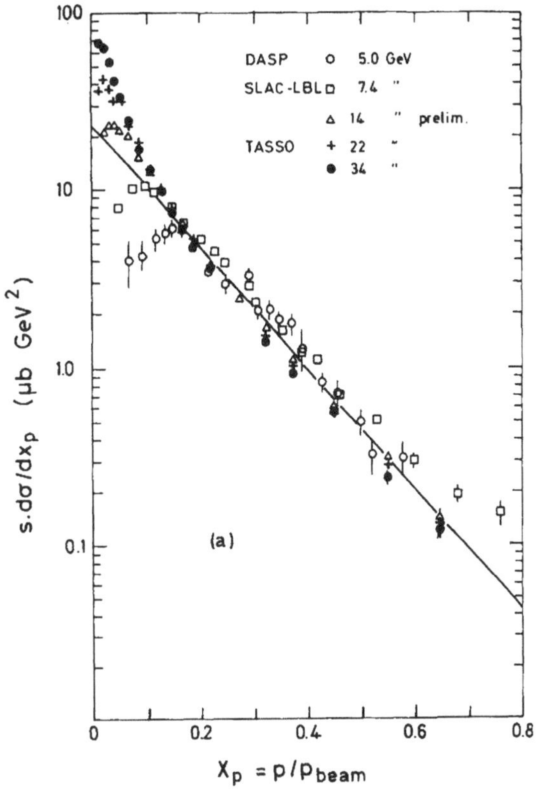

Fig. 64. Inclusive spectrum s.dσ/dx$_p$ of charged particle. (a) shows the data of TASSO together with results from SLAC-LBL and DASP at lower energies.

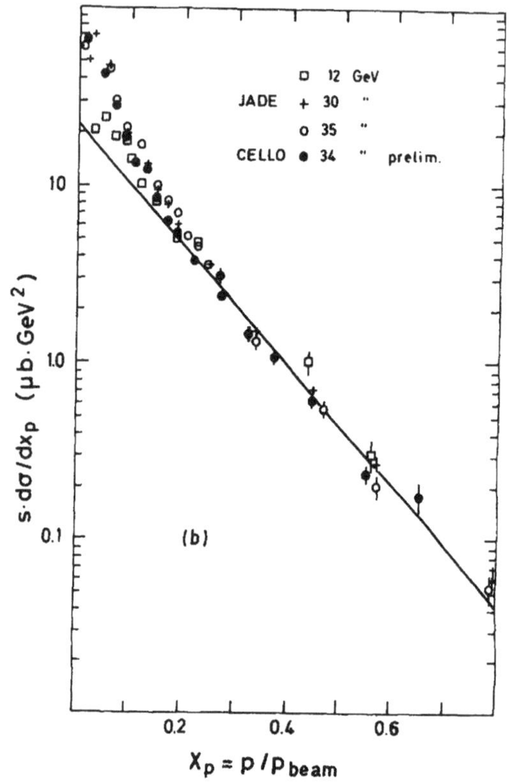

Fig. 64. Inclusive spectrum $s.d\sigma/dx_p$ of charged particle. (b) shows the data from CELLO and JADE. The full lines are drawn to guide the eye and correspond to $s.d\sigma/dx_p = 23\ \exp(-8x_p)\mu B\ GeV^2$. From Felst (1981).

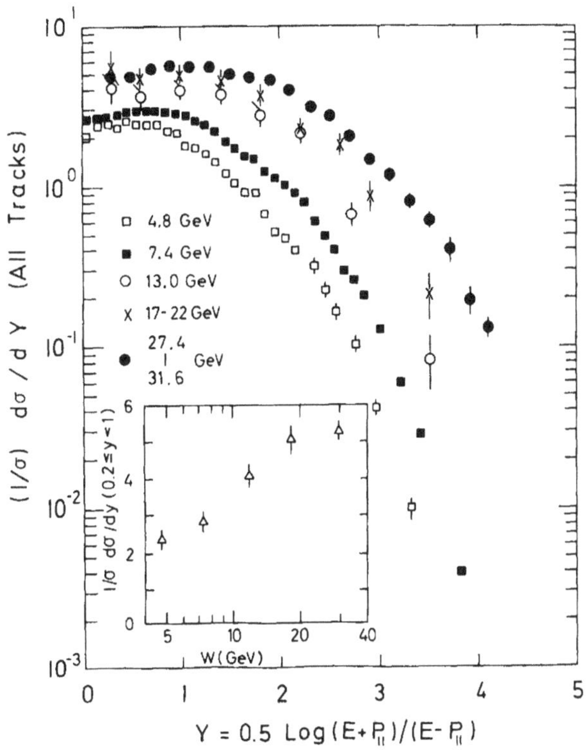

Fig. 65. Rapidity distributions $1/\sigma_{tot}\ d\sigma/dy$ for charged particles (assuming the mass to be the pion mass) measured by SLAC-LBL at 4.8 and 7.4 GeV (Hanson 1978) and by TASSO between 13 and 31.6 GeV (Brandelik et al. 1980a). SLAC-LBL used the sphericity axis, TASSO the thrust axis. The insert shows $1/\sigma_{tot} d\sigma/dy$ averaged over $0.2 < y < 1$ as a function of the c.m. energy.

rises with increasing energy (see insert in Fig. 65). In the fragmentation region (y close to $y_{max} \simeq 0.5 \ln(s/m^2)$) the particle yield is a steeply descending function of y. The width of the fragmentation region is roughly one unit, which is similar to what has been observed in hadron-hadron scattering.

The energy dependences of the plateau and fragmentation regions show that the faster than logarithmic growth of the average multiplicity at high energies is correlated with a rise of the plateau which is due to low energy particles. The same effect was observed in pp collisions (see e.g. Giacomelli & Jacob 1979).

4. Particle Separated Cross Sections

Inclusive cross sections were measured for a variety of particles: $\pi^{\pm}, \pi^o, K^{\pm}, K^o, p, \Lambda$. Table 7 lists the techniques used and the momentum range covered by various experiments.

We start with a discussion of the differential cross sections $d\sigma/dp$. Fig. 66 summarizes the cross sections near W = 30 GeV for the 6 particle species. The curves are drawn to guide the reader. Within errors there are twice as many $\pi^{\pm}$ as π^o. Pions are by far the most abundant type of particle which does not come as a surprise. At momenta below 1 GeV/c one finds large differences between the π yields and those for the heavier particles. As the momentum increases these differences become gradually smaller. The ratio $\pi^{\pm} : K^o, \bar{K}^o : \Lambda, \bar{\Lambda}$ is of the order 100 : 10 : 1 near p = 0.5 GeV/c and 4 : 2 : 1 at 10 GeV/c. All cross sections rise at low momentum, go through a maximum and drop off towards higher momenta. The position of the maximum increases with the particle mass; it appears to be below 0.4 GeV/c for pions, around 0.7 GeV/c for kaons and near 1 GeV/c for lambdas. Most of this turn-on phenomenon at low momenta is an artifact of phase space. Consider the Lorentz-invariant phase space cross section $(E/4\pi p^2)\, d\sigma/dp$ and assume some momentum dependence C(P) for it:

$$\frac{E}{p^2}\frac{d\sigma}{dp} = CP \tag{50}$$

or

$$\frac{d\sigma}{dp} = \frac{p^2}{E}\, C(P) \simeq \frac{p^2}{m}\, C(p) \quad \text{for } p \ll m \tag{51}$$

The low c.m. energy data (W = 4 - 5 GeV, Brandelik et al 1979a) are well described by an exponential,

Table 7. Experiments measuring particle separated cross sections

type of particle	experiment	technique	momentum range (GeV/c)	remark
$\pi^{\pm}$	TASSO[1]	TOF Cerenkov	} 0.4 - 10	
	JADE	dE/dx	< 0.7, 2-7	preliminary
π^{o}	TASSO[2]	liquid Argon	0.5 - 4	
$K^{\pm}$	TASSO[1]	TOF	< 1.1	
	JADE	dE/dx	< 0.7	preliminary
$K^{o}, \bar{K}^{o}$	TASSO[3]	$K^{o}_{s} \to \pi^{+}\pi^{-}$	all p	
	PLUTO[4]	"	"	
	JADE	"	"	preliminary
	MARK II	"	"	preliminary
$p, \bar{p}$	TASSO[1]	TOF	0.5 - 2.2	
	JADE	dE/dx	0.3 - 0.9	preliminary
	MARK II	TOF	0.5 - 2.0	preliminary
$\Lambda, \bar{\Lambda}$	JADE[5]	$\bar{\Lambda} \to \bar{p}\pi^{+}$ + dE/dx	0.4 - 1.1	
	TASSO[6]	$\Lambda \to p\pi^{-}$	1 - 10	
	MARK II	"		preliminary

1 Brandelik et al. (1980d) and (1981d)

2 Brandelik et al. (1981e)

3 Brandelik et al (1980e) and (1981f)

4 Berger et al (1981c)

5 Bartel et al. (1981c)

6 Brandelik et al. (1981f)

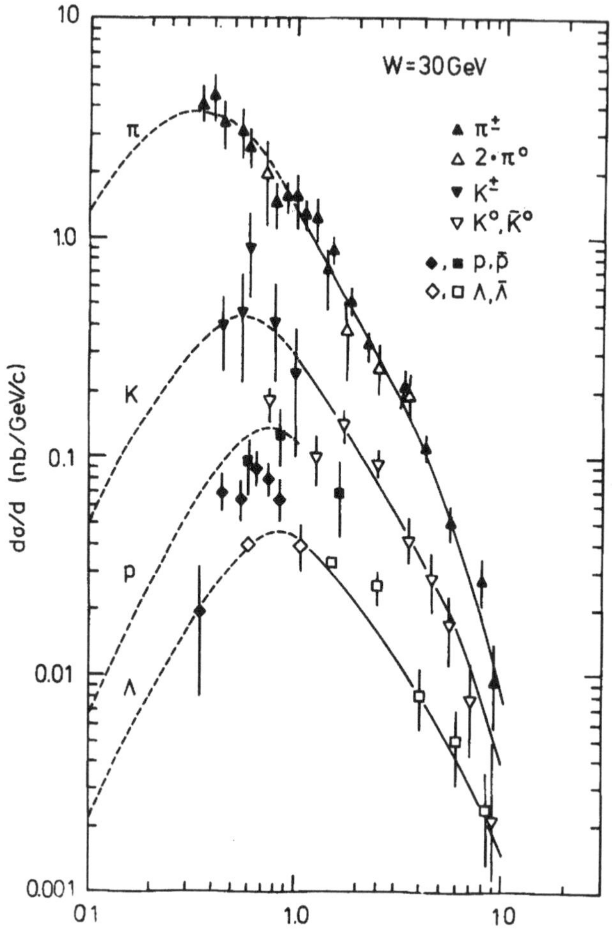

Fig. 66. Differential cross sections $d\sigma/dp$ for inclusive particle production $e^+e^- \to hX$ for $\pi^+ + \pi^-$ (TASSO, preliminary), $2\pi^o$ (TASSO), $K^+ + K^-$ (TASSO), $K^o + \bar{K}^o$ (TASSO), $p + \bar{p}$ (JADE, TASSO), $\Lambda + \bar{\Lambda}$ (JADE, TASSO). The dashed curves are of the form $E/p^2\ d\sigma/dp \sim \exp(-bE)$ with $b = 3.6\ GeV^{-1}$. The solid curves are handdrawn to follow the data.

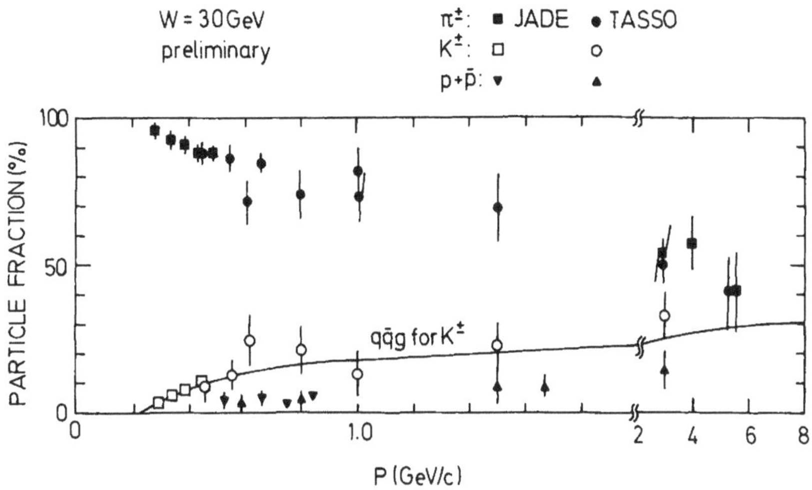

Fig. 67. Charged particle fractions as a function of momentum. From Pandoulas (1980).

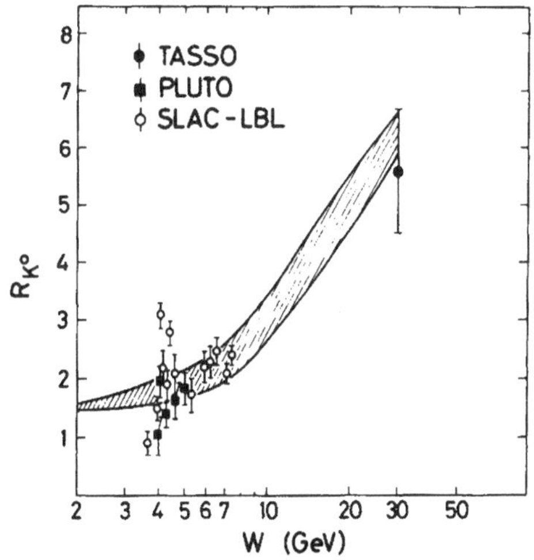

Fig. 68. The total cross section for inclusive $K^o + \bar{K}^o$ production relative to the μ pair cross section, $R_{K^o} = (\sigma(K^oX) + \sigma(\bar{K}^oX))/\sigma_{\mu\mu}$

$$\frac{E}{p^2}\frac{d\sigma}{dp} = C\ \exp(-bE)$$

with b around 5 GeV^{-1}. This form describes also the high energy data up to particle energies around 1 GeV if b = 3.6 GeV is chosen (see dashed curves in Fig. 66). The relative fractions of $\pi^\pm$, $K^\pm$ and p, $\bar{p}$ as a function of particle momentum p at W = 30 GeV are shown in Fig. 67. An average event at this energy has approximately 11 $\pi^\pm$, 5.5 π^0, 1.4 $K^0-\bar{K}^0$, 1.4 $K^\pm$, ≥ 0.4 p,$\bar{p}$ and 0.3 $\Lambda,\bar{\Lambda}$ in the final state. The number of $K^0,\bar{K}^0$ is a factor of 3 larger than observed

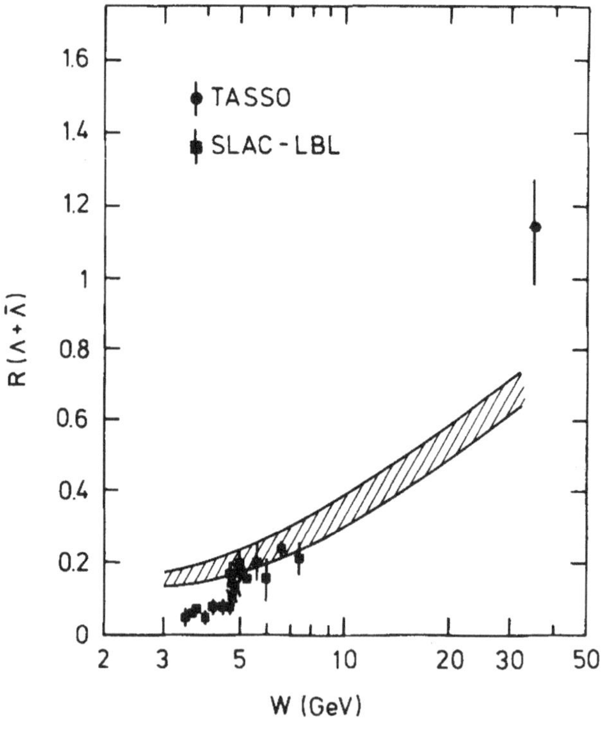

Fig. 69. The total cross section for inclusive $\Lambda + \bar{\Lambda}$ production relative to the μ pair cross section, $R_\Lambda = (\sigma(\Lambda X) + \sigma(\bar{\Lambda}X))/\sigma_{\mu\mu}$

in pp final states: at a c.m. energy of 27.6 GeV there are 0.46 ± 0.02 K^o, $\bar{K}^o$ per event (Kichimi et al. 1979). Whether the excess of kaons in e^+e^- annihilation is due to the c and b quark contributions is not clear (see below).

The W dependence of the inclusive cross sections for K^o and Λ production are shown in Figs. 68 and 69. The K^o yield rises from W = 5 to W = 30 GeV at about the same rate as the average charge multiplicity does while for lambdas the rise could even be faster (see curves).

5. The Scaled Particle Cross Sections

The scaled cross sections $s/\beta\ d\sigma/dx$ ($x = 2E/W$) for π, K, p, Λ are shown in Figs. 70 - 72. All particles exhibit similar features: the cross section falls exponentially with a break in slope around x = 0.1. For $\pi^\pm$, K^o, $\bar{K}^o$ and Λ, $\bar{\Lambda}$ which were all measured up to large x values the same x dependence is observed for $x \gtrsim 0.15$, $s/\beta\ d\sigma/dx \sim \exp(-8x)$ (see Fig. 72). For the π^o data the cross sections measured at energies below W = 8 GeV are systematically higher than those measured above 12 GeV which suggests the presence of scale breaking effects. Qualitatively, they seem to be of the magnitude predicted by QCD (Baier et al. 1979, Frazer & Gunion 1979) but a quantitative comparison has not yet been made.

We turn now to the large K yield observed in e^+e^- annihilation as compared e.g. to that for pp collisions. At first sight it would seem natural to assume the strange quark in the kaons to be either primary produced or to originate from the weak decay of c or b quark. By generating Monte Carlo events according to $e^+e^- \to q\bar{q}$, $q\bar{q}g$ and using a Field-Feynman type fragmentation one can study where the kaons are coming from (Holder 1981). Fig. 73 shows the comparison between the calculation and the data. The curves labelled s, c + b show the amount of K^o production by primary s, c or b quarks. They are seen bo account only for about one third of the produced kaons. For the majority of the kaons the s quark is picked up from the sea. It is surprising to see that this is true event at x values as large as 0.4.

6. Where are all the Baryons from ?

The most surprising result of the inclusive measurements is the large yield of baryons* and their x dependence which is very similar

---

* A probably related phenomenon is the large yield of forward produced protons in μp scattering observed by EMC (Aubert et al. 1981c).

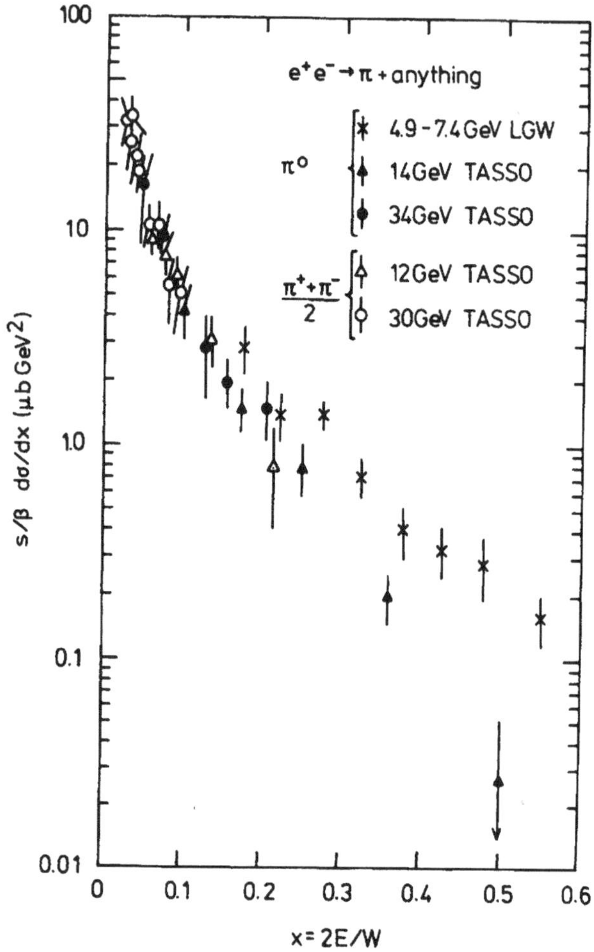

Fig. 70. The scaled cross section $s/\beta\ d\sigma/dx$ for π^0 and $(\pi^+ + \pi^-)/2$ production. From Brandelik et al (1981c)

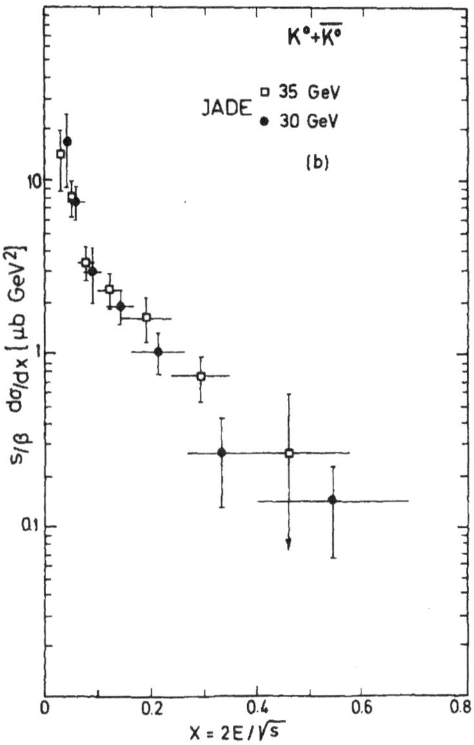

Fig. 71. The scaled cross sections for K^o, $\bar{K}^o$ production

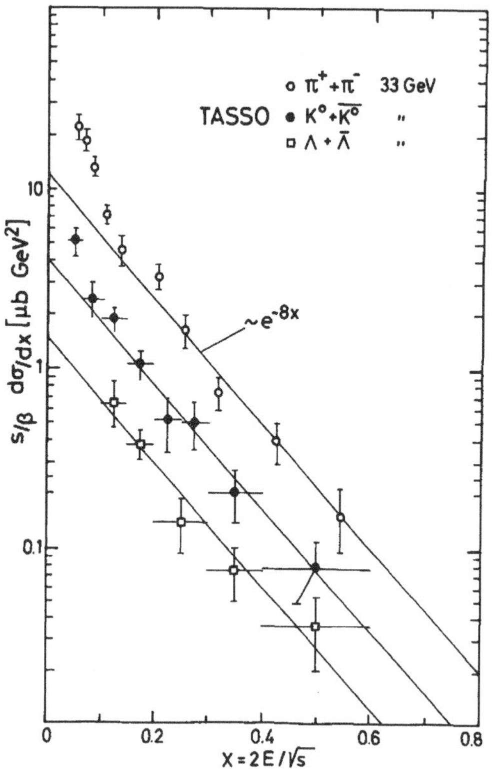

Fig. 72. The scaled cross section for $\pi^+ + \pi^-$, $K^o + \bar{K}^o$ and $\Lambda + \bar{\Lambda}$ production

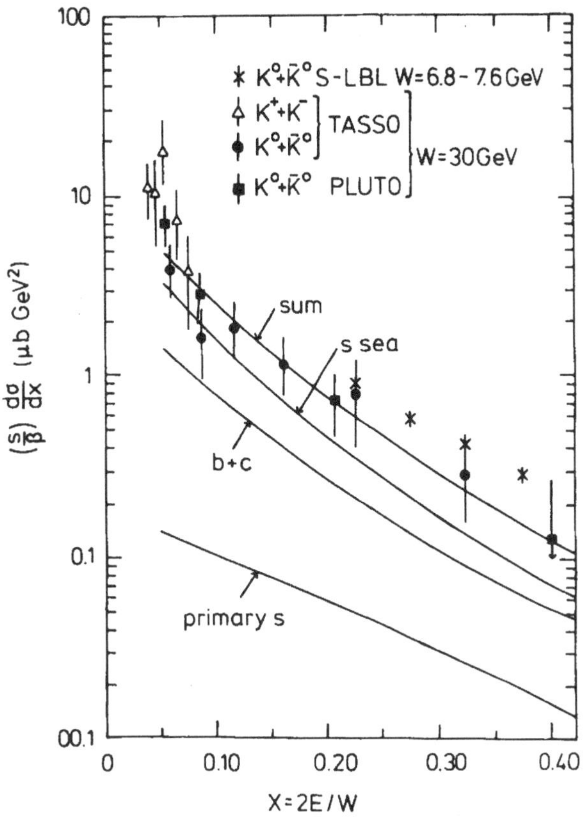

Fig. 73. K^0 production cross sections. The lines show the origin of the strange quark in K^0 according to a cascade model. The curves labelled s and b + c show the contributions from primary s, c and b quarks. The curve labelled sea shows the amount of K^0 production with s quarks from the sea. From Holder (1981)

to that for mesons. Very roughly, every other event contains a baryon-antibaryon pair. There is a long standing prediction by Drell and coworkers (1969) that nucleon production will not vanish at high energies, i.e.

$$R_{(p+\bar{p})} \sim W^{-2}$$

(although the cross section for the exclusive channel, $e^+e^- \to p\bar{p}$, disappears presumably like W^{-10} as a result of the dipole behaviour of the nucleon form factor). The prediction is based on the assumption that if scaling holds proton production in e^+e^- annihilation,

$$e^+e^- \to \bar{p}X \qquad (52)$$

can be related by crossing with inelastic electron proton scattering,

$$e\ p \to eX \qquad (53)$$

Strictly speaking, the crossing relation holds only at $x \equiv 2E/W = 1$ (Gatto and coworkers 1972) but the hope was that it could also be applied in the region close to $x = 1$. Gribov and Lipatov (1971) related the structure functions for the two processes using a field theoretic model. This permits to express the cross section for the annihilation process in terms of the proton structure functions F_1 and F_2 determined from (53) at $\omega = 1/x$.

$$\frac{s}{\beta}\ \frac{d\sigma}{dx}\ (e^+e^- \to \bar{p}X) = \frac{4\pi\alpha^2}{x}\ \{xF_1(\omega = \frac{1}{x}) - \frac{1}{6}\ \beta^2 F_2(\omega = \frac{1}{x})\} \qquad (54)$$

Fig. 74 compares the Gribov Lipatov prediction (shaded bands) with the data at W = 12 and 30 GeV. Like a lower energies (W = 4 - 4.5 GeV, Brandelik et al. 1979a) the theoretical prediction is below the data. Part of the discrepancy - if not all - may have to be attributed to contributions of the type

$$e^+e^- \to hX,\ h = \underset{\hookrightarrow \bar{p}\pi}{\bar{\Lambda}}\ ,\ \underset{\hookrightarrow \bar{p}x}{\bar{\Sigma}}\ ,\ \text{etc}$$

which should be excluded from the e^+e^- data before making the comparison (Schierholz & Schmidt 1975).

Several models were recently proposed which incorporate baryon production in the fragmentation (see e.g. Ilgenfritz et al. 1978,

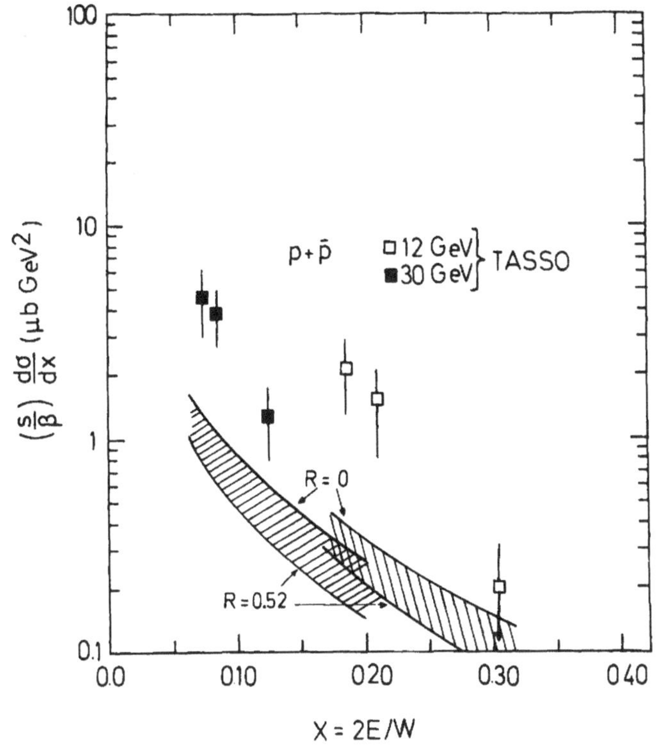

Fig. 74. The scaled cross sections for p + $\bar{p}$ production at W = 12 and 30 GeV compared to the Gribov-Lipatov prediction

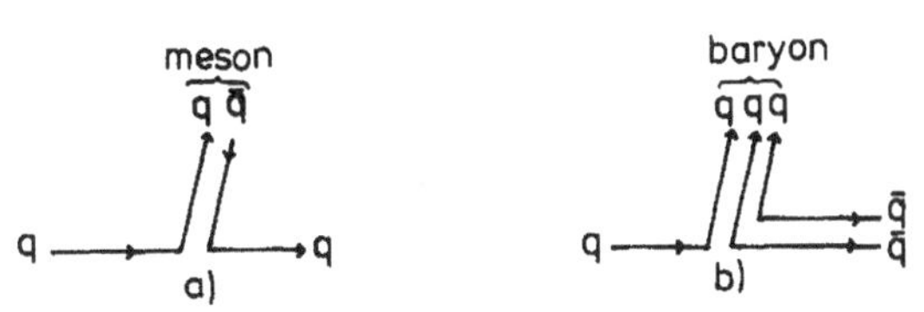

Fig. 75. Quark fragmentation into a meson (a) and a baryon (b)

Casher et al. 1979, Ritter & Ranft 1980, Andersson et al. 1981, Hofmann 1981, Meyer 1981, Schierholz & Teper 1981, Bowler 1981). The simplest approach assumes that a quark cannot only pull a $q\bar{q}$ but also a diquark-diantiquark (qq, $\overline{qq}$) out of the vacuum (see Fig. 75).

A good fit to the p, $\bar{p}$ and Λ, $\bar{\Lambda}$ data is obtained if the relative probability for picking up a diquark, P(qq)/P(q) is determined from the data and all other fragmentation parameters for baryons are assumed to be same as for mesons (T. Meyer 1981). Fig. 76 shows a comparison with the p, $\bar{p}$ and Λ, $\bar{\Lambda}$ data for P(qq)/P(q) = 0.075. Both data sets are represented well by the model. A more detailed description of baryon production, starting also from diquarks, was developed in the string model (Andersson et al. 1981, Bowler 1981). The dashed curves in Fig. 76 show the result of the first group. The p data which had been used to fix the parameters of the model are reproduced. The model fails for the Λ data where the predicted cross sections are a factor of 2 - 3 too small. The model appears to be flexible enough to resolve this discrepancy.

The requirement of diquark- diantiquark states in the vacuum runs into a problem with R: if diquarks are pointlike R will receive a large contribution from diquark pair production, $e^+e^- \to (qq)(\overline{qq})$. This is excluded by the data. A wayout is to assume the diquarks to be extended objects with a size of order GeV^{-1}.

In conclusion, the question asked at the beginning on how baryons are formed, is still open. It remains to be seen whether the diquark concept can be substantiated by other data. A far less exciting alternative is to assume that a quark pulls only $q\bar{q}$ states out of the vacuum; however, quarks and antiquarks from several of the fragmentation steps are so close in phase space that they rearrange into mesons or baryons in a statistical manner.

7. Gluon Fragmentation

Three-jet events interpreted as hard-gluon bremsstrahlung offer the possibility to compare directly quark and gluon fragmentation. It is generally expected that gluon jets will yield a higher multiplicity and a softer hadron spectrum as well as larger jet cone angles than quark jets of the same characteristic momentum (see e.g. Konishi et al. 1979). The reason is the larger colour charge of gluons which leads to a larger parton multiplicity in their evolution.

In gluon bremsstrahlung ($e^+e^- \to q\bar{q}g$) the gluon has on average, a lower energy than quark or antiquark. However, in practice the identification of the gluon amongst the three jets so far has been

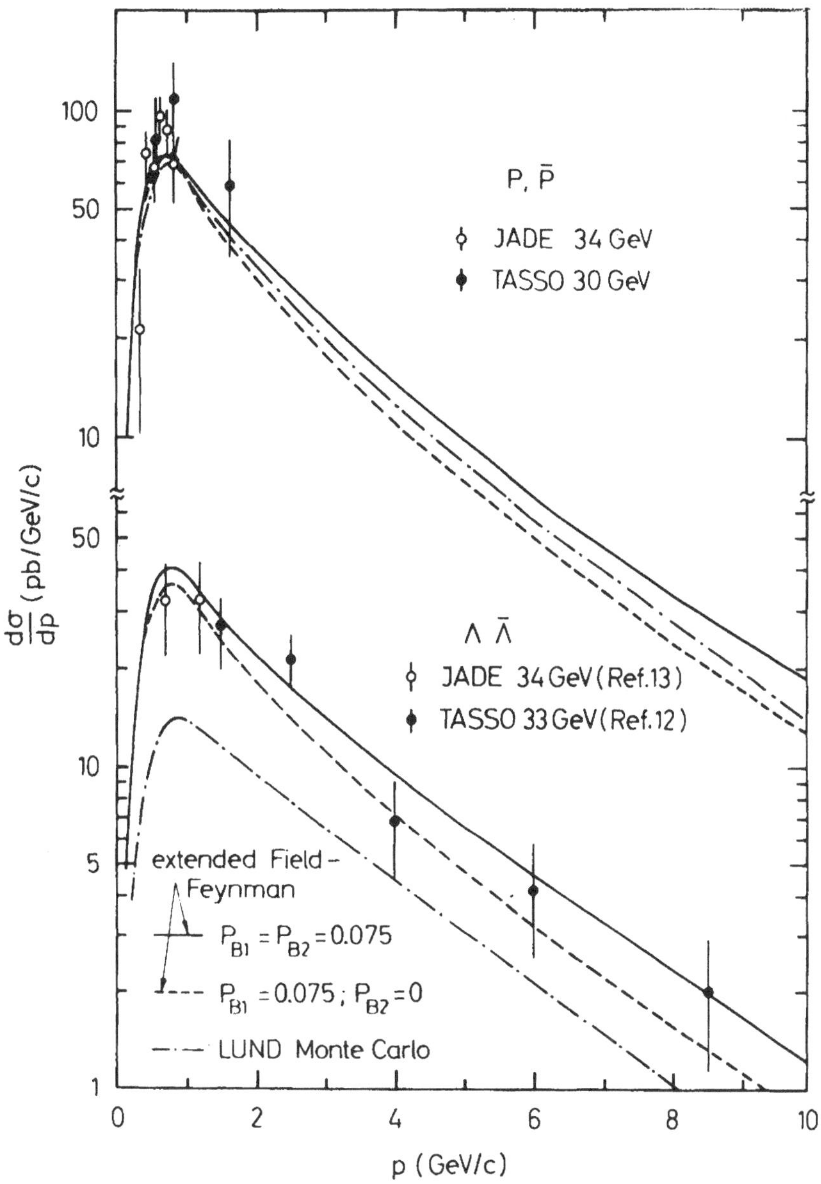

Fig. 76. Inclusive production of $p + \bar{p}$ and $\Lambda + \bar{\Lambda}$ at W = 30 - 34 GeV. The solid and dashed curves show the prediction of the model by Meyer. The dashed-dotted curves show the prediction of the Lund model. From S.L.Wu (1981)

possible on a statistical basis only. An attempt to compare gluon and quark fragmentation has been presented by the JADE group (Bartel et al. 1981). In planar events the slim jet (q) and the two subjets which together form the broad jet were identified. The subjet with the smaller angle relative to the slim jet was called the "gluon" jet (g), the other the "quark" jet ($\bar{q}$). Monte Carlo studies indicated that in 50 % of the events the gluon is correctly identified . The particle yield is plotted in Fig. 77 as a function of the fractional angle θ/θ_{max} between the particle and the q direction (a) for the region between q and $\bar{q}$ and (b) between q and g. One finds that in

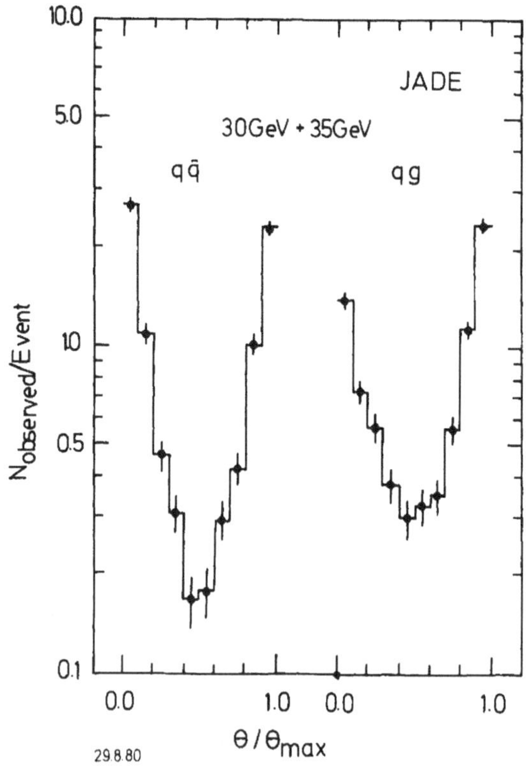

Fig. 77. Angular distribution θ/θ_{max} of particles in the $q\bar{q}$ and qg regions (defined in the text) as measured by the JADE group at W = 30 - 35 GeV.

the center region, $\theta/\theta_{max} \approx 0.5$, the particle angular density in the region between q and g is considerably larger than between q and $\bar{q}$. It is unclear whether this difference is due to higher order effects in QCD or whether it indicates a difference between quark and gluon fragmentation. The string model approach to $q\bar{q}g$ has predicted such a difference (Andersson et al. 1980).

ACKNOWLEDGEMENTS

It is a pleasure to acknowledge the stimulating atmosphere provided by the organizers of the schools, in particular by Profs. D. Speiser, J. Weyers and M. Nikolic. I have profitted from many discussions with my colleagues from TASSO, in particular Dr. W. Koch, and Profs. P. Söding and S.L. Wu. I am very grateful to Mrs. E. Hell for her special efforts to get the manuscript ready and to Mrs. H. Siegner for her help with the drawings.

LITERATURE CITED

Ali, A. et al. 1979, Z. Physik C2:33; Phys. Lett. 83B: 375

Ali, A. et al. 1980. Phys. Lett. 93B: 155; Nucl. Phys. B168: 409

Ali, A. et al. 1981. DESY Report 81/59

Andersson, B. Gustafson, G., Sjöstrand, T. 1980, Phys. Lett 94B, 211 and references quoted therein.

Andersson, B. Gustafson, G., Sjöstrand, T. 1981, Lund Report TP 81-3 TP 81-6, TP 81-7

Aubert, J.J. et al. 1981, Phys. Lett. 103B, 388

Barber, D.P. et al. 1979, Phys. Rev. Lett. 43; 830

Barber, D.P. et al. 1980a, MIT Report LNS 113

Barber, D.P. et al. 1981a, Phys. Rev. Lett. 46, 1663

Barber, D.P. et al. 1981b, MIT LNS Report 115

Bartel, W. et al. 1980a, Z. Phys. C6, 295

Bartel, W. et al. 1980b, Phys. Lett. 91B, 142

Bartel, W. et al. 1981a, DESY Report, 81/72

Bartel, W. et al. 1981b, Phys. Lett. 101B, 361

Bartel, W. et al. 1981c, Phys. Lett. 104B, 325

Bartel, W. et al. 1981d, Phys. Lett.

Behrend, H.J. et al. 1981, DESY Report 81/80

Berends, F.,Kleiss, R. 1981, Nucl. Phys.B178, 141

Berends, F., Gaemers, K.F.J., Gastmans, R. 1973, Nucl. Phys. B63, 381

Berger, Ch. et al. 1979, Phys. Lett. 86B, 418

Berger, Ch. et al. 1980a, DESY Report 80/117

Berger, Ch. et al. 1980b, Phys. Lett. 97B, 459

Berger, Ch. et al. 1981a, Phys. Lett. 99B, 489

Berger, Ch. et al. 1981b, DESY Report, 81/54

Berger, Ch. et al. 1981c, Phys. Lett. 104B, 79

Berger, Ch. 1981, EPS Int. Conf. on High Energy Physics, Lisbon

Bjorken, J.D., Brodsky, S.J. 1970, Phys. Rev. D1; 1416

Bjorken, J.D., Llewellyn Smith, Ch. 1973, Phys. Rev. D7, 887

Bletzacker, F., Nieh, H.T. 1977, Phys. Rev. D16, 2115

Bowler, M. 1981, Oxford Univ. Report 76-81

Brandelik, R. et al. 1979, Phys. Lett. 86B, 243

Brandelik, R. et al. 1979, Nucl. Phys. B148, 189

Brandelik, R. et al. 1980a, Phys. Lett. 92B, 199

Brandelik, R. et al. 1980b, Phys. Lett. 94B, 437

Brandelik, R. et al. 1980c, Phys. Lett. 97B, 453

Brandelik, R. et al. 1980d, Phys. Lett. 94B, 444

Brandelik, R. et al. 1980e, Phys. Lett. 94B, 91

Brandelik, R. et al. 1981a, Phys. Lett. 99B, 163

Brandelik, R. et al. 1981b, Phys. Lett. 100B, 357

Brandelik, R. et al. 1981c, DESY Report

Brandelik, R. et al. 1981d, DESY Report

Brandelik, R. et al. 1981e, DESY Report, 81/69

Brandelik, R. et al. 1981f, Phys. Lett. 105B, 75

Brandt, S. et al. 1964, Phys. Lett. 12B, 57

Brandt, S. 1979, Proc. Int. Conf. High Energy Physics, Geneva, P. 338

Branson, J.G. 1981. Int. Symp. on Lepton and Photon Interactions at High Energies, Bonn
And DESY Report 81/73

Braunschweig, W. 1981, Int. Symp. On Lepton and Photon Interactions at High Energies, Bonn

Bürger, J. 1981, In. Symp. on Lepton and Photons Interactions at High Energies, Bonn

Buras, A.J. 1981, Int. Symp. on Lepton and Photon Interactions at High Energies, Bonn
And Fermilab Report 81/69

Casher, A., Neuberger, H., Nussinov, S. 1979, Phys. Rev. D20, 179

Celmaster, W., Gonsalves, R.J. 1979, Phys. Rev. Lett. 44, 560

Chetyrkin, K.G., Kataev, A.L., Tkachev, F.V. 1979, Phys. Lett. 85B, 277

Cords, D. 1980, Proc. XXth Int. Conf. on High Energy Physics, Madison Wisc.;
DESY Report 80/92

Criegee, L., Knies, G. 1981, DESY Report 81/44 and Physics Reports

de Grand, T.A., Ng, Y.G., Tye, S.H.H. 1977, Phys. Rev. D16, 3251

Dine, M., Sapirstein, J. 1979, Phys. Rev. Lett. 43, 668

Dittmann, P., Hepp, V. 1981, Z. Physik C10, 283

Drell, S.D., Levy, D., Yan, T.M. 1969, Phys. Rev. 187, 2159; 1970 Phys. Rev. D1, 1035, 1617, 2402

Drijard, D. et al. 1979, Nucl. Phys. B155, 269; 1980, Nucl. Phys. B166, 233

Duinker, P. 1981, Int. Conf. on High Energy Physics, Lisbon

Eidelman, S.I., Kurdadze, L.M., Vainshtein, A.I., 1979, Phys. Lett. 82B, 278

Ellis, J., Gaillard, M.K., Ross, G.G. 1976, Nucl. Phys. B111, 253, (E.B130, 516)

Ellis, J. 1980, Private Communication

Ellis, R.K., Ross, D.A., Terrano, A.E. 1980, Phys. Rev. Lett. 45, 1226 1981, Nucl. Phys. B178, 421

Fabricius, K. et al. 1980, Phys. Lett. 97B, 431;
DESY Report 81/35

Farhi, E. 1977, Phys. Rev. Lett. 39, 1587

Farrar, G. 1978, Proc. Int. School of Subnuclear Physics, Erice

Farrar, G., Fayet, P. 1980, Phys. Lett. 89B, 191

Fayet, P. Ferrara, S. 1977, Phys. Reports 32C, 249

Felst, R. 1981, Int. Symp. on Lepton and Photon Interactions at High Energies, Bonn

Field, R.D., Feynman, R.P. 1978, Nucl. Phys. B136, 1

Fournier, D. 1981, Int. Symp. on Lepton and Photon Interactions at High Energies, Bonn

Fritzsch, H., Gell-Mann, M. 1972, Proc. XVI Int. Conf. on High Energy Physics, Chicago, Vol. 2, P. 135

Fritzsch, H., Gell-Mann, M., Leutwyler, H. 1973, Phys. Lett. 47B, 365

Furmanski, W., Petronzio, R., Pokorski, S. 1979, Nucl. Phys. B155, 253

Gatto, R., Menotti, P., Vendramin, I. 1972, Nuovo Cim. Lett. 4, 79

Gatto, R., Preparata, G. 1972, Nucl. Phys. B47, 313

Georgi, H., Politzer, H.D. 1974, Phys. Rev. D9, 416

Giacomelli, G., Jacob, M. 1979, Physics Reports, 55, 1

Glashow, S.L. 1961, Nucl. Phys. 22, 579; 1980, Rev. Mod. Phys. 52, 539

Gol'fan, Yu. A., Likhtman, E.P. 1971, JETP Letters 13, 323

Gribov, V.N., Lipatov, L.N. 1971, Phys. Lett. 37B, 78

Gross, D.J., Wilczek, F. 1973, Phys. Rev. Lett. 30, 1343; Phys. Rev. D8, 3633, also 1974, Phys. Rev. D9, 980

Hanson, G.G. et al. 1975, Phys. Rev. Lett. 35, 1609

Harari, H. 1977, Weizmann Inst. Report WIS 77/56 and 1978, Physics Reports 42, 235

Hofman, W. 1981, Z. Physik C10, 351

Holder, M. 1981, Int. Conf. on High Energy Physics, Lisbon

Hollebeek, R. J. 1981, Int. Symp. on Lepton and Photon Interactions at High Energies, Bonn

Hoyer, P. et al. 1979, Nucl. Phys. B161, 349

Ilgenfritz, E.M., Kripfganz, J., Schiller, A. 1978, Acta Phys. Pol. B9, 881

Kichimi et al. 1979

Koba,Z., Nielsen, H.B., Olesen, P. 1972, Nucl. Phys. B40, 317

Kobayashi, M., Maskawa, T. 1973, Progr. Theor. Phys. 49, 652

Kogut, J.B., Susskind, L. 1974, Phys. Rev. D9, 697; D9, 3391

Kramer, G., Schierholz, G., Willrodt, J. 1978, Phys. Lett. 79B, 249

Kramer, G., Schierholz, G. 1979, Phys. Lett. 82B, 102

Kramer, G., Schierholz, G., Willrodt, J. 1980, Z. Physik C4, 149

Litke, A. 1981, Int. Symp. on Lepton and Photon Interactions at High Energies, Bonn

Marshall, R. 1981, EPS Int. Conf. on High Energy Physics, Lisbon

Meyer, T. 1981, DESY Report 81/46

Nambu, Y. 1966, Preludes in Theoretical Physics, (Amsterdam), p. 133

Niczyporuk, B. et al. 1981, DESY Report 81/008

Pati, J.C., Salam, A. 1978, Nucl. Phys. B144, 445

Polyakov, A.M., 1975, Proc. VIIth Int. Symp. Lepton and Photon Interactions at High Energies, Stanford, p. 855

Preparata, G. 1981, Univ. of Bari Report, GT, 81-13

Renard, F. 1981, Basics of Electron - Positron Collisions, Editions Frontières, Vol. 32

Ritter, S.,Ranft, J. 1980, Acta Phys. Pol. B11, 259

Salam, A. 1962, Phys. Rev. 127;
1980, Rev. Mod. Phys. 52, 525

Scharre, S. 1981, Berkeley Report LBL-13018

Schierholz, G., Teper, M. 1981, DESY Report, 81/41

Schwitters, R.F. et al. 1975, Phys. Rev. Lett. 35, 1320

Söding, P., Wolf, G. 1981, DESY Report 81/13 and Ann. Rev. Nucl. Sci. Vol. 31

Söding, P. 1981, DESY Report 81/70

Spitzer, H. 1980, Proc. XVth Rencontre de Moriond, 1980, Vol. II, p. 157 and DESY Report 80/43

Sterman, E., Weinberg, S. 1977, Phys. Rev. Lett. 39, 1436

Vermaeseren, J.A.M., Gaemers, K.J.F., Oldham, S.J. 1981, Nucl. Phys. B187, 301

Walsh, T.F., Zerwas, P.M. 1980, Phys. Lett. 93B, 53

Wedemeyer, R. 1981, Int. Symp. on Lepton and Photon Interactions at High Energies, Bonn

Weinberg, S. 1967, Phys. Rev. Lett. 19, 1264
1980, Rev. Mod. Phys. 52, 515

Weinberg, S. 1973, Phys. Rev. Lett. 31, 494

Weiss, J.M. et al. 1981, Phys. Lett. 101B, 439

Wess, J., Zumino, B. 1974, Nucl. Phys. B70, 39

Wiik, B.H., Wolf, G. 1979, Electron-Positron Interactions, Springer Tracts in Modern Physics, Vol. 86

Wolf, G. 1980, DESY Report 80/13

Woodworth, Ph. 1981, EPS Int. Conf. on High Energy Physics, Lisbon

Wu, S.L., DESY Report, 81/71

Yamada, S. 1980, Proc. XXth Int. Conf. on High Energy Physics, Madison, Wisconsin; Univ. Tokyo Rep.

e^+e^- COLLISIONS AT CESR

Karl Berkelman

Laboratory of Nuclear Studies
Cornell University
Ithaca, N.Y. 14853

I. CESR, CLEO and CUSB

The experimental results from CESR - the Cornell Electron Storage Ring - concern the physics of the b-quark. The spectroscopy of the bound states of the $b\bar{b}$ system - the $\Upsilon(1S)$, $\Upsilon(2S)$, and $\Upsilon(3S)$ enables us to test QCD, while the decays of the b-flavored mesons produced at the $\Upsilon(4S)$ provide tests of the standard six-quark model of weak interactions.

CESR (see table 1) fills an energy gap between the 7 and 10 GeV maximum center-of-mass energies of SPEAR and DORIS and the lowest energies at which useful luminosity is obtained at PETRA and PEP.

Table 1 Current CESR specifications

Average ring radius	130 m
Bunches per beam	1
Interaction points	2
Free space between interaction quads	7 m
Maximum c.m. energy	11.6 GeV
C.m. energy resolution, r.m.s.	4.1 MeV x $(W/10\ \mathrm{GeV})^2$
Maximum colliding beam current	12 mA
Peak luminosity	$3 \times 10^{30}\ \mathrm{cm}^{-2}\ \mathrm{sec}^{-1}$
Average luminosity	100 nb^{-1}/day
Time to fill	40-60 min
Luminosity lifetime	4 hrs

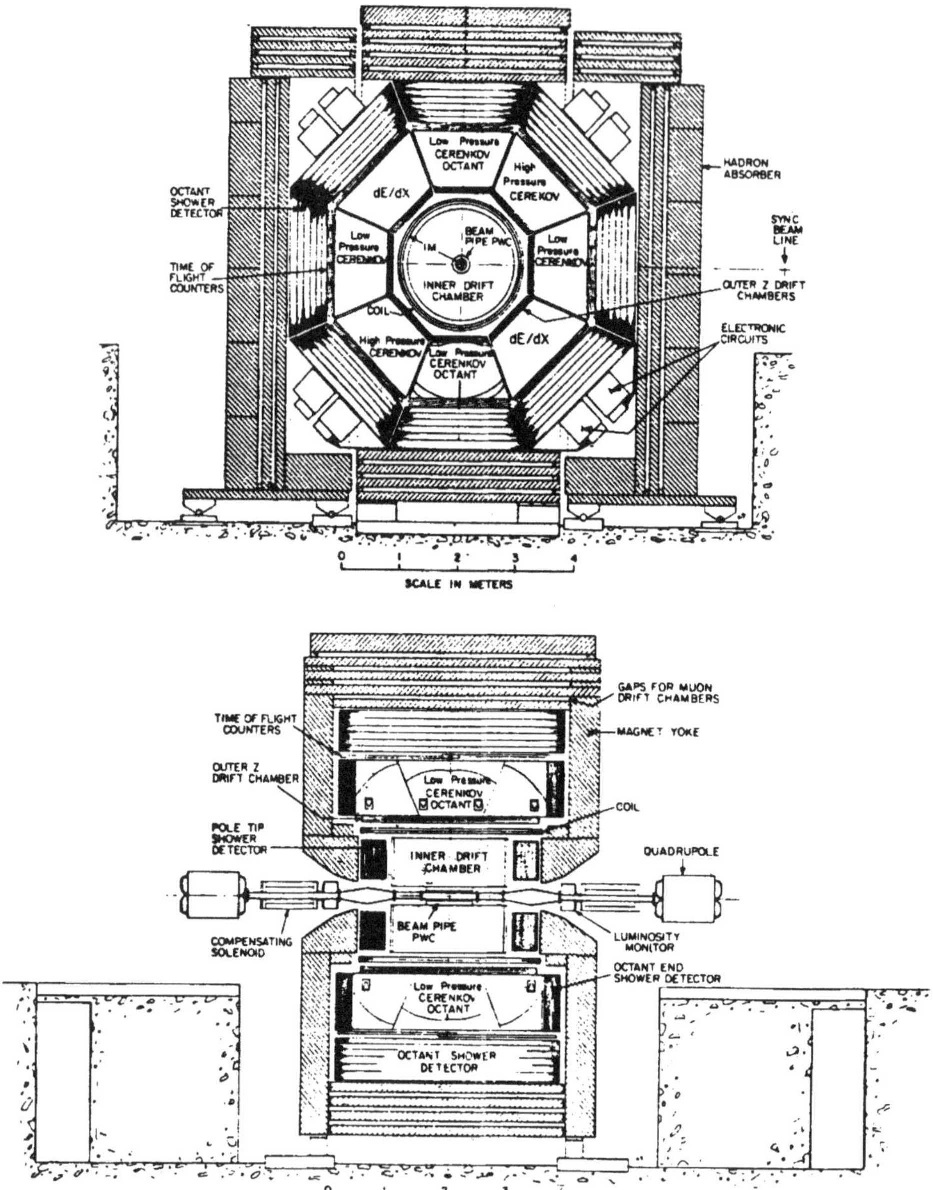

Fig.1. End and side views of the CLEO detector

The single bunch of positrons and of electrons in CESR collide at two diametrically opposite points in the ring. The south interaction region is occupied by a big general purpose magnetic detector CLEO (Fig.1), run by a large Cornell, Harvard, Rochester, Rutgers, Syracuse, Vanderbilt collaboration. Over most of the solid angle the CLEO apparatus detects and measures momenta or energies of charged particles and photons (see table 2), and identifies electrons by their showers and muons by their penetration through iron. Over about half the solid angle and in restricted momentum ranges the CLEO detector also identifies charged hadrons (π, K, p) by time of flight and ionization or Cerenkov light. Like the Mark II detector at SLAC and the PLUTO, TASSO, JADE, and CELLO detectors at DESY, CLEO is designed to measure as wide a range of collision processes as possible.

Table 2. Current CLEO specifications

Dimensions : l		5 m
w		8 m
h		8 m
Total weight		1600 T
Solenoid radius		1 m
length		3 m
field		4.3 kG
Charged particles $\Omega/4\pi$		0.94
$\delta p/p$, r.m.s.		0.02 + 0.03 p
Showering particles $\Omega/4\pi$		0.64
$\delta E/E$, r.m.s.		$\leqslant 0.25\ E^{-1/2}$
Muon identification $\Omega/4\pi$		0.85
Hadron Identification	Time	dE/dx
e − π	< 0.5 GeV/c	> 0.2 GeV/c
π − K	< 1 GeV/c	< 0.7, > 2 GeV/c
K − p	< 1.7 GeV/c	< 1.2, > 4 GeV/c
$\Omega/4\pi$	0.56	0.56

The CUSB detector (Fig.2) in the north experimental area is optimized instead for detection of photons and electrons, with much better energy resolution than is feasible in a general purpose detector (see table 3). Nonshowering charged particles are detected and their directions measured, but in general no information is available on their energies. CUSB is run by a Columbia, Stony Brook, Louisiana State, MPI-Munich collaboration.

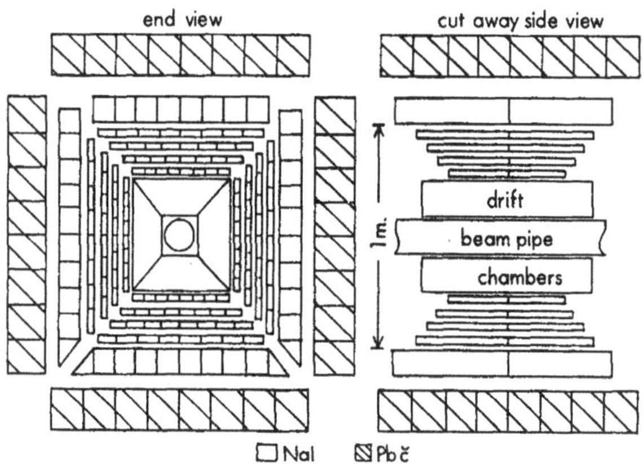

Fig.2. End and side views of the CUSB detector.

Table 3 Current CUSB specifications

Na I blocks : number	332
depth in radiation lengths	8
$\Omega/4\pi$	0.60
Lead glass blocks number	256
Charged particle tracking : $\Omega/4\pi$	0.80

II. UPSILON BOUND-STATE SPECTROSCOPY AND TESTS OF QCD

A. Masses and Potential Models

The upsilon resonances were first observed in 1977 at Fermilab[1] as three overlapping peaks at about 10 GeV in the mass spectrum of $\mu^+\mu^-$ pairs produced in proton-nucleus collisions. The first two states were then seen at DORIS[2-4] as narrow peaks in the cross section for e^+e^- annihilation into hadrons. In the first months of CESR operation in late 1979 both CLEO and CUSB obtained data on the three resonances (Figs. 3 and 4), now called the $\Upsilon(1S)$, $\Upsilon(2S)$ and $\Upsilon(3S)$. The measured mass of the $\Upsilon(1S)$ (Table 4) is limited in accuracy by systematic errors in the storage ring energy calibration from the beam orbit geometry and the magnetic fields. For now the DORIS calibration is probably the more accurate one. However , the precision in the mass differences between the successive upsilon states depends on the linearity of the magnetic field as a function of magnet current, which is better at CESR.

Table 4. Measured masses and mass differences in MeV, compared with predictions of two potential models

Quantity	Measurement	Expt	RBGT[11]	Martin[12]
M(1S)	9462 ± 10	DORIS[2-4]	--	--
	9433 ± 28	CESR[5-6]		
M(2S) - M(1S)	553 ± 10	DORIS[3-4]	560*	560*
	560 ± 3	CESR[5-6]		
M(3S) - M(1S)	889 ± 4	CESR[5-6]	890	890*
M(4S) - M(1S)	1114 ± 5	CESR[7-8]	1160	1120
$M(\chi)$**- $M(\psi)$	426 ± 4	SPEAR[9]	425	425
$M(\psi') - M(\psi)$	589.07 ± 0.10	VEPP4[10]	589	589

* Assumed in fit
** Center of Gravity of $^3P_{1,2,3}$ states

The apparent width of the peaks is a consequence of the energy spread in the colliding beams, caused by synchrotron radiation. It is narrower at CESR (δW_{rms} = (4.1 MeV)(W/10 GeV)^2) than at Doris

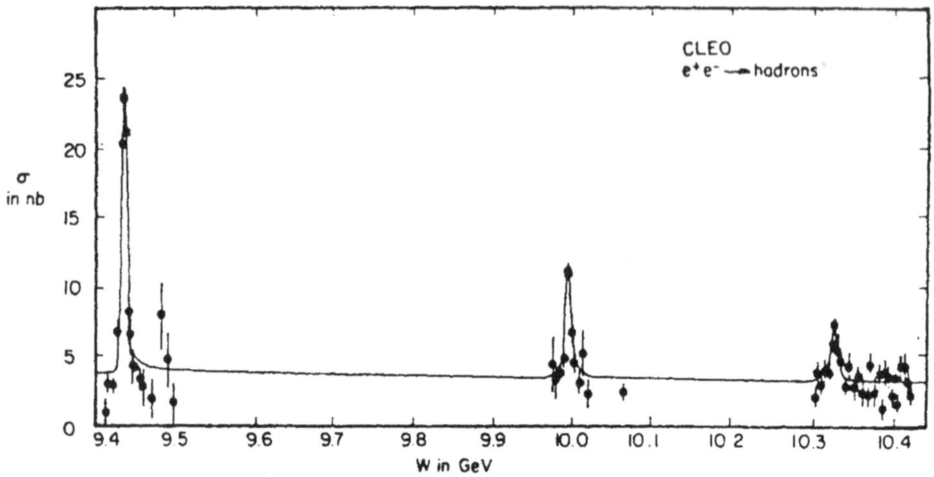

Fig.3. CLEO data[5] for the cross section for $e^+e^- \to$ hadrons as a function of center-of-mass energy from 9.4 to 10.4 GeV, showing the $\Upsilon(1S)$, $\Upsilon(2S)$, $\Upsilon(3S)$.

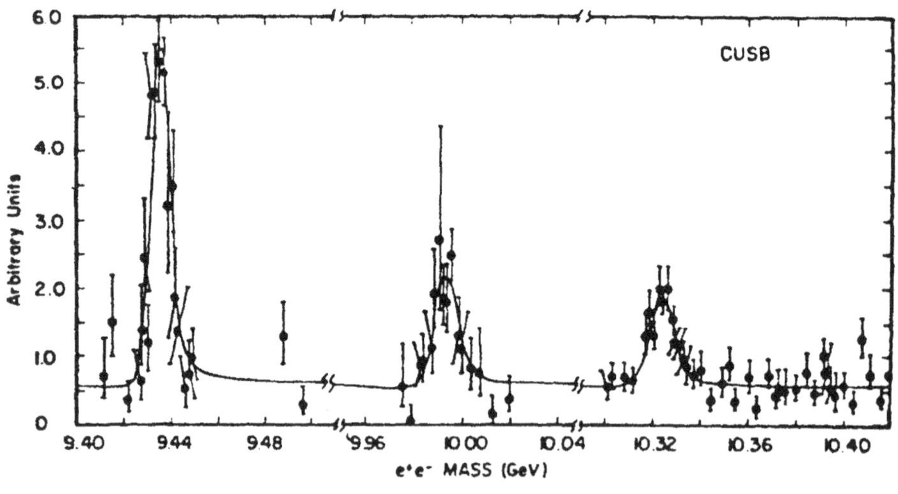

Fig.4. CUSB data[6] for the uncorrected cross section for $e^+e^- \to$ hadrons as a function of center-of-mass energy showing the $\Upsilon(1S)$, $\Upsilon(2S)$ and $\Upsilon(3S)$.

(δW_{rms} = (11 MeV)(W/10 GeV)^2), because of the larger radius of the CESR ring. As in the case of the ψ family, the very narrow intrinsic width of the resonances is taken as evidence that they are bound states of a new heavy quark b and its antiquark $\bar{b}$. Strong decay into a pair of B mesons, each containing the b quark and a light quark are energetically forbidden, and decays into final states not containing the b are suppressed by the Zweig rule.

At about the time of the ψ discovery it became clear[13] that the heavy quark-antiquark spectroscopy might be understood in terms of the nonrelativistic Schrödinger equation with a more-or-less empirical potential function. The first serious attempt at fitting the ψ mass spectrum used a potential of the form [14] $V = A/r + Br$, the coulomb form at short distances suggested by lowest order QCD, the linear form at large distances suggested by the string or flux-tube picture of quark confinement. In a later work[15]) the short-distance coulomb and long-distance linear potentials were pieced together with a logarithmic form at intermediate distances. The strength of the coulomb part was given a logarithmic dependence on the quark mass consistent with the running coupling constant of QCD[16], in order to fit the ψ and Υ systems simultaneously.

The potential model which owes the most to QCD is that of Richardson[17], improved by Buchmüller, Grunberg and Tye[11]. In this model, which I will call the RBGT potential, one starts with a Fourier transformed potential of the form

$$\tilde{V}(Q^2) = -(16\pi/3)(\alpha_s(Q^2)/Q^2).$$

One first has to decide what to use for the running coupling constant $\alpha_s(Q^2)$. The perturbative high-Q^2 form obviously will not do, since it must also make sense at low Q^2.
Consider first the β-function

$$\beta(\rho) = Q^2(d\rho/dQ^2),$$

which specifies the dimensionless logarithmic slope of the coupling $\rho = \alpha_s/4\pi$, expressed as a function of ρ. In the perturbative small-distance (high-Q^2, small-ρ) limit second-order QCD implies

$$\beta(\rho) \sim -b_0\rho^2 - b_1\rho^3 + \ldots$$

where $b_0 = 11 + (2/3)N_f$ and $b_1 = 102 - (38/3)N_f$ (N_f is the number of active quark flavors). In the large-distance (low Q^2, large-ρ) limit we require that the potential be linear : $V \sim (1/2\pi\alpha')r$ (α' is the Regge trajectory slope).
This implies that

$$\beta(\rho) \sim -\rho + \ldots$$

A function which smoothly interpolates between these two limits is given by

$$\frac{1}{\beta(\rho)} = -\frac{1}{b_0\rho^2(1 - e^{-1/b_0\rho})} + \frac{b_1}{b_0^2\rho} e^{-24\rho}$$

The first term, due to Richardson, incorporates lowest-order QCD. The second, by Buchmüller et al., includes two-gluon contributions. The exponential (the factor of 24 is empirical) is supposed to take care of higher orders.

Now we have to fix the magnitude of the coupling. We can either do it at small ρ (small distances) by fixing the Λ in the second order approximation for α_s,

$$\frac{\alpha_s^{(2)}}{4\pi} = \frac{1}{b_0 \ln Q^2/\Lambda^2} - \frac{b_1}{b_0^3}\frac{\ln\ln Q^2/\Lambda^2}{\ln^2 Q^2/\Lambda^2}$$

or we can do it at large ρ (large distances) by fixing the Regge slope α'. We do not have the freedom to do both, since the scale of α_s can be fixed only once; that is, $\alpha' \Lambda^2_{\overline{ms}}$ is already given in terms of integrals over the β function :

$$\alpha' \Lambda_{\overline{ms}} = -\ln(16\pi^2/3) + (b_1/b_0^2)(\gamma_E + \ln 24/b_0)$$
$$- (31/3 - 10N_f/9)/b_0 + \ln b_0$$

If we take $\alpha' = 1\ \text{GeV}^{-2}$ to match the experimental Regge slope, we need $\Lambda_{\overline{ms}} \sim 500$ MeV.

$\alpha_s(Q^2)$ is now completely determined. We then Fourier transform $\tilde{V}(Q^2)$ to $V(r)$, insert it into the Schrödinger equation with an adjustable quark mass, and solve numberically for the energy levels. The agreement with the experimental ψ and Υ mass spacings is remarkably good (table 4). Acceptable fits can still be made with $\Lambda_{\overline{ms}}$ values down to about 100 MeV[11].

Fig.5. shows the r dependence of the RBGT potential compared with other models. Note also the average radii of the various ψ and Υ wave functions. Where they concentrate all potentials are very nearly the same, and the behaviour is neither coulombic nor linear but nearly logarithmic. In fact, a completely empirical power law potential[12], $V = A + Br^n$, with a small power n, fits the mass data equally well (table 4). It is clear then that the ψ and Υ mass measurements are not sensitive to the small-distance singular part of the potential derivable from QCD or even to the large-distance linear part. However, it cannot be an accident that a smooth interpolation between the small-distance behavior fixed by QCD, with the Λ value derived from other phenomena, and the large distance linear behaviour fixed by the Regge trajectories predicts the correct mass differences with no more adjustable input than the quark masses.

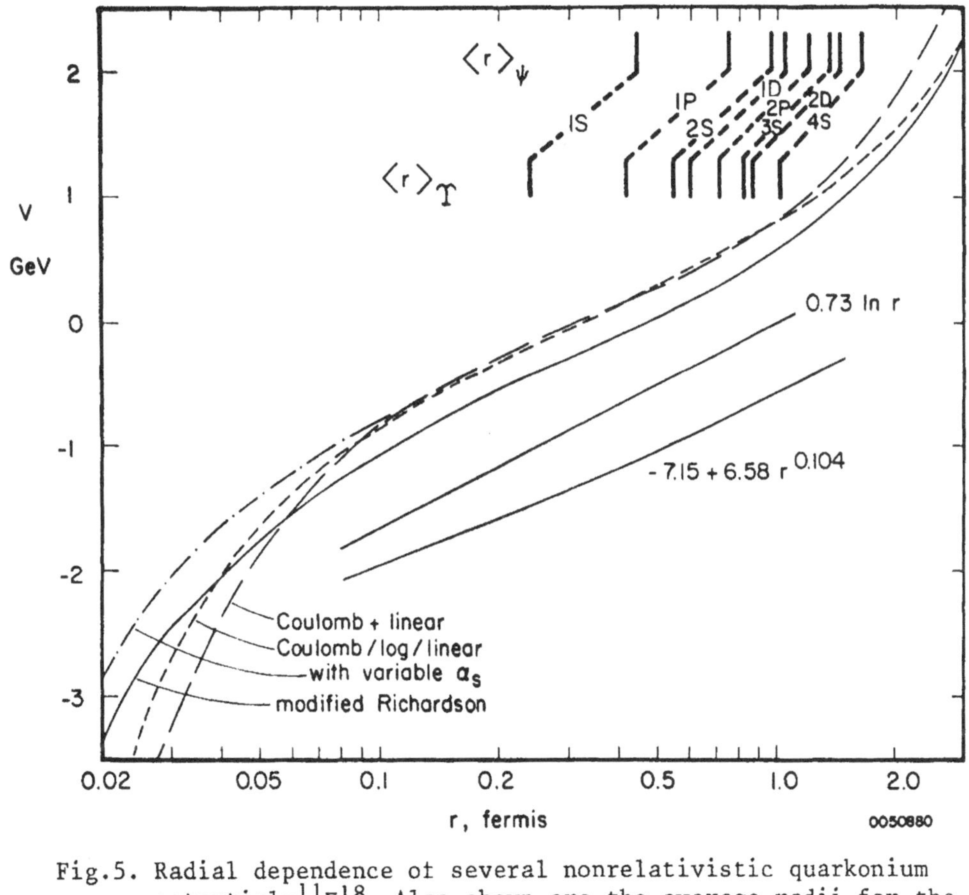

Fig.5. Radial dependence of several nonrelativistic quarkonium potentials[11-18]. Also shown are the average radii for the wave functions in various ψ and ϒ states.

Even if we cannot count it as a crucial test of QCD or make a definitive determination of $\Lambda_{\overline{ms}}$, it must still be considered a very successful consistency check. Indeed, independent of any particular potential model, we have learned that the binding is indeed flavor independent, and that the use of a nonrelativistic potential is justified. Using the RBGT wave functions we get $\langle v^2/c^2 \rangle = 0.23$ in the ψ and 0.077 in the Υ.

B. Leptonic Widths

When we measure the beam energy dependence of the e^+e^- annihilation cross section in the vicinity of a resonance, we determine not only the mass of the state but also the strength of its coupling to the e^+e^- channel. Since the height of the peak is inversely proportional to the machine-dependent resolution width the integrated area (after continuum subtraction and radiative corrections) is given only in terms of the intrinsic resonance mass and decay widths :

$$\int \sigma_{had}\, dW = (6\pi^2/M^2)(\Gamma_{ee}\Gamma_{had}/\Gamma).$$

In practice, the hadronic branching ratio Γ_{had}/Γ is almost 100%, so that the uncertainty in the small electromagnetic branching ratio is not very important in the determination of Γ_{ee} from the measured cross section integral. There is for the $\Upsilon(1S)$, however, a 10 to 20% systematic uncertainty arising from the difficulty in estimating the trigger and event selection efficiencies for the detector. In the case of the higher resonances we prefer to quote ratios $\Gamma_{ee}/\Gamma_{ee}(1S)$, for which the acceptance uncertainty cancels. The DORIS and CESR data are given in table 5.

The Van Royen - Weisskopf prediction[19] with the lowest order QCD correction[20] is

$$\Gamma_{ee} = (16\alpha^2 q_b^2/M^2)\ |\psi(0)|^2 (1 - 16\alpha_s/3\pi)$$

in terms of the $b\bar{b}$ wave function at zero separation.

This provides another test of the nonrelativistic quarkonium models, more sensitive to the short-distance behavior of the potential.

The QCD inspired RBGT potential does quite well for the $\Upsilon(1S)$ and even for the ψ, where the effect of QCD correction is rather large. Nonsingular potentials, such as $A + Br^n$, tend to give lower values for $|\psi(0)|^2$ and hence Γ_{ee}. The ratios for higher states seem to be rather insensitive to the details of the potential as well as to the QCD correction. Therefore, all models which successfully fit the masses give about the same predictions for the Γ_{ee} ratios.

Table 5. Measured and predicted leptonic pair widths and ratios

Quantity	Measurement	Expt	RBGT[11]	Martin[12]
$\Gamma_{ee}(1S)$, keV	$1.29 \pm 0.09 \pm 0.13$	DORIS[2-4]	1.1	-
	$1.02 \pm 0.07 \pm 0.15$	CESR[3-4]		
$\Gamma_{ee}(2S)/\Gamma_{ee}(1S)$	$0.45 \pm 0.06 \pm 0.02$	DORIS[5-6]	0.45	0.43
	$0.45 \pm 0.03 \pm 0.04$	CESR[5-6]		
$\Gamma_{ee}(3S)/\Gamma_{ee}(1S)$	$0.32 \pm 0.03 \pm 0.03$	CESR[7-8]	0.32	0.28
$\Gamma_{ee}(4S)/\Gamma_{ee}(1S)$	$0.24 \pm 0.02 \pm 0.03$	CESR	0.26	0.20

It should be noted that the agreement between theory and experiment confirms the charge assignment $|q_b| = 1/3$ for the b quark. If the b were an up quark, Γ_{ee} would be four times larger.

C. Evidence for the Decay of the Υ(1S) into three Gluons

The only energetically allowed hadronic decay of the Υ(1S) requires the b and $\bar{b}$ to annihilate into lighter quarks. In QCD this must proceed through a three-gluon intermediate state. If the matrix element favors the case in which the three gluon directions are well separated - the "Mercedes Star" configuration - and the transverse momenta of the hadrons produced in the fragmentation of each of the gluons is not too large, then one should be able to see three jets in the Υ(1S) final states. Unfortunately, the two conditions are not well satisfied for the Υ(1S) - it is even worse for the ψ. One must search for subtle effects in the distribution of events with respect to various measures of the event shape. Figs 6, 7 and 8 show three typical distributions, using the variables

Thrust = $\max \Sigma|p_L|/\Sigma p$
(maximum over all choices of axis),
Triplicity = $\max \Sigma p_L(\text{nearest axis})/\Sigma p$
(maximum over all choices of three axes)
Triple sphericity = $\min \Sigma p_T^2(\text{nearest axis})/\Sigma p^2$
(minimum over all choices of three axes).

These are typical of many distributions. In each case the distribution for Υ(1S) hadronic decays (the appropriately weighted distribution for the two jet continuum background and photon-mediated decays is subtracted) are intermediate between the distributions characteristic of $q\bar{q}$ two-jet events and of random phase space. This is in fact what one gets from a three-gluon Monte Carlo simulation. One would however see the same experimental distributions if the Υ were somehow to decay according to phase space about half the time and into two jets half the time.

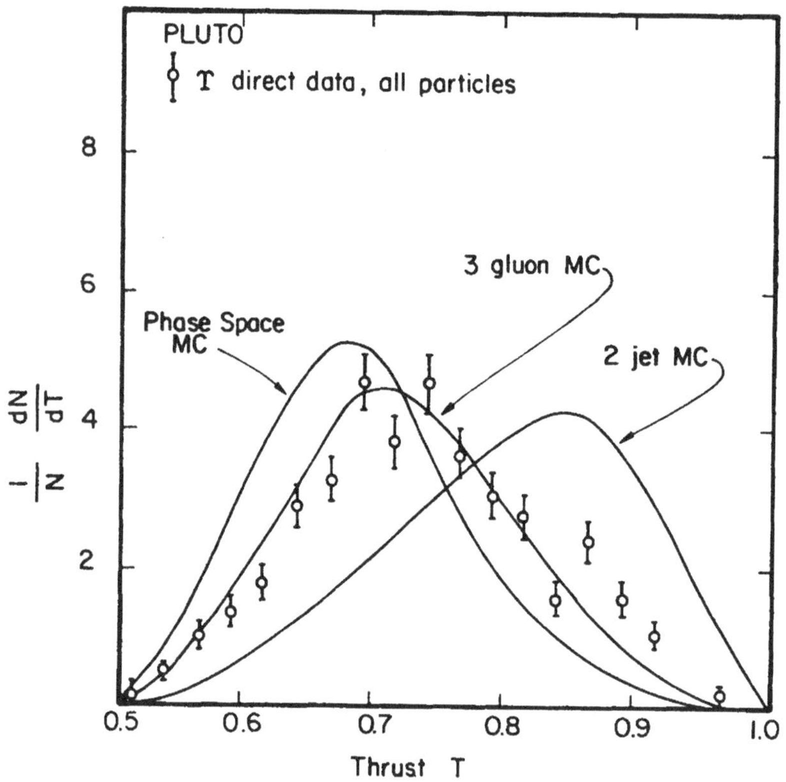

Fig.6. The thrust distribution for Υ(1S) direct decays[21], compared with phase space, three gluon, and two-jet Monte-Carlo simulations.

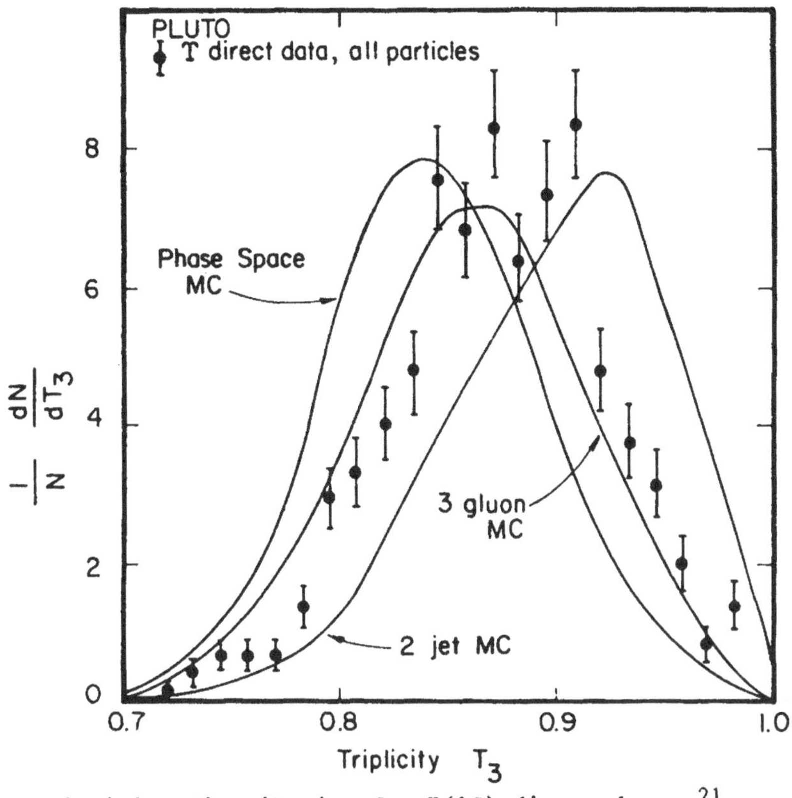

Fig.7. Triplicity distribution for Υ(1S) direct decays[21], compared with phase space, three gluon, and two-jet Monte-Carlo simulations.

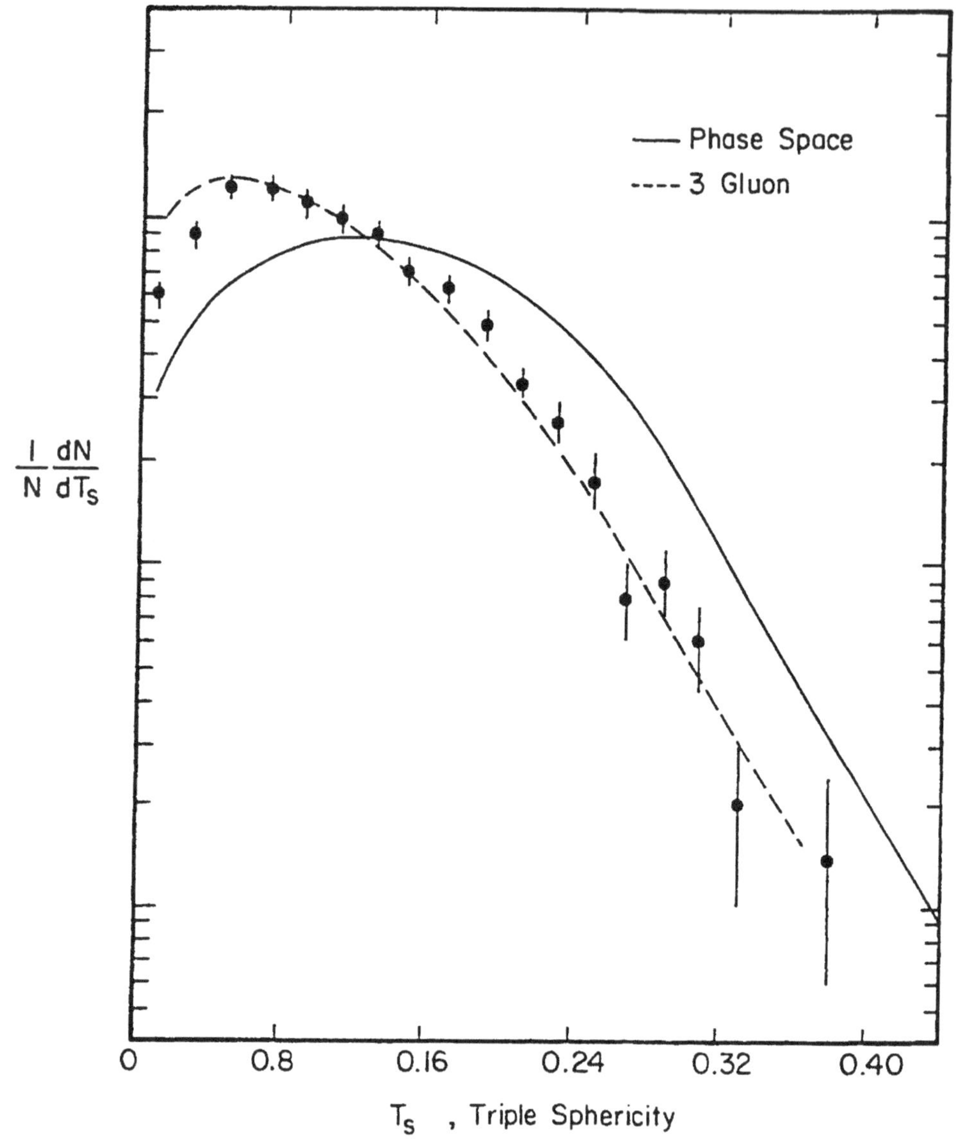

Fig. 8. CLEO triple sphericity distribution for Υ(IS) direct decays, compared with phase space and three gluon Monte-Carlo simulations.

One way to distinguish the three-gluon decay from the phase space two-jet mixture would be to look for the plane defined by the three gluon directions. The triplicity calculation automatically determines for each event the plane within which the hadron momentum components are maximized. QCD makes two predictions. First, the direction of the most energetic gluon in the plane (essentially the thrust axis of the event) should have the polar angle distribution $1 + 0.39 \cos^2\theta$ with respect to the beam axis[22]. And second, the normal to the event plane should have the polar angle distribution $1 - 0.33 \cos^2\beta$.

Table 6 shows the experimental values. The most accurate results, from CLEO, are marginally consistent with the QCD predictions, but are also not far from the completely random correlation between θ and β distributions which one would get from mixed phase-space and two-jet events in which the plane is determined merely by fluctuations in the fragmentation process.

Table 6. Measured and predicted coefficients in the beam polar angle distributions for the thrust axis $\sim 1 + a_T\cos^2\theta$, and the direction of the normal to the event plane $\sim 1 + a_N\cos^2\beta$, in ϒ(1S) direct decay.

Coefficient	Value	Source
a_T	0.39	QCD (vector gluon)[22]
	- 0.995	scalar gluon [22]
	0.83 ± 0.23	PLUTO [21]
	0.7 ± 0.3	LENA [23]
	0.5 ± 0.2	CLEO
a_N	- 0.33	QCD [22]
	- 0.3 ± 0.1	CLEO

I think we have to conclude that although the QCD three-gluon decay is the only theoretically motivated model which is consistent with event topologies in ϒ(1S) hadronic decays, there is really no way yet to convince a skeptic that we have seen direct evidence for three gluons.

D. Rate for ϒ(1S) Decay into three Gluons

QCD, however, also predicts the ϒ(1S) to three gluons partial decay width Γ_{ggg} ; and we can check that against experiment. The total width Γ should be the sum of three terms, diagrammed in Figs. 9a, 9b, 9c :

$$\Gamma = \Gamma_{ggg} + \Gamma_{em} + \Gamma_{\gamma gg}.$$

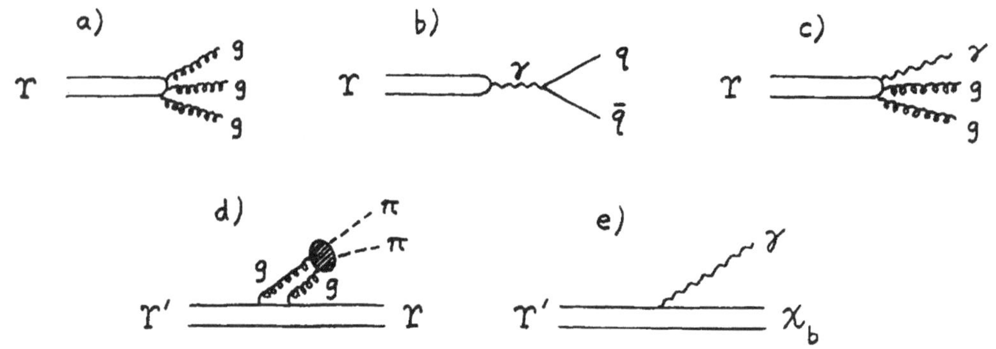

Fig. 9. Diagrams of various processes in upsilon decays.

Dividing by Γ_{ee} and rearranging we have

$$\Gamma_{ggg}/\Gamma_{ee} = 1/B_{ee} - (3 + R) - \Gamma_{\gamma gg}/\Gamma_{ee}.$$

The $\Gamma_{em}/\Gamma_{ee} = 3 + R$ term is about 6.5. The last term is relatively small and has been estimated at 0.8. So to measure Γ_{ggg}/Γ_{ee} we have only to measure the lepton pair branching ratio $B_{ee} = B_{\mu\mu}$ of the $\Upsilon(1S)$.

This has been done in two different ways. At DORIS[24-27] they observe the cross section for e^+e^- annihilation into muon pairs as a function of energy in the vicinity of the $\Upsilon(1S)$ resonance. The continuum cross section is just the QED cross section $\sigma_{\mu\mu}$. On the resonance it increases by $B_{ee}R_{\Upsilon,had}\sigma_{\mu\mu}$, where $R_{\Upsilon,had}$ is the peak hadronic cross section in units of $\sigma_{\mu\mu}$ ($R_{\Upsilon,had} \sim 10$ at DORIS). The increase, about a 30% effect, gives B_{ee}, but with large statistical errors. The DORIS value, combining data from three experimental groups is $B_{ee} = 0.030 \pm 0.008$ (table 7).

Table 7. Measurements of B_{ee}, the branching ratio for $\Upsilon(1S) \to e^+e^-$ (assumed to be equal to that for $\Upsilon(1S) \to \mu^+\mu^-$), and of $B_{\pi\pi}$ the branching ratio for $\Upsilon(2S) \to \Upsilon(1S)\pi^+\pi^-$

	Measurements, %	Expt.
a) B_{ee}	2.2 ± 2.0	PLUTO[24]
	5.1 ± 3.0	PLUTO[25]
	2.9 ± 1.3 ± 0.5	DASP2[26]
	3.5 ± 1.4 ± 0.4	LENA[27]
	3.0 ± 0.8	Av. from (a)
	3.4 ± 0.8	Av. from (b)(c)
	3.3 ± 0.5	World av.
b) $B_{\pi\pi}$	26 ± 13	LENA[28]
	19.1 ± 3.1	CLEO[29]
	19.6 ± 3.0	Av. from (b)
	21 ± 6	Av. from (a), (c)
	20 ± 3	World av.
c) $B_{\pi\pi}B_{ee}$	0.61 ± 0.26	LENA[28]
	0.63 ± 0.13	CUSB[30]
	0.68 ± 0.17	CLEO[29]
	0.64 ± 0.10	Av. from (c)

At CESR we measure B_{ee} from the $\Upsilon(1S)$ by running at the $\Upsilon(2S)$ resonance. We observe the cascade decay $\Upsilon(2S) \to \Upsilon(1S)\pi^+\pi^-$ followed by $\Upsilon(1S) \to e^+e^-$ or $\mu^+\mu^-$. I will give the experimental details when I discuss the $\Upsilon(2S)$ data. This decay sequence is particularly easy to observe and involves no background subtraction. The average CLEO[29] and CUSB[30] result (table 7) for the branching ratio product is

$$B[\Upsilon(2S) \to \Upsilon(1S)\pi^+\pi^-] \cdot B_{ee} = 0.0064 \pm 0.0010.$$

CLEO[29] has also measured the first term in the product by observing a peak at the $\Upsilon(1S)$ mass in the spectrum of missing masses M_X in the process $\Upsilon(2S) \rightarrow \pi^+\pi^-X$. The result is

$$B[\Upsilon(2S) \rightarrow \Upsilon(1S)\pi^+\pi^-] = 0.19 \pm 0.03,$$

which together with the product above implies $B_{ee} = 0.034 \pm 0.008$. The DORIS and CESR methods thus give consistent results with similar errors. The overall average is

$$B_{ee} = 0.033 \pm 0.005.$$

Together with the measurement $\Gamma_{ee} = 1.1$ keV (table 5), this gives us $\Gamma_{ggg} = 26 \pm 8$ keV and $\Gamma = 34 \pm 8$ keV.

The lowest order QCD prediction[31] for Γ_{ggg} contains three powers of α_s:

$$\Gamma_{ggg} = [160(\pi^2 - 9)/81M^2]\ \alpha_s^3\ |\psi(0)|^2$$

We can remove the dependence on the wave function by dividing by the Van Royen-Weisskopf prediction for Γ_{ee}. Although the next higher order QCD correction for Γ_{ee} has been known for some time, it is only very recently that Mackenzie and LePage[32] have succeeded in calculating the next higher order for Γ_{ggg}. The second order result for the ratio is

$$\frac{\Gamma_{ggg}}{\Gamma_{ee}} = \frac{10(\pi^2 - 9)}{81\pi(1/3)^2}\ \frac{\alpha_s^3(M_\Upsilon)}{\alpha^2}\ [1 + 9.1\alpha_s(M_\Upsilon)/\pi + \ldots]_{\overline{ms}}$$

Equating this to the experimental value and solving for α_s gives

$$\alpha_s(M_\Upsilon) = 0.141\ {}^{+\ 0.009}_{-\ 0.008}$$

Note that the effect of the 15% experimental error in B_{ee} is greatly reduced in the α_s result because of the cubic dependence. Using the second order formula for α_s (above in section IIA), we get for the scale parameter

$$\Lambda_{\overline{ms}} = 121\ {}^{+\ 44}_{-\ 31}\ \text{MeV}.$$

Since the second order correction to Γ_{ggg}/Γ_{ee} is about 40%, one might wonder whether the even higher terms might be important. We can try to minimize them by picking the energy at which α_s is evaluated such that the second order correction is exactly zero. If we had a correct formula to all orders the mass choice would not matter, but in finite order it makes a difference. With this choice we have

$$\alpha_s(.48M_\Upsilon) = 0.158 \begin{array}{l} +\ 0.012 \\ -\ 0.010 \end{array} \quad \text{and} \quad \Lambda_{\overline{ms}} = 100 \begin{array}{l} +\ 34 \\ -\ 25 \end{array} \text{MeV.}$$

Until higher orders are calculated (!) this is the best one can do.

We can check the Υ(1S) result by repeating the analysis for the ψ, although one does not expect the perturbation expansion to converge as fast at lower energy. The result is

$$\Lambda_{\overline{ms}} = 53 \begin{array}{l} +\ 16 \\ -\ 12 \end{array} \text{MeV.}$$

If instead we attribute the discrepancy to some unknown relativistic effect, we can assume that the correction is linear in $< v^2/c^2 >$ and use the ψ and Υ(1S) data to determine $\Lambda_{\overline{ms}}$ and the coefficient k in the ad hoc factor $(1 + k < v^2/c^2 >)$ multiplying our previous expression for Γ_{ggg}/Γ_{ee}. The result is $k = -2.6$ and $\Lambda_{\overline{ms}} = 136$ MeV. This is probably not a defensible way to determine $\Lambda_{\overline{ms}}$, but it may give us an idea of the possible uncertainty in the theory. The conclusion is that we have determined $\Lambda_{\overline{ms}}$ from the Γ_{ggg}/Γ_{ee} ratio for the Υ(1S) to be 100 MeV with an experimental uncertainty of about 30 MeV and a similar theoretical uncertainty. This result compares well with recent determinations from other experiments. As far as I know, it is the most accurate determination of $\Lambda_{\overline{ms}}$. The fact that similar values of α_s are derived from deep inelastic nucleon structure functions, gluon bremsstrahlung, quarkonium potential models, and ψ and Υ decay rates is a striking success of QCD.

The Γ_{ggg}/Γ_{ee} check for the Υ(1S) satisfies all the criteria for an ideal test of QCD:

(a) the three gluon decay is a pure QCD effect; it does not even exist in any previous model, such as the quark parton model.

(b) the prediction does not involve any poorly understood features of QCD, such as bound state wave functions or hadron fragmentation distributions.

(c) the higher order corrections are calculated.

(d) the experimental result is well measured, in fact, by two independent methods in five different detectors.

E. Particle Yields in Υ(1S) Decays

If the Υ(1S) decays predominantly into three gluons, and nonresonant annihilations occur primarily through quark-antiquark pairs, we can look for differences between gluon jets and quark jets by comparing Υ(1S) and nonresonant final states. In the CLEO detector charged kaons and protons (or antiprotons) are identified by time-of-flight in scintillation counters (Fig. 11a), provided the momenta

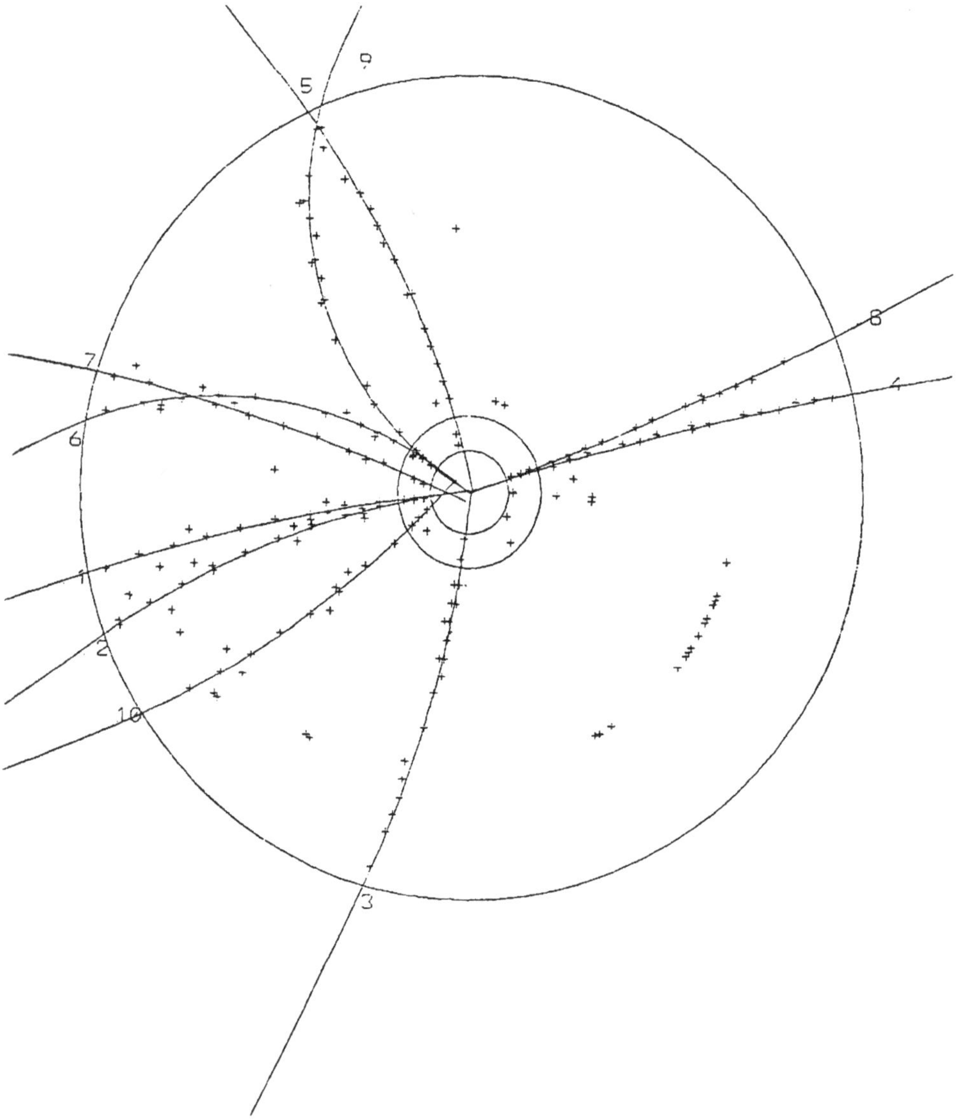

Fig. 10 Computer display of the CLEO tracking chamber showing an event containing a $K_s^o \rightarrow \pi^+\pi^-$ decay (tracks 7 and 10).

are high enough to penetrate the solenoid coil and low enough to separate the flight times. A neutral kaon or lambda (or antilambda) is seen in the tracking chamber as a pair of charged particles forming a separate vertex (see Fig. 10, for example). The identification is made by reconstructing the invariant mass (see Fig. 11b).

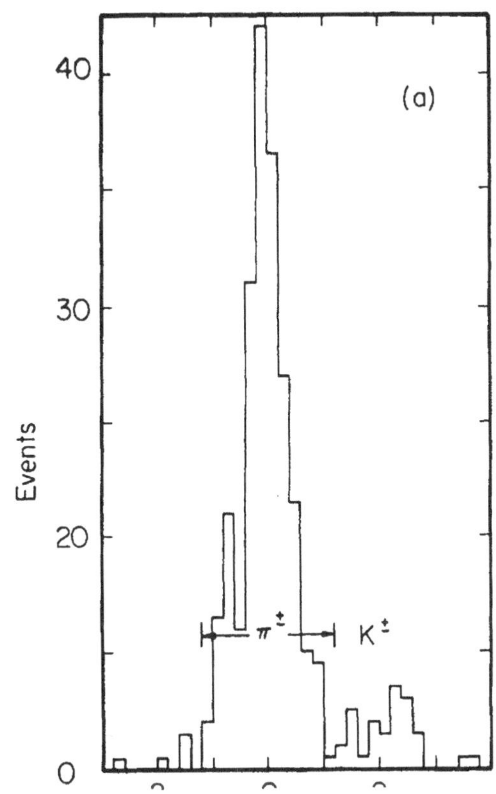

Table 8 shows preliminary CLEO data on the average number of kaons and baryons per event in the direct Υ(1S) decays and in the continuum. Both kaons and especially baryons seem to be significantly more abundant in gluon fragmentation than in quark fragmentation. I don't know of any convincing explanation. The large effect in baryon production is especially puzzling.

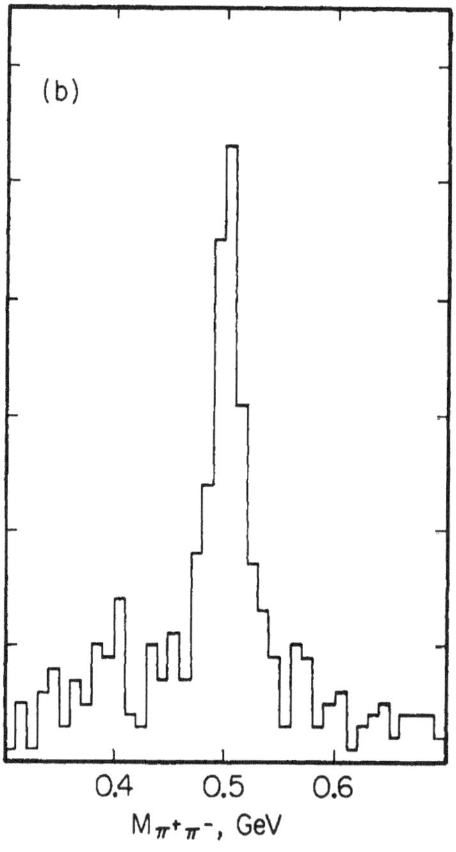

Fig. 11b Mass spectrum of K^0_S candidates.

Table 8. Preliminary CLEO data on inclusive particle rates in momentum ranges within which the particles can be identified, for Υ(1S) direct decays and nonresonant hadron production.

Particle	Momentum GeV/c	Average number per event Υ(1S) direct	nonresonant
K^+K^-	0.5 - 1.0	0.57 ± 0.05	0.34 ± 0.04
$K^o\bar{K}^o$	> 0.3	1.10 ± 0.09	0.78 ± 0.09
$p,\bar{p}$	0.7 - 1.8	0.46 ± 0.07	0.26 ± 0.04
$\Lambda,\bar{\Lambda}$	> 0.4	0.32 ± 0.06	0.12 ± 0.03

F. Hadronic Transitions

The decay sequence $\Upsilon(2S) \to \Upsilon(1S)\pi^+\pi^-$ followed by $\Upsilon(1S) \to e^+e^-$ or $\mu^+\mu^-$ is easily identifiable in any detector (see Figs. 12, 13).

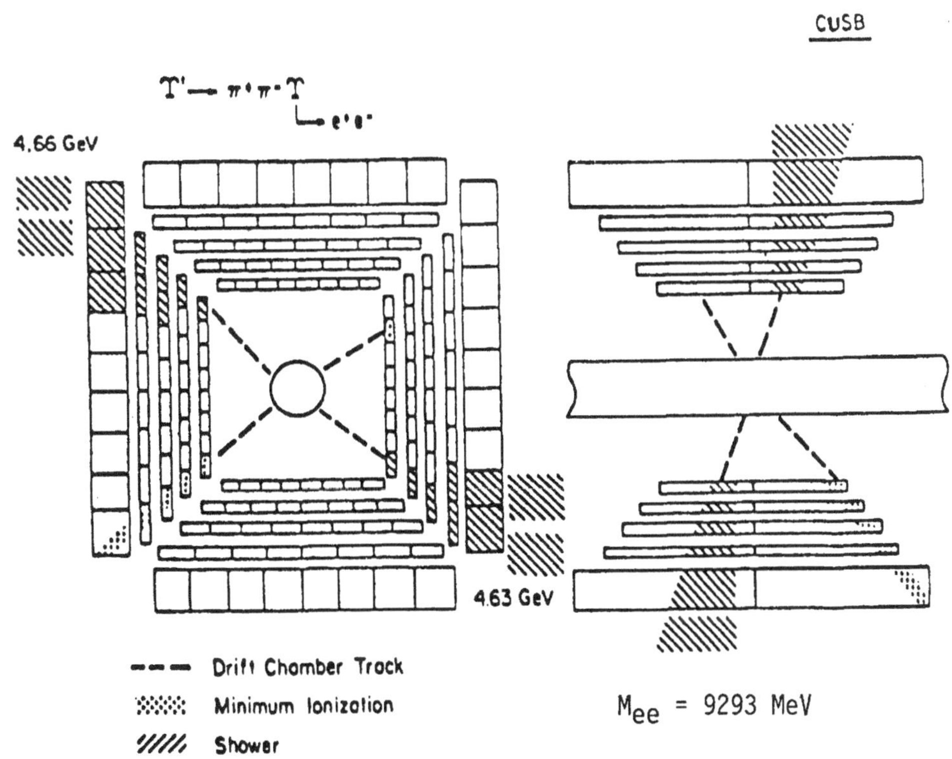

Fig. 12 Example of an event $\Upsilon(2S) \to \Upsilon(1S)\pi^+\pi^-$, $\Upsilon(1S) \to e^+e^-$ in the CUSB detector[30].

The product of the two branching ratios, however, turns out to be rather small (table 7), so there are only a total of 55 events reported so far by CLEO, CUSB and LENA. The $\Upsilon(2S) \to \Upsilon(1S)\pi^+\pi^-$ branching ratio has been measured by CLEO without any requirement on the $\Upsilon(1S)$ decay, by observing the missing M_X peak in $\Upsilon(2S) \to \pi^+\pi^- X$ (Fig. 14). The CLEO branching ratio agrees with the less accurate result derived by dividing the product of branching ratios by the directly measured B_{ee} for the $\Upsilon(1S)$. The overall average is

$$B[\Upsilon(2S) \to \Upsilon(1S)\pi^+\pi^-] = 0.20 \pm 0.03.$$

If the $\Upsilon(2S)$ and $\Upsilon(1S)$ are isospin singlets, there should be half as many $\pi^o\pi^o$ transitions as $\pi^+\pi^-$; that is

$$B[\Upsilon(2S) \to \Upsilon(1S)\pi\pi] = 0.30 \pm 0.04.$$

In order to compare this result with QCD predictions, we need to reexpress it as a width $\Gamma_{\Upsilon\pi\pi}$. To do this we need the total width Γ of the $\Upsilon(2S)$ which is not yet measured. With a theoretical input, however, we can estimate it. The total width can be expressed as a sum over the processes diagramed in Figs. 9 a,...e:

$$\Gamma = \Gamma_{ggg} + \Gamma_{em} + \Gamma_{\gamma gg} + \Gamma_{\Upsilon\pi\pi} + \Gamma_{\chi\gamma}.$$

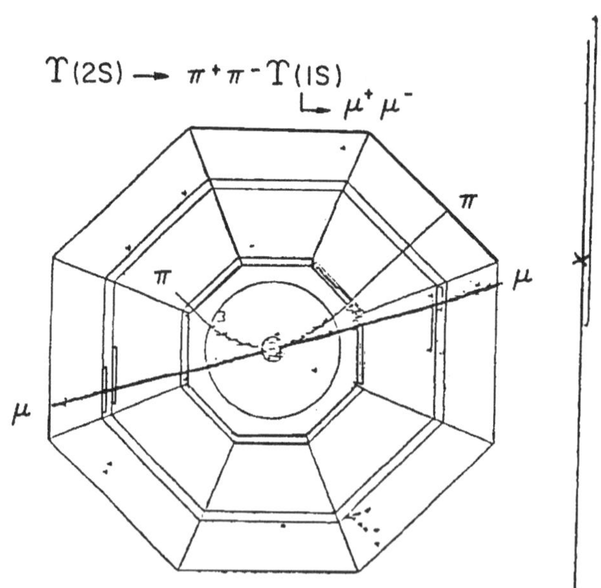

Fig. 13 Display of an event $\Upsilon(2S) \to \Upsilon(1S)\pi^+\pi^-$, $\Upsilon(1S) \to \mu^+\mu^-$ in the CLEO detector[29].

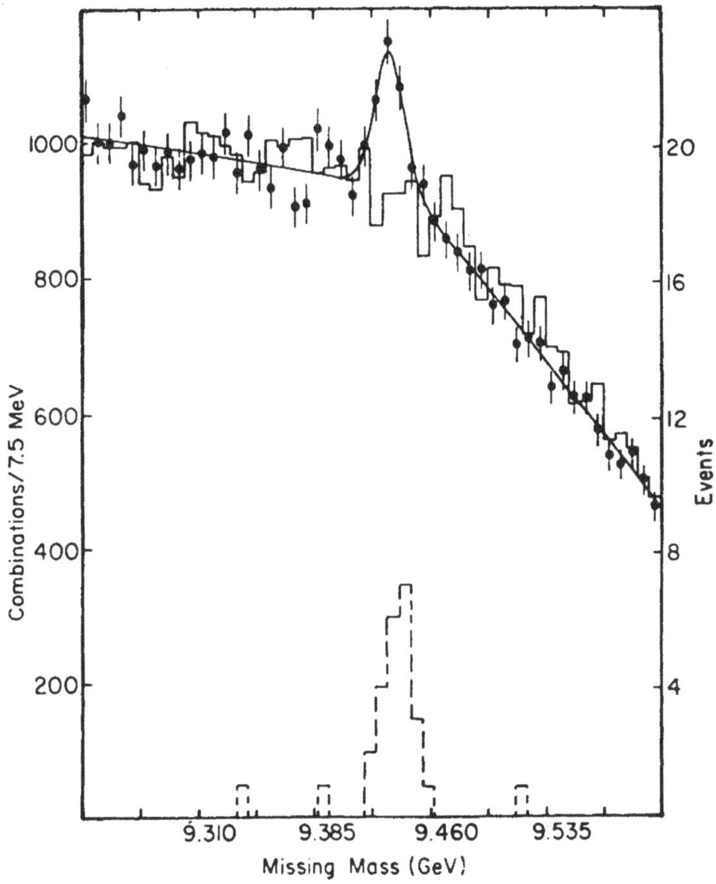

Fig.14 CLEO data[29] for the spectrum of missing masses from $\Upsilon(2S) \to \pi\pi X$. The points are for opposite sign dipions, the histogram for like sign. The curve is a fit to the background plus $\Upsilon(2S) \to \Upsilon(1S)\pi^+\pi^-$. The dashed histogram is for events of the type $\Upsilon(2S) \to \pi^+\pi^-\ell^+\ell^-$

As in the case of the $\Upsilon(1S)$, $\Gamma_{\gamma gg}$ is small, probably about 0.4 keV. $\Gamma_{\chi\gamma}$ can be estimated from the corresponding width in charmonium:

$$\Gamma_{\chi\gamma} = (q_b^2 m_c / q_c^2 m_b)\ \Gamma(\psi' \to \chi\gamma) \simeq 4\ \text{keV}.$$

Γ_{em} is known: $\Gamma_{em} = (3 + R)\Gamma_{ee} = 3.3$ keV. The combination $\Gamma_{ggg} + \Gamma_{em}$ should be proportional to $|\psi(0)|^2$ with the same proportionality factor as in the case of the $\Upsilon(1S)$. We therefore have for the $\Upsilon(2S)$

$$\Gamma_{ggg} + \Gamma_{em} = (\Gamma_{ee}/\Gamma_{ee}^{1S}) \cdot (\Gamma_{ggg} + \Gamma_{em})^{1S}$$

$$= \Gamma_{ee}/B_{ee}^{1S}$$

$$= 16 \pm 4\ \text{keV}.$$

Now inserting $B_{\Upsilon\pi\pi}\Gamma$ for $\Gamma_{\Upsilon\pi\pi}$ and solving for Γ, we have

$$\Gamma = 27\ {}^{+10}_{-7}\ \text{keV}, \qquad \text{and} \qquad \Gamma_{\Upsilon\pi\pi} = 8\ {}^{+3}_{-2}\ \text{keV}.$$

In QCD the transition takes place through the emission of two gluons, which then form the two pions. Since the total energy of the two pions is only 560 MeV, we cannot expect perturbative methods to apply. However, since the energy is very small relative to the Υ masses, we can treat the transition as the radiation of a long wavelength field by a pointlike source; that is, we can use the first term of a multipole expansion in the gluon field[33]. The result is a prediction for $\Gamma_{\Upsilon\pi\pi}$ in terms of the corresponding ψ' width and the relative wave function spreads for the $\Upsilon(2S)$ and ψ':

$$\Gamma_{\Upsilon\pi\pi} = \left(\frac{< r^2 >_{\Upsilon'}}{< r^2 >_{\psi'}} \right)^2 \Gamma_{\psi\pi\pi} \simeq \frac{1}{10}\ (108\ \text{keV}) \simeq 11\ \text{keV},$$

which is consistent with our measurement. This is actually not an insignificant test for QCD. For example, if the gluon had spin 0 instead of 1, the prediction for $\Gamma_{\Upsilon\pi\pi}$ would have been an order of magnitude larger. Using soft pion techniques, one also successfully predicts[33] the shape of the dipion mass spectrum (Fig. 15).

The corresponding transition from the next resonance, $\Upsilon(3S) \to \Upsilon(1S)\pi^+\pi^-$, followed by $\Upsilon(1S) \to \ell^+\ell^-$, has also been seen by CLEO and by CUSB with a branching ratio for the $\pi\pi$ transition consistent with 5 to 10%.

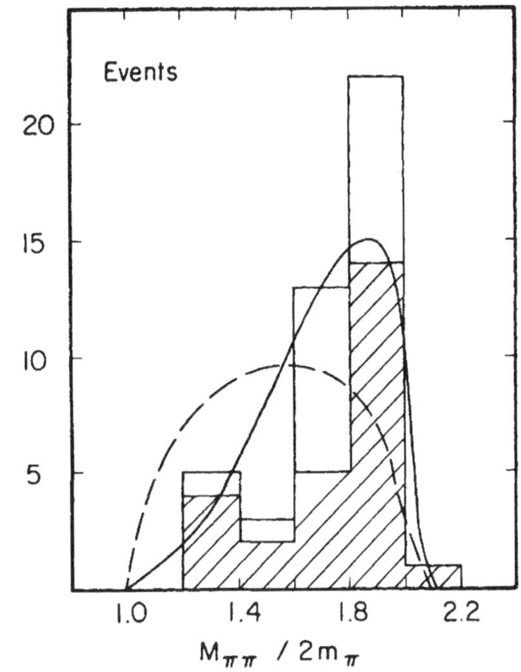

Fig. 15 CUSB[30] and CLEO[29] (shaded) data for the $M_{\pi\pi}$ spectrum from $\Upsilon(2S) \to \Upsilon(1S)\pi^+\pi^-$, $\Upsilon(1S) \to \ell^+\ell^-$. The solid curve is from ref. 33 and the dashed curve is phase space.

III. B MESON DECAYS AND TESTS OF THE STANDARD WEAK INTERACTION

A. Production

The fourth upsilon resonance, discovered at CESR[7,8] at a total energy of 10.55 GeV (Figs 16, 17), has an excitation energy (table 4) and lepton pair width (table 5) which make it the logical choice for the 4^3S state of $b\bar{b}$, and so we call it the $\Upsilon(4S)$. The width of the peak is significantly larger than for the three lower resonances. After unfolding the machine energy resolution, the CLEO and CUSB data imply an intrinsic width Γ_{4S} = 20 ± 2 MeV. The fact that this is

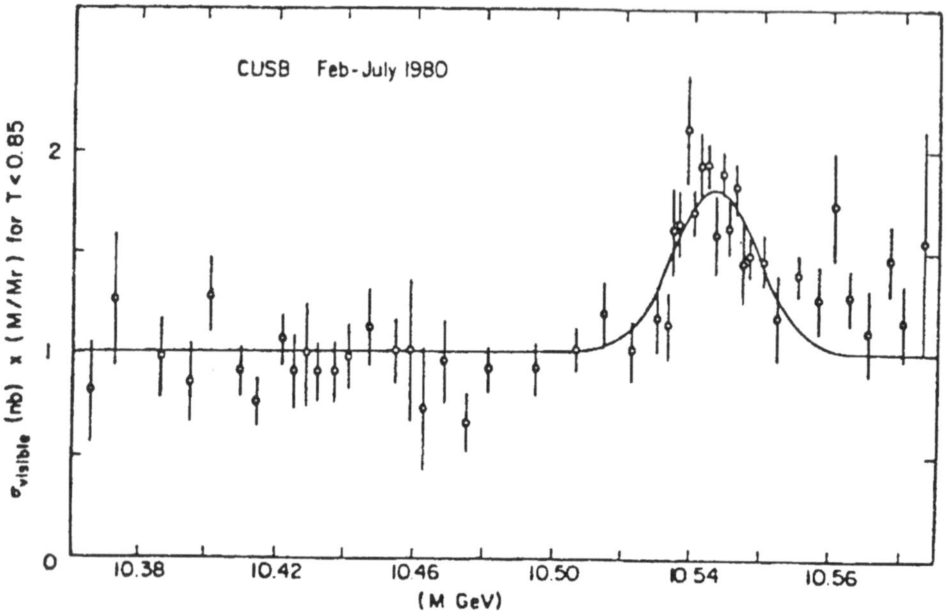

Fig. 16 CUSB data[8] for the uncorrected total cross section for $e^+e^- \rightarrow$ hadrons as a function of center of mass energy, showing the Υ(4S).

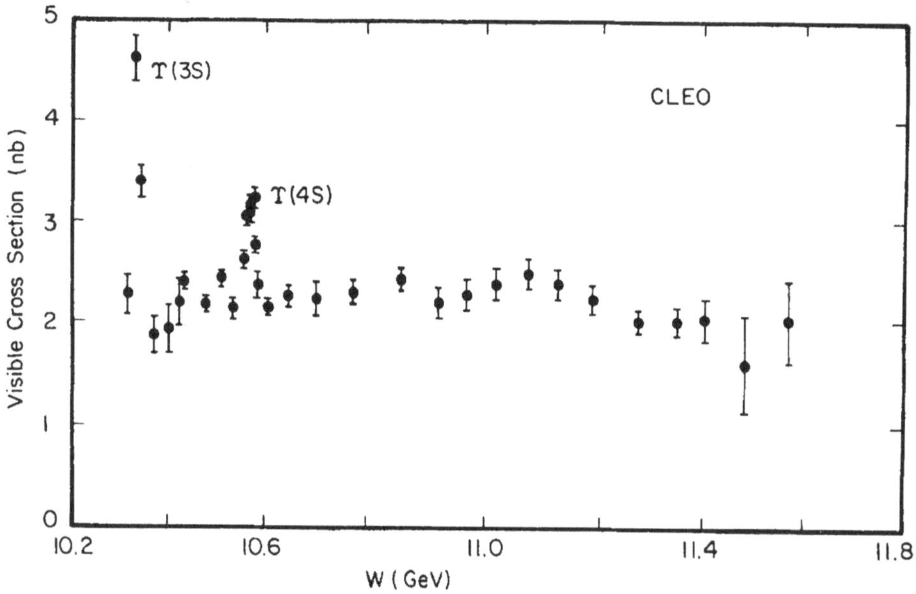

Fig. 17 CLEO data[7] for the cross section for $e^+e^- \rightarrow$ hadrons, showing the Υ(3S), the Υ(4S), and higher energies.

about three orders of magnitude greater than the intrinsic widths of the ϒ(1S), ϒ(2S) and ϒ(3S) suggests that there is above the ϒ(3S) a threshold for a new strong decay channel. This is also consistent with the tendency for the final decay products of the ϒ(4S) to be rather spherically distributed (Fig. 18), as if they came from the isotropic decay of heavy particles created nearly at rest.

We expect the lightest meson states containing a single b quark to be the pseudoscalar $B^+ = \bar{b}u$ and $B^0 = \bar{b}d$, and their charge conjugates. A reasonable guess[14] for the mass would be

$$M_B = M_D + m_b - m_c + (3/4)(1 - m_c/m_b)(M_D{*} - M_D) = 5.2 \text{ GeV},$$

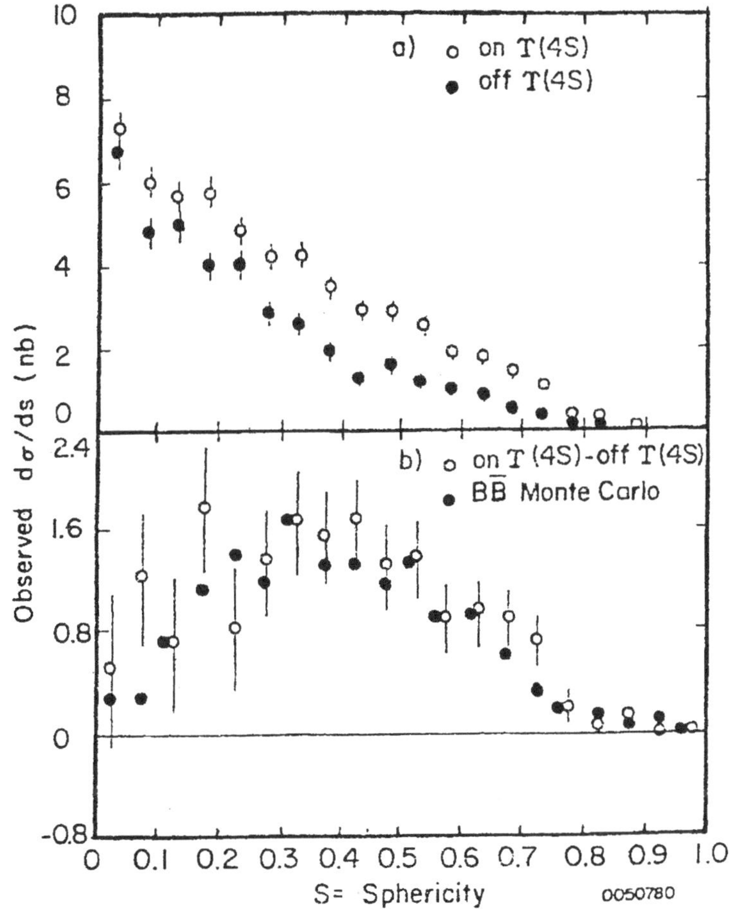

Fig. 18 Sphericity distributions in the vicinity of ϒ(4S), from CLEO

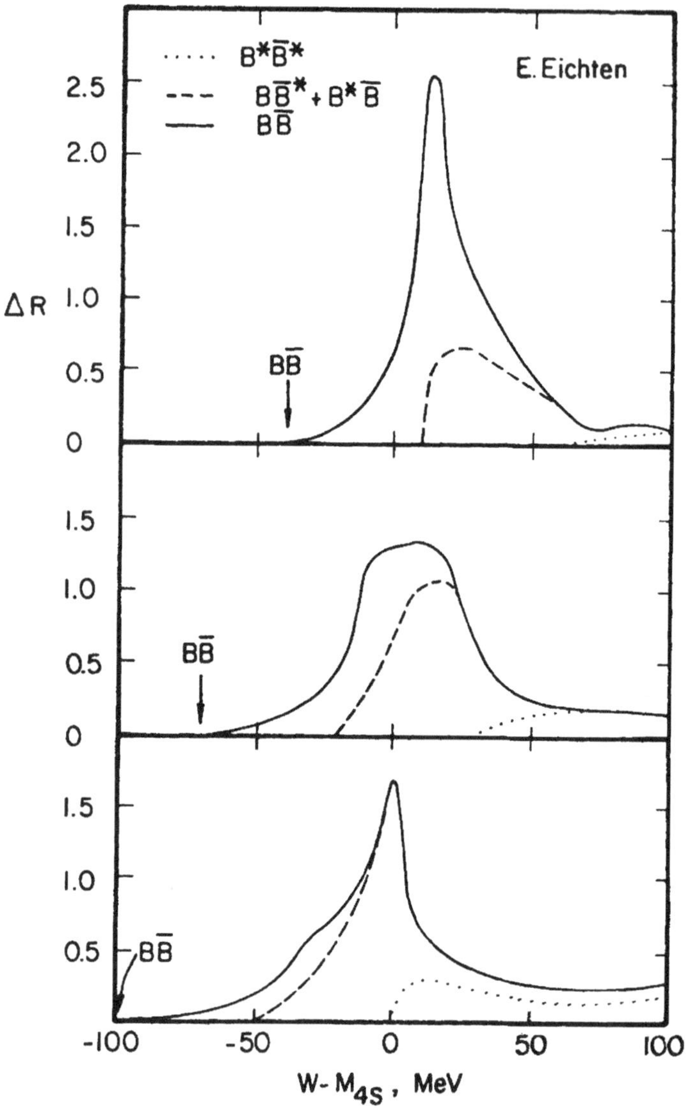

Fig.19. Predicted cross section[34] for production of b-flavored mesons using several B mass values.

which implies a threshold for $e^+e^- \to B\bar{B}$ at about 10.4 GeV. For the next higher meson, the vector B*, we expect

$$M_{B*} - M_B = (m_c/m_b)(M_{D*} - M_D) \simeq 50 \text{ MeV}.$$

One expects the Υ(4S) width to depend on the energy of the resonance relative to the thresholds for $B\bar{B}$, $B\bar{B}^*$ or $B^*\bar{B}$, and $B^*\bar{B}^*$. Making the same kind of coupled-channel calculation used in the charmonium case, Eichten has predicted[34] the shape of the Υ(4S) peak for three assumed values of $\Delta = M_{4S} - 2M_B$ (Fig. 19). The width increases with Δ, the energy above threshold, but not monotonically (Fig. 20). A calculation by Ono[35], based on a quark pair creation model, is qualitatively similar, but predicts a much more rapidly increasing width (Fig. 20). The discrepancy in the models prevents us from using Γ_{4S} to measure M_B. All we can say is that the threshold at $2M_B$ lies between M_{3S} and M_{4S}.

The presence or absence of B^* mesons in the Υ(4S) decays would help to determine Δ and limit the mass of the B. If the photon in

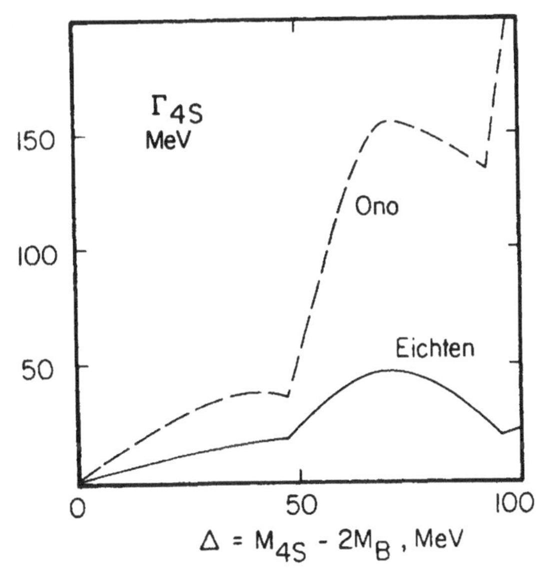

Fig. 20 Total width of the Υ(4S)[34,35] as a function of the assumed mass above $B\bar{B}$ threshold.

$B^* \to B\gamma$ indeed has an energy around 50 MeV, it should show up as a peak above π^0 decay background in the preliminary inclusive single photon spectrum observed by CUSB. There is no convincing evidence for such a peak (Fig. 21), but perhaps the data could accomodate a B^* in as much as one quarter of the ϒ(4S) decays.

The total cross section for $e^+e^- \to$ hadrons above the ϒ(4S) has been measured at CESR up to W = 11.5 GeV (Fig. 17). It shows no surprises, no more prominent peaks. The average level of $R = \sigma_{had}/\sigma_{\mu\mu}$ is 0.3 ± 0.1 higher at energies above the ϒ(4S) than below. Asymptotically one expects $\Delta R = 3q_b^2 = 1/3$ for the $b\bar{b}$ increment.

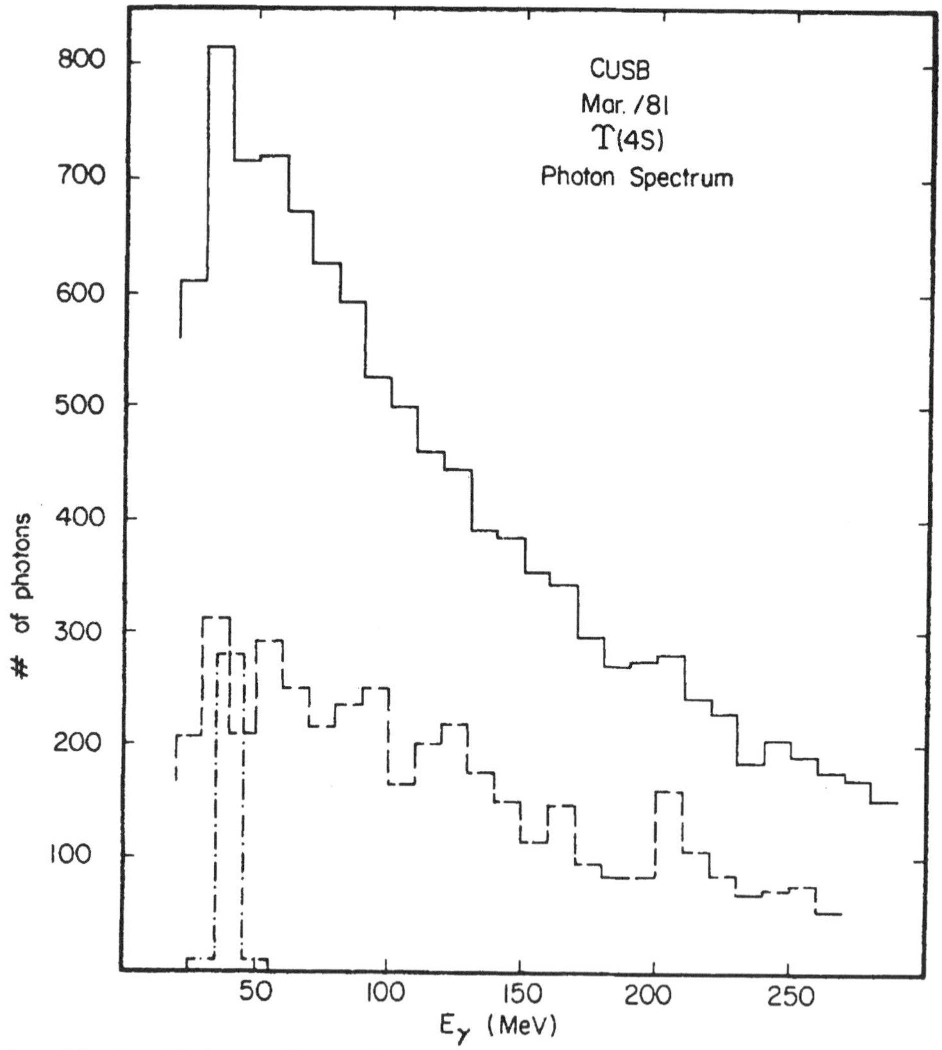

Fig. 21 Preliminary inclusive photon spectrum observed at the ϒ(4S) in the CUSB detector. The dashed histogram is after nonresonant background subtraction. The dash-dot histogram is the prediction for one $B^* \to B\gamma$ per event.

B. The Spectator Model

Assume for now that there are no flavor changing neutral weak currents. In the standard six-quark model then, the only allowed couplings are between the W^- and each of the following weak doublets: $e^-\bar{\nu}_e$, $\mu^-\bar{\nu}_\mu$, $\tau^-\bar{\nu}_\tau$, $d'\bar{u}$, $s'\bar{c}$, $b'\bar{t}$ (or W^+ and the antidoublets). The d', s', b' are the rotated mixtures of d, s, b. Since the t is not energetically accessible, the b quark has to decay to u or c through its mixing in the d' or s'.

Experimentally, we do not observe the b quark decay directly, but the decay of the B meson. Three ways the B meson could decay are diagramed in Fig. 22. In the spectator decay (Fig. 22a) only the b quark participates in the weak interaction. If this were the only contributing process, the lifetimes of the charged and neutral B mesons would be the same and so would the leptonic branching ratios. The diagram in which the b and the light quark exchange a W (Fig. 22b) can only contribute to neutral B decay and cannot produce leptons.

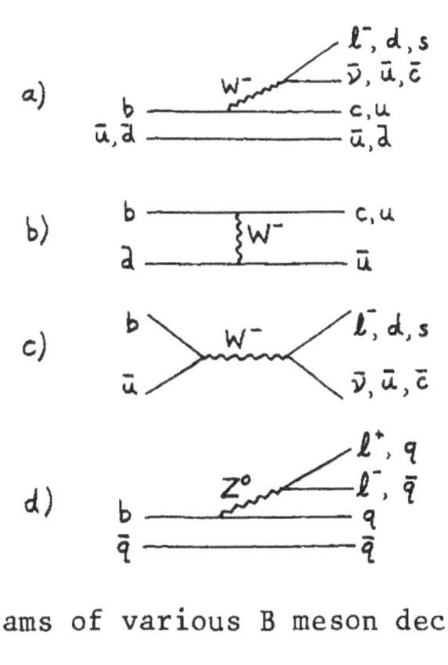

Fig. 22 Diagrams of various B meson decay processes

The annihilation diagram (Fig. 22c) contributes only to charged B decay. In the cases of D decay and K decay the spectator diagram is certainly not dominant. However, one naively expects the exchange and annihilation contributions to be helicity suppressed by a factor m_Q^{-2}, so they might be less important in B decay. The naive expectation can be complicated, however, by the effect of gluons. Leveille (ref. 36) has made a model, including gluon effects, which reproduces the D and K data and predicts the following numbers for the nonspectator contributions to B decay rates for two assumptions about the relative rates of $b \to u$ and $b \to c$:

	u << c	u ≈ c
$B^{\pm}$	0	16%
B^o, $\bar{B}^o$	38%	30%

For now, we have no data on relative lifetimes or leptonic branching ratios for charged and neutral B mssons, so we have no way of checking for the dominance of the spectator process. For simplicity in what follows we will just assume that it dominates, and the B meson decay directly reflects the properties of the b quark decay.

C. Leptonic Decays

Just taking color into account, the b should decay (as in Fig. 22a) into c or u, plus one of the following - $e^- + \bar{\nu}_e$, $\mu^- + \bar{\nu}_\mu$, $\tau^- + \bar{\nu}_\tau$, $d' + \bar{u}$, $s' + \bar{c}$ - in the ratio 1, 1, 1, 3, 3. Putting in the appropriate phase space factors for the heavy leptons and quarks, this becomes 1, 1, 0.2 (0.4), 3, 0.6 (1.4) for the $b \to c$ (u) cases. We would then predict a branching ratio for $B \to e\nu X$ (same as for $B \to \mu\nu X$) of 17% (15%). With Leveille's estimates[36] for the gluon enhancement of the hadronic modes and nonspectator contributions the prediction becomes 12% (10%).

Both of the CESR groups have measured the yield of inclusive electrons as a function of beam energy W. In the CUSB detector[37] electrons are recognized as showering charged particles. CLEO[38] imposes the additional requirements of a signal in a gas Cerenkov counter and a appropriate match between the measured momentum and shower counter pulse height. Fig. 23 shows the CLEO observed rate of electrons with momentum above 1 GeV/c, as a function of W, and the momentum spectrum. The CUSB data look quite similar. In both experiments the electron rate (P_e > 1 GeV/c) increases at the $\Upsilon(4S)$ by a much larger factor than does the total hadronic cross section (Figs. 16, 17). The low electron yield off the $\Upsilon(4S)$ is in fact consistent with the expected rate from charm and tau decays, QED processes, and misidentified hadrons. The rate of high momentum

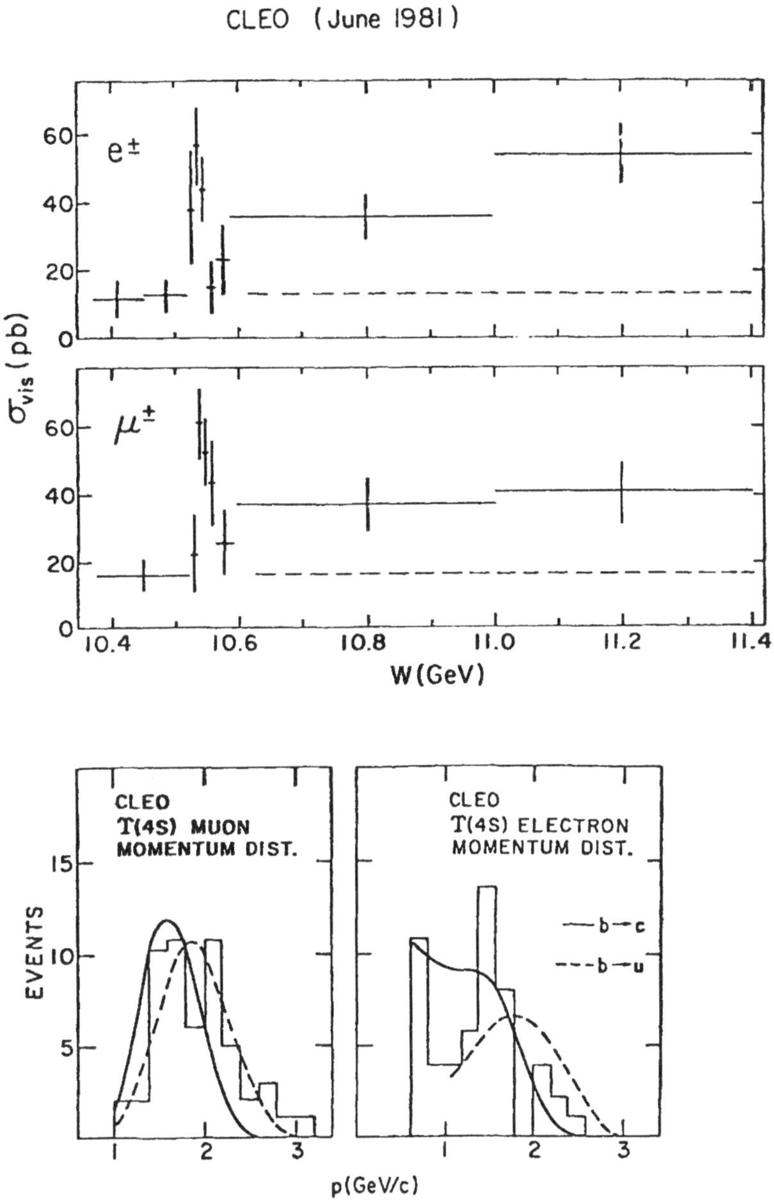

Fig. 23 CLEO observed inclusive electron[38] and muon[39] cross sections as functions of center-of-mass energy , for $p > 1$ GeV/c; and observed momentum spectra of electrons and muons at the Υ(4S).

electrons at the $\Upsilon(4S)$ implies that the B has a significant semileptonic branching fraction, and is good evidence that mesons containing a new quantum number conserved in strong interactions are being produced. Taking the acceptance and electron identification efficiency into account the two groups report the following branching ratio values for $B \to e\nu X$:

$13.6 \pm 2.5 \pm 3.0$ % CUSB[37]

$13 \pm 3 \pm 3$ % CLEO[38]

Only leptons produced in the first generation of the decay are included in these numbers. Also, if the charged and neutral B have different branching ratios, the quoted values apply to the average over charge states produced at the $\Upsilon(4S)$.

The CLEO group has made a similar search for muons[39], identifying them with drift chambers outside of a 60 cm thick iron shield surrounding the detector. The minimum momentum which can penetrate the shield is 1.0 to 1.6 GeV/c depending on the direction. The beam energy dependence and the momentum spectrum (Fig. 23) both behave similarly to those for the electrons. The derived branching ratio for $B \to \mu\nu X$ is:

9.4 ± 3.6 % CLEO[39],

consistent within errors with the electron branching ratio. The average of the electron and muon branching ratio measurements is

12 ± 2 % ,

in good agreement with the standard model prediction.

D. Neutral Currents and Other Unorthodox B Decay Models

Suppose the b quark can change to an s or d by emitting a Z^o (Fig. 22d). Then among the B decays we should expect some with an $\ell^+\ell^-$ pair of high energy leptons. Multilepton events, however, can come from other sources. There are two B's in every event, each one of which can contribute a single lepton, and if a D is produced in the decay, it can decay leptonically too. Listed below are the dilepton ($p > 1$ GeV/c) events found in the first 720 $B\bar{B}$ events seen by CLEO, along with estimates of the number expected from non-neutral-current sources.

	observed	$D \to \ell\nu X$	$B,\bar{B} \to \ell\nu X$
$\mu\mu$	0	< 0.1	0.6 - 1.2
ee	2	< 0.1	0.2 - 0.5
μe	1	< 0.1	0.7 - 1.5

The one μe pair cannot of course come from a single B. Of the two ee pairs, one is consistent with a ψ and the other includes an electron with too high a momentum to be likely for a B decay (probably a continuum $\tau^+\tau^-$ event). So there are so far no likely candidates for dilepton B decays. This implies an upper limit for the $B \to \ell^+\ell^- X$ branching ratio of about 1.3 % at 90% confidence, or a ratio $(B \to \ell^+\ell^- X)/(B \to \ell\nu X) < 8\%$.

There are many non-standard models of the weak interactions, some inspired by the lack of evidence for a t quark. Suppose that the b is either a weak singlet[40] or is in a weak doublet[41], but has a quantum number which forbids $b \to uW^-$ or cW^-. Then either the b is stable, or $b \to \tau\bar{\nu}_\tau X$ or $\bar{q}\bar{q}\ell$ or $\bar{q}\bar{q}\bar{\ell}$. That is, all b decays would contain a lepton and all decays would contain either a τ or an antibaryon (baryon for $\bar{b}$ decay). The data completely rule out this possibility.

Suppose instead that the b is a weak singlet or one of a doublet of charge − 1/3 quarks, and can mix with s and/or d[42]. There is then no way to suppress completely flavor changing neutral currents, and at least 25% of b decays will be through the Z^0. One can derive[43] a lower limit on the dilepton branching ratio of the b:

$$(b \to \ell^+\ell^- X)/(b \to \ell\bar{\nu}X) > 1/8.$$

The CLEO experimental upper limit is on the verge of excluding this hypothesis.

Suppose that we put the b in a right-handed doublet with the c[44]. This model has natural suppression of flavor changing neutral currents. Its predictions can be tailored to be very close to those of the standard six-quark model, so it will be very hard to rule out. We may have to wait for the discovery of the t-quark to know for sure.

Finally it is possible with charged Higgs bosons of mass near 5 GeV to concoct a model in which b decays via a Higgs instead of a W. This would show up as a preference for high mass fermions, τ and c, in the decay products. With the present data one cannot, I believe, rule this one out, but the good agreement between the standard model and the measured e and μ branching ratios makes it rather unlikely.

In short, there are no surprises. The standard model is quite adequate and most of the models without t quark are untenable.

E. The Ratio of $b \to uW^-$ and $b \to cW^-$ Rates

The mixing of the d, s and b is parametrized by the Kobayashi-Maskawa[46] matrix, involving three generalized Cabibbo angles and a phase δ. In terms of the sines and cosines of θ_1, θ_2, and θ_3 we define

$$\begin{pmatrix} d' \\ s' \\ b' \end{pmatrix} = \begin{pmatrix} c_1 & s_1c_3 & s_1s_3 \\ -s_1c_2 & c_1c_2c_3 + s_2s_3e^{i\delta} & c_1c_2s_3 - s_2c_3e^{i\delta} \\ -s_1s_2 & c_1s_2c_3 - c_2s_3e^{i\delta} & c_1s_2s_3 + c_2c_3e^{i\delta} \end{pmatrix} \begin{pmatrix} d \\ s \\ b \end{pmatrix}$$

From strangeness conserving beta decay (proportional to c_1) we know that $s_1^2 = 0.05$. Strangeness changing decays (proportional to s_1c_3) tell us that $0 < s_3 < 0.5$[47]. If CP violation is attributed to the phase δ, then $|c_\delta| \simeq 1$ (letting $c_\delta = \cos\delta$). Finally, from the K^0 mass difference we have (using the bag model) $s_2 = 0.2 \pm 0.1 + (c_\delta/2 + 1/4)s_3$. The uncertainty in the K-M angles is thus reduced to an uncertainty in s_3 (0 to 0.5) and in the sign of $\cos\delta$ (± 1)[46].

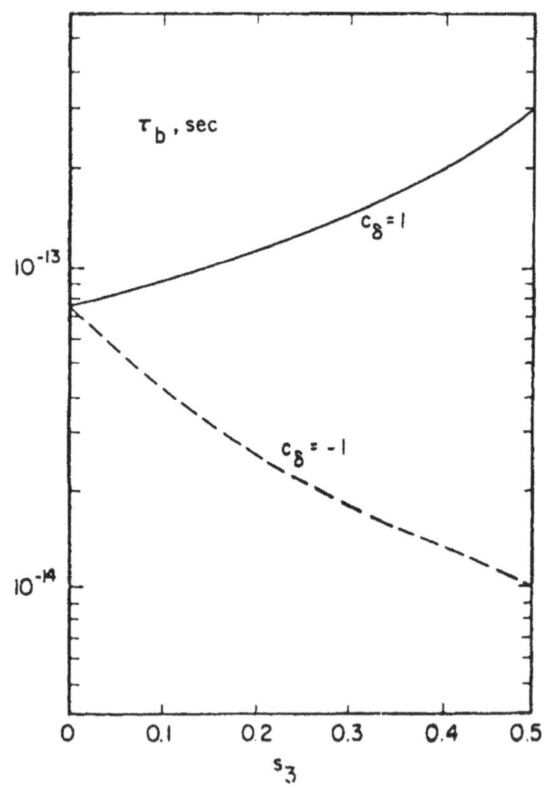

Fig. 24 Dependence of B meson lifetime on $\sin\theta_3$ for the two allowed values of $\cos\delta$[48]

The B meson lifetime depends on the unknown K-M angles (Fig. 24) but the present experimental upper limit[49], $\tau_B < 5\ 10^{-2}$ sec, is too high to be useful.

The other possibility of fixing s_3 and c_δ is a measurement of $(b \to uW^-)/(b \to cW^-)$ (Fig. 25). One indicator of the presence of charm in B decays is the average number of kaons per event. The most naive calculation goes as follows. If $b \to uW^-$, you get two kaons when W^- goes to $s\bar{c}$ (which it does 1/3 of the time by simple counting) otherwise none; so you expect an average of 2/3 kaon per B decay. If $b \to cW^-$, you always get one kaon from the c decay products and again an average of 2/3 from the W^-; so you expect 5/3 kaons per B decay. The naive calculation has to be corrected for phase space, hadronic enhancements, nonspectator contributions, and for kaons generated from the sea in the later hadronization stages. The predictions become very model dependent. Our best guess is that pure $b \to uW^-$ would give about 0.9 kaons per B decay and pure $b \to cW^-$ would yield about 1.6. The measured rate, combining the CLEO rates

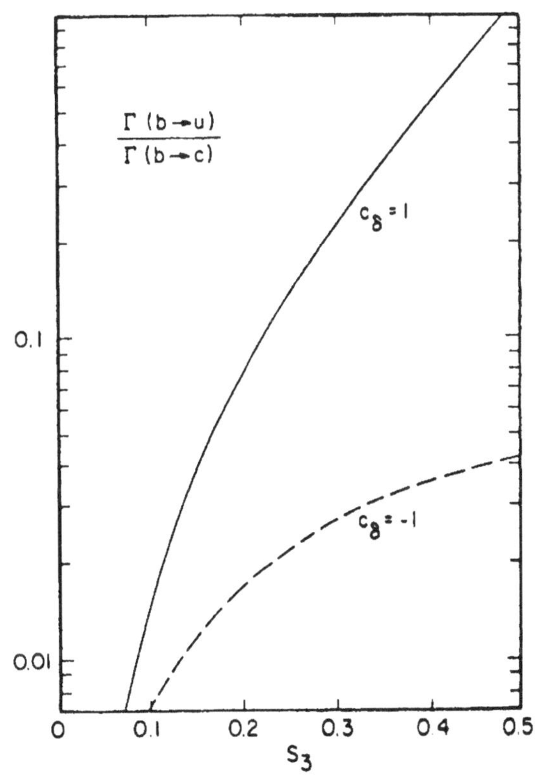

Fig 25 Ratio of rates $(b \to uW^-)/(b \to cW^-)$ as a function of $\sin\theta_3$ for the two allowed values of $\cos\delta$ [48].

for charged and neutral kaons and correcting for the undetected portions of the momentum spectra (Fig. 26), is 1.6 ± 0.6 ± 0.3 per B decay. This agrees with the expected rate for pure $b \to cW^-$, but because of the large errors, it is also marginally consistent with the pure $b \to uW^-$ hypothesis.

Another charm indicator is the lepton momentum spectrum (Fig. 23). The electrons and muons above 1 GeV/c come mainly from the first generation $b \to c\ell^-\bar{\nu}$ or $b \to u\ell^-\bar{\nu}$. The end point of the spectrum depends on the effective mass of the recoiling products of the c or u quark and the spectator quark. Since the recoiling mass is likely to be smaller in the $b \to u$ case, we expect the lepton spectrum to have a higher end point. Until we know more about how the recoiling mass is formed though, we cannot make reliable predictions. The CUSB and CLEO data are consistent so far with either $b \to c\ell^-\bar{\nu}$ or $b \to u\ell^-\bar{\nu}$, probably prefering $b \to c\ell^-\bar{\nu}$.

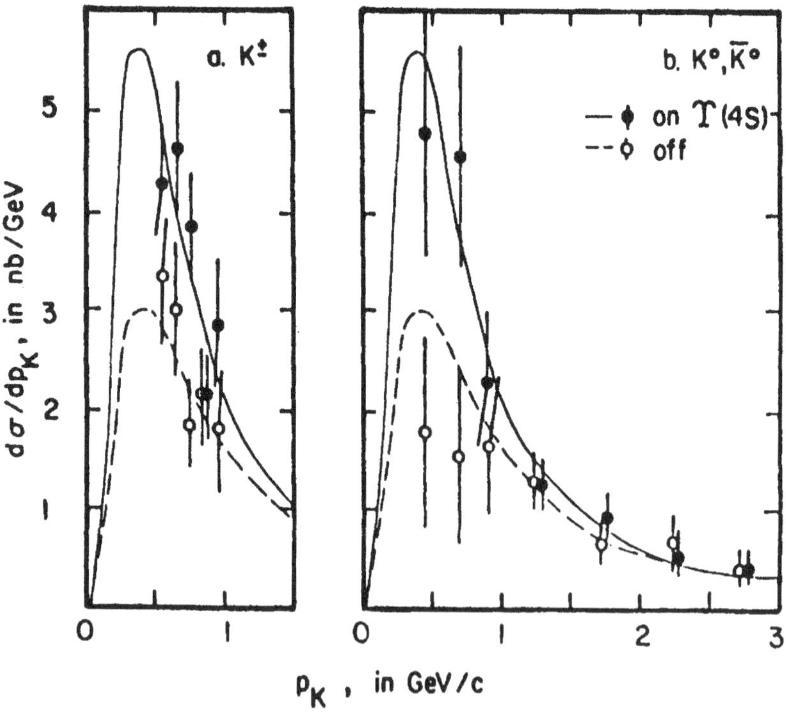

Fig.26. Charged and neutral kaon momentum spectra observed by CLEO on the ϒ(4S) and off resonance. The curves are from a Monte Carlo simulation assuming pure $b \to cW^-$.

Low momentum leptons come from later stages in the decay chain, that is, from D decays if the b decays to c. Again there are large uncertainties in the predictions; for instance, charged and neutral D's have very different leptonic branching ratios, and we do not know how many of each should be produced in B decays. Fig. 23 shows typical predictions. The contrast between $b \to c\ell^-\bar{\nu}$ and $b \to u\ell^-\bar{\nu}$ makes this potentially the best indicator of relative rates. The decay through charm is certainly favored by the CLEO electron data (Fig. 23), but the model dependence prevents us from quoting a value for $(b \to uW^-)/(b \to cW^-)$.

F. Exclusive B Decays

Completely reconstructed final states will give us the B mass as well as more detailed information on decay modes. In CLEO it will require more events and better momentum resolution. More events will

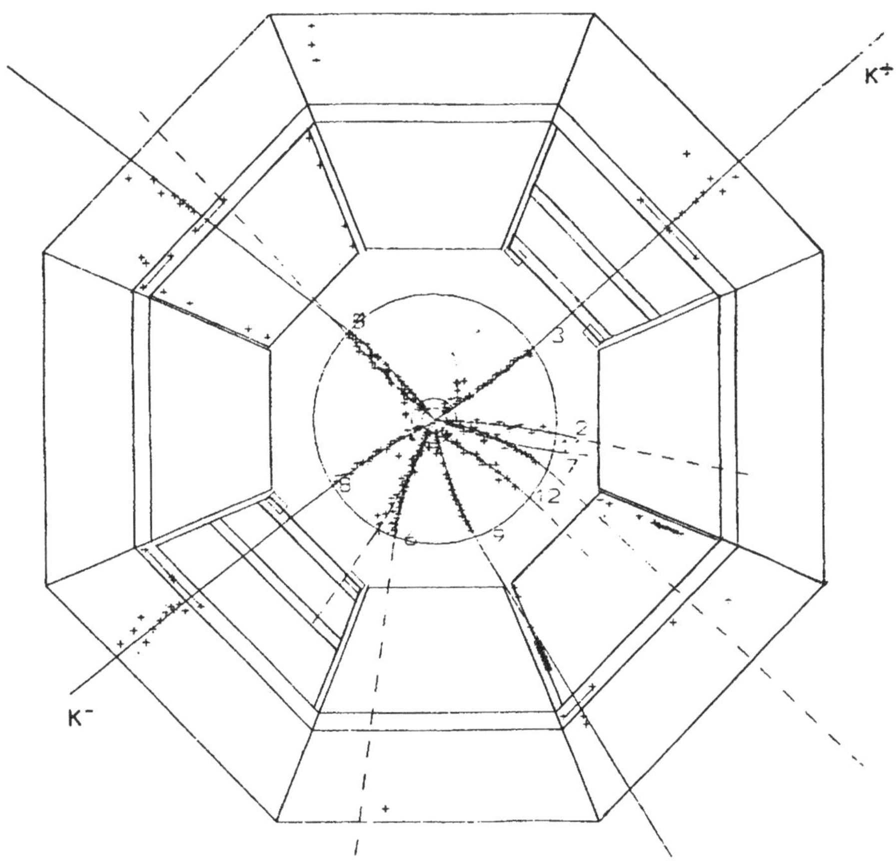

Fig. 27 Display of a CLEO event consistent with the hypothesis $e^+e^- \to B^0\bar{B}^0$, $B^0 \to D^-\pi^+\pi^+\pi^-$, $D^- \to K^+\pi^-\pi^-$, $\bar{B}^0 \to D^0\pi^+\pi^-$, $D^0 \to K^-\pi^+\pi^+\pi^-$.

come with time and increased luminosity; the resolution will be improved soon when the conventional solenoid is replaced with a high-field superconducting solenoid. Meanwhile, we have already some candidates for completely reconstructed $B\bar{B}$ events. Fig. 27 shows one, a possible

$$e^+e^- \rightarrow B^o(\rightarrow D^-(\rightarrow K^+\pi^-\pi^-)\pi^+\pi^-\pi^-)\bar{B}^o(\rightarrow D^o(\rightarrow K^-\pi^+\pi^+\pi^-)\pi^+\pi^-).$$

REFERENCES

1. S.W. Herb et al, Phys. Rev. Lett. 39 , 252 (1977)
2. C. Berger et al (PLUTO), Phys. Lett. 76B , 243 (1978)
3. C.W. Darden et al (DASP2), Phys. Lett. 76B , 246 (1978), and 78B , 364 (1978)
4. J.K. Bienlein et al, Phys. Lett. 78B , 360 (1978)
   P. Boch et al, Z. Phys. C6 , 125 (1980)
5. D. Andrews et al (CLEO), Phys. Rev. Lett. 44 , 1108 (1980)
6. T. Böhringer et al (CUSB), Phys. Rev. Lett. 44 , 1111 (1980)
7. D. Andrews et al (CLEO), Phys. Rev. Lett. 45 , 219 (1980)
8. G. Finocchiaro et al (CUSB), Phys. Rev. Lett. 45 , 222 (1980)
9. M. Oreglia, preprint SLAC-PUB-2529 (1980);
   T.H. Burnett, in High Energy Physics - 1980 (L. Durand and L.G. Pondrom ed., American Institute of Physics, New York, 1981) p. 664.
10. A.A. Zholentz et al (VEPP4), Phys. Lett. 96B , 214 (1980)
11. W. Buchmüller, G. Grunberg and S.-H.H. Tye, Phys. Rev. Lett. 45 103 (1980)
12. A. Martin, High Energy Physics - 1980 (L. Durand and L.G. Pondrom ed., American Institute of Physics, New York, 1981) p. 715.
13. T. Appelquist and H.D. Politzer, Phys. Rev. Lett. 34 , 43 (1975)
14. E. Eichten et al, Phys. Rev. D17 , 3090 (1978) and D21 , 203 (1980)
15. G. Bhanot and S. Rudaz, Phys. Lett. 78B , 119 (1978)
16. H. Krasemann and S. Ono, Nucl. Phys. B154 , 283 (1979)
17. J.L. Richardson, Phys. Lett. 82B , 272 (1979)
18. C. Quigg and J.L. Rosner, Phys. Lett. 71B , 153 (1977)
19. R. Van Royen and V.F. Weisskopf, Nuovo Cimento 50A , 617 (1967)

20. R. Barbieri et al, Phys. Lett. 57B , 455 (1975);
W. Celmaster, Phys. Rev. D19 , 1517 (1979)

21. C. Berger et al (PLUTO), preprint DESY 80/117

22. K. Koller and T. Walsh, Nucl. Phys. B140 , 449 (1978)

23. K.W. Chen et al (LENA), preprint DESY LENA 81/008

24. C. Berger et al (PLUTO), Z. Phys. C1 , 343 (1979)

25. C. Berger et al (PLUTO), Phys. Lett. 93B , 497 (1980)

26. H. Albrecht et al (DASP2), Phys. Lett. 93B , 500 (1980)

27. B. Niczyporuk et al (LENA), Phys. Rev. Lett. 46 , 92 (1981)

28. B. Niczyporuk et al (LENA), Phys. Lett. 100B , 95 (1981)

29. J.J. Mueller et al (CLEO), Phys. Rev. Lett. 46 , 1181 (1981)

30. G. Mageras et al (CUSB), Phys. Rev. Lett. 46 , 1115 (1981)

31. T. Appelquist and H.D. Politzer, Phys. Rev. D12 , 1404 (1975)
R. Barbieri, R. Gatto and R. Kögerler, Phys. Lett. 60B , 183 (1976)
R. Barbieri, R. Gatto and E. Remiddi, Phys. Lett. 61B , 465 (1976)

32. P. Mackenzie and G.P. LePage, Phys. Rev. Lett. (in press)

33. T.-M. Yan, Phys. Rev. D22 , 21 (1980)
K. Gottfried, Phys. Rev. Lett. 40 , 598 (1978)

34. E. Eichten, Harvard preprint HUTP-80/A027 (1980)

35. S. Ono, Aachen preprint PITHA 80/2 (1980)

36. J.P. Leveille, U. Mich. preprint UMHE 81-18

37. L.J. Spencer et al (CUSB), Phys. Rev. Lett. (in press)

38. C. Bebek et al (CLEO), Phys. Rev. Lett. 46 , 84 (1981)

39. K. Chadwick et al (CLEO), Phys. Rev. Lett. 46 , 88 (1981)

40. H. Georgi and S.L. Glashow, Nucl. Phys. B167 , 173 (1980)

41. E. Derman, Phys. Rev. D19 , 133 (1979)

42. V. Barger and S. Pakvasa, Phys. Lett. 81B , 195 (1979)

43. M. Peskin, Cornell internal report CBX 80-65 .

44. M. Peskin and S.-H.H. Tye, Cornell preprint CLNS 81-485, p. 405;
V. Barger, W.Y. Keung and R.J.N. Phillips, Wisconsin preprint DOE-ER/00081-195 (1981)

46. M. Kobayashi and T. Maskawa, Prog. Theor. Phys. 49 , 652 (1973)

47. R. Schrock, S.B. Treiman and L.L. Wang, Phys. Rev. Lett. 42 , 1589 (1979)

48. S.-H.H. Tye, Cornell internal report CBX 81-9
S. Pakvasa, S.F. Tuan and J.J. Sakurai, Phys. Rev. D23 , 2799 (1981)

49. W. Bartel et al (JADE), Z. Phys. C6 , 295 (1980) and private communication.

e^+e^- PHYSICS AT VERY LARGE ENERGIES

M. Davier

LAL
Université Paris-Sud
91405 Orsay, France

0. ABSTRACT

In these lectures we present the motivations to study e^+e^- annihilation at energies $\gtrsim$ 100 GeV. The occurrence of large effects from the interference between weak neutral and electromagnetic currents is a central part of our discussion. Z^o decays provide a plentiful source for new fermion types - quarks or leptons - or even unconventional point-like objects. A detailed study of the structure of the electroweak theory can be performed including tests of the gauge couplings. Finally symmetry breaking can be investigated through the Higgs spectrum or particles related to specific dynamical breaking.

I. INTRODUCTION

1. "Low Energy" e^+e^- phenomenology

Essentially all of e^+e^- annihilation physics studied at SPEAR, DORIS, PETRA and PEP can be described by the fundamental process of pair production of point-like fermions (Fig. 1a), when f stands for any fermion : charged lepton (e^-,μ^-,τ^-) or any quark q (the electron case is further complicated by the existence of a t channel diagram which plays an important role in practice in the forward direction).

When $4m_f^2 \ll s \ll M_Z^2$ we have

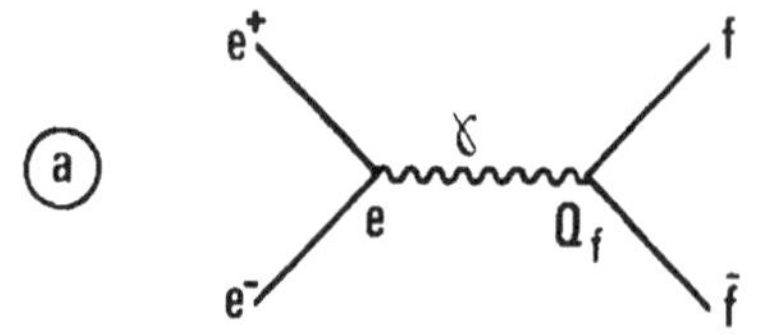

Fig. 1a. e^+e^- annihilation into a pair of point-like fermions

$$\sigma_f = \sigma(e^+e^- \to f\bar{f}) = Q_f^2\sigma_{pt} = Q_f^2 \frac{4\pi\alpha^2}{3s} = Q_f^2 \frac{87\ \text{nb}}{s(\text{GeV}^2)}$$

where Q_f is the fermion charge expressed in units of e.

The total annihilation cross-section can then be scaled from the standard point cross-section:

$$R_T = \frac{\sum_{f\neq e} \sigma_f}{\sigma_{pt}} = N_\ell - 1 + 3 \sum_i Q_i^2$$

where each quark contributes with 3 colours.

This picture has been now thoroughly tested. As far as quarks are concerned, the final state consists of hadron jets (Fig. 1b)

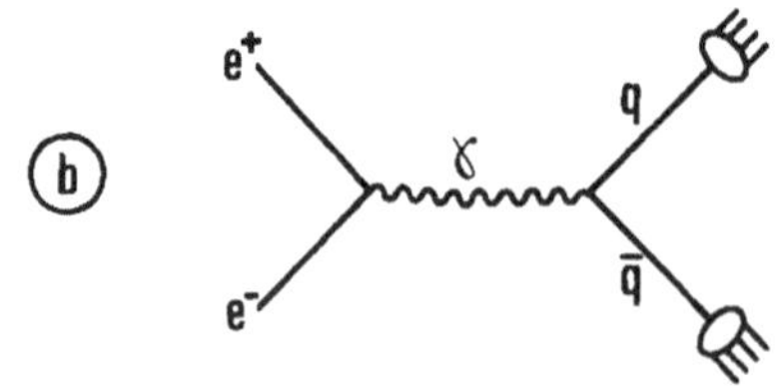

Fig. 1b. $e^+e^- \to q\bar{q} \to 2$ jets

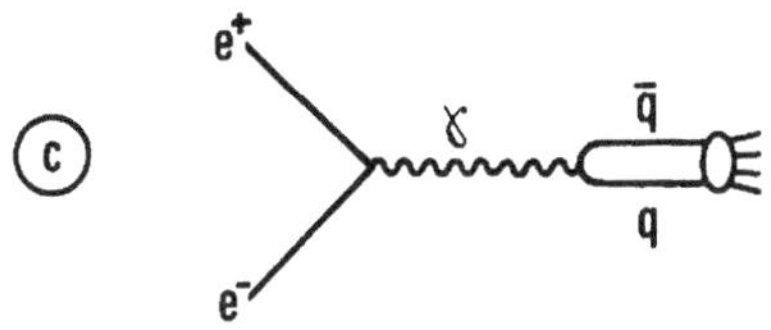

Fig. 1c. $e^+e^- \rightarrow q\bar{q}$ bound state

although sometimes a $q\bar{q}$ bound state can be excited (Fig. 1c).

At short distances, gluons can be hard-radiated from quarks, giving 3-jet final states (Fig. 1d).

Examples of 2 and 3-jet production can be seen in Fig. 2.

So far e^+e^- annihilation has been the most efficient way to produce and analyze new fermions: the most striking example is the τ lepton which has still to be seen in another interaction some six years after its discovery at SPEAR. Its production at PETRA is seen in a straightforward manner since its decay yields low multiplicity events (Fig. 3).

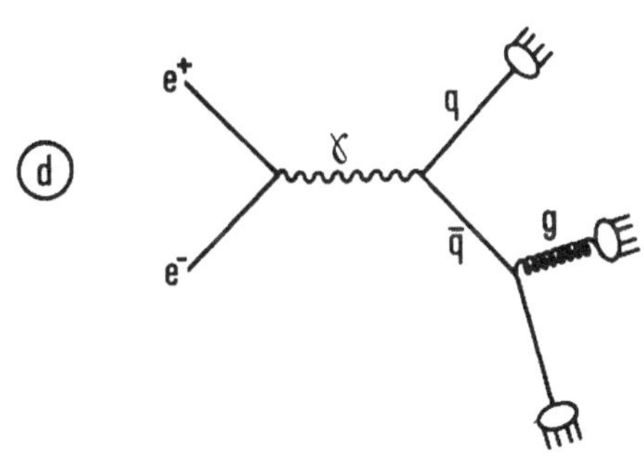

Fig. 1d. Gluon emission in $e^+e^- \rightarrow q\bar{q}g$

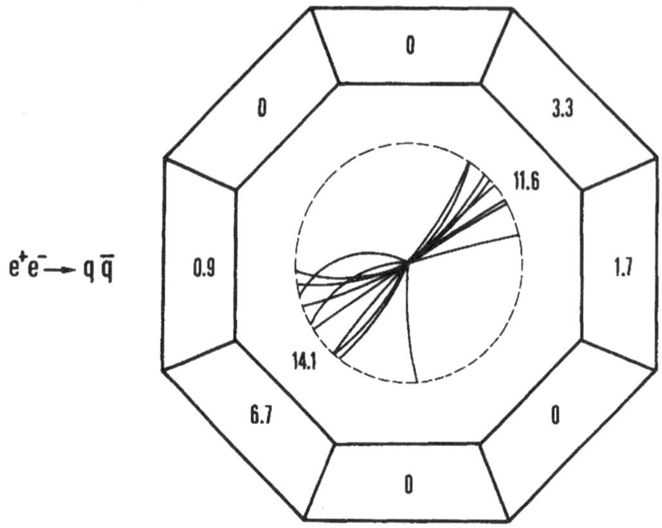

Fig. 2a. Example of 2-jet event in the CELLO detector at PETRA. Neutral energy seen in calorimeter follows the jet pattern of charged particles.

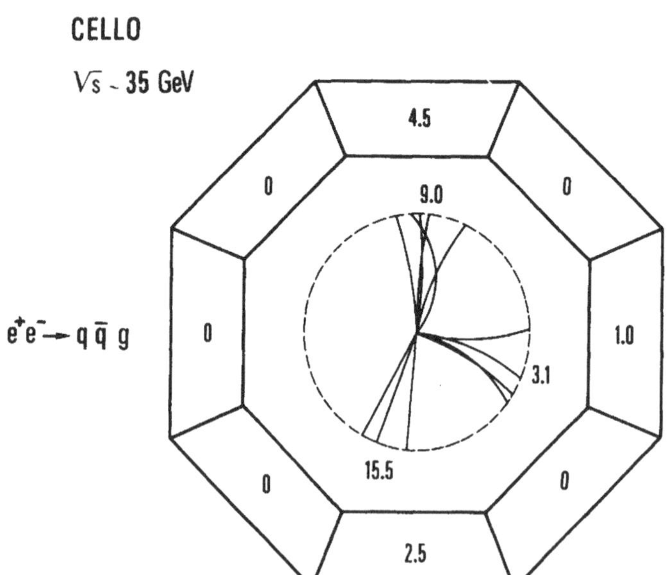

Fig. 2b. Example of 3-jet event in the CELLO detector.

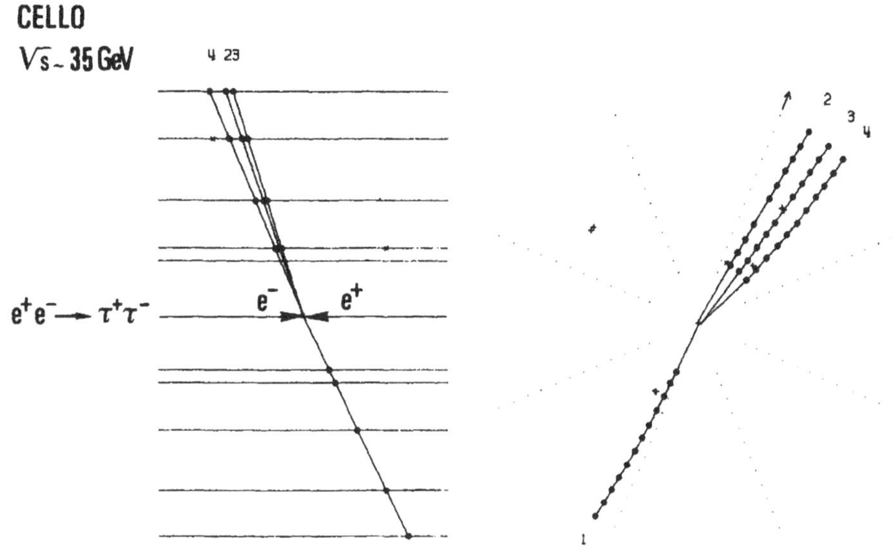

Fig. 3. Event of $e^+e^- \to \tau^+\tau^-$ process observed in the CELLO detector

Finally results presented this summer from PEP and PETRA give us an early glimpse of the weak-electromagnetic interference [1]. To study in detail this important effect, higher energies are needed: it will be our next purpose to determine how much higher we need to go.

2. Energy Scale to Study the Weak Interaction

In general both the electromagnetic current and weak neutral currents will contribute to the most general annihilation process (Fig. 4).

$$\sigma_f = |\gamma + Z|^2$$

Our present experimental knowledge of the weak amplitude from $e^+e^- \to \mu^+\mu^-$ and neutrinos experiments [2], is consistent with a Fermi interaction i.e. infinite M_Z. In that case:

$$|Z|^2 \sim s \qquad \text{while} \qquad |\gamma|^2 \sim \frac{1}{s}$$

At what energy do electromagnetic and weak amplitudes have comparable strength ? As an example let us consider the process $e^+e^- \to \mu^+\mu^-$: Fig. 5 shows that it occurs for $\sqrt{s} \sim 150$ GeV. This provides a natural scale to study experimentally the weak interaction: either the weak force exceeds the electromagnetic one or something else happens. Now in standard SU(2) x U(1) [3] we know what should happen: the energy range discussed encompasses vector boson production:

$$e^+e^- \to Z^o \qquad \sqrt{s} \sim 90 \text{ GeV}$$

$$e^+e^- \to W^+W^- \qquad \sqrt{s} \gtrsim 160 \text{ GeV}$$

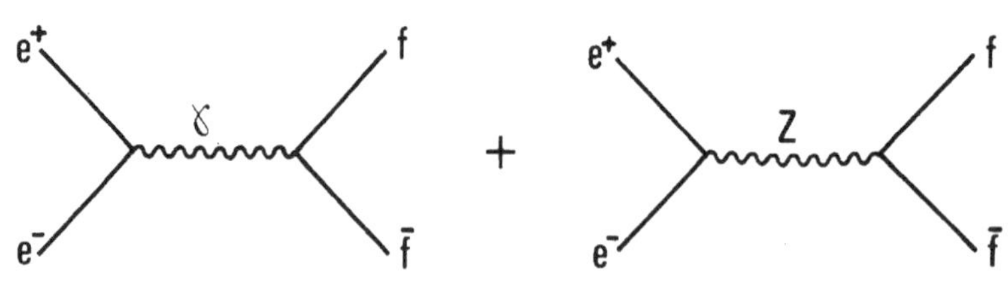

Fig. 4. Electromagnetic and neutral weak currents in e^+e^- annihilation

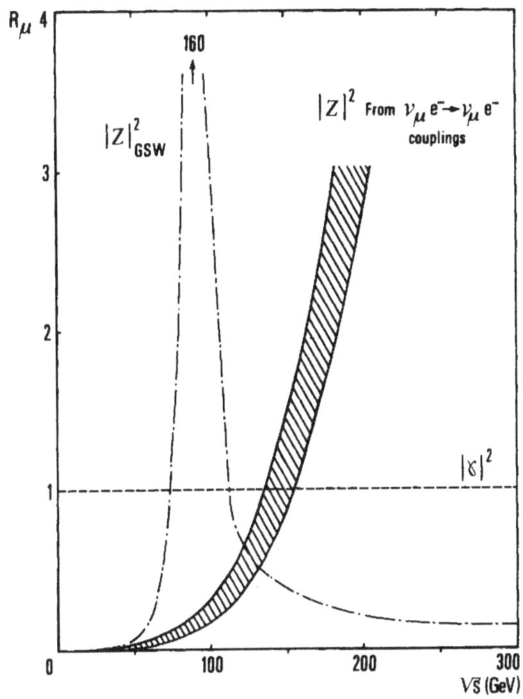

Fig. 5. The ratio $R_\mu = \sigma(e^+e^- \to \mu^+\mu^-)/\sigma_{pt}$ from the $|\gamma|^2$ and $|Z|^2$ terms, the latter being computed from ν-e measurements [2] and assuming a point interaction. The dashed curved corresponds to $|Z|^2$ in the standard GSW model [3]

3. How to Get There ?

The classic way to achieve large s has been e^+e^- storage rings. However synchroton radiation has to be overcome by RF power: for a beam energy E_B and a bending radius ρ , the power radiated is

$$P \sim \frac{E_B^4}{\rho}$$

When E_B increases a cost optimization [4] yields a scaling law

$$\rho \sim s$$

$$\text{cost} \sim s$$

Extrapolating from PETRA with $E_B \sim 20$ GeV and $\rho \sim 0.2$ km we then expect that a 100 GeV ring will require $\rho \sim 5$ km. This is indeed how the LEP project [4] at CERN is optimized (Fig. 6). A first phase of the construction, with reduced RF power, will cover up to 50 GeV in 1987, going to the full 100 GeV a few years later. It is still possible to consider superconducting RF cavities to increase E_B even further in a third phase ($\sim$ 130 GeV ?). Extrapolating from the operation of present-day rings like PETRA or PEP leads one to expect peak luminosities of around 10^{31} $cm^{-2}s^{-1}$ or slightly higher. LEP Phase I will cover the expected Z^o peak with a wealth of physics we shall describe below.

Competition to LEP could occur from two US projects: a 2 x 50 GeV ring at Cornell [5] using superconducting cavities and the single-pass collider [6] at SLAC (Fig. 7). This last case is an interesting development where the SLAC $e^\pm$ beams boosted to 50 GeV are brought in two arcs and clashed once. A luminosity in excess of 10^{30} $cm^{-2}s^{-1}$ is obtained by extreme focussing of the beams down to a $\sim 1\mu$ size. Since polarized electrons can be accelerated in the linac, it is therefore easy in this set-up to collide left or right-handed electrons with unpolarized positrons.

As a final remark in this introduction, we consider the rate for e^+e^- annihilation into hadrons in the minimal Glashow-Salam-Weinberg (GSW) model with $\sqrt{s}$ in the range of a few hundred GeV (Fig. 8). This rate is not corrected for radiative effects (i.e. it represents the yield after radiative events have been removed) - which are large above the Z^o peak. Two points can be made :

(1) the rate is important at the Z^o peak;

(2) outside the Z^o resonance the rate is expected to be small and a luminosity of at least 10^{31} $cm^{-2}s^{-1}$ is required for efficient experimentation.

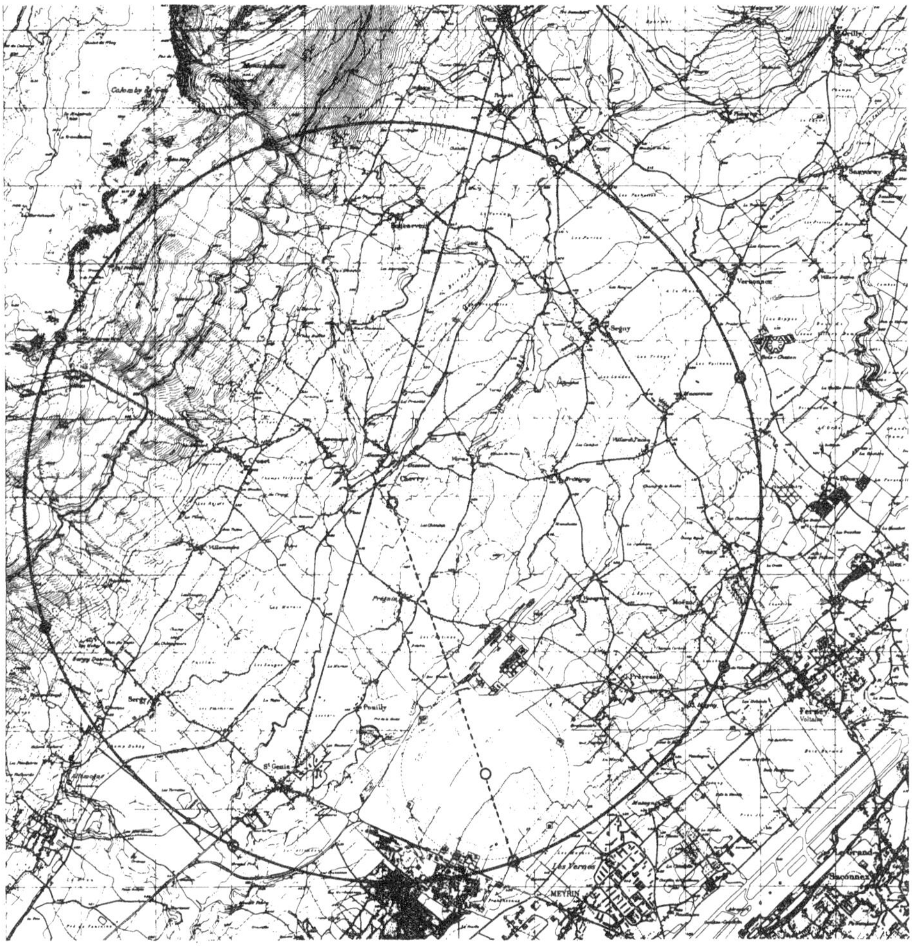

Fig. 6. Lay-out of the LEP project at CERN.

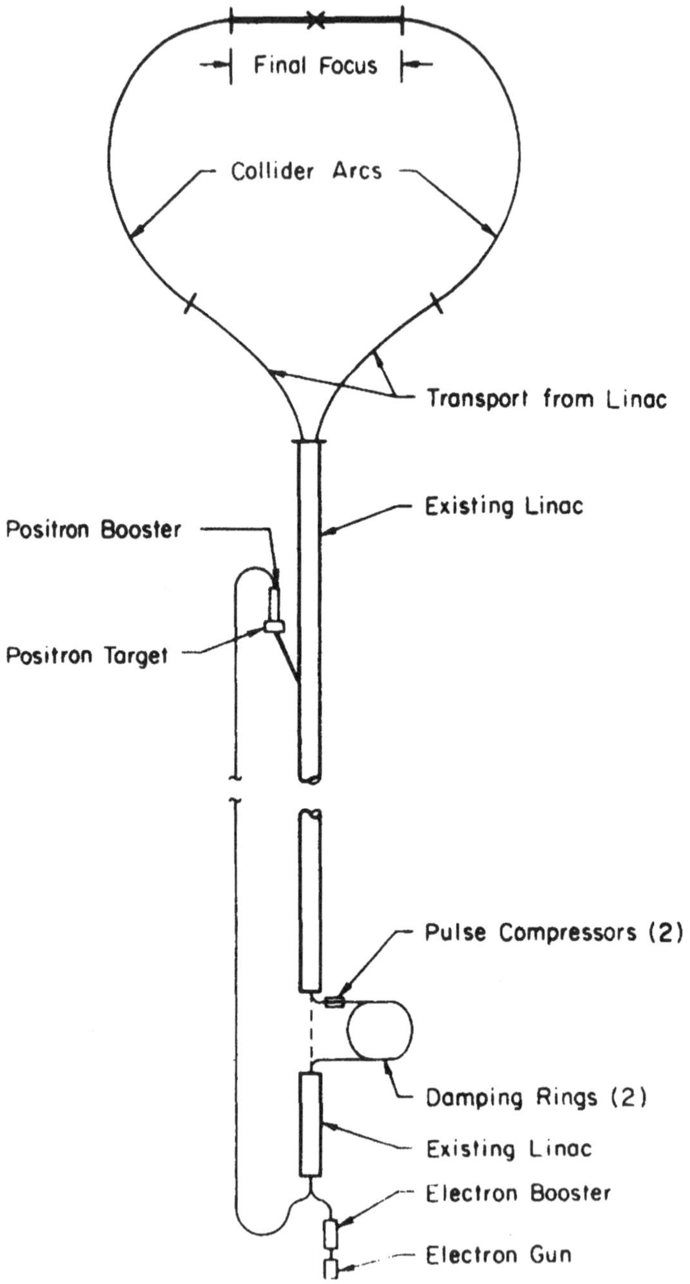

Fig. 7. Schematic lay-out of the SLC facility projected at SLAC.

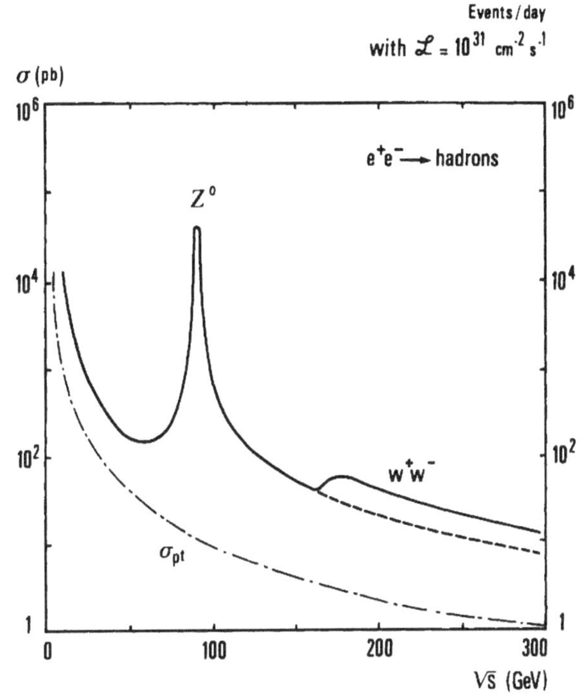

Fig. 8. Total e^+e^- annihilation rate up to $\sqrt{s} \sim 300$ GeV expected in the minimal GSW model after correcting for radiative effects

II. INTERFERENCE BETWEEN ELECTROMAGNETIC AND NEUTRAL WEAK CURRENTS

1. Neutral Weak Couplings

We assume that the elementary interaction between the neutral weak current and a fermion pair (Fig. 4) involves only vector and axial couplings:

$$\bar{f}\gamma_\mu(V_f + A_f\gamma_5)f$$

while the usual vector electromagnetic interaction is:

$$e\bar{f}\gamma_\mu f$$

In the minimal SU(2) x U(1) fermions are arranged in left-handed doublets (weak isospin) and right-handed singlets:

$$\begin{pmatrix} \nu_e \\ e^- \end{pmatrix}_L \quad \begin{pmatrix} \nu_\mu \\ \mu^- \end{pmatrix}_L$$

$$\begin{pmatrix} u \\ d \end{pmatrix}_L \quad \begin{pmatrix} c \\ s \end{pmatrix}_L \quad \ldots$$

$$e^-_R \quad \mu^-_R$$

$$u_R \quad c_R$$

$$d_R \quad s_R$$

Boson masses are generated by the Higgs-Kibble mechanism which implies, in the minimal scheme, the existence of one neutral scalar boson (the Higgs boson H^0). The following relations are then obtained:

$$\mathcal{L}_{Zf\bar{f}} = \frac{e}{2\sin^2\theta_W} \bar{f}\gamma_\mu(v_f - a_f\gamma_5)fZ^\mu$$

$$M_W = \frac{1}{4\sqrt{g}} \frac{1}{\sin\theta_W}$$

$$M_Z = \frac{M_W}{\cos\theta_W}$$

where

$$g = \frac{G_F}{2\sqrt{2}\, e^2} = 4.5\ 10^{-5}\ \mathrm{GeV}^{-2}$$

For $\sin^2\theta_W = .22$ we get $M_W = 79.7$ GeV and $M_Z = 90.3$ GeV.

The neutral couplings v_f and a_f are obtained from the charge Q_f and weak isospin of the corresponding fermions by the relations:

$$a_f = 2(I_3^L - I_3^R)_f$$

$$v_f = 2(I_3^L + I_3^R)_f - 4Q_f^2 \sin^2\theta_W$$

yielding the following expressions depending only on $\sin^2\theta_W$ values

f	a_f	v_f
ν_e	1	1
e^-	-1	$-1 + 4\sin^2\theta_W$
u	1	$1 - \frac{8}{3}\sin^2\theta_W$
d	-1	$-1 + \frac{4}{3}\sin^2\theta_W$

The same couplings should apply to all fermion families.

If $\sin^2\theta_W$ is equal to 1/4 the vector coupling of charged leptons v_ℓ vanishes. Since this is close to the observed value [2)], we expect a small value for v_ℓ which has important consequences for e^+e^- annihilation since all vector weak amplitude interfering with the electromagnetic current will be proportional to the small number v_e. More explicitly , for $\sin^2\theta_W = 0.22$

$$v_e = -\ 0.12$$

$v_u = 0.41$

$v_d = -0.71$

2. Observables of the Reaction $e^+e^- \to f\bar{f}$

a. Angular distribution. In the center of mass with the following definition of the scattering angle θ shown in Fig. 9, we can write the differential cross-section for unpolarized electrons and summing over fermion spins:

$$\frac{d\sigma_f}{d\cos\theta} = \frac{3}{8}\sigma_{pt}\,\{Q_f^2(1+\cos^2\theta) - \frac{2Q_f\, gsM_Z^2(s-M_Z^2)}{(s-M_Z^2)^2+M_Z^2\Gamma_Z^2}\,.$$

$$[v_e v_f(1+\cos^2\theta) + 2a_e a_f\cos\theta] + \frac{g^2s^2M_Z^4}{(s-M_Z^2)^2+M_Z^2\Gamma_Z^2}\,.$$

$$[(a_e^2+v_e^2)(a_f^2+v_f^2)(1+\cos^2\theta) + 8a_e v_e a_f v_f\cos\theta]\}$$

where the three terms correspond to $|\gamma|^2$, interference and $|Z|^2$ terms.

For $|s - M_Z^2| \ll M_Z\Gamma_Z$ the Z width can be neglected (we shall see below that Γ_Z is of the order of a few GeV) and it is useful to introduce a dimensionless variable

$$\chi = \frac{gsM_Z^2}{s - M_Z^2}$$

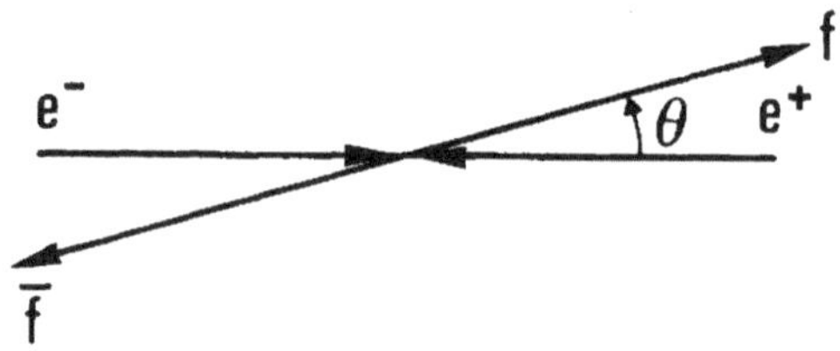

Fig. 9. Definition of scattering angle θ in $e^+e^- \to f\bar{f}$

$$\frac{d\sigma_f}{d\cos\theta} = \frac{3}{8}\sigma_{pt}\{Q_f^2(1+\cos^2\theta) - 2Q_f\chi[v_e v_f(1+\cos^2\theta)$$

$$+ 2a_e a_f \cos\theta] + \chi^2 [(a_e^2 + v_e^2)(a_f^2 + v_f^2)(1+\cos^2\theta)$$

$$+ 8a_e v_e a_f v_f \cos\theta]\}$$

The interference is responsible for terms proportional to $\cos\theta$ which are visible in the angular distribution. More conveniently one can define an angular asymmetry:

$$A_f(\theta) = \frac{d\sigma_f(\theta) - d\sigma_f(\pi - \theta)}{d\sigma_f(\theta) + d\sigma_f(\pi - \theta)}$$

$$= \frac{4\chi a_e a_f(-Q_f + 2\chi v_e v_f)\cos\theta}{[Q_f^2 - 2Q_f\chi v_e v_f + \chi^2(a_e^2 + v_e^2)(a_f^2 + v_f^2)]\,(1+\cos^2\theta)}$$

More globally there is a forwark-backwark asymmetry:

$$\langle A_f\rangle = \frac{\int_o^1 d\sigma_f - \int_{-1}^{o} d\sigma_f}{\int_{-1}^{1} d\sigma_f} = \frac{N_F - N_B}{N_F + N_B}$$

$$= \frac{\frac{3}{2}\chi a_e a_f(-Q_f + 2\chi v_e v_f)}{Q_f^2 - 2Q_f\chi v_e v_f + \chi^2(a_e^2 + v_e^2)(a_f^2 + v_f^2)}$$

It is worth noticing that such an asymmetry is not a violation of parity: it only comes about because of the interference between waves of opposite parities. In fact such an asymmetry can be generated by the electromagnetic interaction alone through higher orders: it is necessary to correct for this effect which is completely calculable to reach the weak-electromagnetic interference.

Due to the effect of χ, $\langle A_f\rangle$ strongly varies near the Z pole (Fig. 10) and follows approximately the real part of the Z Breit-Wigner amplitude since v_e is small. Large asymmetries($|\langle A_f\rangle| \lesssim 3/4$) produce big effects in the angular distribution (Fig. 11).

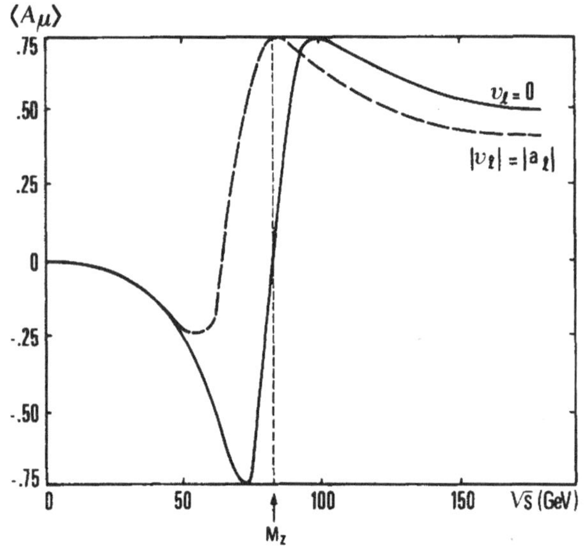

Fig. 10. Forward-backward asymmetry $\langle A_\mu \rangle$ for $e^+e^- \to \mu^+\mu^-$

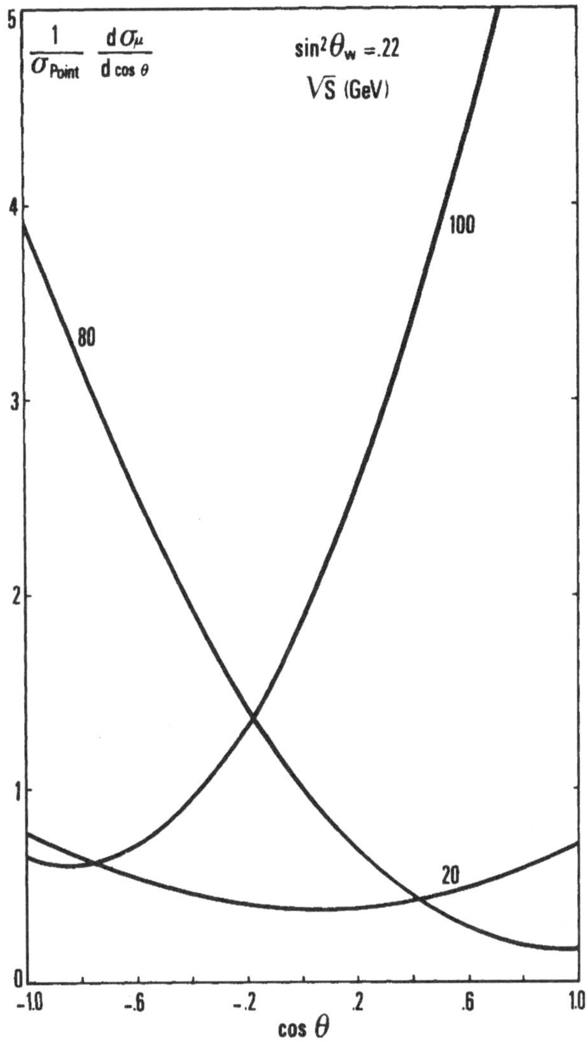

Fig. 11. Angular distribution of $e^+e^- \to \mu^+\mu^-$ at different center-of-mass energies

b. The total cross-section. Upon integration on cos θ, the axial terms disappear and we are left with:

$$R_f = \frac{\sigma_f}{\sigma_{pt}} = Q_f^2 - 2sQ_f v_e v_f g \frac{M_Z^2(s - M_Z^2)}{(s - M_Z^2)^2 + M_Z^2\Gamma_Z^2}$$

$$+ s^2 g^2 (a_e^2 + v_e^2)(a_f^2 + v_f^2) \frac{M_Z^4}{(s - M_Z^2)^2 + M_Z^2\Gamma_Z^2}$$

$$\underset{s \neq M_Z^2}{\simeq} Q_f^2 - 2Q_f v_e v_f \chi + (a_e^2 + v_e^2)(a_f^2 + v_f^2)\chi^2$$

Since v_e is small, the interference is small (Fig. 12).

c. Polarizations

α) unpolarized $e^{\pm}$ beams : final fermion polarization

$$\frac{d\sigma_f}{d\cos\theta}(\lambda_f) = \sigma_1 + \lambda_f \sigma_2$$

$$\lambda_f = \text{fermion helicity}$$

where

$$\sigma_1 = \frac{1}{2} \Sigma \left(\frac{d\sigma_f}{d\cos\theta}\right)(\lambda_f)$$

$$\sigma_2 = \frac{3}{8} \sigma_{pt} \chi \{a_f[v_e a_f(1 + \cos^2\theta) + 2a_e v_f \cos\theta] -$$

$$- \chi[v_f a_f(a_e^2 + v_e^2)(1 + \cos^2\theta) + 2a_e v_e(a_f^2 + v_f^2)\cos\theta]\}$$

The fermion polarization is given by:

$$P_f(\theta) = \langle\lambda_f\rangle = \frac{\sigma_2}{\sigma_1}$$

with a cosθ average value:

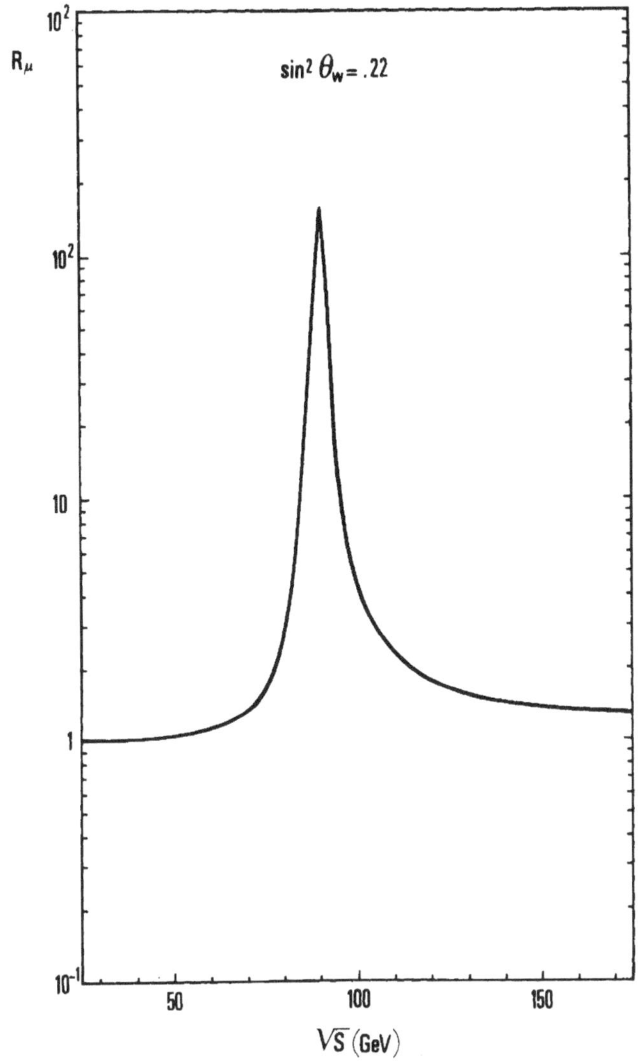

Fig. 12. The ratio R_μ for $\sin^2\theta_W = 0.22$

$$\langle P_f \rangle = \frac{\int_{-1}^{1} \sigma_2 d\cos\theta}{\int_{-1}^{1} \sigma_1 d\cos\theta}$$

$$= \frac{2\chi a_f [Q_f v_e - \chi v_f (a_e^2 + v_e^2)]}{Q_f^2 - 2Q_f v_e v_f \chi + (a_e^2 + v_e^2)(a_f^2 + v_f^2)\chi^2}$$

As an example, $\langle P_\mu \rangle$ is shown in Fig. 13.

β) Polarized $e^\pm$ beams

We can write down the differential cross-section for $e^+e^- \rightarrow f\bar{f}$ with incident electrons (helicity λ_-) and positrons (helicity λ_+), final fermions helicities being summed over,

$$\left(\frac{d\sigma_f}{d\cos\theta}\right)_{\lambda_+,\lambda_-} = (1 - \lambda_+\lambda_-)\sigma_1 + (\lambda_- - \lambda_+)\sigma_3$$

with

$$\sigma_3 = \frac{3}{8}\sigma_{pt}\ \chi\{ Q_f[v_f a_e(1 + \cos^2\theta) + 2a_f v_e \cos\theta]$$

$$- \chi[v_e a_e(a_f^2 + v_f^2)(1 + \cos^2\theta) + 2a_f v_f(a_e^2 + v_e^2)\cos\theta]\}$$

We see that σ_3 can be obtained from σ_2 if we exchange e and f couplings: it means that $\sigma_3 = \sigma_2$ for the process $e^+e^- \rightarrow \ell^+\ell^-$ if charged leptons have universal couplings as in the SU(2) x U(1) model.

With longitudinally polarized beams it is therefore possible to measure the ratio σ_3/σ_1. If we note $(e_L^- e_R^+)$ the initial state with $(\lambda_- = -1,\ \lambda_+ = +1)$ we can define

$$H_f(\theta) = \frac{d\sigma_f(e_R^- e_L^+) - d\sigma_f(e_L^- e_R^+)}{d\sigma_f(e_R^- e_L^+) + d\sigma_f(e_L^- e_R^+)} = \frac{\sigma_3}{\sigma_1}$$

The states $(e_R^- e_R^+)$ and $(e_L^- e_L^+)$ do not contribute at high energies since they are suppressed by a factor m_e^2/s with respect to the "allowed" transitions.

Such measurements are possible in principle since beams can be

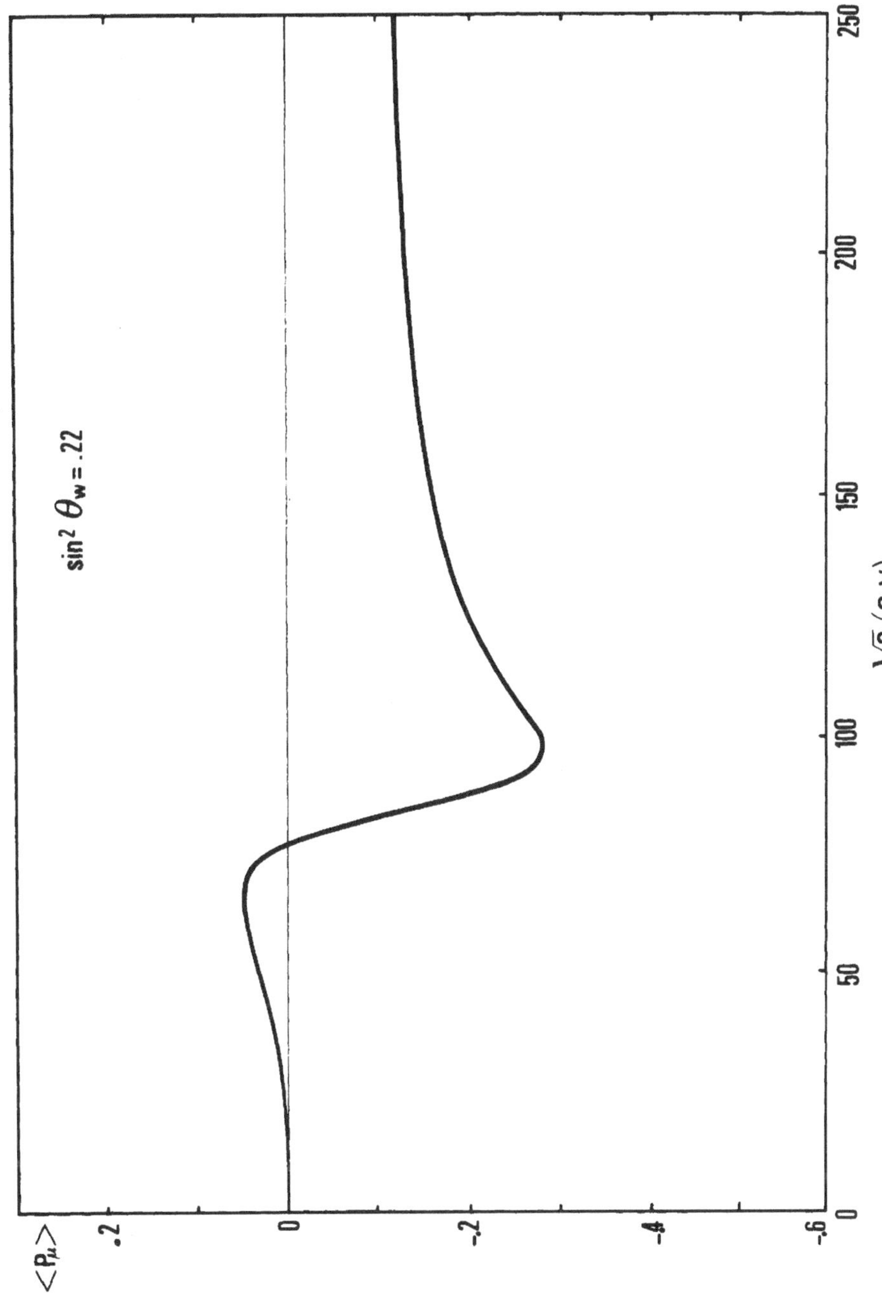

Fig. 13. The average muon polarization $<P_\mu>$ in $e^+e^- \rightarrow \mu^+\mu^-$

transversally polarized in a storage ring due to synchrotron radiation. The transverse polarization can be rotated into a longitudinal polarization with usual spin precession in a magnetic field. However since both e^+ and e^- beams move in opposite directions in the same magnetic field and since their initial transverse polarizations are opposite, the only configurations which are possible in single ring are the "forbidden" ones $e^-_R e^+_R$ or $e^-_L e^+_L$. It is nevertheless possible to turn this difficulty around by selectively depolarizing one beam, keeping the other intact. In that case, one still retains the possibility to measure σ_3:

$$H^e_f(\theta) = \frac{d\sigma_f(e^-_R) - d\sigma_f(e^-_L)}{d\sigma_f(e^-_R) + d\sigma_f(e^-_L)} = \frac{\sigma_3}{\sigma_1}$$

Integrating over solid angle, one gets the total rate asymmetry:

$$\langle H_f\rangle = \frac{2\chi a_e Q_f v_f - \chi v_e(a_f^2 + v_f^2)}{Q_f^2 - 2Q_f v_e v_f \chi + (a_e^2 + v_e^2)(a_f^2 + v_f^2)\chi^2}$$

The measurement of $\langle H_f\rangle$ is important for those cases where $\langle P_f\rangle$ is inaccessible (quarks). Fig. 14 shows $\langle H_\mu\rangle$, $\langle H_u\rangle$ and $\langle H_d\rangle$ as a function of energy. At $s = M_Z^2$ all such quantities are equal since only e couplings are involved.

γ) Final fermion observables with polarized electron beam

Forward-backward asymmetries and final fermion polarizations can be calculated for polarized electrons striking unpolarized positrons. Asymmetries can be quite large, especially for quarks [7] (Fig. 15). At the Z^0 peak

$$\langle A_f\rangle_L = -\frac{3}{2}\frac{a_f v_f}{a_f^2 + v_f^2} \qquad \text{for } e^-_L \text{ beam}$$

$$\langle A_f\rangle_R = \frac{3}{2}\frac{a_f v_f}{a_f^2 + v_f^2} \qquad \text{for } e^-_R \text{ beam}$$

while for an unpolarized beam

$$\langle A_f\rangle = \frac{3\, a_e v_e a_f v_f}{(a_e^2 + v_e^2)(a_f^2 + v_f^2)}$$

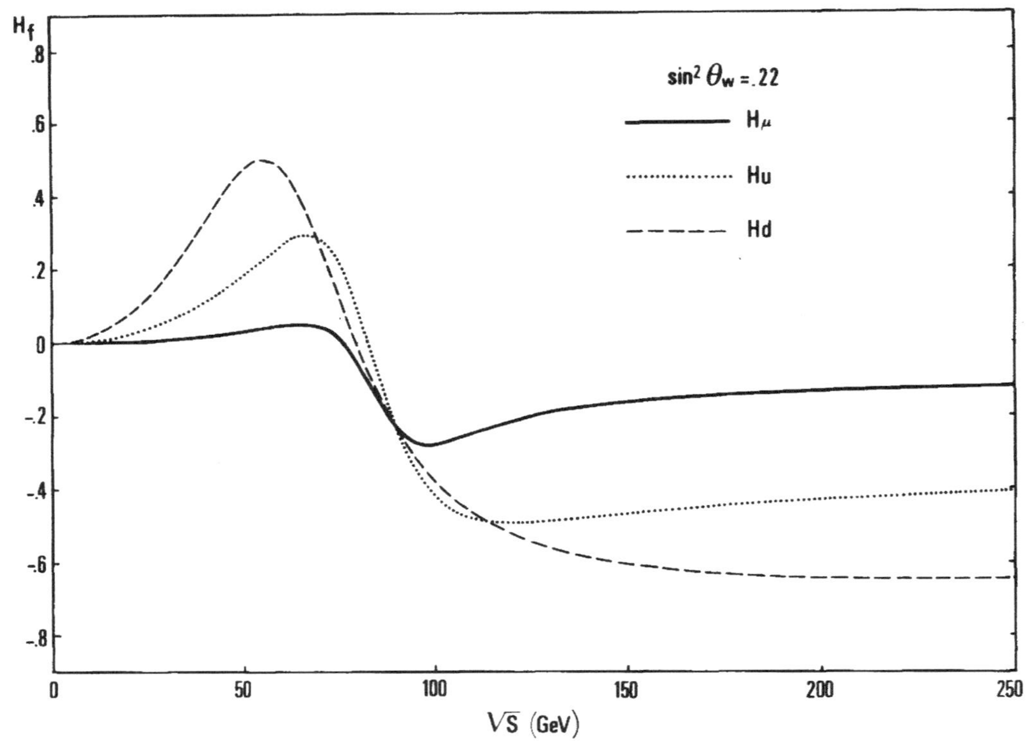

Fig. 14. Polarized-beam asymmetries for $e^+e^- \to \mu^+\mu^-$ (H_μ), $e^+e^- \to u\bar{u}$ (H_u), and $e^+e^- \to d\bar{d}$ (H_d) with $\sin^2\theta_W = 0.22$

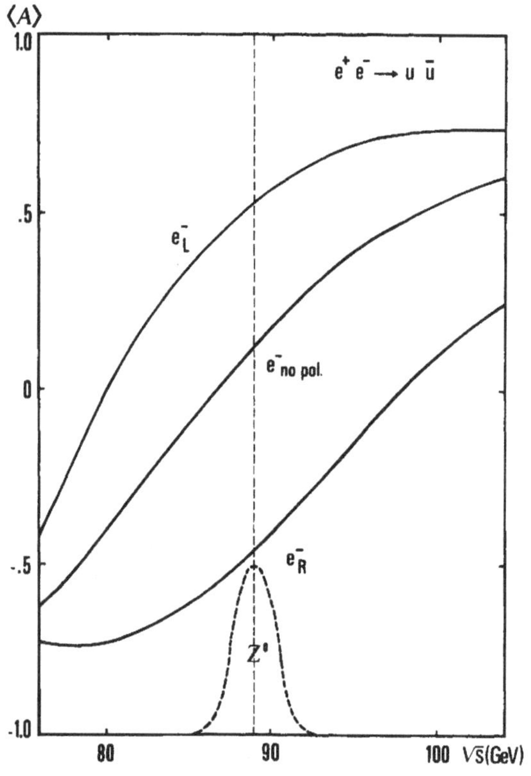

Fig. 15. Forward-backward asymmetries for polarized electrons colliding with unpolarized positrons [7]: (a) $e^+e^- \to u\bar{u}$

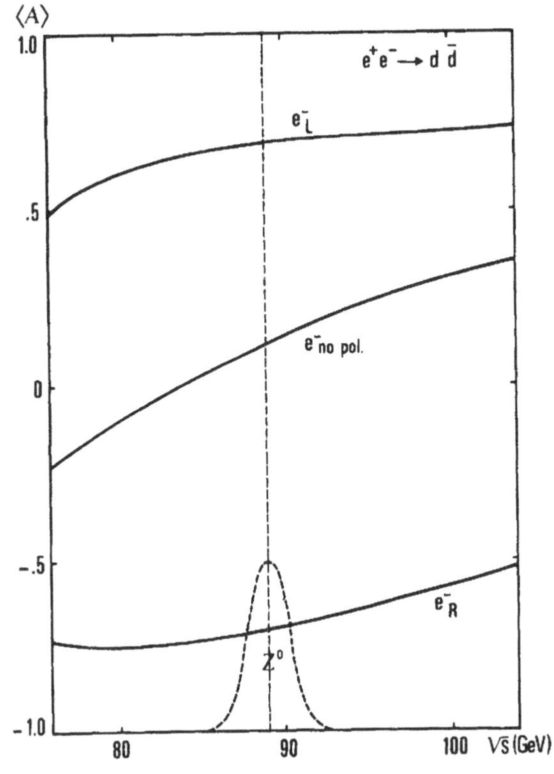

Fig. 15. Forward-backward asymmetries for polarized electrons colliding with unpolarized positrons [7]: (b) $e^+e^- \rightarrow d\bar{d}$

which is strongly reduced because of the small v_e value. Similar effects occur for fermion polarization (Fig. 16).

3. Two Particular Cases

a. $s \ll M_Z^2$ (PETRA, PEP). Here χ is small. For $\sqrt{s}$ = 35 GeV, which is so far about the highest energy used at PETRA with high luminosity $\chi \sim -6\ 10^{-2}$ and therefore $\chi^2 \ll |\chi|$.

$R_f \sim Q_f^2$ with deviations of at most 2%

$$\langle A_f \rangle = -\frac{3}{2}\frac{\chi a_e a_f}{Q_f}$$

giving

$$\langle A_\mu \rangle = -\ 9.4\ \%$$

$$\langle A_d \rangle = -\ 28\ \%$$

The lepton asymmetries are on the verge of becoming significant[1], while quark asymmetries are still not tackled.

$$\langle P_f \rangle = \frac{2\chi a_f v_e}{Q_f} \quad \text{giving } 1.5\ \% \text{ for muons}$$

Although interference effects are still small below $\sqrt{s} \sim 40$ GeV such measurements are helpful in constraining values for the couplings. In particular the vector lepton coupling comes out small as expected in the standard model.

The scheduled increase at PETRA to $\sqrt{s} \sim 45$ GeV should be precious to enhance our understanding of fermion weak couplings. It will however be hard to detect the effect of the Z^o propagator (the χ variation).

b. $s = M_Z^2$. Owing to the large event rate at the Z^o peak it is useful to examine what information can be obtained through interference for this particular energy.

$$R_f = \left(\frac{g M_Z^3}{\Gamma_Z}\right)^2 (a_e^2 + v_e^2)(a_f^2 + v_f^2)$$

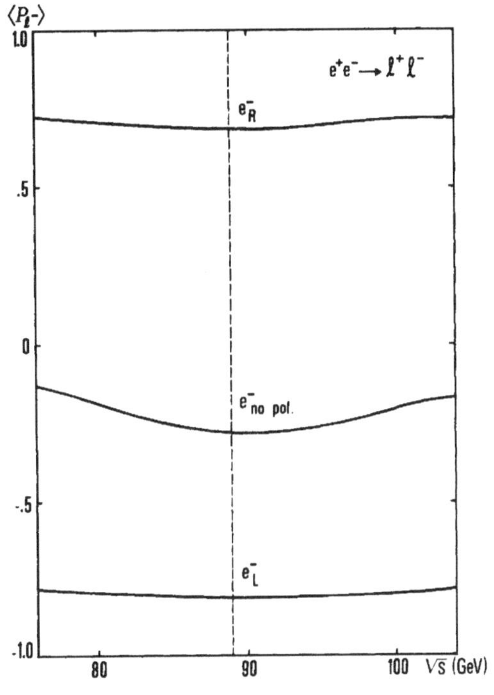

Fig. 16. Fermion polarization for polarized electrons colliding with unpolarized positrons [7].

$$\langle A_f \rangle = \frac{3\, a_e v_e a_f v_f}{(a_e^2 + v_e^2)(a_f^2 + v_f^2)}$$

$$\langle P_f \rangle = - \frac{2 a_f v_f}{a_f^2 + v_f^2}$$

$$\langle H_f \rangle = - \frac{2 a_e v_e}{a_e^2 + v_e^2} \quad \text{independent of f.}$$

Measurements of R_e (t channel exchange must be included), A_e and $\langle H_{tot} \rangle$ determine the electron couplings completely. It is interesting to notice that in principle polarized beams are not necessary if a final state fermion polarization can be measured; indeed

$$\frac{\langle A_f \rangle}{\langle P_f \rangle} = - \frac{3}{2} \frac{a_e v_e}{a_e^2 + v_e^2}$$

which is the same information as $\langle H_f \rangle$.

The obvious interest in these measurements is to see if fermion couplings are really universal and given by a common parameter $\sin^2\theta_W$, which determines lepton and quark couplings as well, regardless of their family assignment.

We are going to examine next some typical situations where this experimental program can be put on work [8].

4. Examples

a. $e^+e^- \to \mu^+\mu^-$. This is the simplest reaction allowing all measurements: R_μ, A_μ and P_μ. The latter implies to stop 45 GeV muons to observe their decay and consequently measure their polarization: this is awkward (or monstruous) for any θ, but it could be handled in the forward direction where the polarization is the highest

$$P_\mu(0) \sim 2\langle P_\mu \rangle$$

b. $e^+e^- \to \tau^+\tau^-$. The $\tau^+\tau^-$ final state is easy to identify at high energy because of its small multiplicity and characteristic leptonic channels. A clear advantage is that parity-violating τ decays are a good analyzer of the τ polarization. This can be achieved in practice using the leptonic decays [9)]

$$\tau^- \to \nu_\tau \ell^- \bar{\nu}_\ell$$

by observing the lepton spectrum shape which is sensitive to P_τ or the 2-body semi-leptonic decays [10)]

$$\tau^- \to \nu_\tau \pi^-$$

$$\tau^- \to \nu_\tau \rho^-$$

where the hadron energy spectrum measures P_τ.

More explicitly in the $\pi\nu_\tau$ case, the π angular distribution in the τ center of mass

$$\frac{dW^\pm}{d\cos\theta^*} \sim 1 \mp \lambda_{\tau^\pm} \cos\theta^*$$

translates into a π energy distribution in the laboratory frame since

$$E_\pi \simeq \frac{\sqrt{s}}{4} (1 + \cos\theta^*)$$

$$\frac{dN}{dE_{\pi^\pm}} \sim 1 \mp P_{\tau^\pm} \left(\frac{4E_{\pi^\pm}}{\sqrt{s}} - 1\right)$$

Therefore a measurement of the slope of the E_π spectrum (irrespective of the π charge) yields the τ polarization. Since at the Z^0 peak

$$\langle P_\tau \rangle = - \frac{2a_\tau v_\tau}{a_\tau^2 + v_\tau^2}$$

the τ couplings can be derived accurately when combined with R_τ

c. Experiments with polarized electrons. We mentioned previously that polarization of both electron and positron beams is unpractical and anyway irrelevant since only one polarized beam is needed.

In storage rings like LEP, transverse polarization is obtained thanks to a tiny spin dependence of synchroton radiation [11]. Theoretically a polarization of about 92% can be achieved but several depolarizing effects come into play to limit the realistic final value. The particular spin resonances (to be avoided to keep polarization) are spaced every 440 MeV while the expected energy spread of the beams will be in the range 50 - 100 MeV at the Z^o resonance; therefore it is very hard to avoid beam depolarization and the prospects for physics with polarized $e^{\pm}$ are still in doubt. Also a non-zero transverse polarization has to be rotated to achieve a longitudinal component: this operation involves strong bending magnets next to the beam crossing and the experiment - a situation which is likely to produce synchroton radiation bachground. (Fig. 17).

The situation is quite different for the SLC project where polarized electrons are produced and accelerated without loss of polarization in the SLAC linac [6),7)]. The spin will process in the bending arcs and their longitudinal polarization P will oscillate according to the beam energy (Fig. 18). Two maxima of P are separated by 880 MeV so that it suitably covers the Z^o peak with several measurements of opposite polarization. Therefore $<H_{tot}>$ can be measured straightforwardly. To cancel systemactic effects the electron beam polarization can be reversed at a fast rate and opposite P data can be recorded almost simultaneously. A significant advantage of polarization is the possibility to measure large asymmetries and polarizations at the Z^o peak (recall Fig. 12 and 13) where large amounts of data can be collected and selected to try to separate different fermions, while similar experiments with unpolarized beams will often have to cope with measurements of small quantities.

d. $e^+e^- \to q\bar{q}$. The study of weak-electromagnetic interference in the production of $q\bar{q}$ pairs in e^+e^- annihilation is a unique situation. All quarks are democratically produced, provided it is kinematically possible and therefore weak couplings of all quarks

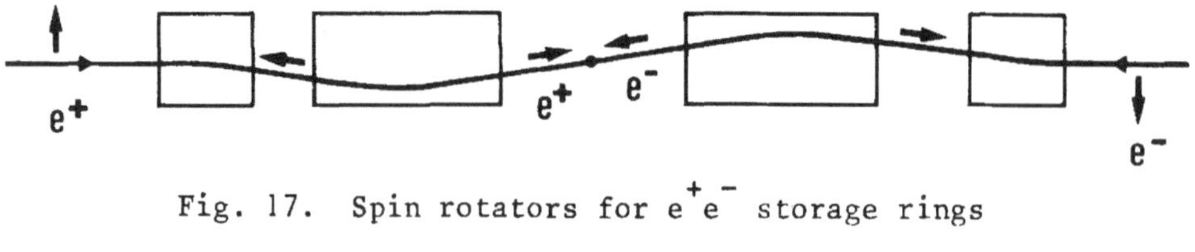

Fig. 17. Spin rotators for e^+e^- storage rings

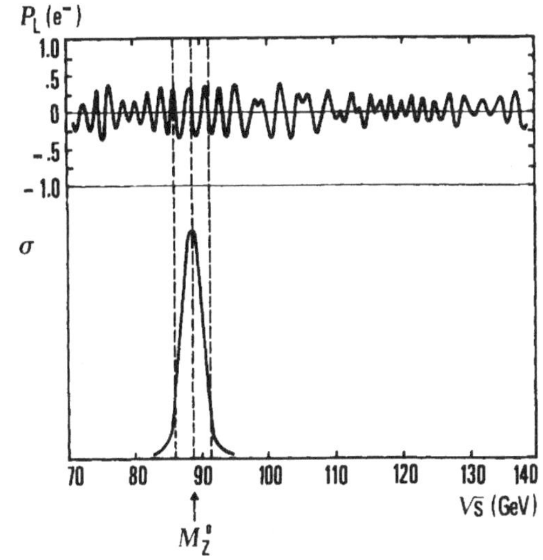

Fig. 18. Electron spin precession in the SLC arc and Z^0 production (schematic)

can be determined in principle. The price to pay is of phenomenological and experimental nature: how to separate different quarks and and how to tell a q jet from a $\bar{q}$ jet ? These are difficult problems which may become easier once sufficient information and understanding are gathered on quark fragmentation and jet properties obtained at PETRA and PEP.

To get a taste of what can be observed, we can look back at Fig. 14 ($\langle H_f \rangle$ with polarized electrons) and Fig. 19 showing pronounced differences in the behaviour of u and d quarks, themselves different from charged leptons.

A hadronic jet must contain information about the primordial quark: this information has been smeared in the random hadronization process but we can appeal to some observables to retrieve part

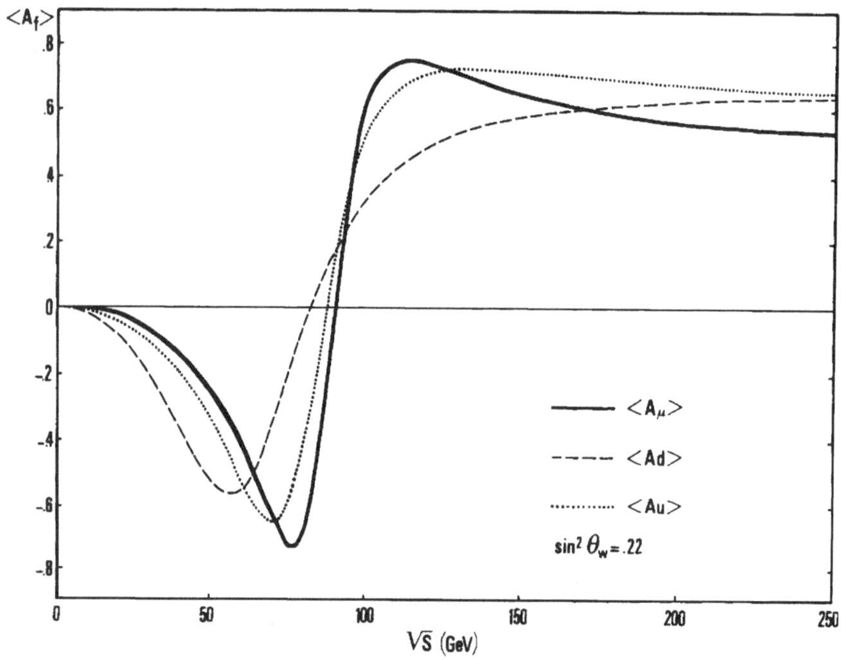

Fig. 19. Forward-backward asymmetries for $e^+e^- \rightarrow \mu^+\mu^-$, $u\bar{u}$ and $d\bar{d}$

of it. To get some feeling we consider the phenomenological fragmentation model of Field and Feynman [12]: in this model the initial quark q_o picks quark pairs $q_1\bar{q}_1$ from the vacuum giving birth to a $q_o\bar{q}_1$ meson and the cascade goes on with the left over quark q_1. The 2-body step $q_o \rightarrow q_1 + (q_o\bar{q}_1)$ is characterized by a probability depending on the fraction z_1 of the q_o momentum taken by the hadron h_1 $(q_o\bar{q}_1)$ (Fig. 20). Rather sophisticated Monte Carlo generators have been constructed following this model; they include resonance production and decay, different quark flavours, ...

α) jet charge. Fig. 21 shows the distribution of the total charge of a jet initiated by a u quark ($Q_u = 2/3$): it is centered near $Q = 2/3$ with a small shift we do not discuss here. However the distribution is quite broad with a dispersion $\sigma_Q \sim 1$: it is therefore impossible on an event-by-event basis to tell a u quark from a d, $\bar{d}$ or even $\bar{u}$ quark for example. Despite this intrinsic difficulty one can get some separation at a statistical level: keeping all jets with a net positive charge we get the following efficiencies for different initial quarks:

$Q > 0$	.63	u
	.52	$\bar{d}$
	.11	d
	.08	$\bar{u}$

Since the efficiencies are close for u and $\bar{d}$ on one hand, and d and $\bar{u}$ on the other hand, the charge asymmetries of $Q > 0$ jets are related simply to the up and down quark (of any multiplet) asymmetries. Approximately we have [13]

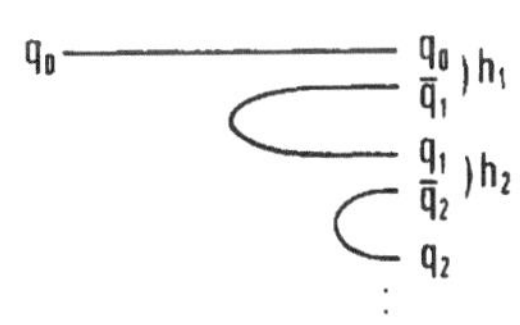

Fig. 20. Feynman-Field model for quark fragmentation [12]

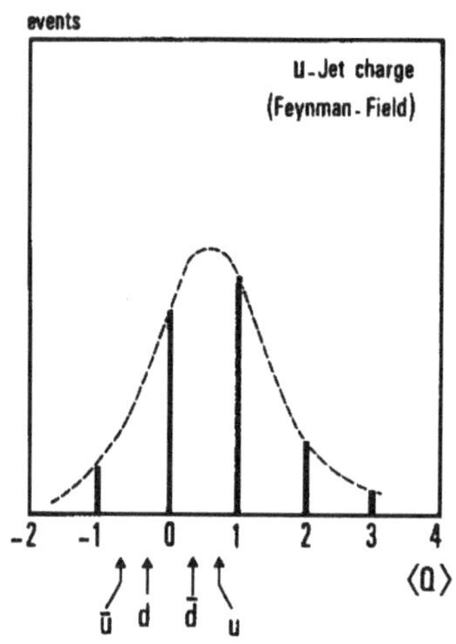

Fig. 21. Distribution of the total charge of a u jet in the Feynman Field fragmentation model [12)]

$$A_Q \simeq \frac{N_u A_u - N_d A_d}{N_u + N_d}$$

where N_u and N_d are the number of produced $u\bar{u}$ and $d\bar{d}$ events.

Such a measurement is undoubtedly very hard to perform and may be even harder to trust, although it is sensitive to all quark couplings at once and would therefore provide a global test of the theory.

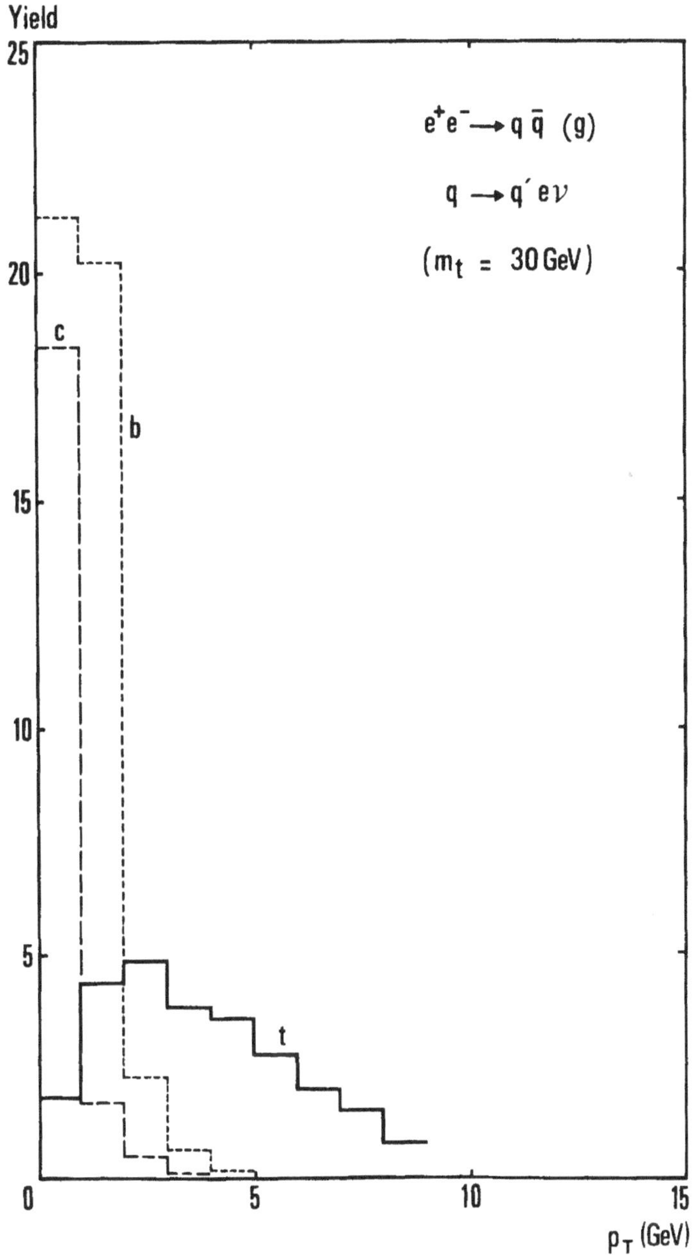

Fig. 22. Transverse momentum distribution (with respect to jet axis) of leptons produced in the semi-leptonic decays of heavy quarks [35])

β) leading particles. A hadron at large z (fraction of total longitudinal momentum) has a high probability to contain the initial quark.

Large z pions are not so interesting: they do not tell u from $\bar{d}$, they are frequent decay products of resonances and furthermore we know already about u and d couplings from neutrino scattering experiments.

The situation is more favourable for $s\bar{s}$ jets with identified charged K at large z. Two oppositely charged kaons at $z > 0.4$ is a good signature for $s\bar{s}$ jets ($c\bar{c}$ jets can in principle be removed using the jet charge since $c \to K^0$ while $s \to K^-$): this configuration should retain about 1.5% of all produced $s\bar{s}$ pairs in the Feynman-Field model.

For $c\bar{c}$ events it may not seem unreasonable to use reconstructed D decays at large z: $D \to K\pi$, $K\pi\pi$.

γ) leptons. Semi-leptonic decays constitute a good signature for heavy quarks:

$$c \to s l^+ \nu$$

$$b \to c l^- \bar{\nu}$$
$$\quad \hookrightarrow s$$

$$t \to b l^+ \nu$$
$$\quad \hookrightarrow c$$
$$\qquad \hookrightarrow s$$

Branching ratios can be estimated to be ~ 0.1 giving an appreciable rate for leptons at the Z^0 peak. It is possible to separate different contributions because of the wide mass spectrum involved: because of the large quark mass difference in the decays, leptons are observed at large transverse momentum measured with respect to the jet axis (Fig. 22). For $p_T \gtrsim 0.9$ GeV leptons come from b and heavier quark decay and it is possible to use thrust or sphericity to disentangle b and t quarks since we know that $m_t > 18$ GeV giving substantially broader jets even at the Z^0 mass.

This is probably the cleaner way to measure the weak couplings of heavy quarks.

5. Conclusions

Weak-electromagnetic interference in the Z^0 mass range will test with precision the couplings of quarks and leptons to the weak neutral current. In a wide range of situations one can expect precisions on $\sin^2\theta_W$ ranging from 10^{-3} to a few 10^{-2}: this experimental program will be a strong consistency test of the standard theory and it may well reveal some discrepancies not visible with cruder measurements of only e, u and d couplings.

It may also happen that the electro-weak gauge group is not $SU(2)_L \times U(1)$ but something bigger. As an example let us consider the left-right symmetric group [14] $SU(2)_L \times SU(2)_R \times U(1)$ with two Z^o bosons (Z^o_L and Z^o_H) and couplings arranged to satisfy all "low energy" data. It is interesting to note that, even if the mass of the heavier boson Z^o_H is kinematically inaccessible, sizeable modifications of the interference patterns can be expected at energies well below $M_{Z^o_H}$. Fig. 23 displays the muon asymmetry $\langle A_\mu \rangle$ in this model [15] showing the long-range sensitivity of interference phenomena. At $\sqrt{s} \sim 200$ GeV it seems possible to have sensitivity to a heavier Z^o of mass up to 400 GeV.

III. Z^o DECAYS

1. Z^o Production

The reaction $e^+e^- \to Z^o$ will be an intense source of Z^o bosons (compared to the expected production off-resonance since in absolute terms the cross section is not that large: to give a useful scale to an experimentalist it corresponds to only about twice the peak cross-section at the ψ'' resonance - the $D\bar{D}$ factory at SPEAR, hidden in the ψ' radiative tail - with an expected similar luminosity).

The resonance shape will be strongly altered by initial state radiative effects. For $s > M_Z^2$ the bare cross section falls very rapidly but the probability to radiate a photon is strongly enhanced by the Z^o pole: this effect leads to a very asymmetric shape (Fig. 24) which has to be corrected to measure the bare Breit-Wigner distribution. The peak cross-section is also reduced.

For 3 fermion multiplets (number of lepton doublets $N_L = 3$, number of quark doublets $N_Q = 3$, each in 3 colours) the rate will be

$$Z^o \text{ rate} \sim 3.10^4/\text{day} \times \frac{\bar{\mathcal{L}}(\text{cm}^{-2}\text{s}^{-1})}{10^{31}}$$

giving in excess of $2.10^6 Z^o$ decays/year.

The branching ratios to quark and lepton species can be calculated from the a_f and v_f couplings:

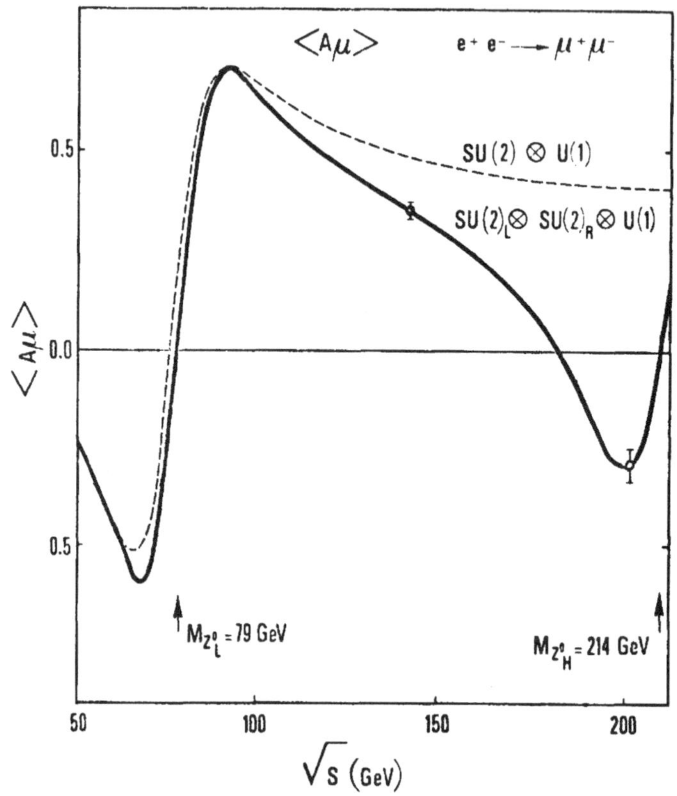

Fig. 23. Forward-backward muon asymmetries in $SU(2)_L \times SU(2)_R \times U(1)$ model with 2 Z^o bosons [15].

$$B(Z^o \rightarrow f\bar{f}) = \frac{a_f^2 + v_f^2}{\sum_i (a_i^2 + v_i^2)}$$

with an expected total width decomposition

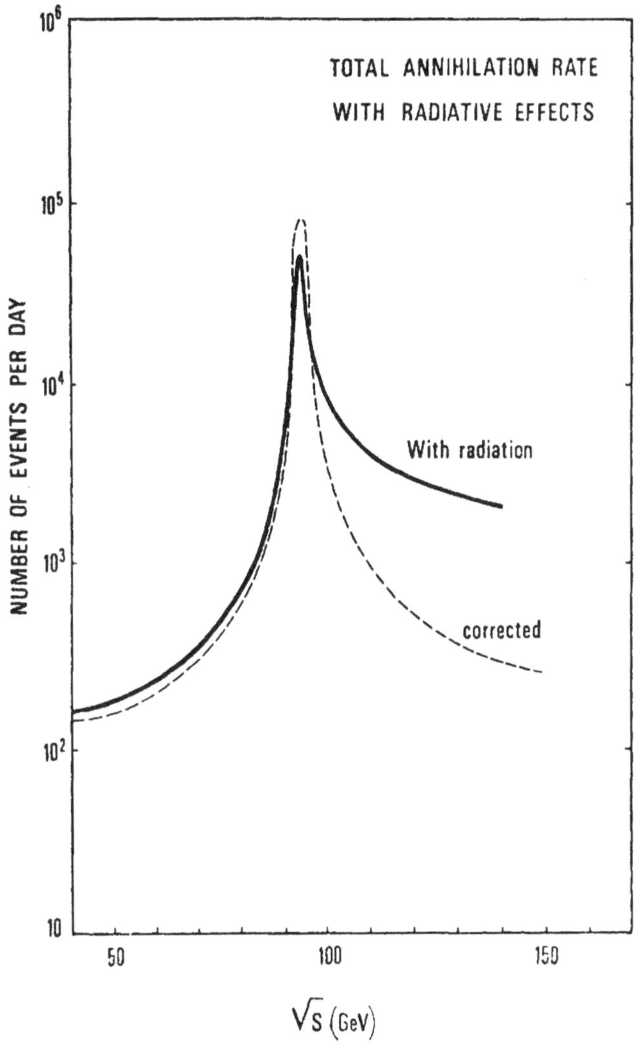

Fig. 24. Effect of initial $e^{\pm}$ radiation on Z^{0} resonance shape.

$$\Gamma_Z(\text{GeV}) = .17\ N_\nu + .09\ N_\ell + .32\ N_u + .40\ N_d$$

where the coefficients in front of N_u and N_d incorporate the colour factor. If all $N_i = 3$ the total width is expected to be 2.9 GeV ($\sin^2\theta_W = 0.20$).

The large rate of Z^o decays can be a good source for new point-like objects - quarks and leptons - up to a mass $M < M_Z/2 \sim 45$ GeV.

2. Production of New Charged Leptons

Charged leptons of mass $M_L < 45$ GeV can be produced and easily identified. Branching ratios can be studied in detail. Even the purely leptonic final state

$$Z^o \to L^+L^- \to \mu^\pm e^\mp 4\nu$$

can be used with a branching ratio $B_\ell^2 \sim .04$. Heavier states can be separated from lighter ones (τ) through the acoplanarity distribution between the two leptons (higher masses will lead to larger acoplanarities). If the gap between two leptons is large, (it will be the case if only one more lepton is produced) this separation will be easy.

The most interesting modes will be the mixed leptonic - hadronic states

$$L^+L^- \to (\ell\nu\nu)(\text{hadrons}\ \nu)$$

where the hadrons are expected to show jet behaviour owing to the large M_L mass. There will be some smearing from the missing ν_L spectrum but the mass of the hadrons can be measured directly (Fig. 25a).

Such a configuration would permit a study of the charged weak current coupling to quarks and leptons in a way similar to e^+e^- annihilation for the electromagnetic current (Fig. 25b).

The complete L^+L^- final state is expected to be:

2 leptons (e,μ,τ) + $4\nu \sim 10\%$

1 lepton + hadrons (2 jets) + $3\nu \sim 40\%$

hadrons (4 jets) + $2\nu \sim 50\%$

3. Production of Heavy Quarks

It is easier to identify new quarks Q near their production

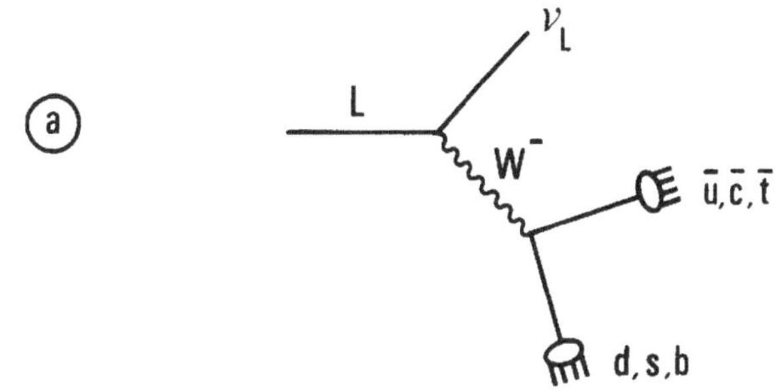

Fig. 25a. Decay of a sequential heavy lepton L into 2 hadron jets.

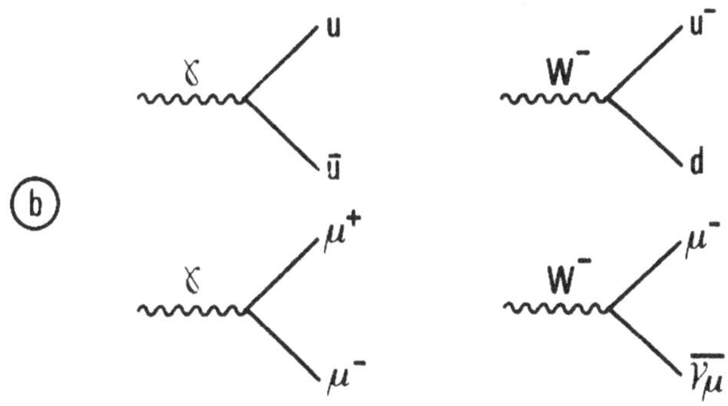

Fig. 25b. Comparison of 2-jet production in γ or W channels, as studied in e^+e^- annihilation and heavy lepton decays.

threshold: sphericity increase just above threshold (due to pair production of heavy $Q\bar{q}$ mesons), increase in R by 3 Q_Q^2, narrow bound states just below threshold.

At high energies where the continuum cross-section is not so large, it is not practical to search directly for thresholds. On the other hand one can look for heavy $Q\bar{Q}$ production at the Z^0 peak where large sphericity 2-jet events should signal their existence. It is also useful to look for large p_T leptons as we have previously discussed.

It will be possible to study the weak interaction of these heavy quarks through the study of their decay to lighter quarks (generalized mixing angles [16]).

4. Production of $\nu\bar{\nu}$ Pairs

Stable neutral leptons (neutrinos) can play an important role in Z^0 phenomenology. The known neutrinos have a mass much smaller that the corresponding charged lepton mass (for example the experimental limit on the ν_τ mass is 250 MeV) allowing even a zero mass value. Under these conditions the "observation" of other low-mass neutrinos could signal the existence of multiplets of quarks and leptons of very high mass (even exceeding largely the Z^0 mass). All these neutrinos would contribute to the total Z^0 rate and be detectable through the Z^0 total width although they would contribute to an invisible final state.

It is possible to determine the number of these neutrinos by measuring precisely the Z^0 width

$$\Gamma_Z = \sum_\nu \Gamma_{\nu\bar{\nu}} + \sum_\ell \Gamma_{\ell\bar{\ell}} + \sum_q \Gamma_{q\bar{q}} = N_\nu \Gamma_{\nu\bar{\nu}} + \Gamma_{visible}$$

The visible width can be derived in the standard way:

$$\int \sigma_{visible}\, d\sqrt{s} \sim \frac{\Gamma_{ee}\,\Gamma_{visible}}{\Gamma_Z}$$

$$\int \sigma_{\mu\mu}\, d\sqrt{s} \sim \frac{\Gamma_{ee}^2}{\Gamma_Z}$$

while Γ_Z is measured directly from the Z^0 line shape. The difference $\Gamma_Z - \Gamma_{visible}$ is a direct measure of the number of neutrinos N_ν of mass less than $M_Z/2$. Since each neutrino contributes $\sim$ 170 MeV to the total width it is necessary to measure Γ_Z and $\Gamma_{visible}$ to an accuracy of $\sim$ 100 MeV to be able to count neutrinos one by one: such a program needs a detailed treatment of radiative corrections, but appears feasible.

Another method to measure N_ν is the measurement of the reaction

$$e^+e^- \rightarrow \gamma Z^o$$
$$\quad\quad\quad \hookrightarrow \text{"nothing"}$$

above the Z^o peak ($\sqrt{s} > M_Z$). Through the detection of only one high energy photon, it is possible to reconstruct the Z^o mass and make sure that no other particle was produced. Form the calculated yield for this reaction one can deduce B_ν, hence N_ν.

One can pursue the idea of a larger number of low-mass neutrinos to the extent that they would dominate Z^o decays. This situation would be most unfortunate since the visible cross-section - the experimentalist's horn of plenty - would essentially vanish. In Fig. 26 such a case is considered: about $\sim 10^3$ neutrinos are needed to wipe out the Z^o peak! Even though the Z^o exists, it becomes invisible! This is however not the last word since an interference measurement like $<A_\mu>$ can still measure the vanishing weak transition (Fig. 27). So even in this catastrophic situation there is a good chance to locate the Z^o pole and measure weak couplings although counting rates will be miserable.

What do we know about N_ν beforehand ?

a. Helium abundance in the universe can be related through standard early universe cosmology to the number of neutrino species less massive than the electron [17]: the limit is $N_\nu \lesssim 4$.

b. A limit of 6.10^{-7} on the $K \rightarrow \pi\nu\bar{\nu}$ branching ratio results in a 95% confidence level limit of 10^4 neutrinos! [18]

c. From purely theoretical arguments one can also try to bound the number of neutrinos or equivalently the number of charged fermions. In the minimal GSW model there exists a relation between M_Z, M_W and θ_W:

$$\rho_o = \frac{M_W}{M_Z \cos\theta_W} = 1$$

following from the existence of only Higgs field doublets. This relation is subject to radiative corrections from fermion loops [19] (see Fig. 28).

$$\rho^2 = \frac{M_W^2}{M_Z^2 \cos^2\theta_W} = 1 + \sum_{m_1 m_2} \eta \frac{G_F}{8\pi^2\sqrt{2}} \left(\frac{2m_1^2 m_2^2}{m_1^2 - m_2^2} \ln \frac{m_2^2}{m_1^2} + m_1^2 + m_2^2\right)$$

$\eta = 1$ for lepton loop
$\eta = 3$ for quark loop

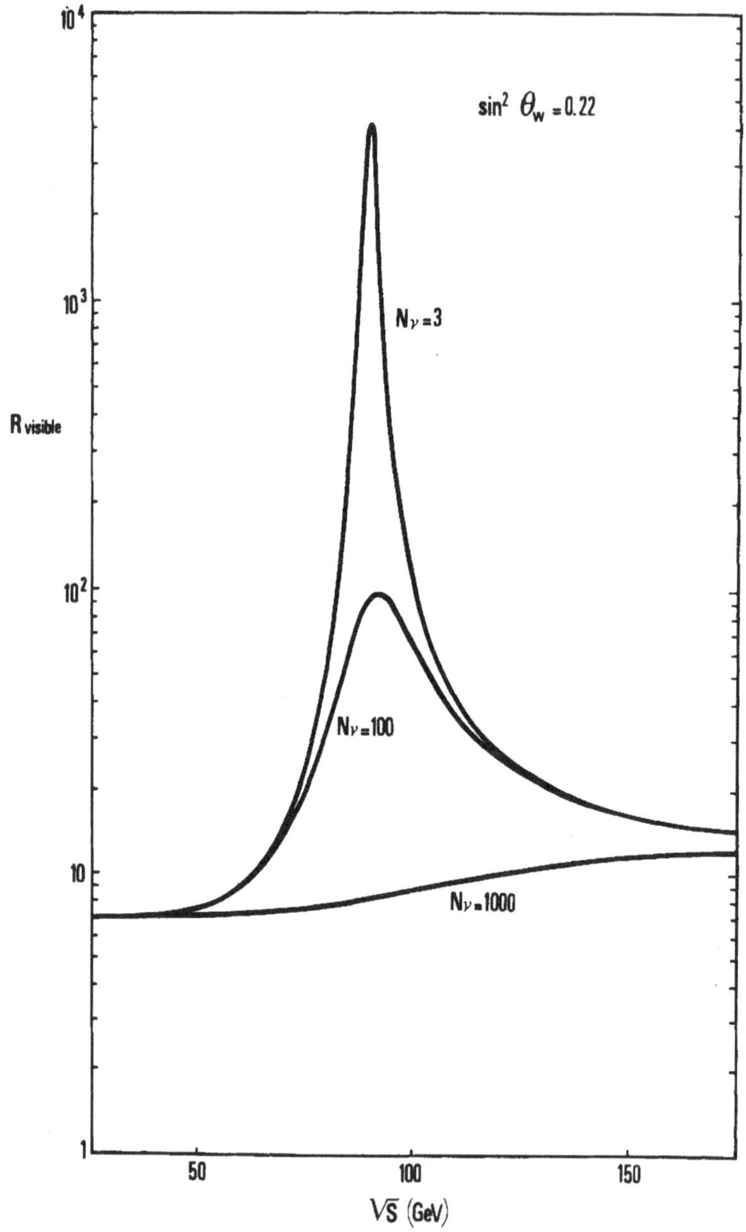

Fig. 26. $R_{visible} = 2R_L + 9R_u + 9R_d$ around Z^o peak as a function of the number N_ν of neutrinos.

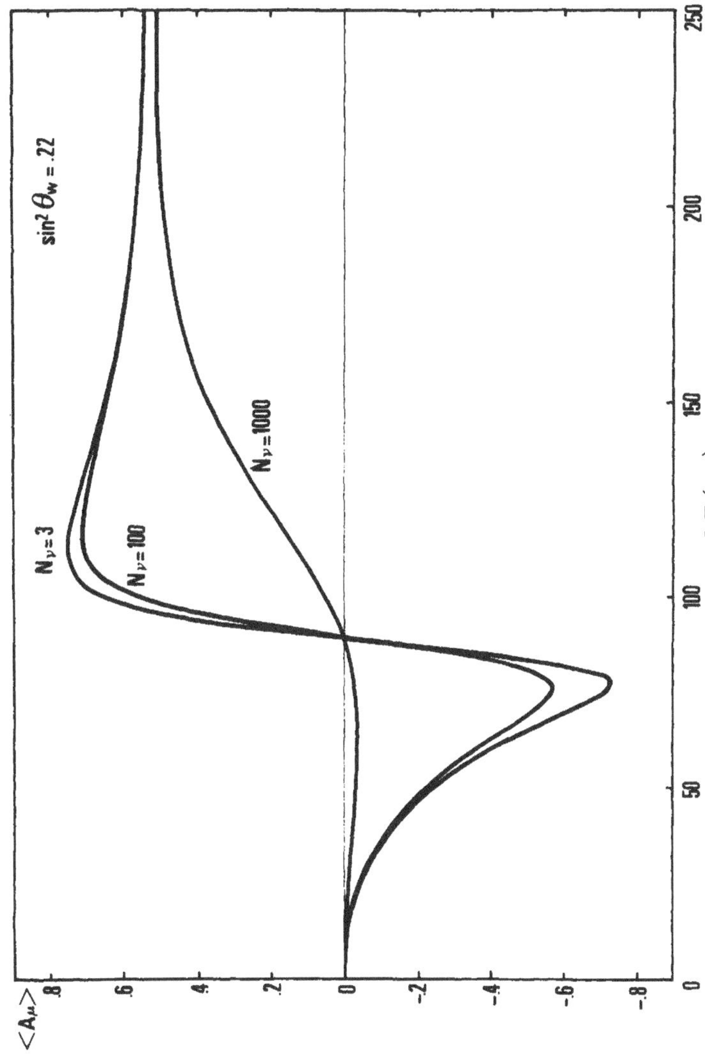

Fig. 27. Forward-backward muon asymmetry $\langle A_\mu \rangle$ as a function of N_ν

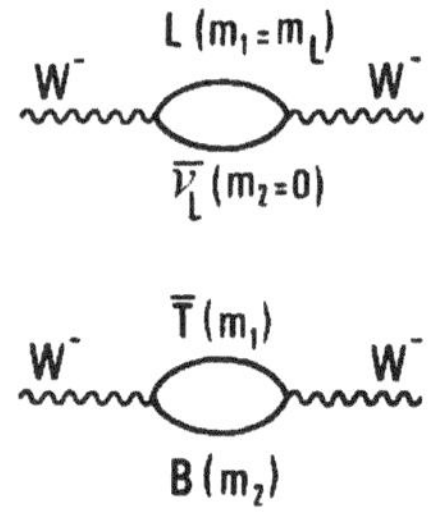

Fig. 28. Higher order corrections for the W propagator

Within the standard model phenomenology, the agreement on cross sections for neutral and charged currents gives a value $\rho = 1.02 \pm .05$. From this 5% possible deviation it is possible to set limits on the occurrence of heavy mass fermions

	Δ_ρ	
lepton loop ($\bar{\nu}_L$ L)	$\sim 10^{-7} m_L^2$	$m_L < 700$ GeV
quark loop (B $\bar{T}$)	$\sim 1.5\ 10^{-6}\ m_1^2$	$m_2 < 540$ GeV
(assuming $m_2 \sim 3m_1$)		

Therefore, with the present fermion spectrum, it seems unplausible to concoct a large number of multiplets before the previous limits are violated.

d. $\nu\bar{\nu}$ decays of heavy quarkonia can also place bounds on N_ν [18] (Fig. 29)

$$\frac{\Gamma(V \to \nu\bar{\nu})}{\Gamma(V \to e^+e^-)} = \frac{G_F^2}{64\pi^2\alpha^2} \frac{m_V^4}{Q_Q^2} [1 - 4|Q_Q|\sin^2\theta_W]^2 \left(\frac{M_Z^2}{m_V^2 - M_Z^2}\right)^2$$

$$= 1.5\ 10^{-9}\ m_V^4 \left(\frac{M_Z^2}{m_V^2 - M_Z^2}\right)^2 \qquad m_V \text{ in GeV}$$

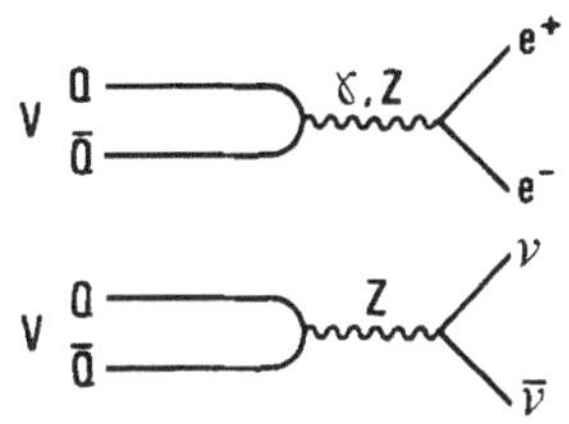

Fig. 29. Diagrams for decays of heavy quarkonium into e^+e^- and $\nu\bar{\nu}$ pairs.

for a top quark (Q_Q = 2/3) and for each neutrino type. We obtain the following branching ratio/neutrino type:

	m_V(GeV)	$B(V \rightarrow \nu_i\bar{\nu}_i)$
$\psi(c\bar{c})$	3.1	9.10^{-9}
$\Upsilon(b\bar{b})$	9.5	4.10^{-6}
$V(t\bar{t})$	40?	5.10^{-4}

Experimentally these invisible transitions can be reached through cascade decays from radially excited quarkonia:

$$V'(Q\bar{Q}) \rightarrow V(Q\bar{Q})\pi^+\pi^-$$
$$\hookrightarrow \text{nothing}$$

If a toponium exists around 40 GeV it will allow to test the effect of neutrinos if their number exceeds ∿ 20.

5. Other Fundamental Objects ?

Z^o decays could be a source of other point-like objects if they exist. To illustrate this point, let us consider the spectrum of fundamental particles predicted by supersymmetric theories [20]. In this scheme spin-raising or lowering transformations generate a group with representations involving particles of different spins. To the spin 1 gauge bosons are associated spin 1/2 and spin 0 partners; to the well-known spin 1/2 constituents are associated spin 0 constituents. Such theories offer the ambitious possibility to connect electroweak and strong forces to the gravity force.

In the simplest scheme, the following types of particles are expected

J = 1	J = 1/2	J = 0
γ	λ_γ	
g	λ_g	
$W^\pm, Z^o$	L	H
	ℓ	s_ℓ, t_ℓ
	q	s_q, t_q

Mass relations are satisfied such as:

$$m^2(t_f) - m^2(f) = m^2(s_f) - m^2(f) = \frac{1}{4}[m^2(W) - m^2(L)]$$

The s and t degenerate states correspond to the 2 degrees of freedom of the fermion field they are transformed from.

It is clear from this mass relation that as $m(s_f)$ increases m(L) must decrease. We know from PETRA experiments [21] that s_e, s_μ and s_τ have masses in excess of $\sim$ 16 GeV since the process (Fig. 30)

$$e^+e^- \rightarrow s_\ell \bar{s}_\ell$$

followed by $s_\ell \rightarrow \ell\lambda_\gamma$ (neutrino-like) is not observed.

Search for scalar leptons (or quarks) in the Z^o decays will give a decisive answer on at least the simplest models of supersymmetry: if they are too heavy and cannot be found, one way out is to look for the leptons L corresponding to the weak gauge bosons.

IV. W^+W^- PRODUCTION

1. Single W production

A single W can be produced in the process

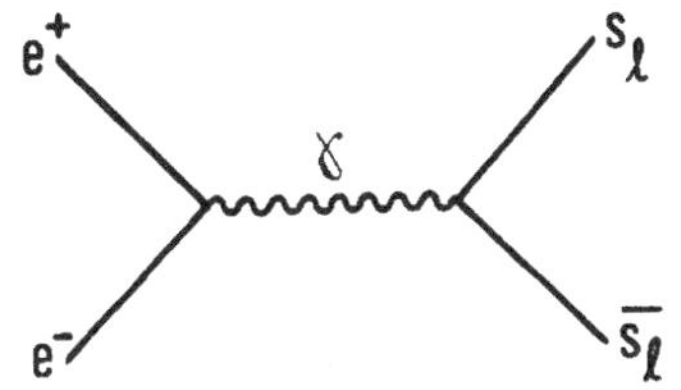

Fig. 30. Pair production of scalar leptons

$$e^+e^- \to W^\pm e^\mp \nu$$

corresponding to a diagram analogous to a γγ collision (Fig. 31). This process can be calculated in the Weizäcker-Williams approximation [22] and yields a cross-section

$$\sigma(e^+e^- \to \nu_e W^- e^+) = \frac{\alpha^2 G_F}{2\pi\sqrt{2}} \ln \frac{s}{m_e^2} \; F\left(\frac{M_W^2}{s}\right)$$

where F(x) is a decreasing function of x equal to 1 for x ∿ 0.15. Small values are obtained for the cross-sections:

$$\sigma(W^+) + \sigma(W^-) = 4.5\ 10^{-37}\mathrm{cm}^2 = 0.1\ \sigma_{pt} \quad \text{at } \sqrt{s} = 140 \text{ GeV}$$

$$1.2\ 10^{-36}\mathrm{cm}^2 = 0.5\ \sigma_{pt} \quad \text{at } \sqrt{s} = 200 \text{ GeV}$$

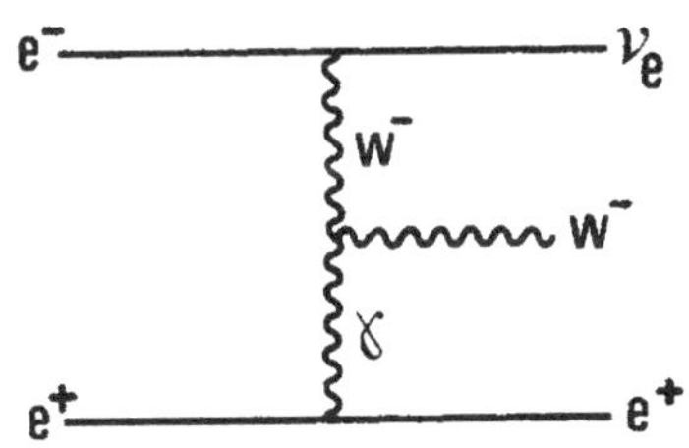

Fig. 31. Diagram for the process $e^+e^- \to e^+\nu_e W^-$

Since $\gamma\gamma$ reactions producing 2 jets and a small angle electron

$$ee \to (e)eq\bar{q}$$

are expected to produce an appreciable background, it leaves us with little hope for this mode of W production. It may however play a role if one wants to test the γW^+W^- coupling as we shall discuss later.

2. W^+W^- production

a. Need for a gauge theory. The process $e^+e^- \to W^+W^-$ plays a key role in the qualitative understanding of why we need a gauge theory.

A W pair can be produced through the electromagnetic current (Fig. 32). Since W is a massive spin 1 object without hadronic structure, the amplitude to produce W with helicity $\lambda_W = 0$ diverges as s increases. In fact depending on the magnetic dipole moment μ and the electric quadrupolar moment ε the s dependence can change significantly [23]

$$\sigma_\gamma \sim \frac{A}{s} + B(\mu,\varepsilon) + sC(\mu,\varepsilon)$$

To this effect, a gauge theory determines the values for μ and ε but it does not eliminate the divergence.

The diverging part coming from the initial $e_L^-e_R^+$ state can be eliminated by considering the t-channel ν exchange diagram in Fig. 32.

In order to cancel the divergence of the amplitude from a $e_R^-e_L^+$ state, a new diagram must be introduced: in gauge theories this is provided by neutral weak boson exchange. In the standard model we have the third diagram in Fig. 32.

One therefore understands qualitatively how a gauge theory with

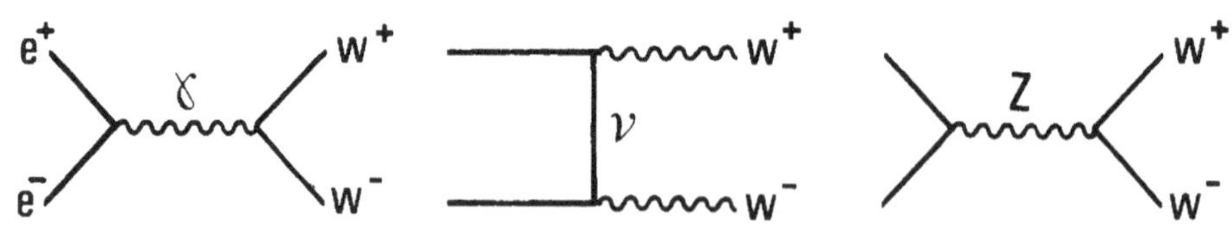

Fig. 32. Pair production of W bosons in e^+e^- annihilation

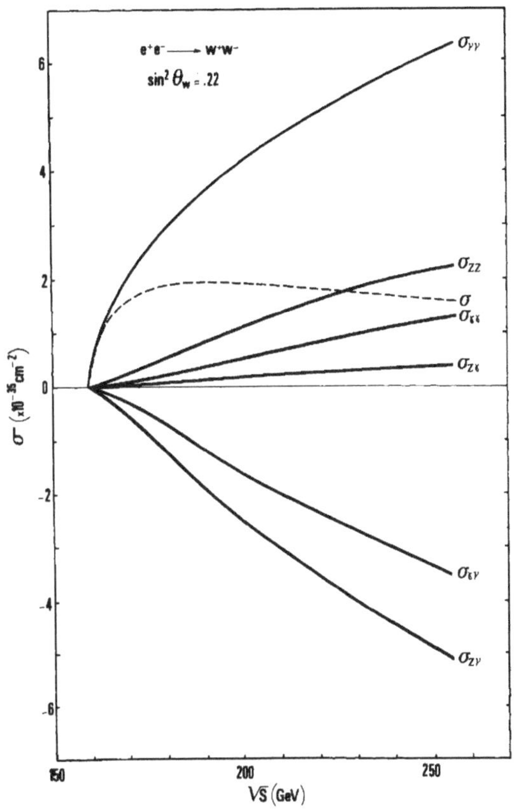

Fig. 33. The different contributions $\sigma_{ij}(i,j = \nu,\gamma, Z)$ to the total cross section of the process $e^+e^- \to W^+W^-$

all its ingredients (bosons: γ, $W^{\pm}$, Z^o; couplings γW^+W^-, $Z^oW^+W^-$; specified magnetic and quadrupolar moments) can result in a divergence-free calculation. As a consequence the simple measurement of a process like $e^+e^- \to W^+W^-$ is a powerful test of the theory.

b. Calculation in the GSW theory. The γW^+W^- coupling fixes $\mu = 1$ (in units of e/M_W) and $\varepsilon = 0$; on the other hand

$$(Z^oW^+W^-) = \frac{(\gamma W^+W^-)}{\tan\theta_W}$$

The calculated cross-section [24] is shown in Fig. 33 where the behaviour of the different terms is explicit:

$$\sigma(e^+e^- \to W^+W^-) = \sum_{ij} \sigma_{ij}$$

$$i,j = \gamma,\nu,W \text{ amplitudes}$$

It is clear from Fig. 33 that each diagram yields a diverging rate. Large cancellations are obvious and the total rate remains finite: it shows a rapid rise above threshold to a R value ~ 10 and the sensitivity to $\sin^2\theta_W$ is shown in Fig. 34.

If the theory was different, important modifications could occur: for example, turning off the $Z^oW^+W^-$ coupling would result in an increase in rate about twofold.

Polarized beams would be very helpful to sort out the different amplitudes. The $e_L^-e_R^+$ state contributes to all diagrams, while the $e_R^-e_L^+$ contribution is absent in ν exchange. How to use this experimentally appears extremely remote at this time.

c. Final states. Weak decays of $W^{\pm}$ involve pairs of left-handed fermions of opposite weak isospin. Leptonic channels are not mixed if neutrinos are massless; on the other hand hadronic channels are formed by linear superposition of $q\bar{q}'$ states. With 3 multiplets, the most general transition current has the form [16]

$$J_\mu \sim (\bar{u}\ \bar{c}\ \bar{t})\ \gamma_\mu\ (1-\gamma_5)\ U \begin{pmatrix} d \\ s \\ b \end{pmatrix}$$

where U is a 3 x 3 unitary matrix depending on 3 mixing angles and a phase δ related to CP violation .

Branching ratios can be estimated

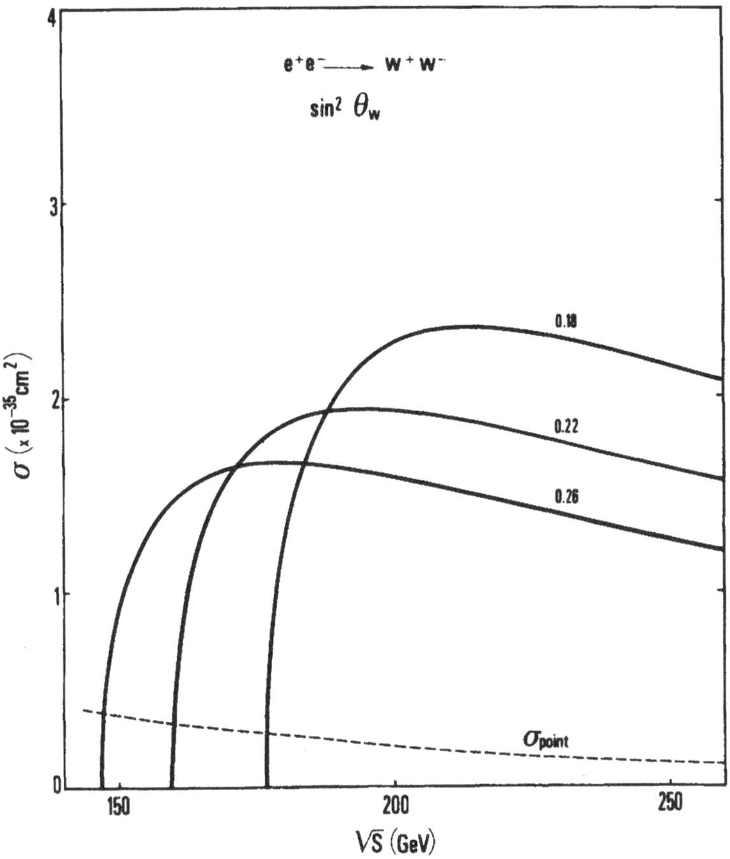

Fig. 34. The total cross-section for $e^+e^- \rightarrow W^+W^-$ for different values of $\sin^2\theta_W$

$$B(W^+ \rightarrow e^+\nu_e) \sim 0.10$$

(similarly for μ,τ)

$$B(W^+ \rightarrow \text{hadrons}) \sim 0.70$$

with the hadronic channel mostly composed of $u\bar{d}, c\bar{s}$ and hopefully $t\bar{b}$. The configuration will be that of 2 tight jets owing to the large W mass. Hence the following topologies:

$e^+e^- \to W^+W^- \to$	4 jets	49%
	2 jets + e,μ,τ	42%
	2 leptons	9%

In $\sim$ 90% of the cases, one or two of the W masses can be reconstructed experimentally: it is easy to do since the 2 corresponding low-mass jets can be associated and their effective mass calculated. In conclusion the reaction can be identified clearly considering the fact that it contributes to about half of the total production rate at $\sqrt{s} \sim 200$ GeV. In numbers:

$$W^+W^- \text{ rate} \sim 20 \text{ events/day} \times \frac{\bar{\mathcal{L}}(\text{cm}^{-2}\text{s}^{-1})}{10^{31}}$$

3. Further Tests of Gauge Couplings

So far we have examined the interactions of gauge bosons with fermions (a_f and v_f couplings in particular), the physical existence of Z^o and $W^{\pm}$ bosons, the trilinear coupling (γW^+W^-, $Z^oW^+W^-$). In the non-abelian gauge theories we also find quadri-linear couplings and it would be interesting to also test this aspect of the Lagrangian.

The 4-boson coupling is in principle accessible to experiment. Above the W^+W^- threshold the process

$$e^+e^- \to W^+W^-\gamma$$

can be considered with 5 diagram types given in Fig. 35. Diagrams (4) and (5) are not interesting since they correspond to initial state radiation of the $e^+e^- \to W^+W^-$ reactions. Tri-linear couplings appear in diagrams (2) and (3) while the quadri-linear coupling governs the behaviour of diagram (1).

Explicit calculations [25)] of the diagrams shows disappointing results (Fig. 36): the overall cross-section is small and the effect of diagram (1) negligible. The measurement of this reaction is still useful since the (small) magnitude of the rate is a test of the theory in itself.

Another possibility exists through the reaction:

$$e^+e^- \to e^+e^-W^+W^-$$

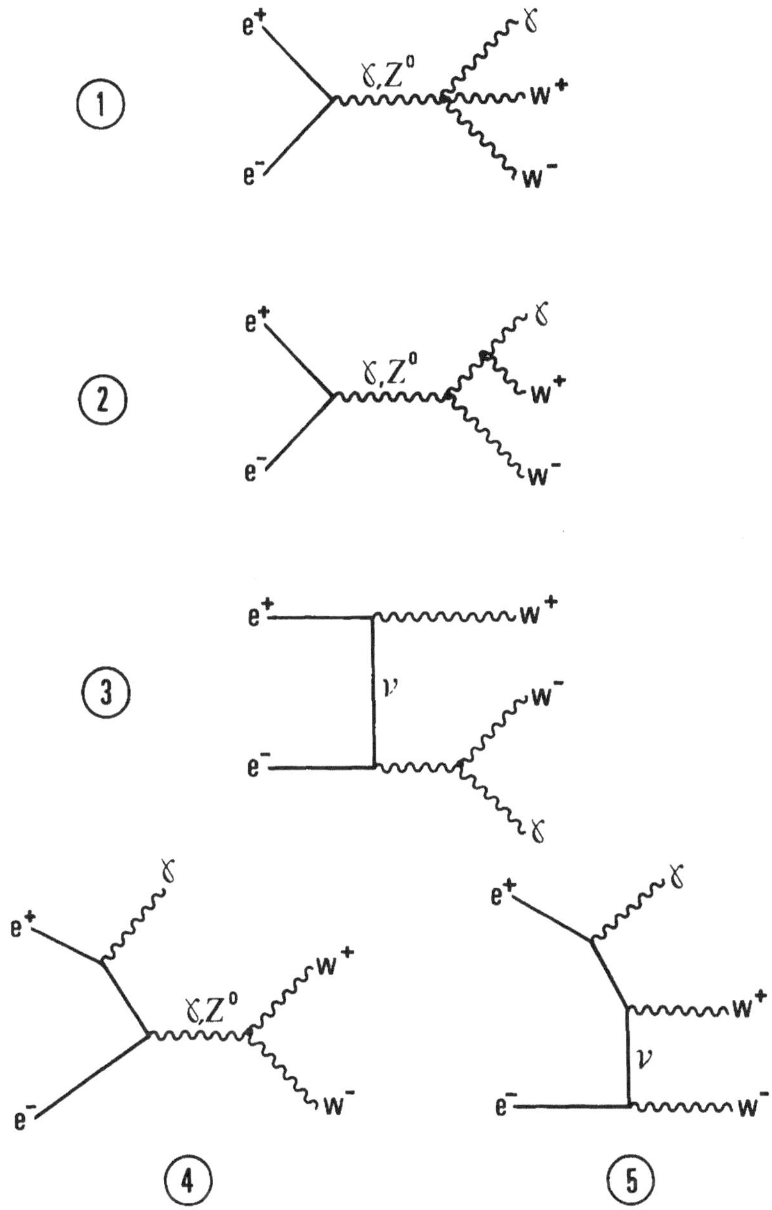

Fig. 35. Diagrams responsible for $e^+e^- \rightarrow W^+W^-\gamma$

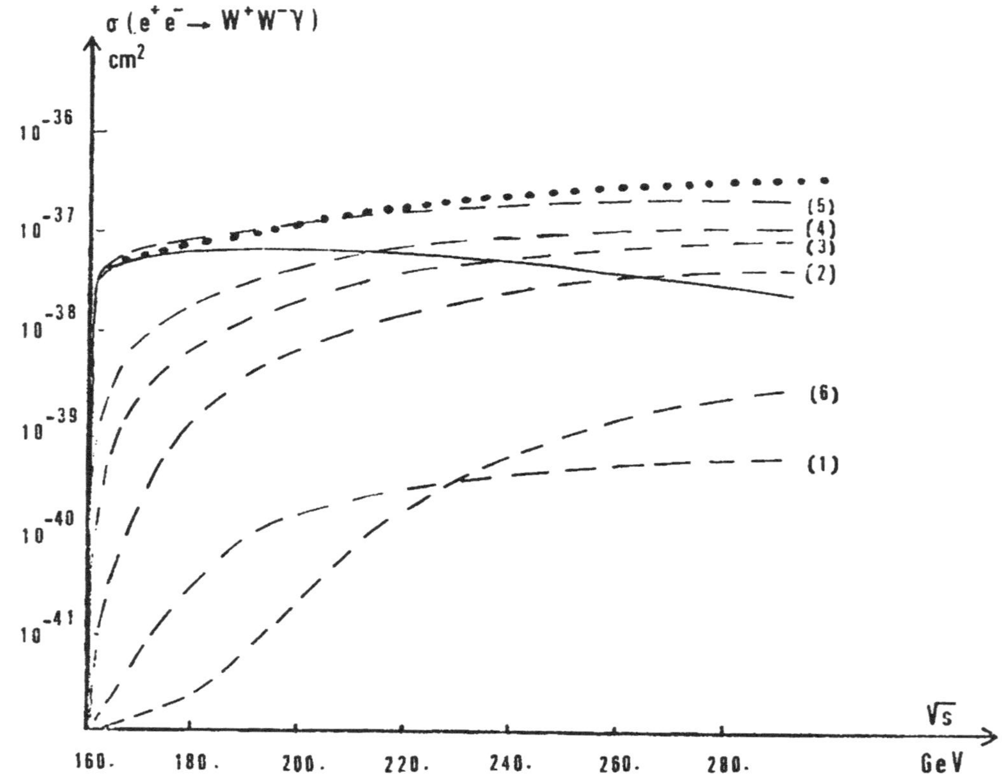

Fig. 36. Total cross section for $e^+e^- \to W^+W^-\gamma$ with $E_\gamma > 2$ GeV as a function of energy [25]. The numbers refer to the diagram types in Fig. 35.

involving the sub-process $\gamma\gamma \to W^+W^-$ (Fig. 37). The cross-section is even more discouraging [25], $\sim 5.10^{-39}cm^2$ at $\sqrt{s}$ = 240 GeV.

In conclusion, while it is difficult, but nevertheless possible, to reach the 3-boson couplings as predicted by non-abelian gauge theories, one does not see at the present time a realistic way to be sensitive to the 4-boson couplings which should also be there.

V. HIGGS BOSONS

1. Minimal Model

Higgs fields are introduced to break the symmetry and give mass to the gauge bosons. While this is not the most aesthetic aspect of the GSW theory, it is however a spectacular feat of the model to predict Z^o and $W^{\pm}$ masses: these masses follow from the simplest assumption of a single doublet of scalar complex fields corresponding to a single neutral physical Higgs boson H^o (the other degrees of freedom disappear in the longitudinal states of massive bosons)

In this simplest case, the couplings of H^o to fermions and bosons go as follows:

$$g_{Hf\bar{f}} \sim \sqrt{G_F}\, m_f$$

$$g_{HWW} \sim \sqrt{G_F}\, M_W^2$$

one of the ugliest features is that no constraints are put on the Higgs mass. Some consistency arguments produce lower limits [26] while for masses around 1 TeV the perturbative approach breaks down.

Because of their particular couplings, the Higgs decays will involve only the heaviest objects which can be reached kinematically: for example $H^o \to c\bar{c}, \tau^+\tau^-$ for a $\sim$ 10 GeV mass, $H^o \to b\bar{b}$ almost essentially much above 10 GeV.

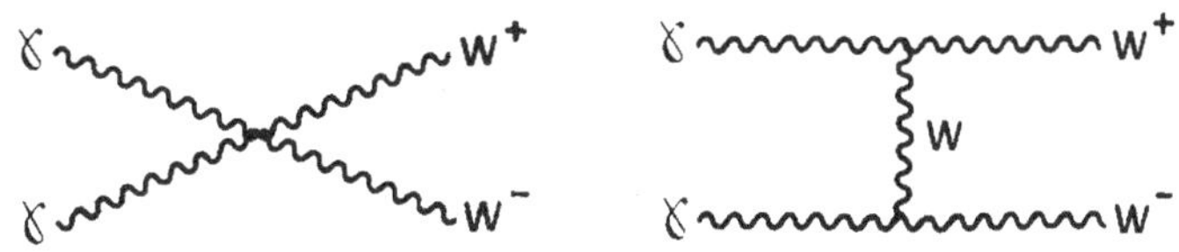

Fig. 37. Diagrams for the process $\gamma\gamma \to W^+W^-$

The small coupling to electrons (m_e small) precludes to see H^o directly in e^+e^- annihilation. The only realistic schemes will involve heavy particles so that H^o couplings can be large: they can be heavy quarks or W,Z bosons. Also clean production modes with detection independent of Higgs decays will be clearly preferred.

2. Minimal Higgs Production

a. Heavy quarkonium decays. The process [27)]

$$V(Q\bar{Q}) \rightarrow H^o\gamma$$

goes through a Q loop and couples H^o and heavy quark (Fig. 38).

The calculated rate

$$\frac{\Gamma(V \rightarrow H^o\gamma)}{\Gamma(V \rightarrow e^+e^-)} \approx \frac{G_F m_V^2}{4\sqrt{2}\pi\alpha} \left(1 - \frac{m_H^2}{m_V^2}\right)$$

increases rapidly wiht m_V. It is negligible for ψ and Υ states, while for a 40 GeV toponium and 10 GeV Higgs one gets

$$\frac{\Gamma(V \rightarrow H\gamma)}{\Gamma_V} \sim 1\%$$

The Higgs can be seen as a peak in the mass recoiling against a photon (Fig. 39). Background from $\pi^o \rightarrow \gamma\gamma$ is most severe for a large mass Higgs, while radiative γgg events (also around ~ 1 % rate but over a wide spectrum) are in the way for a low mass Higgs.

Once a peak is identified, H^o decays can be studied independenly to check for the dominance of heavy fermions.

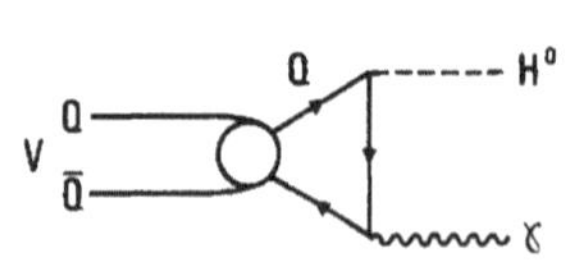

Fig. 38. The Wilczek mechanism for a quarkonium decay into $H^o + \gamma$

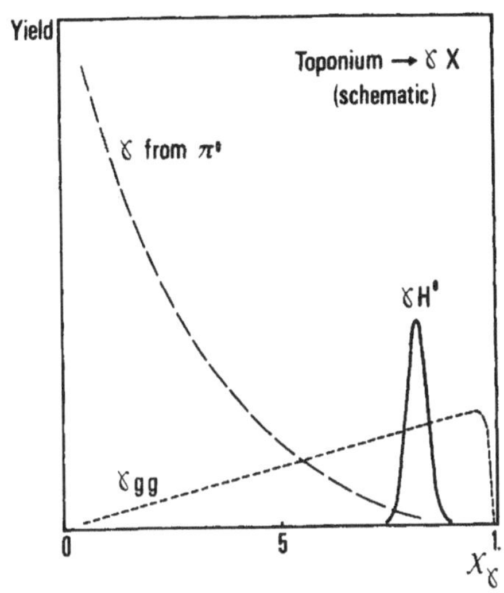

Fig. 39. Photon spectrum from quarkonium decays including a possible contribution from $H^o\gamma$ decay

b. Z^o decays. Despite the large Z rate, nature is not kind since the only possible channels have small branching fractions.

α) $Z^o \rightarrow H^o\gamma$. The direct $Z^oH^o\gamma$ unfortunately vanishes because of the massless photon and one should consider again a loop diagram (as in Fig. 40a.) where a heavy fermion or a W runs around the loop. With a W loop [28] one gets a small branching ratio

$$B(Z \rightarrow H\gamma) \sim 10^{-6}$$

for $m_H/M_Z < 0.4$ as shown in Fig. 41. This is very small, especially

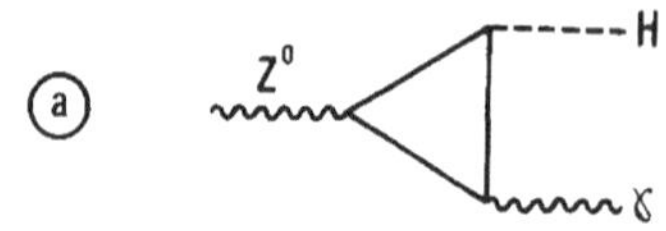

Fig. 40a. Z^o decay into $H^o + \gamma$ through a heavy-particle loop

considering possible backgrounds. However if the Higgs boson is found independently, this process could be a good test of the gauge couplings in the theory ($Z^oW^+W^-$, $W^+W^-H^o$, γW^+W^-)

β) $Z^o \to H^o \ell^+ \ell^-$. Again the $H^oZ^oZ^o$ coupling can be used [29], one of the Z^o being virtual (Fig. 40b.)

The situation is more favourable (Fig. 28) with a branching fraction

$$B(Z \to H\ell^+\ell^-) \sim 10^{-5} - 10^{-6}$$

for

$$0.2 < \frac{m_H}{m_Z} < 0.5$$

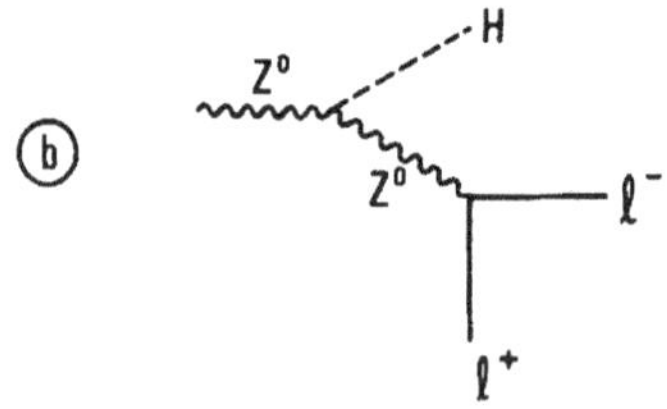

Fig. 40b. Z^o decay into $H^o \ell^+ \ell^-$

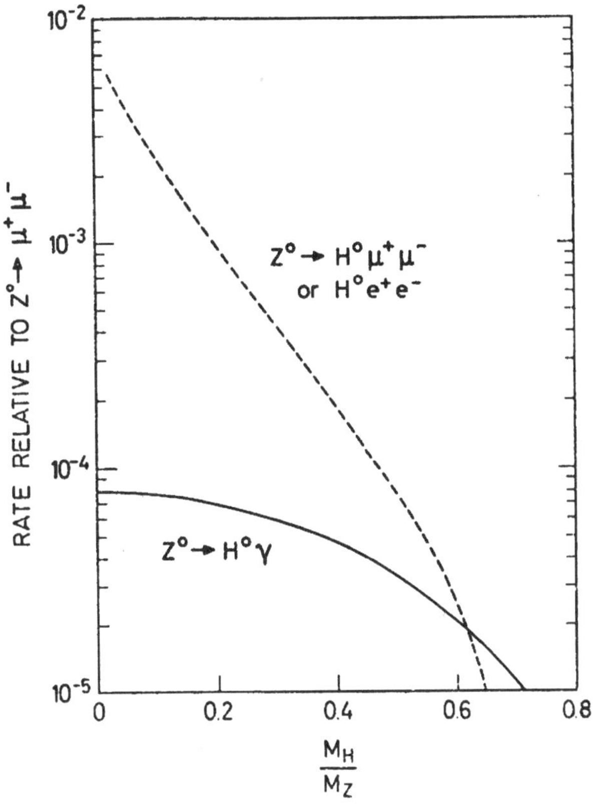

Fig. 41. Branching ratios for $Z^o \rightarrow H^o\gamma$ and $H^o \ell^+ \ell^-$ relative to $Z^o \rightarrow \ell^+\ell^-$ 26)

Background from semi-leptonic quark decays do not appear to be overwhelming since they tend to give small $\ell^+ \ell^-$ mass in contrary to Higgs decays where the virtual Z^o forces the $\ell^+ \ell^-$ mass to be as large as kinematically possible.

In conclusion, however, the Z^o resonance is a poor source of Higgs bosons.

c. H^o production at higher energies.

α) $e^+e^- \to Z^oH^o$. The same $Z^oZ^oH^o$ coupling can be used at higher energies where the final Z^o is real and decays into e^+e^-, $\mu^+\mu^-$ or event $q\bar{q}$ pairs. Again H^o can be reconstructed through the missing mass.

The cross-section [30] reaches a maximum for $\sqrt{s} \sim M_Z + \sqrt{2}\, m_H$

$$\left(\frac{\sigma(e^+e^- \to Z^oH^o)}{\sigma_{pt}}\right)_{max} \sim \frac{50}{m_H} \qquad (m_H \text{ in GeV})$$

for $\sin^2\theta_W = .22$. If all Z^o decays can be used (calorimeter measurement of jets would be extremely useful especially at the highest energies where the jets will be isolated and collimated) this process can test large Higgs masses: at $\sqrt{s} = 230$ GeV, the maximum cross-section occurs for $m_H \sim 100$ GeV with a counting rate of ~ 1 event/day for an average luminosity of 10^{31} cm^{-2}s^{-1}.

β) $e^+e^- \to H^of\bar{f}$. Bremsstrahlung of H^o bosons can be envisaged with, as usual, a preference for heavy fermions (Fig. 42). Other diagrams can occur where H^o is emitted along a t-channel exchange of a heavy boson. These processes can generate large mass Higgs bosons but the rates are small and the final state less constrained. As an example [31], the rate for

$$e^+e^- \to t\bar{t}H^o$$

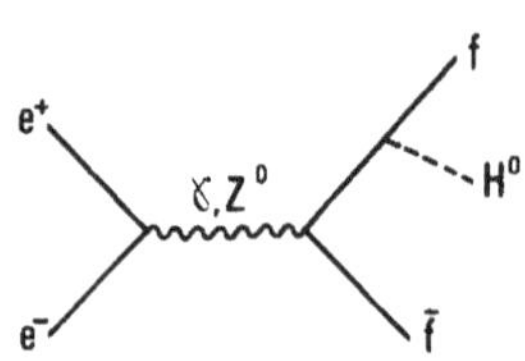

Fig. 42. Bremsstrahlung of Higgs bosons from heavy quarks.

would $\gtrsim$ 0.1 event/day at $\sqrt{s}$ = 240 GeV for $m_H \lesssim$ 140 GeV assuming, as usual, $\mathcal{L} = 10^{31} cm^{-2} s^{-1}$.

To conclude this section, let us restate that it seems possible to perform a clean Higgs search up to a mass $\sim$ 100 GeV if e^+e^- annihilations are realized up to a center-of-mass energy of $\sim$ 200 GeV.

3. More Complicated Higgs Sector

At the present time we have only limited information to constrain the Higgs sector. All the predictions of the standard GSW model which can be experimentally tested have been verified successfully. But most of these predictions do not depend explicitly on the symmetry-breaking mechanism. While we are waiting for the results of the W and Z^0 searches at the CERN $\bar{p}p$ collider we can still speculate on more complicated Higgs structure. As an example, 2 Higgs field doublets (leaving ρ = 1 at the lowest order) lead to 5 physical Higgs particles - 3 are neutral, 2 are charged. Higher (I > 1/2) multiplets could even produce Higgs bosons with charges larger than 1, which would be spectacular experimentally! It is therefore not totally unreasonable to speculate on the existence of charged Higgs bosons which could exist even in the framework of the SU(2) x U(1) model; other groups could obviously have an even more complicated Higgs sector.

The next simplest case to the minimal model - with 2 Higgs doublets - leaves us with already too much flexibility: in particular we would lose in general the feature that Higgs couplings to fermions are proportional to the fermion mass. Let us nevertheless examine some relevant production modes of charged $H^\pm$ bosons.

a. $e^+e^- \to H^+H^-$. It is the simplest process to produce $H^\pm$ as shown in Fig. 43a. At low energy $\sqrt{s} \ll M_Z$ the electromagnetic current dominates with a cross-section typical of a pair of spin-0 particles:

$$\frac{d\sigma}{d\cos\theta} = \frac{\pi\alpha^2}{4s}\,\beta^3\sin^2\theta$$

with $\beta = p_H/E_H$. The rise of the total cross-section is slow as compared to what happens in the production of a fermion pair (Fig.44) and the "stop" is only ΔR = 1/4. The absence of a steep rise makes it rather difficult to be detected experimentally: in fact it is not ruled out that such particles exist even at low masses. As an example let us consider a state with $m_H \sim$ 1 GeV decaying into $e^+\nu_e$ or $\mu^+\nu_\mu$; using the data obtained at SPEAR [32] just below the $\tau^+\tau^-$ threshold one can give a limit

$$B_e^2 + 2B_eB_\mu \lesssim 5.10^{-2}$$

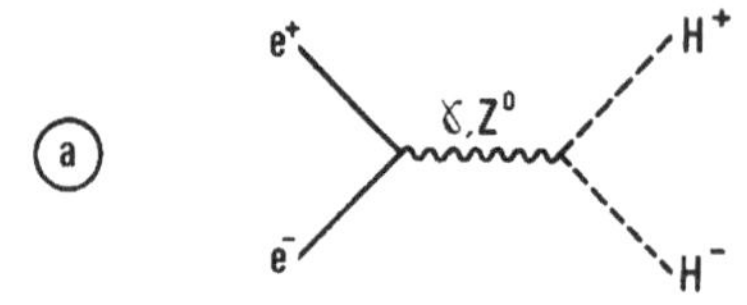

Fig. 43a. Pair production of charged Higgs bosons.

considering only the channels $e^+e^- \to e^\pm$ + 1 prong. For higher $H^\pm$ masses the limit would be correspondingly higher. Because of our ignorance of its decay modes, it is hard to rule out the existence of such a particle.

· Finally we expect $Z^o \to H^+H^-$ - if kinematically possible - at a rate determined by the (vector) coupling v_H of Z^o to H^+H^-. If $|v_H| < 1$, like other fields, the Z^o pole is less favourable (except for rate) than the continuum where the signal/background will be larger.

b. $e^+e^- \to W^\pm H^\mp$. Assuming that the $Z^oW^+H^-$ coupling is of the same magnitude as the usual $W^+W^-H^o$ coupling in the minimal model, the graph in Fig. 43b can be computed and yields a large cross-section [26]: for $m_H \sim 10$ GeV and $\sqrt{s} \sim 100$ GeV we get $R(e^+e^- \to WH) \sim 500$ which would overwhelm everything else being produced. With this large rate the corresponding threshold would not be hard to detect up to $H^\pm$ masses in excess of 100 GeV ($R \sim 5$ at $\sqrt{s}$ = 200 GeV). This reaction is obviously the best tool to investigate the $H^\pm$ spectrum

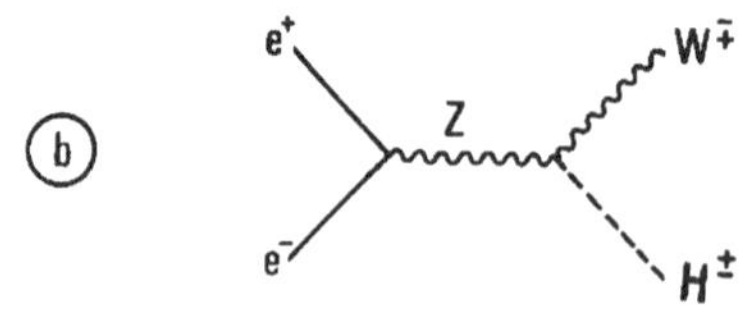

Fig. 43b. Diagram for $e^+e^- \to W^\pm H^\mp$

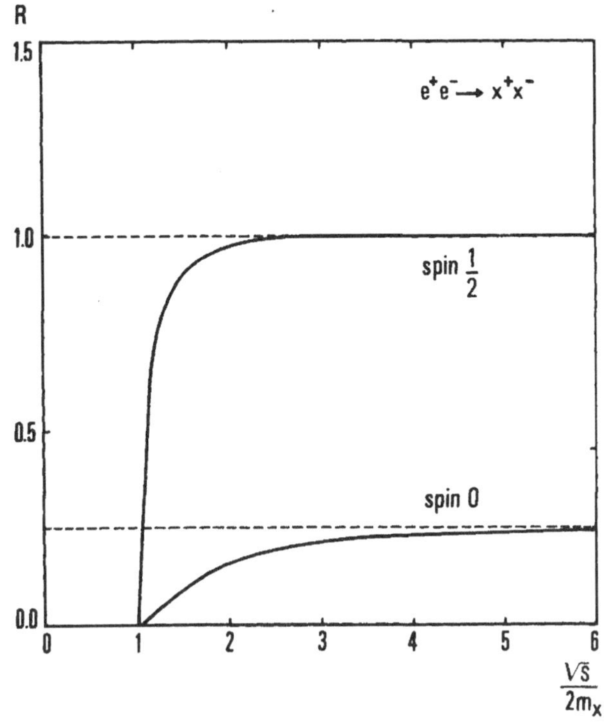

Fig. 44. Threshold behaviour of spin 0 and spin 1/2 particle production in $e^+e^- \rightarrow X^+X^-$

over a wide mass range: experimentally $W \rightarrow 2$ jets can be reconstructed and the $H^\pm$ mass consequently derived.

4. Higgs versus Hypercolour

Dynamical models involving heavy fermions ($\gtrsim$ 1 TeV) have been proposed [33] instead of the "conventional" spontaneous symmetry breaking with the Higgs mechanism. In these schemes low-mass scalar particles appear (pseudo-Goldstone bosons, often called hyperpions): these particles are essentially massless on the scale of the hyperfermions and acquire a small mass which can be computed in a given model.

As for the minimal Higgs scheme, these models have couplings proportional to masses which enhance decays into the heaviest conventional fermions. But the magnitude of the coupling, in particular the ratio of the lepton-to-quark couplings, will be different and can also be different for several hypercolour models. If we remember that for Higgs decays:

$$\frac{H^o \to \ell^+\ell^-}{H^o \to q\bar{q}} = \frac{m_\ell^2}{3m_q^2} \times \text{phase space}$$

the corresponding ratio in hypercolour schemes will be different: for example the colour factor 1/3 is replaced by 3 or 1/27 in specific models [34].

It seems more difficult to distinguish between Higgs bosons and hyperpions when they are charged since in either case the phenomenological constraints on their decay modes are scarce.

VI. CONCLUSIONS

The real conclusions will be drawn when the first results on e^+e^- annihilation around 100-200 GeV are presented. Let us list here a few concluding remarks to these lectures: e^+e^- physics up to $\sqrt{s} \sim 200$ GeV is a great tool to study

- weak-electromagnetic interference in the simplest possible reactions $e^+e^- \to f\bar{f}$; precise measurements of v_f and a_f will permit a severe consistency check of the minimal electroweak model or show deviations.

- the fermion spectrum (quarks, charged leptons, neutrinos) up to a mass $\sim$ 100 GeV; the number of low-mass neutrinos can be measured and will be a clue to possibly heavier multiplets.

- weak gauge bosons; apart from the expected Z^o and $W^\pm$ bosons, additional higher-mass Z^o's can be indirectly detected up to a mass of $\sim$ 400 GeV.

- non-abelian gauge structure through the couplings of 3 gauge bosons, particularly in $e^+e^- \to W^+W^-$.

- Higgs bosons: they can be searched for up to $\sim$ 100 GeV mass in a very clean way; more complicated schemes can be tested, producing charged bosons (more than one Higgs doublet) or dynamical symmetry breaking as in hypercolour models).

- possibly other types of constituents as for example scalar elementary fields like those predicted by supersymmetric theories.

Although not covered in these lectures, we may also add:

- QCD at $q^2 \sim 10^4$ GeV^2, through $Z^o \to q\bar{q}$, $q\bar{q}g$, $q\bar{q}gg$, $q\bar{q}q\bar{q}$ for different quark types (flavour independence ?).

- $\gamma\gamma$ interactions, especially γ structure functions as measured in deep inelastic $e\gamma$ scattering.

- and ... the unexpected!

It is a pleasure for me to acknowledge MM. Lévy, D. Speiser and the organizers of the 1981 Cargèse School who created, with the help of M. F. Hanseler, a very stimulating atmosphere in enjoyable surroundings.

REFERENCES

1. G. Wolf, lectures given at this school
2. G. Barbiellini, Bonn Conference (1981)
3. S.L. Glashow, Nucl. Phys. 22, 579 (1961)
   S. Weinberg, Phys. Rev. Lett. 19, 1264 (1967)
   A. Salam, Proceedings of 8th Nobel Symposium (Almqvist and Wiksell, Stockholm 1968)
4. The LEP Project: CERN/ISR-LEP 79-33 (1979)
5. The Cornell 2 x 50 GeV Project
6. The single-pass collider (SLC) at SLAC: SLAC Report 229 (1980)
7. C.Y. Prescott, SLAC-PUB-2649 (1980)
8. M. Davier, LEP Summer Study, Les Houches, CERN 79-01 (1979)
   K. Winter, id.
9. G. Goggi, ECFA/LEP/2 Note (1978)
10. J.E. Augustin, ECFA/LEP/29 Note (1978)
11. See for example: E. Picasso, ECFA/LEP Report (Sept. 1979)
12. R.P. Feynman and R.D. Field, Nucl. Phys. B136, 1 (1978)
13. M. Davier and F. Richard, ECFA/LEP/34 Note (1978)
14. A. De Rujula et al., Annals of Phys. 109, 242 (1977)

15. B. Degrange, ECFA/LEP/30 Note (1978)

16. M. Kobayashi and K. Maskawa Progr. Theor. Phys. 49, 652 (1973)

17. S. Schramm, Enrico Fermi Institute 78-25 (1978)

18. J. Ellis, SLAC-PUB-2177 (1978)

19. M. Veltman, Nucl. Phys. B123, 89 (1977)
M. Chanowitz et al. Phys. Lett. 78B, 285 (1978)

20. P. Fayet, CERN-TH 2864 (1980)
G. Farrar and P. Fayet, Phys. Lett 89B, 191 (1980)

21. Preliminary results from the CELLO collaboration, to be published.

22. J. Ellis and M.K. Gaillard, CERN 76-18 (1976)

23. N. Cabibbo and R. Gatto, Phys. Rev. 124, 1377 (1961)

24. W. Alles et al., Nucl. Phys. B119, 125 (1977)

25. F.M. Renard, Univ. de Montpellier, PM 78-11 (1978)

26. See for example: G. Barbiellini et al., DESY 79/27 (1979)

27. F. Wilczek, Phys. Rev. Lett. 39, 1304 (1977)

28. R. Cahn et al., LBL 8495 (1978)

29. J.D. Bjorken, SLAC-PUB-1866 (1977)

30. J. Ellis et al., Nucl. Phys. B106, 292 (1976)
B.W. Lee et al., Phys. Rev. D16, 1519 (1977)
S.L. Glashow et al., Phys. Rev. D18, 1724 (1978)

31. D. Jones and J. Petcov, CERN TH-2661 (1979)

32. W. Bacino et al., SLAC-PUB-2113 (1978)

33. S. Weinberg, Phys. Rev. D19, 1277 (1979)
L. Susskind, Phys. Rev. D20, 2619 (1979)

34. J. Ellis et al., Nucl. Phys. B182, 529 (1981)

35. S.L. Wu, private communication.

THEORETICAL ASPECTS IN PERTURBATIVE QCD

G. Altarelli

Istituto di Fisica dell'Università - Roma
Istituto Nazionale di Fisica Nucleare - Sezione di Roma

I. INTRODUCTION

The aim of this set of lectures is to provide an impressionistic view of the current research activity in the field of perturbative QCD [1)] by discussing a selection of topics which in a way or another are of particular interest at present. In spite of the increased technicism of the language spoken by experts, it will be possible, I hope, to convey the meaning and physical significance of what is being done in relatively simple terms, full accessible to all of you who certainly are already familiar with the more elementary aspects of the subjet.

I shall start by summarizing the general framework of the improved version of the parton model which emerges in QCD at the leading logarithmic approximation (LLA) and introduce the different sources of non leading corrections. As a particularly interesting example of the simplicity and physical transparency of the LLA, its application to the photon structure functions will be considered in detail. This is not a particularly recent set of results but the interest on this issue has been revived now by the appearance of the first few data at PETRA and the prospects of more to come soon.

I shall then discuss some problems connected with going beyond the LLA. Precise definitions of the running coupling , of parton densities and so on have to be introduced for this purpose. The sources of prescription dependence and the ways to go around them will be discussed. In particular, I shall consider in detail the separation of what is physical and what is matter of conventions in the form of the running coupling in the large scale region, i.e. the meaning and ways of determining the famous energy scale Λ. Important examples of calculations of next to the leading corrections will be illustrated, including those to the evolution equations which determine the scaling violations in both the spacelike and timelike regions of Q^2. The structure of the second order kernels is analysed and their interpretation, together with some physical implications of their computed explicit forms, is considered. The somewhat related problem of the Q^2 dependence of multiplicities in parton jets, on which some new interesting results have been recently obtained, is also briefly covered.

In the second part of these lectures I shall go to problems related to the resummation of doubly logarithmic terms. The simplest, although somewhat unphysical, example in this class of problems, is the determination of the asymptotic behaviour of the Sudakov form factor of partons, which is best suited as a prototype case. More physical applications are met in the study of transverse momentum distributions and in the resummation of some large non leading corrections as those appearing in Drell-Yan process. Thus I shall also briefly discuss the status of these areas of QCD phenomenology.

II. THE LLA AND BEYOND

In general the naive parton model formulae for deep inelastic processes are also asymptotically valid in QCD within the LLA provided that they are reexpressed in terms of scale dependent effective parton densities and fragmentation functions. This is true provided that one neglects not only terms down by inverse powers of the energy scale, but also non leading corrections which vanish asymptotically with powers of the running coupling constant. The factorization theorem[2)] guarantees that, for a given target or a given final state hadron, the same universal parton densities or fragmentation functions appear in all processes where that target or that hadron enter. For example we have the following list of simple processes at most quadratic in the parton densities and/or the decay functions :

- Inclusive Leptoproduction[3)] : $e + N \to e' + X$

$$\mathcal{F}_a(x,Q^2) \sim e^2 q(x,Q^2) + O[\alpha(Q^2)] \tag{2.1}$$

where $\mathcal{F}_a \equiv (2F_1, F_2/x, F_3)$ and $q(x,Q^2)$ is the (generic) effective quark density in the target, as seen by a virtual probe γ^* of virtual mass mass $-Q^2$. Sums over quarks and antiquarks of different flavours are omitted for simplicity.

- Inclusive $e^+ e^-$ Annihilation [4)] : $e^+ e^- \to$ hadrons

$$R = \frac{\sigma_{hadrons}}{\sigma_{\mu^+\mu^-}} = \frac{4\pi\alpha_{em}^2}{3Q^2} \sum_{flavours} e_q^2 \{1 + O[\alpha(Q^2)]\} \tag{2.2}$$

- One Hadron Inclusive $e^+ e^-$ Annihilation [5)]: $e^+e^- \to H + X$

$$\overline{F}_a(z,Q^2) \sim e^2 D_q^H(z,Q^2) + O[\alpha(Q^2)] \qquad (2.3)$$

where $\overline{F}_a$ are the structure functions analogous to F_a in leptoproduction and D_q^H is the effective fragmentation (or decay) function, i.e. the effective number density of the hadron H in the jet of the quark q.

- Drell-Yan processes [6)]: $H_1 + H_2 \to \ell^+\ell^- + X$

$$\left.\frac{d\sigma}{dQ^2 dy}\right|_{y=0} \sim e^2 \,[\, q^{H_1}(\sqrt{\tau},Q^2)\bar{q}^{H_2}(\sqrt{\tau},Q^2) + 1 \leftrightarrow 2\,] + O[\,\alpha(Q^2)\,] \qquad (2.4)$$

where y is the rapidity and $\sqrt{\tau} = Q/\sqrt{s}$ is the ratio of the lepton pair and incoming hadron invariant masses.

- One Hadron Inclusive Leptoproduction [5), 7)]:
$e + H \to e' + H + X$

$$F_a(x,z,Q^2) \sim e^2 q(x,Q^2) D_q^H(z,Q^2) + O[\alpha(Q^2)] \qquad (2.5)$$

- Two Hadron Inclusive $e^+ e^-$ Annihilation [5)]:
$e^+ e^- \to H_1 + H_2 + X$

$$\overline{F}_a(z_1,z_2,Q^2) \sim e^2[\,D_q^{H_1}(z_1,Q^2)D_{\bar{q}}^{H_2}(z_2,Q^2) + 1 \leftrightarrow 2] + O[\alpha(Q^2)] \qquad (2.6)$$

There are of course also processes of great interest which have no naive parton model analogue in the sense that the relevant rates vanish in the limit of $\alpha(Q^2) \to 0$. Examples which are of particular practical interest are :

- Hard transverse momentum distributions[10]:

$$< p_\perp > \sim \alpha(Q^2)\sqrt{s}\, f(\tau,\alpha(Q^2)) + O[\alpha(Q^2)] \qquad (2.8)$$

where in particular the last formula refers to the average transverse momentum in Drell-Yan processes.

- Heavy Quarkonium Decays [11] :

In particular for the ground state (ortho-quarkonium) :

$$\Gamma(0^{-+} \rightarrow 2G) = \frac{8}{3}\,\alpha^2(M^2)\,\frac{|R(0)|^2}{M^2}\,\{1 + O[\alpha(M^2)]\} \qquad (2.9)$$

and similarly for other decay modes and higher quarkonium states J^{PC}.

The next to the leading corrections to all these processes are by now explicitly known, from the computations quoted in each case. The corrections to transverse momentum distributions are not yet complete and some controversy exists on the non leading terms for three jet distributions in e^+e^- [9),11)]. Such corrections to the leading formulae are important in many respects. In particular we recall that the LLA is only well defined when all possible energy scales in the process are large and of the same order, so that it does not really make a difference which is chosen among them. However near the boundary of phase space ratios of a-priori equivalent energy scales become large and the LLA in general breaks down. The non leading corrections, which of course depend on the choice of scale in the leading term, describe where and how this breaking is to be expected. Thus they provide an order by order criterium for choosing a scale in the leading terms in order to somewhat extend the range of validity of the LLA. More in general it is only after computation of the nth order correction (in the running coupling) that one can estimate the error which affects the $(n-1)^{th}$ term and thus attach a figure of merit to it. The actual size of the corrections turns out to depend very much on the process and for some processes they are conspicuous at relevant energies. An example of large corrections is provided by the relation between quark densities as measured in leptoproduction and Drell-Yan processes [6].

The quoted formulae for processes with naive parton analogue relate different processes at the same energy scale Q^2 once $q(x,Q^2)$ and $D_q^H(z,Q^2)$ (and the like for gluons) are given. A different but related problem is the connection between parton densities and decay functions at different energy scales. This connection is provided by the evolution equations [12].

With the notation :

$$A \otimes B \equiv \int_x^1 \frac{dy}{y} A(y) B(\frac{x}{y}) = \int_x^1 \frac{dy}{y} A(\frac{x}{y}) B(y)$$

they read for parton densities, in the LLA :

$$Q^2 \frac{d}{dQ^2} q_i = \frac{\alpha(Q^2)}{2\pi} [q_i \otimes P_{qq} + G \otimes P_{qG}] + 0[\alpha^2(Q^2)]$$

$$Q^2 \frac{d}{dQ^2} G = \frac{\alpha(Q^2)}{2\pi} [\sum_i q_i \otimes P_{Gq} + G \otimes P_{GG}] + 0[\alpha^2(Q^2)] \qquad (2.10)$$

while for the decay functions one has [13] :

$$Q^2 \frac{d}{dQ^2} D_{q_i} = \frac{\alpha(Q^2)}{2\pi} [D_{q_i} \otimes P_{qq} + D_G \otimes P_{Gq}] + 0[\alpha^2(Q^2)]$$

$$Q^2 \frac{d}{dQ^2} D_G = \frac{\alpha(Q^2)}{2\pi} [\sum_i D_{q_i} \otimes P_{qG} + D_G \otimes P_{GG}] + 0[\alpha^2(Q^2)] \qquad (2.11)$$

Note the transposition $P_{qG} \leftrightarrow P_{Gq}$ in going from the spacelike case to the timelike case. This arise because while in the spacelike case the hadron target is there before the interaction with the probe γ^*, in the timelike case the produced hadron comes out at the end of the production event started by γ^*.

The corrections to the LLA of order $\alpha^2(Q^2)$ for the evolution equations are also known [14),15)]. These are important for a precise comparison of the same process at two different energy scales, i.e. for a more accurate evaluation of scaling violations. Needless to say if one wants to relate with precision two different processes at two different scales, then both the next to the leading corrections to the parton formulae and to the evolution equations are both needed and enter on an equal footing.

As explained in detail later, we mention that the form of the correction depends on the exact definitions which are introduced for the quantities appearing in the leading terms. Beyond the LLA the definition of running coupling, of parton densities and decay functions can be given in infinitely many ways. For example take a quark density. In the naive parton model this quantity is meant to be the number density of quarks in the target (in the infinite momentum frame). The Q^2 dependent effective quark density differs from the previous one by inclusion of a factor, extracted from the parton-current cross section, which contains the leading logarithmic series of terms and the related mass singularities. This statement is enough to make the definition unambiguous in the LLA but beyond it one could include extra non singular, non leading logarithmic terms in the factorization mentioned above, without affecting the LLA but modifying the next to leading (and higher) terms.

This discussion will be resumed later in all details. But first we consider the photon structure functions as a highly interesting and physically very important application of the LLA formalism.

III. THE PHOTON STRUCTURE FUNCTIONS

The interest of deep inelastic scattering on a photon target stems from the fact, first discovered by Witten [16] that the photon structure functions can be absolutely predicted in the asymptotic limit, and yet the result does not coincide with the free field prediction.

The photon structure functions $2F_1^\gamma$, F_2^γ/x are determined as usual once the quark (and gluon) densities in the photon q_i^γ (and G^γ) are known. In the LLA these densities can be determined as follows. One starts by observing that the evolution equations are modified in this case by the presence of another parton species : the photon itself. They read [17] :

$$Q^2\frac{d}{dQ^2}\, q_i^\gamma(x,Q^2) = \frac{\alpha(Q^2)}{2\pi}\,[q_i^\gamma \otimes P_{qq} + G^\gamma \otimes P_{qG}] + \frac{\alpha_{em}}{2\pi}\,\gamma^\gamma \otimes P_{q\gamma} + \dots \tag{3.1}$$

$$Q^2\frac{d}{dQ^2}\, G^\gamma(x,Q^2) = \frac{\alpha(Q^2)}{2\pi}\,[\Sigma^\gamma \otimes P_{Gq} + G^\gamma \otimes P_{GG}] + \dots \tag{3.2}$$

The last term in eq. (3.1), proportional to the photon density in a photon γ^γ, is of order $\alpha_{em} \simeq 1/137$ and cannot be neglected in this case because q_i^γ and G^γ are themselves of order α_{em}. We shall always work in first order in α_{em}. The photon density starts at zeroth order in α_{em} and we can make the identification :

$$\gamma^\gamma(x,Q^2) \simeq \delta(1-x) \tag{3.3}$$

In fact a factor of α_{em} is already brought in by the probability $(\alpha_{em}/2\pi)P_{q\gamma}$ of finding a quark in the photon. $P_{q\gamma}$ and P_{qG} only differ by a colour factor (T/f is replace by tr(I) = N, the number of colour replicas for each quark) :

$$P_{q\gamma}(x) = N[x^2 + (1-x)^2] \tag{3.4}$$

In eq. (3.2) Σ^γ is defined as

$$\sum_{a=1}^{f} (q_a^\gamma + \bar{q}_a^\gamma)$$

and there is no term in $P_{G\gamma}$ because there is no vertex in the theory directly connecting a photon with a gluon.

In the unphysical limit of completely neglecting the strong interactions one would obtain from eqs. (3.1, 3.3) the naive parton model result :

$$q_i^\gamma(x,Q^2)_{naive} = \frac{\alpha_{em}}{2\pi} e_i^2 \, N[x^2 + (1-x)^2] \ln \frac{Q^2}{\Lambda_\gamma^2} + \ldots \tag{3.5}$$

where the dots stand for non logarithmic terms which can be asymptotically neglected. The energy Λ_γ is a convenient scale that, on a purely conventional basis, can be set equal to the coupling constant scale : $\Lambda_\gamma \sim \Lambda$. Eq. (3.5) correspond to the absorptive part of the box diagram contribution to the $\gamma^*\gamma$ forward amplitude (fig. 1a)

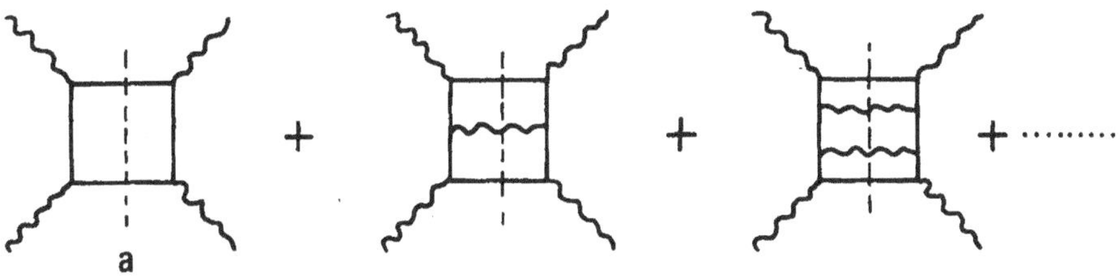

Fig. 1. A sequence of ladder diagrams (with gluons as rungs) for deep inelastic scattering on a photon target. Labelled with (a) in the box diagram corresponding to the naive parton model.

In order to proceed in studying the effect of strong interactions we separate as usual the non singlet and the singlet quark evolution equations :

$$Q^2 \frac{d}{dQ^2} (q_i^\gamma)_{NS} = \frac{\alpha(Q^2)}{2\pi} (q_i^\gamma)_{NS} \otimes P_{qq} + \frac{\alpha_{em}}{2\pi} (e_i^2)_{NS} P_{q\gamma} + \ldots \tag{3.6}$$

$$Q^2\frac{d}{dQ^2}\,\Sigma^\gamma = \frac{\alpha(Q^2)}{2\pi}[\Sigma^\gamma \otimes P_{qq} + G^\gamma \otimes 2fP_{qG}] + \frac{\alpha_{em}}{2\pi}\,2f <e^2> P_{q\gamma} + \ldots \tag{3.7}$$

where:

$$(q_i^\gamma)_{NS} = q_i^\gamma - \frac{1}{2f}\,\Sigma^\gamma \tag{3.8}$$

$$< e^k > = \frac{1}{f}\sum_{i=1}^{f} e_i^k \tag{3.9}$$

$$(e_i^2)_{NS} = e_i^2 - < e^2 > \tag{3.10}$$

We first consifder the non singlet equation which contains all the interesting features in simplest form. By dropping all inessential labels and going to moments, the relevant equation is of the form:

$$Q^2\frac{d}{dQ^2}\,q_n(Q^2) = \alpha(Q^2)q_n(Q^2)bd_n^{qq} + \alpha_{em}e^2bd_n^{q\gamma} \tag{3.11}$$

where d_n are the logarithmic exponents defined by:

$$d_n = \frac{1}{2\pi b}\int_0^1 dx\, x^{n-1}\, P(x) \tag{3.12}$$

with $\alpha(Q^2) \simeq 1/b\ln(Q^2/\Lambda^2)$. The general solution of this inhomogeneous equation is given by the general solution of the associated homogeneous equation plus a particular solution of the complete equation. One obtains:

$$q_n(Q^2) = q_n^0\left[\ln\frac{Q^2}{\Lambda^2}\right]^{d_n^{qq}} + \frac{\alpha_{em}e^2 d_n^{q\gamma}}{(1-d_n^{qq})}\,\frac{1}{\alpha(Q^2)} \tag{3.13}$$

The particular solution is given by the second term and it would exactly coincide with the naive parton result (see eq. (3.5)) if not for the factor $(1-d_n^{qq})^{-1}$. This rescaling factor is smaller than 1 for $n > 1$, because $d_n^{qq} < 0$ in this range. The particular solution is asymptotically dominant over the first term which contains the whole hadronic structure of the photon (for example the vector meson dominance contributions). Thus the asymptotic behaviour of $q_n(Q^2)$ is an absolute prediction.

The way the asymptotic solution is built up in perturbation theory is elucidated by a diagrammatic analysis in terms of dressed ladders in physical gauges [18] (fig. 1). The ladder structure (the dressing replaces α with $\alpha(Q^2)$) leads to the recursive structure :

$$q_n(Q^2) = \alpha_{em} e^2 b d_n^{q\gamma} \Big[\ln\frac{Q^2}{\Lambda^2} + \int^{Q^2} \frac{dQ_1^2}{Q_1^2} \alpha(Q_1^2) b d_n^{qq} \ln\frac{Q_1^2}{\Lambda^2}$$

$$+ \int^{Q^2} \frac{dQ_1^2}{Q_1^2} \alpha(Q_1^2) b d_n^{qq} \int^{Q_1^2} \frac{dQ_2^2}{Q_2^2} b d_n^{qq} \ln\frac{Q_2^2}{\Lambda^2} + \ldots \Big] \quad (3.14)$$

All orders in the recursion contribute with a linear term in $\log Q^2/\Lambda^2$ because $\alpha(Q^2)\, b \log Q^2/\Lambda^2 \simeq 1$:

$$q_n(Q^2) = \alpha_{em} e^2 b q_n^{q\gamma} \ln\frac{Q^2}{\Lambda^2} \{1 + d_n^{qq} + (d_n^{qq})^2 + \ldots\}$$

$$= \frac{\alpha_{em} e^2 b d_n^{q\gamma} \ln Q^2/\Lambda^2}{(1 - d_n^{qq})} \quad (3.15)$$

Note that the free limit corresponds to $b \to \infty$, i.e. $d_n^{qq} = 0$ and $b d_n^{q\gamma}$ finite.

It is straightforward to also solve the singlet equations by following quite similar lines. The complete asymptotic solutions can then be written as :

$$(q_i^\gamma)_n = \frac{\alpha_{em} d_n^{q\gamma}}{\alpha(Q^2)} \Big\{ \frac{e_i^2 - \langle e^2\rangle}{1 - d_n^{qq}} + \frac{\langle e^2\rangle(1 - d_n^{GG})}{(1-d_n^+)(1-d_n^-)} \Big\}$$

$$(G^\gamma)_n = \frac{\alpha_{em} d_n^{q\gamma}\, 2f\langle e^2\rangle\, d_n^{Gq}}{\alpha(Q^2)(1-d_n^+)(1-d_n^-)} \quad (3.16)$$

$d_n^\pm$ are the eigenvalues of the singlet logarithmic exponent matrix.

Note that

$$(\Sigma^\gamma)_n/(G^\gamma)_n = (1 - d_n^{GG})/d_n^{Gq} \quad (3.17)$$

In particular for n = 2 the ratio of the momentum fractions carried by quarks and gluons in a photon turns out to be independent of the number of quark flavours f (and of Q^2) [19] :

$$\frac{(\Sigma^\gamma)_{n=2}}{(G^\gamma)_{n=2}} = \frac{11C_n}{8C_F} \underset{QCD}{=} \frac{99}{32} \tag{3.18}$$

It is now easy to reconstruct the structure functions. From

$$F_2(x,Q^2) \simeq 2xF_1(x,Q^2) \simeq x \sum_{i=1}^{2f} e_i^2 q_i(x,Q^2)$$

by taking moments, we immediately find (recall eq.(3.9) :

$$(2F_1^\gamma)_n \simeq (F_2^\gamma/x)_n$$

$$\simeq \frac{\alpha_{em}}{\alpha(Q^2)} \frac{2f\, d_n^{q\gamma}}{} \{\frac{\langle e^4\rangle - \langle e^2\rangle^2}{1 - d_n^{qq}} + \frac{\langle e^2\rangle^2 (1 - d_n^{GG})}{(1-d_n^+)(1-d_n^-)}\} \tag{3.19}$$

With respect to the naive parton model result the effect of QCD produces a pronounced depletion of the structure function at large x because of the gluon radiation that softens the quark distributions.

The contributions of q_i^γ and G^γ to the longitudinal structure function of the photon F_L^γ, $F_L^\gamma = F_2^\gamma - 2xF_1^\gamma$, is directly obtained as usual (see eq. (5.16)). In addition there is also a contribution from the photon density, eq. (3.3), which can immediately be obtained from the gluon term by a simple coupling and colour factor rescaling. One finally has :

$$F_L^\gamma(x) = \frac{\alpha_{em}}{2\pi} \; 24f\langle e^4\rangle x^2(1-x) + x^2 \int_x^1 \frac{du}{u^3} \{\frac{8}{3}\frac{\alpha(Q^2)}{2\pi} \; F_2^\gamma(u,Q^2)$$

$$+ 4f\langle e^2\rangle(\frac{\alpha(Q^2)}{2\pi} \; uG^\gamma(u,Q^2))(1 - \frac{x}{u})\} \tag{3.20}$$

Note that the second term is independent of Q^2 as is the first one because both F_2^γ and G_2^γ are proportional to $1/\alpha(Q^2)$. Thus F_L^γ scales in the LLA. In terms of moments :

$$\int_0^1 dx\, x^{n-2} F_L^\gamma(x) = 2f \frac{\alpha_{em}}{2\pi} \{ \frac{12<e^4>}{(n+1)(n+2)} + d_n^{q\gamma} [\frac{8}{3(n+1)} (\frac{<e^4> - <e^2>^2}{1 - d_n^{qq}} + \frac{<e^2>^2 (1-d_n^{GG})}{(1-d_n^+)(1-d_n^-)}) + \frac{4f<e^2>^2 d_n^{Gq}}{(n+1)(n+2)(1-d_n^+)(1-d_n^-)}] \} \quad (3.21)$$

The next to leading leading corrections to F_2^γ and $2F_1^\gamma$ are also absolutely computable for $n > 2$. In fact at subleading accuracy the moments with $n > 2$ are of the form :

$$(F_2^\gamma/x)_n = \alpha_{em} [\frac{E_n}{\alpha(Q^2)} + H_n^2] \quad (3.22)$$

where E_n can be read from eq. (3.19) and $\alpha(Q^2)$ is to be taken here in the improved from (eq. (4.19)). The important point is that the hadronic components do not enter in the next to leading corrections $H_n^{1,2}$ because all the logarithmic exponents d_n^{qq}, $d_n^{\pm}$ are negative for $n > 2$. The moment with $n = 2$ is exceptional :

$$\int_0^1 dx\, F_2^\gamma(x,Q^2) = \alpha_{em} \{ \frac{E_2}{\alpha(Q^2)} + E_2' \ln\alpha(Q^2) + H_2^2 \} \quad (3.23)$$

Here E_2' is computable while H_2^2 is affected by the unknown matrix elements of the hadronic energy momentum tensor (whose anomalous exponent d_2^+ vanishes) between photon states. The values of H_n for $n > 2$ and of E_2' were computed by Bardeen and Buras[20]. For $n > 2$ non leading corrections (in the $\overline{MS}$ prescription for α) turn out to be sizeable at practical values of Q^2 and their sign is in the direction of increasing the difference with the naive parton model.

IV. PRESCRIPTION AND SCALE DEPENDENCE FOR THE RUNNING COUPLING

As a very important preliminary step before the discussion of the non leading corrections to several processes of physical interest we first recollect here a number of important points on the QCD coupling (fixed and running) especially on the definition and relevance of the scale parameter Λ which appears in the asymptotic expansion of the running coupling at large values of the energy scale.

Even in the case of massless QCD, where no parameter with dimension of mass appears in the bare lagrangian, a mass scale is necessarily introduced by the process of regularization and renormalization of the theory. A mass scale is needed for stating a proper definition of renormalized quantities, in particular of the renormalized coupling constant. The renormalized coupling can be defined in infinitely many ways, all of them implying, in some way or another, the introduction of a reference mass μ. We mention here two typical, vastly different, practical ways of defining the coupling : the momentum subtraction schemes and the minimal subtraction schemes (in dimensional regularization).

In momentum subtraction schemes the renormalized coupling $\alpha = g^2/4\pi$ is defined by some three (or four) point Green function at specified values of the external momenta. For example in QCD one can take the renormalized (one particle irreductible) three gluon vertex funtion $\Gamma_{3G}(p^2,q^2,r^2)$ at given symmetric values of p^2, q^2 and r^2 (the virtual squared masses of the external gluons) as the quantity defining the renormalized coupling g. A possible choice is $p^2 = q^2 = r^2 = -\mu^2$. The corresponding coupling α is thus a function of μ : $\alpha \equiv \alpha(\mu^2)$. Note that $\mu^2 = 0$ is not allowed because of the infrared singularities of Γ_{3G} for on shell massless gluons. In this respect the interplay of infrared and ultraviolet singularities is crucial and in fact it plays an important role in fixing the asymptotic behaviour of Green functions. Similarly the definition of the scale for the renormalized fields (after factorization of the wave function renormalization factors) requires to consider the inverse propagator at $p^2 = -\mu^2 \neq 0$.

In dimensional regularization the reference mass μ enters in a quite different way, as a scale for the coupling g which in $d < 4$ spacetime dimensions is no more adimensional :

$$g_{dim} = \mu^{\varepsilon} g_{adim} \equiv \mu^{\varepsilon} g \tag{4.1}$$

with $2\varepsilon = 4 - d$. Consider the same three gluon vertex function as before with gluons off shell by an amount p^2. The unrenormalized quantity, Γ_U, obtained directly from the computation of one loop Feyman diagrams in d dimensions, leads to a result of the form :

$$\Gamma_U = g\ \{1 + \alpha\left(\frac{-\mu^2}{p^2}\right)^{\varepsilon} [\frac{B}{\varepsilon} + A + O(\varepsilon)\ ]\} \tag{4.2}$$

The factor $\mu^{2\varepsilon}$ corresponds to eq. (4.1) while the factor $p^{-2\varepsilon}$, forced by dimensions, follows from the increased degree of infrared singularity at $\varepsilon > 0$. Renormalization by minimal subtraction [21] (MS) consists at this order (and similarly in higher order) in the prescription of subtracting only the pole part in $1/\varepsilon$ with no finite terms :

$$\Gamma_R = g\{1 + \alpha[(-\frac{\mu^2}{p^2})^{\varepsilon}(\frac{B}{\varepsilon} + A) - \frac{B}{\varepsilon}]\} = 1 + \alpha[B\ln\frac{\mu^2}{-p^2} + A] \tag{4.3}$$

Renormalization by modified minimal subtraction[3] ($\overline{MS}$) prescribes to only subtract the pole part in $1/\hat{\varepsilon}$, where :

$$\frac{1}{\hat{\varepsilon}} = \frac{1}{\varepsilon} + \ln 4\pi - \gamma_E \tag{4.4}$$

where γ_E is the Euler-Mascheroni constant ($\gamma_E \simeq 0.5772...$). This superficially curious definition is motivated by the systematic appearance of this combination of constants from all loop integrals in d dimensions. Note that the coefficient of $\ln\mu^2$ is independent of these details which only affect the finite terms. In principe Γ_R can be compared, at least indirectly, to experiment, thereby fixing α. It is clear that α_{MOM} defined by a given momentum subtraction prescription α_{MS} or $\alpha_{\overline{MS}}$ are all different. Roughly speaking the relation between α_{MOM} and α_{MS} is obtained by considering that

$$\Gamma_R(p^2, q^2, r^2)\Big|_{p^2=q^2=r^2=-\mu^2} \equiv g_{MOM} = g_{MS}(1 + \alpha_{MS}A + ...)$$

In general the relation between any two different prescriptions will be up the form :

$$\alpha = \alpha'(1 + k^{(1)}\alpha' + ...) \tag{4.5}$$

provided that the two couplings coincide in trivial order as always the case.

Although the value of any physical quantity is obviously unaffected by a change of prescription the functional form of any observable in terms of different couplings is clearly different. In particular all terms in the perturbative expansion of any physical quantity, except for the first non trivial term, depend on the definition adopted for the coupling. For example if $\rho(\alpha)$ is measurable and its perturbative expansion in terms of α is given by

$$\rho(\alpha) = \rho^{(0)} + \rho^{(1)}\alpha + \rho^{(2)}\alpha^2 + ... \tag{4.6}$$

then its expansion in terms of α', by inserting eq. (4.5) results to be :

$$\rho(\alpha') = \rho^{(0)} + \rho^{(1)}\alpha' + (\rho^{(2)} + k^{(1)}\rho^{(1)})\alpha'^2 + ... \tag{4.7}$$

The running of α , i.e. the dependence of α on the scale μ, is ruled by the beta function :

$$Q^2 \frac{d}{dQ^2} \alpha(Q^2) = \beta\{\alpha(Q^2)\} \qquad \alpha \equiv \alpha(\mu^2) \tag{4.8}$$

with

$$\beta(\alpha) = -b\alpha^2(1 + b'\alpha + b''\alpha^2 + \dots) \tag{4.9}$$

where b is positive in QCD as in all asymptotically free theories [1].

In view of the above discussion it is remarkable that in the expansion of $\beta(\alpha)$ both b and b' are independent of the renormalization prescription for α. This is of course due to the close relation between $\beta(\alpha)$ and α which follows from eq. (4.8). In fact on one hand by replacing eq. (4.5) into eq. (4.9) one finds :

$$\beta(\alpha) = -b\alpha^2(1 + b'\alpha + \dots) = -b\alpha'^2(1 + b'\alpha' + \dots)(1 + 2k^{(1)}\alpha' + \dots) \tag{4.10}$$

On the other hand from the definition of $\beta(\alpha)$ in eq. (4.8) one has :

$$\beta(\alpha) = \frac{\partial \alpha}{\partial \ln\mu^2} = \frac{\partial \alpha'}{\partial \ln\mu^2} \frac{\partial\alpha}{\partial\alpha'} = \beta(\alpha')(1 + 2k^{(1)}\alpha' + \dots) \tag{4.11}$$

By comparison one concludes that the first two terms of $\beta(\alpha)$, and only these, are indeed independent of the definition of α in the massless theory. In particular in QCD we have [22] :

$$b = \frac{33 - 2f}{12\pi} \qquad b' = \frac{153 - 19f}{2\pi(33 - 2f)} \tag{4.12}$$

where f is the number of excited flavours. The value of b" is also known in the MS/$\overline{\text{MS}}$ prescriptions (the two are the same : $b''_{MS} = b''_{\overline{MS}}$), but this last quantity has no physical meaning in itself (ref.23).

The solution of the differencial equation (4.8) with $\beta(\alpha)$ given by eq. (4.9) and the stated boundary condition leads to the relation :

$$\frac{\alpha}{\alpha(t)} = 1 + b\alpha t + b'\alpha \ln\frac{\alpha}{\alpha(t)} + \dots \tag{4.13}$$

where

$$t = \ln Q^2/\mu^2 \tag{4.14}$$

or equivalently :

$$\alpha(t) = \frac{\alpha}{1 + b\alpha t}\left[1 - \frac{b'\alpha}{1 + b\alpha t}\ln\frac{\alpha}{\alpha(t)} + \ldots\right] \qquad (4.15)$$

This in turn can be solved by iteration : first one obtains in the LLA :

$$\alpha(t)_{LLA} = \frac{\alpha}{1 + b\alpha t} \equiv \frac{1}{b\ln Q^2/\Lambda^2_{LLA}} \qquad (4.16)$$

with

$$\ln\frac{\mu^2}{\Lambda^2_{LLA}} = \frac{1}{b\alpha} \qquad (4.17)$$

In the second iteration one replaces the expression so obtained from the LLA into the second term :

$$\alpha(t) = \alpha(t)_{LLA}\left[1 - b'\alpha(t)_{LLA}\ln\frac{\alpha}{\alpha(t)_{LLA}} + \ldots\right]$$

$$= \frac{1}{b\ln\frac{Q^2}{\Lambda^2_{LLA}}}\left[1 - \frac{b'}{b\ln\frac{Q^2}{\Lambda^2_{LLA}}}\left(\ln\ln\frac{Q^2}{\Lambda^2_{LLA}} - \ln\ln\frac{\mu^2}{\Lambda^2_{LLA}}\right) + \ldots\right] \qquad (4.18)$$

Actually at this point one may decide, for sake of elegance, to change the definition of Λ in such a way as to cancel the last term. In fact one can set :

$$\alpha(t) = \frac{1}{b\ln Q^2/\Lambda^2}\left[1 - \frac{b'\ln\ln Q^2/\Lambda^2}{b\ln Q^2/\Lambda^2} + \ldots\right] \qquad (4.19)$$

provided that

$$\ln\mu^2/\Lambda^2 = \frac{1}{b\alpha} + \frac{b'}{b}\ln b\alpha \qquad (4.20)$$

to be compared with eq. (4.17). It is quite important to realize that the difference between Λ_{LLA} and Λ is not small in any sense. In other words at large Q^2 a (reasonably) large change of Λ amounts to a small change of $\alpha(t)$. A consequence of this fact is also that a small experimental error in $\alpha(t)$ at large Q^2 corresponds to a large error on Λ.

Since b and b' are independent of renormalization prescriptions the asymptotic form of $\alpha(Q^2)$ in eq. (4.19) can be adopted in all cases with b and b' given by eq. (4.12). Therefore a change of definition of α reflects itself in a change of the corresponding Λ. This means that, when mentioning Λ, the prescription must be specified which one refers to. It is however important to recall that, in view of the previous discussion, saying $\Lambda_{\overline{MS}}$ is not precise enough. One should also quote the exact relation this $\Lambda_{\overline{MS}}$ derives from. For example whether it is $(\Lambda_{\overline{MS}})_{LLA}$ defined from eq. (4.16) or eq. (4.18) or it is $\Lambda_{\overline{MS}}$ from eq. (4.19) or still something else. If we take into account this rather complicated set of specifications to be given in order to fix the meaning of Λ and the above remarks on the experimental errors on Λ at large energies we are led to recommend that experiments should be phrased in terms of $\alpha(Q_0^2)_{\overline{MS}}$ at some convenient Q_0^2, rather than in terms of $\Lambda_{\overline{MS}}$.

In order to further discuss these matters let us consider the non leading corrections to the ratio R in e^+e^- annihilation and to the rates of heavy quarkonium decays. For R one knows that :

$$R = R_0 \{1 + \frac{\alpha(Q^2)}{\pi} + C_2 \frac{\alpha^2(Q^2)}{\pi^2} + \dots\} \tag{4.21}$$

with [24] (for α in the $\overline{MS}$ prescription) :

$$C_2 = \frac{C_F}{16}\{\frac{123}{2} C_A - \frac{3}{2} C_F - 11f - 48\pi b\zeta(3)\} \underset{QCD}{=} 1.986 - 0.115f \tag{4.22}$$

where $\zeta(3) = 1.2021 \dots$ If we go from α to α' related as in eq. (4.5), then clearly :

$$C_2 \to C_2 + \pi k^{(1)} \tag{4.23}$$

C_2 not only depends on the definition of α but also on the scale Q^2 which is chosen for $\alpha(Q^2)$. In fact from

$$\alpha(Q^2) \simeq \alpha(\eta Q^2) \{1 + b\ln\eta\; \alpha(\eta Q^2) + \dots\} \tag{4.24}$$

with any reasonable η(i.e. $b\ln\eta\; \alpha(\eta Q^2) \ll 1$), it follows that

$$C_2 \to C_2 + \pi b\ln\eta \tag{4.25}$$

Thus C_2 is only determined after fixing a prescription and a scale for α. Taken by itself, the value of the calculation of C_2 is to show that measuring R at the energy Q^2 in a sensible quantitative way (up to terms of order α^3) of measuring $\alpha_{\overline{MS}}(Q^2)$. Also it serves in a sense to completely qualify the previous term in the expansion. For example it confirms that the prediction of an approach from above of the asymptotic limit is expected to hold at present energies.

We now consider more than one single process and take as an example the valuable set of results which are known for heavy quarkonium decays (moments of structure functions in leptoproduction or annihilation will be considered in next section). Let us define

$$B(J^{PC}) = \Gamma(J^{PC} \to 2G)/\Gamma(J^{PC} \to 2\gamma) \tag{4.26}$$

then for the ground state quarkonium we have[11] :

$$B(0^{-+}) = \frac{2}{9e_Q^4}\left(\frac{\alpha(\mu^2)}{\alpha_{em}}\right)^2 \{1 + \frac{\alpha(\mu^2)}{\pi}[\frac{61}{2} - \frac{13}{8}\pi^2 - \frac{2}{3}\ln 2 + \frac{25}{3}\ln(\frac{\mu}{2m_Q})] + \ldots\} \tag{4.27}$$

where the dependence on the scale μ chosen for the running coupling was explicitly shown. By putting numbers one finds :

$$\begin{aligned} \{\ \} &= (1 + 14\,\frac{\alpha}{\pi}) \quad \text{for} \quad \mu = 2m_Q \\ &= (1 + 8.2\,\frac{\alpha}{\pi}) \quad \text{for}\ \mu = m_Q \end{aligned} \tag{4.28}$$

We see that a smaller scale is favoured. This is reasonable because $2m_Q$ is the largest momentum transfer one can think of for a gluon in this process and not the average one (in a sense specified by dynamics). However even for $\mu = m_Q$ (m_Q is the heavy quark mass) the correction is still large. Of course one can think of changing the definition of α in order to make the coefficient smaller by force. Since the leading term is proportional to α^2 the coefficient of α/π in the non leading term is rather sensitive to a change of α. Thus one can make this coefficient small without spoiling the apparently good convergence of R (and of the first few moments in deep inelastic scattering). However this playing around is not very much of use. In fact let us now consider other 2G or 2γ decays and express the corresponding results in the form of ratios (all with the same scale) so that all prescription dependence drops away [11] :

$$\frac{B(0^{++})}{B(0^{-+})} = (1 + 0.9\,\frac{\alpha}{\pi} + \ldots)_\psi \; ; \; (1 + 2.1\,\frac{\alpha}{\pi} + \ldots)_\Upsilon \tag{4.29}$$

$$\frac{B(0^{++})}{B(2^{++})} = \left(1 + 6.5\,\frac{\alpha}{\pi} + \ldots\right)_{\psi} \;;\; \left(1 + 4.0\,\frac{\alpha}{\pi} + \ldots\right)_{T} \tag{4.30}$$

$$\frac{\Gamma(0^{++} \to 2\gamma)}{\Gamma(2^{++} \to 2\gamma)} = \frac{15}{4}\left(1 + 5.5\,\frac{\alpha}{\pi} + \ldots\right) \tag{4.31}$$

$$\frac{\Gamma(0^{++} \to 2G)}{\Gamma(2^{++} \to 2G)} = \frac{15}{4}\left(1 + 12\,\frac{\alpha}{\pi} + \ldots\right) \tag{4.32}$$

where separate values for the J/ψ and T families were given (when distinct). We see that no single choice of prescription can make the corrections to all processes small. Clearly the possibility of different ad hoc scales for these different processes of quite similar topology is not appealing either.

In conclusion we have that all prescription dependence is eliminated when data are compared with data. Actually all prescription dependent formulae are to be considered as useful intermediate steps, before one picks up a given measurement to replace α. Also, the non leading terms of each process can be adjusted by a variation of scale in the leading term. For example it has been proposed to pick up that scale which minimizes the scale dependence [53)] (to the given order). I think that the change of scale is a sensible procedure (i.e. likely to be supported in still higher orders) only if the needed (not too large) scale difference can be a-priori justified by some physical argument.

Before closing this section we mention that quite recently the calculation was completed [25)] of the next to leading corrections to the total width of ortho-quarkonium ($J^{PC} = 1^{--}$) which in lowest order starts as α^3 corresponding to three gluons in the final state. This important result is reported in detail in Berkelman's course.

V. SPACELIKE AND TIMELIKE STRUCTURE FUNCTIONS BEYOND THE LLA

In this section we consider the corrections to the LLA in leptoproduction. We have seen that the naive parton model results are formally reproduced in the LLA in terms of Q^2 dependent effective parton densities which satisfy evolution equations with known kernels. Corrections of order $\alpha(Q^2)$ to the parton formulae and of order $\alpha^2(Q^2)$ to the evolution equations have been simply indicated in section II. Here we first consider in succession the non leading corrections to the parton formulae and then we go in more detail to the evolution equations.

For a compact description of all independent structure functions in leptoproduction it is convenient to introduce the notation :

$$\mathcal{F}_a \equiv (\mathcal{F}_1, \mathcal{F}_2, \mathcal{F}_3) \equiv (2F_1, F_2/x, F_3) \tag{5.1}$$

In order to simplify the writing we temporarily restrict the discussion to the non singlet projections of $\mathcal{F}_a$. We start from the general expression of $\mathcal{F}_a$ as a convolution of an effective parton density and of a reduced cross section after extraction of a factor containing the leading logarithmic sequence :

$$\mathcal{F}_a(x,Q^2) = \int_x^1 \frac{dy}{y}\, q(y,Q^2)e^2 \{\delta(\frac{x}{y} - 1) + \frac{\alpha(Q^2)}{2\pi} f^a(\frac{x}{y}) + \dots\}$$

$$\equiv q \otimes e^2 \{ \delta + \frac{\alpha(Q^2)}{2\pi} f^a + \dots \} \tag{5.2}$$

with q(x, t) obeying the evolution eq. (2.10) (here the sum over q and $\bar{q}$ flavours was omitted). On the other hand from general arguments of light cone dominance and renormalization group techniques one can write for moments of $\mathcal{F}_a$:

$$M_n^a(Q^2) = C_n^a\,[\alpha(Q^2)]\, Q_n(\alpha) \exp \int_\alpha^{\alpha(Q^2)} \frac{\gamma_n(\alpha)}{\beta(\alpha)}\, d\alpha \tag{5.3}$$

the correspondence between the two formulations might be put as follows :

$$\int_0^1 dx\, x^{n-1}\, q(x,Q^2) = Q_n(\alpha) \exp \int_\alpha^{\alpha(Q^2)} \frac{\gamma_n(\alpha)}{\beta(\alpha)}\, d\alpha$$

$$\int_0^1 dx\, x^{n-1}\, e^2 \{\delta(1-x) + \frac{\alpha(Q^2)}{2\pi} f^a(x) + \dots\} = C_n^a\, [\alpha(Q^2)] \tag{5.4}$$

However neither $C_n^a(\alpha)$ nor $\gamma_n(\alpha)$ are separately well defined beyond the LLA. Correspondingly we have already mentioned that the set of $f^a(x)$ is not unambiguous and in fact these non leading terms depend on the regularization adopted for infrared singularities and on the renormalization prescription (through virtual diagrams). This is an indication that the effective parton densities have been so far completely specified only within the accuracy of the LLA. In the naive parton model one thought to have an intrinsic, measurable definition of parton densities. But in going from naive to QCD improved parton densities the original definition was modified by inclusion of a universal (convolution) factor, extracted from the parton-current cross section and containing the leading logarithmic series of terms with all related mass singularities. Beyond the LLA the definition becomes ambiguous because one could clearly also include an additional universal non leading logarithmic factor without altering the results of the LLA. A change of the definition of q(x, t) and G(x, t) beyond the LLA corresponds to a change of the reduced coefficients $C_n^a(\alpha)$, of the anomalous dimension functions $\gamma_n(\alpha)$ and to a variation of the non leading corrections to individual processes and also to the evolution equations for parton densities.

A precise definition of parton densities beyond the LLA can be fixed in many different ways. Depending on the definition adopted the charge and momentum sum rules can be either preserved to all orders in the running coupling or receive computable corrections. One possibility is to exactly specify the computation method and the renormalization prescription and then use the first eq. (5.4) for defining q(x,t). A second possibility, which we prefer, is to link the definition of parton densities beyond the LLA to some suitable processes acting as standard "partonometers".

According to this idea quark densities which exactly satisfy the valence charge sum rules i.e. without corrections of order $\alpha(Q^2)$, can be defined from the structure functions F_2 for all possible beams on a given target [6]. That is one defines quark densities by setting to all orders in $\alpha(t)$:

$$F_2(x,Q^2) \equiv \sum_i e_i^2 q_i(x,Q^2) \tag{5.5}$$

or equivalently

$$\int_0^1 dx\, x^{n-1} \sum_i e_i^2\, q_i(x,t) = C_2[\,\alpha(t)\,]\, Q_n(\alpha) \exp \int_\alpha^{\alpha(t)} \frac{\gamma_n(\alpha)}{\beta(\alpha)}\, d\alpha \tag{5.6}$$

where e_i^2 are the appropriate naive parton model couplings squared. Clearly the r. h. s. is now independent of prescriptions. F_2 is given a special role in that it is connected with the Adler sum rule. This sum rule which follows from the algebra of charges is exact and not only asymptotically true. In particular when all powers of m^2/Q^2 are neglected it can be written in the form (to all orders in $\alpha(Q^2)$) :

$$\int_0^1 \frac{dx}{x} \{F_2^{\nu T}(x,Q^2) - F_2^{\bar{\nu} T}(x,Q^2)\} = A_0 \tag{5.7}$$

where T is the target and A_0 is a fixed number depending on T and the structure of weak couplings. Note that $(F_2^{\nu T} - F_2^{\bar{\nu}T})/x$ is a combination of differences $\Sigma C_\ell(q_\ell - \bar{q}_\ell)$ between quarks and antiquarks of the same flavour, with $\Sigma C_\ell = 0$. Thus by defining all quark densities from F_2 we automatically ensure that the valence charge sum rules are preserved from corrections of order $\alpha(Q^2)$.

With the definition of $q_i(x,t)$ from F_2 the LLA formulae, which for each term in the current are schematically given by

$$F_1 \simeq F_2 \sim q + \bar{q} \qquad ; \quad F_3 \sim -q + \bar{q} \tag{5.8}$$

in next to leading accuracy are modified into:

$$F_2 \sim q + \bar{q} \tag{5.9}$$

$$F_1 \sim (q+\bar{q}) \otimes \{\delta + \frac{\alpha(Q^2)}{2\pi}(f_q^1 - f_q^2)\} + 2\frac{\alpha(Q^2)}{2\pi} G \otimes (f_G^1 - f_G^2) + O(\alpha^2(Q^2)) \tag{5.10}$$

$$F_3 \sim (-q+\bar{q}) \otimes \{\delta + \frac{\alpha(Q^2)}{2\pi}(f_q^3 - f_q^2)\} + O(\alpha^2(Q^2)) \tag{5.11}$$

The non singlet restriction has been dropped here. The differences $(f^a_{q,G} - f^2_{q,G})$ are completely well defined. This can easily be understood because the leading logarithmic series cancels in the differences $(F_a - F_2)$ and so does their ambiguous fall-out on the $f_{q,G}(x)$ terms. An explicit computation leads to [3),6)]:

$$f_q^2(x) - f_q^1(x) \equiv f_q^L(x) = C_F\ 2x \tag{5.12}$$

$$f_G^2(x) - f_G^1(x) \equiv f_G^L(x) = \frac{T}{f}\ 4x(1-x) \tag{5.13}$$

$$f_q^2(x) - f_q^3(x) \equiv C_F(1+x) \tag{5.14}$$

Note that only real gluon emission diagrams contribute to eqs. (5.12 to 14). We see that the correction terms to the naive parton model formulae for the structure functions in leptoproduction are extremely well behaved at all x and even near x = 0,1.

Of particular importance is the expression of the longitudinal structure function:

$$F_L(x,Q^2) = F_2(x,Q^2) - 2xF_1(x,Q^2) \qquad (5.15)$$

One immediately obtains from eqs (5.12, 5.13) with N = 3, T = f/2 :

$$F_L(x,Q^2) = \frac{\alpha(Q^2)}{2\pi} x^2 \int_x^1 \frac{dy}{y^2}\{\frac{8}{3} F_2(y,Q^2) + 2(\sum_{i=1}^{2f} e_i^2)(1 - \frac{x}{y})yG(y,Q^2)\}$$
$$+ 0[\alpha^2(Q^2)] \qquad (5.16)$$

$$\sigma_L(Q^2) \equiv \int_0^1 dxF_L(x,Q^2)$$

$$= \frac{\alpha(Q^2)}{2\pi} \{\frac{8}{9} \int_0^1 dxF_2(x,Q^2) + \frac{1}{6}(\sum_{i=1}^{2f} e_i^2) \int_0^1 dxxG(x,Q^2)\} + 0[\alpha^2(Q^2)] \qquad (5.17)$$

where (Σe_i^2) is the sum of all coefficients of q and $\bar{q}$ in the naive parton model expression for F_2/x (with f = 4 it is 20/9 in electroproduction and 8 in ν or $\bar{\nu}$ scattering from charged currents). It would be of great importance to experimentally verify these expressions in detail. The present data only allow a very vague consistency statement (for a complete list of references to the data see for example ref. 26). Note that measure of F_L would allow a determination of the shape of the gluon density which otherwise in totally inclusive leptoproduction is to be extracted from observed scaling violations of F_2.

We know consider the non leading corrections to the evolution equations. For each structure function the convolution structure of the equations is kept in higher orders but the kernels start depending on the particular structure function which is considered and on the definition adopted for parton densities. This is seen for the non singlet case by going back to eq. (5.3) and writing in the form ($t = \ln Q^2/\mu^2$) :

$$\frac{d}{dt}M_n^a(t,\alpha) = \beta[\alpha(t)]\frac{d}{d\alpha(t)}M_n^a(t,\alpha)$$

$$= \{\gamma_n[\alpha(t)] + \beta[\alpha(t)][\frac{d}{d\alpha(t)}C_n^a[\alpha(t)]]C_n^{-1\,a}[\alpha(t)]\}M_n^a(t,\alpha)$$

$$= \{\alpha(t)\gamma_n^{(1)} + \alpha^2(t)[\gamma_n^{(2)} - bC_n^{(1)a}] + \dots\}M_n^a(t,\alpha) \qquad (5.18)$$

where eq.(4.18) and expansions of γ_n and C_n^a($\gamma_n = \gamma_n^{(1)}\alpha + \gamma_n^{(2)}\alpha^2 + \ldots$; etc.) were used. Since products in moment space are equivalent to convolutions in density space, one has :

$$\frac{d}{dt} F_a(x,t) = \int_x^1 \frac{dy}{y} F_a(y,t)\, P_a(\frac{x}{y},\alpha(t)) \equiv F_a \otimes P_a \qquad (5.19)$$

with

$$\int_0^1 dx\, x^{n-1} P_a(x,\alpha) = \alpha\gamma_n^{(1)} + \alpha^2 [\gamma_n^{(2)} - bC_n^{(1)a}] + \ldots \qquad (5.20)$$

Consider quark densities defined in terms of F_2 from eq. (5.5). Then the terms of order $\alpha^2(Q^2)$ in the non singlet evolution equations are completely determined. For an extension to the singlet sector a definition of the gluon density beyond the LLA is also necessary. This has not been necessary for the next to leading corrections to the parton formulae because gluons start interacting with electroweak currents only at order $\alpha(Q^2)$. Thus shifting the gluon density by terms of order $\alpha(Q^2)$ only affects structure functions at order $\alpha^2(Q^2)$. On the other hand the gluon density is already present in the evolution equations in the LLA and consequently the singlet kernels are sensitive at order $\alpha^2(Q^2)$ to the definition of the gluon density. We are not interested now in specifying a particular definition for the gluon density because all explicit examples will be restricted to the non singlet sector. We denote by P the general kernels for the parton densities in a given definition and in particular in the non singlet sector, we shall always take :

$$P(x,\alpha) \equiv P_2(x,\alpha) \qquad \text{(non singlet)} \qquad (5.21)$$

Quite in general we can then write for the evolution of parton densities :

$$\frac{d}{dt} q_i = \sum_{j=1}^{f} q_j \otimes P_{q_i q_j} + \sum_{j=1}^{f} \bar{q}_j \otimes P_{q_i \bar{q}_j} + G \otimes P_{q_i G} \qquad (5.22)$$

$$\frac{d}{dt} G = \sum_{j=1}^{f} q_j \otimes P_{G q_j} + \sum_{j=1}^{f} \bar{q}_j \otimes P_{G \bar{q}_j} + G \otimes P_{GG} \qquad (5.23)$$

By charge conjugation we have :

$$P_{q_j q_j} = P_{\bar{q}_i \bar{q}_j} \quad ; \quad P_{q_i \bar{q}_j} = P_{\bar{q}_i q_j}$$

$$P_{q_iG} = P_{\bar{q}_iG} \quad ; \quad P_{Gq_i} = P_{G\bar{q}_i} \tag{5.25}$$

The equivalence of all flavours in the massless theory implies that in the quark sector one only needs to distinguish flavour conserving kernels from flavour non conserving kernels :

$$P_{q_iG} = P_{qG} \quad ; \quad P_{Gq_i} = P_{Gq} \tag{5.26}$$

$$P_{q_iq_j} = \begin{matrix} P^D_{qq} & i = j \\ P^{ND}_{qq} & i \neq j \end{matrix} \tag{5.27}$$

$$P_{\bar{q}_iq_j} = \begin{matrix} P^D_{\bar{q}q} & i = j \\ P^{ND}_{\bar{q}q} & i \neq j \end{matrix} \tag{5.28}$$

D = DIAGONAL
ND = NON DIAGONAL

For example

$$P_{uu} = P^D_{qq},\ P_{du} = P^{ND}_{qq},\ P_{\bar{u}u} = P^D_{\bar{q}q}$$

Note that P^{ND}_{qq}, $P^{D,ND}_{\bar{q}q}$ all start at order α^2.

The structure of the gluon equation is not changed with respect to eq. (2.10). We therefore concentrate on the quark sector. From eqs (5.22, 5.26 to 28) we obtain :

$$\frac{d}{dt}q_i = q_i \otimes Q_{qq} + \bar{q}_i \otimes Q_{\bar{q}q} + \Sigma_q \otimes P^{ND}_{qq} + \Sigma_{\bar{q}} \otimes P^{ND}_{\bar{q}q} + G \otimes P_{qG} \tag{5.29}$$

where

$$Q \equiv P^D - P^{ND} \tag{5.30}$$

and Σ_q, $\Sigma_{\bar{q}}$ are the sums of all quark and all antiquark densities respectively.

As first observed by Ross, Sachrajda [27] $P^D_{\bar{q}q}$ and $P^{ND}_{\bar{q}q}$ are different already at order $\alpha^2(Q^2)$. For example $P_{\bar{u}u} \neq P_{\bar{d}u}$. The difference at this order is due to the presence in the final state of two identical quarks in the $\bar{u}u$ case but not in the $\bar{d}u$ case, as illustrated in fig. 2.

Fig. 2. Lowest order diagrams for the probabilities of finding a $\bar{u}$ or $\bar{d}$ antiquark in a quark u.

On the other hand at order α^2 the equality holds :

$$P^{ND}_{qq} = P^{ND}_{\bar{q}q} \qquad \text{(to the order } \alpha^2\text{)} \tag{5.31}$$

For example $P_{du} = P_{\bar{d}u}$. Eq. (5.31) is true to this order because all flavour changing transitions arise through one gluon exchange, as in fig. 2. We shall limit the following discussion to the next to leading approximation where eq. (5.31) is valid. Then eq. (5.28) becomes :

$$\frac{d}{dt} q_i = q_i \otimes Q_{qq} + \bar{q}_i \otimes Q_{\bar{q}q} + \Sigma \otimes P^{ND}_{qq} + G \otimes P_{qG} \tag{5.32}$$

By taking differences in order to restrict to the non singlet sector one finds :

$$\frac{d}{dt} (q_1 - q_m) = (q_1 - q_m) \otimes Q_{qq} + (\bar{q}_1 - \bar{q}_m) \otimes Q_{\bar{q}q} \tag{5.33}$$

and similarly for a difference of antiquarks :

$$\frac{d}{dt} (\bar{q}_1 - \bar{q}_m) = (\bar{q}_1 - \bar{q}_m) \otimes Q_{qq} + (q_1 - q_m) \otimes Q_{\bar{q}q} \tag{5.34}$$

or of a quark and an antiquark :

$$\frac{d}{dt} (q_1 - \bar{q}_m) = (q_1 - \bar{q}_m) \otimes Q_{qq} + (\bar{q}_1 - q_m) \otimes Q_{\bar{q}q} \tag{5.35}$$

Note, for example, that eq. (5.34) implies that the difference $(\bar{u}-\bar{d})$ cannot be 0 at all Q^2 in a proton.

The previous results show the existence of two independent non singlet kernels at order α^2. In particular one predicts a different evolution for, say, $(F_2^{eP} - F_2^{eN}) = 1/3\ (u + \bar{u} - d - \bar{d})$ and $F_2^{\nu P} - F_2^{\bar{\nu}P} = 2\ (u - \bar{u} - d + \bar{d} - s + \bar{s} + c - \bar{c})$ where these expressions are valid (for f = 4) because of our definition (eq. (5.5)) of quark densities. In fact the two evolution equations are :

$$\frac{d}{dt}(F_2^{eP} - F_2^{eN}) = (F_2^{eP} - F_2^{eN}) \otimes (Q_{qq} + Q_{\bar{q}q})$$

$$\frac{d}{dt}(F_2^{\nu P} - F_2^{\bar{\nu}P}) = (F_2^{\nu P} - F_2^{\bar{\nu}P}) \otimes (Q_{qq} - Q_{\bar{q}q}) \qquad (5.36)$$

The validity of the Adler sum rule implies that :

$$\int_0^1 dx[\ Q_{qq}(x,\alpha) - Q_{\bar{q}q}(x,\alpha)\ ] = 0 \qquad (5.37)$$

while the corresponding moment of $(Q_{qq} + Q_{\bar{q}q})$ is not expected to vanish, so that the charge moment of $(F_2^{eP} - F_2^{eN})$ starts depending on Q^2 at order α^2.

$(Q_{qq} \pm Q_{\bar{q}q})$ are the non singlet kernels for the structure functions F_2. The non singlet kernels of other structure functions can easily be found by using eqs (5.9 to 11 and 5.18 to 20). As examples consider $(F_3^{\bar{\nu}P} \pm F_3^{\nu P})$. In the naive parton model or in the LLA in QCD one has $(F_3^{\bar{\nu}P} \pm F_3^{\nu P}) = 2\ (-u \pm \bar{u} \mp d + \bar{d} \mp s + \bar{s} - c \pm \bar{c})$. It then follows that the corresponding evolution equations are :

$$\frac{d}{dt}(F_3^{\bar{\nu}P} \pm F_3^{\nu P}) = (F_3^{\bar{\nu}P} \pm F_3^{\nu P}) \otimes \{(Q_{qq} \mp Q_{\bar{q}q}) + \alpha^2(t)\ \frac{b}{2\pi}\ (f_q^2 - f_q^3)\} \qquad (5.38)$$

where the difference $(f_q^2 - f_q^3)$ was given in eq. (5.14)

The explicit form of the non singlet kernels in the next to leading accuracy was first computed in moment form by Floratos, Ross, Sachrajda [14]. The moment expression was later considerably simplified [28]. Independent recomputations have recently been done [15] leading to the following compact form for Q_{qq} in density space ($\alpha \equiv \overline{MS}$) :

$$Q_{qq}(x,\alpha) = \frac{\alpha}{2\pi} C_F\left(\frac{1+x^2}{1-x}\right)_+ + \left(\frac{\alpha}{2\pi}\right)^2 \{C_F^2[-2\left(\frac{1+x^2}{1-x}\right)\ln x \ln(1-x) - 5(1-x)$$

$$- \left(\frac{3}{1-x} + 2x\right)\ln x - \frac{1}{2}(1-x)\ln^2 x] + \frac{1}{2}C_F C_A\, [\left(\frac{1+x^2}{1-x}\right)\cdot$$

$$\left(\ln^2 x - \frac{11}{3}\ln\frac{1-x}{x^2} + \frac{367}{16} - \frac{\pi^2}{3}\right) + 2(1+x)\ln x + \frac{61}{12} - \frac{215}{12}x]$$

$$+ \frac{2C_F T}{3}\, [\left(\frac{1+x^2}{1-x}\right)\left(\ln\frac{1-x}{x^2} - \frac{29}{12}\right) + \frac{1}{4} + \frac{13}{4}x]\}_+$$

$$+ \delta(1-x) \int_0^1 dx\, Q_{\bar{q}q}(x,\alpha) \tag{5.39}$$

where the "+" distribution is defined by :

$$\int_0^1 dx\, \{\}_+\, \phi(x) \equiv \int_0^1 dx\, \{\}\, [\phi(x) - \phi(1)] \tag{5.40}$$

Similarly for $Q_{\bar{q}q}$ one finds :

$$Q_{\bar{q}q}(x,\alpha) = \left(\frac{\alpha}{2\pi}\right)^2 \left(C_F - \frac{1}{2}C_A\right) C_F\, [2(1+x)\ln x + 4(1-x)$$

$$+ \frac{1+x^2}{1-x}\left(\ln^2 x - 4\ln x \ln(1+x) - 4Li_2(-x) - \frac{\pi^2}{3}\right)] \tag{5.41}$$

with

$$Li_2(x) = - \int_0^x \frac{dy}{y}\ln(1-y) \tag{5.42}$$

In the case of QCD :

$$\int_0^1 dx\, Q_{\bar{q}q}(x,\alpha) \simeq - 0.1598 \left(\frac{\alpha}{2\pi}\right)^2 + \ldots \tag{5.43}$$

Note that $Q_{\bar{q}q}$ is regular near $x = 1$ so that no "+" redefinition was needed in this case as opposite to that of Q_{qq}. The expression of Q_{qq} is completely independent of prescriptions once the definition of the coupling constant is specified, which explains the explicit mention to the modified minimal subtraction in eq. (5.39). We recall from eq. (4.7) that a change in the definition of the coupling induces a variation of the second order term proportional to the lowest order result $P_{qq}(x)$.

In the limit $x \to 1$ the dominant terms in Q_{qq} are given by :

$$Q_{qq}(x,\alpha) \xrightarrow[x\to 1]{} \frac{\alpha}{2\pi} P_{qq}(x) - \left(\frac{\alpha}{2\pi}\right)^2 2\pi b C_F \frac{1+x^2}{1-x} \ln(1-x) + \ldots \quad (5.44)$$

while the behaviour of $Q_{\bar{q}q}$ is very damped near $x = 1$:

$$Q_{\bar{q}q}(x,\alpha) \xrightarrow[x\to 1]{} \left(\frac{\alpha}{2\pi}\right)^2 C_F\left(C_F - \frac{1}{2}C_A\right) \frac{(1-x)^5}{10} + \ldots \quad (5.45)$$

The behaviour of Q_{qq} is equivalent to $[\alpha \ln n + \alpha^2 a(\log n)^2]$ (a being a constant) for moments at large n. It is interesting to observe that the most singular term near $x = 1$ at order α^2 in Q_{qq} can be reabsorbed into the leading term of order α by a change of scale $Q^2 \to Q^2(1-x)$. In fact for $\alpha(Q^2)$ given by eq. (4.19) one has :

$$\alpha[Q^2(1-x)] \sim \alpha(Q^2)\,[1 - b\alpha(Q^2)\ln(1-x)] \quad (5.46)$$

The physical origin of this rescaling is easily traced back to a phase space effect. The log Q^2 is actually a log $(k_\perp^2)_{MAX}$ where $k_\perp$ is the transverse momentum of the produced parton and its maximum value is $k_\perp^2 \sim Q^2(1-x)$ near $x = 1$. This phase space effect is present for each rung of the ladder and can be controlled to all orders. In fact the region $(1 - x) \sim 0$ corresponds to the soft gluon region (recall that $(1 - x)$ is the energy lost by the quark). One can therefore take advantage of the techniques of soft gluon resummation. As a result, by diagrammatic [29] or by coherent state methods [30] for the non singlet moments one obtains the exponentiation :

$$q_n(Q^2) = q_n^0 \exp \int_0^1 dx\,(x^{n-1} - 1) C_F \frac{1+x^2}{1-x} \int_{\mu^2}^{Q^2(1-x)} \frac{dk_\perp^2}{k_\perp^2} \frac{\alpha(k_\perp^2)}{2\pi} \quad (5.47)$$

Equivalently the corresponding evolution equation can be written, at leading accuracy in the expression for $\alpha(Q^2)$, in the form [31] :

$$\frac{d}{dt} q(x,Q^2) = \int_x^1 \frac{dy}{y} q(y,Q^2) \left\{P_{qq}\left(\frac{x}{y}\right) \frac{\alpha[Q^2(1-\frac{x}{y})]}{2\pi}\right\}_+ \quad (5.48)$$

This result is interesting as a nice illustration of a source of non uniformity near $x = 1$. Arguments have been given [32] to prove that to all orders the phase space effect provides the leading sub-dominant corrections near $x = 1$. Phenomenological applications of eq. (5.48) have been recently discussed [33]. Its importance is in practice severely limited by ignorance of power behaved scaling violations (mass effects, higher twists ans so on).

A relatively simple analytical form for the even (odd) moments of $Q_{qq} + Q_{\bar{q}q}$ $(Q_{qq} - Q_{\bar{q}q})$ has been given [28]. Incidentally the reason why two distinct non singlet kernels arise from a unique non singlet leading twist operator with definite anomalous dimension is that [27] the light cone formalism only predicts even (odd) moments of crossing even (odd) structure functions. $Q_{qq} \pm Q_{\bar{q}q}$ are obtained as separate inversions from even/odd moments. It is clear on the other hand that the formalism in x space eqs (5.39, 5.41) also allow to derive odd (even) moments of crossing even (odd) structure functions.

Moments of structure functions at non leading accuracy are of the form :

$$M_n^a(Q^2) \sim \left[\frac{\alpha}{\alpha(Q^2)}\right]^{d_n} \{1 + e_n^a \frac{\alpha(Q^2)}{\pi} + \ldots\} \tag{5.49}$$

where

$$d_n = \frac{\gamma_n^{(1)}}{b} \; ; \; e_n^a = - \frac{\pi}{b} \{\gamma_n^{(2)} - bC_n^{(1)a} - b'\gamma_n^{(1)}\} \tag{5.50}$$

with $\gamma_n^{(1)}$ and $\gamma_n^{(2)}$ being the one and two loop anomalous dimensions and $C_n^{(1)a}$ the one loop reduced coefficient. Note that

$$\int_0^1 dx\, x^{n-1} \{Q_{qq}(x,\alpha) \pm Q_{\bar{q}q}(x,\alpha)\} \simeq \gamma_n^{(1)}\alpha + (\gamma_n^{(2)} - bC_n^{(1)2})\alpha^2 + \ldots \tag{5.51}$$

Also note that in eq. (5.49) $\alpha(Q^2)$ is to be taken in its improved form, including the b' term, as in eq. (4.19)

Clearly e_n^a depend on the definition of α, because under a change of α as in eq. (4.5), $\gamma_n^{(2)}$ is modified according to :

$$\gamma_n^{(2)} \to \gamma_n^{(2)} + k^{(1)}\gamma_n^{(1)} \tag{5.52}$$

Once more this dependence on the definition of α is to be eliminated by comparing data with data. For example :

$$M_n^a(Q^2) = [M_{n_o}^{a_o}(Q^2)]^{\frac{d_n}{d_{n_o}}} \{1 + (\frac{e_n^a}{d_n} - \frac{e_{n_o}^{a_o}}{d_{n_o}}) \frac{\alpha(Q^2)}{\pi} + \ldots\} \quad (5.53)$$

or

$$\frac{d\ln M_n^a(Q^2)}{d\ln M_k^a(Q^2)} = \frac{d_n}{d_k} \{1 + (\frac{e_n^a}{d_n} - \frac{e_k^a}{d_k}) \frac{\alpha(Q^2)}{\pi} + \ldots\} \quad (5.54)$$

are relations where all coefficients are independent of the definition of α.

Fig. 3 shows a plot of the coefficients e_n^a in the $\overline{MS}$ prescription. The computed values are of order unity for the first few values of n. The differences among the various structure functions are small. In particular the difference between crossing even and odd structure functions is barely visible even on an expanded scale and drops away very fast with n as a consequence of the rapid decrease of $Q_{\bar{q}q}$ near $x \simeq 1$, eq. (5.45). This fact a posteriori justifies the neglect of this difference which is so often found in the literature.

Thus the numerical evaluation of the non leading corrections to all leptoproduction structure functions shows that, for reasonable values of n, the convergence of the perturbative expansion (in the $\overline{MS}$ prescription) appears as quite good.

We now consider the comparison between the non leading corrections in the spacelike region and in the timelike region (that is those for the structure functions of one hadron inclusive e^+e^- annihilation). The non singlet evolution equations are in this case :

$$Q^2 \frac{d}{dQ^2} (z\bar{F}_a)_{NS} = (z\bar{F}_a)_{NS} \otimes \bar{P}_a \quad (5.55)$$

where the bar always refers to annihilation. The kernels P_a can be split into a "regular" part and a delta function term at x = 1 :

$$P_a = P_a^{reg} + P_a^{sing}\,\delta(1-x) \quad (5.56)$$

In the LLA the spacelike and the timelike kernels are connected by the analytic continuation rule

$$-xP_a^{reg}(1/x,\alpha) = \bar{P}_a^{reg}(x,\alpha) \quad (5.57)$$

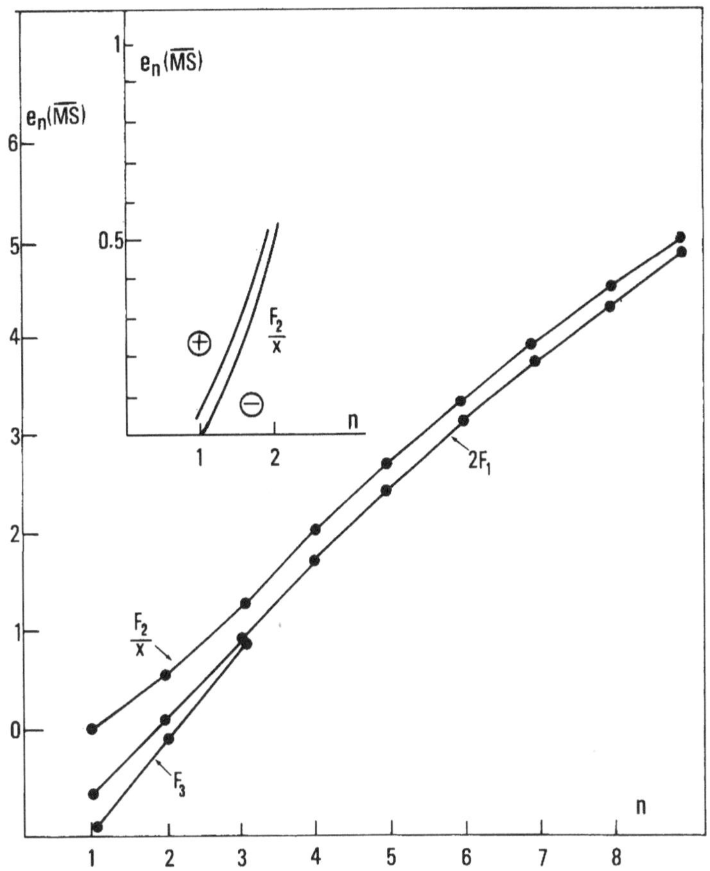

Fig. 3. Plot of e_n^a, defined in eqs (5.49), (5.50), for a = 1,2,3, corresponding to (non singlet) moments of $2F_1$, F_2/x and F_3 respectively in deep inelastic scattering. The $\overline{MS}$ prescription for α was adopted. The differences for crossing even and crossing odd structure functions is negligible but for small n where it is barely noticeable (see expanded scale).

For example :

$$P_{qq}^{reg}(x) \sim \frac{1+x^2}{1-x} \to -x \frac{1 + (1/x)^2}{1 - (1/x)} = \frac{1+x^2}{1-x} \simeq P_{qq}^{reg}(x)$$

i.e. $P_{qq}^{reg}(x)$ is self-conjugate under this transformation. Also when all colour and spin factors are ignored.

$$P_{Gq}(x) \sim \frac{1 + (1-x)}{x} \to x \, \frac{1 + (1-1/x)^2}{1/x} \sim - [\, x^2 + (1-x)^2] \sim P_{qG}(x) \tag{5.58}$$

(an extra minus sign is present in the non diagonal case with respect to eq. (5.57) valid in the non singlet case.

The analytic continuation rule was found [34] to hold for P_a^{reg} and $\bar{P}_a^{reg}$ in the next to the leading approximation as well (with natural conventions like interpreting all ln(1-x) as ln|1-x|) although P_a^{reg} is no more selfconjugate as was the case for P_{qq}^{reg}. However as a consequence of the continuation from q^2 negative in leptoproduction to q^2 positive in annihilation, one finds that [34] :

$$\bar{P}_a^{sing} = P_a^{sing} - \frac{b}{2\pi} \, \alpha^2(Q^2) C_F \pi^2 \tag{5.59}$$

We recall that the moments of P_a correspond to $\alpha^2(\gamma_n^{(2)} - bC_n^{(1)a})$ (see eq. (5.19)). The additional term proportional to π^2 in eq. (5.59) arises from the reduced coefficient $C_n^{(1)a}$, and in fact is proportional to b. Precisely it is obtained from the virtual diagram contribution (which is only present at x = 1 in the structure functions; see section VII). This contribution is formally identical both in the spacelike and the timelike regions when expressed in terms of q^2. However in terms of $Q^2 = |q^2|$, we have $-Q^2 = q^2$ for F_a and $Q^2 = q^2$ for $\bar{F}_a$. In the language of dimensional regularization because of the double pole $1/\varepsilon^2$ the real part of the relative factor $(-1)^\varepsilon = 1 + i\pi\varepsilon - 1/2\, \pi^2\varepsilon^2 + \ldots$ leads to a difference between F_a and $\bar{F}_a$:

$$(\bar{F}_a - F_a)_{virtual} = \frac{\alpha}{2\pi} \, C_F \pi^2 \delta(1-x) \tag{5.60}$$

which exactly corresponds to eq. (5.59). We see that the π^2 term is associated with $\log^2 q^2$ in the vertex correction. The leading term $\log^2 Q^2$ is cancelled by the real diagrams but the change of sign of q^2 is felt by the non leading terms. As a result one can state that the analytic continuation rule is preserved at order α^2 only if both x and q^2 are continued [34]. The possibility of factoring out in exponential form this particular effect will be discussed later in connection with a similar situation in Drell-Yan processes.

If the moments of $z\bar{F}_2$ (non singlet), which correspond to those of F_2/x in the spacelike region, are parametrized as (compare with eq. (5.49)) :

$$\bar{M}_n^2(Q^2) \simeq \left[\frac{\alpha}{\alpha(Q^2)}\right]^{d_n} \{1 + \bar{e}_n^2 \, \frac{\alpha(Q^2)}{\pi} + \ldots \} \tag{5.61}$$

then it turns out that the numerical values of $\bar{e}_n^2$ (in the $\overline{MS}$ prescription for α) are substancially larger than those of e_n^2, i.e. the corresponding quantities in the spacelike region, as shown in fig. 4.

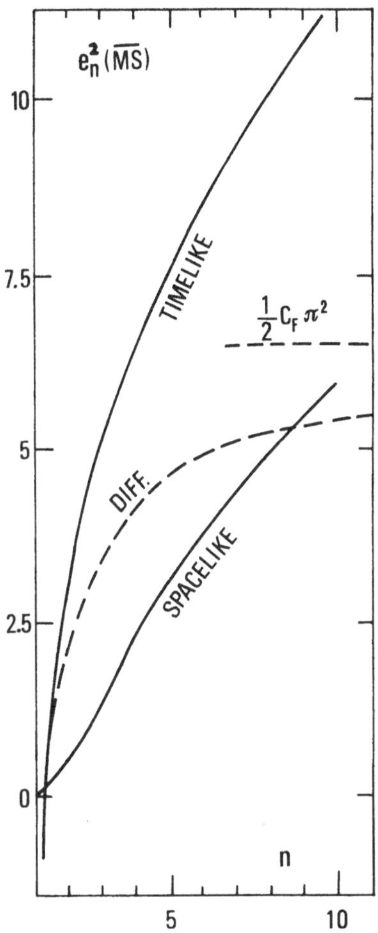

Fig. 4. A comparison of the values of e_n^2 (see eqs (5.49), (5.50)) for the non singlet moments of F_2/x in the spacelike region and of $\bar{e}_n^2$ (see eq. (5.61)) for the non singlet moments of $z\bar{F}_2$ (in the MS prescripion). The difference is also shown. At large n the main contribution to the difference arises from the continuation of the quark form factor from negative to positive q^2 (the dashed term $1/2\ C_F\pi^2$).

At large value of n the difference is indeed dominated by the additional term in eq. (5.59). Recall that in the LLA the scaling violations are the same for non singlet structure functions in leptoproduction and annihilation (this remark concerns the exponent d_n in eq. (5.61)). Thus it is not that the scaling violations are expected to be larger in the timelike region, but rather that it appears that one needs more energy before the LLA becomes equally accurate. It is an interesting empirical fact that the perturbative expansion appears better behaved for light cone or short distance dominated phenomena rather than for other processes (an example of short distance effect is the effective weak non leptonic hamiltonian which has been recently studied at next to the leading accuracy [35]).

VI. AN ASYMPTOTIC FORMULA FOR MULTIPLICITIES

The study of non leading terms in the evolution equations was also extended to the singlet sector both in the spacelike and the timelike regions [14),15]. Of course the non singlet evolution equations, being homogenuous are more predictive than those for the singlet sector. Moreover at the next to the leading accuracy a more precise definition of the gluon density is also to be specified in order to be able to unambiguously write down the singlet kernels. For the singlet equations to become predictive one should have an independent way to measure the parton densities, in particular the gluon density, to a sufficient degree of accuracy. But actually the scaling violations are the only practical possibility of studying the gluon density to some accuracy. Thus the main practical advantage of the important theoretical effort made on the next to the leading corrections of order α^2 to the singlet kernels is to provide the possibility for a precise measurement of the gluon density (and the singlet quark density as well).

There is one interesting point which is worth mentioning about the singlet sector. In leading order P_{GG} and P_{Gq} both show a simple pole near x = 0 which corresponds to a pole near n = 1 in the moments. For example

$$\bar{\gamma}_n^{GG}(\alpha) = \frac{\alpha}{2\pi} A_n^{GG} \sim \frac{\alpha C_A}{\pi(n-1)}$$

From the explicit calculations in two loops the leading triple pole at n = 1 in the timelike region was found to be consistent with the branch point structure for $\bar{\gamma}_n^{GG}(\alpha)$ given by :

$$\gamma_n^{GG}(\alpha) \simeq \frac{1}{4}\left[- (n-1) + \sqrt{(n-1)^2 + \frac{8\alpha C_A}{\pi}}\ \right] \qquad (6.1)$$

Recently this conjecture was further supported by a three loop calculation of the leading term near n = 1 [36]. The removal of the pole singularity makes it possible to evaluate the Q^2 behaviour of the average multiplicity $\bar{n}(Q^2)$ of gluons, which is proportional to the moment with n = 1 of the corresponding fragmentation function. For a gluon jet eq. (6.1) leads to :

$$\overline{n}(Q^2) \sim \exp \int_{\alpha}^{\alpha(Q^2)} d\alpha \frac{\overline{\gamma}_1^{GG}(\alpha)}{\beta(\alpha)} = \exp\sqrt{\frac{2C_A}{\pi b} \ln \frac{Q^2}{\Lambda^2}} \tag{6.2}$$

This remarkable result was first obtained, with a slight difference in the numerical factor at exponent in refs (31), (37).

The gluon component of singlet fragmentation functions is expected to be dominant near x = 0. Furthermore since all Q^2 dependence arises from the partonic cascade in the jet the prediction in eq. (6.2) should presumably be relevant for hadrons as well. This matter is currently under study and it may be that more will be learnt on this issue soon.

VII. THE SUDAKOV FORM FACTOR OF PARTONS

The behaviour of the quark form factor in the LLA and beyond is a very important and delicate problem in QCD. Its importance arises from the fact that this is the simplest problem involving the resummation of a series of doubly logarithmic terms. A complete understanding of this problem would allow the control of many similar doubly logarithmic effects of great relevance in phenomenology. Some of them will be discussed in the following sections. Thus an intense research activity on this area is currently under way.

Consider the one loop calculation in fig. 5 for the quark form factor. We take an on shell massless quark and regularize both infrared and ultraviolet singularities by dimensional regularization. It is convenient to work in the Landau gauge where the quark self energy diagrams vanish and the vertex diagram of fig. 5 leads by itself to the gauge invariant result. One obtains for the coefficient of $e\gamma_\mu$ (see for example the first of ref. 6 in N.P. B157) :

$$\Gamma(q^2) = 1 + \frac{\alpha}{4\pi} C_F \left(\frac{4\pi\mu^2}{-q^2}\right)^{\varepsilon} \frac{\Gamma(1+\varepsilon)\Gamma^2(1-\varepsilon)}{\Gamma(1-2\varepsilon)} \left\{-\frac{2}{\varepsilon^2} - \frac{3}{\varepsilon} - 8 + O(\varepsilon)\right\} \tag{7.1}$$

where $2\varepsilon = 4 - d$ and q is the momentum of the virtual photon γ^*. The double pole in ε is of infrared origin. At fixed ε and large $Q^2 = |q^2|$ it corresponds to :

$$\Gamma(Q^2) \simeq 1 - \frac{\alpha}{4\pi} C_F \ln^2 \frac{Q^2}{\mu^2} \tag{7.2}$$

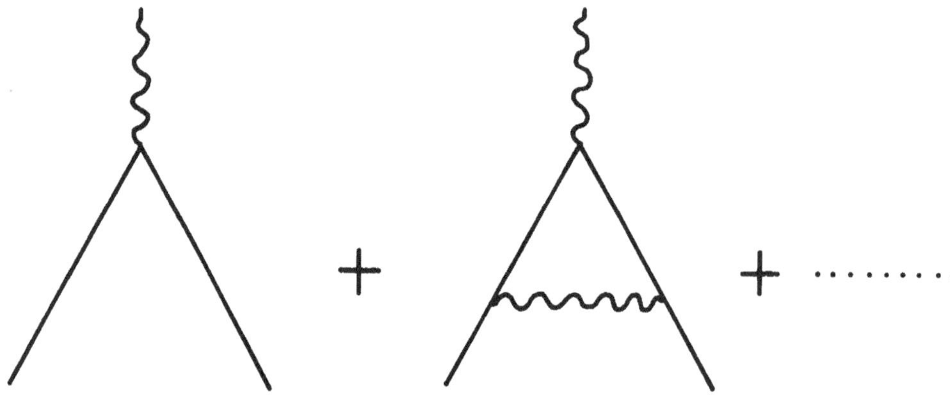

Fig. 5. Diagrams for the quark form factor.

The presence of these infrared doubly logarithmic terms is a well known feature. At order n in α they are of the form $(\alpha t^2)^n$. They cancel when one adds the real emission diagrams in order to construct some measurable quantity (all charges being necessarily accompanied by soft radiation).

For example if one extends the calculation in the same regularization to the real diagrams of relevance for leptoproduction then one obtains for the structure function F_2 of the quark :

$$F_2(x,Q^2)_{real} = \frac{\alpha}{2\pi} C_F \left(\frac{4\pi\mu^2}{Q^2}\right)^{\varepsilon} \frac{\Gamma(1-\varepsilon)}{\Gamma(1-2\varepsilon)} \left\{\frac{2}{\varepsilon^2}\,\delta(1-x) - \frac{1+x^2}{(1-x)_+}\,\frac{1}{\varepsilon}\right.$$

$$+ \frac{3}{2\varepsilon}\,\delta(1-x) + \left[1 + x^2\left(\frac{\ln(1-x)}{(1-x)}\right)_+ - \frac{3}{2(1-x)_+}\right.$$

$$\left.\left. - \frac{1+x^2}{1-x}\,\ln x + 3 + 2x + \frac{7}{2}\,\delta(1-x)\right]\right\} \qquad (7.3)$$

One easily checks the cancellation of the double poles after multiplication of eq. (7.1) by $\delta(1-x)$ and also that the total residue of the single poles is indeed $P_{qq}(x)$.

The contribution to $\Gamma(Q^2)$ of the leading double logarithmic series of terms $(\alpha t^2)^n$ was proven long ago [38] to exponentiate in QED

$$\Gamma(Q^2) \simeq \exp\left(-\frac{\alpha}{4\pi}\ln^2\frac{Q^2}{\mu^2}\right) \qquad \text{(QED, mass shell)} \quad (7.4)$$

The corresponding equation for QCD is simply obtained by inserting a colour factor :

$$\Gamma(Q^2) \simeq \exp\left(-\frac{\alpha}{4\pi} C_F \ln^2 \frac{Q^2}{\mu^2}\right) \qquad \text{(QCD, mass shell)} \qquad (7.5)$$

This extension is non trivial because of the gluon self couplings which are absent in the Abelian case. In QCD the exponentiation has been explicitly verifyed to three loops [39] and proven to all orders in ref. (40). Similar results also hold for other form factors of partons, for example the form factor of the gluon.

The physical meaning of eqs (7.4, 7.5) appears to be that the probability amplitude for a (either Abelian or non Abelian) charge to to be deflected with high momentum transfer without radiating soft quanta is exponentially small (in the cut-off theory).

The detailed form of the exponent is dependent on the infrared regularization procedure. It is well known that for an off shell quark the exponent differs by a factor of 2. For example :

$$\Gamma(Q^2) \simeq \exp\left(-\frac{\alpha}{2\pi} \ln^2 \frac{Q^2}{\mu^2}\right) \qquad \text{(QED, off shell)} \qquad (7.6)$$

Up to this point we have been working in the leading doubly logarithmic approximation and in particular with a fixed coupling. We now consider the problem, very relevant in QCD, of how to make the coupling to run. In this case this has to do with non leading terms because, according to eq. (4.16), $(\alpha-\alpha(t))\, t^2 = b\alpha^2 t^3$, which is non leading with respect to terms of order $\alpha^2 t^4$. On the other hand the control of non leading terms is important because of the exponential damping of the leading sequence of terms.

This problem has been studied in ref. (41). If one translates into QCD the results obtained for QED with negative α, a non unitary but asymptotically free theory, then one obtains :

$$\Gamma(Q^2) \simeq \exp\left(-C_F \int^{Q^2} \frac{dk^2}{k^2} \frac{\alpha(k^2)}{2\pi} \ln\frac{Q^2}{k^2}\right) \qquad \text{(QCD, on shell)} \qquad (7.7)$$

For $\alpha(k^2)$ in the LLA as given by eq. (3.28) the above formula was proven in non Abelian theories by Kinoshita, Ukawa [40]. Within the stated accuracy, they in fact proved the validity for the form factor of a modified RGE proposed by Korthals-Altes, de Rafael [42] with a t dependent anomalous dimension term. This modified RGE leads to eq. (7.7) as a solution. For fixed α eq. (7.5) is reobtained. For running $\alpha(k^2)$ the asymptotic exponent is computed by observing that

$$\int^{Q^2} \frac{dk^2}{k^2} \frac{\alpha(k^2)}{2\pi} \ln \frac{Q^2}{k^2} = \frac{1}{2\pi b} \int^{t} \frac{d\ln k^2}{\ln k^2} (\ln \frac{Q^2}{\Lambda^2} - \ln \frac{k^2}{\Lambda^2})$$

$$\sim \frac{1}{2\pi b} \ln \frac{Q^2}{\Lambda^2} \ln\ln \frac{Q^2}{\Lambda^2} + \ldots \qquad (7.8)$$

This implies by using eq. (4.12) for b :

$$\Gamma(Q^2) \simeq \text{const. } \exp \left(\frac{-6C_F}{11C_A-4T} \ln \frac{Q^2}{\Lambda^2} \ln\ln \frac{Q^2}{\Lambda^2}\right) \text{ (QCD, on shell)} \qquad (7.9)$$

Note that, because $\alpha(k^2)$ is only known at large k^2, the scale of the Sudakov form factor is not predictable. In fact the constant in eq. (7.9) arises from the low k^2 part of the integration range.

Eq. (7.9) is thought to be the correct asymptotic behaviour when all terms down by logarithms are included and all terms down by powers are neglected. However it cannot be excluded that terms that order by order are down by powers finally dominate the asymptotic behaviour.

VIII. EXAMPLES OF DOUBLY LOGARITHMIC EFFECTS OF PHYSICAL INTEREST

We stated in the previous section that the understanding of the asymptotic behaviour of the Sudakov form factor is of importance for a number of physical problems. Here we consider two examples of particular interest. The first deals with transverse momentum distributions, while the second provides an illustration of how the Sudakov result may be useful in some cases to alleviate problems caused by too large next to leading corrections.

In order to fix the ideas consider the structure of the final state in leptoproduction. The discussion can (and shall) easily be extended to other processes as well.

In the naive parton model all deep inelastic events in leptoproduction are expected to appear as collinear within an angular accuracy $p_\perp/E$, where $p_\perp$ is an "intrinsic" (Q^2 independent) transverse momentum of the partons in the target and /or of the hadrons in the parton jets. The collinear events consist of two jets, the target fragmentation jet and the struck parton jet, at opposite sides of the rapidity interval, smoothly connected by a soft hadron plateau.

Correspondingly the longitudinal cross section σ_L vanishes as a power of m^2/Q^2 or $p_\perp^2/Q^2$. This follows because in the frame where γ^* carries no energy the hit parton simply reverses its momentum. In the massless limit the quark helicity is conserved in the interaction with γ^* and in the collinear limit, this implies a spin flip for the hit parton and a unit of spin component for γ^*. Thus σ_L, which corresponds to a γ^* of zero helicity, can only arise from mass and acollinearity effects.

In QCD this picture is invalidated because of the probability of order $\alpha(Q^2)$ of emitting, for example, a gluon of large $k_\perp \sim Q$ from the quark line. This produces a hard tail in the $k_\perp$ distribution of the final parton.

This high $k_\perp$ tail behaves as $1/k_\perp^2$ and extends up to $k_\perp^2 = W_p^2/4$ where W_p is the total energy squared of the parton-current system in its center of mass : $k_\perp^2 \leqslant Q^2(1-x)/4x$. This small (because of order $\alpha(Q^2)$) but long tail is responsible of scaling violations, of a contribution of order $\alpha(Q^2)$ to σ_L and of a linear rise with energy of the average $k_\perp$ of partons :

$$Q^2 \frac{d\sigma_T}{dQ^2} = Q^2 \frac{d}{dQ^2} \int^{Q^2} \frac{dk_\perp^2}{k_\perp^2} \alpha(k_\perp^2) \simeq \alpha(Q^2) + \ldots \tag{8.1}$$

$$\sigma_L/\sigma_T = \int^{Q^2} \frac{dk_\perp^2}{k_\perp^2} \alpha(k_\perp^2)\left(\frac{k_\perp^2}{Q^2}\right) \simeq \alpha(Q^2) + \ldots \tag{8.2}$$

$$< k_\perp^2 > \simeq \int^{Q^2} \frac{dk_\perp^2}{k_\perp^2} \alpha(k_\perp^2)\, k_\perp^2 \simeq \alpha(Q^2)\, Q^2 + \ldots \tag{8.3}$$

In the last equation for $<k_\perp^2>$ the dots indicate terms which are constant in Q^2, apart from logarithms, and cannot be computed in perturbation theory. In fact these terms are affected by the non perturbative intrinsic $p_\perp$ distribution. On the other hand the slope of the linearly rising term is determined by values of $k_\perp \sim Q$ on the perturbative tail and is computable.

The leading term at large Q (or W) for the average moments of the $k_\perp$ distribution, which amount to reconstruct the tail at $k_\perp \sim Q$ of the transverse momentum distribution, are dominated at large energies by the renormalization group improved lowest order contributions (order α) in perturbation theory and have been computed since a long time by now [43]. Recently sophisticated computations of non leading effects of order α^2 have been completed in some cases [44] (for example the non singlet Drell-Yan processes, i.e. $\pi^+N - \pi^-N$). A further improvement is possible when Q is sufficiently large that it makes sense to study the transverse momentum distribution in a region of $k_\perp$ such that :

$$\mu^2 \ll k_\perp^2 \ll Q^2 \tag{8.4}$$

In this regime terms of order $\ln Q^2/k_\perp^2$ are no more negligible with respect to terms of order $k_\perp^2/\mu^2$ as was instead true for $k_\perp \sim Q$. We shall see that this leads to consider a problem of resummation for a sequence of doubly logarithmic terms in close relation with the Sudakov exponentiation considered in section VII. This problem was first formulated and discussed by Dokshitzer, Dyakonov, Troyan [45]. The solution that they proposed was modified by Parisi, Petronzio [46] who established the direct connection of this problem with the Sudakov form factor. Explicit calculations to two loops have since then supported their solution [47].

The origin of the double logarithms can be understood as follows. Consider the quantity :

$$\Sigma(k_\perp^2, Q^2) = \frac{1}{\sigma(Q^2)} \int^{k_\perp^2} \frac{d\sigma}{dp_\perp^2}\, dp_\perp^2 \tag{8.5}$$

where $\sigma(Q^2)$ is the total cross section. Σ is the generating function (by differentiation with respect to $k_\perp^2$) of $1/\sigma \;\; d\sigma/dk_\perp^2$. We write down the expression for $\Sigma(k_\perp^2,Q^2)$ to order $\alpha(k_\perp^2)$ and for a parton quark target in the form of 1 minus the probability of hard radiation

$$\Sigma(k_\perp^2, Q^2) \simeq 1 - \int_{k_\perp^2}^{Q^2} \frac{dp_\perp^2}{p_\perp^2} \frac{\alpha(p_\perp^2)}{2\pi} \int_0^{1-p_\perp^2/Q^2} dx P_{qq}(x) + \ldots \tag{8.6}$$

the upper limit of integration in x arises because of the condition $p_\perp^2 < W_p/4$ on the maximum allowed value of $p_\perp^2$ at fixed x (we dropped a factor of 4 because Q^2 or $Q^2/4$ is the same in the LLA) :

$$p_\perp^2 \leqslant Q(\frac{1-x}{x}) \;, \text{ i.e. } \; x \leqslant \frac{1}{1+p_\perp^2/Q^2} \sim 1 - \frac{p_\perp^2}{Q^2} \tag{8.7}$$

For the sake of this argument one can take the expression of $P_{qq}(x)$ at $x \leqslant 1$. The double logarithms are produced from the soft singularity of $P_{qq}(x)$ near $x = 1$ (which we know is typical of vector gluons)

$$\Sigma(k_\perp^2,Q^2) \simeq 1 - \int_{k_\perp^2}^{Q^2} \frac{dp_\perp^2}{p_\perp^2} \frac{\alpha(p_\perp^2)}{2\pi} \int^{1-p_\perp^2/Q^2} dx C_F \left(\frac{1+x^2}{1-x}\right) + \ldots$$

$$\simeq 1 - \int_{k_\perp^2}^{Q^2} \frac{dp_\perp^2}{p_\perp^2} \frac{\alpha(p_\perp^2)}{2\pi} \left[2C_F \ln \frac{Q^2}{p_\perp^2} + \ldots \right] \qquad (8.8)$$

Consider first the case of QED with α fixed and $C_F = 1$. Then :

$$\Sigma(k_\perp^2,Q^2) = 1 - \frac{\alpha}{2\pi} \ln^2 \frac{Q^2}{k_\perp^2} + \ldots \qquad \text{(QED)} \qquad (8.9)$$

which corresponds to

$$\left(\frac{1}{\sigma} \frac{d\sigma}{dk_\perp^2}\right)_{\text{order}\,\alpha} \underset{k_\perp^2>0}{=} \frac{\alpha}{\pi} \frac{1}{k_\perp^2} \ln \frac{Q^2}{k_\perp^2} + \ldots = \alpha\nu(k_\perp) + \ldots \qquad (8.10)$$

In the region $\alpha/\pi \log^2 Q^2/k_\perp^2 \simeq 1$ the leading expression for Σ or $1/\sigma\ d\sigma/dk_\perp^2$ is obtained by taking double logarithms into account to all orders. The double logarithms arise from the emission of photons (gluons) with momenta much smaller than Q^2. In this soft region the cross section for production of n photons (gluons) factorizes into independent emissions :

$$\frac{1}{\sigma} d\sigma_n \simeq \frac{\alpha^n}{n!} d^2\vec{k}_\perp^1 d^2\vec{k}_\perp^2 \ldots d^2\vec{k}_\perp^n \frac{\nu}{\pi}(k_\perp^1) \ldots \frac{\nu}{\pi}(k_\perp^n) \qquad (8.11)$$

In the leading bilogarithmic approximation this corresponds to the exponential result :

$$\Sigma(k_\perp^2,Q^2) \simeq \exp\left(-\frac{\alpha}{2\pi} \ln^2 \frac{Q^2}{k_\perp^2}\right) \simeq \Gamma^2\left(\frac{Q^2}{k_\perp^2}\right) \text{ (QED)} \qquad (8.12)$$

where we made explicit the connection with the square of the on shell Sudakov form factor in eq. (7.4). Equivalently :

$$\frac{1}{\sigma} \frac{d\sigma}{dk_\perp^2} \underset{k_\perp^2>0}{\simeq} \frac{\alpha}{\pi} \frac{1}{k_\perp^2} \ln \frac{Q^2}{k_\perp^2} \exp\left(-\frac{\alpha}{2\pi} \ln^2 \frac{Q^2}{k_\perp^2}\right) \text{ (QED)} \qquad (8.13)$$

More in general one can write :

$$\frac{1}{\sigma}\frac{d\sigma}{d^2\vec{k}_\perp} = \sum_{n=0}^{\infty}\frac{\alpha^n}{n!}\prod_{i=1}^{n}\int d^2k_\perp^i\,\frac{\nu(k_\perp^i)}{\pi}\,\delta^{(2)}(\vec{k}_\perp - \sum_i \vec{k}_\perp^i)\ \text{(QED)} \tag{8.14}$$

The δ function term can be cast in a convenient form by the familiar impact parameter formalism :

$$\delta^{(2)}(\vec{k}_\perp - \sum_i \vec{k}_\perp^i) = \frac{1}{(2\pi)^2}\int d^2\vec{b}\, e^{-i\vec{b}(\vec{k}_\perp - \sum_i \vec{k}_\perp^i)} \tag{8.15}$$

One then obtains the eikonal result :

$$\frac{1}{\sigma}\frac{d\sigma}{d^2\vec{k}_\perp} \simeq \frac{1}{(2\pi)^2}\int d^2b\, e^{-i\vec{b}\vec{k}_\perp}\exp\{\frac{\alpha}{\pi}\int d^2\vec{p}_\perp \nu(p_\perp)\,[\,e^{i\vec{b}.\vec{p}_\perp}-1\,]\,\}\ \text{(QED)} \tag{8.16}$$

The factors $(e^{i\vec{b}.\vec{p}_\perp}-1)$ arise from interpreting $\nu(p_\perp)$ as a distribution of the "+" type (near $p_\perp \simeq 0$).

The extension to the non Abelian case is done by following the same discussion as for the Sudakov form factor. One obtains :

$$\Sigma(k_\perp^2,Q^2) \simeq \exp\left(-C_F\int_{k_\perp^2}^{Q^2}\frac{dp_\perp^2}{p_\perp^2}\frac{\alpha(p_\perp^2)}{\pi}\ln\frac{Q^2}{p_\perp^2}\right) = \Gamma^2\left(\frac{Q^2}{k_\perp^2}\right) \tag{8.17}$$

where the square of the Sudakov form factor is again reproduced (see eq. (7.7)). Alternatively one could exponentiate the whole lowest order differential distribution :

$$\Sigma(k_\perp^2,Q^2) \simeq \exp\left(-\int_{k_\perp^2}^{Q^2} dp_\perp^2\left(\frac{1}{\sigma}\frac{d\sigma}{dp_\perp^2}\right)_{\text{order }\alpha}\right) \tag{8.18}$$

This is an equivalent procedure at the leading doubly logarithmic approximation with the additional advantage of being automatically in agreement with the calculation at order $\alpha(Q^2)$ for $k_\perp \sim Q$.

By inserting the LLA for the running coupling one obtains the asymptotic expansion :

$$\Sigma(k_\perp^2,Q^2) \simeq \exp\left\{\frac{-4C_F}{11C_A-4T}\left(\ln\frac{Q^2}{\Lambda^2}\ln\frac{\ln Q^2/\Lambda^2}{\ln k_\perp^2/\Lambda^2} - \ln\frac{Q^2}{k_\perp^2}\right)\right\} \tag{8.19}$$

In analogy with eq. (8.16) one finds :

$$\frac{1}{\sigma}\frac{d\sigma}{d^2\vec{k}_\perp} = \frac{1}{(2\pi)^2}\int d^2\vec{b}\, e^{-i\vec{b}.\vec{k}_\perp} \exp\left[\frac{1}{\pi}\int\frac{d^2p_\perp}{p_\perp^2}\left(e^{i\vec{b}.\vec{p}_\perp}-1\right)\frac{\alpha(p_\perp^2)}{\pi} C_F \ln\frac{Q^2}{p_\perp^2}\right] \tag{8.20}$$

(In this connection see also ref. (48)). It is due to Parisi, Petronzio [46] the observation that the previous approach can be extended down to $k_\perp = 0$ provided that Q^2 is large enough. Physically this means that at sufficiently large Q^2 the dominant contribution to the cross section at $k_\perp = 0$ arises from the cancellation of several hard emissions each with large transverse momentum.

The generating function Σ for the actual hadronic process can be obtained by folding the partonic Σ with the effective parton densities as usual. The only new point is that the effective scale will be $k_\perp^2$ in this case rather than Q^2 because what matters is the maximum allowed transverse momentum, as we have seen many times by now.

The relevant formulae can readily be adapted to the case of Drell-Yan processes where the total transverse momentum of the charged lepton pair is a direct measure of the total transverse momentum of the annihilating partons. If we denote by $q_\perp$ the γ^* transverse momentum, after convolution of Σ with the quark parton densities and differentiation one obtains at zero rapidity [45),46)] :

$$\left.\frac{d\sigma}{dQ^2 dq_\perp^2 dy}\right|_{y=0} = \frac{4\pi\alpha_{em}^2}{9Q^2 s}\frac{d}{dq_\perp^2}\{\Gamma^2(\frac{Q^2}{q_\perp^2})[\sum_h e_h^2 (q_h^{(1)}(\sqrt{\tau},q_\perp^2) q_h^{(2)}(\sqrt{\tau},q_\perp^2) + 1 \leftrightarrow 2)]\} \tag{8.21}$$

This is often referred to as the DDT formula. Once again, a further improvement of this formula is obtained by transformation to the impact parameter space, where the conservation of transverse momentum in the otherwise independent soft multiple emissions can most simply be taken into account. In analogy with eq. (8.20), by using :

$$\int_0^{2\pi} d\phi\, e^{\pm iz\cos\phi} = 2\pi J_0(z) \tag{8.22}$$

one finds :

$$\left.\frac{d\sigma}{dQ^2 dq_\perp^2 dy}\right|_{y=0} = \frac{1}{2}\int_0^\infty b\,db\,J_0(bq_\perp)\frac{4\pi\alpha_{em}^2}{9Q^2 s} B(\tau,b,Q^2) \tag{8.23}$$

where $B(\tau,b,Q^2)$ is :

$$B(\tau,b,Q^2) = \sum_h e_h^2 \{q_h^{(1)}(\sqrt{\tau},\frac{1}{b^2})\, q_h^{(2)}(\sqrt{\tau},\frac{1}{b^2}) + 1 \leftrightarrow 2\}.$$

$$\exp \frac{1}{\pi} \int \frac{d^2\vec{p}_\perp}{p_\perp^2} \frac{\alpha(p_\perp^2)}{\pi} C_F \ln \frac{Q^2}{p_\perp^2} [\exp(i\vec{b}.\vec{p}_\perp) - 1]$$

These developments are of great theoretical interest and still the subject of active research. Recently, for example, a systematic expansion method for the subleading correctionsto the DDT formula has been discussed [48]. However, since the kinematical inequalities in eq. (8.4) are to be interpreted in a logarithmic sense, the practical usefulness of the two scale approach is still limited. Presumably by merging the results from the explicit calculation [44]of the terms of order $\alpha^2(Q^2)$ in the $q_\perp$ distributions together with the mentioned expansion of non leading corrections to the DDT formula (the former fixing the unknown parameters of the latter) it will be possible to further sharpen the QCD predictions. The same technique can be applied in e^+e^- annihilation to the energy-energy correlation function defined as [49] :

$$\frac{dW}{d\cos\theta} = \sum_{a,b} \int x_a x_b \frac{1}{\sigma} \frac{d\sigma}{dx_a dx_b d\cos\theta} dx_a dx_b$$

where x_a and x_b are energy fractions carried by two hadrons and θ is the angle between their directions such that $\theta = 0$ in the back to back configuration. This quantity is independent of fragmentation functions, which drop away as a consequence of the energy conservation sum rule and is infrared safe in the massless theory. Since for small θ one has $\theta^2 \sim k_\perp^2/Q^2$ the previous approach can be extended to this case [50].

Besides the scaling violations and the increase with energy of the average $q_\perp$ a third effect is also of great interest for establishing the relevance of QCD in Drell-Yan processes. This is the issue of normalization for the cross section. We have seen that the parton formulae in terms of effective parton densities are valid in QCD within the LLA. In order to evaluate the first order corrections it is preliminarly necessary to precisely specify what is meant by quark densities beyond the LLA. As discussed in Section V, this is conveniently done by specifying that the quark densities are to be measured from the structure function F_2 in leptoproduction at the same absolute value of the virtual photon mass. The relation between the effective parton densities so defined and the bare densities which appear in the expression for the Drell-Yan cross section is given by :

$$q(x,t) = \int_x^1 \frac{dy}{y} \{q_0(y)[\ \delta(1-\frac{x}{y}) + \frac{\alpha}{2\pi}(tP_{qq}(\frac{x}{y}) + f_q^2(\frac{x}{y}))]$$

$$+ G_0(y)\frac{\alpha}{2\pi}(tP_{qG}(\frac{x}{y}) + f_G^2(\frac{x}{y}))\} + \dots \qquad (8.24)$$

By replacement in the Drell-Yan cross sections of the bare with the effective densities, in analogy with eqs. (5.10, 5.11), one finds :

$$\frac{d\sigma}{dQ^2} \simeq \frac{4\pi\alpha_{em}^2}{9Q^2 s}\int_0^1 \frac{dx_1}{x_1}\int_0^1 \frac{dx_2}{x_2}\{[\sum_h e_h^2 q_h^{(1)}(x_1,t)\bar{q}_h^{(2)}(x_2,t) + 1 \leftrightarrow 2].$$

$$[\delta(1-z) + \theta(1-z)\frac{\alpha(Q^2)}{2\pi}(f_q^{DY}(z) - 2f_q^2(z))]$$

$$+ [\sum_h e_h^2(q_h^{(1)}(x_1,t) + \bar{q}_h^{(1)}(x_1,t))G^{(2)}(x_2,t) + 1 \leftrightarrow 2].$$

$$\theta(1-z)\frac{\alpha(Q^2)}{2\pi}(f_G^{DY}(z) - f_G^2(z))\ \} \qquad (8.25)$$

While the function $f_{q,G}^{DY}$ and $f_{q,G}^2$ have no physical meaning by themselves and in fact depend on the details of the regularization method, the differnces in eq. (8.25) are completely will defined. This is a direct consequence of the factorization theorem that makes the leading logarithmic series to disappear when one process is referred to the other. It is also evident from the fact that these differences appear in a relation between measurable quantities. In fact to the accuracy required the gluon density is only needed in the LLA and to that level it also is well defined and measurable, for example from the longitudinal structure function in leptoproduction (eq. (5.16)). An explicit computation leads to [6) :

$$\frac{\alpha}{2\pi}(f_q^{DY}(z)-2f_q^2(z)) = \frac{\alpha}{2\pi}C_F\{(1 + \frac{4\pi^2}{3})\delta(1-z) + 2(1+z^2)(\frac{\ln(1-z)}{1-z})_+$$

$$+ \frac{3}{(1-z)_+} - 6 - 4z\ \} \qquad (8.26)$$

$$\frac{\alpha}{2\pi}(f_G^{DY}(z)-f_G^2(z)) = \frac{\alpha}{2\pi}\frac{T}{f}\{(z^2 + (1-z)^2)\ln(1-z) + \frac{9}{2}z^2 - 5z + \frac{3}{2}\ \} \qquad (8.27)$$

Had we chosen to define quark densities by a different combination of structure functions rather than by F_2 only the polynomial terms in the previous equations would have been modified. This is because of eqs. (5.12 to 5.14) relating the various structure functions.

It turns out that the gluon terms, eq. (8.27), is of normal size. It can be neglected at present energies and not too large values of τ in all processes where the quark leading term is of valence-valence type, while it makes a non negligible contribution to valence-sea processes. On the other hand the quark correction is quite large at present energies with currently accepted values of $\alpha(Q^2)$. The physical origin of these large corrections is quite clear : it can be traced back to the continuation of q^2 from the spacelike region in leptoproduction to the timelike region in lepton pair production and to the difference in phase space between the two processes (the heavy photon is in the initial state in one case and in the final state in the other case).

In fact the large corrections in eq. (8.26) are due to the δ function and to the logarithmic term. Most of the contribution to the δ function term arises from the continuation of the vertex diagram from $q^2 < 0$ to $q^2 > 0$. Precisely a term $\alpha/2\pi\ C_F\pi^2\delta(1-z)$ is due to this effect as was already derived in eq. (5.60). The origin of the logarithmic term is immediately traced back to the expression of the maximum allowed transverse momentum in deep inelastic scattering and in Drell-Yan processes (recall the discussion leading to eq. (5.47)) :

$$(k_\perp^2)^{DIS}_{MAX} = \frac{s}{4} = Q^2\,\frac{(1-z)}{4z} \qquad (z = \frac{x}{y}) \tag{8.28}$$

$$(q_\perp^2)^{DY}_{MAX} = \frac{(s-Q^2)^2}{4s} = Q^2\,\frac{(1-z)^2}{4z} \qquad (z = \frac{\tau}{x_1x_2}) \tag{8.29}$$

Detailed studies of the first order correction show that it is approximately a constant in the limited τ range covered by the data and in most of the rapidity range. Taken at face value the effect of the first order correction should amount at current energies and τ values to a rescaling of the cross section upward by an approximately constant factor of about two (called the K factor) with respect to parton model prediction.

Note that all typical tests of the Drell-Yan mechanism are not affected by the above results. This is immediately clear for the predictions of a linear dependence on the atomic number, of approximate scaling and of an approximately linear rise of $< q_\perp >$ with energy at fixed τ (although the value of the slope is affected). As for the intensity rules they are essentially left unaltered because the whole quark correction is homogeneous in $q\bar{q}$ in the same way as the leading term. Finally the dominance of the transverse component in the angular distribution is not altered because the vertex diagram and the leading logarithms are not present in the longitudinal component.

The presence of a nearly universal and constant factor of the right magnitude and in the right direction is impressively confirmed by the data in all the measured channels (see for example ref. (51)). What is most impressive, at first sight, is that the experiments appear to closely reproduce all the quantitative features of the first order result, although its magnitude should correspond to a breakdown of the perturbative expansion in $\alpha(Q^2)$. The answer to this puzzle is to some extend provided by the observation that the largest contributions to this effect can be resummed and exponentiated [52)]. The terms arising from the continuation of the form factor from spacelike to timelike values of q^2 can be treated to all orders according to eq. (7.5) :

$$\left|\frac{\Gamma(-q^2)_{DY}}{\Gamma(-q^2)_{DIS}}\right|^2 \sim \exp(C_F \frac{\alpha}{2\pi} \pi^2) \simeq 1 + C_F \frac{\alpha}{2\pi} \pi^2 + \ldots \tag{8.30}$$

Similarly the exponentiation of the logarithms arising from the phase space effect follows from arguments as those leading to eq. (5.47). For moments in the τ variable one obtains a factor :

$$\exp \int_0^1 dx(x^{n-1} - 1)C_F \frac{1+x^2}{1-x} \int_{Q^2 \frac{(1-x)}{4x}}^{Q^2 \frac{(1-x)^2}{4x}} \frac{dk_\perp^2}{k_\perp^2} \frac{\alpha(k_\perp^2)}{2\pi} \tag{8.31}$$

While the possibility of factoring out these sequences of terms to all orders is quite well founded at the moment, it certainly remains a conjecture that the residual expansion, which starts with small terms in lowest order, is also well behaved in higher orders. An obvious task for the theory is to put these arguments on a more solid basis in order to understand the K-factor more completely. On the experimental side it is important to study this effect in more detail, for example by detecting its dependence on the scaling variables and on Q^2.

ACKNOWLEDGEMENT

I am really grateful to Prof. M. Levy and D. Speiser, as well as to the whole organizing Committee for their kind invitation to this School. I am also in debt with Miss M.F. Hanseler for her efficient and nice assistance in Cargèse.

REFERENCES

1. For recent reviews on perturbative QCD see :
A.J. Buras; Rev. Mod. Phys. 52, 199 (1980)
Yu. L. Dorkshitzer, D.I. Dyakonov, S.I. Troyan; Phys. Rep. 58, 269 (1980)
E. Reya; Phys. Rep. (1981)
A.H. Mueller; Phys. Rep. (1981)
G. Altarelli; Phys. Rep. (to appear)

2. D. Amati, R. Petronzio, G. Veneziano; Nucl. Phys. B140, 54 (1978); B146, 29 (1981)
A.V. Efremov, A.V. Radyushkin; Serpukhov E2-11535 (1978)
S. Lilly, G. Sterman; Phys. Lett. 78B, 618 (1978); Phys. Rev. D18, 3252, 4737 (1978)
A.H. Mueller; Phys. Rev. D18, 3705 (1978)
S. Gupta, A.H. Mueller; Phys. Rev. D20, 118 (1979)
R.K. Ellis, H. Georgi, M. Machacek, H.D. Politzer, G.G. Ross; Phys. Lett. 78B, 281 (1978); Nucl. Phys. B152, 285 (1979)

3. W.A. Bardeen, A.J. Buras, D.W. Duke, T. Muta; Phys. Rev. D18, 3998 (1978)

4. T. Applequist, H. Georgi; Phys. Rev. D8, 4000 (1973)

5. R. Baier, K. Fey; Bielefefd BI-TP 79/10 (1979)
G. Altarelli, R.K. Ellis, G. Martinelli, S.Y. Pi; Nucl. Phys. B160, 301 (1979)

6. G. Altarelli, R.K. Ellis, G. Martinelli; Nucl. Phys. B143, 521 (1978); E B146, 544 (1978); B157, 461 (1979)
J. Kubar-André, F. Paige; Phys. Rev. D19, 221 (1979)
J. Kubar, M. Le Bellac, J.L. Meunier, G. Plaut; Nice NTH 80/8 (1980)

7. N. Sakai; Phys. Lett. 58B, 67 (1979)

8. R.K. Ellis, D.A. Ross, A.E. Terrano; Nucl. Phys. B178, 421 (1981);
J. Vermaseren, K. Gaemers, S. Oldham CERN-TH 3002 (1981)
R.K. Ellis, D.A. Ross; CERN-TH 3131 (1981)

9. K. Fabricius, H. Kramer, G. Schierholz, I. Schmitt; DESY 81/035 (1981)

10. R.K. Ellis, G. Martinelli, R. Petronzio; CERN-TH (1981)

11. R. Barbieri, G. Curci, E. D'Emilio, R. Remiddi; Nucl. Phys. B154, 535 (1979)
R. Barbieri, M. Caffo, R. Gatto, R. Remiddi; Phys. Lett. 95B, 93 (1980); CERN-TH 3071 (1981)

12. G. Altarelli, G. Parisi; Nucl. Phys. B126, 298 (1977)

13. H. Georgi; H.D. Politzer; Nucl. Phys. B136, 445 (1978)
T. Uematsu; Phys. Lett. 79B, 97 (1978)
J.F. Owens; Phys. Lett. 76B, 85 (1978)

14. E.G. Floratos, D.A. Ross, C.T. Sachrajda; Nucl. Phys. B129, 66 (1977) ; E B139, 545 (1978); B152, 493 (1979)

15. G. Curci, W Furmanski, R. Petronzio; Nucl. Phys. B175, 27 (1980)
W. Furmanski, R. Petronzio; Phys. Lett. 97B, 437 (1980); CERN-TH 3046 (1981)
E.G. Floratos, R. Lacaze, C. Kounnas; Phys. Lett. 98B, 89, 285 (1981); LPTENS 81/3 (ENS-Paris) (1981)
J. Kalinowski, K. Konishi, T.R. Taylor; CERN-TH 2902 (1980) and (with P.N. Scharbach) CERN-TH 2917 (1980)
T. Munehisa, H. Okada, K. Kudoh, K. Kitani; TIT-HEP-64 (Tokyo) (1981)

16. E. Witten; Nucl Phys. B120, 189 (1977)

17. R.J. De Witt, L.M. Jones, J.D. Sullivan, D.E. Willen, H.W. Wyld Jr; Phys. Rev. D19, 2046 (1979)
A. Nicolaidis; Nucl. Phys. B163, 156 (1980)

18. C.H. Llewellyn-Smith; Phys. Lett. 79B, 83 (1978)

19. W.R. Frazer, G.F. Gunion; Phys. Rev. D20, 147 (1979)

20. W.A. Bardeen, A.J. Buras; Phys. Rev. D20; 166 (1979); see also D.W. Duke, J.F. Owens; Florida State Univ. FSU-HEP 802401 (1980)

21. G. 't Hooft; Nucl. Phys. B61, 455 (1973)

22. W. Caswell; Phys. Rev. Lett. 33, 244 (1974)
D.R.T. Jones; Nucl. Phys. B75, 531 (1974)

23. O.V. Tarasov, A.A. Vladimirov, A. Yu. Zharkov; Phys. Lett. 93B, 429 (1980)

24. K.G. Chetyrkin, A.L. Kataev, F.V. Tkachov; Phys. Lett 85B, 277 (1979)
M. Dine, J. Sapirstein; Phys. Rev. Lett. 43, 688 (1979)
W. Celmaster, R.J. Gonsalves; Phys. Rev. Lett. 44, 560 (1979) Phys. Rev. D21, 3772 (1980)

25. P.B. Mackenzie, G.P. Lepage; Cornell CLNS/81-498 (1981)

26. F. Sciulli; Rapporteur talk, Proceedings of the Madison Conference, (1980), page 1279

27. D.A. Ross, C.T. Sachrajda; Nucl. Phys. B149, 497 (1979)

28. A. Gonzales-Arroyo, C. Lopez, F.J. Yndurain; Nucl. Phys. B153, 161 (1979)

29. V.N. Gribov, L.N. Lipatov; Sov. Jour. Nucl. Phys. 15, 438, 675 (1972)

Yu. L. Dokshitzer; Sov. Phys. JETP 46, 641 (1977)
K. Konishi, A. Ukawa, G. Veneziano; Nucl. Phys. B157, 45 (1979)

30. M. Greco, F. Palumbo, G. Pancheri, Y. Srivastava; Phys. Lett. 77B, 282 (1978)
D.R. Butler, C.A. Nelson; Phys. Rev. D18, 1196 (1978)
G. Curci, M. Greco; Phys. Lett. 79B, 406 (1978)
G. Curci, M. Greco, Y. Srivastava; Phys. Rev. Lett. 43, 834 (1979); Nucl Phys. B159, 451 (1979)
H.D. Dahmen, F. Steiner; DESY 80/37 (1980)
C.A. Nelson; Fermilab 80/82 THY (1980)

31. D. Amati, A. Bassetto, M. Ciafaloni, G.Marchesini; G. Venziano; Nucl. Phys. B173, 492 (1980)

32. M. Ciafaloni; Phys. Lett. 95B, 113 (1980)
M. Ciafaloni, G. Curci; Scuola Norm. Pisa SNS 1 (1981)

33. R.M. Barnett, D. Schlatter, L. Trentadue; SLAC-PUB-2714 (1981)

34. See the first paper of ref. (15)

35. G. Altarelli, G. Curci, G. Martinelli, S. Petrarca; Phys. Lett 99B; 141 (1981); CERN-TH 3003 (1980)

36. A.H. Mueller; Columbia Univ. CV-TP-197 (1981)

37. A. Bassetto, M. Ciafaloni, G. Marchesini; Nucl. Phys. B163; 477 (1980)
W. Furmanski, R. Petronzio, S. Pokorski; Nucl Phys. B155, 253 (1979)

38. V. Sudakov; Sov. Phys. JETP 3 , 65 (1956)

39. J. Cornwall, G. Tiktopoulos; Phys. Rev. Lett. 35, 338 (1975) Phys. Rev. D13, 3370 (1976);
J.J. Carazzone, E.R. Poggio, H.R. Quinn; Phys. Rev. D11; 2286 (1975) E D12, 3368 (1975)

40. T. Kinoshita, A. Ukawa; Phys. Rev. D13, 1573 (1976); D15, 1596 (1977); D16, 332 (1977)

41. A.H. Mueller; Phys. Rev. D20, 2037 (1979)

42. C.P. Korthals-Altes, E. de Rafael; Nucl Phys. B106, 237 (1976); B125, 275 (1977)

43. For leptoproduction :
E.G. Floratos; Nuovo Cim. 43A; 241 (1978)
G. Altarelli, G. Martinelli; Phys. Lett. 76B, 89 (1978)

44. R.K. Ellis, G. Martinelli, R. Petronzio; CERN-TH (1981)

45. Yu. L; Dokshitzer, D.L. Dyakonov, S.I. Troyan; Phys. Lett. 78B, 290 (1978)

46. G. Parisi, R. Petronzio; Nucl. Phys. B154, 427 (1979)

47. C.Y. Lo, J.D. Sullivan; Phys. Lett. 86B,327, (1979)
H.F. Jones, J.Wyndham; Nucl. Phys. B176, 466 (1980)
S.D. Ellis, W.J. Stirling; Phys. Rev. D23, 214 (1981)

48. J.C. Collins, D.E. Soper; Oregon Univ. OITS-155 (1981); Proceedings of the Moriond Workshop on Lepton Pair Production (1981), ed. by J. Tran Thanh Van.

49. C.L. Basham, L.S. Brown, S.D. Ellis, S.T. Love; Phys. Rev. D19, 2018 (1979); Phys. Lett. 85B, 287 (1979)

50. R. Baier, K.Fey; Bielefeld BI-TP 80/10 (1980)

51. J. Lefrançois; Rapporteur talk, Proceedings of the Madison Conference, (1980), page 1319
G. Matthiae; Riv. Nuovo Cim. 4,1 (1981)

52. G. Parisi; Phys. Lett. 90B, 295 (1980)
G. Curci, M. Greco; Phys. Lett. 92B, 175 (1980); Frascati LNF-81/7 (1981)

53. P.M. Stevenson ; Wisconsin DOE-ER 00881-155 (1980)

DYNAMICAL MASS GENERATION FOR QUARKS AND LEPTONS

P. Sikivie

CERN, Geneva, Switzerland and
Physics Department, University of Florida
Gainesville, Florida, 32611

I. INTRODUCTION

The history of physics can be summarized into the invention of classical physics followed by a progressive reduction of the number of separate concepts - i.e. mass, space, time and temperature - on which classical physics is based. First, statistical mechanics taught us that temperature is nothing but the energy per degree of freedom of a very many body system in thermal equilibrium. And Boltzmann constant k_B is merely the conversion factor between two identical (but previously separately defined) concepts: energy and temperature. Next, special relativity taught us that the concepts of space and time are essentially identical. Indeed space and time are rotated into one another by the Lorentz transformations which are symmetries of the laws of physics. The speed of light c is the conversion factor. Finally, quantum mechanics taught us that there are no separate concepts of energy and time. Indeed, an eigenstate of the energy has a completely determined time development:

$$\psi_E(t) = e^{-\frac{i}{\hbar}Et} \psi_E(0).$$

Planck's constant is the conversion factor.

We are thus at a point where there is only one concept left. It may be called mass or energy or $(\text{time})^{-1}$ or temperature. It would then appear there is one big step left to make: get rid of it. After that is done, all physics will be in terms of pure (dimensionless) numbers.

Now, it appears we already have a very good candidate for a conversion factor between mass and pure numbers, i.e. Planck's mass:

$$M_{Pl} = \frac{1}{\sqrt{G_N}} \tag{1.1}$$

where G_N is Newton's gravitational constant. But here my glibness ends. There is no theory of mass yet, as there are theories of entropy, speed and action. For example, nobody as yet has explained the ratio of the electron mass to M_{Pl}. It is well-known that the standard model of the strong and electroweak interactions [1], although eminently successful in describing and predicting the interactions of the quarks and leptons, the photon and the $W^{\pm}$ and Z^0 vector bosons, is unable in its present form to predict the masses of these particles. The particle masses are arbitrary parameters in the standard model.

If one were to make a guess today as to what a theory of mass might look like, it would be reasonable to think that QCD-like gauge theories and the renormalization group equations might play an important role in it. Indeed, QCD predicts in principle a very large number of quantities with dimension of mass in terms of just one intrinsic scale Λ^c. And Λ^c can be calculated using the renormalization group equations from knowledge of the grand unification coupling constant at the grand unification scale [2].

We briefly review the various mechanisms that have been proposed to dynamically generate and hence to calculate the masses of the $W^{\pm}$ and Z^o vector bosons, and those of the quarks and leptons. The recurring themes throughout are the following:

1. The elementary Higgs scalar of the standard model must be done away with. Indeed it is the arbitrariness in the Higgs self-interactions and in its Yukawa couplings that makes respectively the masses of $W^{\pm}$ and Z^o, and those of the quarks and leptons uncalculable.

2. A new QCD-like gauge interaction with intrinsic scale of order 1/2 TeV is necessary to break $SU(2)_L \times U_Y(1) \to U(1)^{e.m.}$.

3. Yet other gauge interactions characterized by energy scales a few TeV to a few hundred TeV might be needed to generate quark and leptons masses.

In section II, technicolour [3] (TC) and its extension [4] (ETC) are briefly introduced. In section III, I review the various issues and problems which have been raised in the path of a successful ETC model. In section IV, I argue that if quarks and leptons are elementary, there is no gauge interactions other than ETC which can generate quark and lepton masses from the breaking of $SU(2)_L \times U_Y(1)$ by the technicolour interactions. Finally, in section V, the possibility that quarks and leptons are composite is considered. If composite, the quarks and leptons must satisfy 't Hooft's constraints[5] so that they may be massless on the scale of their binding energy. I show that one can envisage three different mechanisms through which massless composite fermions à la 't Hooft can be given a small mass, and give an example of each mechanism. Whether one of these mechanisms can be used to construct a realistic model of quark and lepton mass generation remains to be investigated.

II. TECHNICOLOUR AND ITS EXTENSION

We want to replace the elementary Higgs scalar of the standard model by a composite one, a bound state (technimeson) resulting from a new QCD-like gauge interaction, called Technicolour (TC). Consider thus a theory without elementary scalar and based on the gauge group:

$$SU(2)_L \times U_Y(1) \times SU(3)^C \times G^{TC} \tag{2.1}$$

The fermions in the theory are the usual quarks and leptons, all of which are technicolour singlets, plus a set of technifermions:

$$\begin{pmatrix} U_L \\ D_L \end{pmatrix} = (2,0,1,n)_L$$

$$U_R = (1,1/2,1,n)_R$$

$$D_R = (1,-1/2,1,n)_R \tag{2.2}$$

The technicolour gauge group G^{TC} and the technifermion representation (n) need not be specified at this time. The energy scale Λ^{TC} at which the technicolour gauge coupling becomes strong is of order 1 TeV. At those energies one can in first approximation neglect the electroweak and colour interactions.

Now (2.2) shows that the technicolour interactions are chiral isospin and $SU(2)_L \times SU(2)_R$ invariant. The same is true of QCD with two massless quark flavors u and d. In the case of QCD we know that the colour interactions cause quark-antiquark condensates to form:

$$< \bar{u}u >_o = < \bar{d}d >_o \simeq (\Lambda^c)^3 \tag{2.3}$$

which spontaneously break the chiral isospin symmetry of QCD down to isospin $SU(2)_{L+R}$. The pions are the corresponding Goldstone Bosons.

There is good reason to believe that the technicolour interactions behave in exactly the same way, i.e. the technifermion condensates:

$$< \bar{U}U >_o = < \bar{D}D >_o \simeq (\Lambda^{TC})^3 \tag{2.4}$$

form and break the chiral isospin symmetry $SU(2)_L \times SU(2)_R$ of the technicolour interactions spontaneously down to techni-isospin $SU(2)_{L+R}$. Three massless Goldstone Bosons result which are the technipions.

Let us now turn on the colour and electroweak interactions. The technifermion condensates (2.4) are colour and electrically neutral, but not $SU(2)_L \times U_Y(1)$ invariant. The $SU(2)_L \times U_Y(1) \times SU(3)^C$ gauge group has thus been spontaneously broken down to $U(1)^{e.m.} \times SU(3)^C$ by (2.4). The $W^{\pm}$ and Z^o vector bosons "eat up" the three technipions and acquire a mass. One can show that [3]:

$$M_W = M_Z \cos\theta_W = \frac{g}{2} F_\pi \tag{2.5}$$

Where F_π is the technipion decay constant which is the technicolour analogue of $f_\pi \simeq 95$ MeV. To obtain the correct strength for the weak interactions one must have:

$$F_\pi = \sqrt{\sqrt{2}\, G_{Fermi}} \simeq 250 \text{ GeV} \tag{2.6}$$

which indeed implies that the technicolour scale must be of order 1 TeV. The first equality in (2.5) holds to all orders in the technicolour interactions as a result of the techni-isospin symmetry. This particularly elegant explanation of the ratio of neutral to charged currents in weak interactions is one of the foremost reasons of technicolour's physixappeal. Another is the fact that if G^{TC} is grand unified along with the other gauge groups in eq.(2.1), then Λ^{TC} can be determined using the renormalization group equations, and hence G_{Fermi} becomes calculable.

The trouble with the model as developed so far is that it leaves the quarks and leptons massless. Some way must be found to couple the ordinary fermions (leptons and quarks) to the technifermions so that the former acquire mass when the latter condense. This can be done [4] by unifying the ordinary fermions with the technifermions into multiplets of a larger gauge group G^{ETC}, called "extended technicolour" of which G^{TC} is the unbroken subgroup.

Consider for example a gauge group $SU(6)^{ETC}$ which breaks spontaneously down to $SU(3)^{TC}$ at successive scales μ, μ' and μ'':

$$SU(6)^{ETC} \xrightarrow{\mu} SU(5) \xrightarrow{\mu'} SU(4) \xrightarrow{\mu''} SU(3)^{TC} \qquad (2.7)$$

Consider also six fermions which form a sextet of $SU(6)^{ETC}$. These will break into a triplet (F_i, i = 1,2,3) and three singlets (f,f' and f") of $SU(3)^{TC}$:

$$\begin{array}{c}(\underbrace{f \quad f' \quad f''}_{\text{generation}} \quad \underbrace{F_1 \quad F_2 \quad F_3}_{TC}) \\ \longleftarrow\!\!\longrightarrow \\ ETC\end{array} \qquad (2.8)$$

At some scale Λ^{TC} (below μ,μ' and μ'') the technicolour forces become strong and the technifermions condense as before:

$$< \bar{F}F >_o \simeq (\Lambda^{TC})^3 \qquad (2.9)$$

However, the ordinary fermions f, f' and f" now acquire a small mass. Indeed there are in G^{ETC} heavy vector bosons which couple the ordinary fermions to the technifermions and which mediate four-fermion interactions of the type:

$$-\frac{1}{\mu^2}(\bar{F}\gamma_\alpha f)(f\gamma^\alpha F) - \frac{1}{\mu'^2}(\bar{F}\gamma_\alpha f')(\bar{f}'\gamma^\alpha F) - \frac{1}{\mu''^2}(\bar{F}\gamma_\alpha f'')(\bar{f}''\gamma^\alpha F) \qquad (2.10)$$

This can be Fierz transformed into:

$$(\bar{F}F)(\frac{1}{\mu^2}\bar{f}f + \frac{1}{\mu'^2}\bar{f}'f' + \frac{1}{\mu''^2}\bar{f}''f'') + \ldots \qquad (2.11)$$

This implies that:

$$m_f = \frac{< \bar{F}F >_o}{\mu^2} \quad , \quad m_{f'} = \frac{< \bar{F}F >_o}{\mu'^2} \quad , \quad m_{f''} = \frac{< \bar{F}F >_o}{\mu''^2} \qquad (2.12)$$

To apply this ETC mechanism to the quarks and leptons of the standard model, we introduce a family of technifermions U, D, N and E which have the same quantum numbers as u, d, ν and e under the standard gauge group $SU(2)_L \times U_Y(1) \times SU(3)^C$, but which in addition carry technicolour. We then unify our technifamily with (for example) three ordinary families into ETC multiplets, as in (2.8):

$$\begin{array}{cccccccc} u & c & t & U_1 & U_2 & \cdots & U_n \\ d & s & b & D_1 & D_2 & \cdots & D_n \\ \nu_e & \nu_\mu & \nu_\tau & N_1 & N_2 & \cdots & N_n \\ e & \mu & \tau & E_1 & E_2 & \cdots & E_n \end{array}$$

$\longleftarrow$ generation $\longrightarrow$ $\longleftarrow$ TC $\longrightarrow$

$\longleftarrow$ ETC $\longrightarrow$ (2.13)

where n is the number of technicolours. The techniquarks U and D come in three colours (R, W, B) as do the quarks. The technifermion condensates:

$$< \bar{U}_R U_R >_o = < \bar{U}_W U_W >_o = < \bar{U}_B U_B >_o = < \bar{N}N >_o$$

$$= < \bar{D}_R D_R >_o = < \bar{D}_W D_W >_o = < \bar{D}_B D_B >_o = < \bar{E}E >_o \simeq (\Lambda^{TC})^3 \qquad (2.14)$$

break $SU(2)_L \times U_Y(1) \times SU(3)^c \rightarrow U(1)^{e.m.} \times SU(3)^c$ and assure $M_W = M_Z \cos\theta_W$. Moreover, they give mass to three generations of quarks and leptons through the ETC mechanism described above.

By assigning different ETC representation (4.6) to the charge 2/3 and charge - 1/3 quarks, realistic looking quark mass-matrices with up-down splittings [6] and non-zero Cabibbo angles [7] can be obtained. The left-handed neutrinos are made approximately massless by assigning large masses to their right-handed counterparts [8]. The right-handed technineutrinos can not receive such large masses however since they must participate in the condensates (2.14).

III. ETC ISSUES, ANSWERS AND PROBLEMS

1. Isospin Violation in the Quark and Lepton Mass Matrices.

To obtain:

$$M_W = M_Z \cos\theta_W \qquad (3.1)$$

the technicolour interactions and the technifermion condensates must conserve isospin. Therefore, the isospin violation which is necessary to generate realistic quark and lepton mass matrices must only occur in the ETC sector. Because $G^{TC} \subset G^{ETC}$, the global symmetries of the ETC interactions are always global symmetries of the TC interactions, but not vice-versa. Therefore, it is possible as a rule to have isospin violation in the ETC sector but isospin conservation in the technicolour interactions and condensates. In fact, several models [6), 9), 7)] have been constructed which make use of this and which have both $M_W \simeq M_Z \cos\theta_W$ and $m_u \neq m_d$, $m_c \neq m_s$, ... The isospin violation in the ETC sector will add corrections to (3.1) of order $O(\Lambda^{TC}/\mu^{ETC})^2$. These corrections will be sufficiently small provided the isospin splittings in the quark and lêpton mass matrices are not too large.

2. Hierarchies in the Quark and Lepton Mass Matrices

Eq. (2.11) shows that the ordinary fermion mass-matrix can be calculated from the $(\text{mass})^2$-matrix of the ETC vector bosons. But what determines the latter ? If the $G^{ETC} \to G^{TC}$ breaking requires the introduction of additional QCD-like gauge group (like technicolour but at higher energy scales), the contemplated theory will soon become ugly and lose its predictive power. The "tumbling" hypothesis [10)] might provide us with a satisfactory alternative. It plausibly states that asymptotically free gauge theories with non-real left-handed fermion representation content break themselves through the formation of successive Lorentz scalar but "colour" non-singlet fermion bilinear condensates. It is possible that G^{ETC} breaks itself down to G^{TC} through this mechanism. In that case, the ETC vector boson masses would all be in principle calculable in terms of Λ^{TC}.

3. Vacuum Alignment

As mentioned in section II, the most straightforward way to implement ETC into the standard model is to introduce one family of technifermions:

$$\{F_{ai}\} = \begin{pmatrix} U_R & U_W & U_B & N \\ D_R & D_W & D_B & E \end{pmatrix} \qquad (3.2)$$

(a = 1, ..., 4; i = 1, 2; the technicolour index has been suppressed)

alongside the three ordinary families, as in (2.13). One expects the following condensates to form [11) 6)] :

$$< F_{aiL}^{\dagger} F_{bjR} >_o = (U_L^{\dagger} \not{\pi} U_R)^{ai}{}_{bj} \frac{1}{2} \mu_{TC}^3 \qquad (3.3)$$

where U_L and U_R are arbitrary 8 x 8 special unitary matrices. The degeneracy of the vacuum (3.3) is lifted by the $SU(2)_L \times U_Y(1) \times SU(3)$ gauge interactions, the ETC interactions, and whatever other interactions may be present. These interactions must align the condensates in the direction $U_L = U_R = 1\!\!1$, since only the condensates (2.14) break $SU(2)_L \times U_Y(1) \times SU(3)^C \rightarrow U(1)^{e.m.} \times SU(3)^C$ and assure $M_W = M_Z \cos\theta_W$. Peskin and Preskill have independently shown that the technifermion condensates do indeed align in the direction (2.14) provided the technifermions transform as a complex representation of G^{TC} or a real representation of G^{TC} with a symmetric metric [12]. These two cases are referred to respectively as complex technicolour and real technicolour. If the technifermions transform as a real representation of G^{TC} with an anti-symmetric metric (this case is referred to as pseudo-real technicolour), the technifermion condensates align differently and in fact break $U(1)^{e.m.}$ [12]. Pseudo-real technicolour is thus ruled out.

4. Pseudo-Goldstone Bosons

The technifermion condensates (2.14) spontaneously break the global symmetries of the technicolour interactions: $SU(8)_L \times SU(8)_R \times U_Y(1) \rightarrow SU(8)_V \times U_V(1)$ if technicolour is complex or $SU(16) \rightarrow O(16)$ if technicolour is real. This spontaneous symmetry breakdown produces 63 (Pseudo-) Goldstone Bosons in the complex case, 135 in the real case [13]. Three of these are "eaten up" by $W^\pm$ and Z^o. The remaining ones acquire masses, ranging from a few GeV to 300 GeV, from the $SU(2)_L \times U_Y(1) \times SU(3)^C$ gauge interactions, the ETC interactions and from Pati-Salam unification. Their properties have been studied in detail by a number of authors [14]. One of them called $P^\pm$, of mass 5 to 15 GeV, has recently been searched for by the JADE collaboration [15] at PETRA, with negative results. The data rule out one techni-family ETC models in which the ETC gauge group commutes with the standard $SU(2)_L \times U_Y(1) \times SU(3)^C$ gauge group. The possibility that the ETC vector bosons are triplets of $SU(3)^C$ is not ruled out however.

5. CP

The quantity:

$$\bar{\theta}_{QCD} = \theta_{QCD} - \arg(m_u m_d m_s m_c m_b \dots) \tag{3.5}$$

is constrained to be very small $[< O(10^{-8})]$, for example by the experimental upper limit on the neutron electric dipole moment. In the standard model with one elementary Higgs doublet $\bar{\theta}_{QCD}$ is an arbitrary parameter however. To obtain $\bar{\theta}_{QCD} = 0$ naturally in models with elementary Higgs scalars one can introduce a Peccei-Quinn U(1) symmetry [16] and the concomitant axion [17] which can be made "invisible" [18]. Technicolour models offer an alternative. Because they have no bare fermion mass terms in their Lagrangian, the various θ parameters associated with non-abelian gauge group in technicolour models can as a rule all be set equal to zero. The Lagrangian is then CP conserving. In that case, it is natural that all fermion condensates be CP conserving as well [6]. Hence there is no CP violation at all and $\bar{\theta}_{QCD} = 0$. Of course, upon this state of affairs a small CP violating perturbation must be introduced to explain $K_L \to 2\pi$, etc. A satisfactory way to do this has not yet been agreed upon yet[19], in great part because of the problem with $\delta m_K^2 = m_K^2 - M_{KS}^2$ explained below. It is clear however that the small perturbation responsible for $K_L \to 2\pi$ will in general introduce only a tiny deviation of $\bar{\theta}_{QCD}$ from zero.

6. Flavor Changing Neutral Currents

These are the "bête noire" of scalarless gauge theories which attempt to be realistic. In one techni-family ETC models, the problem arises as follows. Consider the unification into G^{ETC} multiplets of three ordinary families f, f' and f" with one techni-family F, as in (2.8) and (2.13). The gauging of G^{ETC} implies the gauging of the generation group. To avoid trouble with flavor changing neutral currents (FCNC), the vector bosons $\mathcal{E}'$ that gauge the generation group must have masses larger than about g_{ETC} x (1000 TeV). On the other hand, in order that the ordinary fermions f, f', ... acquire masses ranging from a few MeV to several GeV, the masses of the vector bosons $\mathcal{E}$ that couple to the $\bar{F}\gamma_\mu f$ currents must range from g_{ETC} x (a few TeV) to about g_{ETC} x (100 TeV). However, in any reasonable symmetry breaking scheme of G^{ETC} down to G^{TC}, the masses of the $\mathcal{E}$ and $\mathcal{E}'$ vector bosons are of the same order of magnitude. Hence, one expects $\delta m^2{}_K$, $\delta m^2{}_D$ and other FCNC processes to be too large by roughly a factor 100 [20]. It is not excluded however that some FCNC suppression mechanism appropriate to ETC models can be invented. In a recent model [7], $\delta m^2{}_D$ was found to be naturally suppressed by a factor 100 below the naive estimate above and hence consistent with present data. $\delta m^2{}_K$ however was as naively estimated above and hence 100 times too large.

IV. ALTERNATIVES TO ETC ?

Let us take for granted the absence of elementary scalars, the existence of technicolour and technifermions and the breaking of the electroweak gauge symmetry by the technifermion condensates. One might well ask then whether there is another mechanism (other than ETC) in which quark and lepton masses can be generated. The answer appears to be no if quarks and leptons are elementary down to length scales $<1/10^4$ TeV.

Indeed, let us suppose we are not allowed to use the ETC mechanism. This means that we are not allowed to gauge the $\bar{F}\gamma_\mu f$ currents. We can then only gauge the $\bar{f}\gamma_\mu f$ and $\bar{F}\gamma_\mu F$ currents in arbitrary linear combinations that form a Lie group. In that case, there is no Feynman

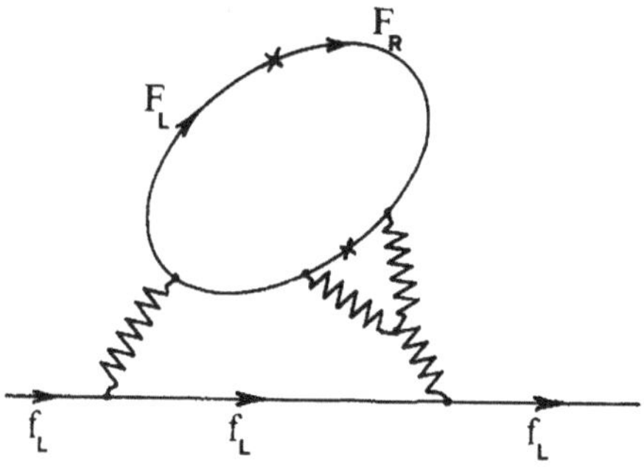

Fig. 1. In perturbation theory, the ordinary fermions f cannot acquire mass from the technifermion condensates (denoted by X) if the $\bar{f}\gamma_\mu F$ currents are not gauged.

diagram that produces a helicity flip of an ordinary fermion f (see Fig. 1). In other words, the overall chiral U(1) symmetry of the ordinary fermions can not be broken perturbatively, unless ETC is used.

It must be pointed out that non-perturbative effects make that U(1) anomalous. For example, a colour instanton emits both techniquarks and ordinary quarks. The techniquarks are absorbed into the techniquark condensates (see Fig. 2). This result into a chiral U(1) breaking effective interaction among ordinary quarks. However, this interaction is not a quark mass term, but rather a many quark interaction, since we know that a colour instanton emits every quark flavour. To state this more precisely and more generally, the non-perturbative effects which make the overall chiral U(1) symmetry anomalous cannot produce quark mass terms because those effects still obey discrete Z_N chiral symmetries (N = # quark flavors) which forbid such mass terms.

At any rate, even if some complicated model were constructed to foil the discrete Z_N symmetries, the contributions from colour instantons to quark masses through the mechanism of Fig. 2 would probably be too small since one naively estimates such contributions to be of order:

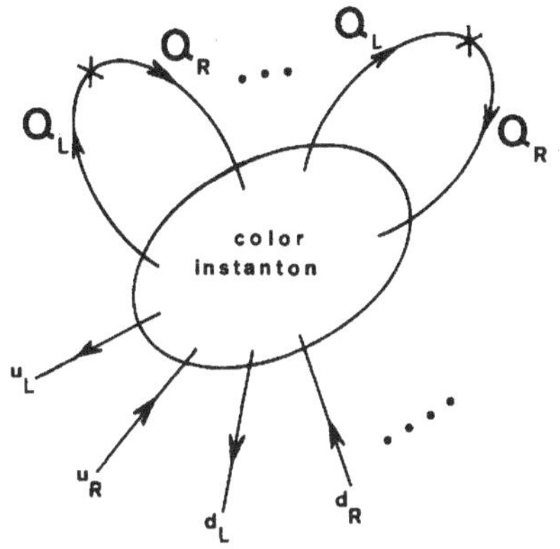

Fig. 2. Colour instantons emit both quarks and techniquarks Q. The techniquarks are absorbed in the condensates X. This produces a $U_A(1)$ breaking interaction among many quarks rather than a quark mass term.

$$\Lambda^{TC} \exp \frac{-2\pi}{\alpha_C(\Lambda^{TC})} = \Lambda^{TC}\left(\frac{\Lambda^C}{\Lambda^{TC}}\right)^{11-\frac{2}{3}N} \tag{4.1}$$

Finally, if the quarks and leptons are composite, the above arguments do not hold since, in that case, they would have residual interactions due to their composite structure. On the other hand, if the composite structure only appears at very high energies, one can neglect it in low energy considerations and the above arguments would hold.

V. COMPOSITE QUARKS AND LEPTONS

Let us now consider the possibility that quarks and leptons are composites of more elementary objects called preons. Two questions arise. First, why are the lepton (and presumably also quark) binding energies so much larger than their masses? Second, what is the origin of the small masses of quarks and leptons ? In answer to the first question, 't Hooft [5)] has given conditions under which composite fermions can be massless. We will briefly review 't Hooft's conditions. Next, in answer to the second question, we will propose mechanisms through which these composite fermions can acquire a small mass.

Consider a QCD-like gauge group G^{MC} with left-handed spin 1/2 fermion representation content:

$$N_1(n_1) + N_2(n_2) + \ldots + N_k(n_k) \tag{5.1}$$

We use the name "metacolour" for the degrees of freedom on which G^{MC} acts, and the name "metaflavour" for the degrees of freedom on which acts the global symmetry group of the metacolour gauge interactions:

$$G^{MF} = SU(N_1) \times \ldots \times SU(N_k) \times U_1(1) \times \ldots \times U_{k-1}(1) \tag{5.2}$$

Also, by convention, all fermions are left-handed spin 1/2 whether preons or composites. Let us consider the theory defined by G^{MC} and the fermion representation content (5.1) at energies well below Λ^{MC}, the scale at which the gauge coupling constant of G^{MC} becomes strong. At those energies, metacolour is confined: only metacolourless (and therefore in general composite) degrees of freedom appear. The global symmetries of physics at those energies form some subgroup H^{MF} of G^{MF}, since in general G^{MF} has been partly broken down spontaneously at energy scales of order Λ^{MC}:

$$G^{MF} \underset{\Lambda^{MC}}{\rightarrow} H^{MF} \tag{5.3}$$

't Hooft [5)] pointed out that composite fermions can be kept massless by chiral symmetries in H^{MF}. Moreover, 't Hooft postulated the following constraints on the H^{MF} multiplets of massless composite fermions that might appear [5)]:

1. $$\mathrm{Tr}_{\Psi}(Q^a \{Q^b, Q^c\}) = \mathrm{Tr}_{\psi}(Q^a \{Q^b, Q^c\}) \tag{5.4}$$

for any triplet of generators Q of H^{MF}. Ψ represents the set of composites, whereas ψ represents the set of preons.

2. For any explicit preon mass term:

$$\psi^T m\sigma_2 \psi + \text{h.c.} \tag{5.5}$$

that breaks $H^{MF} \to H'^{MF}$, H'^{MF} must be such as to allow mass for all composites Ψ that contain massive preons ψ.

Eq (5.4) is thought to follow from general principles and to be a necessary constraint [11), 21)]. The second constraint, as has recently been emphasized [22)], follows from the Appelquist-Carazzone decoupling theorem and the assumption that no phase transition occurs when m is varied from zero to infinity. Since m/Λ^{MC} can be a critical parameter, it is not clear that this assumption is valid.

Solutions to 't Hooft's constraints exist [23)]. And, if the second constraint is dropped, numerous solutions can be found to eq.(5.4) by itself. Let us thus take for granted the existence of massless composite fermions à la 't Hooft and address the question of how to give them small masses.

Obviously this must be done by introducing a small perturbation that breaks the chiral symmetries in H^{MF}. Let us require that this perturbation be introduced without the use of elementary scalars. We wish to consider theories with gauge fields and spin 1/2 fermions only. In such theories, the only interactions present are gauge interactions. We then distinguish three different ways in which composite fermions can acquire small masses [24)]:

1. Chiral symmetries in H^{MF} can be broken explicitly by gauge interactions broken above the scale Λ^{MC}.

2. Chiral symmetries in H^{MF} can be broken explicitly by gauge interactions unbroken above the scale Λ^{MC}.

3. Chiral symmetries in H^{MF} can be broken spontaneously by gauge interactions unbroken above the scale Λ^{MC}.

The fourth possibility of principle, where chiral symmetries in H^{MF} would be broken spontaneously by gauge interactions broken above the scale Λ^{MC}, is presumably impossible in practice. Let us now analyze the three mentioned possibilities in somewhat greater detail.

1. "Unification" : G^{MC} can be the unbroken subgroup of a larger gauge group G:

$$G \xrightarrow[\mu]{} G^{MC} \qquad , \mu > \Lambda^{MC} \tag{5.6}$$

As a rule, G will unify several multiplets (5.1) of G^{MC} into larger multiplets. As a result, H^{MF} will be explicitly broken by interactions mediated by heavy G/G^{MC} vector bosons. One can expect composite fermion masses of order $(\Lambda^{MC})^3/\mu^2$.

As an example, consider G^{MC} = SU(5) and the preons

$$\{\psi^i, \psi_{ij} = -\psi_{ji}\} \sim (\bar{5}) + (10)$$

G^{MF} = U(1) generated by $Q = 3N_{\bar{5}} - N_{10}$. The massless composite

$$\Psi = \varepsilon_{ijklp}(\psi_{ij}{}^{\dagger}\sigma_2\psi_{kl}{}^{\dagger})\psi^P$$

satisfies 't Hooft's constraints . Now assume unification into G = O(10) at some scale $\mu \gg \Lambda^{MC}$. We introduce a SU(5) singlet ψ_o to complete the O(10) multiplet: (16) = ($\bar{5}$) + (10) + (1). The four fermion interactions mediated by the O(10)/SU(5) vector bosons:

$$\frac{-1}{\mu^2} J^{\alpha}_{ij} J^{ij}_{\alpha} \tag{5.7}$$

with

$$J^{\alpha}_{ij} = \psi_o^{\dagger}\sigma^{\alpha}\psi_{ij} + \frac{1}{2}\varepsilon_{ijklp}\psi^{\dagger}_{kl}\sigma^{\alpha}\psi^P \tag{5.8}$$

explicitly breaks the $U(1)^{MF}$ symmetry of the metacolour interactions. After a Fierz transformation, one of the terms in (5.7) becomes:

$$\frac{-2}{\mu^2}\varepsilon_{ijklp}(\psi^{\dagger}_{kl}\sigma_2\psi^{\dagger}_{ij})(\psi_o\sigma_2\psi^P) + \text{h.c.}$$

$$\simeq -\frac{(\Lambda^{MC})^3}{\mu^2}(\Psi^T\sigma_2\psi_o + \text{h.c.}) \tag{5.9}$$

A massive Dirac fermion is obtained, whose left-handed part is composite whereas its right-handed part is elementary [24]. Although this might appear unsettling at first, one should remember that such left-right asymmetries are present in physics, in particular in the weak interactions. In fact, it is possible to interpret the weak interactions as residual interactions due to the composite nature of the left-handed quarks and leptons. This has been proposed recently by several authors [25].

2. Chiral symmetries in H^{MF} can be broken explicitly by gauging some subgroup K^{MF} of G^{MF}. This possibility has been discussed by several authors [26]. The resulting Ψ masses are of order $\Lambda^{MC}e^{-2\pi/\alpha}$ or $\alpha\Lambda^{MC}$ depending on whether or not K^{MF} instantons are necessary to break the relevant chiral symmetries in H^{MF} (α is the K^{MF} gauge coupling constant at the scale Λ^{MC}). We give here an updated version of Weinberg's original example. [26]

Consider QCD with three massless quarks

$$\{u,\ d,\ s\}_{L,R} \sim 3(3)_L + 3(3)_R$$

of $SU(3)^C$. In this example, it will be convenient to use the more familiar four-component notation for spin 1/2 fermions. $G^F = SU(3)_L \times SU(3)_R \times U_V(1)$. Let us consider a scenario where $<\bar{s}s>_o \simeq (\Lambda^C)^3$ but $<\bar{u}u>_o = <\bar{d}d>_o = 0$. In that case,

$$H^F = SU(2)_L \times SU(2)_R \times U_V(1) \times U_V^8(1).$$

The massless composites $p = (u_i\mathbb{C}dj)u_k\varepsilon^{ijk}$ and $n = (u_i\mathbb{C}dj)d_k\varepsilon^{ijk}$ satisfy 't Hooft's constraints. ($\mathbb{C}$ is the 4 x 4 charge conjugation matrix). The other members of the baryon octet (Λ, Σ and Ξ) have masses of order Λ^C.

Now let us gauge $K^F = SU(3)_V$. The vacuum aligns in such a way that

$$SU(3)_V \xrightarrow{\Lambda^C} SU(2)_V \times U_V^8(1)$$

The gauging of $SU(3)_V$ explicitly breaks the axial symmetries in H^F that keep the p and n massless. p and n acquire mass of order $\alpha\Lambda^C$ through self-energy diagrams with a $SU(3)_V/SU(2)_V$ vector boson and a Λ or Σ baryon inside the loop.

3. "Metatechnicolour" : The gauging of a subgroup G^{TC} of H^{MF} can lead to the spontaneous breaking of chiral symmetries in H^{MF}. The mechanism combines the ideas of metacolour and technicolour to produce an alternative to extended technicolour.

As an example, consider $SU(5)^{MC}$ with the preons

$$\{\psi_\alpha^i,\ \psi_{ij}^\rho\} \sim N(\bar{5}) + N(10) \qquad \alpha,\ \rho = 1,\ 2,\ \ldots,\ N;\ i,j = 1,\ldots 5.$$

$$G^{MF} = SU(N)_{\bar{5}} \times SU(N)_{10} \times U(1)$$

where the abelian generator is $Q = 3N_{\bar{5}} - N_{10}$. As a result of a "tumbling" scenario, $H^{MF} = O(N) \times U(1)$ where $O(N)$ is the diagonal subgroup of $O(N)_{\bar{5}} \times O(N)_{10}$. The massless composites

$$\Psi_\alpha = (\psi_{ij}^{\rho\dagger}\sigma_2\psi_{kl}^{\rho\dagger})\psi_\alpha^p\varepsilon_{ijklp} \sim (N)_5 \text{ of } O(N) \times U(1)$$

satisfy 't Hooft's constraints. Now, let us gauge a $O(M)^{TC}$ subgroup of $O(N)^{MF}$. Ψ_r (r = 1, ..., M) are composite techni-fermions; Ψ_a (a = M + 1, ..., N) are composite "ordinary" fermions. We expect the technifermions to condense:

$$< \sum_{r=1}^{M} \Psi_r \sigma_2 \Psi_r >_o \simeq (\Lambda^{TC})^3 \tag{5.10}$$

Thereby spontaneously breaking the $U(1)^{MF}$ symmetry that keeps the "ordinary" fermions massless. The latter acquire mass as a result of residual interactions between metacolour composites (these are analogous to baryon-baryon interactions in QCD). In our model, we expect the O(N) x U(1) symmetric residual interaction:

$$\frac{1}{(\Lambda^{MC})^2} \sum_{\alpha,\beta=1}^{N} (\Psi_\alpha \sigma_2 \Psi_\alpha)(\Psi_\alpha \sigma_2 \Psi_\alpha)^\dagger \tag{5.11}$$

to be present. A mass term for the ordinary fermions results:

$$\frac{(\Lambda^{TC})^3}{(\Lambda^{MC})^2} \sum_{a=M+1}^{N} (\Psi_a \sigma_2 \Psi_a + h.c.) \tag{5.12}$$

We can thus conclude that composite quarks and leptons offer the possibility of several new mass generation mechanisms. Whether one or more of these can be incorporated into a realistic model of the quark and lepton masses and interactions remains to be explored.

Acknowledgement

This work has been supported in part by DOE contract DSR 80136 je 7.

REFERENCES

1. S.L. Glashow, Nucl. Phys. 22 (1961), 579;
S. Weinberg, Phys. Rev. Lett. 19 (1967), 1264;
A. Salam in "Elementary Particle Physics", ed. N. Svartholm (Almqvist and Wiksells, Stockholm, 1968).

2. H. Georgi and S.L. Glashow, Phys. Rev. Lett. 32 (1974), 433;
H. Georgi, H. Quinn and S. Weinberg, Phys. Rev. Lett. 33 (1974) 451.

3. L. Susskind, Phys. Rev. D20 (1979), 2619;
S. Weinberg, Phys. Rev. D19 (1979), 1277.

4. S. Dimopoulos and L. Susskind, Nucl. Phys. B155 (1979) 237;
E. Eichten and K. Lane, Phys. Lett. 90B (1980), 125;
S. Weinberg, unpublished.

5. G. 't Hooft in "Recent Developments in Gauge Theories", ed. 't Hooft et al. (Plenum Press, New York, 1980).

6. P. Sikivie, L. Susskind, M. Voloshin and V. Zakharov, Nucl. Phys. B173 (1980), 189.

7. J. Ellis and P. Sikivie, Phys. Lett. 104B (1981), 141.

8. M. Gell-Mann, P. Ramond and R. Slansky, unpublished;
P. Ramond, Caltech preprint CALT-68-709 (Feb. 1979);
T. Yanagida in Proc. of the Workshop on the Unified Theory and the Baryon Number in the Universe, KEK 1979.

9. S. Dimopoulos, S. Raby and P. Sikivie, Nucl. Phys. B176 (1980), 449.

10. S. Raby, S. Dimopoulos and L. Susskind, Nucl. Phys. B169 (1980) 373.

11. S. Coleman and E. Witten, Phys. Rev. Lett. 45 (1980), 100.

12. M.E. Peskin, Nucl. Phys. B175 (1980), 97;
J. Preskill, Nucl. Phys. B177 (1981), 21.

13. S. Dimopoulos, Nucl. Phys. B168 (1980), 69.

14. For reviews see:
P. Sikivie in the Proceedings of the Rencontres de Moriond, 1981, CERN preprint TH 3083 (May 1981)
J. Ellis in the Proceeding of the SLAC Topical Conference on Particle Physics, 1981, CERN preprint TH 3139 (Aug. 1981).

15. JADE collaboration, data presented by J. Burger, Rapporteur talk at the Bonn Symposium on lepton and photon interactions at High Energies (1981).
For a recent discussion of the $P^{\pm}$ mass see:
P. Binétruy, S. Chadha and P. Sikivie, CERN preprint, Sept. 1981.

16. R. Peccei and H. Quinn, Phys. Rev. Lett. 38 (1977), 1440 and Phys. Rev. D16 (1977), 1791.

17. S. Weinberg, Phys. Rev. Lett. 40 (1978), 223;
F. Wilczek, Phys. Rev. Lett. 40 (1978), 279.

18. M. Dine, W. Fischler and M. Srednicki, Princeton IAS preprint, 1981;
M. Wise, H. Georgi and S.L. Glashow, Phys. Rev. Lett. 47 (1981) 402.

19. For a proposal, see E. Eichten, K. Lane and J. Preskill, Phys. Rev. Lett. 45 (1980), 225.

20. S. Dimopoulos and J. Ellis, Nucl. Phys. B182 (1981), 505.

21. Y. Frishman, A. Schwimmer, T. Banks and S. Yankielowicz,

Nucl. Phys. B177 (1981) 157;
S. Coleman and B. Grossman, to be published.

22. J. Preskill and S. Weinberg, Phys. Rev. D24 (1981) 1059;
S. Dimopoulos and J. Preskill, Harvard preprint HUTP - 81/A045 (Sept. 1981).

23. S. Dimopoulos, S. Raby and L. Susskind, Nucl. Phys. B173 (1980) 208;
T. Banks, S. Yankielowicz and A. Schwimmer, Phys. Lett. 96B (1980), 67;
C. Albright, Max-Planck preprint MPI-PAE/Th 7181;
I. Bars and S. Yankielowicz, Phys. Lett. 101B (1981), 159.

24. P. Sikivie, Phys. Lett. 103B (1981), 437.

25. L. Abbott and E. Farhi, Phys. Lett. 101B (1981), 69 and CERN preprint TH 3057 (1981);
H. Fritzsch and G. Mandelbaum, Phys. Lett. 102B (1981), 319.

26. S. Weinberg, Phys. Rev. D13 (1976), 974;
G. 't Hooft in Ref. 6;
M. Peskin, Cornell preprint CLNS 81/503 (1981);
S. Dimopoulos and L. Susskind, Stanford preprint ITP-681 (1980);
S. Weinberg, Phys. Lett. 102B (1981), 401.

QUARK CONFINEMENT AND LEPTON LIBERATION IN AN ANISOTROPIC SPACE-TIME

G. Preparata

Istituto di Fisica dell'Università - Bari
and
Istituto Nazionale di Fisica Nucleare - Bari

I. INTRODUCTION

The remarkable progress in our understanding of particle physics in the last ten years has convincingly provided a picture of elementary interactions (excluding gravity) based on the "standard model":

$$SU(3)_{colour} \times [SU(2) \times U(1)]_{electroweak}$$

The existence of such symmetries, either exact ($SU(3)_{colour}$) or spontaneously broken ($SU(2) \times U(1)_{ew}$) is brought out by so many experimental facts that it would appear unreasonable to deny their deep relevance in the realm of modern particle physics.

Where doubts may still occur, and are somewhat strongly motivated (see below), is in the precise theoretical framework that should underly the description of such symmetries.

A seemingly reasonable extrapolation from our understanding of local field theories like QED, has led most physicists to the conviction that the symmetries of the "standard model" should find their theoretical implementation in the framework of local Quantum Field Theories of the Yang-Mills type[1]. This attitude has certainly scored an impressive success in the Glashow-Salam-Weinberg (GSW) theory of electroweak interactions [2]; even though the Higgs mechanism for symmetry breaking is admitted to have some perplexing features.

In the domain of hadronic physics, however, Quantum-Chromo-Dynamics (QCD)[3], the gauge theory of coloured quarks and gluons, has struggled without any real success with the difficult and, apparently paradoxical, notion of colour-confinement. In spite of the high hopes based on Asymptotic Freedom (AF) and approximate scaling behaviour at short distance, there exists no really convincing evidence that QCD, in the way we formulate and understand it today, is the correct theory of hadronic phenomena. In particular, the complicated vacuum phenomena [4] that should lead to confinement at long distances, while not affecting the short distance behaviour appear to the unprejudiced eye as unlikely to be able to account for the stunning simplicity of both the quark-model and of the parton behaviour at short distance.

While I do not advocate that one should stop all the valiant efforts that go into understanding the behaviour of QCD in the non-perturbative regime, for instance through the methods of Lattice Gauge Theories [5], I believe it worthwhile to investigate whether it should not be possible to formulate the "standard model" in a different theoretical framework, that could bring out, right from the start, confinement for quarks and liberation for leptons, while at the same time reproducing in a precise way the main features of electroweak and strong interactions.

We shall see that this is indeed possible, provided we are willing to modify our long cherished ideas about the structure of space-time [6], [7]. It will be shown that a simple solution of the most severe problems that prevent us from understanding the role of the quarks and gluons, can be obtained if we postulate a new structure for the event continuum: the Anisotropic Space-Time. But before going into the details I think it useful to pause and ask ourselves whether the hadronic world, that has been revealed to us in countless experiments and observations in the last thirty years, suggests the possible existence of a fundamentally simple dynamics.

It turns out that the physics of a local QFT in 1 + 1 space-time dimensions has some important kinship with the hadronic behaviour familiar to us. This shall give us an important clue. But let us see how.

II. IS HADRON DYNAMICS TWO DIMENSIONAL ?

I should now like to concentrate on a few important facts of hadron dynamics which show a striking resemblance of hadron dynamics with the gauge-dynamics of a two-dimensional (one space, one time) world.

a) Confinement is a natural feature of a two-dimensional world

In fact in a two-dimensional space-time the Coulomb potential between two opposite charges is confining, and is given by

$$V_{1+1}(r) = g^2 r \tag{2.1}$$

where the coupling constant g^2 has the dimension of a mass-squared and can be called "string tension".

The two dimensional situation is in sharp contrast with what happens in 3 + 1 dimension, where the Coulombic potential is given by

$$V_{3+1}(r) = \frac{e^2}{r} \tag{2.2}$$

and vanishes for very large distances. This difference can be appreciated at once by considering the Gauss law in 1- and 3- spatial dimensions (See Fig. 1)

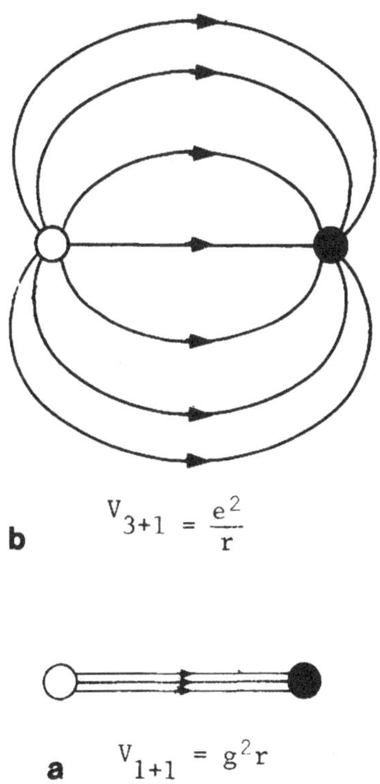

Fig. 1. In two dimensions the electric flux is focussed between the two charges (a), while in 4-dimensions (b) it is spread around them.

It is the lack of focussing of the electric flux between the charges that is responsible for the drastic difference between the 2-dimensional and the 4-dimensional worlds.

b) In two dimensions freedom at short distances holds naturally

We can see this simply by noticing that at short distances $(r \to 0)$

$$V_{1+1}(r) \xrightarrow[r \to 0]{} 0, \tag{2.3}$$

while in 3 + 1 dimensions:

$$V_{3+1}(r) \xrightarrow[r \to 0]{} \infty \tag{2.4}$$

even for the non-Abelian theories.

Deep inelastic data seem to support the "early" freedom of two-dimensional dynámics more strongly that the "asymptotic" freedom of non-Abelian theories in 4-dimensions [8].

c) In two dimensions gluons are fundamentally more elusive than quarks

Indeed, in a two-dimensional world they do not exist even in a confined mode. The simple reason for the disappearence of gluons as independent degrees of freedom lies in the absence of transverse directions. The situation is again drastically different in four dimensions.

The lack of success of the many glueball searches launched in the last few years is, I believe, another very solid support for the two-dimensional character of hadron dynamics.

d) In hadronic physics longitudinal dynamics differs significantly from transverse dynamics

The experimental revelation of jet structures in the most diverse hadronic processes, ranging from hadron-hadron collisions to e^+e^- physics, is another strong hint that hadron dynamics has a subtle way to distinguish one spatial dimension from the other two. This is another piece of supporting evidence for the fundamental bidimensionality of hadronic interactions.

From the preceding arguments it appears that we would be in a more advantageous theoretical position if we could cut two dimensions out of our 4-dimensional world; in other words, if we could, so to speak, "dynamically sterilize" two dimensions of the three dimensional space.

But how can we achieve this apparently impossible operation of surgery ? How can this be done consistently with the countless p pieces of evidence that suggest that we do live in a 4-dimensional world ?

Surprisingly enough there exists a simple answer to these questions, which I shall present and discuss in the next sections.

III. THE ANISOTROPIC SPACE-TIME (AST)

The problem we are confronted with is to formulate a gauge theory in a 4-dimensional space-time, whose dynamical properties resemble very closely those of a two-dimensional theory.

A solution to this problem has been found [6), 7)] by postulating that the elementary physical event is no more a point of a 4-dimensional Minkowskian manifold M_4, but belongs to a seven-dimensional continuum given by the topological product

$$M_4 \times \bar{S}_3,$$

which shall be called the "Anisotropic Space-Time" (AST), where $\bar{S}_3$ is a 3-dimensional pseudosphere of a Minkowskian manifold.

In other words, the elementary physical event is described by

$$E \equiv (x,n) \in M_4 \times \bar{S}_3 \qquad (3.1)$$

with $x \in M_4$ and $n \in \bar{S}_3 (n^\mu n_\mu = -1)$.

The principle of relativity can be expressed by the diagram in Fig. 2, which stipulates that

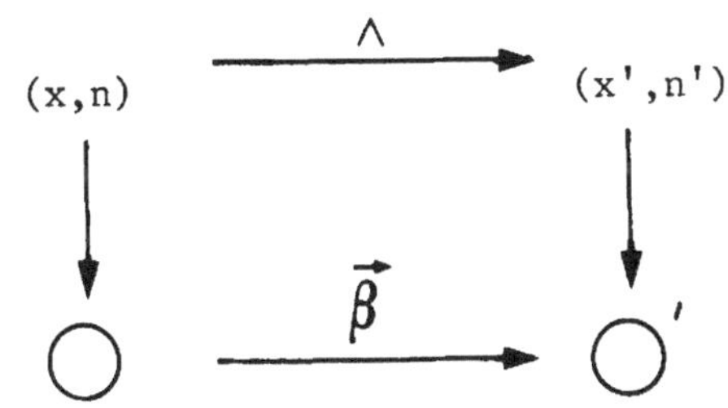

Fig. 2. The operation of the principle of relativity in $M_4 \times \bar{S}_3$

if an observer O assigns the event E the coordinates (x,n) the observer O', moving with respect to O with constant velocity $\vec{\beta}$, assigns the same event E the coordinates (x',n'), with

$$(x',n') = (\Lambda x, \Lambda n), \tag{3.2}$$

where Λ is the homogeneous Lorentz-transformation connecting the two observers. Notice that the same Λ acts on both 4-vectors x_μ and n_μ. On the other hand space-time translations are assumed to act on x_μ only.

The introduction of AST in place of M_4, means that the description of an elementary physical event given by a point x_μ of 4-dimensional Minkowskian continuum is in general incomplete; and that in order to specify an event completely, it is necessary to assign another 4-vector $n_\mu \in \bar{S}_3$.

Let us briefly discuss the geometrical structure of AST. It turns out that it is particularly useful to visualize AST as an infinite collection of Minkowskian sections $M_4(n)$, parametrized by a generic "polarization" vector $n_\mu \in \bar{S}$.

Such sections shall be called "n-sheets". (Fig. 3)

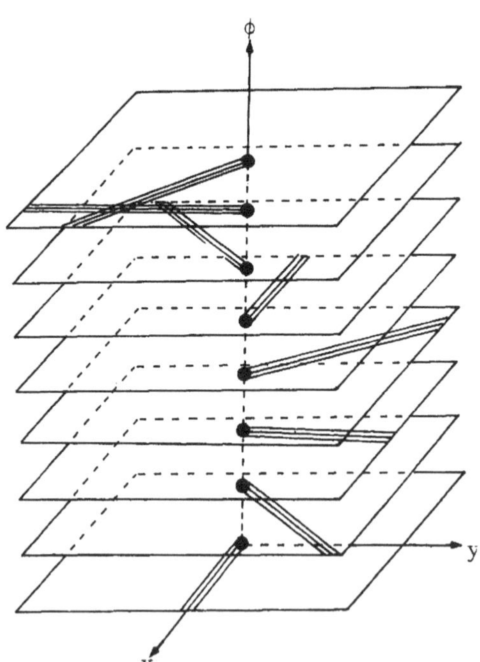

Fig. 3. The (xy;ϕ)-section of AST

As indicated in Fig. 3 the physical inequivalence of the different ϕ-sheets, suggests that it should be possible to construct a theory such that a given "charge", located at the origin = (0,0) of the plane xy, could give rise to "electric flux tubes" pointing in different directions ϕ, according to the particular ϕ-sheet it belongs to. If we did not have the extra-dimensions of the AST, the electric flux leaving the "charge" at the origin would have no other way but to spread isotropically around it. In AST this is no more necessary, because a given point-charge located at one space-time point does not uniquely define the elementary physical event, but there exists an infinity of inequivalent elementary events denoted by the same space-time points but by different directions n_μ.

After having introduced AST, the new "arena of physical events", we shall now take up the task of constructing a gauge theory that, using the new descriptive elements of the event continuum, succeeds in achieving the focussing of flux-tubes which distinguishes a two-dimensional dynamics from a four-dimensional one. This is the aim of the following sections.

IV. THE ANISOTROPIC YANG-MILLS INTERACTIONS[6), 7)]

1. The pure gauge-system

We concentrate first on the Lagrangian of a pure gauge-system. It is a function of the AST-variables (x,n), which we write:

$$L_{AYM}(x,n) = -\frac{1}{4} f^a_{\mu\nu}(x,n)\, F^{\mu\nu}_a(x,n) \tag{4.1}$$

where $\mu,\nu = 0,\ldots, 3$ and $a = 1, \ldots, N^2 - 1$ are the indices of a vector belonging to the regular representation of SU(N).

The "intensity tensor" $F^a_{\mu\nu}(x,n)$ is given by the usual expression:

$$F^a_{\mu\nu}(x,n) = \partial_\mu A^a_\nu(x,n) - \partial_\nu A^a_\mu(x,n) - igf^{abc}A^b_\mu(x,n)A^c_\nu(x,n) \tag{4.2}$$

where the derivatives act on the Minkowskian space-time coordinates x_μ only. The "magnitude tensor" $f_{\mu\nu}(x,n)$, on the other hand, is expressed by

$$f^a_{\mu\nu}(x,n) = \epsilon^{\alpha\beta}_{\mu\nu}(n)F^a_{\alpha\beta}(x,n) \tag{4.3}$$

where the "anisotropy tensor $\epsilon^{\alpha\beta}_{\mu\nu}(n)$ is given by

$$\varepsilon^{\alpha\beta}_{\mu\nu}(n) = -2\,(\delta^{\alpha}_{\mu}n_{\nu}n^{\beta} + \delta^{\beta}_{\nu}n_{\mu}n^{\alpha}). \tag{4.4}$$

With our definitions (4.2)-(4.4), the Lagrangian (4.1) describes a gauge-field in a medium which has the character of an infinite, perfect magnetic superconductor with electric flux lines in the direction n_μ. The "intensity tensor" $F^a_{\mu\nu}$ describes the pair $(\vec{E}^a, \vec{B}^a)$, while the "magnitude tensor" $f^a_{\mu\nu}$ stands for the pair $(\vec{D}^a, \vec{H}^a)$.

The action S_{AYM} of the gauge system is obtained by integrating (4.1) over the AST variables, i.e.

$$S_{AYM} = \int d^4x \int d\mu(n)\; L_{AYM}(x,n) \tag{4.5}$$

where, by parametrizing n_μ as

$$n_\mu \equiv (\sinh\psi,\ \cosh\psi\vec{u}) , \tag{4.6}$$

we have

$$d\mu(n) = d\psi\; d\Omega_{\vec{u}}. \tag{4.7}$$

The Euler-Lagrange equations of the theory are obtained from (4.5) through the Hamilton principle, i.e. by imposing

$$\delta S_{AYM} = 0 \tag{4.8}$$

for infinitesimal variations $\delta A^a_\mu(x,n)$.

We get

$$\partial^\mu f^a_{\mu\nu}(x,n) - igf^{abc}f^b_{\alpha\nu}(x,n)A^{\alpha c}(x,n) = 0 \tag{4.9}$$

These equations are invariant under the infinitesimal gauge-transformations

$$\delta A^a_\mu(x,n) = -\frac{i}{g}\,\partial_\mu\phi^a(x,n) + f^{abc}A^b_\mu(x,n)\phi^c(x,n) \tag{4.10}$$

which are "local" both in x_μ and n_μ.

What kind of theory have we got ?

Let us choose the "axial gauge"

$$n^\mu A^a_\mu(x,n) = A^a_n(x,n) = 0. \tag{4.11}$$

The Euler-Lagrange eqs. (4.9) simplify considerably (we use the notation $v_\mu = v_n n_\mu + \bar{v}_\mu$; $v_n = -(n.v)$)

$$\partial_n^2 \bar{A}_\mu^a = 0 \tag{4.12}$$

$$\partial_n(\bar{\partial}\bar{A}^a - igf^{abc}\partial_n \bar{A}^b A^c) = 0. \tag{4.13}$$

The remarkable feature of (4.12) is the elliptic nature of the differential operator ∂_n^2. Differently from the hyperbolic operator like the d'Alembertian $\Box$, ∂_n^2 does not allow for wave solutions, and if we further require that the fields $\bar{A}_\mu^a$ have good behaviour at spatial infinity, the only solution of (4.12) and (4.13) is

$$\bar{A}_\mu(x,n) = 0. \tag{4.14}$$

As in the case of 2-dimensional gauge theories, in AST the pure-gauge system does not possess any independent dynamics.

Thus we have obtained a theory where

(i) gluons, monopoles, glueballs and other dynamical artifacts of four-dimensional pure-gauge systems disappear;

(ii) the colour-dynamics is essentially two-dimensional;

(iii) there exists no field-matter dualism, in the sense that without fermionic matter no physically meaningful realization of the gauge-fields (even the non-Abelian ones) is possible.

Another remarkable feature of our results is that they do not depend on the Abelian or non-Abelian nature of the gauge-group.

2. Introducing Fermionic Matter

We wish now to introduce fermionic matter in our AST.

We first define a set of fermionic fields:

$$q(x,n)$$

and write the "Dirac" action as ($V = \int d\mu(n)$)

$$S_F = \int \frac{d\mu(n)}{V^{1/2}} \int \frac{d\mu(n')}{V^{1/2}} \int d^4x\, \bar{q}(x,n')\,(i\not{\partial} - M)\, q(x,n), \tag{4.15}$$

while the interaction part is given by:

$$S_I = g \int d\mu(n) \int d^4x \, \bar{q}(x,n) \, A\!\!\!/^a(x,n) \, \frac{\lambda^a}{2} \, q(x,n) \tag{4.16}$$

Thus the full quantum theory is defined by the total action:

$$S = S_F + S_I + S_{AYM} \tag{4.17}$$

where S_F S_I and S_{AYM} are given by (4.15), (4.16) and (4.5) respectively.

A few comments on eq. (4.17) are now in order. While S_I and S_{AYM} are "diagonal" in n_μ, i.e. they do not mix field-variables on different n-sheets, S_F is totally n- non-diagonal. The reason for this seemingly capricious choice is that, in the absence of interaction with the gauge field, fermionic matter should have no knowledge whatsoever of the n-sheeted structure of space-time. A natural way of achieving this is to "average" on the n-degree of freedom of the Fermi-fields in the Dirac-action, as we have done in (4.15). As a result S_F breaks local gauge-invariance in n. The only invariance which survives in the theory is global (x-,n-independent) gauge invariance. We shall see later, (in section VI) however, that the theory does indeed possess effectively the invariances that we now believe physically relevant.

3. The Dynamics of the Full Quantum Theory

Let us now derive the Euler-Lagrange equations of the theory. In order to do so in a more efficient way, we expand q(x,n) in $\bar{S}_3$-spherical harmonics:

$$Y_k(n) \qquad k \equiv (\lambda, l, m) \tag{4.18}$$

where the index k stands for the angular-momentum indices (l,m) and another continuous index λ, which originates from the non-compact nature of $\bar{S}_3$. Taking a "box", $(L \geqslant \psi \geqslant -L)$ and requiring periodic boundary conditions, λ becomes discrete, and in such case k denotes a set of 3 integers.

We thus write:

$$q(x,n) = \sum_k q_k(x) Y_k(n), \tag{4.19}$$

where

$$q_k(x) = \int d\mu(n) Y_k^*(n) q(x,n) \tag{4.20}$$

The Fermi-action S_F can be rewritten as:

$$S_F = \int d^4x\, \bar{q}(x)\, (i\not\partial - m)\, q(x), \tag{4.21}$$

where we have renamed the zero-mode $q_o(x)$, $q(x)$. On the other hand the interaction part of the action takes the form:

$$S_I = g \sum_{kk'} \int d^4x \int d\mu(n) V^{kk'}(n) A^a_\mu(x,n) \bar{q}_{k'}(x) \gamma^\mu \frac{\lambda^a}{2} q_k(x), \tag{4.22}$$

with

$$V^{kk'}(n) = Y^*_{k'}(n) Y_k(n) \tag{4.23}$$

The Dirac equations are obtained by varying the full action S with respect to $q_k(x)$. One obtains for k = 0, the zero-mode,

$$(i\not\partial - M) q(x) + g \sum_k \int d\mu(n) V^{ok}(n) \not A^a(x,n) \frac{\lambda^a}{2} q_k(x) = 0 \tag{4.24}$$

while for $k \neq 0$ a set of "constraint" equations emerges:

$$g \sum_{k'} \int d\mu(n)\, V^{kk'}(n) \not A^a(x,n) \frac{\lambda^a}{2} q_{k'}(x) = 0 \tag{4.25}$$

(4.24) and (4.25) show that only the zero-mode $q(x)$ is a dynamically independent degree of freedom. The fields $q_k(x)$, with $k \neq 0$, become, through the equations (4.25), functionals of $q(x)$ and of the gauge-fields $A_\mu(x,n)$. Finally, by varying with respect to $A^a_\mu(x,n)$, we obtain:

$$\partial^\mu f^a_{\mu\nu}(x,n) - igf^{abc} f^b_{\alpha\nu}(x,n) A^{c\alpha}(x,n) =$$

$$= g \sum_{kk'} V^{kk'}(n) \bar{q}_{k'}(x) \gamma_\nu \frac{\lambda^a}{2} q_k(x) \tag{4.26}$$

In the preceding section we have checked that, before coupling the gauge field to matter the gauge fields possess no dynamical degrees of freedom. The quantum field theory can thus be defined by the Euler-Lagrange equations ((4.22), (4.24), (4.25) and (4.26)) and the canonical quantization procedure on the field $q(x)$ only. All the other fields are functionals of $q(x)$.

V. QUANTUM FLUCTUATIONS AND THE ROLE OF CHIRALITY

If the full Lagrangian were n-diagonal, confinement would be natural and automatic, but at the same time we would be describing a totally unrealistic world. We would in fact be dealing with an infinite set of copies of identical worlds (the n-sheets), each equipped with a two-dimensional dynamics, but with no possibility to merge together and generate the isotropic world that we, uncoloured objects, are familiar with.

This is another important reason why the completely non diagonal Fermi-action (4.15) has been introduced. By virtue of (4.15) fermionic matter can propagate from any n-sheet to any other n'-sheet and through this process those unrealistic, effectively two-dimensional worlds get mixed together, giving us hope that the observable world will come close to what we know. But this hope hides a poisonous nag: the suspicion that the thorough mixing of the n-sheets induced by S_F will at the same time render the AST completely isotropic. If this happens, the confinement mechanism based on the focussing of the colour-fluxes on each n-sheet evaporates.

That this suspicion in indeed a founded one can be perceived at once by looking at the effects of "vacuum polarization" diagrams like in Fig. 4. The gauge-field A_n creates a quark pair $q\bar{q}$ on the sheet n, through the interaction part of the action S_I, S_F now propagates the $q\bar{q}$ pair from the sheet n to the sheet n', where again S_I annihilates the pair producing a gauge field $A_{n'}$. As a result of these processes, whose occurrence is explicitly contained in the Lagrangian of the theory, A_n mixes with every $A_{n'}$ in such a way that the peculiar n-structure of the gauge-fields tends to get washed out.

Thus we expect that <u>if the vacuum gets polarized, AST gets isotropized, and quarks are liberated</u>.

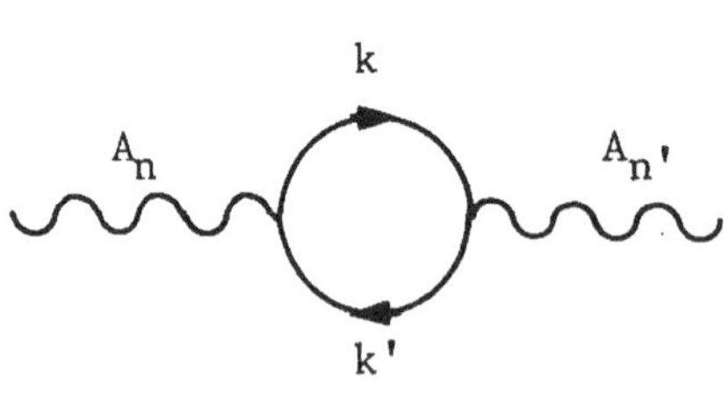

Fig. 4. A typical "vacuum polarization" contribution to the gauge-field propagator.

1. The Survival of Colour Confinement

Remarkably, however this does not happen. Let us see why.

We set up the perturbative expansion by lifting the singularity of the Lagrangian through the addition of a "regularizing" term of the form

$$S' = \varepsilon \int d^4x \sum_{k\neq 0} \bar{q}_k(x)\,(i\not\partial - M)\,q_k(x) \tag{5.1}$$

with the understanding that the limit $\varepsilon \to 0$ should be taken at the end.

The Feynman rules are thus based on the following building blocks:

(i) the "gluon propagator":

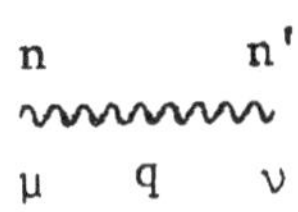

$$= -i(g_{\mu\nu} + n_\mu n_\nu)\frac{\delta(n-n')}{(q.n)^2 + \lambda^2} = +\, iD^i_{\nu\mu}(q;n',n), \tag{5.2}$$

where λ is an infrared cut-off ($\lambda \to 0$ should be taken at the end);

(ii) the fermion propagator:

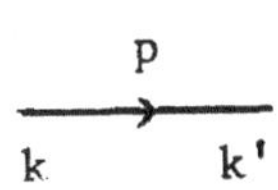

$$= S_{kk'}\,\frac{1}{\varepsilon_k(\not p - M)} \tag{5.3}$$

$$\begin{cases} \varepsilon_k = 1 & k = 0 \\ \varepsilon_k = \varepsilon & k \neq 0; \end{cases}$$

(iii) the "fermion-gluon vertex":

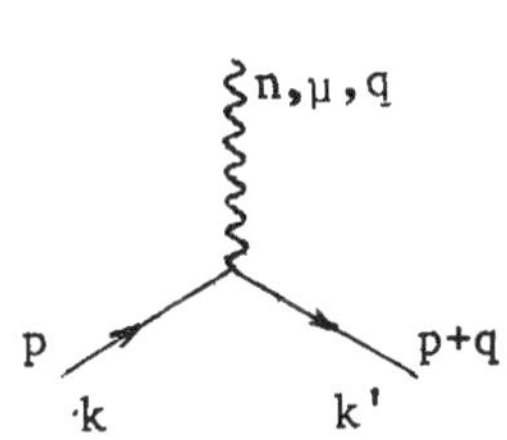

$$= -igV^{k'k}(n)\gamma_\mu \tag{5.4}$$

Note that no "bare" three-and four-gluon coupling has been introduced, due to their vanishing in AST.

The dressed Green's functions are on the other hand:

(a) the full "gluon propagator":

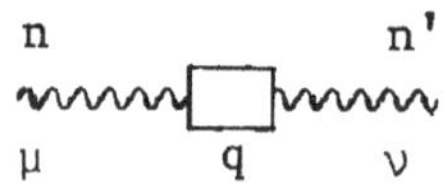

$$= + iD'_{\nu\mu}(q;n,\bar{n}) \ ; \tag{5.5}$$

(b) the full Fermion propagator

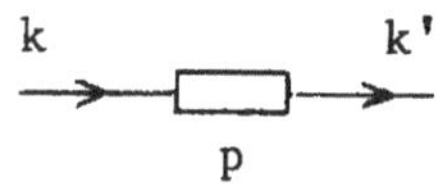

$$= iS'(p)_{k'k}; \tag{5.6}$$

(c) the full fermion-gluon vertex

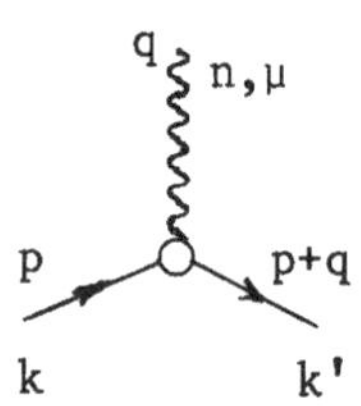

$$= - ig\Gamma_{\mu}^{k'k}(p+q,p;n). \tag{5.7}$$

We should also define the 3- and 4-gluon full vertex but we shall not need them in our discussion.

Finally the following important derived Green's functions should be defined:

(α) the vacuum polarization tensor:

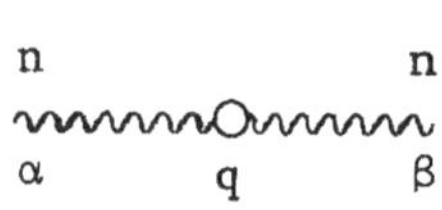

$$= i\Pi_{\beta\alpha}(q;n',n), \tag{5.8}$$

related to the full gluon-propagator through the following equations:

$$D'_{\mu\nu}(q;n',n) = P_{\mu\rho}(n')G_{\rho\nu}(q;n',n), \tag{5.9}$$

where

$$P_{\mu\rho}(n) = (g_{\mu\rho} + n_{\mu}n_{\rho}), \tag{5.10}$$

and

$$(G^{-1})_{\rho\nu} = (q.n)^2\delta(n - n')g_{\rho\nu} - \Pi_{\rho\lambda}(q;n',n)P_{\lambda\nu}(n). \tag{5.11}$$

(β) the fermion self-energy:

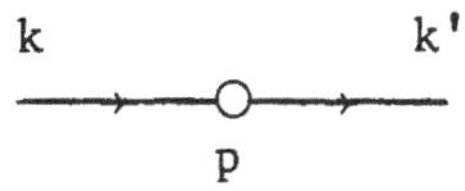

$$= \Sigma_{k'k}(p), \qquad (5.12)$$

related to the full fermion propagator by:

$$S'(p)^{-1}_{k'k} = \delta_{k'k}\varepsilon_k(\not{p} - M) - \Sigma_{k'k}(p). \qquad (5.13)$$

(1) The self-energy equation : diagrammatically it is given by

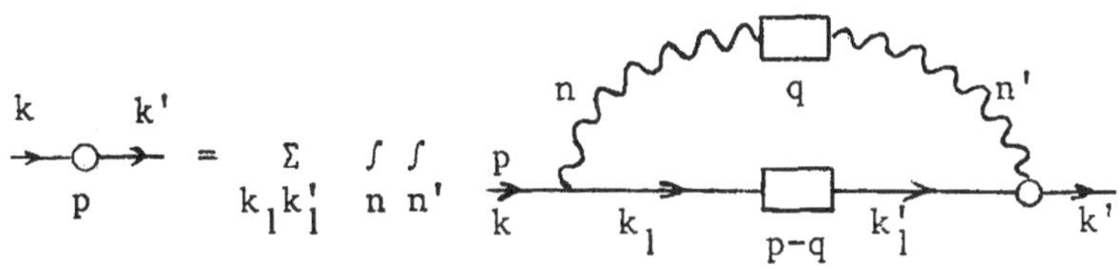

(5.14)

(2) The vacuum polarization equation.

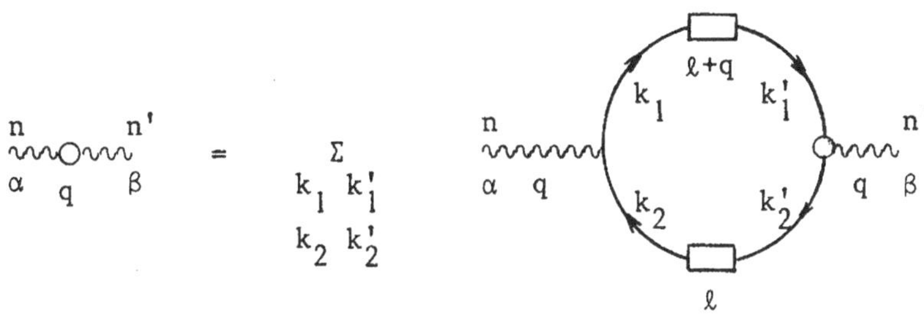

(5.15)

and finally:

(3) The vertex equation:

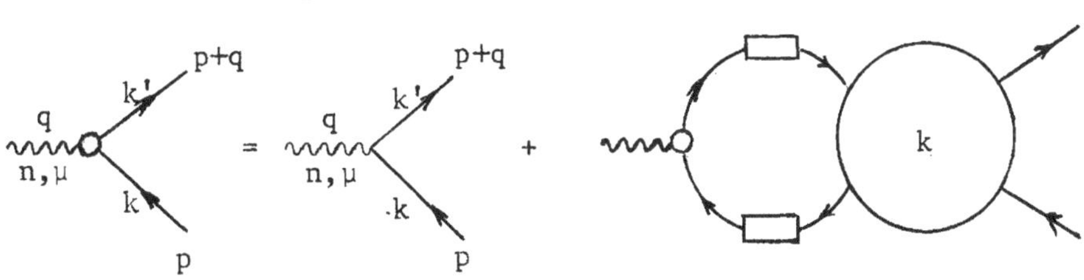

+ 3- and 4-gluon-intermediate states (5.16)

We may now ask the question whether confinement is preserved. Consider the first two equations ((5.14) and (5.15)). We look for solutions which do not mix the different colour fields, i.e. solutions for which

$$\Pi_{\alpha\beta}(q;n',n) \xrightarrow[\lambda\to 0]{} 0. \tag{5.17}$$

Suppose such solutions exist, then the full gluon propagator is given by

$$D'_{\mu\nu}(q;n',n) \xrightarrow[\lambda\to 0]{} P_{\mu\nu}(n) \frac{\delta(n - n')}{(qn)^2 + \lambda^2} \tag{5.18}$$

Inserting such an expression in the self-energy equation (5.14) we check at once that $\Sigma_{kk'}(p)$ develops, in the infrared limit $\lambda \to 0$, a singularity in the "mass" component $[\Sigma_{LR}, \Sigma_{RL}]$. The limit $\varepsilon \to 0$ is now harmless due to the dominance of the Σ-term in the inverse propagator (5.13). It is easy to check that a solution exists for which

$$\Sigma_{k'k}(p) \xrightarrow[\lambda\to 0]{} \frac{C}{\sqrt{\lambda}}\, \delta_{k'k} + \ldots , \tag{5.19}$$

but we must be sure that this solution is consistent with (5.17). This can be ascertained by inserting (5.19) in the vacuum-polarization equation (5.15).

Again a simple calculation shows that (an ultraviolet cut-off is introduced in order to regularize the integral)

$$\Pi_{\alpha\beta}(q;n',n) \xrightarrow[\lambda\to 0]{} (g_{\alpha\beta}q^2 - q_\alpha q_\beta) d_{nn'} \lambda^2 \tag{5.20}$$

Thus the "confining" solution is indeed consistent.

The diagram in fig. 5 depicts the logic of our proof.

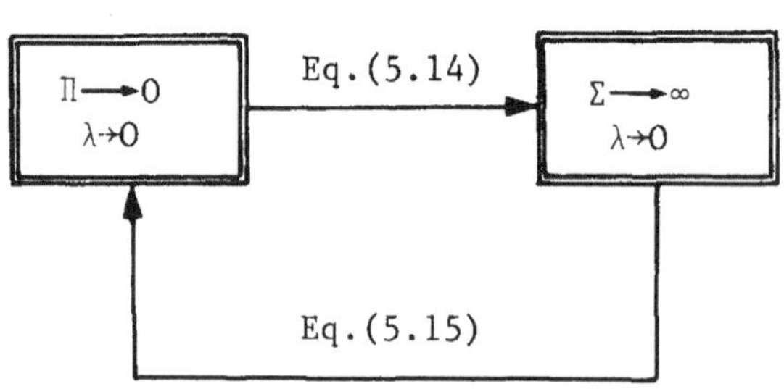

Fig. 5. The consistency diagram of the "confining" solution of the Dyson equations.

What we have found has a particularly transparent physical meaning. Before switching the interaction on the Fermi field hops freely from one n-sheet to another while propagating in usual space-time, and at the same time the gauge fields establish well defined connections among the space-time points of the generic n-sheet. When the interaction is switched on the fermion, which was wandering through the different n-sheets, feels an infinitely strong interaction which supplies it an infinite mass. The fermion gets literally nailed down to the sheet he found itself on when the interaction was switched on.

The Fermi action S_F is now powerless and cannot deploy its dangerous potential for deconfinement.

The link among the gauge fields on different n-sheets is cut-off forever, and on each n-sheet we are left with a dynamics which has a two-dimensional character. But this does not mean that all quark systems cannot wander through the n-sheets and recover our well known isotropic world. We shall see later that quark systems forming colour singlets do not receive from the gauge field an infinite mass, and on them S_F recovers its powers; but at this point confinement has already been secured!

Thus colour-confinement survives the potentially fatal quantum fluctuations.

2. The Liberation of Leptons

If AST is to have a theoretical status which goes beyond mere phenomenology, we should be able to develop a theory of the $SU(2)_L \times U(1)_Y$ electroweak system along the lines discussed in Section IV. But then, doing so, would we not obtain confined leptons ? For, if this were the case, we better go back and think the fundamentality of AST over.

But, almost miraculously, things go very differently for the electroweak system: leptons are liberated and the gauge fields acquire the usual isotropic dynamics.

The reason for such a remarkable dynamical bifurcation between the colour and the electroweak system is in fact quite simple. Going back through the discussion of the preceding paragraphs we notice that a consistent confining solution can only be obtained if the gauge interaction can produce a divergence in the mass component $(\Sigma_{LR}, \Sigma_{RL})$ of the self energy function. A vector interaction (like the colour interaction), we have seen, produces precisely such a divergence. But for a chiral interaction this is not the case.

Let us take for instance the neutrino. In $SU(2)_L \times U(1)_Y$ it is coupled only through its left-handed component. This means that the interaction cannot produce an infrared divergence in the mass-component of the self energy function, but only in its wave function component (Σ_{LL}). Such a divergence, however, can be reabsorbed in the definition of the fermion-gluon vertex, with the result that no consistent solution exists for a confining situation ($\Pi \to 0$).

Unencumbered by an infinite mass, the neutrino can now freely propagate from one sheet to another linking together the gauge fields on different n-sheets, thus inducing a massive rearrangement of the gauge degrees of freedom.

As a result:

(i) we obtain an effective isotropic gauge theory;

(ii) all other leptons, besides the neutrino, get liberated;

(iii) the electroweak angle is determined, to lowest order, independently of the initial coupling constants (g_I, g_Y).

(i) Let us assume that the gauge field $A^a_\mu(x,n)$ becomes n-independent. The Dirac eq. (4.25) implies then that $q_k(x) = 0$ for $k \neq 0$; i.e. all the non-zero modes decouple. With this information we return to the Dyson equation for the vacuum polarization function (5.15). We obtain $[V = \int d\mu(n)]$

$$\Pi_{\alpha\beta}(q;n',n) = \frac{g^2}{V^2}(g_{\alpha\beta}q^2 - q_\alpha q_\beta)\Pi(q^2) \tag{5.21}$$

i.e. the vacuum polarization function becomes, as it should, n-independent. We must now diagonalize the inverse propagator function $(G^{-1})_{\rho\nu}$ (eq. (5.11)). It is straightforward to see that in the limit $V \to \infty$, with (g^2/V) kept finite, the first term in (5.11) is negligible and the diagonalization produces a single non-zero eigenvalue

$$\frac{g^2}{V}q^2\,\Pi(q^2) \tag{5.22}$$

with eigenvector $Y_o(n) = 1/V^{1/2}$. Thus our assumption of the n-independence of the gauge field is indeed a consistent one.

(ii) With an isotropic propagator, the self energy equation is perfectly regular when $\lambda \to 0$, and confinement disappears, not only for ν's but for all leptons.

(iii) From the reshuffling of the theory engendered by the diagonalization of the gauge field propagator, the new, isotropic effective theory is characterized by coupling constants which do not depend (to lowest order) on the initial coupling constants. In fact, the effective interaction is determined to lowest order by the diagram in fig. 6; sticking in the propagator (5.22), and the vertex (5.4) for small g, we find for the amplitude in fig. 6

$$A = \frac{g_i^2}{V^2} (\gamma_\mu)(1 \pm \gamma_5)(\gamma_\nu)(1 \pm \gamma_5)(-g_{\mu\nu}) \frac{1}{q^2 \Pi_i(0) \left(\frac{g_i^2}{V^2} \right)} \tag{5.23}$$

Thus we see that the "effective coupling constant" of the gauge field A_i is nothing but

$$(g_i^2)_{eff} = \frac{1}{\Pi_i(0)} \tag{5.24}$$

where $\Pi_i(0)$ is calculated through the usual vacuum polarization fermion loop. This immediately fixes the value (to lowest order) of the electroweak angle as:

$$\left(\frac{g_Y^2}{g_I^2} \right) = tg^2\theta_W = \frac{\Pi_I(0)}{\Pi_Y(0)} = \frac{\sum_{lept} (I_3)^2}{\sum_{lept} 1/4\ (Y_L^2 + Y_R^2)} = \frac{1}{3} \tag{5.25}$$

leading to the value very close to experiments:

$$\sin^2\theta_W = \frac{1}{4} \tag{5.26}$$

Note that in (5.25) only the leptons need to be considered in

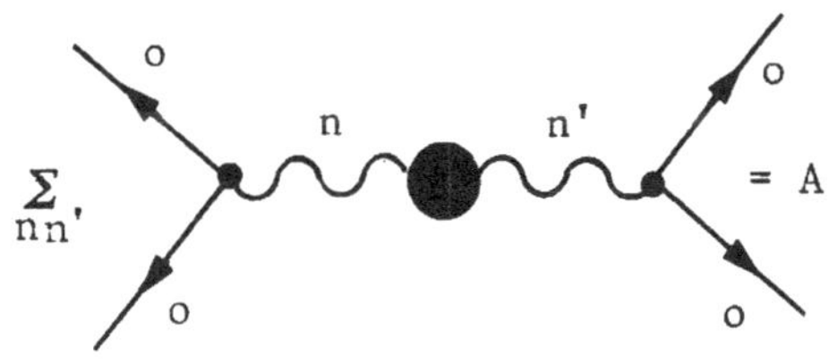

Fig. 6. The lowest order fermion-gauge field interaction.

the summation; the quarks in fact, being confined, do not take part in the isotropization process.

Beyond the remarkable results presented in this subsection the problem still remains of symmetry breaking and mass generation. Consistency seems to require some dynamical mechanism for mass generation, but it should be stressed that AST contains so many novel features, that it is not unconceivable that a slight generalization of the theory described here could lead to a natural and elegant resolution of such a fundamental problem. I end this section by summarizing its main results.

A Yang-Mills interaction scheme in AST has been constructed for the "standard model": $SU(3)_c \times SU(2)_L \times U(1)_Y$. Consistent solutions of the full quantum field theory have been found that confine quarks interacting through $SU(3)_c$ and leave leptons, governed by $SU(2)_L \times U(1)$ only, unconfined. An "effective" gauge theory of the electroweak interactions emerges that is a perfectly standard one with a value of the electroweak angle which is completely determined (to lowest order).

Chirality is seen to demarcate the drastically different behaviours of the $SU(3)_c$ and $SU(2)_L \times U(1)_Y$ interactions.

VI. GAUGE INVARIANCE RESTORED

In section IV the observation has already been made that the peculiar form of S_F (eq. (4.15)) removes from the theory all local invariances. It is a remarkable feature of the "bifurcated" solutions, discussed in the precedind section, that the relevant local gauge invariance is at the end restored.

1. The Colour Interaction

It is a straightforward matter to check that, submitting the quark fields and the gauge fields to a local (in (x,n)) colour gauge transformation, the Lagrangian density in x-space $L^{colour}(x)$ varies as

$$\delta L^{colour}(x) = O^{(8)}(x) \tag{6.1}$$

where $O^{(8)}(x)$ is a local operator transforming as a colour-octet. As we shall see in the following sections in more detail, the physical Hilbert space (spanned by finite energy states) turns out consisting of colour singlets; while non-trivial colour representations exhibit infinite masses and cannot be linked to the physically realizable states by any physical means. Thus on any state of the physical Hilbert space $|PHYS>$

$$\langle PHYS|\ \delta L(x)\ |\ PHYS\rangle = 0, \tag{6.2}$$

and the theory recovers local colour gauge invariance.

2. The Chiral Interaction

For chiral interactions we have seen that the k ≠ 0 modes of the quark field decouple. Thus for the solution considered we can limit ourselves to the Lagrangian

$$L^{chiral}(x) = \bar{q}(x)(i\partial\!\!\!/)q(x) + g/V \int d\mu(n)\ \bar{q}(x)A\!\!\!/^{a}(x,n)\lambda^{a}/2\ q(x)$$
$$+ \int d\mu(n)\ \{-\frac{1}{4} f^{a}_{\mu\nu}(x,n)F^{\mu\nu}_{a}(x,n)\} \tag{6.3}$$

which is obviously invariant under the following local, n-independent gauge transformation:

$$q(x) \to i\,\frac{\lambda^{a}}{2}\ \phi^{a}(x)q(x)$$

$$A^{a}_{\mu}(x,n) \to -\frac{i}{g}\,\partial_{\mu}\phi^{a}(x) + f^{abc}\ A^{b}_{\mu}(x,n)\phi^{c}(x). \tag{6.4}$$

Thus physical gauge invariance gets again reinstated.

VII. THE STRUCTURE OF STRONG INTERACTIONS [9]

We have just seen how the theory of quarks and colour proposed in section IV does lead to quark confinement through the appearance of an infrared divergent quark self energy and the consequent decoupling of gauge fields on different n-sheets. This means that the basic features of the classical theory discussed in a recent paper[6] survive substantially unmodified in the quantum case. The theory described in Ref 6 is defined by a Lagrangian L(x,n)

$$L\ (x,n) = L_{AYM}(x,n) + g\ J^{a}_{\mu}(x,n)A^{\mu}_{a}(x,n) \tag{7.1}$$

where $J^{a}_{\mu}(x,n)$ is a classical current for a set of pointlike particles moving on a given n-sheet, whose expression is given by

$$J^{a}_{\mu}(x,n) = \sum_{q} Q^{a}_{q} \int_{-\infty}^{+\infty} d\tau\,\delta^{4}(x - s^{q}_{\mu}(\tau))\ \frac{ds^{q}_{\mu}}{d\tau}(\tau) \tag{7.2}$$

τ being the proper time, and $s_\mu^q(\tau)$ the world line of the particle q. The main consequence of (7.1) and (7.2) are the following:

(i) An isolated charge has an infinite (infrared divergent) energy, thus it is confined;

(ii) All finite energy, unconfined states must satisfy:

(a) the triality rule:

$$\sum_q Q_q^a \mid \text{COLOUR STATE} > = 0 \tag{7.3}$$

i.e. they must be colour-singlets;

(b) at a given instant the colour charges, comprising a colour singlet state, must find themselves on the same $\vec{u}$-sheet ($n_\mu \equiv (0,\vec{u})$);

(c) on the common $\vec{u}$-sheet the trajectories of the classical charges must have the form:

$$\vec{x}_q(t)_{\vec{u}} = \vec{x}_\perp(t) + x_q(t)\vec{u} \; , \tag{7.4}$$

i.e. they must consist of a common translation $\vec{x}_\perp(t)$, in the direction transverse to $\vec{u}$, and oscillations in the direction $\vec{u}$. For illustration in fig. 7 I report two baryonic (3 quark) configurations of which only one carries finite energy;

(d) the potential energy at large distances is given by

$$\mathcal{E}(\vec{u}) = \mu^2 \sum_{p<q} Q_p^a Q_q^a \left| x^{(p)} - x^{(q)} \right| \tag{7.5}$$

i.e. it increases linearly with the relative distances.

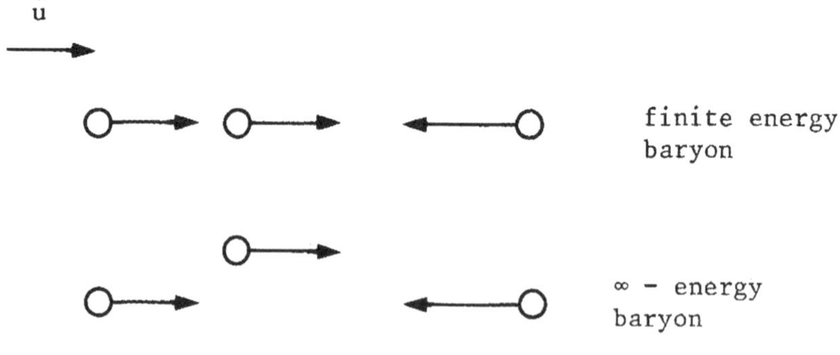

Fig. 7. Only perfectly aligned configurations carry finite energy.

1. The Classical Dynamics

Using the previous qualitative results, we now wish to find a dynamical formulation of our theory in the classical case.

Given N pointlike non relativistic quarks, their classical trajectories in AST can be described by 5N trajectory functions

$$\{\vec{x}_i(t),\ \vec{u}_i(t)\} \qquad i = 1,\ldots,N \tag{7.6}$$

When we try to transcribe our theory (4.17) for this non-relativistic classical problem, we notice at once that the kinetic part does not contain the trajectory functions $\vec{u}_i(t)$.

The Lagrangian

$$L = L\,(\dot{\vec{x}}_i,\ \vec{x}_i,\ \vec{u}_i) \tag{7.7}$$

is in fact a function which does not contain time derivatives of the $\vec{u}_i$. The Lagrange equations for the trajectory functions $\vec{x}_i$ and $\vec{u}_i$ read (i = 1, ... ,N)

$$\frac{d}{dt}\left(\frac{\partial L}{\partial \dot{\vec{x}}_i}\right) - \frac{\partial L}{\partial \vec{x}_i} = 0 \tag{7.8}$$

$$\frac{\partial L}{\partial \vec{u}_i} = 0 \tag{7.9}$$

Eqs. (7.8) are the truly dynamical equations of the problem, while eqs. (7.9) act as constraint equations. We know from the preceding discussion that L contains a divergent term proportional to

$$|\vec{u}_i(t) - \vec{u}(t)|^2$$

(the sheet variables of the different particles at a given instant must coincide). Furthermore defining the N-1 relative variables

$$\vec{y}_{ij}(t) = \vec{x}_i(t) - \vec{x}_j(t),$$

we have another divergent term in the Lagrangian proportional to

$$[\vec{y}_{ij}(t) - y_{ij}(t)\ \vec{u}(t)]^2.$$

Thus the constraint equations (7.9) have the effect to impose the following conditions for finite energy

$$\vec{u}_i(t) = \vec{u}(t) \tag{7.10}$$

and

$$\vec{y}_{ij}(t) = \vec{x}_i(t) - \vec{x}_j(t) = y_{ij}(t)\vec{u}(t). \tag{7.11}$$

Inserting such constraints in (7.7) we obtain the Lagrangian

$$L = L(\dot{y}_{ij}(t),\ y_{ij}(t),\ \dot{\vec{u}}(t),\ \vec{u}(t)), \tag{7.12}$$

which does not depend on the 3(N-1) Lagrangian degrees of freedom of a non-relativistic N-body problem, but only on N+1. Such a dramatic reduction stems from the strict alignment necessary to achieve finite energy. This is a distinctive feature of this theory for all systems but the two-body ($q\bar{q}$) system.

In such a case, in fact, we can choose as the Lagrangian degrees of freedom the relative coordinate $\vec{x}(t) = \vec{x}_1(t) - \vec{x}_2(t)$; the two-body Lagrangian can be written as

$$L(\dot{\vec{x}},\vec{x}) = \frac{m}{2}\dot{\vec{x}}^2 - V(|\vec{x}|), \tag{7.13}$$

with the constraints

$$\begin{cases} \vec{u}_1(t) = \vec{u}_2(t) = \vec{u}(t) \\ \vec{x}(t) = x(t)\ \vec{u}(t) \end{cases} \tag{7.14}$$

and the potential given by (7.5).

2. The Quantum Dynamics

Inspired by the classical analysis we look for a solution of the quantum problem which reduces to the classical solution in the corresponding limit.

We shall construct it in several steps:

a) quantize canonically the zero mode q(x) at t = 0,

$$q(\vec{x},0) = \sum_r \int \frac{d^3k}{(2\pi)^{3/2}} \left(\frac{m}{E_k}\right)^{1/2} \{ b(r,\vec{k}) e^{i\vec{k}\vec{x}} u(r,\vec{k}) + d^+(r,\vec{k}) e^{-i\vec{k}\vec{x}} v(r,\vec{k}) \} \tag{7.15}$$

(b) Assume that the dynamically dependent fields ($k \neq o$ modes) are linear functions of the creation and annihilation operators of the zero-mode quark field;

(c) Write the Hamiltonian

$$H = H_k + H_f \tag{7.16}$$

where H_k stands for the kinetic hamiltonian, and H_f for the colour field hamiltonian;

(d) Split H_f in the no-pair part H_{np} and in the pair-part H_p:

$$H_f = H_{np} + H_p \tag{7.17}$$

where H_{np} contains only the no-pair creation terms b^+b, d^+d, while H_p contains the pair creation terms bd, b^+d^+.

This can be achieved by writing:

$$H_f = \sum_{\vec{u}} \int d^3x \, \Delta_{\vec{u}}(\vec{x}) \, \bar{J}^a_\mu(\tfrac{\vec{x}}{2},0,\vec{u}) \, \bar{J}^\mu_a(-\tfrac{\vec{x}}{2},0,\vec{u}) \tag{7.18}$$

with

$$\Delta_{\vec{u}}(\vec{x}) = \lim_{\lambda \to 0} \int \frac{d^3k}{(2\pi)^3} e^{i\vec{k}\vec{x}} \frac{1}{(\vec{k}.\vec{u})^2 + \lambda^2} \tag{7.19}$$

and separating the colour current in its no-pair and pair creation parts. Thus

$$H_{np} = \sum_{\vec{u}} \int d^3x \, \Delta_{\vec{u}}(x) \, \bar{J}^a_\mu(\vec{x}/2,0,\vec{u})\Big|_{np} \, \bar{J}^\mu_a(-\tfrac{\vec{x}}{2},0,\vec{u})\Big|_{np} \tag{7.20}$$

(e) Write the hamiltonian as

$$H = H_0 + H' \tag{7.21}$$

with

$$H_0 = H_k + H_{np}$$

$$H' = H_p, \qquad (7.22)$$

and treat H' as a perturbation;

(f) diagonalize H_0.

This procedure generates the "primitive world". The physical states are exactly those of the quark model, as a result of the important fact that H_0 commutes with the quark number operator

$$N_q = \int d^3k\ b^+(k)b(k) \qquad (7.23)$$

and antiquark number operator

$$N_{\bar{q}} = \int d^3k\ d^+(k)d(k) \qquad (7.24)$$

separately.

(g) Obtain relativistic equations.

For mesons, introducing the 3-dimensional w.f. in the rest frame,

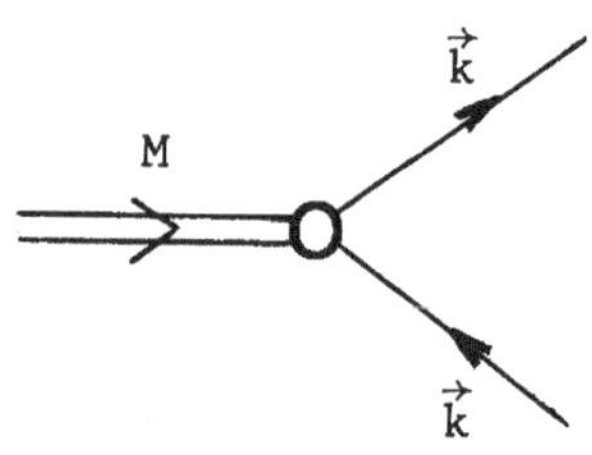

$$= \psi_M(\vec{k}) \qquad (7.25)$$

the diagonalization of H_0 leads to a relativistic Schrödinger-type equation

$$\{\sqrt{\vec{k}^2 + m_1^2} + \sqrt{\vec{k}^2 + m_2^2}\}\psi_M(\vec{k}) + \int\frac{d^3q}{(2\pi)^3} V(\vec{k},\vec{q})\psi_M(\vec{q}) = M\psi_M(\vec{k})$$
(7.26)

with $V(\vec{k},\vec{q})$, computed in a straightforward manner from H_f^{np}.

(h) Calculate corrections by a systematic expansion in H'.

It is easy to see that decay processes, scattering amplitudes, mass shifts etc. are all generated by the pair creations induced by H'.

3. Why can we use Perturbation Theory ?

As all the calculational programme rests on the assumed possibility of isolating a perturbative part in the quark Hamiltonian, we should ask ourselves first whether such a procedure is supported by the structure of the theory and second, whether some vestiges of such a perturbative structure can be recognized in Nature as we know it. The second question has been already positively answered by many authors. In particular the very notion of "primitive world" stands on the definite evidence of a perturbative structure of hadronic interactions, stemming from the validity of the quark model, the simplicity of deep inelastic phenomena, and the remarkable validity of the OZI-rule [10].

In order to answer the first question, we should find out whether a small parameter exists in the theory.

Let us note that apart from quark masses the theory has only one free parameter, g, which gets "dimensionally transmuted" into a "string tension" μ[9].

The field Hamiltonian H_f is proportional to μ^2, but only the H_{np} piece exhibits a singular behaviour at long distances ($\sim \mu^2 r$). The rest is perfectly finite both at long and short distances. Suppose now that μ^2 is a small number then it is clear that, whereas we can treat H' as a perturbation, H_0 can never be treated so (it is sufficient that we choose r large enough).

From a preliminary analysis μ^2 does indeed turn out to be a fairly small (compared with typical hadronic mass parameters) parameter:

$$\mu^2 \simeq .25 \div .30 \text{ GeV}^2. \tag{7.27}$$

And if we furthermore notice that μ^2 is the fundamental hadronic yardstick, we can easily understand the following key hadrodynamical features:

a) the success of the quark model; pair creation in associated typically with small interaction energies of the order of a few hundred MeV, the characteristic size of hadronic widths.

b) The precocity of Regge behaviour; The asymptotic behaviour in hadronic scattering amplitudes sets in when the energies are large compared with level spacing, which is controlled by μ. Thus when

$$s \gg \mu^2$$

we are asymptotic; and the smallness of μ^2 (eq. (7.27)) accounts for precocious Regge behaviour.

c) The precocity of Light Cone behaviour; Like for Regge behaviour we enter the asymptotic regime when $Q^2 \gg \mu^2$.

REFERENCES

1. C.N. Yang and R.L. Mills, Phys. Rev. 96, 191 (1954).
2. S.L. Glashow, Nucl. Phys. 22, 579, (1961)
A. Salam, Proc. 8th Nobel Symposium, N. Svartholm ed. (Almqvist and Wiksell Stockholm, 1968) p. 367
S. Weinberg, Phys. Rev. Lett. 19, 1264 (1967).
3. QCD is a really collective achievement. However, the earliest ideas can be traced to Y. Nambu;
M. Han and Y. Nambu, Phys. Rev. 139B, 1006 (1965) and to O.W. Greenberg, Phys. Rev. Lett. 13, 598 (1964).
4. Cfr for instance G. 't Hooft's lectures at this School.
5. The theoretical formulation of these attempts is due to K.G. Wilson, Quarks and Strings on a Lattice, in New Phenomena in Subnuclear Physics, A. Zichichi ed. (Plenum Press, New York, 1977) p. 69.
6. G. Preparata, Beyond 4-dimensional space-time: a possible answer to the riddle of colour confinement. Phys. Lett. 102B, 327 (1981).
7. G. Preparata, A new approach to the interactions of quarks and leptons in an anisotropic space-time, Preprint BA-GT-80-2; and Anisotropic Yang Mills interactions for quarks and leptons, Preprint BA-GT-80-3.
8. For a theoretical analysis of recent deep inelastic data see P. Castorina, G. Nardulli and G. Preparata, Phys. Rev. Lett. 47, 468 (1981).
9. J.L. Basdevant and G. Preparata, The structure of strong interactions in Anisotropic Chromo-Dynamics, Preprint BA-GT-06 (1981).
10. The relevance of the "primitive world" in the description of hadronic behaviour has been discussed at length in G. Preparata: QCD, problems and alternatives. Proceedings of the 1979 EPS Geneva Conference, p. 442 (CERN, Geneva, 1980).

PHYSICS AT COLLIDER ENERGIES

M. Jacob

CERN

Geneva

I. INTRODUCTION.

By the end of 1981, the first physics results in the range of energy presently referred to as collider energy ($\sqrt{s} \geqslant 500$ Gev), should be available, and the topic of these lectures will no longer have the speculative character which it has had, while being discussed over several years already. The success of the ISR, the first proton-proton collider, prompted design studies for colliding rings of much larger energies. The key motivation for such new machines was the discovery of the weak intermediate bosons which the many successes already scored by the Standard model of the Electro-Weak interactions[1)] lead one to expect in the 80-100 Gev range. In order to produce the weak vector bosons as copiously as possible in accordance with their actual couplings, one needs the corresponding mass-energy at the quark-parton level. This requires hadron center of mass energy typically six times larger, hence the range of energy actually considered. The design of Isabelle, a proton-proton colliding ring system with beams of 400 Gev, has prompted many investigations[2)]. The anticipated physics at Collider energy is indeed described in several Isabelle workshop documents. Prospects are great but experimentation is still remote in time.

The time scale for experimentation could however be much reduced with the idea of using existing proton supersynchrotrons as proton-antiproton colliders, and implementing it through the development of cooling techniques[3)]. According to such a scheme protons and antiprotons are simultaneously accelerated in opposite directions in a synchrotron and kept coasting as bunched beams. Interactions occur at specific points where configuration with anomalously large multiplicity than a confinecessary for injection corresponds to 10^{11} to 10^{12} antiprotons, patiently collected (over one day say) among the particles produced from a high energy proton beam impinging on a target. Such a source has now been built at CERN using stochastic beam cooling techniques[4)]. It is possible to pile up in a dedicated storage ring, the Antiproton Accumulator, many $\bar{p}$ bursts collected with a rather wide momentum spread, in order to insure a reasonable intensity, and cooled to a small momentum spread (small overall convex cross section of the stored beam) in order to allow for piling up within the aperture of the storage ring.

The CERN SPS, which then receives and accelerates beams of protons and antiprotons can keep them coasting, thus acting as a storage ring, but not at its full energy. The magnets cannot indeed be used in a DC mode at full power. The center of mass energy is thus limited in practice to 540 Gev. This is however certainly high enough an energy for W, Z production when one considers the factor 6 already mentionned. The relatively weak point of this ingenious scheme remains however its limited luminosity. Its value is controlled by still not fully understood beam - beam effect[5)]. The design value which we nevertheless hope to reach eventually is 10^{30} $cm^{-2}s^{-1}$. The counting rate for a particular process is given by the luminosity L times the corresponding cross section σ, namely

$$N = L\sigma \tag{1}$$

With 1 even/day set as a reasonable limit, one cannot thus hope to actually study processes with reaction cross sections much less than 10^{-35} cm^2. While ingenious low β techniques could eventually be used to increase the luminosity, it will remain rather limited. If in such a case one may say that ingeniosity (and some money!) is buying the energy, a (more expansive) dedicated collider, such as Isabelle, offers a much larger luminosity. The design value of the original Isabelle design is indeed

$$L = 10^{33} cm^{-2} s^{-1}$$

At present, Fermilab is also developing a $p\bar{p}$ Collider programme. It will use the energy doubler when it is completed. In this case the superconducting magnet can be used at full current intensity, even in the storage ring mode, and the nominal machine energy can be reached. The center of mass energy should then be of the order of 2000 Gev a much appreciable gain over the 540 Gev reached at CERN. The luminosity is however expected to be of the same order, namely 10^{30} - Table 1 puts together the main characteristics of the hadron Colliders presently at the building stage. It also includes the CERN-ISR. The CERN programme is starting in 81, whereas the Fermilab one will probably not start before '83-'84. The completion date of Isabelle is still uncertain, and changes in the project may result in different values than those quoted in the table.

Table 1.1. Hadron colliders and collider energies

Machine	Mode	$\sqrt{s}$ (GeV)	E (TeV)	Δy	$\sqrt{\hat{s}}$ (GeV)	L ($cm^{-2}\ s^{-1}$)
CERN ISR	Two rings pp($p\bar{p}$)	62	2	8	10	$10^{31}(10^{29})$
CERN SPS	One ring $p\bar{p}$	540	150	12.5	90	10^{30}
Isabelle	Two rings pp	800	340	13.3	140	10^{33}
Fermilab Doubler	One ring $p\bar{p}$	2000	2000	15.2	340	10^{30}

$\sqrt{s}$ = maximum centre-of-mass energy; E = equivalent laboratory energy; $\Delta y \sim \ln s$ = rapidity range: $\sqrt{\hat{s}}$ = maximum effective centre of mass energy at the consistent level (requiring larger energies implies diminishing luminosities); L = luminosity. A typical 1000 h period may correspond to an integrated luminosity of 10^6 times the quoted design value.

All new projects meet the condition of reaching a center of mass energy $\geqslant 6$ times the mass of the expected weak vector bosons. The production cross sections are then estimated to be at the level of 10^{33} cm^2, hence the luminosity achieved with $p\bar{p}$ colliders should be enough. The advent of such machines offer extraordinary perspectives for particle physics in the eighties.

Producing W or Z in $p\bar{p}$ (or pp) collisions is of course a far more complicated process than in e^+e^- annihilations. Nevertheless confidence in the quark parton model, and the interpretation of its success in the framework of perturbative quantum chromodynamics[6] is such that one can advance predictions for the cross section which should be valid at least at the order of magnitude level, provided of course that the standard model is correct[1]. The case for such an optimistic stand will be presented in section 2.

One may of course question the validity of a reduction by a mere factor of 6 when considering the energy of quark-antiquark collisions as opposed to proton-antiproton collisions, in view of the fact that the effect of scaling violations could be important. It turns out, however, that for the presently most interesting value of the subenergy ($\sim$ 100 GeV at the constituent level, corresponding to Z and W production), the relevant value of x (the fraction of the hadron momentum taken by the relevant partons) is of the order of 0.2 (SPS) to 0.05 (Fermilab Doubler). This is a range for which the structure functions show only a relatively small variation with the probing momentum, Q^2 [7]. This is shown in Fig.1-1. The differential luminosity introduced in section 2 is thus used in a region where it should be particularly reliable even with present uncertainties about scaling violations.

While the Z and W search is the main motivation behind the collider programme, the result of other investigations are also eagerly awaited. This is in particular the case for jet studies[8], where spectacular effects are expected. It is also clear from present ISR results on hadron production[9], and the hints from Cosmic Ray results[10], that even the study of large cross section processes is going to provide very interesting new results and, may be, important surprises. We should of course hope that unexpected and therefore even more exciting features will appear as one jumps ahead by an order of magnitude in center of mass energy. The foreseen prospects are already very interesting.

This review of present expectations is divided into three main sections. The first one (Section 2) is concerned with W and Z production. Since the dominant production mode is expected to show a close connection with large-mass lepton-pair production, the section starts with a discussion of the lepton-pair continuum, which is used as a reference process to normalize expectations. The production of narrow vector mesons (quarkonia in the present vernacular), that could be discovered through their lepton-pair decay mode, is also discussed. In Section 3 we deal with large cross-sections. We start with a review of the key features of hadron production, as they are now well known from research at the ISR and at Fermilab, and present their extrapolation as the rapidity range, which appears as the

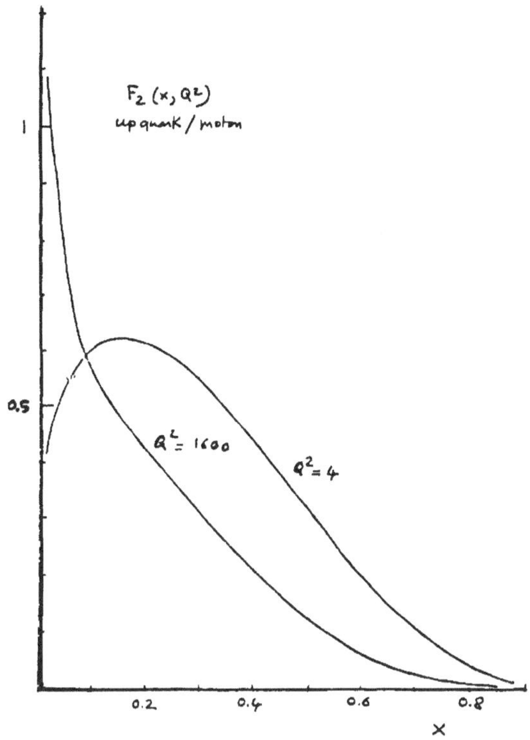

Fig. 1.1. The structure function $F_2(x)$ (the u quark momentum distribution) for two widely different values of Q^2. As a result of scaling violations, the large x quark distribution gets depleted as the low x distribution increases. Nevertheless, the density hardly changes for x ≃ 0.1.

important parameter, is extended by 5 units (from 8 to 13). Information available from cosmic ray research is then reviewed and more particularly the results obtained with emulsion chambers, which cover the relevant energy range. It is stressed that some important changes seem to occur as one reaches energies of the order of 100 TeV[10], which should be well within the range of colliders.

In Section 4 we discuss large p_T and jet phenomena [8]. Hard scattering among hadron constituents is known to result in specific kinematical distributions, with hadronic jets as the key feature. The corresponding yields increase rapidly with energy as the relevant constituents change from rare hard ones (taking a large fraction of the hadron momentum) to numerous soft ones. One therefore expects relatively important jet-production cross-sections at collider energies. As a result, jet studies should be an important research activity using such machines. We review present knowledge about jets in hadron interactions [8] and discuss extrapolations to very high energies. This brings us back to weak vector boson production inasmuch as their two-jet decay modes are expected to dominate. This hadronic background appears however to be overwhelming. Insofar as gluon interactions appear to play a very important role in jet phenomena, this also leads us to consider the production of hitherto unknown objects which could be coupled to colour and hence produced in gluon collisions. If the Higgs mesons of the standard model[1] are condensates of hitherto unknown fermions, also coupled to colour as well as to a force which confines them within very small distances, they should have massive coloured partners which could be within the new energy range considered. This so called Higgs sector, while still speculative, offers very interesting possibilities at Collider energy.

Gluon interactions should in any case result in relatively copious heavy quark production. In the framework of QCD, rates should be scaled down according to quark mass squared only, as opposed to the exponential decrease with hadron mass usually considered in the framework of statistical models successful with low p_T standard hadron production. Semi-leptonic decay modes may provide relatively clear signals and spectacular events with the cascade decay of very heavy quarks can be anticipated.

While this review represents an updated survey of the subject as I can see it today, it borrows much from previous reviews which I have been called to write over the past few years[11]. I apologize for the amount of overlap which one may find between this new review and previous ones. Nevertheless it stands for the conforting fact that predictions which could appear as much speculative a few years ago now seem to be further supported by experimental information, and in particular by the many results on lepton pair production which have been accumulated over the past few years[12], together with the better understanding of hard hadronic processes provided by QCD[6].

A large fraction of the exciting prospects of the Collider programme are shared by LEP (large electron positron storage ring) which also appears as an extremely interesting machine[13].

Before discussing the expected phenomena, we therefore first dwell a little on a comparison between machines, in order to stress the pros and cons of the different approaches. Figure 1-2 shows how the machine radius R varies as a function of the centre-of-mass energy $\sqrt{s}$. There is a linear variation for a proton colliding-beam machine. It can be normalized at $\sqrt{s}$ = 500 GeV to R = 1 km (SPS), though special magnets with improved cooling could allow the energy at fixed radius to be upgraded by a factor about 1.5. The use of super-conducting magnets with present technology, should allow for a gain in energy by an extra factor of about 2.5. We take a second normalization point (according to the Fermilab collider project) with $\sqrt{s}$ = 2000 GeV at R = 1 km. Niobium-tin technology, if it becomes manageable, should allow another factor of two in energy at fixed radius[14]. ($\sqrt{s}$: 20,000 GeV at R = 5 km). Considering the energy alone, proton colliders have a high performance when compared with a fixed-target proton machine (VBA) or to an e^+e^- colliding ring (LEP)[13]. In both latter cases the centre-of-mass energy increases only as the square root of the machine radius. In the case of an e^+e^- machine this results from a cost optimization, taking into account synchrotron radiation power losses, which vary as s^2/R[15]). In Fig.1-2 we have normalized the VBA line at $\sqrt{s}$ = 30 GeV for R = 1 km (standard magnets). However, the use of superconducting magnets would reduce the radius by a factor of 2.5. We have normalized the e^+e^- line at $\sqrt{s}$ = 200 GeV for R = 5 km. This is half way between the standard and superconducting cavity options for LEP[16]. Whilst a VBA is not favoured by such a graph, its prominent feature is of a different nature, namely the existence of secondary beams allowing for a very wide variety of experiments. A proton storage ring looks far more advantageous than an e^+e^- ring, but one should acknowledge the fact that what actually matters in many interesting cases is the centre-of-mass energy at the constituent level, which in hadron-hadron collisions corresponds to typically one sixth of the total centre-of-mass energy (dashed-dotted line in Fig.1-2 corresponding to the pp(standard) line). This factor 1/6 corresponds to the fact that the momentum of the proton is roughly shared by gluons and quarks and that the mean momentum of a quark is roughly half its maximum value.

This is now well documented by deep inelastic lepton scattering results[17], and Figure 1-1 shows one structure function F_2 and its expected evolution as Q^2 varies. While far larger energies than $\sqrt{s}/6$ are found at the constituent level, it is with a very low probability, as later discussed at length. It is still true, however, that a high energy ($\sqrt{s} \geqslant$ 100 GeV), hadron machines retain an advantage over electron machines. Protons are just much easier to accelerate than electrons. One has, however, to face a very large background (often an overwhelming one) for many very interesting processes far better studied with an electron machine, hence the cardinal interest of the LEP project[13], despite the availability of a proton-antiproton collider in 1981.

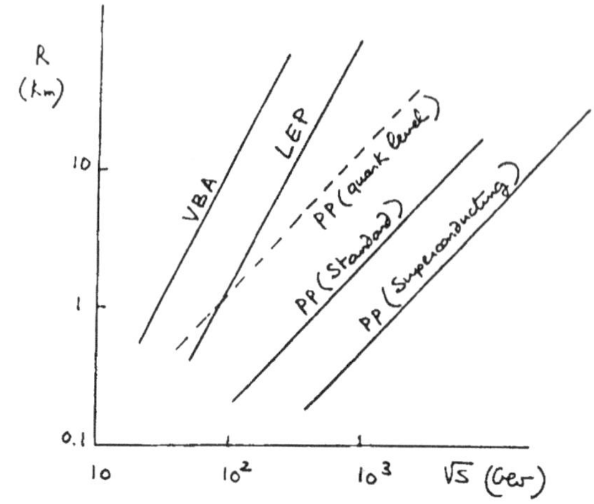

Fig. 1.2. Size to energy relation for different types of circular machines. The centre-of-mass energy $\sqrt{s}$ increases as R for a proton-proton (antiproton) collider and only as $R^{1/2}$ for a fixed target machine (obvious) and e^+e^- colliding-beam machine (the result of cost optimization).

As we previously stressed, energy is very important, but so is luminosity, and now considering rates, one may compare the performances of a pp ($p\bar{p}$) collider with that of an e^+e^- storage ring for Z° production (the cardinal case at present). For LEP running at the Z° energy[13)], one expects close to 10^5 events/day for a luminosity at that energy of about 10^{32}. In pp collisions at Isabelle energy the production cross-section is expected to be of the order of 10^{-33} cm^2 (nb level) and the over-all rate should therefore be very similar ! The experimental conditions offered by LEP are, of course, far better (low background and well-defined kinematics). On the other hand, Isabelle will also produce about twice as many W^+ (and as many W^-) at the same time when pair production will only be sizeable enough at LEP, and once the very high threshold energy is reached, ($\sqrt{s}$ > 160 GeV).

The production of objects coupled to colour (new quarks in particular) will be far more abundant on a pp ($p\bar{p}$) machine than on an e^+e^- machine, namely by a factor of $(\alpha_s/\alpha)^2 \simeq 10^2$, provided that the centre-of-mass energies and the luminosities are comparable at the constituent level (this is the case for Isabelle at 800 GeV and LEP at 140 GeV, looking at quark jets of 70 GeV, say). Nevertheless, such a larger yield is likely to be associated with a very large background which may saturate detectors. In much the same way as LEP, the $p\bar{p}$ Collider appears to be the proper way to answer some very important topical questions. It cannot compete with LEP, however, in its testing in detail of the electroweak theory[1)].

Considering more specifically the present CERN programme, five experiments are nearing the date taking stage[18]. Two operate very large detectors, which are designed in order to conduct a powerful Z-W search. The three others use much smaller and dedicated apparatuses.

Experiment UA1 (Aachen - Annecy - Birmingham - CERN - London - Paris - Riverside - Rome - Rutherford - Saclay - Vienna) uses a very large dipole magnet with electremagnetic and hadronic calorimetry. It spans a very large solid angle and coverage for Z and W search is practically 4π . Located in zone LSS5, this experiment is design to meet almost all present questions, namely :

- Find the intermediate vector mesons (prospects are high)
- Find new narrow vector mesons (prospects are low)
- Study jet phenomena (prospects are high) and also the production of heavy quarks
- See if (and how) lns physics continues (prospects are high as Cosmic Ray results hint at new phenomena).

Experiment UA2 (Bern - CERN - Copenhagen - Orsay - Pavia - Saclay), uses electromagnetic and hadronic calorimetry with a very large solid angle detector (also a practically 4π coverage for Z and W search) and has however only forward (toroïdal) magnets. It is located in zone LSS4. The aims of UA2 are similar to those of UA1 but the techniques used are different. Experiment UA3 (Annecy - CERN) is a monopole search experiment located in zone LSS5. It uses Kapton foils for anomalous ionization search. Experiment UA4 (Amsterdam - CERN - Genova - Naples - Pisa) is designed for an elastic and total cross section measurement. It uses Roman pot detectors. It is located in zone LSS4. Finally, Experiment UA5 (Bonn - Brussels - Cambridge - Stockholm) uses a streamer chamber. Located in zone LSS4 it should provide the first results about the structure of high multiplicity events and rapidly check information from Cosmic Rays in the 100 Tev range.

The energy and the luminosity of the $p\bar{p}$ collider are such that most of the pressing questions which one now has about the weak vector bosons, jet production, the continuation of ln s physics...should be answered to a large extent. As previously said, it would, however, be surprising if new and unexpected things do not occur as the centre-of-mass energy increases by an order of magnitude. Thresholds can be reached. Such surprises will probably better define the second generation of experiments - in particular those demanding higher luminosities or higher energies - than the mere further exploration of present questions.

This is to be kept in mind when extrapolating in energy (Fermilab Doubler) or in luminosity (Isabelle).

II. HUNTING THE WEAK VECTOR BOSONS.

II.1. The Drell-Yan process

The production of the weak bosons in $p\bar{p}$ (pp) collisions has long been considered in the framework of the parton model, using the Drell-Yan process as a guide[19]. The dominant amplitude when calculating the expectation value of a current operator at large $|Q^2|$ is associated with quark-antiquark annihilation (Fig.2-1). The production cross-section of a lepton pair from (electromagnetic) quark-antiquark annihilation is written as the product of a quark-antiquark cross-section by quark and antiquark distribution functions at values fixed by the kinematics. It reads

$$\frac{d^2\sigma}{dx_1 dx_2} = \frac{1}{3}\sum_i e_i^2 \{q_i(x_1)\bar{q}_i(x_2) + q_i(x_2)\bar{q}_i(x_1)\} \frac{4\pi\alpha^2}{3sx_1x_2} \tag{2.1}$$

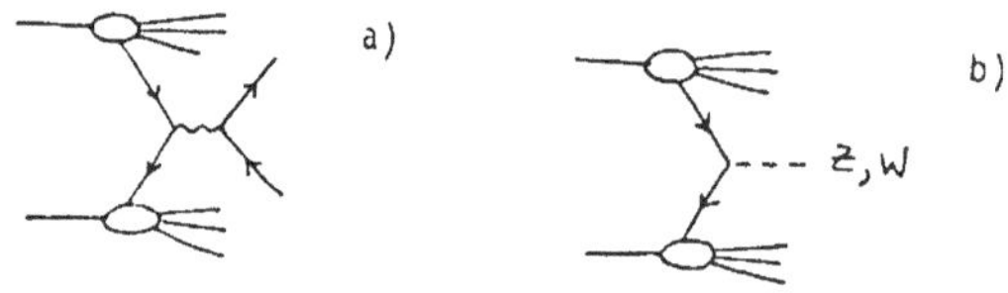

Fig. 2.1. The Drell-Yan process
a) for lepton-pair production
b) for the production of the weak boson

The fraction of the hadron momentum taken by the two constituents is denoted by x_1 and x_2, respectively. This corresponds to a distribution probability of $q_i(x)$ (for quarks of species i and charge e_i) and $\bar{q}_i(x)$ (for antiquarks). One sums over species, weighting according to the quark charge squared $e_i^2 = 4/9$ or 1/9. The factor 1/3 appearing in front corresponds to colour averaging. The last factor in (2.1) is the point-like annihilation cross section. The relevant center-of-mass energy squared (or the lepton mass squared) is

$$M^2 = sx_1x_2 = s\tau \tag{2.2}$$

The lepton pair has a longitudinal momentum

$$p_L = |x_1 - x_2| \frac{\sqrt{s}}{2} = x\frac{\sqrt{s}}{2} \qquad (2.3)$$

The transverse momentum (not explicitly included in (2.1)) is also the sum of the (transverse) momenta of the constituents which, in the approximation of Fig.2.1, are limited to Fermi motion inside the incident hadrons.

Insofar as relation (2.1.) describes the coupling of the electromagnetic current at large $|Q^2|$, one expects the same approximation to apply to the coupling of the weak currents, as graphically described by Fig2-1, a and b, therefore to W, Z production.

Coming back to lepton pair production , one may rewrite the production cross section (2.1) in terms of directly observable quantities, M and x. This reads

$$\frac{d^2\sigma}{dMdx} = \left(\frac{8\pi\alpha^2}{9M^3}\right)\left(\frac{\tau}{\sqrt{x^2 + 4\tau}}\right) \sum_i e_i^2 R_i(x_1,x_2) \qquad (2.4)$$

with

$$R_i(x_1, x_2) = q_i(x_1)\bar{q}_i(x_2) + q_i(x_2)\bar{q}_i(x_1) \qquad (2.5)$$

$$x_1, x_2 = \frac{1}{2} \left(\sqrt{x^2 + 4\tau} \pm x\right) \qquad (2.6)$$

The cross-section is written as the product of the cross-section for annihilation times the probability of finding such constituents in the incident hadrons. This is the parton model expression[20].

From relations (2.4) and (2.5), one readily derives the scaling property

$$M^3 \frac{d^2\sigma}{dMdx} = F(x, \tau) \qquad (2.7)$$

or in terms of the rapidity of the lepton pair $y = \ln \frac{E + p_L}{M}$

$$M^3 \frac{d\sigma}{dMdy} = \sqrt{x^2 + 4\tau}\, F(x,\tau) \qquad (2.8)$$

Such a differential cross-section scales, depending on M^2/s only, and is thus in principle predictable at very high energy and very large mass, provided that the relevant scaling properties are actually verified, and that the cross-section is known for the corresponding values of τ

Before turning to the validity of the Drell-Yan approach, on which predictions for Z and W production are based, we introduce a very useful parameter : the differential luminosity. We write (2.4) and (2.5) as :

$$\frac{d\sigma}{dM} = \frac{8\pi\alpha^2}{9M^3} \int_0^{1-\tau} \frac{\tau}{\sqrt{x^2 + 4\tau}} \sum_i e_i^2 R_i(x_1,x_2)dx \tag{2.9}$$

The upper limit of integration comes from (2.2) and (2.3), and x_1 (x_2) are given by (2.6). Going from variable x to variable x_1 (at fixed τ), one writes

$$\frac{d\sigma}{dM} = \frac{8\pi\alpha^2}{9M^3} \tau \int_\tau^1 \sum_i e_i^2 R_i(x_1,\frac{\tau}{x_1}) \frac{dx_1}{x_1} \tag{2.10}$$

The differential luminosity is then defined as

$$\frac{dL_i}{d\tau} = \int_\tau^1 R_i(x,\frac{\tau}{x}) \frac{dx}{x} \tag{2.11}$$

and the cross-section is written as

$$\frac{d\sigma}{dM} = \frac{8\pi\alpha^2}{9M^3} \sum_i e_i^2 \tau \frac{dL_i}{d\tau} \tag{2.12}$$

$$= \frac{2}{3M} \sigma_o \sum_i e_i^2 \frac{dL_i}{d\tau} \tag{2.13}$$

where σ_0 is the well-known "point-like" cross-section for $e^+e^- \to \mu^+\mu^-$ in the one-photon exchange approximation

$$\sigma_o = \frac{4\pi\alpha^2}{3s} = 10^{-35}(\frac{90}{\sqrt{s}})^2 cm^2 \tag{2.14}$$

with $\sqrt{s}$ in GeV.

The differential luminosity readily gives the lepton-pair production cross-section in the Drell-Yan approximation. It is the probability of finding incident constituents with the proper matching value of τ, within the incident hadrons, namely :

$$\frac{dL_i}{d\tau} = \int R_i\ (x_1, x_2)\ \delta\ (x_1 x_2 - \tau)\ dx_1\ dx_2 \tag{2.15}$$

The quantity $\tau(dL/d\tau)$ is, however, the important parameter in (2.13) and this is the quantity which we shall directly calculate.

The validity of the Drell-Yan approach was neither checked nor understood in any satisfactory way when prediction had first to be ventured, as colliders where first considered in the mid seventies. One is now on a much stronger footing !

Indeed, over the past few years, the validity of the Drell-Yan approach was able to be tested to a surprisingly good accuracy, granting the oversimplified nature of the model. We illustrate this with two figures. Figure 2-2 a gives $M^3(d\sigma/dMdx)$ at $x = 0$ as a function of $\sqrt{\tau}$, combining ISR results and Fermilab results[21]. The scaling behaviour, which was known to hold over the Fermilab energy range (200 to 400 GeV of incident energy) is seen to apply up to 2000 GeV. The value of $F(\tau)$, at $\sqrt{\tau} \sim 0.15$, important for Z and W production at collider energy cannot be reached properly at $\sqrt{s} \sim 25$ GeV (Fermilab and SPS energy), since it corresponds then to J/ψ production. The τ dependence of the scaling curve, which reflects the fall-off of the differential luminosity at large τ, implies an important change of the cross-section with energy. This is displayed explicity in Fig.2-2b, which gives to $d\sigma/dMdy$ at $y = 0$, as a function of M, at Fermilab and ISR energies. Both distributions are, however, to a good approximation compatible with a unique scaling parametrization in the region where they overlap (around $\sqrt{\tau} \sim 0.2$).

Scaling, as implemented in the quark parton model, is thus found to be reasonably well satisfied. Comparing distributions over the full τ range, one already sees however, some, relatively small, deviations from the scaling behaviour. Such deviations are indeed expected in the framework of QCD and can be associated with the variation of the parton distribution with Q^2 (Fig.1-1). Since they are however expected to be relatively small for $\sqrt{\tau} \sim 0.15$, one can at present consider predictions for W and Z production on a much safer ground than 4 years ago.

Very encouraging also over the past few years was the fact that there is now also good evidence for the association of the relevant partons with quarks. This is, in particular, clear for lepton-pair production in π-nucleus collisions away from the prominent J/ψ and Υ resonances. Specific features associated with the quark charges are seen. In $\pi^+(\pi^-)$ induced reactions one indeed expects the dominant term to be $d\bar{d}(u\bar{u})$ annihilation, since quarks dominate over antiquarks in the target nucleus. For isospin zero nuclei this implies a factor of the order of 0.25 between the π^+ and π^- induced cross sections.

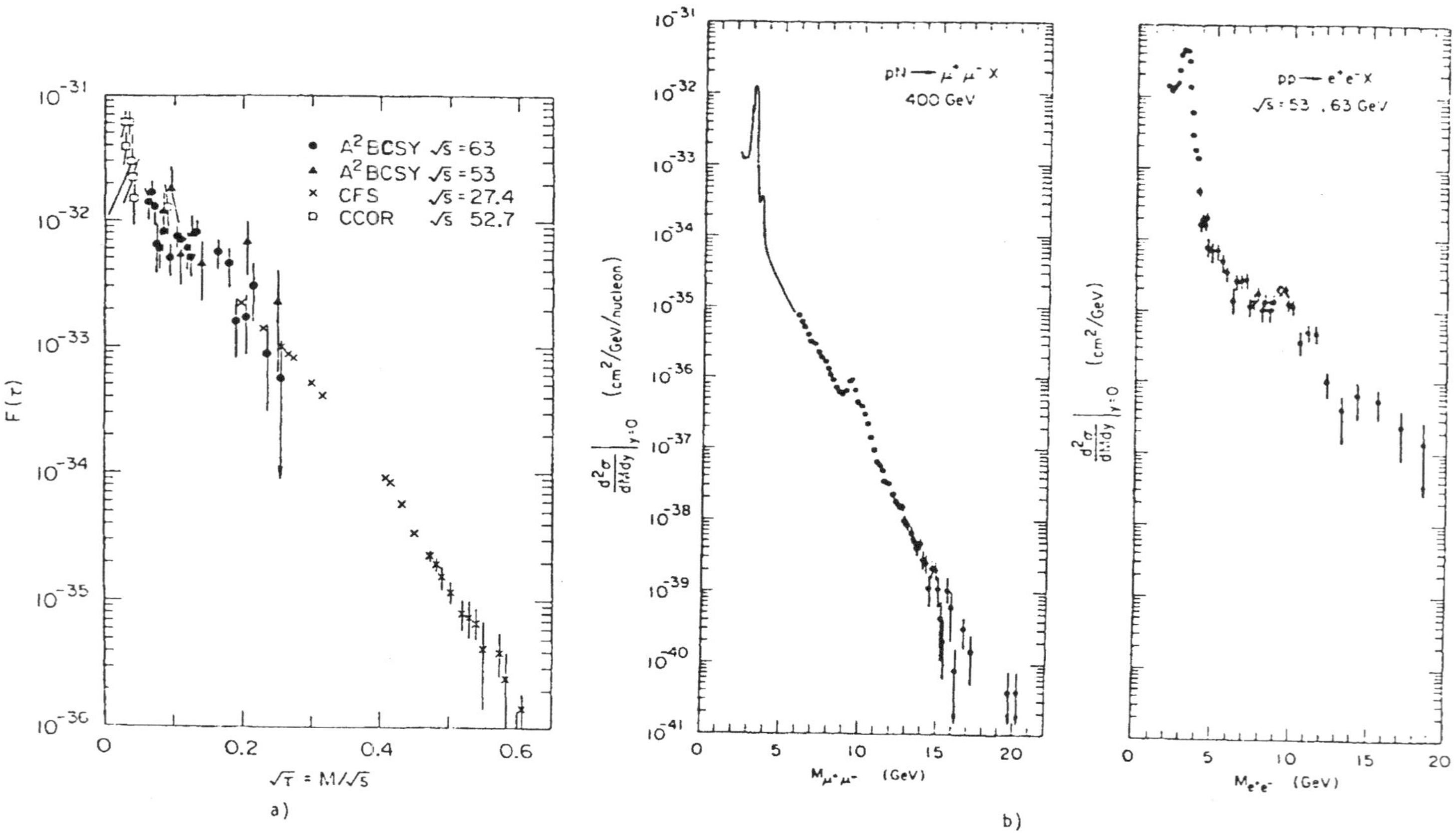

Fig. 2.2. a) lepton-pair differential cross-section times M^3, at x = 0 as a function of the scaling variable $\tau = M/\sqrt{s}$. Data from Fermilab ($\sqrt{s}$ = 27.4 GeV) and ISR ($\sqrt{s}$ = 62 GeV).
b) lepton-pair differential cross-section as a function of the mass at Fermilab and ISR energies.

The actual ratio, taking account of sea quarks as well as valence quarks, is expected to be slightly larger. The expected behaviour is well verified for $M > 4$, as shown in Fig.2-3, which displays a compilation of data from CERN and Fermilab[21-22]. The behaviour for $M \leqslant 3$ GeV actually signals the limit below which the model does not apply and where hadron physics has to be considered in a more conventional (and complicated) way.

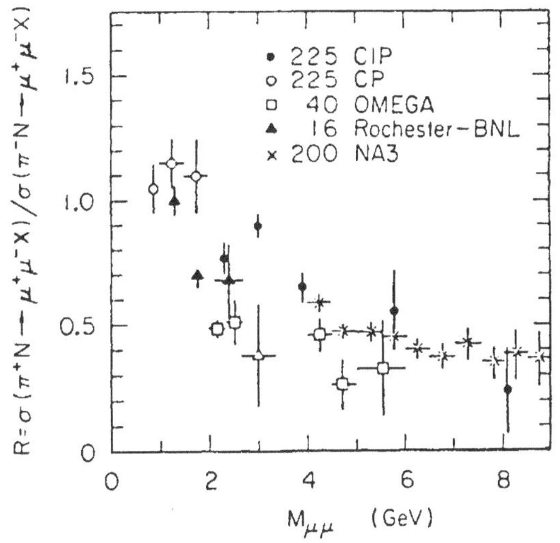

Fig.2-3. Ratio of π^+N to π^-N induced cross-sections as a function of lepton-pair mass

Having gained confidence in the relevance of the key features implied in the approximation (2.1) one now proceeds to the calculation of R (2.5) and of the differential luminosity (2.15).

In principle the value of R can be obtained from quark distributions determined form deep inelastic lepton scattering. For instance, an approximate scale-invariant parametrization of the CDHS neutrino scattering results[23] can be obtained with

$$\begin{aligned} xu(x) &= 2.13\sqrt{x}(1-x)^{2.3} + 0.27(1-x)^{3.1} \\ xd(x) &= 1.26\sqrt{x}(1-x)^{3.8} + 0.27(1-x)^{3.1} \\ x\bar{u}(x) &= x\bar{d}(x) = 0.27(1-x)^{3.1} \\ xs(x) &= x\bar{s}(x) = (1/2)x\bar{u}(x) \end{aligned} \tag{2.16}$$

and such a distribution, which refers to the proton can be used to calculate $dL/d\tau$ in (2.15).

The corresponding value, valid at relatively low $|Q^2|$, can then be determined at any higher $|Q^2|$ using standard parametrization of scaling violation effects, as predicted by QCD[24].

At present there is an encouraging agreement between the order of magnitude and shape of the distribution functions obtained from deep inelastic scattering and from lepton-pair production; or, in other words lepton-pair production can be predicted from deep inelastic scattering. Nevertheless, there appears to be an over-all discrepancy in rate by a factor of the order of two, the lepton-pair production cross-section being larger than that calculated from distribution functions determined from deep inelastic scattering. This is illustrated by Fig.2-4, which puts together the proton F_2 function, as presently determined from neutrino scattering (CDHS) and from muon-pair production (NA3).

We shall take this with optimism, remarking that the order of magnitude of the lepton-pair production cross-section is correctly predicted, and face later some justified pessimism.

The procedure followed is indeed now well justified to lowest order in QCD[6)]. The Q^2 variation of the structure function corresponds to the leading log approximation which takes into account quark evolution as quark branch into soft or collinear gluons and gluons turn into quark-antiquark pairs, as the analysis power Q^2 varies.

Perturbative QCD thus justifies the factorization procedure used in (2.1) to lowest order in a coupling constant which is of the order of $\alpha_s/\pi \sim 0.06$ providing that one takes into account the varying probability of finding a quark (antiquark) in a proton. Since the weak and the electromagnetic coupling are closely related in the standard model, one may (as done later) bravely use a Drell-Yan calculation up to the weak boson masses ($M^2 \sim 10^4$) in order to calculate the production cross-section. Doubts remain of course concerning the validity of the lowest order calculation. We already mentioned the discrepancy of a factor close to 2 when comparing experimental results. Caculations to first order in α_s have been done25) and give a first order term about as large as the zeroth order one. This is however largely due to the fact that a piece in the first order term appears with a large numerical factor ($\sim \pi^2$). Whether higher order corrections exponentiate (and thus remain globally of the same order) or not is still unclear. We take the attitude that the lowest order Drell-Yan approach should give the correct order of magnitude.
Higher order corrections effects are also easily seen in the transverse momentum (p_T) dependence of the lepton pair. It should be limited by the small transverse momentum of the quarks in the parton model. In QCD, $< p_T > \sim \sqrt{|Q^2|}$ as the radiation of hard gluons may provide quarks with large transverse momentum. A large p_T "tail" with a mere p_T^{-2} fall off thus develops beyond the

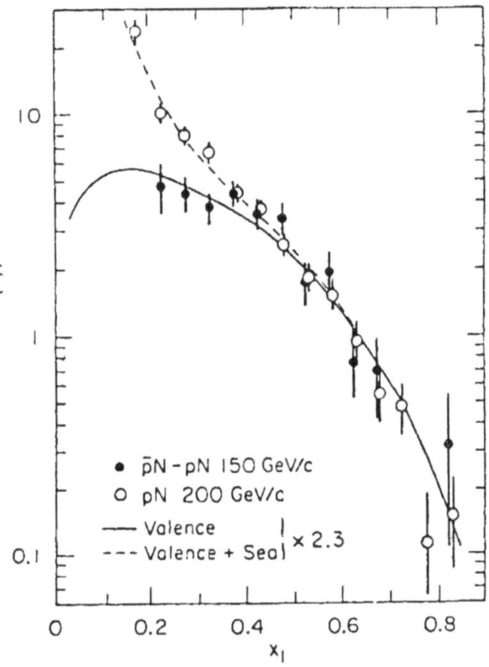

Fig. 2-4 The proton structure function $F_2(x)$ as determined from lepton-pair production results (data points) and from the deep inelastic neutrino scattering curve. The latter results have to be scaled by a factor of 2.3 in order to agree with the former one.

exponential drop found at low p_T ($p_T < 0.5$ GeV) and which is associated with quark motion within the colliding hadrons. This is only poorly displayed by data on $<p_T>$. Nevertheless the related increase of $<p_T>$ as a function of mass with a saturation value increasing with energy is now well documented, as shown by Figure 2-5.

We are thus lead to expect a rather large value of $<p_T>$ when reaching the W-Z mass range ($<p_T> \sim 4$ GeV/c say). Nevertheless one may be fairly confident that the cross section obtained in the leading logarithm approximation is providing the correct order of magnitude. We thus proceed to the calculation of the differential luminosity.

At present there are convenient M^2 (Q^2) parametrizations of the structure functions which can be used for this calculation[24]. This typically leads to a decreasing luminosity at large τ and an increasing one at low τ, as M^2 increases, when compared to values obtained

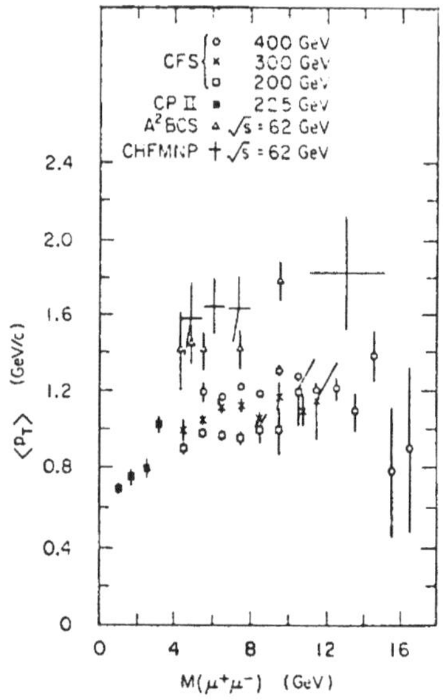

Fig. 2.5. The mean value of the transverse momentum for lepton-pair production in pp collisions. This compilation puts together Fermilab results (200 to 400 GeV) and ISR results ($\sqrt{s}$ = 62 GeV)

without scaling violations. This is illustrated by Fig.2-6a, which gives the $u\bar{u}$ contribution to $\tau(dL/d\tau)$ in $p\bar{p}$ collisions without scaling violation and with scaling violation for $\sqrt{s}$ = 540 GeV. The shape for $\sqrt{s}$ = 2000 is not much changed. Note however the small changes between scaling and non-scaling distributions at $\sqrt{\tau} \sim 0.15$, which are relevant for Z and W production at $\sqrt{s}$ = 540 GeV. The relative merits of pp and $p\bar{p}$ are displayed in Fig.2-6b, which gives the value of $\tau(dL/d\tau)$(with scaling violation) for $u\bar{u}$ in pp (lower curve) and $p\bar{p}$ (upper curve) collisions at the same energy ($\sqrt{s}$ = 540 GeV). For the expected W and Z masses $p\bar{p}$ collisions are only twice as efficient as pp collisions. One may also say that pp collisions at Isabelle energy are as efficient as $p\bar{p}$ collisions at the CERN collider at producing the Z^0. The luminosity curves also show that for $\sqrt{\tau} \leqslant 0.2$, the production rate is essentially determined by the cross-section at the parton level and of course, by the nominal machine luminosity, $\tau(dL/d\tau)$ is of the order of 1. The proton behaves as a wide-band unseparated beam of quarks, antiquarks, and gluons, which are all important in number. As $\tau \geqslant 0.2$, rates drop as needed hard constituents are becoming very few. Figure 2-6 thus motivates the "rule of thumb" of demanding six times the required energy at the constituent level for the hadron centre-of-mass energy.

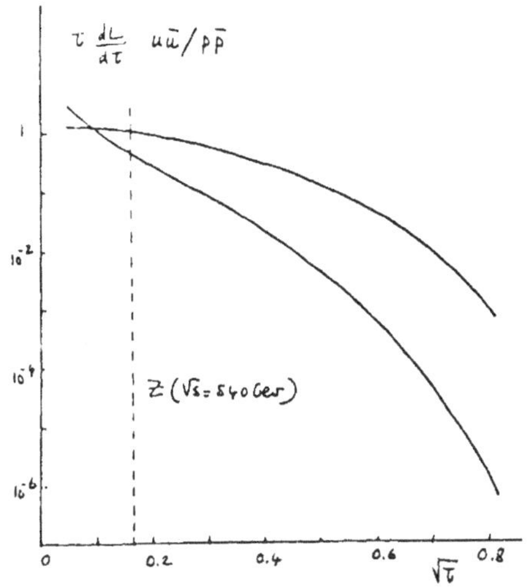

Fig. 2.6.a) Differential luminosity $\tau(dL/d\tau)$ as a function of $\sqrt{\tau}$ for $u\bar{u}$ collisions in $p\bar{p}$ scattering. The upper curve (for $\sqrt{\tau} > 0.1$) is calculated without scaling violation, while the lower one includes scaling violations.

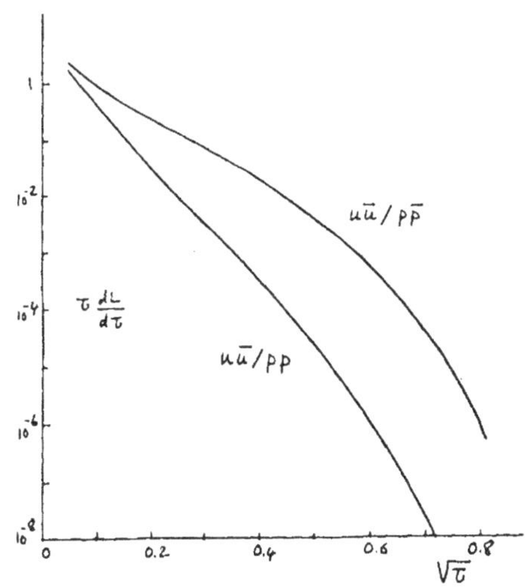

Fig. 2.6.b) Differential luminosity $\tau(dL/d\tau)$ as a function of $\sqrt{\tau}$ for $u\bar{u}$ collisions in $p\bar{p}$ (upper) and pp (lower) collisions for $\sqrt{s}$ = 540 GeV. Scaling violations are included

Figure 2-7 gives the differential cross-section $d\sigma/dM$ for $p\bar{p} \to \mu^+\mu^-$ at $\sqrt{s}$ = 540 GeV as a function of $\sqrt{\tau}$ (lower scale) and M (upper scale), without scaling violation (solid line) and with scaling violation (dashed line). Scaling violations are of little effect in the mass range which can be probed ($M \lesssim 20$ GeV) with a luminosity of 10^{30}. In this range the $\sqrt{\tau}$ dependence shown in Fig.2-7 corresponds essentially to a M^{-3} fall off of $d\sigma/dM$ dominating over the variation associated with the structure functions. Increasing the energy (up to $\sqrt{s}$ = 2000 GeV) would therefore not much increase the cross-section for fixed mass. We may conclude that study of the lepton-pair continuum on the $p\bar{p}$ collider will, in practice, be limited to a range already covered by the PETRA and PEP machines.

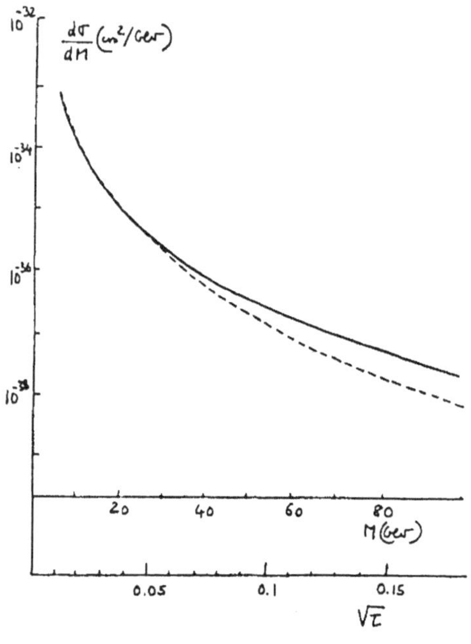

Fig. 2.7. Lepton-pair production cross-section in $p\bar{p}$ collisions at $\sqrt{s}$ = 540 GeV as a function of $\sqrt{\tau}$ (lower scale) and M (upper scale) without scaling violation (solid line) and with scaling violation (dashed line)

Increasing the luminosity by 3 orders of magnitude would allow much higher mass values to be reached even with a pp machine (Figure 2-5). One could then probe the lepton-pair continuum beyond 100 GeV or say cover the mass range available to LEP phase 1 (beam energy up to 50 GeV).

Background from heavy-quark decay has of course to be considered. We shall come back to it later. Given the fact that unsymmetric pairs can be used to subtract the background and that the Drell-Yan contribution is peaked at zero transverse momentum, signal and background could however be easily separated. The limitation comes actually from rate rather than background.

Studying the lepton-pair continuum is interesting in view of its direct relation to the proton structure as analyzed in perturbative QCD. Nevertheless for mere rate reasons we will not be able to push the analysis much beyond 20 GeV (remaining at very low values of τ!). The emphasis given here finds however its main justification in its providing a reference frame for the calculation of production rates for the W and Z mesons.

II.2. The production of the weak bosons

In the framework of the standard model, the dominant production mode is expected to correspond to quark-antiquark fusion in much the same way as it is the dominant term for large mass lepton-pair production. The photon, the Z and the W are indeed coupled to closely related currents. Two quarks of species i and j, annihilate into a weak boson (Fig.2.1b). The cross section is readily obtained. It reads

$$\frac{d\sigma^{ij}}{dx} = \pi \frac{x_1 x_2}{\sqrt{x^2 + 4\tau}} \frac{1}{M^2} g_{ij}^2 R_{ij}(x_1, x_2) \tag{2.17}$$

where x_1 and x_2 are given by (2.6); x and τ are as defined by (2.2), (2.3); R is as defined in (2.5).

In the case of a pure vector coupling, the coupling constant introduced in (2.18) is defined through a Lagrangian density

$$-g\bar{q}\gamma_\mu q V^\mu \tag{2.18}$$

in which quarks of the proper species are used.

For the charged weak bosons $W^\pm$, the corresponding Lagrangian density is written as

$$-g_W\bar{q}\gamma_\mu(1 - \gamma_5)qW^\mu \tag{2.19}$$

with

$$g_W = \sqrt{\frac{G}{\sqrt{2}}}\, M_W \cos\theta_c \tag{2.20}$$

for $W^+(W^-)$ production in $u\bar{d}$ ($\bar{u}d$) collisions. A similar expression with the cosine of the Cabibbo angle replaced by sin θ_c, applies for $u\bar{s}$ ($\bar{u}s$) collisions.

For the neutral weak boson, the standard parametrization of the Lagrangian density is

$$-g_Z\bar{q}\gamma_\mu \frac{v - a\gamma_5}{\sqrt{2}} qZ^\mu \tag{2.21}$$

or

$$-\sqrt{\frac{G}{\sqrt{2}}} M_Z\bar{q}\gamma_\mu \frac{v - a\gamma_5}{\sqrt{2}} qZ^\mu \tag{2.22}$$

$$\begin{aligned} \text{with } v_u &= v_c = v_t = 1 - 8/3 \sin^2\theta_W \simeq 0.37 \\ v_d &= v_s = v_b = -1 + 4/3 \sin^2\theta_W \simeq -0.69 \\ a_u &= a_c = a_t = 1 \\ a_d &= a_s = a_b = -1 \end{aligned} \tag{2.23}$$

The total production cross-section is obtained from (2.18) and expressed in terms of the relevant differential luminosity

$$\sigma^{ij} = \frac{\pi}{M^2} g_{ij}^2 (\tau \frac{dL_{ij}}{d\tau}) \tag{2.24}$$

We use the same transformation as in (2.10). For $W^\pm$ production g_{ij}^2 is replaced by $1/3\ G\sqrt{2}M_W^2 \cos^2\theta_c$, when using the $u\bar{d}$ luminosity, and by $1/3\ G\sqrt{2}M_W^2 \sin^2\theta_c$, when using the (much smaller for $p\bar{p}$ collisions) $u\bar{s}$ luminosity. In both cases the factor 1/3 follows from colour averaging. For Z^o production we replace g_{ij}^2 by $(\sqrt{2}/6)GM_Z^2(v^2 + a^2)$, with the values of v and a corresponding to the initial quark channels $u\bar{u}$, $d\bar{d}$, or $s\bar{s}$ read from (2.23).

While the production cross-section includes a scaling M^{-2} term, the coupling constant squared increases with M^2, the g^2/M^2 ratio being fixed by the Fermi constant. The production cross-section depends upon the weak boson mass only through the differential luminosity, and also through the value of $\sin^2\theta_W$ (1.2) for Z^o production.

The W^+ cross-section is thus given (to a good approximation)

$$\sigma_{W^+} = \frac{G\sqrt{2}\pi}{3} \cos^2\theta_c \ (\tau \frac{dL_{u\bar{d}}}{d\tau}) \tag{2.25}$$

$$\simeq 0.65\ 10^{-32}\ cm^2\ (\tau \frac{dL_{u\bar{d}}}{d\tau}) \tag{2.26}$$

The Z° production cross-section is obtained in a similar way, but using the $u\bar{u}$ and $d\bar{d}$ differential luminosities. One gets

$$\sigma_Z \simeq 0.25 \times 10^{-32} cm^2 (\tau(dL_{u\bar{u}}/d\tau)) \quad (2.27)$$

assuming that, to a good approximation $L_{u\bar{u}} = 4\ L_{d\bar{d}}$ in $p\bar{p}$ collisions (at the relevant τ value).

The differential luminosities can be read from Fig. 2.6.

The values of $\sqrt{\tau}$ corresponding to $\sqrt{s}$ = 540 GeV are taken to be 0.157 (W) and 0.176 (Z°). The order of magnitude of the cross-section is then determined by the value of the Fermi coupling and not by the fact that it is improbable to find the relevant partons in a collision, as would be the case at much lower energy. While some increase in rate should then result from an increase in energy, it should no longer be a prominent fact.

There are, of course, large uncertainties in such estimates. We shall take as a working hypothesis

$$\sigma_{W^+} = \sigma_{W^-} \simeq 3.2 \times 10^{-33} cm^2 \quad (2.28)$$

$$\sigma_Z \simeq 1.6 \times 10^{-33} cm^2 \quad (2.29)$$

With a luminosity of 10^{30}, one therefore expects rates at the level of a few hundred events per day. Scaling violations do not result in an important effect for the pertinent range of τ values and hence calculations done in the framework of the standard parton model[19)] are still valid to a good approximation.

While the expected rates look comfortable, we have to face the fact (section 4) that jet yields which correspond to quark (antiquark) decay weighted by a factor of 3 for colour are much below those expected from hadronic interactions analysed in the framework of perturbative QCD. The dominant decay modes look as if they are almost impossible to extract from a large background. In order to be on the safe side, one should therefore direct attention to the leptonic modes, such as $e\nu$, $\mu\nu$ for W decay (with an expected probability of 8% for each for three quark and lepton families), and e^+e^-, $\mu^+\mu^-$ for Z° decay, with a probability of 3% for each as read from (2.23). Decay rates can be obtained from (2.21) and (2.22). A straightforward calculation gives

$$\Gamma(W^+ \to e^+\nu) \simeq \frac{GM_W^3}{6\pi\sqrt{2}} \quad (2.30)$$

with

$$\Gamma_{e\nu} \simeq \frac{1}{4N}\,\Gamma_{tot} \tag{2.31}$$

where N is the number of families.

Similarly one finds

$$\Gamma(Z^o \to e^+e^-) \simeq \frac{GM_W^3}{24\sqrt{2}\pi}\,(1 + (1 - 4\sin^2\theta_W)^2) \tag{2.32}$$

with

$$\Gamma_{e^+e^-} \simeq \frac{1}{11N}\Gamma_{tot} \tag{2.33}$$

Starting from a few hundred events per day, we are now left with a few events per day in each leptonic mode.

With the production cross-section taken as

$$\sigma(W^+ \to e^+\nu) \simeq 2.7 \times 10^{-34}\ cm^2 \tag{2.34}$$

$$\sigma(Z^\circ \to e^+e^-) \simeq 4.8 \times 10^{-35}\ cm^2$$

an integrated luminosity of the order of 10^{36} cm^{-2} (which can be summed over a good running period of 1000 hours) will provide $\sim$ 300 W events in the $e^+\nu$ mode and $\sim$ 50 Z° events in the e^+e^- mode.

The total width of the W and Z is close to 3 GeV (with three families). This is close to the expected experimental resolution ΔM for a lepton-pair at that mass. The signal-to-lepton-pair background ratio is therefore very large. The Drell-Yan electromagnetic background is obtained from (2.13)

$$Br = \int_{\Delta M} \frac{d\sigma}{dM}\,dM \simeq 10^{-37}\,(\tau\,\frac{dL_{u\bar{u}}}{d\tau})\ cm^2 \tag{2.35}$$

while the e^+e^- signal estimated from (2.33) is

$$Sg \simeq 0.7 \times 10^{-34}\ (\tau\,\frac{dL_{u\bar{u}}}{d\tau})\ cm^2 \tag{2.36}$$

The signal-to-background ratio is therefore close to 3 orders of magnitude! In connection with this a useful relation is

$$\sigma_W \simeq 0.1\ (M^3\,\frac{d\sigma_{e^+e^-}}{dM}) \qquad \text{with M in GeV.} \tag{2.37}$$

Some serious background problem could a priori originate from heavy quark decays into lepton pairs, since the rate for $c\bar{c}$, $b\bar{b}$,... production will be sizeable. This should not be worrisome for the weak bosons. For the leptonic decay mode the limitation comes only from rate.

We may indeed illustrate the two points made about background problems with two figures. Figure 2-8, to which we shall come back in section 4, gives the yields for two jets observed in an approximately back to back configuration. While such a correlation should favour Z and W formation and decay over other sources, one sees that the hadronic background remains overwhelming since tagging of c, b,... jets cannot probably be done very efficiently.

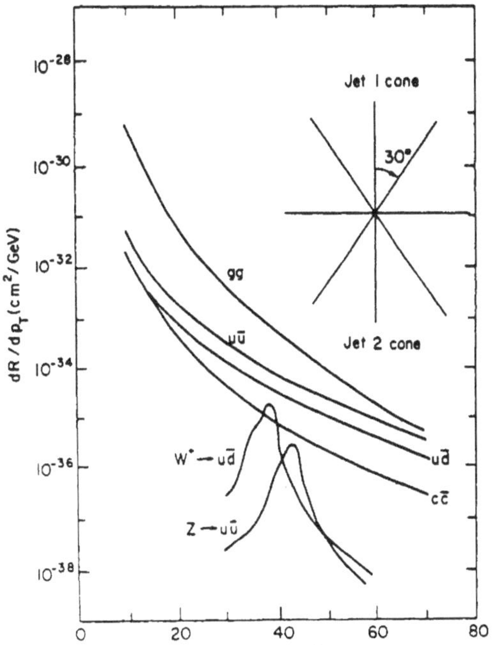

Fig. 2.8. Transverse-momentum distribution of approximately back-to-back jets over the angular interval $60° < \theta < 120°$. Dominant QCD contributions are shown, together with expected signals from $W^{\pm}$ and Z^0 decays.

On the other hand , figure 2-9 shows that using lepton channels there are no important (known) background problem[26]. This is particularly the case for Z° production, where one can look at the lepton pair mass distribution (Fig.2-9a). This is also true for the W when considering the p_T distribution of the μ(e) associated with its leptonic decay. A wide angle detector providing evidence for missing energy corresponding to the production of a large p_T neutrino would of course much enlarge the signal over background ratio, reaching a value comparable to that in figure 2.9.a).

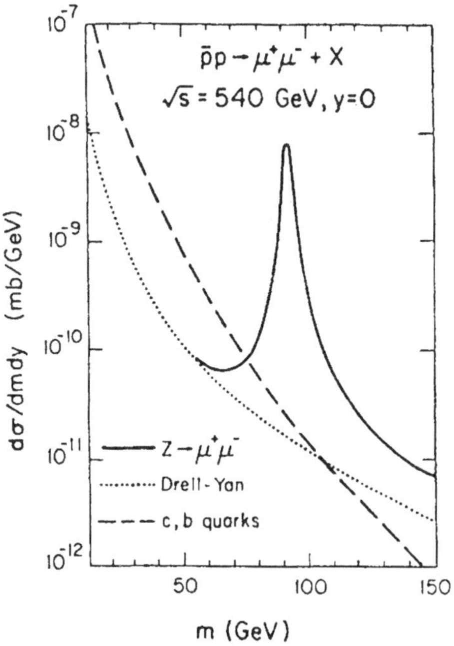

Fig. 2.9.a) The lepton-pair spectrum in p̄p collisions at $\sqrt{s}$ = 540 GeV, y = 0, with different background contributions

One sees in figure 2-9b the Jacobian peak effect associated with production and two body decay of a resonance. The peak in the 90° distribution is at a value of $p_T \simeq M/2$, or occurs as $x_T = 2p_T/M$ approaches 1.

This follows from the fact that the $d\sigma/dx_T$ distribution is equal to the angular distribution in the resonance center of mass system $d\sigma/d\cos\theta$ times a Jacobian term, readily calculated to be ($x_T/\sqrt{1 - x_T^2}$). The singularity at $x_T = 1$ is tempered by the folding of the resonance width and its p_T distribution. A peak remains which gets however eroded away if the resonance transverse momentum spread becomes too large. One hopes that this is not the case for W, Z production, with an expected value for the mean transverse momenta of the order of 4 GeV/c.

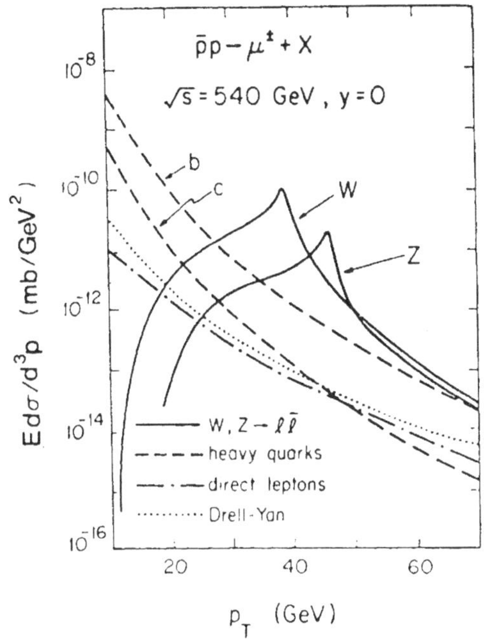

Fig. 2.9.b) Single-lepton invariant cross-section in $\bar{p}p$ collisions at $\sqrt{s}$ = 540 GeV, y = 0, with various background contributions from Drell-Yan processes, almost-real high-p_T photons, and c and b quarks. The separation between two peaks (W,Z) should hardly appear once higher order QCD effects one included.

We now turn to global production rates.

We give in Fig.2-10 the variation of the W^+ production cross-section in $p\bar{p}$ and pp collisions as a function of τ. This is a "historical" document where the calculation is done without scaling violation and the production rate then depends only on τ[19]. Rates do not change much when including scaling violation, provided the W has the expected mass. The production rate, however,

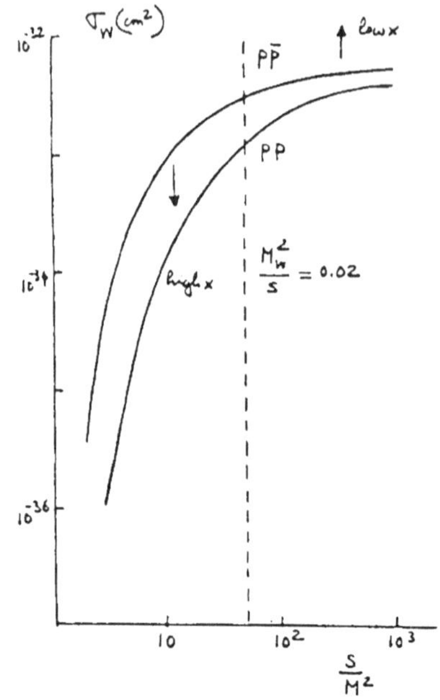

Fig. 2.10. The inclusive production cross-section for W^+ in $p\bar{p}$ and pp collisions. The calculation is done without scaling violations and the cross-section depends only upon the scaling variable τ. Arrows indicate the trends associated with scaling violation

drops faster at larger τ and increases faster at lower τ .

While, as previously said (and seen from fig.2-5), scaling violations do not have a major effect for the relevant range of τ, this would not be the case at significantly lower machine energies (rates would decrease even faster as a result of scaling violations) or at much higher energies (rate would increase slightly faster).

Figure 2-11 gives the x dependence of W^+ production in $p\bar{p}$ and pp collisions. It is then calculated with parametrized scaling violations for $\sqrt{s}$ = 540 GeV and M_W = 80 GeV. Scaling violations affect the x dependence, since they deplete the quark distribution at large x as compared with that at lower Q^2. This is shown to stress the wide angular distribution which is expected, and the importance of covering a large solid angle.

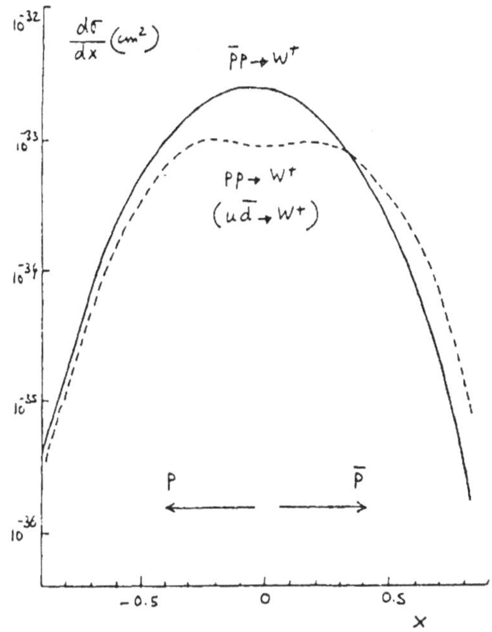

Fig. 2.11. The x dependence of W^+ production in $p\bar{p}$ and pp collisions. The calculation is done with parametrized scaling violations for $\sqrt{s}$ = 540 GeV and M_W = 80 GeV

Parity violation effects lead to specific asymmetries which have already been discussed in great detail in the framework of the parton model[19)]. Everything remains essentially valid at the present level of perturbative QCD with which predictions are made. For instance, the μ^+ from W^+ decay being emitted preferentially along the W^+ spin is therefore thrown backward ($u\bar{d}$ pair moving in the direction of the u quark). The μ^- from W^- decay being emitted preferentially in the direction opposite to the W^- pair is conversely thrown forward ($\bar{u}d$ pair moving in the direction of the d quark).

As a result, the μ^- distribution is wider than the μ^+ in pp collisions[19)]. In $p\bar{p}$ collisions the μ^- distribution peaks towards the proton side, while the μ^+ distribution peaks towards the antiproton side.

Such a charge asymmetry is illustrated by Figure 2-12. It is rather strong for W production and decay should help in detecting the signal. It is very weak for Z° since the value of $\sin^2\theta_W$ is such that one has almost a pure vector coupling the e^+e^- channel.

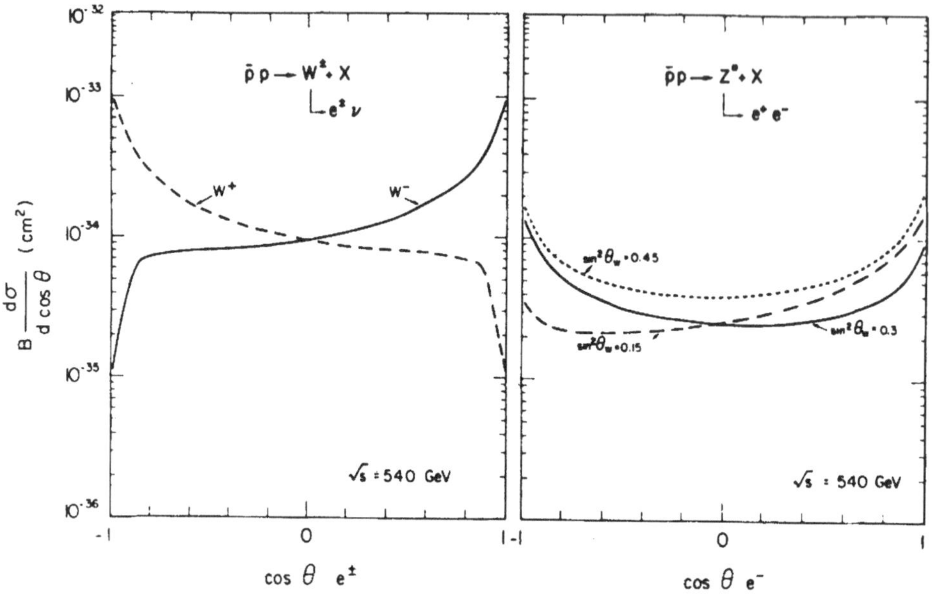

Fig. 2-12. Charge asymmetries for leptonic $W^{\pm}$ and Z° decays from $\bar{p}p$ collisions at $\sqrt{s}$ = 540 GeV

In conclusion, one may say that W and Z production should occur at the level of a few events per day in each leptonic mode for which no background problem should arise, provided that the luminosity reaches the designed value of 10^{30}. Experimental information about high-mass lepton-pair production and theoretical understanding based on perturbative QCD have very much strengthened the predictions of four years ago. They remain essentially valid, with small effects only expected from scaling violations.

Prospects for finding the Z° in its e^+e^- ($\mu^+\mu^-$) decay mode can be considered to be very encouraging. While expected rates are low, taking into account the fact that the designed luminosity will not be reached immediately, the signal is very neat. For the W, rates are higher (by a factor of 5 to 6), but the signal is not as clear. It requires a large p_T lepton at $p_T \simeq M_W/2$ and missing momentum on the away side.

One may hope that by the time these notes appear in print, evidence will have been obtained.

The advantage of a much higher luminosity is clear and as shown by Fig. 2.6, rates for pp and $p\bar{p}$ induced reactions are not expected to differ by a great deal for $\sqrt{\tau} \leqslant 0.2$. This would become very important if there are W and Z mesons but if their mass is larger than that obtained in the standard model. As given by (2.27), the production rate is controlled by the differential luminosity, and an increase by a factor of 10^3 in luminosity while using pp would allow one to reach values of $\sqrt{\tau}$ of the order of 0.4, hence $\sim$300 GeV at $\sqrt{s}$ = 800 GeV. The same applies when increasing much the energy at fixed luminosity. Keeping τ fixed and going from 540 GeV to 2000 GeV also gives masses of the order of 300 GeV, which could be detectable at the same level as the expected W and Z bosons are at the CERN $p\bar{p}$ collider.

The limiting value of $\sim$300 GeV is an interesting one insofar as it represents an energy where weak interactions would become strong if they are not properly damped at high energy, as is the case in the standard model. One has, however, to reckon with the fact that the width of the weak boson increases as M^3 (2.30). Hence a 300 GeV W would be 130 GeV wide and would be even wider if the needed damping mechanism should use a large number of new fermions. There would still be no serious lepton background, but finding a "peak" would be very difficult if luminosity is low.

While our discussion of the production of the weak bosons has been presented in the framework of the standard model, the model is however expected to meet its crucial test : whether or not the W and Z exist with the expected masses. Physics could be different. The safety margin which one has with the collider experiments can be assessed from Figure 2-6. The relatively slow variation of the differential luminosity times τ at $\sqrt{\tau} < 0.3$ say, is such that the Z (W), if unique, could still be discovered up to 150 GeV (optimistically). A situation with several weak bosons, with definitely weaker couplings for the low mass ones, would be more difficult to manage.

While the CERN $p\bar{p}$ collider appears as a suitable machine to discover the Z and W bosons if they are as presently expected, machines with a much higher luminosity (Isabelle), or with a much higher energy (Fermilab Doubler used as a Collider), would be needed if nature, whilst not vicious, is more subtle.

Granting the fact that a very large amount of energy may be available at the quark level, though with relatively low probability, one may hope to also get access to high energy effects involving W, Z - pair-production, and thus test for the presence of specific gauge theory couplings. In the energy range considered, the pair production cross sections are still rising sharply with energy and, going from $\sqrt{s}$ = 540 GeV to $\sqrt{s}$ = 2000 GeV, now represents over an order of magnitude gain in rate ! Rates are however very low and it is hard to see how such a small signal could be extracted from the background. Figure 2-13 gives the variation of the cross sections as a function of energy[27]. At the CERN Collider the pair production cross section is three orders of magnitude below that for a single W or Z ! Prospects of seeing it are extremely low. The same applies to the Higgs meson.

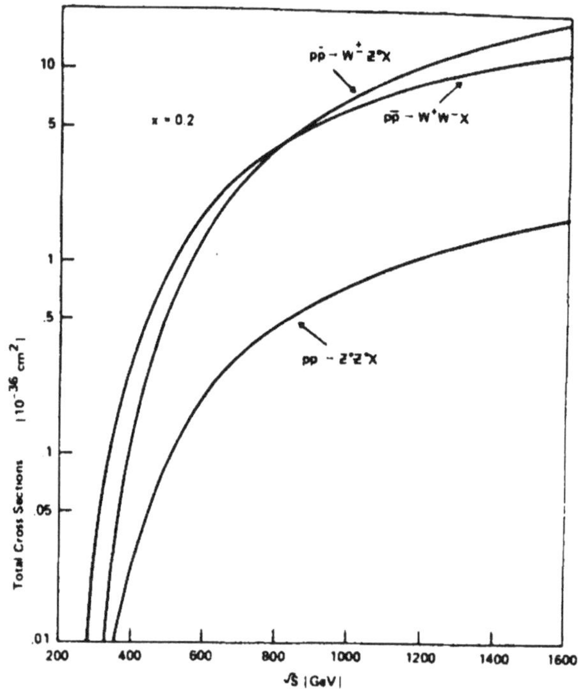

Fig.2-13. Total cross-sections for the production of bosons pairs, W^+W^-, $W^{\pm}Z^0$, and Z^0Z^0, in $\bar{p}p$ collisions ($\sin^2\theta_W = 0.2$)

II.3. Search of Higgs mesons

The Higgs meson is a cardinal ingredient of the standard model[1]. It is the more strongly coupled the higher are the masses of the particles it couples to. Its coupling to a quark-antiquark system is determined to be

$$g_{q\bar{q}H}\, q\bar{q}H \tag{2.38}$$

with

$$g_{q\bar{q}H} = \sqrt{\sqrt{2}\, G}\; m_q \tag{2.39}$$

The coupling constants to the W and Z are given by

$$g_{WWH} = 2\sqrt{\sqrt{2}G}\, M_W^2; \; g_{ZZH} = 2\sqrt{\sqrt{2}G}M_Z^2 \qquad (2.40)$$

There are no constraints on the mass, while one may consider with particular attention a mass of $\sim$ 10 GeV[13], which would originate from radiative corrections alone. The Higgs meson mass may, however, be very large, several times that of the W.

The particular couplings of the Higgs meson (2.38) and (2.39) will lead to rather spectacular decays with several leptons often seen as resulting from cascade decays involving heavy quarks. Nevertheless, they are likely to be lost amongst an overwhelming background in $p\bar{p}$ and pp collisions. It is therefore necessary to look for Higgs mesons as well-defined missing-mass peaks. Two cases thus raise particular interest and have been considered in detail in connection with LEP studies[13]. One is the radiative decay of a heavy enough quarkonium :

$$V \rightarrow H + \gamma \qquad (2.41)$$

The other reaction is the decay of a Z into Z + H, with a Z (seen as a $\mu^+\mu^-$ pair, say) recoiling from a Higgs meson. This is the only one a priori accessible with the collider. One would then for a Z (lepton pair) with a rather large momentum transfer as a signal for the observation of a peculiar cascade on the away side. Since the branching ratio into Z + H decreases rapidly with Higgs mass and the production rate decreases as the intermediate Z becomes more off shell, one is extremely limited by luminosity. It is hopeless with $p\bar{p}$ colliders. With Isabelle it may be marginal for Higgs masses of $m_H \lesssim$ 30 GeV.

This is illustrated by Figure 2-14 which gives the ratio between the reactions rates for pp $\rightarrow$ ZH + ... and pp $\rightarrow$ Z + ... a function of m_H for $\sqrt{s}$ = 800 GeV(this also corresponds to what is expected at $\sqrt{s}$ = 540 GeV in $p\bar{p}$ collisions if the corresponding enhanced production of Z° is taken into account (Fig.2.6)). Even

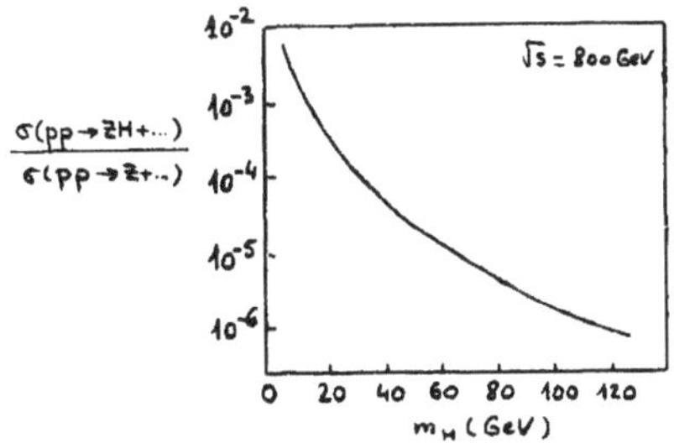

Fig. 2.14. Ratio of the cross-section for pp $\rightarrow$ ZH + ... to pp $\rightarrow$ Z + ... as a function of the Higgs meson mass at $\sqrt{s}$ = 800 GeV

though one may expect 10^4 Z^o/day at Isabelle, the rate for Z + H falls to one a day at $m_H \simeq 30$ GeV and one should keep in mind that only 6% of the decay modes can be used! A search for Higgs mesons, as expected in the standard model, looks therefore to be extremely difficult.

II.4. Search of quarkonia

The production of heavy narrow states and in particular those made out of a new quark and its antiquark, usually referred to as quarkonia, should also be amenable to a Drell-Yan type calculation[28]. The signal looked for is a lepton pair corresponding to one of the important decay modes of the vector meson which has the lowest mass in the family (J/ψ for $c\bar{c}$, Υ for $b\bar{b}$, etc.). Several production channels, have, however, to be considered at the same time. Production can occur from standard quarks in the colliding hadrons. This is expected to be suppressed by the OZI rule, and increasingly so with increasing mass, if running coupling constant effects are important. However, the relevant luminosity is large. Production can also occur through collision among the heavy quarks involved in the structure of the colliding hadrons and which should be present at large enough Q^2. The cross-section is then important, but the relevant luminosity may be very low. One can also consider the collision of two gluons producing a bound system with C = + 1, which eventually decays into the vector meson which is observed through its lepton-pair decay mode. Present information from J/ψ and Υ production shows that such mechanisms are most probably all present[28]. In particular, different rates for Υ production in $\pi^\pm p$ collisions point to the existence of the former mode. The observation of a photon in association with J/ψ production, part of the time, also points to the existence of the latter one.

For the sake of argument, we first consider the former mode and therefore write the production cross-section as (2.24)

$$\sigma = \frac{\pi}{M^2}\,\frac{1}{3}\,\sum_i g_i^2\left(\tau\,\frac{dL_i}{d\tau}\right) \tag{2.42}$$

thus including explicitly the colour averaging factor.

We assume that the decay width is narrower than the mass resolution for a lepton pair at this mass, denoted by ΔM, and thus get a contribution to the lepton-pair differential cross-section averaged over ΔM at the resonance

$$\frac{d\sigma^R}{dM} = \frac{\pi B}{M^2}\,\frac{1}{\Delta M}\,\frac{1}{3}\,\sum_i g_i^2\left(\tau\frac{dL_i}{d\tau}\right) \tag{2.43}$$

where B is the branching ratio for lepton-pair production.

The signal will be visible if this quantity is reasonably large as compared with the Drell-Yan background, given by (2.12) The signal-to-background ratio is of the order of

$$\frac{Sg}{Br} \sim \left(\frac{M}{\Delta M}\right)\left(\frac{Bg^2}{\alpha^2}\right) \qquad (2.44)$$

The differential luminosity essentially cancels out, since we are dealing with similar (light) quarks. We have separated out two factors. The second one is of the order of 1, at least if the quark-jet decay mode corresponds to a sizeable fraction of the dominant hadronic decay. The decay rate into a light quark-jet system and that into a lepton-pair are in a ratio of the order of g^2/α^2.

The first factor is the resolution power of the detector. It may hopefully be large and it actually determines the signal-to-background ratio in this approach.

The same type of reasoning applies in the gluon-gluon production case. The production yield and the hadronic decay rate are both controlled by similar QCD coupling, whilst lepton-pair decay and direct formation are scaled according to α^2. The argument cannot, however, be made as quantitative as it could in a model calculation carried out in detail for the former case. We remark that production through heavy quarks is hard to assess. If dominant it would, however, give a larger signal-to-background ratio than the one obtained in the other two cases.

We shall therefore tentatively guess quarkonia production rates according to the scaling behaviour of the direct lepton-pair production cross-section(Fig.2-7) , but assuming that one benefits from a signal-to-background ratio given by $M/\Delta M$. While mass resolutions are presently at the 20% level, one could consider for future experiments at larger masses values of the order of 2%, hence a factor of 50 for the first term in (2.44). Possible signals can be estimated from Fig.2-2. The mass resolution is, however, still only of the order of 20% in the ISR experiment and an improvement by an order of magnitude could be hoped for. Taking one event/day/GeV as a limiting value to reach any evidence for a narrow vector state, we can take 40 GeV as a reasonable mass limit with a luminosity of 10^{30}, for charge 2/3 (with quarkonia with quark charge - 1/3 the expected signal is lower by a factor of 4). Whilst the search for narrower vector states is thus limited by over-all production rates rather than by background (figure 2-9a), we have to conclude that the $p\bar{p}$ collider will not probe beyond the range already well covered by PETRA. Increasing the energy of the machine would help, but not very much in view of the limited dependence on τ. One could thus extend the probed range from 40 GeV (at $\sqrt{s}$ = 540 GeV) to 50 GeV (at $\sqrt{s}$ = 2000 GeV). Increasing the luminosity by 3 orders of magnitude would on the other hand correspord to a large potential gain and narrow states beyond 100 GeV could then perhaps be discovered. In this case the range covered by LEP phase I would be efficiently covered in one stroke. Near the Z peak interference effects would also strongly boost production rates. This has been discussed in connection with LEP studies [13)].

Our estimate may be too pessimistic, since we focus on only one particular production mode. However, it may be deemed optimistic for other reasons. We have indeed assumed proportionality between the electromagnetic width and the hadronic width and we have parametrized the former in a scaling form, controlled by α^2 alone (one-photon intermediate state), whatever the mass of the quarkonium. It should be acknowledged that the dimensionless phenomenological coupling, introduced in (2.42), may actually decrease with increasing mass. This would imply a smaller signal-to-background ratio at large mass than the one which we have estimated. The argument is the following. For a constant coupling one expects the partial widths (and in particular $\Gamma_{e^+e^-}$) to increase linearly with mass (scaling). In the potential model[29)], which has had many successes with the spectroscopy of the J/ψ and Υ families, the lepton partial width is, however, not only controlled by α^2, but also by the value of the wave function at the origin. It reads

$$\Gamma_{e^+e^-} = 16\pi\alpha^2 e_u^2 \frac{|\Psi(0)|^2}{M^2} \tag{2.42}$$

This is the Van Royen-Weisskopf formula. The partial width increases linearly with M, as it would with a constant phenomenological coupling, only provided that $|\Psi(0)| \sim M^{3/2}$. This should be the case for a Coulomb wave function, implied by the one-gluon exchange part of the potential at small distances. However, it is known that for the J/ψ and even the Υ, the wave functions are still far from being of a Coulomb type as should be the case for very heavy quark mass. The value at the origin is largely determined by the confining part of the potential. With the potential usually considered[29)], one actually finds an interesting regime setting in at the J/ψ and Υ levels, with

$$\Gamma_{e^+e^-} \sim e_i^2 \tag{2.43}$$

in good agreement with experiments, the relation holding empirically for all vector mesons. This is displayed in Fig.2-15.

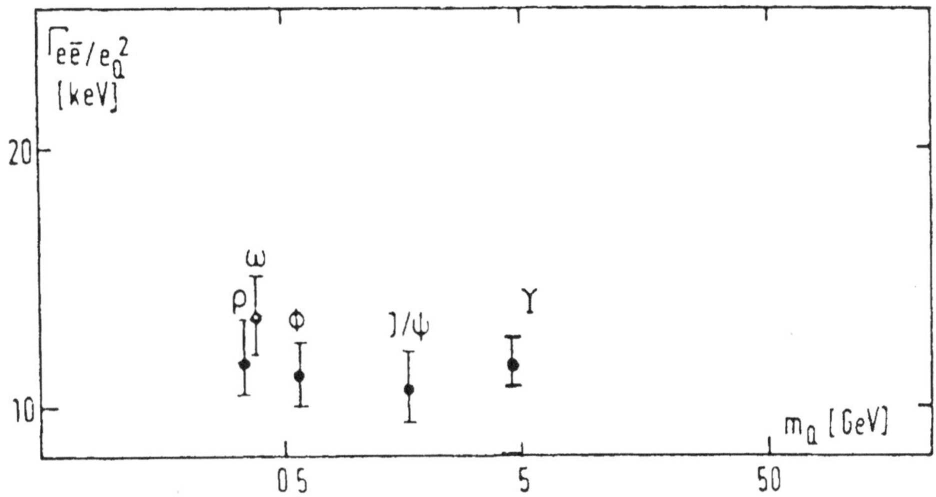

Fig.2-15. Reduced leptonic decay widths.

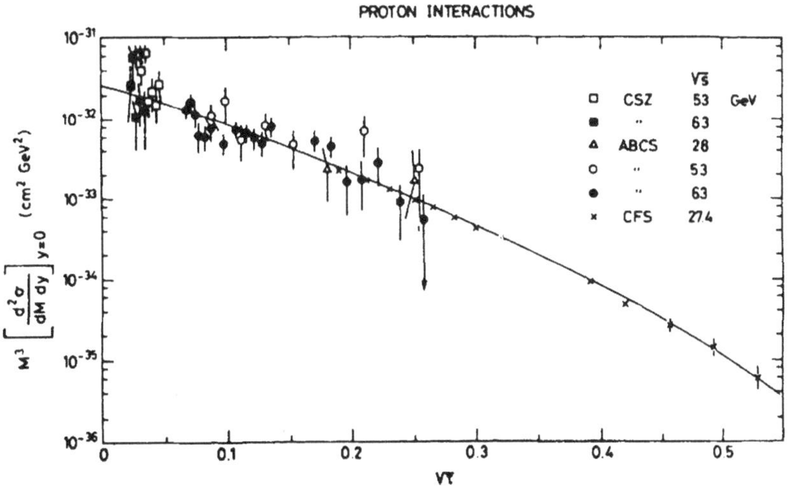

Fig.2-16. Scaling function for vector resonances as a function of $\sqrt{\tau}$. The curve represents the Drell-Yan function $F_{DY}(\tau)$ rescaled by 4×10^6.

If such a behaviour continues as M increases, the production rate will keep falling faster than M^{-2}, to the extent that the value of $\Psi(0)$ controls production as well as decays. The signal-to-background ratio will no longer increase with mass as in (2.44). With increasing quark mass the wave function is in any case going to be more and more "sucked" into the Coulomb region and the partial width should eventually increase accordingly. Nevertheless, this may be only a very belated effect if a running coupling constant with sizeable variations is included in the formulation of the potential. In this case one gets a $(\ln r)^{-1}$ behaviour at small r as opposed to the r^{-1} singularity of the Coulomb potential.

With a decrease of the (running)coupling constant with quark mass, weak decays may also eventually claim a sizeable fraction of the total width,thus decreasing the branching ratio entering into the calculation of the signal-to-background ratio.

There are, of course, some compensating effects associated with the fact that the gluon-gluon process can contribute through a number of bound states with increases with quark mass (as $M^{1/2}$ in the WKB approximation [30]), with already 6 s-states below threshold at $M \simeq 40$ GeV, each of them associated with several states with even C parity.

In view of these uncertainties, one may follow a different approach and remark that a simple scaling law has been verified to hold experimentally. It reads

$$\frac{M^3}{\Gamma_h} \left.\frac{d\sigma}{dy}\right|_{y=0} = F(\tau) \qquad (2.45)$$

and its present success with ϕ, J/ψ, ψ' and Υ can be assessed with figure 2-16. In (2.45) Γ_h is the hadronic width.

Observation through a leptonic decay mode corresponds to a cross section

$$B_\ell \left.\frac{d\sigma}{dy}\right|_{y=0} = \frac{\Gamma_\ell}{M^3} F(\tau) \qquad (2.46)$$

with a value of Γ_ℓ guessed from Figure 2.15 assuming constancy and a value of $F(\tau)$ read from figure 2-16. This approach is perhaps more secure than the previous one insofar as it extrapolates known trends. It is also a little less optimistic as it uses a constant width. Conclusions are similar, namely that observing a quarkonia with mass much larger than 40 GeV is out of question. What matters is the rate, not the background. The Drell-Yan background is indeed close to two orders of magnitude smaller than the signal. The background provided by heavy quark decay is more serious (figure 2-9a) but it is expected to be over a factor two below the signal. In this case considering associated hadrons, subtracting the symmetric charge rate, imposing transverse momentum balance among leptons should very much enhance the signal over background ratio.

While quarkonia could certainly be detected up to 40 geV, we already know that this is too low a value for the toponium. New quarkonia are certainly worth search (high mass resolution being required for Z° search already) in case the toponium is just above the present PETRA range ! Nevertheless one cannot be very optimistic since rates are low.

Studies of known quarkonia at collider energies should benefit from reasonable rates and this should provide very valuable information about production mechanisms. This is already interesting. One should, however, not be very optimistic at using the higher nominal energy to discover states which would not have been already seen either at PETRA (for the CERN $p\bar{p}$ Collider) or with LEP phase 1 (for Isabelle).

Concluding this lepton-production section, one may say that prospects for finding the Z and W, as they appear in the standard model, are quite good at the $p\bar{p}$ collider. For anything else one meets a very strong luminosity limitation.

III. THE DOMINANT HADRON PROCESSES

III.1. Overview

Experimentation at the ISR has provided evidence for the setting in of a régime which was anticipated in the framework of Regge theory, namely the scaling property of production yields and the emergence of a rapidity plateau. These questions have been covered in several review articles[31]. Experimentation at the ISR has also given evidence for deviations from the simplest implementation of the Regge approach, based on the eventual dominance of a Pomeron trajectory with intercept one. The most dramatic feature is, of course, the rise of the total cross-section, which is compatible with a $\ln^2 s$ behaviour, when a Pomeron of intercept one would have led to a constant asymptotic value[32]. Less dramatic, but probably of equal significance, is the slow rise of the inclusive yield for slow secondary particles[33] referred to as the "plateau" density rise, a rapidity plateau appearing only as a first approximation. At present, the main question is whether such departures from a too-simplified (but yet powerful) Regge parametrization are to be interpreted in a more refined version of the Regge approach, such as the one offered by Reggeon field theory[34] or signal the emergence of some new mechanisms in hadronic interactions with no hitherto recognized origin in the analysis of data at lower energies.

While observed rates of change are slow (scaled to a good approximation according to ln s, as opposed to the centre-of-mass energy), the collider offers such a large gain in energy that dramatic changes should result. On the other hand, information from cosmic-ray results hint at more sizeable changes than those expected from the extrapolation of recognized trends and, by incident energy of E $\sim$ 100 TeV, qualitatively new features seem to appear, separate from the observation of very peculiar events such as Centauros[35]. Experimentation at $\sqrt{s}$ = 540 GeV thus offers extremely interesting prospects and it does so even at relatively low luminosity. A first look at event structures is eagerly waited for. This section is organized as follows. We first briefly review key features of hadron interactions at ISR(SPS) energies and discuss their extrapolation. We then turn to a review of cosmic-ray results and, in particular, those covered with emulsion chamber techniques used for energies similar to those covered by the collider.

Even at low luminosity, detectors such as those of experiments UA1 and UA5, first to be at the data taking stage, can provide valuable information.

III.2. Hadron interactions at ISR energies and their extrapolation

The total cross-section should rise by a significant amount. Figure 3-1 shows the pp and $p\bar{p}$ cross sections as presently known, and their common extrapolation according to a $\ln^2 s$ behaviour. At $\sqrt{s}$ = 540 GeV, the total cross-section should have reached a value of 63 mb (a 50% increase over its value at SPS energy). The pp total cross section by itself does not allow one to extrapolate with much confidence. However the combined analysis of σ_{tot} and ρ , the ratio between the real part and the imaginary part of the forward elastic amplitude, does allow a more confident extrapolation to be made. Figure 3-2 gives the energy dependence of parameter ρ . Dispersion relations provide a relation between ρ and σ_{tot}, which is fairly unique if one assumes that the ratio between the $p\bar{p}$ and pp cross-sections converges uniformly to unity[36], with increasing energy. Insofar as ρ probes the behaviour of σ_{tot} at much higher energies, a combined analysis provides bounds on σ and ρ, which give information on the high-energy behaviour of both, but which of course become increasingly vague as the energy becomes very high. This is shown in Figs.3-1 and 3-2, where the shaded area corresponds to the uncertainty attached to a variation by one standard deviation of the data point.

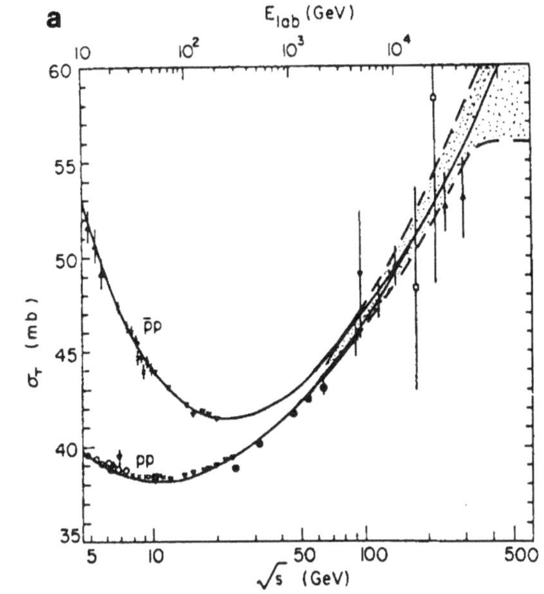

Fig. 3.1. The total cross-section for pp and $p\bar{p}$ collisions and their extrapolation up to energies where large uncertainties appear. Results from the CERN-Rome and Pisa-Stony Brook Collaborations at the ISR.

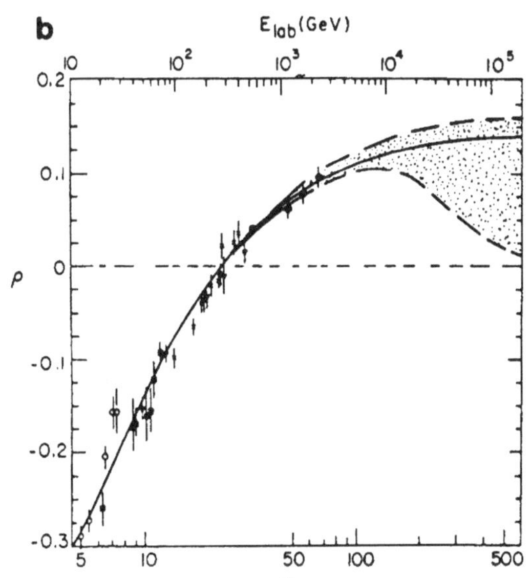

Fig. 3.2. Energy dependence of the parameter ρ for pp collision. The ratio between the real and the imaginary part of the forward scattering amplitude. Uncertainties in the common dispersion relation analysis of ρ and σ_{tot} are shown as shaded areas in this figure and in the preceding one.

The positive value of ρ over the ISR energy range was able to be predicted from the rise of the total cross-section in 1973. At this time this was particulary interesting since ρ was measured to be negative at PS and Serpukhov energies. We may dwell on this point. As is well known, the elastic scattering amplitude is an analytic function of s with branch lines on the positive and negative real axės (unitarity in the direct pp and crossed $p\bar{p}$ channels). This is shown in Fig.3-3. The physical value, obtained on the real axis approached from above (A), is asymptotically related through crossing to the value at -s, on the negative real axis approached from below (A'). This follows from the fact that the scattering amplitude is asymptotically even (a ratio of one between the pp and $p\bar{p}$ cross-sections being imposed by the Pomeranchuk theorem). Since the scattering amplitude is a real analytic function, its discontinuity across the real axis is equal to twice its imaginary part. The value on the negative real axis approached from above (A'') is therefore the complex conjugate of the value at A, whilst continuation from A to A'' in the upper half plane is readily possible (Fig.3-3). It follows that a model amplitude involving a logarithmic dependence (which through the optical theorem will readily appear as required experimentally in the total cross-section) can parametrize this structure only through the term $(\ln s - (i\pi/2))$, which only transforms into its complex conjugate when rotating counterclockwise in the upper half plane, whereas ln s alone does not. We may then write the leading term in the asymptotic forward amplitude as

$$A\,(s,\,0) = is\,(\,\ln s - \frac{i\pi}{2})^{\alpha} \tag{3.1}$$

with $\alpha \lesssim 2$, in order to meet both empirical behaviour and required analytic properties.

It therefore follows that

$$\rho = \frac{\mathrm{Re}\;A(s,\,0)}{\mathrm{Im}\;A(s,\,0)} \underset{s\to\infty}{\sim} \frac{\pi}{\ln s} > 0 \tag{3.2}$$

More precisely, one finds that for any even amplitude

$$\rho(s) = \frac{1}{s\sigma(s)}\,(\mathrm{tg}\,\frac{\pi}{2}\,\frac{\partial}{\partial \ln s})\;s\sigma(s) \tag{3.3}$$

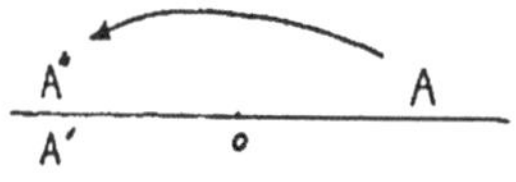

Fig.3-3. The scattering amplitude as defined in the complex s plane, when the right-hand cut is approached from above. Points A, A', and A" are as defined in the text.

With an asymptotically rising cross-section $\rho(s)$ rises and decreases logarithmically after going through a very wide maximum, while remaining positive.

In practice, one has to replace (3.3) by a relation involving also the odd part of the amplitude, which appears to rise only as $s^{1/2}$, as opposed to $s(\ln s)^{\alpha}$ for the even part. It may also be necessary to replace power expansion in the derivative (3.3) by a dispersion integral.

From a combined analysis of ρ and σ one can conclude that the total cross-section should continue to rise, at least up to $\sqrt{s} \cong 300$ GeV, and most probably beyond. It would be a surprise if total cross-section measurements at the collider would not give a value of the order of 60 mb. As indicated in Fig.3-4 this is compatible with information from cosmic-ray results (Maryland and Tien-Shan). Measurements of σ_{tot} at $\sqrt{s} = 540$ GeV (SPS) and measurements of ρ for $p\bar{p}$ collisions at the ISR will allow one to make an important consistency check. Measurement of ρ at collider energy could allow for a sizeable contribution to evidence for an asymptotic logarithmic behaviour with σ and ρ closely related.

While a $\ln^2 s$ extrapolation, fitted to present data , gives an expected cross section at the level of 63 mb, it should be stressed that Reggeon field theory[34] leads to expect different asymptotic behaviours. This is particularly the case for a "critical Pomeron", when the renormalized Pomeron intercept is at 1[34]. In such a case, the asymptotic logarithmic behaviour is controlled by the critical exponents in the theory[35]. An extrapolation from present data then gives an expected value of σ_{tot} between 55 and 60 mb, thus lower than the merely empirical $\ln^2 s$ extrapolation (figure 3-1). It is clear that the Froissart bound is very far from being saturated, the coefficient of the $\ln^2 s$ term being much smaller than the value obtained from the unitary limit.

This brings us to the study of the proton shape.

A striking feature found at ISR energy is that the proton is still fairly transparent. The Fourier analysis of the forward peak, done assuming a purely imaginary amplitude (a good approximation) gives the shape shown in the insert of Fig. 3.5. The unitarity limit is no quite reached at impact parameter 0. The absorption falls rather rapidly with distance. One is very far from a black disk limit. Indeed the second maximum which shows departure from a simple Gaussian shape is 6 orders of magnitude lower than the first one. As the energy increases the proton size increases (the radius increases logarithmically) but the

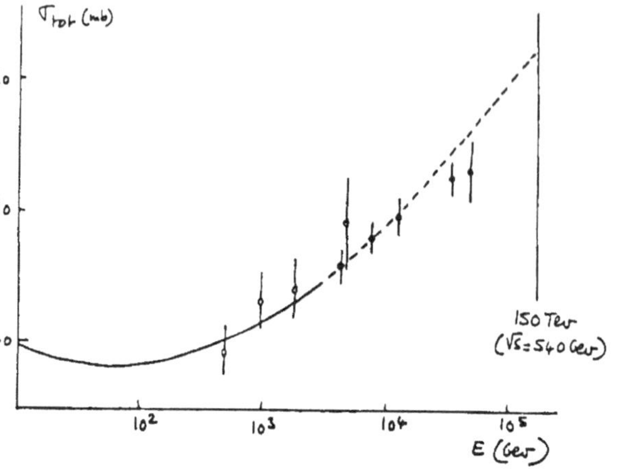

Fig.3-4. Extrapolated pp total cross-section (assumed $\ln^2 s$ behaviour) and cosmic ray points Maryland (open dots) and Tien-Shan (full dots) as obtained by C.B. Yodh from Glauber calculations.

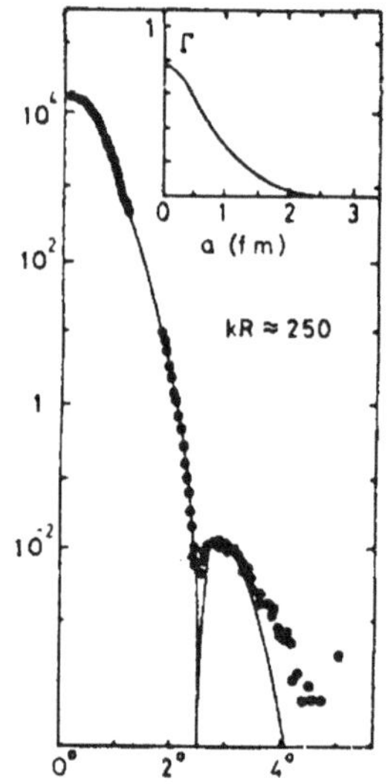

Fig. 3.5. The proton shape as inferred from the diffraction peak. While the radius appears to increase (logarithmically) with energy the shape does not change appreciably over the ISR energy range.

proton shape remains the same. This is referred to as geometrical scaling[36]. As a result σ_{tot} remains proportional to σ_{el}, while being much larger (5 to 6 times) than twice the elastic cross section, as expected for a black disk. Extrapolating the present trend one expects the shrinking of the forward peak to continue, with as a very clear consequence, a displacement of the dip (figure 3-6) towards lower $|t|$ values.

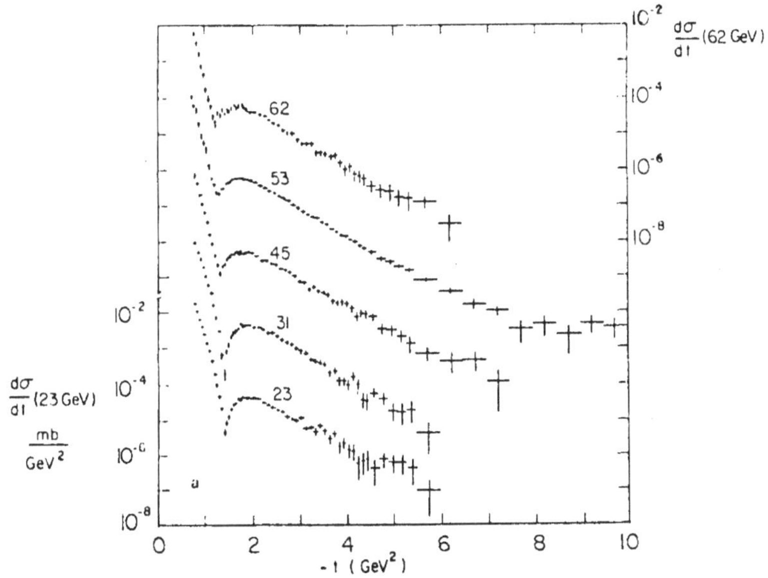

Fig.3-6. Differential cross-section for pp elastic scattering at ISR energies. Data from the CERN-Hamburg-Orsay-Vienna Collaboration at the ISR.

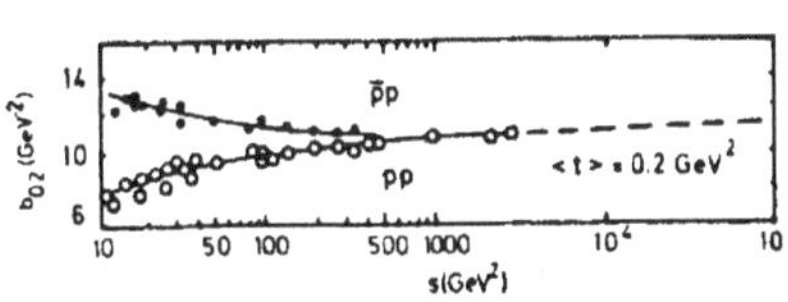

Fig.3-7. The energy dependence of the slope of the pp elastic forward peak. The pp and $p\bar{p}$ slopes are expected to approach the same behaviour.

With the parametrization provided by Reggeon field theory[35] one expects the dip to move from $|t| \simeq 1.35$ at $\sqrt{s} = 50$ GeV down to $|t| \simeq 0.7$ to 0.8 at $\sqrt{s} = 500$ GeV.

We discuss $p\bar{p}$ scattering as if it were the mere extrapolation of pp scattering. One expects it to be in a simple way though surprises could occur.

Over the SPS energy range there is indeed convergence between the pp and $p\bar{p}$ elastic cross-sections. This is true at low $|t|$ (slope parameter)(Figure3-7). This is also true at larger $|t|$ as beautifully displayed in Fig.3-8 which puts side by side pp elastic data at the ISR[37] with, as a prominent feature, a dip

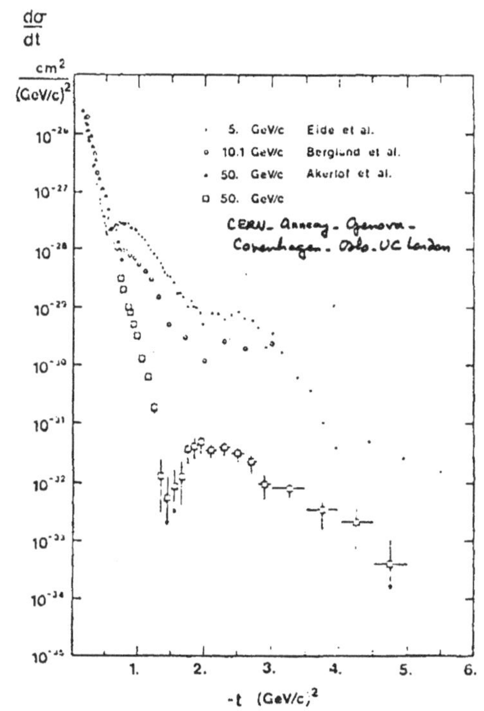

Fig.3-8. Differential cross-section for $p\bar{p}$ elastic scattering at 50 GeV incident energy.

at $|t| \sim 1.6(\text{GeV/c})^2$ which moves in as the forward peak shrinks, and SPS $p\bar{p}$ elastic data which, though at much lower energy, show the same dip appearing[38]. The difference between the pp and $p\bar{p}$ energy dependence can be understood in terms of duality arguments with a real (pp) and an imaginary ($p\bar{p}$) non-diffractive part in the amplitude. At collider energy the diffractive part should be overwhelming in this t range and one expects the pp and the $p\bar{p}$ differential cross-sections to become identical.
The energy jump offered by the $p\bar{p}$ Collider is such that results are eagerly waited for. The total and differential cross-sections could show the expected logarithmic change, with Reggeon field theory suggests. The ISR range may however have provided a misleading view when looking for an asymptotic regime.

It would be interesting to have data on inelastic diffractive scattering as well as on elastic scattering. The signature is of course not as easy to define experimentally. Studying diffractive excitation may require a modification of the magnet-vacuum chamber complex downstream, in order to use a section of the SPS as a forward spectrometer. With a gain in rapidity form 8 to 13, the study of Double Pomeron exchange which is still marginal at ISR energy[9]), could be done in a very convincing way. Evidence could even be obtained for triple Pomeron exchange. These are important issues.

We now turn to particle production. Two key features of hadronic production are displayed in Fig.3-9. It shows the inclusive rapidity distribution for different secondaries as measured at fixed transverse momentum (p_T = 0.4 GeV/c) at different ISR energies. Yields are measured as a function of $y - y_{max}$, where y_{max} is the rapidity of the incident proton. Rapidity is defined as

$$y = \frac{1}{2} \ln \frac{E + p_L}{E - p_L} \tag{3.4}$$

where E and p_L are the energy and the longitudinal momentum, respectively. As follows from the definition (3.4), a Lorentz boost along the incident direction (going, say, from the centre-of-mass system to the rest frame of the target or to that of the

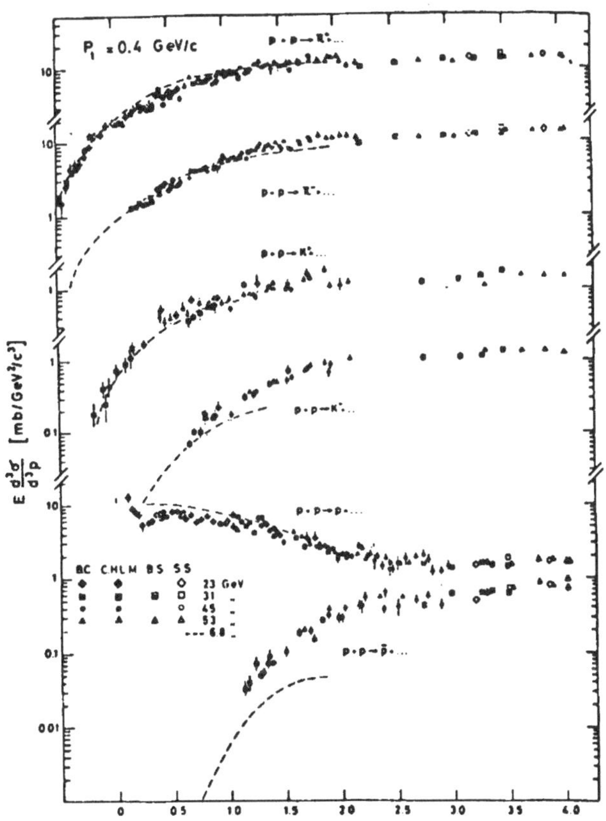

Fig. 3.9. Inclusive rapidity distributions for different particles at p_T = 0.4 GeV/c. Data points cover the full ISR energy range.

incident particle) corresponds to a mere translation in rapidity. Rapidity intervals, related to invariant masses, are not affected The quantity $y - y_{max}$ is the rapidity of a secondary particle as measured in the rest frame of the incident proton. A fixed value of $y - y_{max}$ corresponds (to a good approximation) to a fixed ratio between the longitudinal momentum of the secondary particle and that of the incident proton (the Feynman scaling variable).

Presenting the inclusive distribution as a function of $y - y_{max}$ corresponds to a description in the rest frame of the incident of proton. One sees from Fig.3-9 that the distributions of secondary particles, as seen from the colliding proton, do not change with energy. The proton fragments in the same way, irrespective of the energy with which it has been hit, provided that it is large enough. This is the scaling property. With increasing energy, particles are also produced at large energy in the proton rest frame (they are slow in the centre-of-mass system), and this with a constant density per species. When analysing the same results in the centre-of-mass system, one gets a symmetric picture with an extending central plateau. This is the second important property. The plateau density dN/dy is $\sim$ 2.2 for charged particles at ISR energies.

All these features are well known. They are striking properties which were actually expected in the framework of the Regge pole approach. Nevertheless, with rising cross sections (and a complicated Pomeron) one is lead to expect some (logarithmic) deviations.

In the Mueller-Regge theory[39], the scaling property and the existence of a rapidity plateau follow in a particularly simple way when one assumes that Pomeron exchange (a Regge pole with interpt one) controls the asymptotic behaviour. This is graphically represented in Fig.3-10. The total cross-section for AB collisions is related through the optical theorem to the imaginary part of the elastic AB amplitude, or rather to its discontinuity Δ across the real axis (Fig.3-3). The energy dependence of this discontinuity is asymptotically controlled by the Pomeron (a linear dependence on s of Δ giving a constant value for σ_{tot} as expected for a Regge trajectory with intercept one). One writes

$$\text{Im } A(s, 0) = \Delta(s)/2 = s\sigma_{tot}(s) \tag{3.5}$$

with $$\Delta(s) = \beta s^{\alpha(0)} \tag{3.6}$$

for a Regge pole with intercept $\alpha(0)$.

This is represented by the graph (a) in Fig.3-10. The inclusive cross-section for production of particle C in de AB collisions is similarly related to a specific discontinuity of the elastic $AB\bar{C}$ forward scattering amplitude. If one subenergy (A with respect to BC, say) is large while the BC energy is small, the discontinuity behaves according to Pomeron exchange and the inclusive yield for C, seen as a fragment of B, remains constant (scaling property). If both subenergies $A\bar{C}$ and $B\bar{C}$ are large (C is then distant in rapidity from both A and B), the relevant discontinuity is the product of two terms controlled by the Pomeron. The inclusive yield is constant in energy and rapidity (rapidity plateau).

For the benefit of the reader not familiar with the Regge formalism, we can explicitly display the dependence of the inclusive production cross-section upon large rapidity gaps in the three-body elastic amplitude, the introduction of which generalizes the optical theorem.

One fixes p_T and considers the production cross-section as a function of rapidity y

$$E \frac{d\sigma}{d^3 p} = \frac{1}{\pi} \frac{d\sigma}{dy dp_T^2} \qquad (3.7)$$

When one subenergy ($S_{A\bar{C}}$ say) is large, while the other ($S_{B\bar{C}}$) remains limited, as well as p_T there is only one large rapidity gap justifying a Regge approximation for the three-body elastic amplitude and hence the particular discontinuity associated with the inclusive production cross section (Fig. 3.9b). The dependence on $S_{B\bar{C}}$ and on p_T cannot be predicted. One finds

$$\frac{d\sigma}{dy dp_T^2} = \frac{1}{s} F(p_T, S_{B\bar{C}}) \, S_{A\bar{C}}^{\alpha(0)} \qquad (3.8)$$

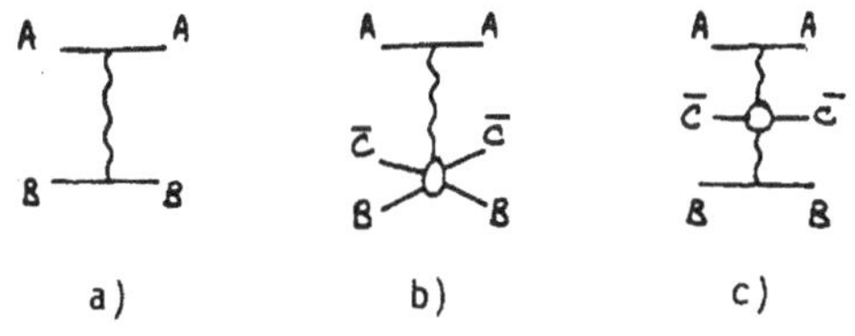

Fig.3-10. a) The Mueller-Regge approach to total cross-section.
b) Inclusive production in the fragmentation region.
c) Inclusive production in the central region. The wiggly line corresponds to a Pomeron exchange approximation made whenever there is a large rapidity gap for forward elastic two-body AB and three $AB\bar{C}$ amplitudes.

where $\alpha(0)$ is the intercept of the leading trajectory.

If one first remains in the framework of standard Reggeism, the Pomeron with $\alpha(0) = 1$ should asymptotically dominate. Fixing the values of $S_{B\bar{C}}$ and p_T implies that $S_{A\bar{C}} \sim s$ and all s dependence disappears from (3.8). An easy calculation shows that $S_{B\bar{C}}$ is related to the Feynman scaling variable (C is fragment of B) or to the rapidity of C measured in the rest frame of B ($y - y_{max}$ as previously considered). A Pomeron with intercept one thus implies scaling as an asymptotic property.

When both subenergies $B\bar{C}$ and $A\bar{C}$ are large (particle C is distant in rapidity from both particle A and particle B), a Regge behaviour can be introduced for the dependence on both $S_{A\bar{C}}$ and $S_{B\bar{C}}$. In practice, one finds that Pomeron dominance applies with 3 units or rapidity gap. The inclusive distribution reads

$$\frac{d\sigma}{dydp_T^2} = \frac{1}{s} G(p_T) S_{A\bar{C}}^{\alpha(0)} S_{B\bar{C}}^{\alpha(0)} \tag{3.9}$$

when an easy calculation shows that, asymptotically

$$S_{A\bar{C}} S_{B\bar{C}} = sh(p_T) \tag{3.10}$$

A Pomeron with intercept one now imposes an energy-independent (as before) but also constant, rapidity distribution (the rapidity plateau).

The Mueller-Regge approach actually translates into simple asymptotic expressions ideas developed in the framework of the multiperipheral model[40] . It predicts both key features of particle production (the scaling property and the rapidity plateau) relating then to a constant asymptotic behaviour for the cross-section. While such a relation is conceptually very interesting, we should remember that we are only dealing with approximations.

Although the validity of this description is a beautiful success of the Regge approach, the discovery of rising cross-sections implied that scaling and fixed plateau densities could also be but approximate properties.

Over the ISR energy range there is no appreciable violation of the scaling property for fast pion fragments, for which it should best apply. There is however a noticeable rise of the central plateau (albeit by 40% only over the full ISR range). This is shown in Figure 3-11[33].

Extrapolating from present data, one expects the particle

density to increase by a sizeable amount when reaching collider energies. The important question is whether such a rise can be related to a more sophisticated Regge behaviour, incorporating a rising total cross-section[41], or if it signals the emergence of new features outside the coherent aspects of hadron interactions covered by Regge theory.

As a matter of fact, the rise of the central density may be thought to be in agreement with hydrodynamical models applied to central collisions[42]. The central density then increases as $s^{1/4}/\ln s$.

We see immediately that the central particle density at Collider energy is an important issue. One expects a rise. With a simple logarithmic extrapolation it would typically double between ISR and Collider energies.

An important and relatively new feature with respect to the central density is that the same density value is found in e^+e^- annihilation, where an extending rapidity plateau also appears[43]. This is shown in Figure 3-12 where the distribution is folded, with a resulting value thus twice as large as the value to be compared with the one found in hadron collisions (figure 3-9).

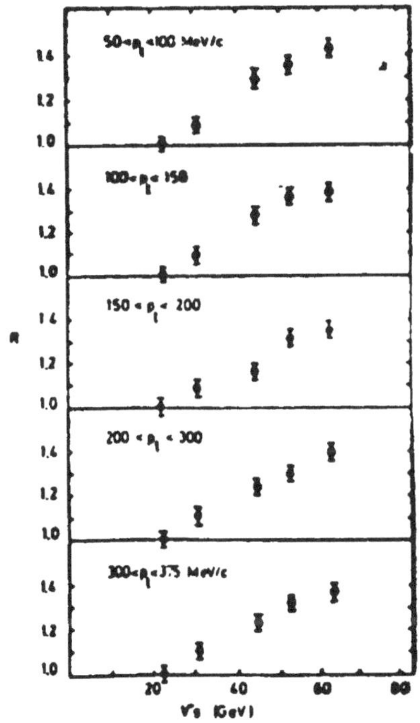

Fig. 3.11. Energy behaviour of the central density for charged particles. Results from the British-Scandinavian-MIT Collaboration at the ISR.

When one takes away the energy carried by the leading baryons, the remaining center of mass energy, effective for particle production, is on the average only half the nominal center of mass energy. It is thus indeed natural to compare ISR results ($\sqrt{s} \sim 60$ Gev) to PETRA results ($\sqrt{s} \sim 30$ GeV). The central densities could however have been very different. In the framework of QCD, one associates central hadronization with a $3-\bar{3}$ flux tube appearing in both cases, either as a result of the violent encounter of two protons, or as the result of a $q\bar{q}$ pair formation.

In e^+e^- annihilation perturbative QCD predicts a rise of the central plateau resulting from jet evolution before hadronization[44]. In hadron collisions the production of jets from relatively low x partons could also lead to an increase in central density which would also result from perturbative QCD, though in a different way. In both cases the increase in central density should be associated with an increase of $\langle p_T \rangle$. Is it the case ?

It would be surprising if there was not a substancial rise of the central density. Whether it is (perturbative QCD) or not (simple Regge extrapolation) associated with an increase of $\langle p_T \rangle$ is an important issue.

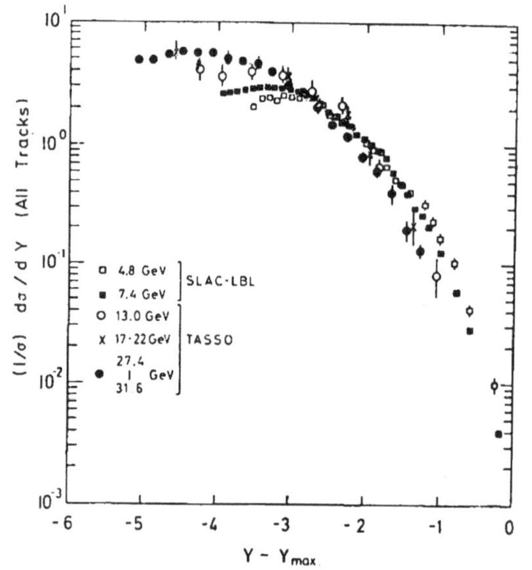

Fig. 3.12. Inclusive rapidity distribution for charged particles produced in e^+e^- annihilation. The distribution is folded. The actual plateau density is therefore half the one read on the Fig. When one takes into account the fact that two leading protons take on the average half of the available energy, PETRA and ISR (effective) energies are comparable.

One should also insist on the fact that a test of scaling in the fragmentation region is also very important. This could be easily done at the $p\bar{p}$ collider comparing the response of a forward calorimeter to beam-beam and beam-gas events. If scaling applies the two angular distributions should be in a fixed ratio corresponding to the relative value of the total cross sections.

The test would be similar (though it should be done in a more precise way) to a recent experiment at the ISR where the fragmentation of a 31 GeV proton into fast secondaries was studied in pp collisions (against a proton of 31 GeV) and $p\alpha$ collisions (against an α particle with 15 Gev/nucleon). The proton is here found to fragment irrespectively of the way it has been hit[45)]. This is shown in figure 3-13.

If scaling holds in the fragmentation region, the mean multiplicity should follow the behaviour of the plateau density.

With a constant plateau density, the mean multiplicity would increase according to the rapidity interval

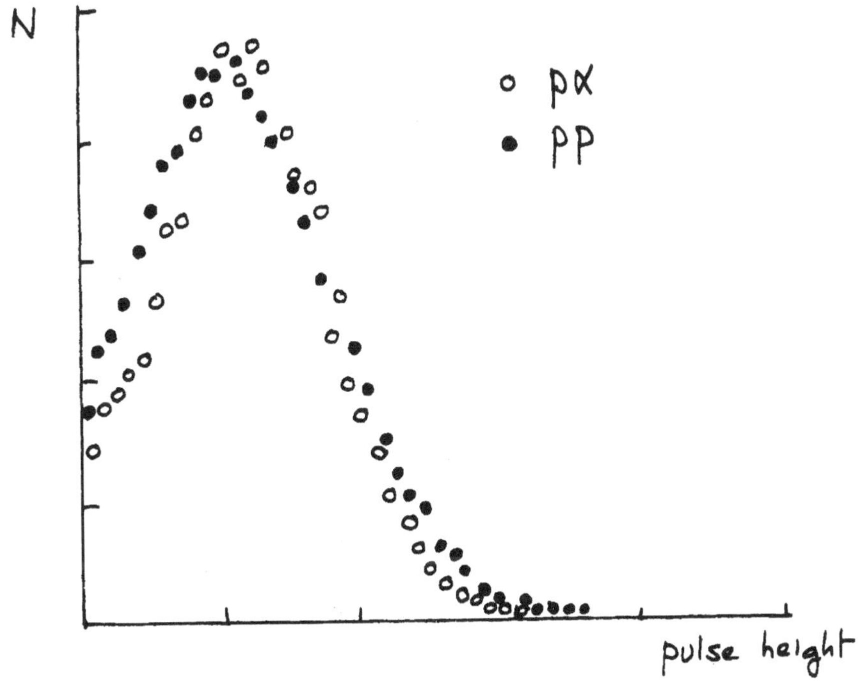

Fig.3-13. Distributions of pulse heights in a forward calorimeter for pp and $p\alpha$ collisions

[$<n> = 1/\sigma \int (d\sigma/dydp_T^2)dp_T^2dy \sim \ln s$] . With a rising density it should increase faster. Indeed recent ISR measurements of the total charged multiplicity favour a steeper than logarithmic increase. This is shown in Fig. 3.14.

An interesting issue is the validity of the Wroblenski's rule, which states that the dispersion $\Delta = (<n^2> - <n>^2)^{1/2}$ is a linear function of $<n>$[47].

If for instance there are different and exclusive reaction mechanisms at work at collider energies, as suggested by Cosmic Ray results, one would expect Δ to increase faster with $<n>$. This could be quickly tested with particles produced in the Central region ($\Delta y \sim 2$ say).

A measurement of the central particle density of charged particles should be relatively quickly obtained at the collider. It should carry important clues about the continuation of ln s physics.

Another key feature of particle production is displayed in Fig.3-15. This is the short-range nature of correlations among secondary particles. In this figure the two-particle correlation function is presented. It is defined as

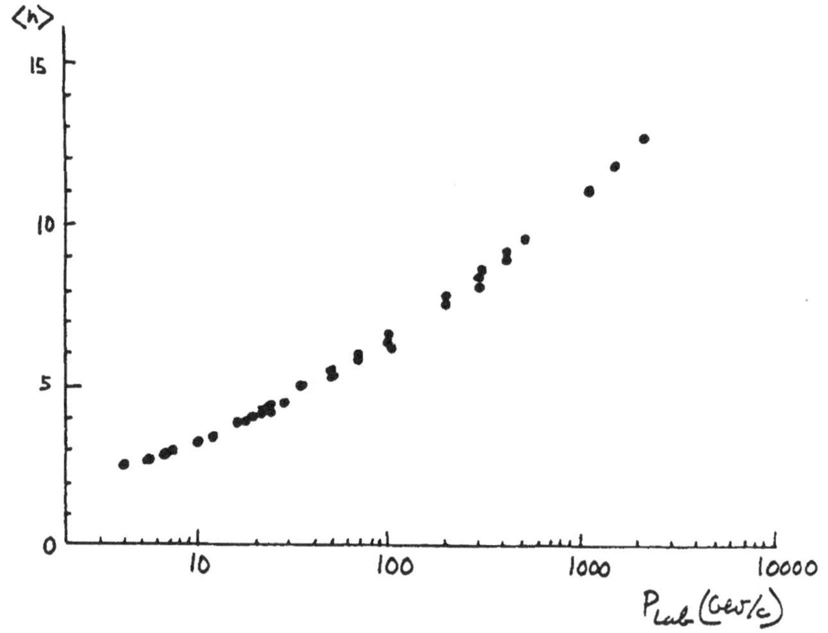

Fig. 3.14. Mean charged multiplicity as a function of energy for pp collisions. The ISR results are from the CERN-Munich Collaboration.

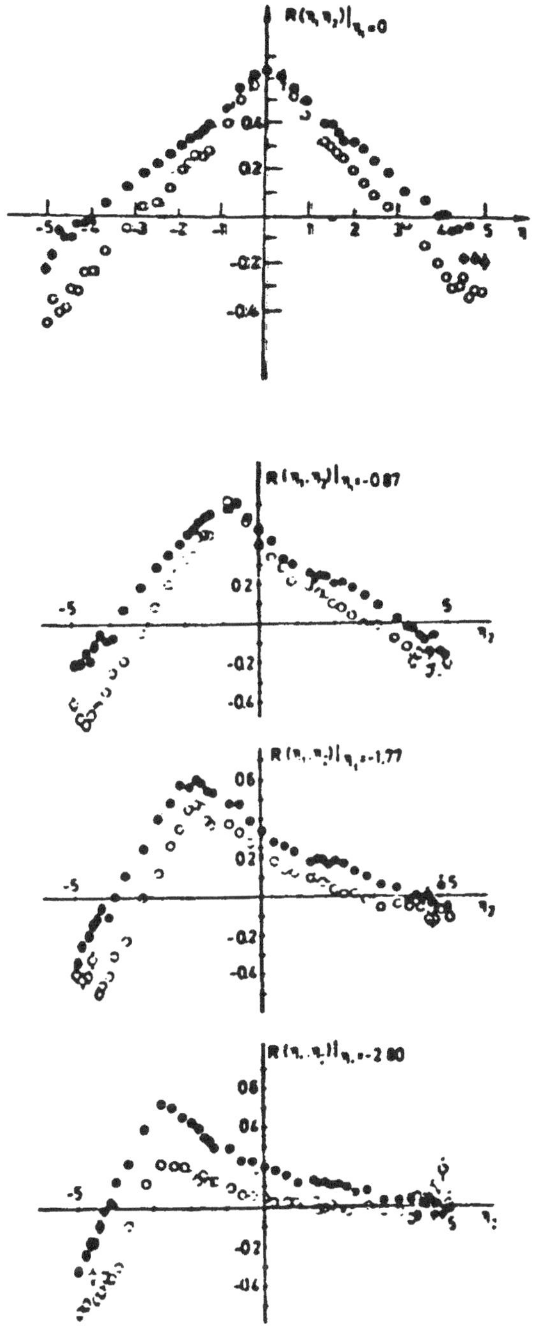

Fig. 3.15. Rapidity correlations for two charged particles at the two extreme ISR energies 62 GeV (full dots) and 22 GeV (open dots). Results from the Pisa-Stony Brook Collaboration at the ISR.

$$R(\eta_1,\eta_2) = \frac{1}{\sigma_{in}} \frac{d^2\sigma}{d\eta_1 d\eta_2} - \frac{1}{\sigma_{in}^2} \frac{d\sigma}{d\eta_1} \frac{d\sigma}{d\eta_2} \tag{3.11}$$

where η_1 and η_2 are the pseudorapidities of the two particles. The correlation function should be zero if particles are not correlated. Clustering among secondary particles gives a component in R which decreases with $|\eta_1 - \eta_2|$.

Pseudorapidity is defined as

$$\eta = \frac{1}{2} \ln \frac{p + p_L}{p - p_L} = - \ln \mathrm{tg} \frac{\theta}{2} \tag{3.12}$$

where θ is the production angle in the centre-of-mass system, hence its practical interest. It often provides a good enough approximation for the actual rapidity (3.4).

The shifting peak in Fig.3-15 shows that particles have a tendency to cluster at the same rapidity. The almost constant value of the correlation function (3.11) shows that, as one covers the whole ISR energy range, the plateau extends, but fluctuations in density, as well as the mean density, hardly change. This is parametrized in terms of cluster formation and fragmentation[48]. An increase in energy allows for the production of a larger number of clusters, but the cluster properties remain fixed. Extrapolating from present results one would not expect much change in cluster properties. This is, however, an area where early results at the collider could display qualitatively new features. Such pseudorapidity correlations should be relatively easy to measure.

One can look at Cluster formation remarking that the attempted separation of a colour $3\bar{3}$ system represents a stored energy of $\sim$ 1 Gev/f. Since pionization occurs over a mean linear distance of 1.5 to 2 f (guessed from the value of $\langle p_T \rangle$) one is lead to expect clusters with an average energy of 1.5 - 2 Gev. This is the value (with an associated multiplicity of 3 - 4 particles) which one is lead to consider when analyzing short range correlations[49]. With such a picture in mind, one expects little change in the short range part of the correlations when reaching Collider energies. However, the existence of different mechanisms would increase the correlation value. This is an important point to check. It could easily be done with the angular distributions available from the Streamer Chamber (UA5).

Particle production at the ISR is mainly a pion affair, since about 85% of the secondary particles are pions. It should be stressed that little is known about quantum number excitation and, in particular, there are practically no data on correlations involving heavy particles (kaons, baryons, antibaryons, etc., outside of resonance production). This is a topic where many questions are still open and where surprises could occur. Extrapolating from Fig.3.9, thus extending all known "plateau" behaviour to higher energies, one may conclude that there will be only a relatively small variation among particle ratios. Surprises may, however, be in store.

While we have only been able to present here a brief survey of the many features of hadron production, we hope to have shown that, even in a domain which may not seem very topical to many, challenging questions remain and great expectations arise with the $p\bar{p}$ collider. We have a zeroth order-approximation ; it consists of an extending rapidity plateau with a charged particle density of about two and fixed correlations among dominant pions, involving rather low mass clusters. As an expected departure we have a rise of all particle densities. It would be surprising if a rise of the total cross-section by a factor of 1.5 would not be associated with more quantitative and qualitative changes.

Considering the excitation of new heavy flavours, two mechanisms seem to be relevant[50)]. The first one corresponds to $q\bar{q}$ production through gluon interactions. It should be important at very high energy but yields can be reliably estimated only at large p_T. We shall come back to it in section 4. This is the relevant mechanism when discussing background level for lepton pair production, as done in section 2.

The second one corresponds to the production of a forward baryon. A leading diquark in the incident proton may pick up a heavy flavour quark, thus transforming itself into a leading Λ, Λ_c, Λ_b,... Cross sections are supposed to be in the ratios of the quark mass squared. This mechanism seems to be responsible for the fairly large cross section (σ >100 μb) for Λ_c observed at the ISR[51)].

Either way one should expect heavy flavour particles to be rather copiously produced at Collider energy. In the later case a heavy flavour meson is produced at wide angle with a resulting large p_T lepton part of the time.

While combinatorial background may be a very important question for hadron decays, sequential leptonic decay may give very particular reactions.

III.3. Some information from cosmic-ray results

There is now a large amount of data in the 10 to 10^3 TeV range[52,53)], which will be that covered by $p\bar{p}$ colliders (table1.1). In particular, the region of $\sim 10^2$ TeV, corresponding to the typical running energy of the CERN collider, is already well covered by emulsion chambers. Our discussion borrows much from a recent review of such results by Lattes, Fujimoto and Hasegawa[10), to which the reader is refered for more information,

An emulsion chamber detector has typically two decks, each consisting of alternating layers of lead, emulsion and X-ray film, separated by a wide air gap with a target layer (typically carbon) placed just below the upper deck. The detector is sensitive to electromagnetic energy (π^o's, γ's and electrons) and to track detection. Some events with a shower seen in the lower deck can be traced to an origin in the carbon layer. They are referred to as C jets[10)]. While there is a relatively high threshold on the energy of the registered secondaries ($\sim$ 2 TeV), such C jets can be analysed with enough detail to give a relatively good picture of the reaction process. Carbon jets are now available in the hundreds with electromagnetic total energy seen in excess of 20 TeV and hence an estimated total energy of the order of 100 TeV. At lower energy (E $\sim$ 10 TeV) information is usually available from smaller balloon-borne detectors. At much higher energies, observed events usually correspond to a more extended shower with its origin any place in the atmosphere above the detector. Such events are referred to as A jets. They correspond to most of the available information about E $\sim$ 1000 TeV reactions. The estimated reaction energies are of the order of three to five times the observed electromagnetic energy ($E_\gamma = k_\gamma E$). The detector is often only sensitive to energetic particles and thus covers essentially the fragmentation of the projectile, only reaching the central plateau region if any.

With such limitations in mind, one can test the scaling property considering the distribution dN/dF_γ , with $F_\gamma = E_\gamma/\Sigma E_\gamma$ where E_γ is the electromagnetic energy measured in association with each track.

Figure 3-16 shows such a distribution, combining 80 C jets with $\Sigma E_\gamma > 20$ TeV, and corresponding to a mean total energy of 130 TeV (full dots), and nine C jets, obtained in a balloon flight, with $2 < \Sigma E_\gamma < 5$ TeV and a mean total energy of 7 TeV. The high energy jets were obtained in the mountain exposure of the Brazil-Japan collaboration at Mt. Chacaltaya[10)]. Scaling holds, at least to a good approximation. Simulation calculations using accelerator data can reproduce these data reasonably well[10)]. The fact that $E/\Sigma E$ is measured, as opposed to the actual fragmentation ratio $x = 2E/\sqrt{s}$, or $y - y_{max}$, limits the sensitivity of the scaling test. One may, however, expect that at collider energies the scaling property of the distribution (fast particles take a fixed fraction of the incident momentum) should be approximately verified. There is also no dramatic change in the p_T distribution, as shown in Figure 3-17.

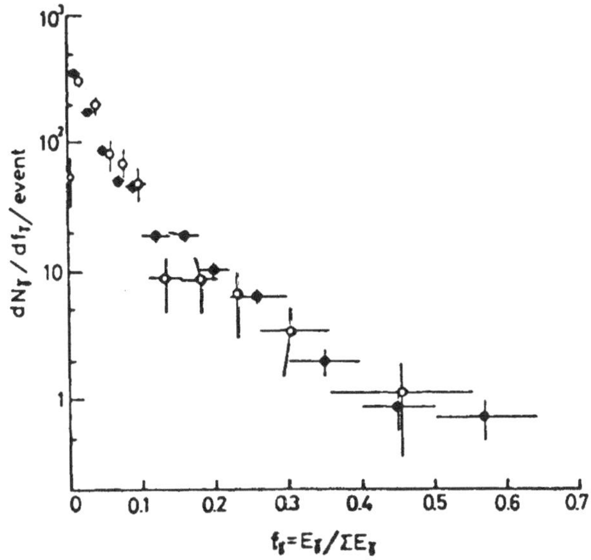

Fig.3-16. Scaling distribution for cosmic-ray jets observed in an emulsion chamber (E $\sim$ 100 TeV) shown as full dots and in balloon-borne detectors (E $\sim$ 10 TeV) shown as open dots.

Such comparisons are important, as they refer to similar techniques and similar selection criteria. One may conclude that the inclusive distribution of the fast proton fragments scales over the 10-100 TeV range.

There seems, however, to be a sizeable change when considering secondaries which could be attributed to the plateau region. Emulsion chambers have been exposed at Fermilab and simulation calculations can be made to compare their results with those observed with C jets. This is shown in Fig.3-18. The pseudorapidity distribution measured with C jets (ΣE_γ > 20 TeV), plotted as full dots, is compared with the pseudorapidity distribution, measured at Fermilab, plotted as open dots. The pseudorapidity in the projectile rest frame is defined by

$$\eta = \ln \frac{\Sigma E_\gamma}{k_\gamma m} \operatorname{tg}\theta \tag{3.13}$$

where k_γ is the correction factor between observed electromagnetic energy and total energy ($\langle k_\gamma \rangle \sim 0.3$). As a simple calculation shows, it provides a reasonable approximation to the rapidity ($y - y_{max}$ in Fig.3-9). The distribution is, however, presented in terms of tg θ as opposed to ln tg θ, which is proportionnal to rapidity, hence its very different shape!

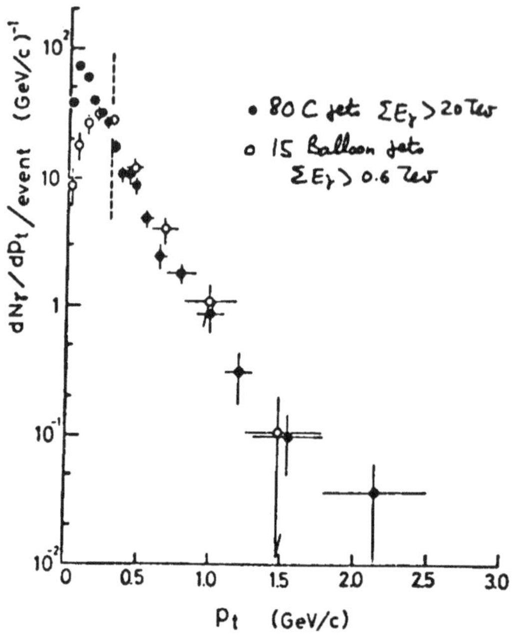

Fig.3-17 Transverse momentum distributions for forward jet particles in the 100 TeV range (full dots) and 10 TeV range (open dots). The discrepancy at low P_T corresponds to an as yet uncorrected detector cut-off. The P_T distributions can be considered as identical.

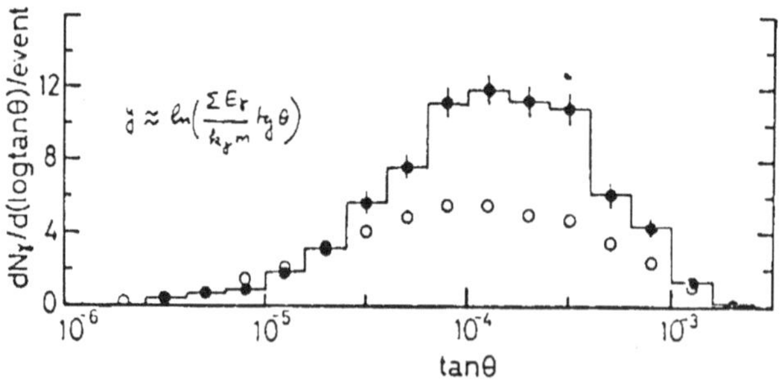

Fig.3-18 Inclusive distributions as measured for C jets in emulsion chambers (E $\sim$ 100 TeV), shown as full dots, and in emulsion at Fermilab, show as open dots. There is a very sizeable rise.

One notices the scaling behaviour in the left-hand tail of the distribution, but the "plateau value" is definitely increasing with energy and this by a very large amount : a factor of 2.5! For reasons previously mentioned, the C jet experiment suffers a cut-off at a large value of rapidity as measured in the projectile rest frame. Particles that are too slow in the laboratory are not seen. The distribution would otherwise extend much to the right, beyond the Fermilab distribution.

This rise in central density seems to be a strong feature of the emulsion chamber data. It seems however, to be already present when comparing Fermilab results with balloon results at $\sim$ 10 TeV[10]. With full confidence in the results, one may thus conclude that there is a sharp rise of the plateau density between 1 and 10 TeV and then but a small variation. One may, however, question the fact that the comparison between the accelerator data and the emulsion chamber data is free of any systematic bias. Such an energy behaviour is, of course, a very challenging question. As a second piece of information from cosmic rays, one may thus say that the central density will be significantly larger at the collider than that observed at the ISR, with a mean rapidity density of about five for charged particles. The energy dependence of the plateau density is a very important issue. It is also claimed that the p_T distribution becomes wider, with a change already observed between 0.4 and 10 TeV.

The analysis of the Chacaltaya experiment goes further and raises the possibility that such changes could be due to the emergence of new types of interactions. The claim made is rather clearly displayed in the plot of Fig.3-19. For each of the $\Sigma_\gamma E_\gamma >$ 20 TeV C jets collected, one plots the mean γ-ray density against the mean transverse momentum. One may then postulate that the events fall into two categories. One of them centres on values of $<n_\gamma> = 2$ and of $<p_T> \doteq 150$ MeV/c, while the other centres on values of $<n_\gamma> \simeq 7$ and of $<p_T> \simeq 250$ MeV/c. The separation into two clusters which may still sound somewhat arbitrary, is strengthened by the fact that simulation calculations, based on scaling and using Fermilab data as input, reproduce only the former batch.
The latter one would then correspond to a new type of configuration appearing at very high energy. The two configurations would be about equally probable at E $\sim$ 100 TeV. These two configurations are referred to as Mirin and Açu, in the vernacular of the Chacaltaya Collaboration [10]. These names encode the size of the fireball which is associated with the forward moving cluster. While for Mirin events small fireballs are found, with a typical cluster mass of 2-3 GeV, for Açu events large fireballs appear, with a typical cluster mass of 15-30 GeV.

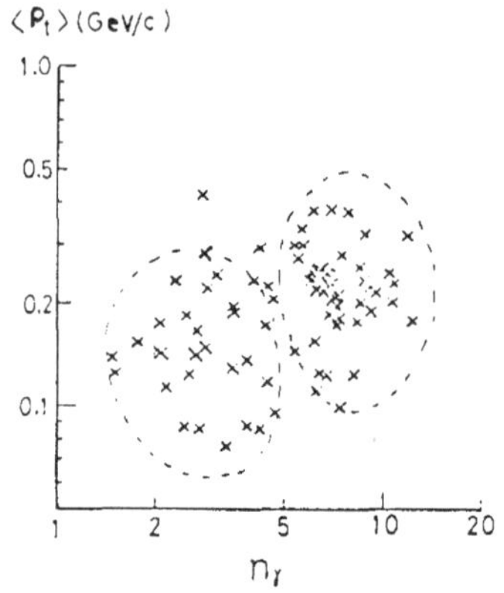

Fig.3-19 Scatter plots for C jet events showing the mean transverse momentum and the rapidity density. Separation into two populations is indicated to guide the eye.

Such as possible separation between two types of events is a challenging question at collider energy. It would be evidence for a new kind of dynamics (incoherent interactions among low x partons perhaps) and would go against an interpretation of the energy dependence in terms of a Regge behaviour successful at lower energies. If such a separation exists (the reaction is of one type or of the other type), important effects will occur among correlations, and in particular among the two-body correlations in the central region, so stable over the ISR energy range (Fig.3-15). Such a dichotomic separation may also be only an approximation.

Early results at the collider should settle these questions. Measurements of two body correlations should clearly show whether a new type of configuration is present. 510

While the occurrence of Açu configurations corresponds to a large increase in multiplicity, events with very large multiplicity may also occur with a sizeable cross-section. Such events have been seen and are sometimes referred to as Guaçu (very big fireball) events. While they are typically seen as A jets, at $\Sigma_\gamma E_\gamma \sim 1000$ TeV, the sample of 80 C jets considered in the analysis referred to so far can be described as containing 40 Mirin, 39 Açu, and 1 Guaçu configurations. Their respective p_T distributions show spectacular differences. This is shown in Fig.3-20. The occurrence of Guaçu type events at the rate of $\sim$ 1000/s would be an immediate signal of new physics! It is among such large multiplicity events that the peculiar Centauro configuration occurs. Before we turn to this, we should leave the emulsion chamber data to mention another

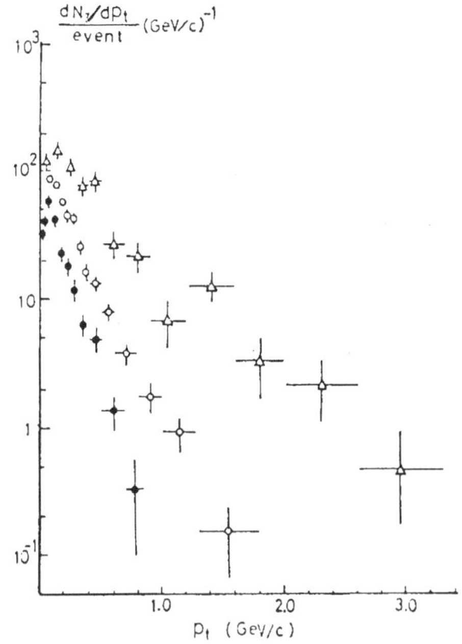

Fig.3-20 Transverse momentum distribution for Mirin (full dots), Açu (open dots) and Guaçu (triangles) events as sorted out among 80 C jets in the 100 TeV range and corresponding to 40 Mirin, 39 Açu and 1 Guaçu.

peculiar effect which is of relevance at collider energy. Measurements have been made for absorption of the cascade, as a function of hadron energy, using deep calorimeters with depths greater than five interaction lengths. The attenuation length increases slowly with energy at $\sim$ 10 Tev. The Tien-Shan experiment reports, however, a sudden increase of the attenuation length at E $\sim$ 100 TeV, and this by a factor close to 2[53]. Such an increase would signify the onset of the copious production of some penetrating componet. The absorption length as a function of primary energy is shown in figure 3-21. The collider energy is just where an important change occurs.

III.4. The Centauro events

A Centauro event corresponds to a large multiplicity (N $\sim$ 100) with a staggering absence of neutrals. There is a very clean event, shown Fig.3-22. It is clean insofar as its

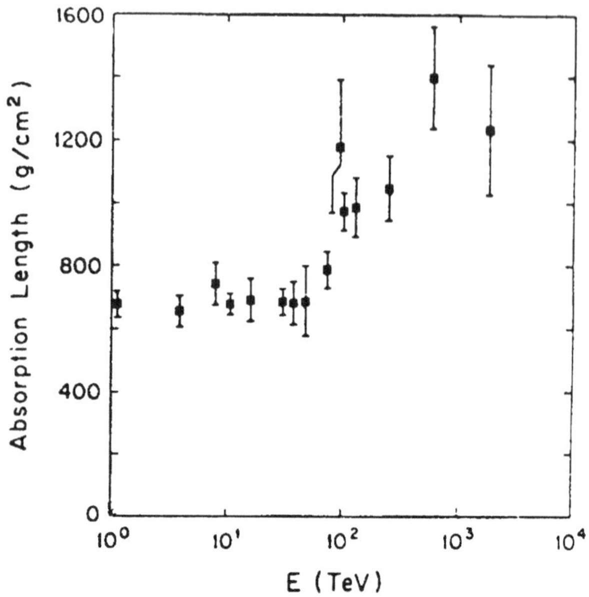

Fig.3-21 The Tien-Shan anomaly in the penetration of cosmic-ray showers in a calorimeter as a function of the measured total energy.

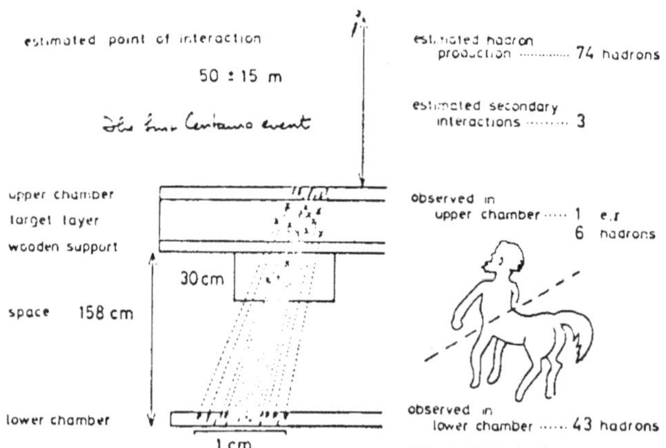

Fig.3-22 A Centauro event. A strong shower registered in the lower part of the emulsion chamber is not matched by a significant amount of energy deposited in the upper part, as if the particle associated with the event did not include neutrals. This mismatch between the responses of the 2 chambers explains the name Centauro.

origin can be traced to a relatively well defined point 50 ± 15 m above the chamber. There are very few tracks seen in the upper chamber and then a very large number are seen in the lower chamber. The latter set originates from hadron interactions in the carbon layer and the smallness of the former set bears witness to the fact that there were practically no neutrals (one seen, which could have well originated from secondary interactions in the air above the chamber) among the fast hadrons reaching the chamber. This is very peculiar. Indeed, the name Centauro comes from the impossibility of guessing a priori, what happened in one chamber when knowing what happened in the other, and being right. In the same way, reaching Mt. Olympus with only a knowledge of ordinary things and catching a glimpse of a human torso or a horse's rump, no one could guess that the torso was not attached to human legs and the rump to the rest of a horse.

Several events with some similarity to the one already mentioned have now been collected. They correspond to a small amount of prompt electromagnetic energy together with a much larger amount of hadronic energy as seen through the associated electromagnetic showers, or to a small number of electrons (photons) as compared to a much larger number of hadrons. However, with none of them is it possible to get the good reconstruction of the original event shown in Fig.3-22. Plots of $\Sigma E_h/\Sigma E_{e,\gamma}$ and $N_H/N_{e,\gamma}$ are shown in Figs.3-23a and 3.23b respectively. They correspond to events with total observed energy in excess of 100 TeV. Also shown, as a dashed line, is an interpolation of the result of a simulation calculation. Data from three different experiments (Mt. Chacaltaya, Mt. Fuji, Pamir) are put together.

One sees how Centauro configurations stand out and cannot easily be claimed as fluctuations. They may correspond to a new type of interaction, but one is at a loss trying to understand why no π^o's appear. Do they correspond to droplets of a new kind of nuclear matter which has very high density in order to be able to penetrate down to the atmospheric depth at which they are observed ?

The energy at which they are reported may appear as too high for the collider. Nevertheless, mini-Centauro events have also been reported, with lower typical multiplicity but still the same relative absence of π^o's, and they also apparently occur at lower energies E $\sim$ 100 TeV[10].

A very large amount of work has gone into analysing such peculiar events. They may be evidence for a new mode of interaction which could also be indicated by the occurrence of very large multiplicities (Guaçu). There are other types of peculiar events (Andromeda, Geminion etc.). They, however, all appear beyond the range of energy covered by the CERN $p\bar{p}$ Collider.

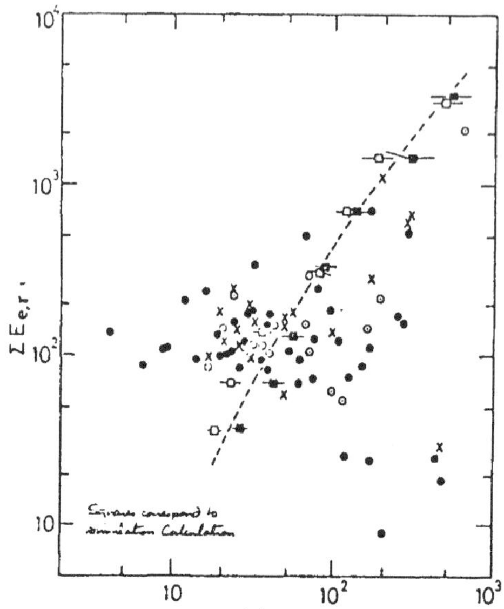

a

Fig. 3.23a. Compilation of events with energy larger than 100 TeV showing the total electromagnetic energy and the total hadronic energy. Centauro events stand at the lower right corner.

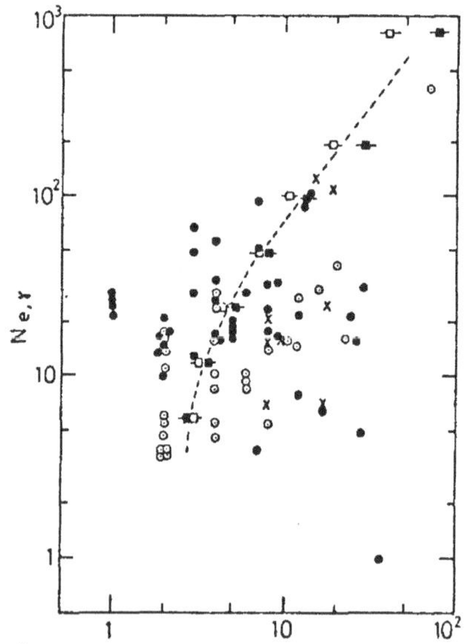

b

Fig. 3.23b. Compilation of events with energy lower than 100 TeV showing the total multiplicities for electrons and photons, and for hadrons.

While extrapolation from present data already leads to many interesting questions, information from cosmic-rays allows one to anticipate a very new and exciting venture, and this already at relatively low luminosity.

Of particular interest are the Guaçu events which would give configuration with very large multiplicity at the 1% level. They should immediately stand as a striking feature.

IV.JET PHENOMENA

IV.1. Large p_T production and jets

Whilst Section 3 was concerned with relatively large cross-sections, we now turn to a particular class of processes with relatively small cross-sections, but where the parton structure of the colliding hadrons is more clearly tested. Such processes correspond to the appearance of secondary particles with large transverse momentum p_T[53]. They are described in terms of the scattering of hadron constituents, acting to a large extent incoherently. Whilst quark and gluon constituents are not observed as such, they manifest themselves as jets[8,54], thus giving a very special structure to the hadron configuration in the final state. Evidence for a jet structure in reactions triggered by a large p_T particle has already been much discussed[8]. The hard scattering models which are considered when studying such processes[55] [56] have also been much discussed at the same time. We shall therefore limit ourselves to a brief overwise.

In the parton model, one considers the scattering of two constituents, with fractional momenta x_1 and x_2, respectively, and with differential cross-section $d\sigma/d\hat{t}$, where $\hat{t}$ is the momentum transfer squared at the constituent level. The centre-of-mass energy squared of the relevant constituents is $\hat{s} = sx_1x_2$. One of the scattered constituents fragments inclusively into the observed hadron at large p_T, giving it a fraction z of its momentum. This is schematically displayed in Fig.4-1. The inclusive yield reads :

$$E\frac{d\sigma_{abc}}{d^3p} = \sum_{ijk}\int F_i^a(x_1)F_j^b(x_2)\frac{1}{\pi}\frac{1}{z}D_k^c(z)\frac{d\tilde{\sigma}_{ijk}}{d\hat{t}}dx_1dx_2 \qquad (4.1)$$

The summation extends over different types of constituents (ij) and D_k^c is the fragmentation function for constituent k into hadron c. All kinematical variables are easily expressed in terms of x_1 and x_2, the fractional momenta of the two initial constituents[57] namely :

$$z = -\frac{1}{s}\left(\frac{t}{x_1} + \frac{u}{x_2}\right) \qquad (4.2)$$

$$t = (p_c - p_a)^2 \;;\; u = (p_c - p_b)^2 \qquad (4.3)$$

$$\hat{t} = \frac{x_1 t}{z} \quad ; \quad \hat{s} = x_1 x_2 s \qquad (4.4)$$

The momenta of the two incident hadrons and that of the observed hadron at large p_T are denoted by p_a, p_b, and p_c, respectively.

Such a parton model approach is justified in the leading-log approximation in perturbative QCD provided that the structure functions (F's and D's) are considered as varying with the probing momentum squared (typically p_T^2)[56]. In this case, the differential cross-section to lowest order in the running coupling constant reads:

$$d\sigma_{ijk}/d\hat{t} = \pi\alpha_s^2(p_T^2) \quad |A_{ijk}|^2/\hat{s}^2 \qquad (4.5)$$

The amplitudes A are dimensionless expressions, averages over spin and colour, which vary according to the particular subprocess considered[58].

In much the same way as with the Drell-Yan process considered in detail in Section 2, one may question the smallness of higher order effects in α_s. We shall here take the same attitude, namely assuming that a lowest order QCD calculation gives the correct order of magnitude, provided that the probing momentum (in practice p_T^2) is large enough. We may also expect, that whilst rates may be in reasonable agreement with experiment, departures from coplanarity should be much larger than expected from the parton model.

While one should not trust yields calculated in the leading log approximation to better than providing the correct order of magnitude, the expected effects remain spectacular. Actually some rates become so large that one may consider the new Cosmic Ray configurations as possibly due to incoherent effects among partons, typical of those responsible for large p_T production at present machine energy. The distinction between sections 3 and 4 may then partially disappear as the energy increases.

From (4.1) and (4.5) one easily finds that at 90° (where the kinematics simplifies), the particle yield then obeys a simple scaling law, namely[8]

$$E\, d\sigma/d^3p = 1/p_T^4\, f(x_T) \qquad (4.6)$$

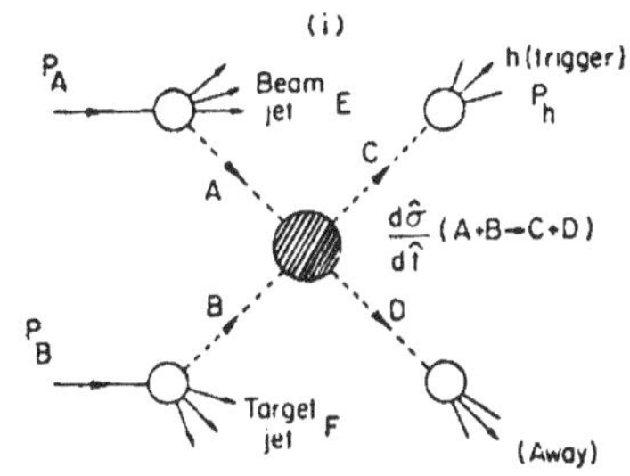

Fig.4.1 Parton model description of a reaction process with large transverse momentum particles. Two constituents scatter. Fragments of the scattered constituents appear as hadronic jets at wide angle.

with

$$x_T = 2p_T/\sqrt{s} \tag{4.7}$$

It should be valid up to some (logarithmic) dependence associated with scaling violations.

At lower p_T (p_T < 10 GeV/c say), terms associated with specific higher order perturbative effects, and usually referred to as "higher twist effects", could be of some importance, even though lowest order QCD may already be a reasonable approximation. These terms may correspond to the scattering of constituents with some structure (a two-quark system for instance). Their relative importance should decrease as inverse powers with p_T. They could however result in an important deviation from the scaling form (4.6) on top of the log effects.

If one adds to this the fact that any primordial (transverse) motion of the colliding constituents will result in a steeper p_T dependence at lower p_T than that read from (4.1) and (4.5), one may reconcile the asymptotic form (4.6) with the empirical behaviour found to hold for pions over a rather large range $2 < p_T < 8$ GeV/c, namely

$$E\frac{d\sigma}{d^3p} \simeq \frac{1}{p_T^8}\, g(x_T) \tag{4.8}$$

At 90°, where (4.8) applies, $x_T \simeq x_1 \simeq x_2 \simeq \sqrt{\tau}$ (section 2) It therefore follows that much of the variation with x_T corresponds to the x dependence of the structure functions which fall sharply as x approaches one. Increasing the energy at fixed p_T corresponds to decreasing x_T and hence to a rise of the inclusive yields as one switches from relevant rare constituents at large x to far more abundant ones at low x. There is, accordingly, a sharp rise of the large p_T yield and the more so the larger the initial x_T. This should be valid irrespective of the details of the basic scattering process.

This brief introduction is but an overview of the present approach to the physics of large p_T processes. We illustrate it with two figures. Figure 4-2a shows the inclusive π^0 yields as observed over the 2 to 7 GeV/c range at the ISR. The solid line is the exponential extrapolation of the low p_T distribution ($p_Y < 1$ Gev/c. One notices a very sizeable excess in production yield associated with the importance of hard scattering effects at large p_T and its increase with energy (as x_T decreases). Whilst these data are very well reproduced by relation (4.8), with the corresponding fit shown as dashed curves in Fig. 4.2b, one notices with the more recent results a clear departure from this parametrization. This signals a change of behaviour which should eventually dominate at large enough p_T [59].

At present there are still too many uncertainties about the gluon distribution and fragmentation functions and one can claim neither full success nor clear failure for perturbative QCD. One

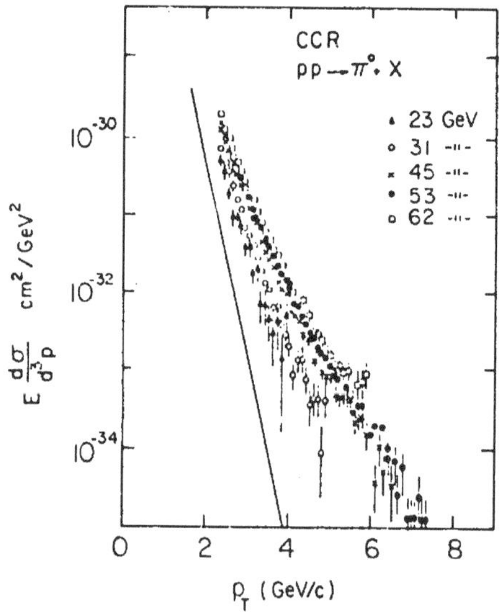

Fig. 4.2a. Inclusive π^0 yield at intermediate p_T values. The solid line corresponds to the extrapolation of the exponential behaviour valid up to $p_T \simeq 1$ GeV/c. Observed yields far exceed such an extrapolation and the more so at high p_T (an effect increasing with $\sqrt{s}$). Results from the CERN-Columbia-Rockefeller Collaboration at the ISR.

may, however, claim some reasonable agreement[56,60]. With some confidence, therefore, we may use perturbative QCD in its leading-log approximation to extrapolate yields to collider energy. Two features then stand out. The first one is that yields are expected to increase greatly at fixed p_T While very large p_T reactions by present standards (p_T > 10 GeV, say) imply constituents of typically x $\gtrsim$ 0.3 at ISR energy, which have only a low probability of being found (and probably the more so when one considers gluons), the relevant values of x will fall by an order of magnitude when considering collider energies.

We already discussed this question in terms of the quantity $\tau(dL/d\tau)$ in Section 2, when we considered lepton-pair production. Figure 4-3 shows $\tau(dL/d\tau)$ as a function of $\sqrt{\tau}$ for $u\bar{u}$ and gg reactions at $\sqrt{s}$ = 540 GeV. One readily notices the importance of moving from $\sqrt{\tau} \simeq x \simeq 0.3$ down to $\sqrt{\tau} \simeq x \simeq 0.05$. The luminosity is calculated here not including colour. Proper colour averaging and summation are usually introduced in the expression for the differential cross-sections (4.5)[58].

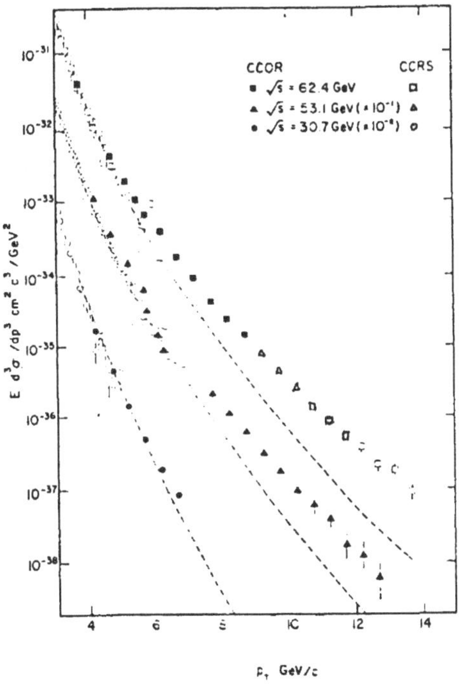

Fig. 2.4b. Inclusive π^0 yields at higher p_T values. The dashed curves correspond to a very good parametrization of the production cross-section at lower p_T (preceding figure). Observed yields depart from above as expected in perturbative QCD. Results from the CERN-Columbia-Rockefeller-Oxford Collaboration at the ISR.

The calculation of a large p_T pion yield, according to the factorization property implied by figure 4-1 goes through the calculation of the inclusive cross-section for a constituent, e.g. a jet. At fixed p_T a jet cross section, collecting momentum flow within a limited solid angle around the direction attached to the constituent, should thus be much larger than the single particle cross section.

This expected feature has been verified and the proportionality factor between jet and single pion is of the order of 10^2 to 10^3. Since the large collider detectors will respond to jet (measuring the amount of energy emitted within a particular solid angle) rather than to single particles (as usually done so far with spectrometers at the ISR) the yield of registered large p_T events will enormously increase from what one is presently used to at ISR energy (figure 4-2).

One can therefore envisage very high counting rates ($>$ 1/s say) up to relatively large p_T (corresponding to PETRA and PEP jets say). One can therefore expect a very important development in jet physics, and the study of jet phenomena will certainly be an important topic at collider energy.

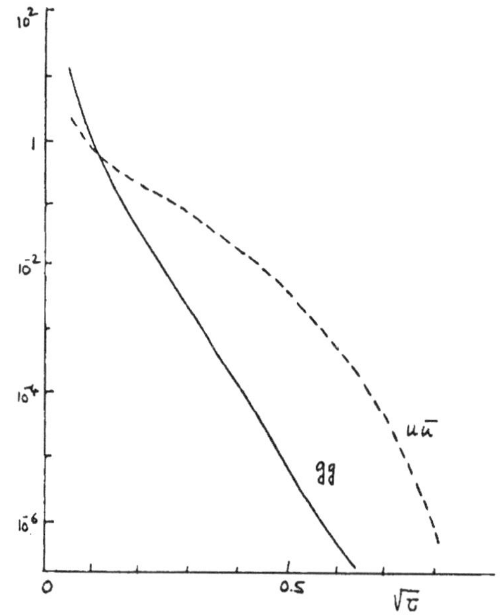

Fig. 4.3. Differential luminosities. Variable τ is equal to $\hat{s}/s$. Scaling violations are included and the calculation done for $\sqrt{s}$ = 540 GeV. The solid (dashed) curve corresponds to gg and $u\bar{u}$ collisions (for $p\bar{p}$ interactions).

In e^+e^- annihilation one may assign all produced hadrons to a two jet system : one may then use global parameters such as thrust and spherocity, to measure the "jettiness" of each configuration.

With hadron collisions it is far more difficult since together with the jet fragment proper there are a large number of low p_T background particles. A jet of a few GeV will have only few fragments standing above background by their transverse momenta. Jet analysis of the ISR thus centered on correlation studies among particles which one could relatively safely consider above background (p_T < 1 GeV/c say). One then finds the angular correlations which are expected from fragments of a jet and eventually checks that these angular correlations correspond to but limited transverse momenta for all fragments with respect to the jet axis [8]. This is now well documented. The last point is illustrated by figure 4.4. The transverse momentum distribution of hadronic jet compares well with that observed in e^+e^- annihilation at similar global jet momentum.

The coplanarity structure, expected for a process described in terms of figure 4-1 is well satisfied as shown for instance by figure 4-5 which combines data with the same transverse energy, calculated taking into account whatever has to balance the transverse momentum of the two observed π^0's. This is done in order to avoid

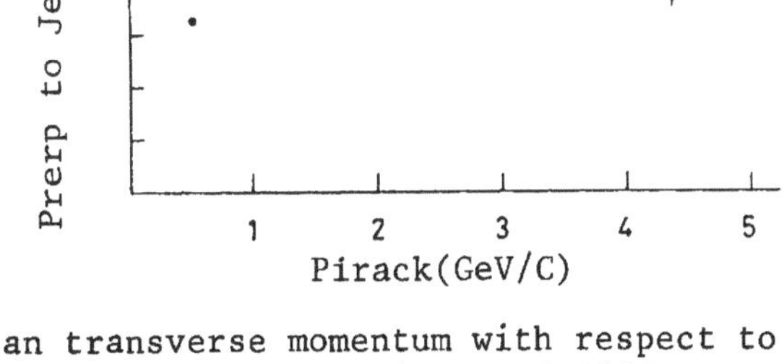

Fig. 4.4. Mean transverse momentum with respect to the reconstructed jet axis for particles of different p_T associated with trigger particles of different p_T. Results from the CERN-Columbia-Oxford-Rockefeller Collaboration at the ISR.

an obvious bias favouring coplanar events because these minimize the total transverse energy[61].

Coplanarity is not as perfect as expected in the "naive" parton model. Deviations actually reflect higher order QCD effects and agree with calculations, as shown by figure 4-5, which gives momentum distributions out of the reaction plane for charged particles associated with a large p_T π^o the direction of which defines the reaction plane.

An important property expected from jet fragmentation is scaling. Jet fragments take, on the average, a fixed fraction of the jet momentum. Figure 4-7 illustrates the point with distributions in variable x_e, the fraction of the trigger particle transverse momentum balanced by a large p_T particle seen on the away side. This particle is a fragment of the opposite-side jet.

The x_e distribution can be written explicitly in a simplified model where one would assume that the large p_T yield is given by an inverse power behaviour

$$d\sigma/dp_T \sim 1/p_T^N \tag{4.9}$$

This is good approximation to (4.8) since x_T covers in practice a limited range. The x_T dependence can then be absorbed into an effective power N [58]. With the jet fragmentation determined by a scaling function F(z), one easily gets

$$1/N\ dN/dx_e = \int_0^1 z^N F(z)F(zx_e)dz \Big/ \int_0^1 z^{N-1}F(z)dz \tag{4.10}$$

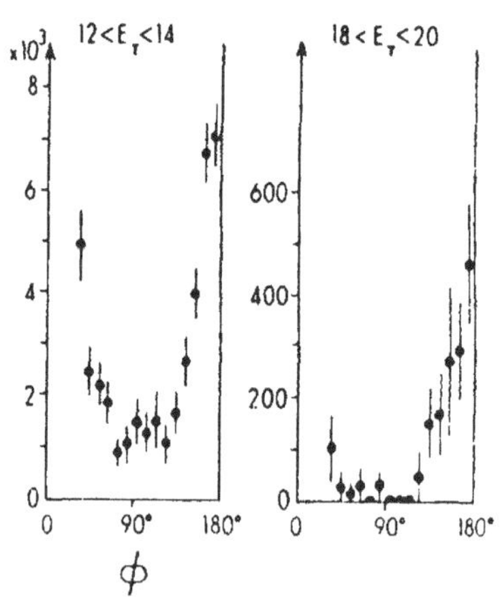

Fig. 4.5. Coplanarity structure for two large p_T π^o's in configurations with the same total transverse energy. Results from the Athens-Brookhaven-CERN-Syracuse Collaboration at the ISR.

with no p_T dependence left.

Sacling readily leads to the trigger bias effect[8]. This is also easily illustrated with the same simplifying assumption. The single particle (trigger) yield is then related to the jet cross-section by

$$\frac{d\sigma}{dp_T} = \frac{d\sigma_J}{dp_T} \int_0^1 z^{N-1} F(z) dz \tag{4.11}$$

The differential cross-section for a jet (summing over jet members to define a total jet momentum) is written as $d\sigma J/dp_T$. Relation (4.11) shows two things. The first one is the "parent-child" relationship, which takes a particularly simple form for inverse power behaviour (4.9): the single-particle distribution is proportionnal to the jet distribution. The second one is that the coefficient of proportionality is very small. Indeed F(z) decreases with increasing z when the high z part of the distribution is highly weighted with, in practice, $N \simeq 9$ to 10 (4.9). It therefore follows, that, while the now well-studied large p_T yields (fig.4-2) provide evidence for hard scattering, they are very small as compared with jet yields, which are expected to be two orders of magnitude larger.

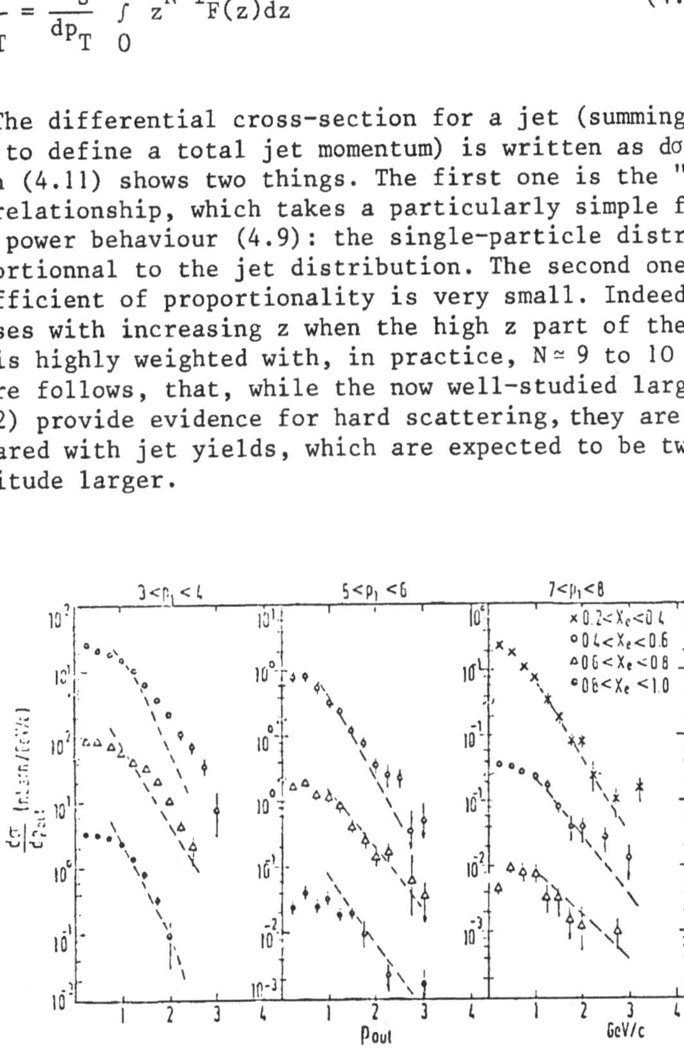

Fig. 4.6. The coplanarity distribution for particles of different x_e associated with two different trigger momenta. Results of the Athens-Brookhaven-CERN-Syracuse Collaboration at the ISR. The dashed lines (calculations by Z. Kunszt et al.) correspond to QCD calculations including next-to-leading-order effects.

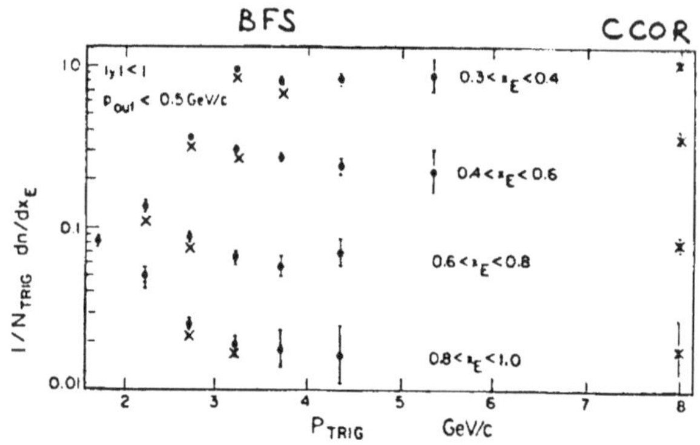

Fig. 4.7. Scaling in jet fragmentation. Particle yields for fixed x_e (the fraction of the trigger transverse momentum compensated by the observed particle) for different trigger momenta. Results from the Athena-Brookhaven-CERN-Syracuse Collaboration at the ISR.

Figure 4-8 shows a comparison between the jet yield and the π meson yield at large p_T, as respectively measured in two different experiments at Fermilab[63]. The ratio (typically two orders of magnitude) increases with x_T. This is due to two reasons. The first one read from (4.11), is due to the fact that the effective power N increases slowly with p_T The second one is due to scaling violation effects since F(z) decreases slowly with p_T at fixed and large z values which are relevant for high p_T trigger particles.

While single-particle yields are very small (fig.4-2), jet yields are much larger. They are both expected to increase sharply

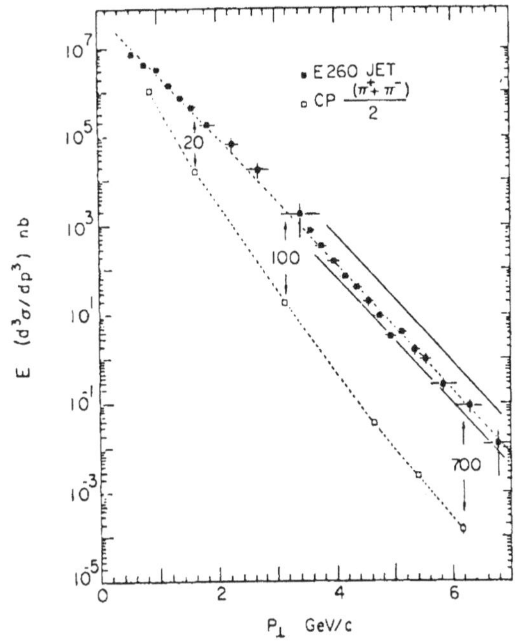

Fig. 4.8. Jet yield and single-pion yield at large p_T. Results from the Caltech-UCLA-Fermilab-Illinois Collaboration at Fermilab (data points for jets) and from the Chicago-Princeton Collaboration at Fermilab (solid curve for pions). The two dashed lines bracket QCD predictions (Feynman-Field parametrization). The beam energy is 200 GeV.

at fixed p_T, as one goes from ISR (Fermilab)energy to collider energy. Also shown in Fig.4-8 is a QCD calculation of the jet yield. This calculation takes into account the processes of Fig.4-1 with structure function varying (albeit slowly) with probing momentum (in practice p_T^2), and treats the differential cross-section in lowest order perturbation theory. This corresponds to the leading log approximation. It gives a jet yield where the jet is a quark or a gluon jet, not a hadronic jet, as actually observed. It therefore makes a difference whether one identifies the parton (quark or gluon) momentum (or energy) with the momentum or the actual energy of the jet. Two such identifications correspond to two different p_T values differing typically by 1 GeV/c, and hence to a difference in yield at fixed p_T which is illustrated in Fig.4-8.

We may conclude this bird's eye view of large p_T and jet phenomena by saying that there is now ample evidence for a jet structure in events triggered by a large p_T particle and that the over-all jet yield agrees to a reasonable approximation with QCD calculations in the leading-log approximation. This may then be used to venture extrapolations for collider energy with, as previously said, much increased yields. One should, however, stress that agreement such as that displayed in Fig.4.8, or obtained for symmetric pair triggers[64], or claimed with correlation studies, always strongly relies on an assumed gluon distribution. In perturbative QCD gluon-gluon scattering appears as the dominant process for jet production for $x_T \leqslant 0.3$[65], which covers the p_T range probed so far. One may, however, argue that, in this p_T range, a better analysis should use a different gluon parametri-zation, but include contributions from higher twist effects. Their contributions to the high p_T yield become relatively negligible only at very high p_T values. Predictions for very high p_T jets observable at collider energy would then be different. While one may say that to some extent agreement with present data is the result of a particular parametrization of the gluon distribution, experimentation at collider energy appears as a very efficient way to put it to a thorough test, together with its assumed scaling violation properties[66].

Since the first observation of an anomalous large p_T yield, evidence for hadronic jet has become well documented. While even jet production cross-sections are relatively small at present machine energy (figure 4-8) one sees why they should become much larger at collider energy. Fixing p_T, one requires a value of x_T decreasing as $\sqrt{s}$ and therefore relevant partons ($x \simeq x_T$) which are far more likely to be found.

Perturbative QCD justifies parton model calculations to a good approximation and while no precise agreement can be claimed in view of the uncertainties involved, no worrisome contradiction can be pointed out. Higher order effects are likely to be importance. We shall consider estimates as valid at the order of magnitude level.

It should be said that the measurement of jet cross section is still in a difficult situation. Some controversy has arisen in connection with the results of Experiment NA5 at CERN[67], which studies yields obtained with a 2π calorimeter and $\Delta\phi = \pi/2$ modules thereof. It is found that the response of the calorimeter exceeds (by a factor S or so) estimates based on jet production according to the jet yield of figure 4-8. This is shown in figure 4-9, where the expected jet yield corresponds to the full curve (2π) and dashed curve (back to back) configurations, respectively. I think that this question has been overemphasized. Triggering on transverse energy at SPS energy, one is more likely to obtain a configuration with anomalously large multiplicity than a configuration with a jet structure. Both results in a large transverse energy. He registered multiplicity should show large variations as one goes from one configuration to the other. Jettish configurations should be found only when the multiplicity is on the low side or when a particle with medium p_T ($p_T \sim 2$ GeV/C for a trigger set at 10 GeV) is seen. Figure 4-10 gives the topological cross-sections measured at Fermilab[68]. It shows how important cross-sections for rather large multiplicities still are.

Increasing the energy while keeping the transverse energy threshold fixed, the cross section for large p_T jets increases much faster than that for large multiplicities. A jet trigger set at $p_T \sim 20$ GeV, say at Collider energy should therefore not be much affected by the presence of configurations with multiplicities large enough to give globally the same amount of transverse energy while including only typical low p_T particles.

We therefore expect Calorimeter triggering to work reasonably well at Collider energy. We then turn to the estimate of rates.

IV.2. Jet yields at collider energy

The inclusive jet yield is given by [8,66]

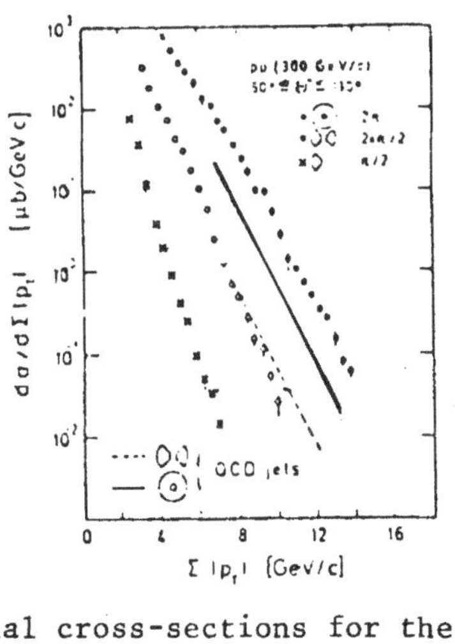

Fig. 4.9. Differential cross-sections for the observation of transverse momentum flow using calorimeters of different azimuthal coverage. Results from experiments NA5 at the SPS.

$$E\frac{d\sigma_k}{d^3p} = \sum_{ij}\int F_i(x_1)F_j(x_2)\ \frac{\hat{s}}{\pi}\ \delta(\hat{s}+\hat{t}+\hat{u})\frac{d\tilde{\sigma}_{ijk}}{d\hat{t}}\ dx_1dx_2 \qquad (4.12)$$

This is the standard parton model result, assuming massless partons and integrating over the limited transverse momentum distributions of the incident partons (quarks and gluons). This is a perturbative QCD result using distributions with a built-in scaling violation behaviour, as parametrized using deep inelastic scattering. At present it is expected to correspond

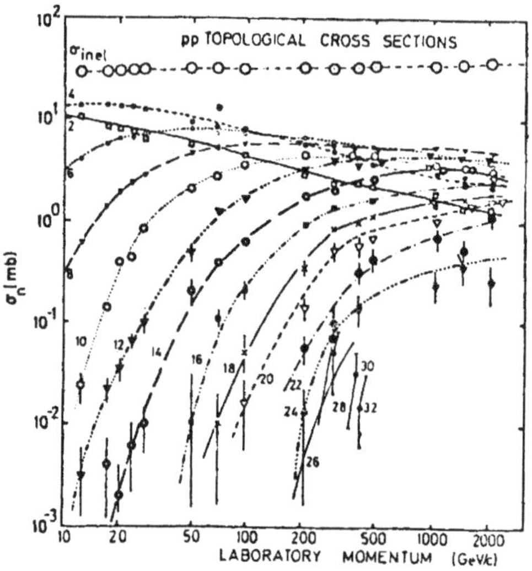

Fig. 4.10. Topological cross-sections for charged particles observed in pp collisions (Fermilab-Bubble-Chamber results). To a good approximation the Koba-Nielsen-Olesen scaling distribution is satisfied namely $\sigma_n/\sigma_{in} = 1/\langle n\rangle 4(n/\langle n\rangle)$.

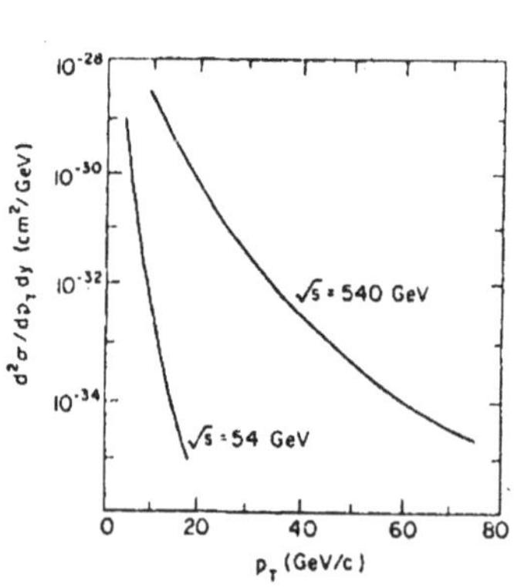

Fig.4-11 Jet yields at collider and at ISR energies as calculated in the leading-log approximation for the momentum distribution of the scattered constituent.

to a reasonable approximation at large enough p_T(p_T > 10 GeV/c, say). Higher order effects could be numerically important, as they are found to be for large-mass lepton pairs (Section 2). We consider (4.2) as providing the correct order of magnitude of the jet yield.

Figure 4-11 gives the inclusive jet yield (summing over k) thus expected at $\sqrt{s}$ = 540 GeV and, as a matter of comparison, for ISR energy, $\sqrt{s}$ = 62 GeV. The differential cross-section at 90°, $d^2\sigma/dp_T dy$ is shown as function of jet momentum. Spectacular effects are thus expected to occur going from ISR energy to collider energy. Yields at $\sim$ 10 GeV/c are 3 orders of magnitude above those expected at the ISR (and compatible with present fragmentary information about jets). Yields at $\sim$ 20 GeV/c are still extremely high ($\sim$1 event with a luminosity of 10^{30}). The jet yield could be detected up to $p_T \sim$ 70 GeV/c with such a luminosity.

The large value of the jet yield reflects the fact that, while p_T becomes large, x_T never becomes very large ($x_T \leqslant 0.25$). The large gluon component at low x can thus be used to the full. By the same token the p_T dependence (in practice $d\sigma/dp_T \sim p_T^{-9}$) corresponds to a rapidly falling x_T dependence, as the differential luminosity drops (fig.4-3), more than to an intrinsic p_T dependence (4.6). The magnitude of the jet yield and its strong p_T dependence appear therefore to provide very good tests of present ideas about the behaviour of gluon interactions and gluon distributions.

Figure 4-11 corresponds, of course, to a "bare" (optimistic) view as far as rates are concerned. One has first to acknowledge the fact that the p_T of the quark (or gluon) jet will be smaller than the value obtained in lowest order perturbation. Some jet evolution has to be included[69]. While the mean value of the thrust should approach unity asymptotically, one should expect, in practice, a lower value (0.9 say, at 20 GeV/c). While this corresponds to the energy flow associated with a quark or gluon jet, the hadronic energy flow will be smaller still, as some energy is taken away in mass and transverse motion by the relatively slow particles in the jet (in practice about 1 GeV). In any case, the jet energy distribution has to be fed into a Monte Carlo calculation giving the actually expected response of the calorimeter detector being used. This alone should be compared with the data.
When considering the actual response of a calorimeter the effective cross section should thus be read from the graph after some shift to the right by a few GeV.

It remains that very large jet yields are expected even if actual counting rates should be definitely smaller than those read from Fig.4-11. Expected global jet yields in pp and $p\bar{p}$ reactions are very similar (quark-antiquark annihilation contributes relatively little). Another important point, which has already been stressed, is that gluon jets are expected to dominate. The gluon luminosity is expected to decrease faster with $\sqrt{\tau}$ than the quark luminosity. Nevertheless, the relative importances of the gluon-gluon and gluon-quark cross-sections are such that gluons maintain a leading role up to rather large values of x_T ($x_T \lesssim 0.3$), as shown in Fig.4-12. The gluon yield is therefore expected to remain dominant over the full p_T range covered by the collider at a luminosity of 10^{30}. This applies to 90° production. Rapidity distributions, shown in Fig.4-13 for two different p_T ranges, $10 < p_T < 15$ GeV/c and $40 < p_T < 45$ GeV/c, respectively, display the leading role which valence quarks eventually get at small enough angles. It is clear that over the whole wideangle range ($|\Delta y| \leqslant 1$) gluon jets dominate. In Fig 4-12 and 4-13 gluon jet yields are distinguished from u and d jet yields (in $p\bar{p}$ collisions the $\bar{u}(\bar{d})$ jet yields are obtained from the u(d) jet yields by reflecting the u(d) curves in y = 0 (fig.4-13). Thus quark and antiquark yields are equal at 90° (Fig.4-12).

The strange quark yield corresponds to an SU(3) symmetric sea. The charm quark yield is calculated assuming no charm quark in the sea distribution. It therefore corresponds to the excitation of any new flavour (b, t, etc .) in a regime where the relevant quark mass can be neglected (in practice $p_T \gg m_q$).

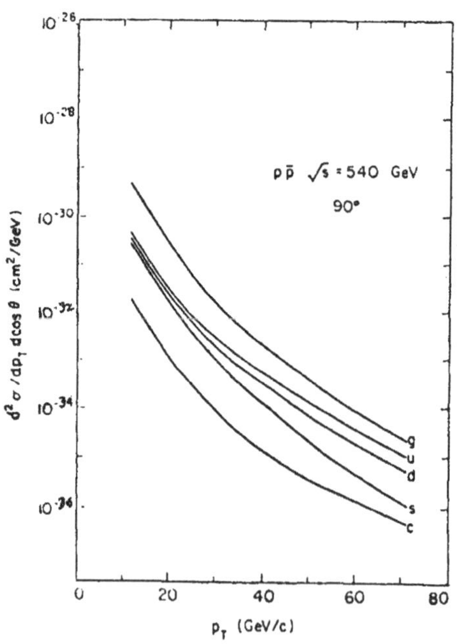

Fig. 4.12a. 90° large p_T yields for different constituents. $p\bar{p}$ collisions at $\sqrt{s}$ = 540 GeV.

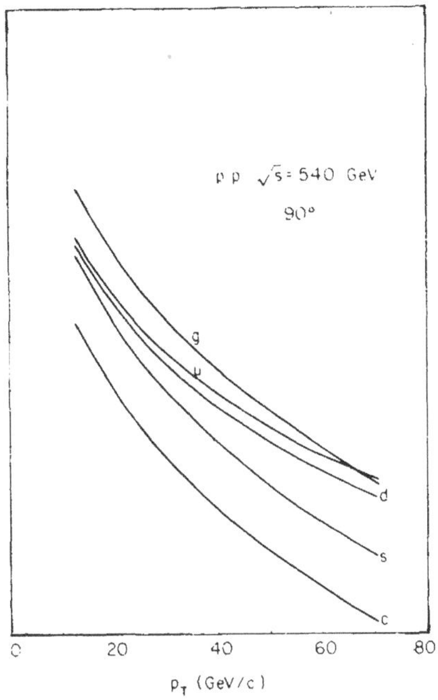

Fig. 4.12b. 90° large P_T yields for different constituents, pp collisions at $\sqrt{s}$ = 540 GeV. In both cases the charm quark yield corresponds to the excitation of a quantum number not present in the proton sea.

This is a regime where one can relatively safely estimate heavy flavour excitation, as opposed to low p_T production where soft hadronic effect imply a far more complicated picture.

Figures 4-12a and 4.12b distinguish jet yields in $p\bar{p}$ and pp collisions at $\sqrt{s}$ = 540 GeV. While, as previously said, the global yields are very similar, important differences occur when considering jets of specific nature. The main difference is here with charmed jets. This is due to the particular role of the one-gluon annihilation term, present in $p\bar{p}$ reactions, and absent in pp reactions. Charm (and other flavours) can originate in this model from gluon-gluon collisions and quark-antiquark annihilation. One also notices the more important role of the u jet in pp collisions at large p_T. This, however, merely the result of quark counting among valence quarks.

It is difficult to distinguish jet species. Nevertheless, this can to some extent be done[70]. As later discussed, charmed (and b and t) jets should signal themselves by high p_T leptons.

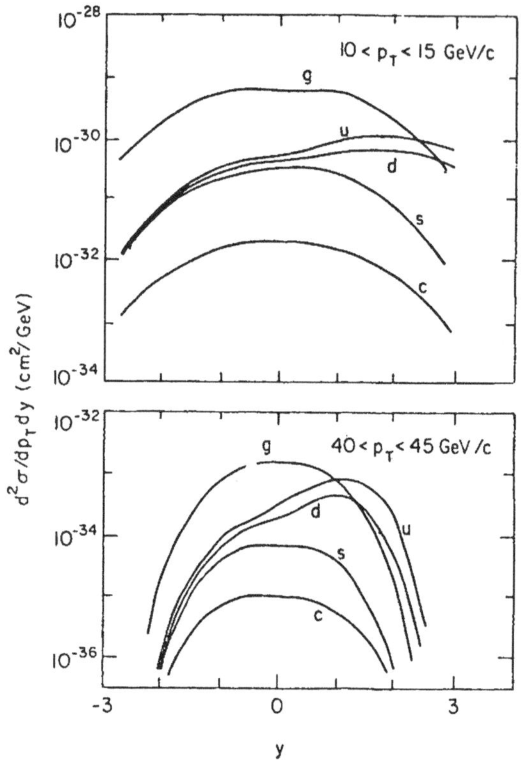

Fig. 4-13 Rapidity distributions for different constituents for $10 < p_T < 15$ GeV/c and for $40 < p_T < 45$ GeV/c. $p\bar{p}$ collisions at $\sqrt{s} = 540$ GeV.

In Fig.4-12, jet distributions are arbitrarily cut off at $p_T = 10$ GeV/c, since one may certainly question the validity of perturbative QCD at lower values. One may, however (optimistically) consider yields calculated as such down to lower values on the grounds that, for $p_T > 2$ GeV/c say, particles should still originate from incoherent parton interactions. The rather large yields thus obtained have indeed to be associated with the numerous partons at low x which can interact incoherently at very high energy even though they certainly do not at lower energy. As the number of "operative" partons becomes large, non-linear contributions should also be taken into account. They are associated with multiscattering among partons. An interaction enhances the probability of having other partons in the same hadrons interacting, in view of the fact that they are nearby in space. If σ_1 is the cross-section (assumed to be small) for parton-parton interaction, the cross-section for an interaction involving double scattering will be of the order of $\sigma_1(\sigma_1/\sigma_0)$, where σ_0 is the hadron-hadron geometrical cross-section. Such effects may be responsible for the larger than A increase of inclusive large p_T yields off nuclei with atomic number A. They shoul1 then also be responsible for the relatively high medium p_T ($p_T \simeq$ 5 GeV/c) yields observed in $\alpha\alpha$ collisions at the ISR[45])

Such multiscattering processes should result in several small jets giving eventually large central multiplicities. They may be linked to the Açu behaviour discussed in Section 3.

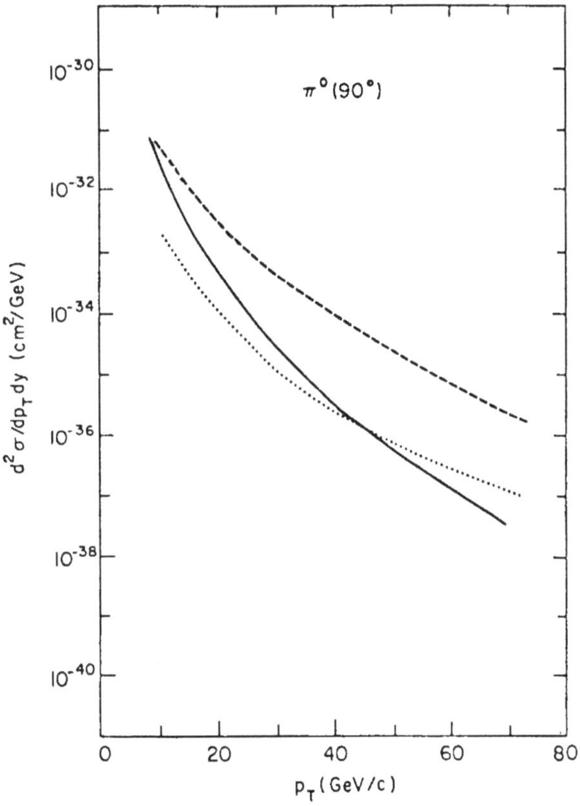

Fig. 4.14 Neutral-pion yield at 90° for $p\bar{p}$ collisions at $\sqrt{s}$ = 540 GeV. The production rate obtained with scaling violations (solid curve) is much smaller than the one calculated without scaling violations (standard parton model: dashed curve). Also shown is the expected yield for prompt gammas (dashed dotted curve).

The single-particle yields are much lower than the jet yields at the same p_T, but still much larger than at ISR energy, and probably measurable up to values of the order of 40 GeV/c (Fig.4-14) with a luminosity of 10^{30} . In this case scaling violation effects in QCD have a more important role, since they appear both in the structure functions and in the fragmentation functions. In Fig.4-14 , the standard parton model yield, with fixed structure and fragmentation functions, is shown as dashed curve. The particle yields and their p_T dependence are strongly modified by scaling violation effects.

If present ideas about the respective role of quark and gluon jets are correct, specific correlation effects should be found and could be tested measuring charge ratios among hard jet components. Figure 4-15 illustrates such an effect. Here a π^+ is used as a trigger particle ($20 < p_T < 25$ GeV/c). This results in a bias in favour of a u jet. A hard pion favours a quark jet as opposed to a gluon jet, in view of an expected softer fragmentation function. Observing a π^+, expected to take a large fraction of the jet momentum when used as a trigger particle, is then a strong bias in favour of a u quark jet. The nature of the recoiling jet depends on the angle at which it is observed. As shown in fig. 4.15, a forward recoil ($30^\circ<\theta<60^\circ$) will include u jets almost as frequently as gluon jets and the more so when x_e increases ($x_e > 1$ by construction). There will be far more u jets (associated with valence quarks) than $\bar{u}$ jets (associated with $u\bar{u}$ pairs). The situation is very different at wide angles ($60^\circ < \theta < 120^\circ$). Now gluon jets dominate (qg scattering) and $\bar{u}$ dominate (albeit by a small amount) over u, as a result of the role of valence quarks in the $\bar{p}$, as well as $u\bar{u}$ pair production.

Having emphasized the importance of the jet-production cross-section at collider energy, and the important counting rates which should allow one to efficiently push jet studies beyond the momentum range covered at PETRA and PEP (from 30 to 60 GeV/c, say), we may discuss the continuation of these studies increasing the energy or increasing the luminosity.

In the former case expected gains can be rather large. Rates fall with increasing p_T, but eventually more so with increasing x_T($\sim\sqrt{\tau}$ at wide angles). Going to lower values of $\sqrt{\tau}$ by a factor of 3 to 4 helps a great deal. This can be assessed from Fig.4-16, which gives the quantity $\tau(dL/d\tau)$ for $u\bar{u}$ and gg at $\sqrt{s}$ = 540 GeV (solid curves) and at $\sqrt{s}$ = 2000 GeV (dashed curves). There is by then little dependence of L on $\sqrt{s}$, and rates at fixed (large) p_T ($p_T \sim 50$ GeV/c) increase very much going from 540 to 2000 GeV. While one can thus probe jet production up to much larger p_T values, one will, however, remain within the realm of gluon jets, dominant at $\sqrt{s}$= 540 GeV over the whole probed range,

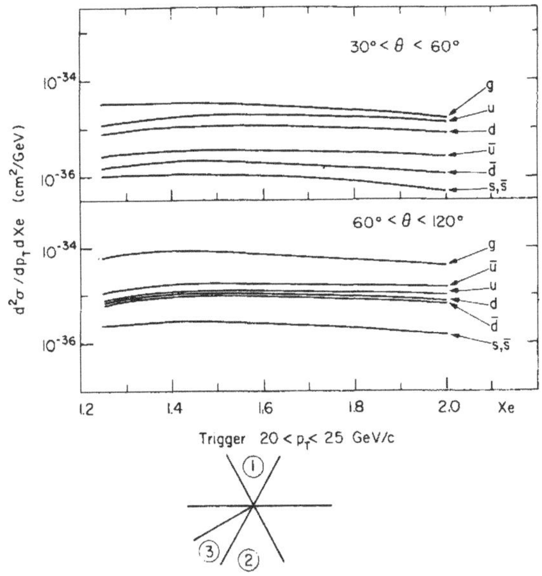

Fig. 4.15. Quantum number correlations. Different jet yields expected when a π^+ with $20 < p_T < 25$ GeV/c is first observed at wide angles. Jet yields in the opposite hemisphere are calculated for two different angular ranges.

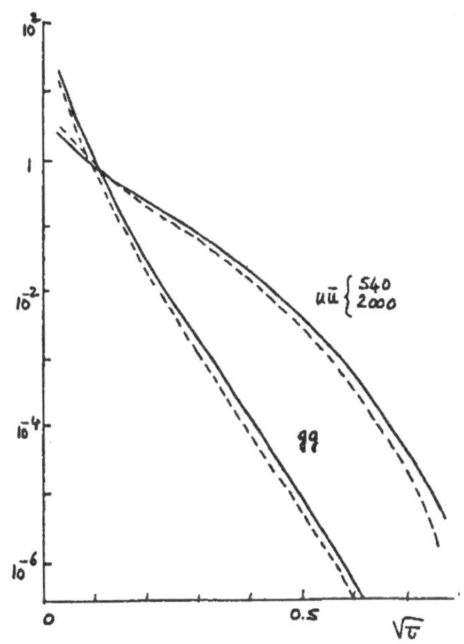

Fig. 4.16. Differential luminosities for $u\bar{u}$ and gg subprocesses. Scaling violations imply only small changes between $\sqrt{s}$ = 540 GeV (solid curves) and $\sqrt{s}$ = 2000 GeV (dashed curves).

and still very much dominant at $\sqrt{s}$ = 2000 GeV at the same luminosity, since the highest experimentally accessible value of x_T will drop when increasing the energy by a factor of 4 while keeping the same luminosity. Significant gains, may, however, be obtained with heavy quarks, since the condition $p_T \gg m_q$ has to hold in order to get reasonably high yields and the p_T range will be much extended. Heavy quarks should give rather spectacular jets, with a multicore structure issuing from weak decays.

In the latter case (Isabelle), a gain in luminosity by 3 orders of magnitude allows one to reach much larger values of x_T ($x_T \simeq 0.6$, say), while still exceeding the (limiting) production rate expected at $x_T \simeq 0.25$ with the $p\bar{p}$ collider. While this is of course a very significant gain for heavy quark studies, it also allows one to study a greater variety of jets, as quark jets become eventually dominant over gluon jets (for $x_T \geqslant 0.3$ at wide angles).

While the $p\bar{p}$ colliders will mainly look at gluon effects, Isabelle will study quark-quark scattering at very large momentum transfers, putting high thresholds on jet energy triggers while achieving a practical counting rate.

It is however, certainly premature to discuss the future of jet physics. The CERN $p\bar{p}$ collider offers already a unique and great opportunity to study jet production. This should provide a much better understanding of the relevant processes than the one on which present predictions are ventured.

IV. 3 Jet yields and weak boson decay

Large jet yields open up a very large range of topical studies. They, however, provide a very large background for the dominant (two-jet) decay modes of the Z and W mesons. This is why so much emphasis was put on their relatively rare leptonic decay modes in Section 2.

One can calculate the jet production cross section associated with W or Z formation and decay. All couplings are known (section 2). In the inclusive distribution at 90°, there is as well known a Jacobian peak at $p_T \simeq M/2$. Such peaks are however found to be much below the hadronic background previously calculated (figure 4-12).

The corresponding effect is shown in Fig.4-18, which presents together the gluon and quark jets expected from hard scattering (Section IV.3), and the u quark yield calculated from W and Z formation and decay. The u quark yield corresponds to $u\bar{d}$

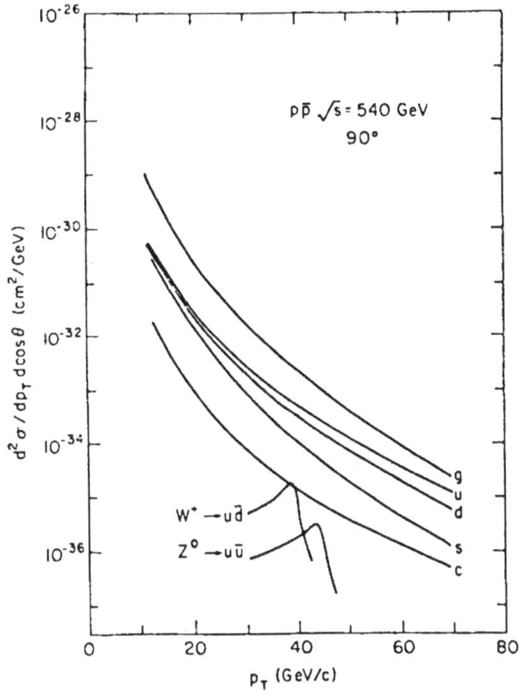

Fig.4-18 Jet yields at large p_T associated with W^+ and Z° formation and decay (with a u jet produced) and as resulting from constituent scattering. The background appears a overwhelming for jets.

and $u\bar{u}$ configurations, respectively. We have taken M_W = 78 GeV, M_Z = 88 GeV, and total widths of 3 GeV in both cases. One sees the Jacobian peak appearing at $p_T \simeq M/2$. It is calculated here with negligible transverse momentum for the intermediate boson. A wide transverse momentum distribution would erode it away. As clear from Fig.4-18, the signal is rather specific, but the background is overwhelming. It is clear that one should not consider any of these estimates (weak boson production and hard scattering jet yields) to be more reliable than at the order of magnitude level. Nevertheless, one may conclude that jet yields as expected from W and Z, are likely to be lost in the background. There are three ways out. One is to consider leptonic yields. They are smaller than those associated with specific jets (the e^+ yield is a third of the u yield for W^+ and $\sim$ 1/8 of the u yield for Z^0). Nevertheless, they should stand out over the corresponding background and this by a large amount. Leptons are expected from the Drell-Yan process (Section 2). However, when selecting a particular mass the signal-to-background ratio is then close to 3 orders of magnitude. Leptons can also come from the weak decay of heavy quarks. Since the branching ratios for leptonic modes are of the order of 10^{-1} for the heavy mesons obtained from c, b, and t jets, and that these jets are relatively infrequent, the signal should appear against background even for an inclusive distribution (Fig. 4-18), and very much so for the lepton-pair decay of the Z^o. This was mentioned in section 2. The lepton way appears therefore as a safe way. Nevertheless, it selects only a minority mode (a few percent) when luminosity considerations are so important.

The wide angular distribution expected from jets originating from weak boson formation and decay, which results from their rather wide distribution in longitudinal phase space, is such that using angular dependence and asymmetry effects associated with non-invariance under parity is unlikely to be reliable.

The second possibility is to focus on jets associated with heavy quarks. The $W^+ \to u\bar{d}$ and $c\bar{s}$ rates are the same. The $Z^o \to u\bar{u}$ and $c\bar{c}$ rates are also the same! In such cases signal and background are comparable (Fig. 4-18). One may, however, have to lose an order of magnitude in yield, defining an unambiguous signature for heavy quarks. Interest in heavy quark production is, however, such that much effort will be needed in that direction and some structure may then appear in the corresponding distributions.

One possibility is to look for peaks in the invariant mass distributions of same sign dileptons, thus reaching configurations where leptonic decays occur at different steps in the decay chain but where the leptons take most of the available momentum. The signal is poor but the background is very low.

The third way considers global jet yields, but uses a refined signal based on correlation measurements. It, however, does not seem to be very promising. Figure 4.19 shows how strong such correlations are expected to be. One takes the attitude here that a jet observed at wide angle ($60° < \theta < 120°$) is likely to be compensated for by a jet in the other hemisphere, also at wide angle ($60° < \theta < 120°$), and in the same plane. The differential production cross-sections for such configurations, with the same p_T on both sides, are presented for $u\bar{d}$ (W^+) and $u\bar{u}$ ($Z°$) jets as a function of p_T, with a clear peak in either case at $p_T \approx M/2$. The QCD background expected with the same selection criterion is also shown. The background remains very strong. The relatively small gain thus achieved is of course due to the weak angular correlation among jets, whatever their production modes are. One may, of course, go further and apply stringent invariant mass cuts. The effect is shown in Fig. 4.20. The differential cross-section $d^2\sigma/dp_T dM$, integrated over a momentum resolution of 8 GeV, is shown as a function of p_T, for ranges which favour W^+ (72 to 84 GeV) and $Z°$ (82 to 94 GeV) production. This is done in the case of $u\bar{d}$ and $u\bar{u}$ jet production, respectively. The hard scattering jet background remains very strong (gluon-gluon configurations

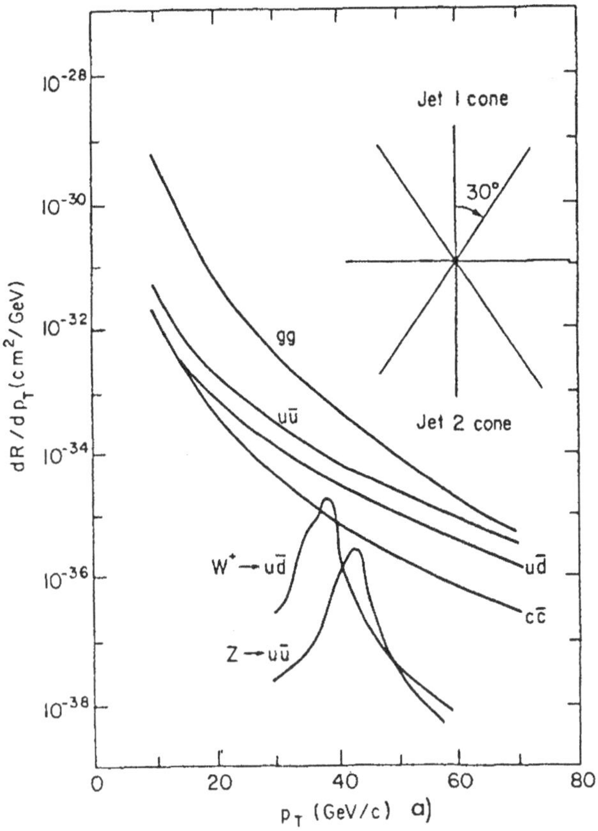

Fig. 4.19. Yields for jet pairs in opposite cones with an opening angle of 30° around 90°. The observation of such correlated jets clearly favours W and Z production. Nevertheless the background remains overwhelming.

are more important still, as shown in Fig. 4.19). One may then conclude that it is probably hopeless to get a signal out.

There is, of course, a rather large amount of uncertainty in all these predictions and, in particular, in those involving gluon jets. Nevertheless, one may conclude that, despite its relatively low yield, the lepton signal appears as the only reliable one (lepton pair for Z^0 and the missing energy of the lepton for W). One may be lucky with others. It is, however, not possible to put any emphasis on them at present.

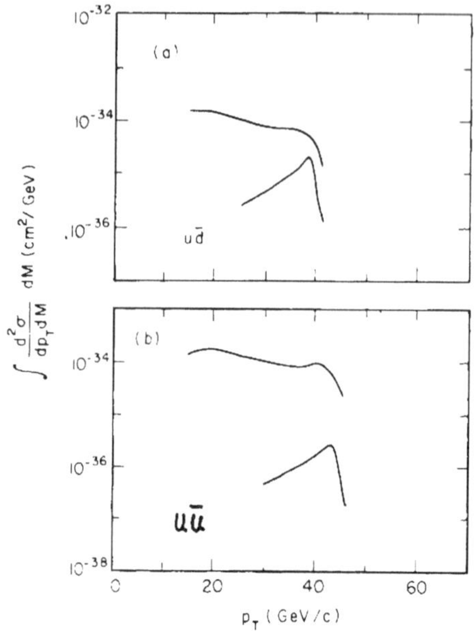

Fig. 4.20. $u\bar{d}$ (for W^+) and $u\bar{u}$ (for Z°) yields. Comparing the signal expected from weak boson formation and decay, and the background associci ted with constituent scattering applying stringent invariant mass cuts.

IV. 4. Heavy-quark production

While the charmed jet yield is relatively small (altogether a few per cent of the global jet yield), it is sizeable. Furthermore, considering $p_T \geqslant 10$ GeV, jets of b and c quarks should be on the same level. With increasing knowledge about such jets coming from PETRA and PEP, good enough signatures will certainly be found.

Since we arbitrarily set the charm quark content of the sea at zero, our charmed jet yield (Fig.4-12) applies for any quark yield, provided threshold effects are negligible. This is certainly the case for $p_T \gg m_q$, since $p_T^2 \sim \hat{s}$. Conversely, the charmed jet yield at high p_T (Fig.4-12) provides an order of magnitude for the production cross-section for any new heavy quark. It is typically given by the production rate for $p_T > m_q$.

Perturbative calculations can be done more thoroughly, introducing masses in the relevant reaction amplitudes (4.5), and integrating with the corresponding kinematical limits[71].

Since semi leptonic decays offer a relatively good signature for heavy quarks, one can calculate the lepton yield associated with c and b quark production. The result is shown in figure 4-21. One obtains similarly the background curves considered in section 2 when discussing W, Z, and quarkonia production, as observed through leptonic modes.

As it was previously mentioned however, the dominant mode for heavy flavour production should be associated production with a leading baryon and a central meson. This is what one is tempted to infer from ISR results[51]. Rates for heavy flavour excitation should be rather large. This would result in a sizeable prompt lepton yield at medium p_T (a few GeV/c).

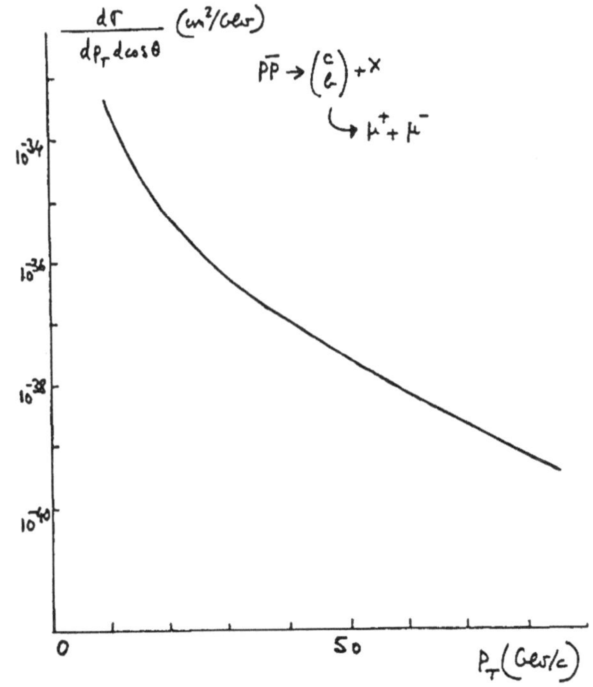

Fig. 4.21. Lepton yield (μ^+ and μ^- or e^+ and e^-) expected from c and b quark leptonic decays, when produced in $p\bar{p}$ collisions at $\sqrt{s}$ = 540 GeV.

IV. 5. An intense colour source

Very high energy protons and antiprotons have been repeatedly considered as broad-band unseparated beams of quarks and gluons. In particular, the gluon differential luminosity is fairly high, provided that $\sqrt{\tau} \lesssim 0.15$ (Fig.4-16). A $p\bar{p}$ (pp) collider appears therefore as an efficient instrument to produce anything coupled to colour, provided that the mass is not too large ($M = \sqrt{\tau s} \lesssim M_Z$, say, at $\sqrt{s} = 540$ GeV). We already saw that for heavy quarks the limitation comes from background rather than rate for $M_q < 40$ GeV. Besides quarks, one may also look for hitherto unknown objects which could be coupled to colour as are quarks and gluons.

At present, there is an interesting possibility which is being considered under the general heading of hypercolour[72] and we shall briefly discuss here a possible aspect of this new scheme which could be of some relevance at collider energy. One could have fermions of a new kind Q, which would be coupled to colour as quarks are, but also through another force referred to as hypercolour, confining them at extremely small distances, and in particular within $Q\bar{Q}$ mesons, which would appear to be point-like up to probing momentum transfers of the order of 0.5 TeV, as opposed to 0.5 GeV for typical hadronic states associated with colour confinement. Such $Q\bar{Q}$ bound states could be neutral under colour. Others could have colour and therefore be confined, but only at typical hadronic distances.

The colourless mesons are at the origin of the idea. Evidence for the coloured ones would provide experimental support. The idea is indeed to associate such colourless scalar mesons with the Higgs mesons of the standard model. Some of these mesons could have a zero-mass resulting from a dynamical breaking of the chiral symmetry imposed at the beginning[73]. They could provide the required Higgs field without one having to depart from a theory with spin-1/2 fermions and gauge fields as the only fundamental entities. Bound states with colour, which should then also exist, would, on the other hand, acquire a mass through their coupling to gluons. These are the ones which one would hope to observe at collider energy. Being coupled to the gluon field, the neutral mesons could be singly produced in gluon-gluon collisions, in much the same way as the W and Z mesons are expected to be produced in quark-antiquark collisions.

Within such a scheme, a new and rich spectroscopy appears in the mass range 10-1000 GeV[72,74]. While an e^+e^- machine would be better suited to search for the charged particles (pair produced), the collider could do just the same (an intermediate gluon replacing an intermediate photon) and also produce the neutral ones singly, thus being able to extend the probed mass range by a large amount.

Insofar as some bound states are identified with Higgs mesons, the game is rather contrived. We may illustrate this with a few relations providing the order of magnitude, at the level of which one has to stay at present.

The Higgs fields are considered as being $Q\bar{Q}$ consendates, built out of the new fermion fields Q, with a non-zero vacuum expectation value $<\phi>$. Since they are Higgs fields and thus responsible for the Z meson mass, one should have

$$M_Z \sim e <\phi> \tag{4.13}$$

The mass of the Z results from the spontaneous breaking and e is a coupling constant of the standard model (we remain at the order-of-magnitude level and do not distinguish between couplings). The Fermi coupling is related to e by

$$\sqrt{G} \sim e/M_Z \tag{4.14}$$

Finally, since the Higgs field is a condensate, the vacuum expectation value $<\phi>$ is equal to the weak-decay coupling F_π , which plays the role of the well-known pion-decay coupling f_π , defined as

$$< 0|A_\mu(0)|\pi(a)> = f_\pi q_\mu \tag{4.15}$$

but now for the $Q\bar{Q}$ condensate. It follows from the definition of F_π that

$$<\phi> = F_\pi \tag{4.16}$$

Combining (4.13) (4.14) and (4.16), one can then write

$$\sqrt{G} \sim (F_\pi)^{-1} \tag{4.17}$$

We may now consider the formation of a coloured partner in gluon-gluon collisions. The coupling, which can be calculated in the one-Q fermion loop approximation, is written as :

$$g \sim \frac{\alpha_s M}{F_\pi} \tag{4.18}$$

with a production cross-section given by

$$\sigma \sim \frac{\alpha_s^2}{F_\pi^2} (\tau \frac{dL_{gg}}{d\tau}) \tag{4.19}$$

The calculation is almost identical to that the π^0 lifetime in the one-quark loop approximation. The coupling introduced, g, is dimensionless and M is the coloured meson mass. One can assume that $(F_\pi)^{-1}$, which is related to the size of the heavy fermion bound system should be of the same order of magnitude whether the system is coloured or not. Using (4.17), the production cross-section can now be compared with that for the Z (Section 2)

$$\sigma_Z \sim G(\tau \frac{dL_{u\bar{u}}}{d\tau}) \qquad (4.20)$$

One may conclude by saying that the production rate for such a new object, which should exist if Higgs mesons are bound states of coloured fermions interacting through the highly confining hypercolour force, should be scaled down, as compared to that for the weak bosons by a factor of $\alpha_s^2 \sim 0.01$, and scaled up or down according to the ratio of the gg and the $u\bar{u}$ differential luminosities. These luminosities are read from Fig.4-15. A mass comparable to that of the Z meson would then correspond to the limit set for a detectable production rate at $\sqrt{s}$ = 540 GeV, for example, of the order of a few per day. A mass of the order of 100 GeV may, however, be too restrictive, since present expectations correspond rather to the range 100 to 1000 GeV. Probing higher masses can be done in two ways, either by increasing the energy (keeping $\sqrt{\tau}$ at a limited value), or by increasing the luminosity. As read from Fig.4-15, going from $\sqrt{s}$ = 540 GeV to $\sqrt{s}$ = 2000 GeV, or increasing the luminosity by 3 orders of magnitude would have similar results. One could thus reach masses up to 300 GeV with a production rate at the level of a few events per day.

Once such a massive meson is produced and fragments (it is confined by colour), it has to be looked for among an overwhelming background. It is hard to venture any safe signature. Nevertheless, the γg decay mode (down by a factor of the order of 10 only, as compared to the dominant gg mode) appears as a likely signature. It would provide γ-rays of $p_T \approx M/2$ with a Jacobian peak, as already discussed in connection with the production of the weak bosons. While such considerations are still very speculative, and while the estimate made may be much in error, it remains worth while to look for structure in the high p_T π^o or γ spectrum, which could probably be seen up to p_T of 40 to 50 GeV at a luminosity of 10^{30} (fig.4-14). This is meant merely to illustrate the type of entirely new physics which could be reached with hadron-hadron interactions at extremely high energies, the so-scalled "Higgs sector" covering a rich and new domain[72,74].

CONCLUSION

The $p\bar{p}$ collider programme has started already through successful test runs with luminosity at the $\sim 10^{25}$ level.

One may have that the luminosity achieved will rapidly increase and that running conditions will soon become stable. One should soon leave speculations based on present knowledge but applied to an energy range an order of magnitude above the one already explored, and enter with awe this new frontier of particle physics.

If the weak vector bosons exist as expected in the present Electroweak theory, they should be found. The research programme as it appears at present is extraordinarily rich. Much should have been learned already when these notes appear in print.

I would like to thank the organizers of the Cargèse ASI to have given me the opportunity to try to summarize once again expectations on the eve of actual experimentation.

REFERENCES

1. S. Weinberg, Phys. Rev. Lett. 19, 1264 (67)
A. Salam, in Elementary Particle Theory, 8th Nobel Symposium (ed. N. Svartholm) (68)
For a review see: E. Abers and B. Lee, Phys. Rep. 9, 1 (73)
P. Langacker, Phys. Rep. 72, 4 (81)

2. Brookhaven Isabelle Report BNL 50648 (77) and BNL 50885 (78)

3. C. Rubbia played a keep role in such an endeavour

4. D. Möhl, G. Petrucci, L. Thorndahl and S. V.D.Meer, Phys. Rep. 58, 2 (80)
A description of the CERN source is given in G. Goggi CERN EP/81-08, Proceedings of the IX International Winter Meeting on Fundamental Physics, Siguenza, Spain (81). This paper is also a general and very good reference on Physics at Collider Energy

5. M. Month, Proceedings of the St Croix ASI (80), Plenum Press

6. Perturbative Quantum Chromodynamics, Phys. Rep. reprint, 5 (82)
AIP Conference Proceedings 53 La Jolla Institute (78)

7. A.J. Buras and K.J. Gaemers, Nucl. Phys. B130 , 249 (78)

8. M. Jacob and P.V. Landshoff, Phys. Rep. 48, 4 (78)
Proceedings Symposium on Jets in High Energy Collisions, Copenhagen (78), Phys. Scr. 19, 69 (79)

9. G. Giacomelli and M. Jacob, Phys. Rep. 55, 1 (79)

10. C. Lattes, Y. Fujimoto and S. Hasegawa, Phys. Rep. 65, 2 (80)

11. ECFA invited talk (77), Proceedings of the Fermilab Aspen Workshop (77)

CERN School of Physics, Malente, Germany (80), CERN 81-04

12. R. Stroynowski, Phys. Rep. 71, 1 (81)

13. For a detailed discussion of Physics at LEP see CERN 79-01; Proceedings of the Les Houches Summer Study (78)

14. R.R. Wilson, Rev. Mod. Phys. 51, 259 (79)

15. B. Richter, Nucl. Inst. Meth. 136, 47 (76)

16. CERN/ISR-LEP/79-33 (79)

17. Gauge Theories and Neutrino Physics, Phys. Rep. reprint 2 (78)

18. Experiments at CERN in 1980 (CERN report 80)

19. S.D. Drell and T. Yuan, Ann. Phys. 66, 578 (71)
The application of the Drell-Yan model to weak vector boson production, in the framework of the parton model is discussed by
C. Quigg, Rev. Mod. Phys. 49, 297 (77)
R. Peirerls, L. Trueman and L.L. Wang, Phys. Rev. D16, 1397 (1977)

20. R.P. Feynman, Photon-hadron interactions (Benjamin NY, 72)

21. J.H. Cobb et al. Phys. Lett. 72B, 273 (77)
C. Kourkoumelis et al. Phys. Lett. 91B, 475 (80)
J.K.Yols et al. Phys. Rev. Lett. 41, 684 (78)
A.L.S. Angelis et al. CERN EP/79-112
D. Antreasyan et al. CERN EP/80-75

22. C. Kluberg et al. Phys. Rev. Lett. 37, 1451 (76)
J. Badier et al. Phys. Lett. 89B, 145 (79)

23. J.G.H. de Groot et al. Z. Phys. C 1, 143 (79),Phys. Lett. 82B, 292 (79)

24. G. Altarelli, Partons in Quantum Chromodynamics, Phys. Rep. to be published (81)
A.J. Buras and K.J. Gaemers, Nucl. Phys. B130, 249 (78)
J.F. Owen, E. Reya and M. Gluck, Phys. Rev. D18, 1501 (78)

25. J. Kubar-Andre and F.E. Paige, Phys. Rev. D19, 221 (78)
G. Altarelli, R.K.Ellis and G. Martinelli, Nucl. Phys. B143, 521 (78), Nucl. Phys. B146, 544 (78)

26. D.M. Scott, Topical Workshop on the production of new particles in super high energy collisions, Madison (79)

27. R.W. Brown and K.O. Michaelian, Phys. Rev. D19, 922 (79)
R.W. Brown, D. Sadhev and K.O. Michaelian, Phys. Rev. D20, 1164 (79)

28. A. Donnachie and P.V. Landshoff, Nucl. Phys. B112, 233 (76); Z. Phys. C 4, 231 (80)
N.S. Craigie, Phys. Rep. 47, 1 (78)

29. For a detailed study of quarkonia decay modes see M. Kramer and H. Kraseman, Proc. Schladming School (79), Acta Phys. Austr. Supp. 21, 259 (79)

30. C. Quigg and J. Rosner, Phys. Rep. 56, 4 (79)
H. Grosse and A. Martin , Phys. Rep. 60, 6 (79)

31. For a review of basic concepts in particle production see:
L. van Hove, Phys. Rep. 1, 7 (71)
W. Frazer et al. Rev. Mod. Phys. 44, 284 (72)
E. Feinberg, Phys. Rep. 5, 5 (72)

32. U. Amaldi et al. Phys. Lett. 44B, 112 (73)
S.R. Amendolia et al. Phys. Lett. 44B, 119 (73)
For a review see G. Giacomelli, Phys. Rep. 23, 2 (76)
U. Amaldi et al. Nucl. Phys. B145, 367 (78)

33. K. Guettler et al. Phys. Lett. 64B, 111 (76)

34. H. Abarbanel et al. Phys. Rep. 21, 3 (76)
M. Baker and K. Ter-Martirosyan, Phys. Rep. 28, 1 (77)
M. Moshe, Phys. Rep. 37, 3 (78)

35. For a review see C. Lattes, Y. Fujimoto and S. Hasegawa, Phys. Rep. 65, 3 (80)
D. Goulianos and A. White, Phys. Rep. in preparation
J. Baumel et al. Technion PH-81/28

36. U. Amaldi et al. Ann. Rev. Nucl. Sci. 26, 187 (76)
J. Fischer, Scattering laws at very high energy and small momentum transfer, Phys. Rep. to be published (81)
A.J. Buras and J. Dias de Deus, Nucl. Phys. B71, 481 (74)
P.V. Landshoff, Rapporteur talk, 17th Int. Conf. on High Energy Physics, London (74)

37. E. Nagy et al. Nucl. Phys. B150, 221 (79)

38. CERN-Annecy-Genova-Copenhagen-Oslo-UCLondon; Contribution to the Madison Conference (80)

39. A. Mueller, Phys. Rev. D2, 2963 (70)

40. S. Fubini, Proc. 4th Scottish Universities Summer School (1963)

41. S.N. Ganguli and D.P. Roy, Phys. Rep. 67, 2 (80)

42. P. Carruthers et al. Phys. Lett. 41B, 597 (72)

43. G. Wolf, DESY 80/13 (80)

44. K. Konishi, A. Ukawa and G. Veneziano, Phys. Lett. 78B, 243 (78)
D. Amati and G. Veneziano, Phys. Lett. 83B, 87 (79)

45. $\alpha\alpha$ and pp interactions at ISR energies, M. Jacob, 5th High Energy Heavy Ion Study, Berkeley (81). The results are from experiment R807 at the ISR

46. W. Thome et al. Nucl. Phys. B129, 365 (77)

47. A general reference on multiplicity distributions is Z. Koba, CERN 73-12

48. D. Horn, Phys. Rep. 4, 1 (72)
R. Slansky, Phys. Rep. 11, 3 (74)
I. M. Dremin and A.M. Dunaevskii, Phys. Rep. 18, 3 (74)

49. L. Foa, Phys. Rep. 22, 1 (75)

50. M. Jacob, Blacksburg Conference on Physics in Collisions, Blacksburg, Virginia (81)

51. Excitation of New Flavour, CERN report ISR Discussion Meeting 2-4 (M. Alhow and M. Jacob), 1981

52. See for instance the Proceedings of the 14th (Munich), 15th (Plovdiv) and 16th (Kyoto) Int. Cosmic Ray Conferences

53. T. Gaiser and G. Yodh, Ann. Rev. Nucl. Sci. 30, 475 (80)
First evidence came from three experiments at the ISR:
M. Banner et al. Phys. Lett. 44B, 537 (73)
F.W. Büsser et al. Phys. Lett. 46B, 471 (73)
B. Alper et al. Phys. Lett. 44B 521 (73)

54. M. Jacob, Jet Phenomena ASI St Croix, Plenum press (81)

55. D.I. Sivers, R. Blankenbecker and S.J. Brodsky, Phys. Rep. 23, 1 (76)
S.D. Ellis and R. Stroynowski, Rev. Mod. Phys. 49, 753 (77)

56. R. Field, La Jolla Institute (78), in Quantum Chromodynamics AIP Conf. Proc. 53, 97 (79)

57. For an explicit calculation of yields in the parton model see S. Ellis and M. Kisslinger, Phys. Rev. D9, 2027 (74)

58. R. Combridge, J. Kripfganz and J. Ranft, Phys. Lett. 70B, 234 (77)

59. A.L.S. Angelis et al. Phys. Lett. 79B, 505 (78)
A.G. Clark et al. Phys. Lett. 74B, 267 (78)
C. Kourkoumelis et al. Phys. Lett. 84B, 271 (79)

60. M. Jacob, Rapporteur talk EPS Conference, Geneva (79)
P. Daniulat, Ann. Rev. Nucl. Sci. 30, 159 (80)

61. J.H. Cobb et al. Phys. Rev. Lett. 40, 1420 (78)

62. Z. Kunszt and E. Pietarinen DESY 79/34 (79)

63. C. Bromberg et al. Phys. Rev. Lett. 38, 1447 (77); Nucl. Phys. B134, 189 (78)
J. Rohlf, Caltech Thesis (80)

64. H. Jöstlein et al. Phys. Rev. Lett. 42, 146 (79)

65. W. Furmanski and S. Pokorski, CERN TH 2665 (79)

66. R. Horgan et al. Nucl. Phys. B179, 441 (81)

67. J. Seyboth et al. Rencontre de Moriond (80) and XXth Int. Conf. on High Energy Physics, Madison (80)

68. J. Whitmore, Phys. Rep. 10, 15 (74); 27, 5 (76)

69. See for instance F. Paige and S.D. Protopopescu, ISAJET, BNL report DD19 (80)

70. See for instance the discussion given in CERN 79-01, Les Houches Summer Study for jets expected at LEP.

71. B.L. Combridge CERN TH 2574 (79)

72. L. Susskind, Phys. Rev. D20, 2619 (79)
L. Susskind and E. Fahri, Technicolour and techniquarks, Phys. Rep. to be published (81)

73. S. Weinberg, Phys. Rev. D13, 974 (76); D19, 1277 (79)

74. S. Dimopoulos and L. Susskind, Nucl. Phys. B155, 237 (79)
S. Dimopoulos, Nucl. Phys. B168, 69 (80)

PROTON LIFETIME EXPERIMENTS

D. Treille

C.E.R.N.

Geneva, Switzerland

The possibility that the nucleon could be unstable has long been considered. However, it is clear that the advent of Grand Unified Theories (GUTs) [1] has given a new impetus and clear motivations to this line of research. I would therefore like to start with a brief (and somewhat superficial) review of what GUTs predict.

I. THEORETICAL PREAMBLE

The mechanism leading to nucleon decay is readily understood by looking at the SU(5) multiplets (Fig. 1). In the matrix of the gauge bosons, we recognize the gluon octet, the SU(2) x U(1) sector, and new non-diagonal parts. When sandwiched between fermion multiplets, in which quarks and leptons co-exist, we see clearly that these non-diagonal parts will induce "cross-talk" between the quark and the leptonic world, making possible processes such as that shown in Fig. 1c. The X and Y are the leptoquarks mediating nucleon decay. They are supposed to be very heavy. If M_X is their mass (in SU(5) there is one single mass to be considered), the expression of the propagator leads to a proton lifetime varying as M_X^4. So the first task is to evaluate M_X [2].

This mass is the grand unification mass, at which the three evolving effective coupling constants -- of the strong and electroweak sectors -- are supposed to merge (Fig. 2a). The reason for the difference in the evolution of the three couplings is shown in Fig. 2b. This difference, naively stated, is simply due to the fact that to probe small structures -- or equivalently to feel the effect of loops with high masses -- we need to have a sufficient resolution. Each interaction has its own trajectory, and the aim is

$$A_\mu = \begin{vmatrix} V+\sqrt{\frac{2}{15}}U & V & V & X^r & Y^r \\ V & V+\sqrt{\frac{2}{15}}U & V & X^w & Y^w \\ V & V & V+\sqrt{\frac{2}{15}}U & X^b & Y^b \\ \bar{X}^r & \bar{X}^w & \bar{X}^b & \frac{W_0}{\sqrt{2}}-\sqrt{\frac{3}{10}}U & W^+ \\ \bar{Y}^r & \bar{Y}^w & \bar{Y}^b & W^- & -\frac{W_0}{\sqrt{2}}-\sqrt{\frac{3}{10}}U \end{vmatrix} \quad \text{a)}$$

$$\psi_5 = \begin{vmatrix} d^r \\ d^w \\ d^b \\ e^+ \\ \bar{\nu} \end{vmatrix}_R \qquad \psi_{10} = \begin{vmatrix} 0 & u_c^b & -u_c^w & -u^r & -d^r \\ -u_c^b & 0 & u_c^r & -u^w & -d^w \\ u_c^w & -u_c^r & 0 & -u^b & -d^b \\ u^r & u^w & u^b & 0 & e^- \\ d^r & d^w & d^b & -e^- & 0 \end{vmatrix}_L \quad \text{b)}$$

c)

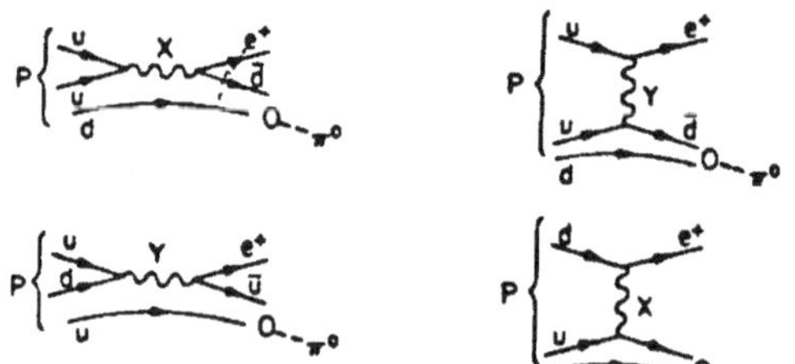

Fig. 1.

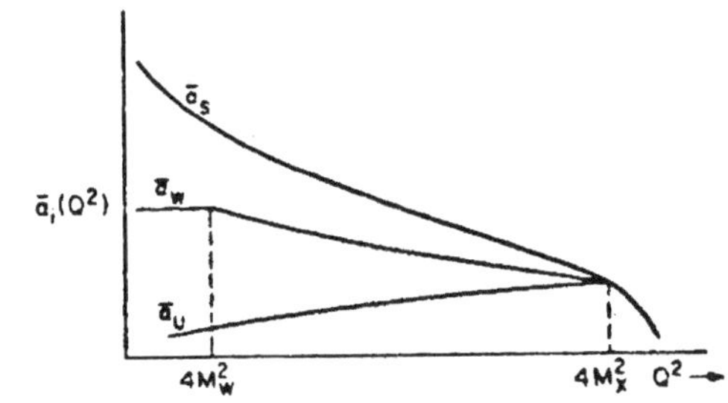

Fig. 2.

i) to project them as accurately as possible;

ii) to match the low-energy couplings with experimental values and the Glashow-Weinberg-Salam (GWS) model.

By doing this, we are left with a single equation:

$$\frac{1}{\alpha_{em}} - \frac{8}{3\alpha_s(M_W^2)} = \frac{11}{2\pi} \ln\left(\frac{M_X^2}{M_W^2}\right)$$

Knowing the "running" strong coupling α_s, the electromagnetic coupling, and the W mass, we can calculate M_X. Taking $\alpha_s(M_W^2) = 0.14$, which corresponds to $\Lambda = 400$ MeV in the usual, but perhaps irrelevant, expression for α_s:

$$\alpha_s(Q^2) = \frac{12\pi}{(33-2f)\ \ln(Q^2/\Lambda^2)} ,$$

the result is

$$M_X = 2 \times 10^{16} \text{ GeV},$$

which is a total disaster, since the corresponding nucleon lifetime, τ_N, would be several orders of magnitude beyond reach.

However, the naive calculation involves several approximations which represent large modifications of τ_N because one computes the logarithm of the mass and then raises M_X to the fourth power. When these approximations are corrected for [2] (see Table 1), the disaster turns into a triumph: M_X decreases by more than a factor of 50 and becomes $\sim 4 \times 10^{14}$ GeV. The major step came simply from renormalizing α from the Thomson limit to the correct value at M_W. From M_X, the computation of the nucleon lifetime (Fig. 3) still involves a few delicate steps, such as the evaluation of the wave function at the origin, $|\psi(0)|^2$, which governs the annihilation rate of the two quarks. The results obtained by various authors within SU(5) fall within the region

$$\tau_p = 10^{31\pm 2} \text{ years}.$$

Table 1.

Physical phenomenon	Effect on M_X
1) Correct treatment of the fermion mass threshold	x 2
2) α_{em} at M_W^2	x 1/10
3) Correct treatment of gauge boson mass thresholds (including the right choice of gauge)	x 2 x 1/1.7
4) Two-loop contributions	x 1/4
5) Effect of the Higgs sector	x 1/1.7
6) x,y are felt below their mass	x 1/2
Total	1/58

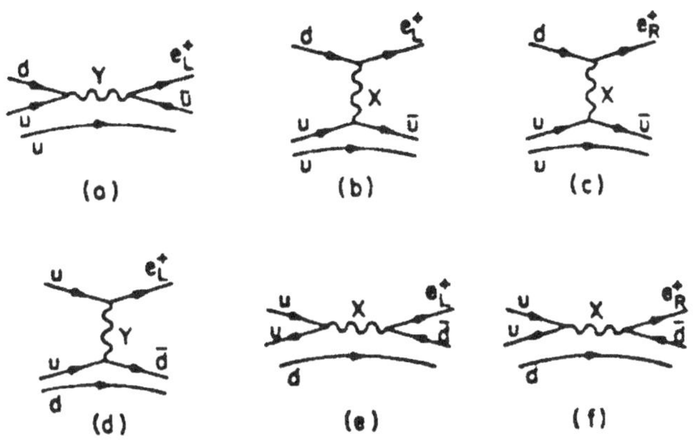

Fig. 3.

The error in the exponent includes the dispersion of the different results as well as the safety margin generally quoted by each group.

Regarding the branching ratios into two-body final states (Table 2), there are a few disagreements. One (which is difficult to understand -- at least by me) concerns the rate into μ's: some authors [3] expect ∿ 20 %, whilst others [4] expect almost nothing. This may be due to the treatment of phase space. Another disagreement concerns the nature of the meson accompanying the positron: generally a pion is favoured, but one model [5] claims that vector mesons will dominate. Moreover, some authors [6] argue that three-body decay modes can be as important as two-body modes.

What we said about SU(5) is in fact valid for all models satisfying the "Desert" hypothesis. General arguments [7], [8] involving dimensional counting for baryon-violating terms, lead to the retention of only those operators which give the following selection rules:

i) $\Delta B = \Delta L$ $\quad p \to e^+\pi^o$

$n \to e^+\pi^-$

$n \not\to e^-\pi^+$

Table 2a. Antilepton + meson branching ratios (from Ref. 32)

Mode		SU(6)					Bag model	
		a)	b)	Static model c)	Recoil model c)	Relativistic model c)	d)	e)
$e^+\pi^{-0}$	p	0.33	0.37	0.36	0.40	0.38	0.08	0.31
	n	0.58	0.74	0.79	0.72	0.68	0.22	0.45
$e^+\rho^{-0}$	p	0.17	0.02	0.02	0.07	0.10	0.18	0.21
	n	0.30	0.04	0.05	0.12	0.18	0.49	0.30
$e^+\eta$	p	0.12	0.07	0.07	0.01	~ 0	0.03	0.05
$\bar{\nu}_e\eta$	n	0.02	0.01	0.01	~ 0	~ 0	0.01	0.03
$e^+\omega$	p	0.21	0.18	0.21	0.25	0.26	0.47	0.19
$\nu_e\omega$	n	0.04	0.04	0.05	0.05	0.05	0.13	0.15
$\bar{\nu}_e\pi^{+0}$	p	0.09	0.15	0.14	0.16	0.15	0.03	0.11
	n	0.04	0.07	0.08	0.07	0.07	0.02	0.04
$\bar{\nu}_e\rho^{+0}$	p	0.04	0.01	0.09	0.03	0.04	0.07	0.08
	n	0.02	0.005	0.01	0.01	0.02	0.05	0.03
μ^+K^o	p	0.005	0.19	0.18	0.08	0.05	0.14	0.005
$\nu_\mu K^o$	n	0.004	0.10	0.01	0.03	0.01	0.10	0.002
τ_n/τ_p		0.88	1.0	1.1	0.9	0.9	1.35	0.75

a) M. Machacek, Ref. 4.

b) M.B. Gavela et al., Ref. 3.

c) G. Kane and G. Karl, Ref. 3.

d) J.F. Donoghue, Ref. 5.

e) A. Din et al., Ref. 4.

Table 2b. Current theoretical predictions for proton decay

$<\tau B>$ years	Lifetime range	Decay modes	Author and model
5×10^{30}	$10^{30\pm2}$	$e^+ + (\pi,\omega,\rho,\eta)$	Ellis, Gaillard, Rudaz, Nanopoulos [SU(5)]
10^{31}	$10^{31\pm2}$	μ decay suppressed	Wilczek, SU(5), SO(10)
10^{32}	-		Georgi, Quinn, Weinberg
$\sim 10^{31}$	$10^{31\pm2}$	e^+ decays	Goldmann Ross, Marciano [SU(5)]
8×10^{30}	-	MIT bag Mode $2+e^+$dec.	Donoghue [SU(5)]
-	-	Do not forget muonic decays	Weinberg, Glashow
10^{30}	10^{29}-10^{32}	$3\ \nu\pi,\ \nu\pi$	

ii) $\Delta S/\Delta B \lesssim 0$ $\quad p \to K^{+-}\nu$

$n \not\to K^- e^+$

iii) $\Delta I = 1/2$ $\quad \Gamma(p \to e^+\pi^o) = \frac{1}{2}\,\Gamma(n \to e^+\pi^-)$

etc.

Furthermore, the exchange of heavy vector bosons leads to the existence of a universal antilepton polarization. This is illustrated for O(10) [9] in Fig. 4.

However, the existence of this desert is simply an assumption. Left/right symmetric models [10], such as SU(16), having spontaneous violations of baryonic B and leptonic L numbers, would permit other types of nucleon decays which may be classified according to the degree of B - L violation:

i) $\Delta(B - L) = -4$ $\quad p \to 3\nu + \pi^+$

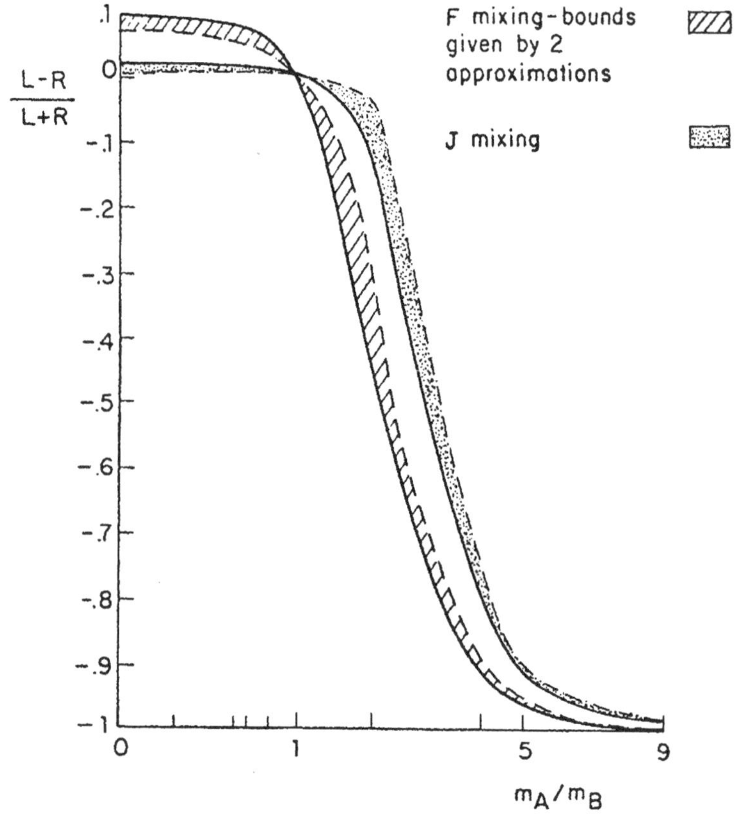

Fig. 4. μ^+ polarization from proton decay for final state in which pions are not distinguished from kaons in the final state. The horizontal scale is linear in $(m_A/m_B)1/2$. The cross-hatched (dotted) band corresponds to F(J) flavour mixing schemes in O(10), where the bounds on each band correspond to various approximation schemes for hadronic matrix elements.

ii) $\Delta(B - L) = -2$ $\quad$ $\underline{p \to e^- \pi^+ \pi^+}$

iii) $\Delta(B - L) = 0$ $\quad$ see above

iv) $\Delta(B - L) = 2$ $\quad$ $\underline{p \to 3\bar{\nu} + \pi^+}$

We see that class (iii) is the only one which was predicted by "desert" models. These "abnormal" proton decays imply the existence of intermediate mass scales. We may ask which mass scale would, for a given class, lead to an observable nucleon lifetime in the 10^{30}-10^{33} years range [11]: for $N \to 3\nu + X$, for instance, the answer is 10^3-10^4 GeV. Also relevant to the existence of such intermediate mass scales is the phenomenon of n-$\bar{n}$ oscillations [12], which will be described later.

To summarize, what would we like to know ?

1) Independently of all subtleties, it is clear that the irrefutable demonstration of nucleon instability would be a fantastic adventure. Obtaining such a clear result implies that huge problems linked to the rate and the backgrounds should be solved, and discussing them will be my main objective.

2) Then, in order to increase audacity, one would like to:

- know the sign of the lepton in order to check the selection rules;
- find its polarization;
- obtain the exact conservation laws;
- observe the nature of the accompanying hadronic final state [13];
- Check whether free quarks are liberated; people cannot help playing with big numbers, and it has already been noticed[14] that the fraction of nucleons which could have decayed since the beginning of the Universe, assuming for the proton lifetime the value found above, is of the same order of magnitude as the fraction of non-integer charges found in the Stanford experiment [15];
- check whether Poincaré and Lorentz invariance holds for processes occurring at such a wild scale of space time (10^{-28} cm).

II. THE EXPERIMENTAL PROBLEM

1. The General Idea

Nothing is more simple in principle. We have just to observe enough nucleons during enough time: 10^3 tons contain 6×10^{32} nucleons. If $\tau = 6 \times 10^{30}$ years, we would get 100 decays/year in such a mass.

The problem is to avoid or reject backgrounds which could simulate the signal.

2. The Signal

We are looking for a decay which in its ideal configuration is easy to imagine: a lepton with momentum balanced by a π or a vector meson and $\sim$ 1 GeV energy deposit. However things are not so clear.

A major problem is linked to the fact that the decaying nucleon is bound inside a nucleus. First, Fermi momentum ruins the back-to-back configuration and apparent momentum balance of the products, but the total energy is unchanged. Figure 5 shows, for instance, the resulting plot for $p \to e^+\pi^o$ in the (acollinearity versus fractional momentum) plane [17]. This spread is obviously important, because it allows background to creep in.

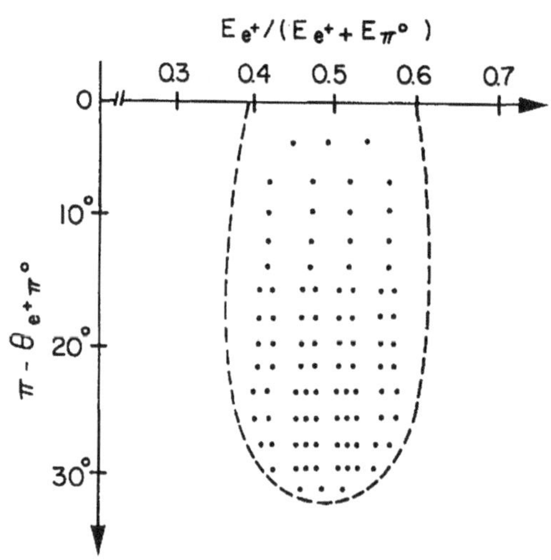

Fig. 5.

Then the decay products have to escape from the nucleus. Figure 6 gives the probability of non-reinteraction for pions in various nuclei as a function of energy. η and ω escape before decaying. The ρ does not, and should be treated as two pions. Summarizing various measurements and estimates, we can say that for nucleon decay in iron, one pion escapes from the nucleus for every two produced. If we further require that no energy loss occurs, then only one pion in three survive. More details can be found elsewhere [18]

Then, can one get the sign of the lepton ? Implementing a sufficient magnetic field on such a volume seems to discourage the richest physicists since a respectable field (a few kilogauss) is needed. Using the delayed μ-decay is in principle the solution.

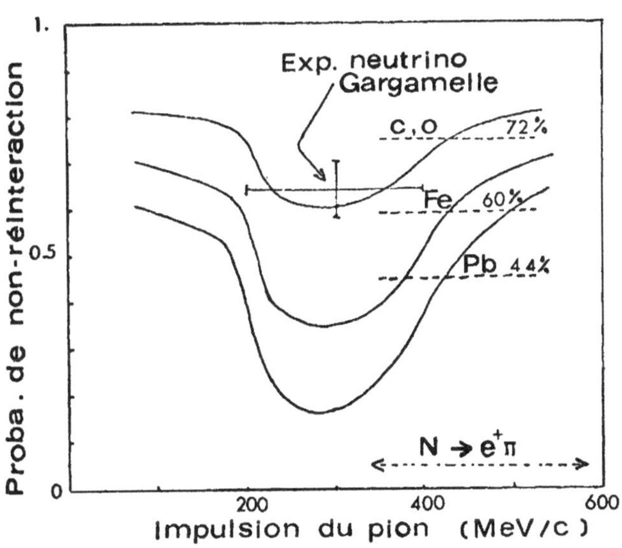

Fig. 6.

However, this does not work for water since the μ^+ and μ^- have the same lifetimes because the nuclei are too small to cause much capture of μ^- [19], whereas for iron, the mechanism is indeed viable: $\tau^-/\tau^+ \simeq 0.07$. The problem here may be linked to the structure of the calorimeter: the low-energy positron ($\sim$ 30 MeV) has to cross iron and find its way to detectors (tubes) which are "alive", i.e. which have not been already fired by the decay itself and set into dead time.

Another possibility is to look, not at the direct μ (which according to some authors does not exist anyway), but at the μ from the π decay, for instance in $n \to e^+\pi^-$ ($\pi^- \to \mu^-$). What is relevant now is the difference in the behaviour of π^+ and π^- due to absorption of the latter indicated by the absence of μ's.

Finally, can we measure the polarization of the lepton ? This question is relevant only for μ's, again through their $\mu^+ \to \ldots$decay. First we need a non depolarizing medium, so that the stopping μ keeps its polarization. Water is all right but not iron. If x is the fraction of the μ momentum carried by the electron and P the muon polarization, the electron angular distribution goes as

i) $1 - P \cos\theta$ if $x \simeq 1$

ii) $1 - 1/3\, P \cos\theta$ if all x are retained.

The problems are evident:

a) the nucleon decay to $\mu^+ + X$ may have a small branching ratio;

b) we will be forced to keep all x's so that the effect is small;

c) measuring the angle between the μ and the decaying electron, even with poor resolution, is problematic. Čerenkov techniques (see later) claim that an angular resolution of 25° can be achieved.

3. Muon Background

These are μ's originating from atmospheric showers. The incident spectrum has a power law in energy. At a shallow depth, energy loss by ionization maintains this power law. However, at greater depths, only very energetic muons are present: for these, other types of energy losses (bremsstrahlung, etc.) which are energy-dependent have been at work, and the energy spectrum has become an exponential. Vertically it is $\sim A \exp(-h/\lambda)$. At an angle θ, there are two competing effects: in the atmosphere the π's have a longer decay path, whereas in earth the shielding is increased.

The real flux through a given underground set-up [20] depends on the exact thickness of the overlying rock. Figure 7 shows how different this can be for a flat surface or for a hemispherical one.

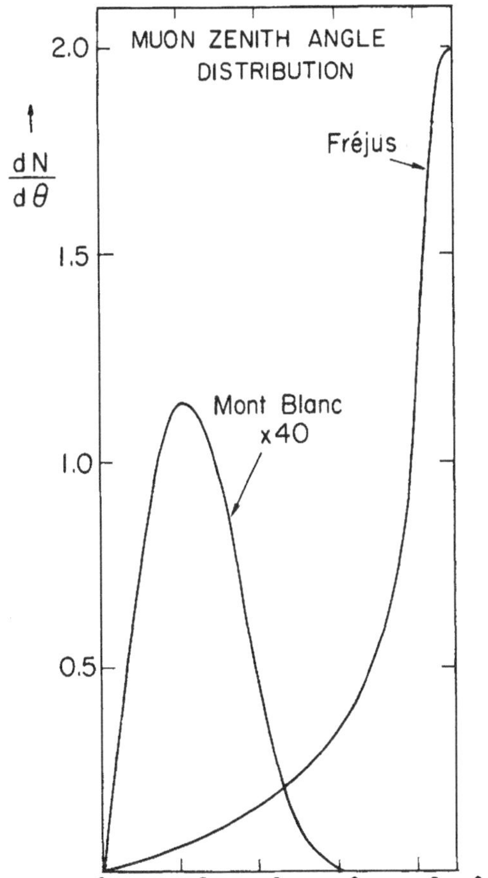
MUON ZENITH ANGLE DISTRIBUTION
2.0
1.5
1.0
0.5
dN/dθ
Fréjus
Mont Blanc x40
0°
20°
40°
60°
80°
90°
Zenith angle θ

Fig. 7.

In this calculation [20] such rock layers were assumed for the Mont Blanc and Fréjus tunnels, respectively. Experimentally the observed distributions do not reproduce the predictions well: it is thus demonstrated that the Mont Blanc region is not flat and the Alps at Fréjus are not hemispherical. Figure 8 shows the μ flux in the various set-ups we will consider. Note the two components of the μ flux, with the ν-induced muons dominating at great depths.

The μ's themselves are not so dangerous: it is relatively easy to protect a detector against entering μ's (see later). The problem is due to μ's which do not penetrate the detector but send to it, for instance, an isolated neutron originating from a hadronic cascade. To give a dangerous background, the n should deposit $\sim$ 1 GeV and no other hadron should penetrate. This has been computed by Grant for a $(10\ m)^3$ detector [21]. Clearly the problem occurs on the edge of the detector. The full rejection, obtained with good energy resolution and perfect detection of eventual accompanying hadrons, gives an event rate which corresponds to $\sim 2 \times 10^{-5}$ of the number of μ's through the detector.

If you are deep enough, this background is so low that you can forget about it. If not, you have to protect the detector on its sides so as to veto such events. Here comes the important notion of fiducial volume. One can define the fiducial region F as the inner part of the detector, which satisfies the two conditions :

i) all products of a nucleon decay occurring in F are contained and measured in the full detector;
ii) no incoming neutral can reach F.

4. Neutrino Background

Neutrinos have the same origin as muons, but they are not shielded. The angular distribution is therefore nearly isotropic, slightly peaked horizontally. The ν and $\bar{\nu}$ fluxes are equal. In the absence of ν oscillations (see later) there should be twice as many ν_μ as ν_e. The rates are given in Table 3 [22].

Considering $p \to e^+\pi^o$, the two dangerous modes are:

$(\nu_e,\bar{\nu}_e)\ N \to e\pi N,$ $\qquad \nu_e$: charged current

$(\nu,\bar{\nu})\ N \to (\nu,\ \bar{\nu})\ (\begin{smallmatrix}\pi \\ \pi\ \pi\end{smallmatrix})\ N$ $\qquad \nu$: neutral current.

With $\Delta E/E = \pm$ 20 %, each of these represents 10^{-2} events per year per ton. So if only an energy cut is applied, the equality of the signal from N decay and the ν background is obtained for:

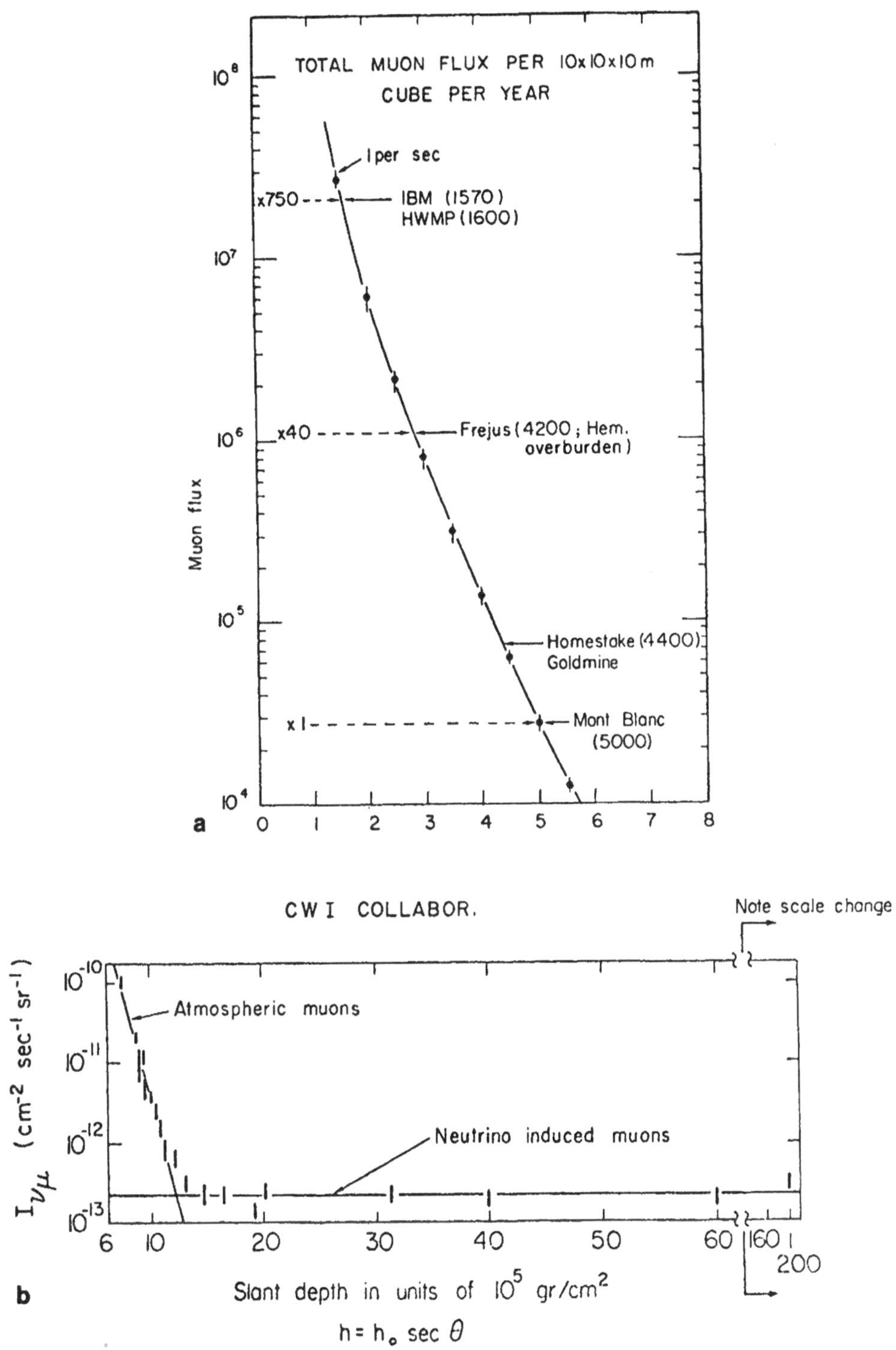
TOTAL MUON FLUX PER 10x10x10 m
CUBE PER YEAR
1 per sec
x750
IBM (1570)
HWMP (1600)
x40
Frejus (4200; Hem. overburden)
Homestake (4400)
Goldmine
x1
Mont Blanc
(5000)
Muon flux
a
CWI COLLABOR.
Note scale change
Atmospheric muons
Neutrino induced muons
$I_{\nu\mu}$ (cm^{-2} sec^{-1} sr^{-1})
b
Slant depth in units of 10^5 gr/cm^2
h = h$_o$ sec θ

Fig. 8.

Table 3. Neutrino events/year/100 t

Visible energy GeV	Charged current $\nu_\mu, \bar{\nu}_\mu$	Charged current $\nu_e, \bar{\nu}_e$	Neutral current	Total
0.4 - 0.6	2.06	0.94	1.02	3.96
0.6 - 0.8	1.34	0.64	0.51	2.49
0.8 - 1.0	1.13	0.47	0.31	1.91
1.0 - 1.2	0.89	0.38	0.23	1.51
1.2 - 1.4	0.74	0.31	0.10	1.22

$$\tau/B_{e^+\pi^o} \sim 2 \times 10^{31} \text{ years.}$$

Further rejection can be obtained from topological requirements. In principle, a decay at rest and a fixed target ν interaction are different. We have seen that owing to Fermi momentum the topology of n decay is somewhat confused. In Fig. 5 we can see the domain to be considered. It has been claimed that nevertheless a gain of two orders of magnitude in background rejection can be expected (in water): this may be somewhat optimistic. Nuclear scattering and energy loss in the nucleus will cause good events to leave the domain, and should be considered as a source of loss.

Depending on the degree of optimism, we can say that the ultimate limit on $\tau/B_{e\pi}$ will be somewhere between 2×10^{32} and 2×10^{33} years (Fig. 9). To improve this, we must turn to drastic solutions, such as performing the experiment in a cave beneath the surface of the moon!

5. Other Backgrounds

Radioactivity may be a problem at the trigger level for calorimeters but not further. Unknown backgrounds (see later, the Kolar events) could well be present, but on the other hand could introduce new physics.

III. STATUS OF N DECAY STUDIES

Figure 10 gives as a function of time the experimental limit on the N lifetime[23]). Most of the time the values quoted are model-dependent.

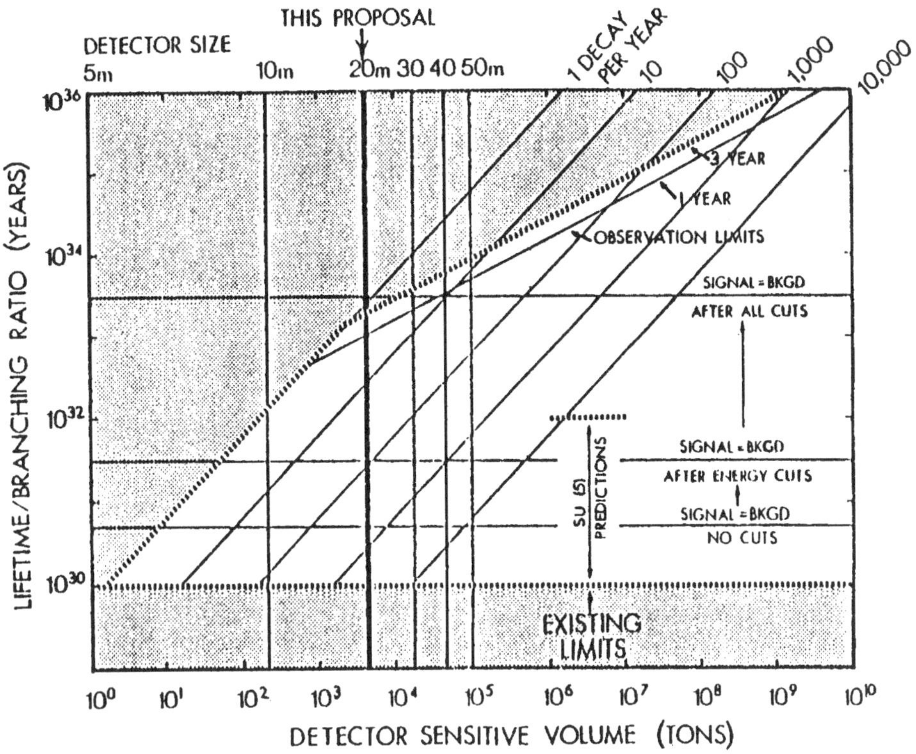

Fig. 9

Radiochemical methods have been employed [24]. We use the fact that N decay inside the K nucleus will produce argon, which can be detected. For instance:

$$^{39}_{19}K \xrightarrow{(p)} {}^{38}_{18}Ar \xrightarrow{n} {}^{37}_{18}Ar .$$

However, neutrinos set a limit on the sensitivity of this method to $\sim 10^{30}$ years.

Direct detection of N decay is mostly considered nowadays.

Figure 11 [25] shows the present detectors, in operation or planned. They are either calorimeters or Čerenkov set-ups.

The only results available come from the "small" calorimeters in the bottom left-hand corner. One sees that their fiducial mass, as previously defined, is rather modest; it is even more exact to say that this first generation does not really satisfy the criterion of the fiducial volume.

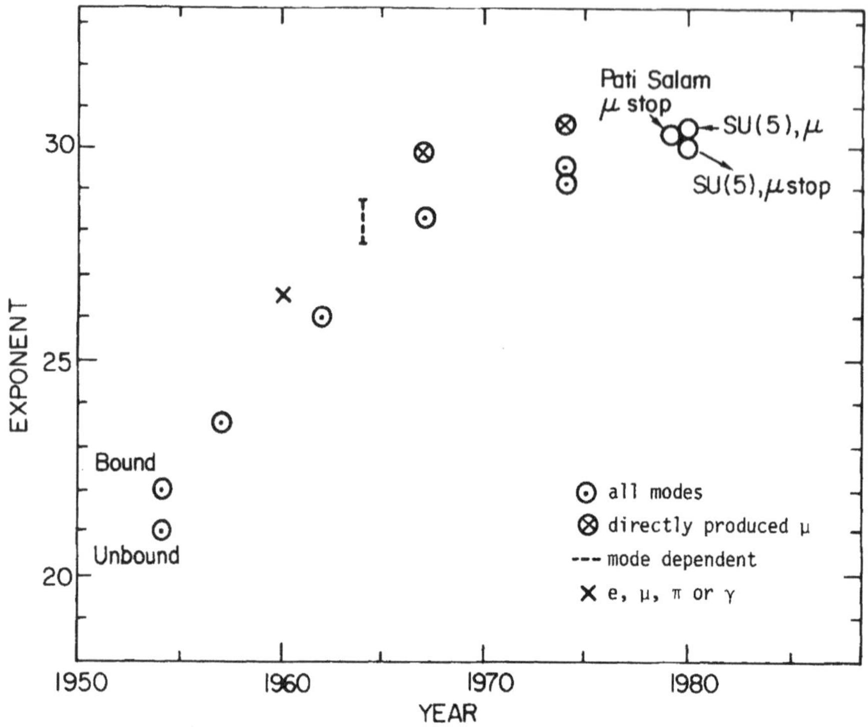

Fig. 10

1. Three Recent Results

1. Homestake Mine (USA) [26)] The depth is 4400 mwe. The detector (Fig. 12) is made of water Čerenkov modules, piled up, with a wavelength shifter; the total mass is 180 t. Above and below, scintillation vetos are used against penetrating particles. The only signal which is looked for corresponds to the delayed sequence

$$N \to \mu + \ldots$$
$$\quad \hookrightarrow e + \ldots,$$

where the μ can be prompt or originating from a π. The electron threshold is 10 - 15 MeV.

Candidates are those which satisfy the following requirements:

i) no veto was recorded;

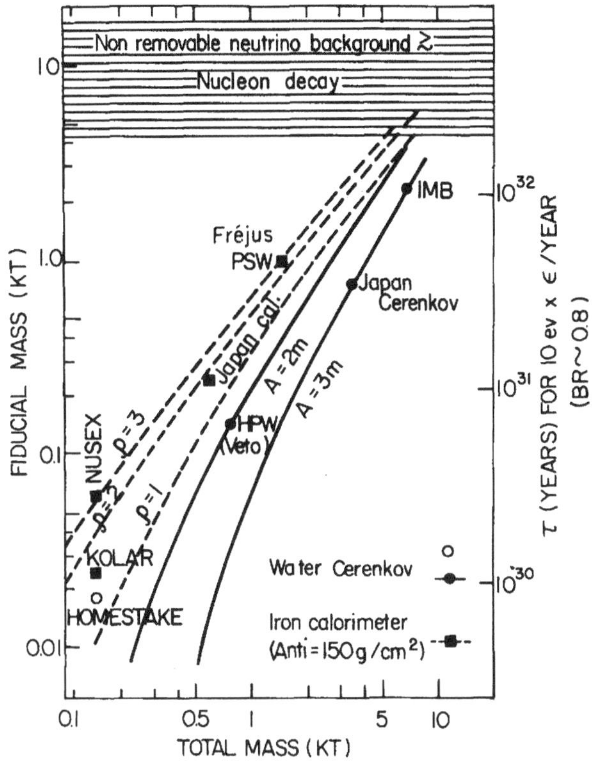

Non removable neutrino background ≳
Nucleon decay
10
1.0
0.1
0.01
FIDUCIAL MASS (KT)
Fréjus
PSW
IMB
Japan Cerenkov
Japan cal.
A = 2m
A = 3m
HPW (Veto)
NUSEX
ρ = 3
ρ = 2
ρ = 1
KOLAR
HOMESTAKE
Water Cerenkov
Iron calorimeter (Anti = 150g/cm²)
10^32
10^31
10^30
τ (YEARS) FOR 10 ev x ε /YEAR (BR ~ 0.8)
0.1
0.5
1
5
10
TOTAL MASS (KT)

Fig. 11

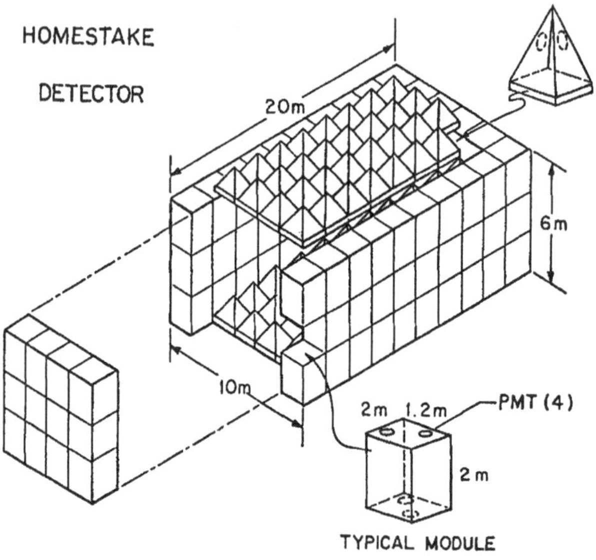

Fig. 12

ii) the event is not downward, i.e. the delayed electron signal is not geometrically recorded below the first signal;

iii) not more than two water tanks ($2.3 \times 1.5 \times 1.5\ m^3$) are concerned: this has been shown to guarantee confinement with a good efficiency;

iv) 50-600 MeV of energy are deposited.

The important mode, $e^+\pi^o$, does not contribute. Averaging over the contributing ones within SU(5) and putting in detection efficiencies, one predicts for water a net muon number per proton of 0.2 - 0.38. The signal obtained in 127 days of exposure is: $\leqslant 2$ possible candidates. Interpreted as background, this sets a limit:

$$\tau \geqslant 1 - 2 \times 10^{30} \text{ years.}$$

2. Case-Western-Irvine (CWI) [27]) This exposure represents 67 tons per year. The detector consists of liquid scintillator and flash tubes at a depth of 9000 mwe in a South African mine.

One limit comes from μ stops *), again through the delayed pulse. The result is $\tau/B_\mu > 6 \times 10^{30}$ years with 90 % CL. For SU(5) this means

$$\tau_N \geqslant 10^{30} \text{ years.}$$

For the Pati-Salam model, this would be $\tau_N \geqslant 2 \times 10^{30}$ years. Other, less restrictive limits are obtained from throughgoing particles.

This set-up is in fact used as a detector for decays which can occur in the surrounding rock; the philosophy is therefore the opposite of the one which relies on a fiducial volume.

3. The Kolar Gold Field experiment (KGF) [28]), India Here the authors claim that they observe a positive result.

The detector is a calorimeter (Fig. 13) with 140 tons of iron plates (2 cm thick) and layers of proportional tubes ($10 \times 10\ cm^2$ of cross-section) situated at 7600 mwe. The mean density is therefore quite small, less than that of the surrounding rock.

The μ's are going through the detector demonstrate that the tubes have a good efficiency. Muons from ν interactions are observed

*) Stopping μ's mostly come from the pions of the hadronic cascades induced by primary muons.

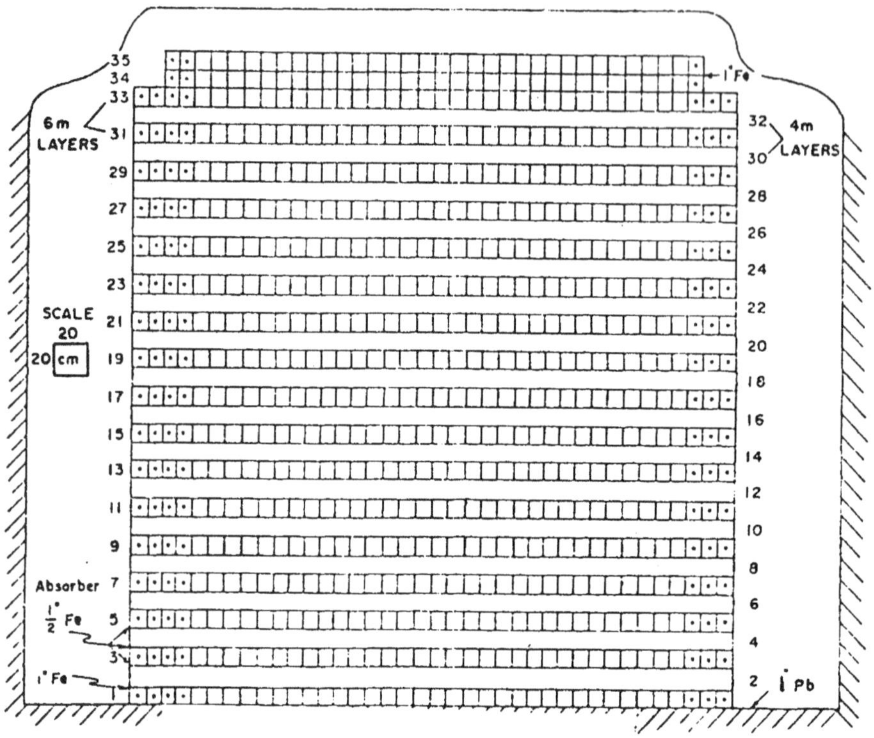

Fig. 13

at very large angles. A Monte Carlo simulated e of 460 MeV is shown in Fig. 14d. Note that owing to the large size of the tubes, it appears to be rectilinear with a good fraction of the tubes not being fired.

The three candidates are shown in Fig. 15. All three vertices are close to the edge. Each time, there is a track leaving (or entering), but according to the authors it does not look like a μ or a π because several tubes have not fired. Energy deposit does not contradict nucleon decay, but it does not tell us much either.

The authors claim that these three events cannot be explained by the usual interactions (Table 4) and that they demonstrate the instability of the proton with

$$\tau_p \simeq 8 \times 10^{30} \text{ years.}$$

However, we are allowed to be sceptical. The events are never contained; the surrounding rock, within a couple of interaction lengths, represents much more weight than the calorimeter itself. There is in this experiment another class of unexplained events, the so-called

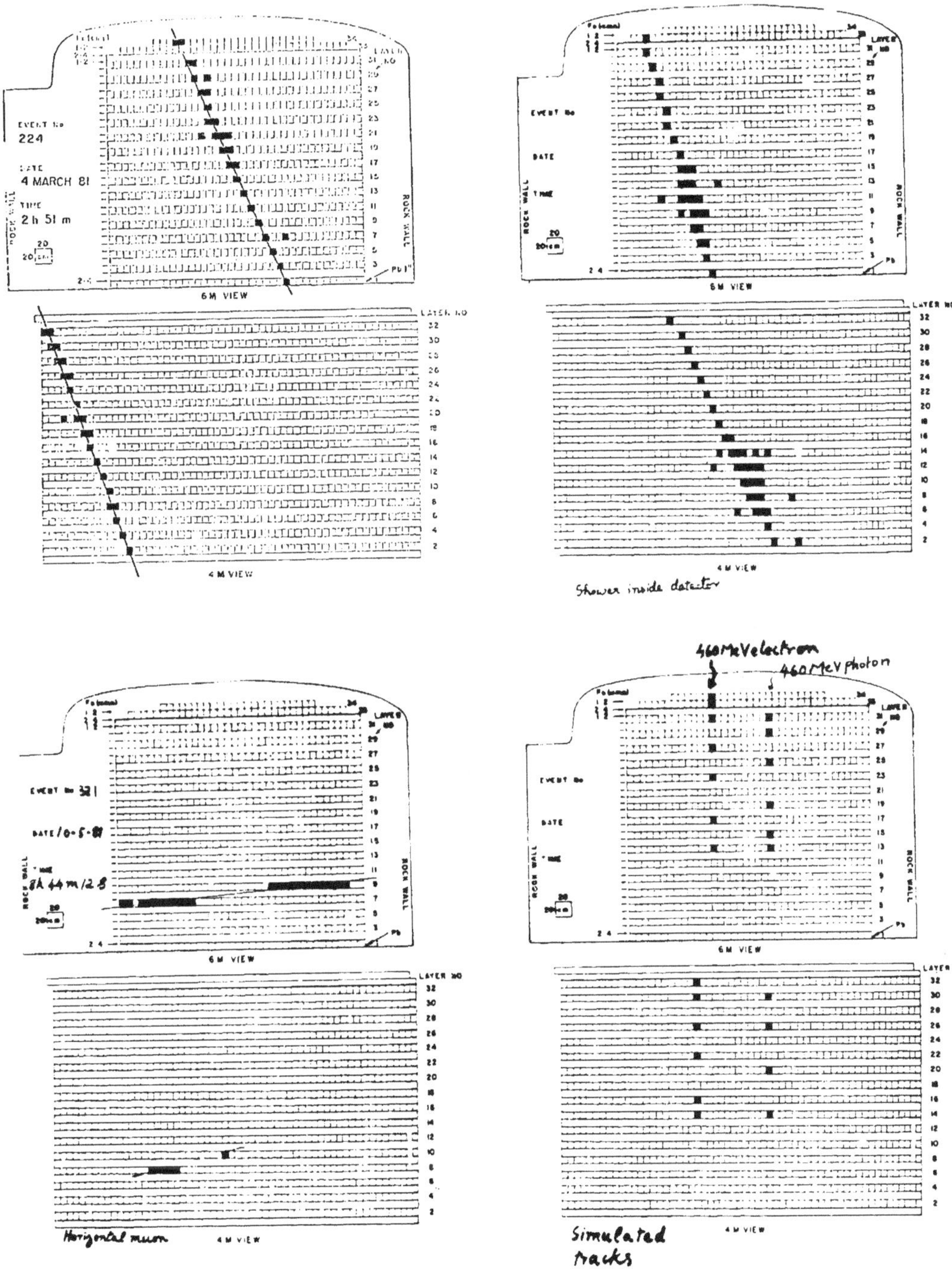
EVENT No
224
DATE
4 MARCH 81
TIME
2 h 51 m
ROCK WALL
LAYER NO
6 M VIEW
4 M VIEW
Shower inside detector
460MeV electron
460MeV photon
EVENT No 321
Horizontal muon
Simulated tracks

Fig. 14

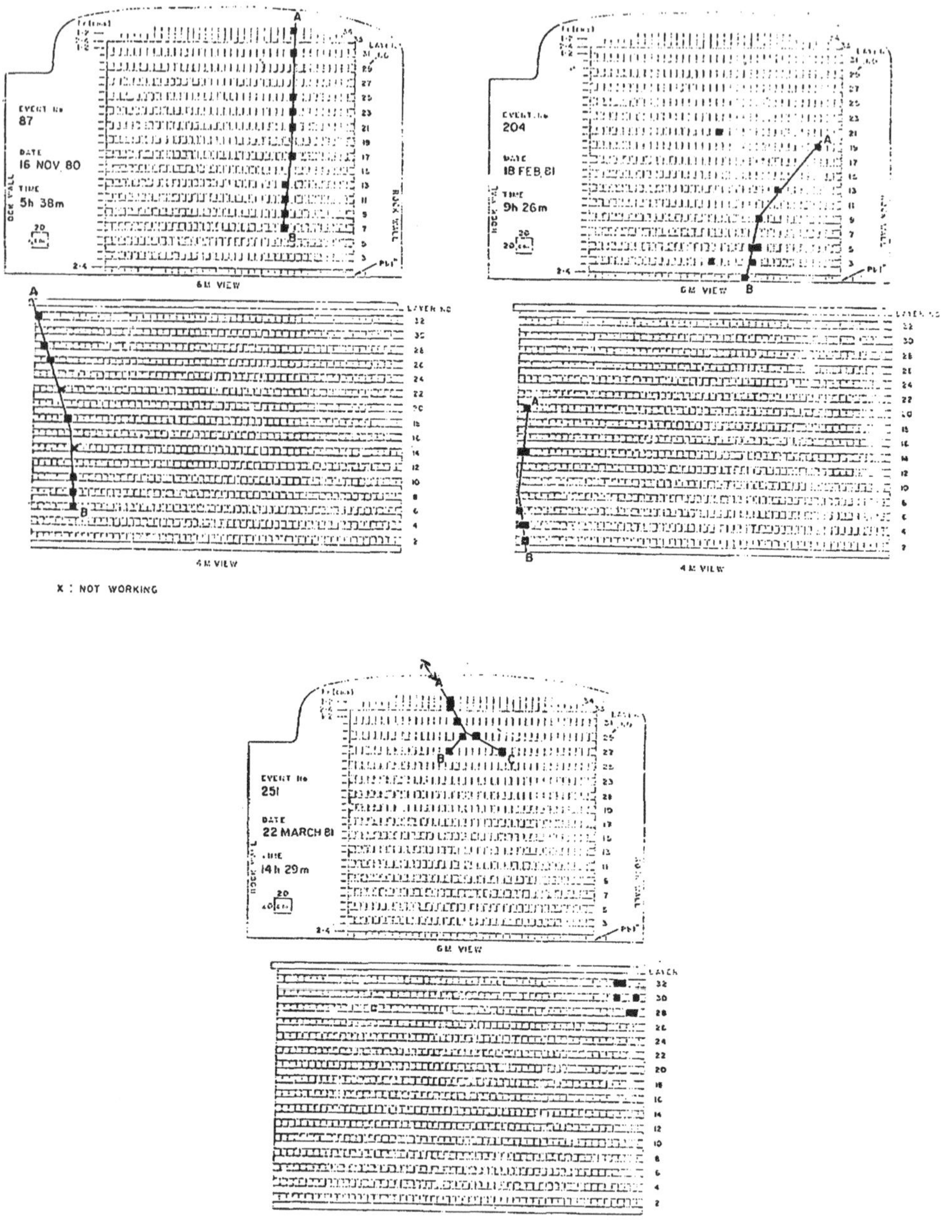
EVENT No
87
DATE
16 NOV 80
TIME
5h 38m
6 M VIEW
4 M VIEW
X : NOT WORKING
EVENT No
204
DATE
18 FEB 81
TIME
9h 26m
6 M VIEW
4 M VIEW
EVENT No
251
DATE
22 MARCH 81
14h 29m
6 M VIEW
4 M VIEW

Fig. 15

Table 4. Basic data: detector volume : 6 m x 4 m x 3.7 m
total weight : 140 tons (iron)
life-time : 205 days
total Nr of events: 388

Category of events	Observed nr.	Estimated nr.
1. Single muons penetrating the detector (θ = 0-55°)	360	350 ± 50
2. Side showers	2	-
3. Parallel muons (E_{μ} ∿ 15 TeV multiplicity ⩾ 2)	2	-
4. Stopping muons	2 *	-
5. ν-interactions in rock θ = 0-55°	Grouped under category (1)	∿ 8
θ = 56-90°	13	13.2 ± 2.5
6. ν-interactions inside the detector	3	6 ± 2
7. Kolar events (Multiprong-wide angle-air vertex events)	1(?)	(?)
8. Unusual events (Nucleon decay candidates)	3	(?)

* A fraction of these events would be ν-induced muons or pions that stop inside the detector.

Kolar events [28)]. In a previous version of the experiment they appeared as decays in the air; but since now the detector is compact, they appear to come from dense matter. As they are defined, the deposited energy is in principle larger than m_p but only if they are correctly contained.

Without proving that the result is wrong, all these facts indicate that second-generation experiments with a large and true fiducial mass will really be welcome. Personally I will wait for their results before making up my mind about proton stability at the 10^{31} years level.

IV. THE FUTURE

1. The Very Near Future

The Mont Blanc detector [29] is a 150 t fine-grain digital calorimeter: 1 cm thick iron plates, $(1\ \text{cm})^2$ streamer tubes with resistive cathodes and X - Y strip read-out (Fig. 16). A test module has been exposed to various beams at CERN [30]. The response to e^- indicates that for the $p \rightarrow e^+\pi^o$ decay, one may expect an average of $\sigma_E/E \sim 20\ \%$. The response to π^- gives

$\sim$ 85 % for π identification,

$\sim$ 10 % for $\pi\ \mu$ ambiguity,

$\sim$ 5 % for π-e confusion.

The exposure to the CERN Proton Synchroton ν beam is very instructive because the ν energy spectrum has a similar shape to that of the atmospheric neutrinos. But a test run of 3.5×10^{18} protons and 2 tons fiducial volume gives the same number of events ($\sim$ 600) as a one-year exposure of 2500 tons to atmospheric neutrinos.

A preliminary analysis of 180 events from the test reveals that several of them have a structure which resembles the Kolar candidates (Fig. 17): ν interactions inside or outside the module can simulate N decay candidates. After normalization to the Kolar ν luminosity, the claim is that one should expect, in the Indian-Japanese set-up, only 0.3 such background, whilst three candidates are observed. However, such an extrapolation from one experiment to the other looks to me to be rather tricky because of the very different structures of the detectors and their surroundings. Hopefully one will soon get results from the Mont Blanc detector (NUSEX) *in situ* and see whether candidates of the Kolar type are also observed.

A 31 t detector, Soudan 1, at 1750 mwe, made of layers of taconite concrete (taconite: a high iron-content flint-like rock, valuable as iron ore) and layers of gas proportional tubes (1" diameter) (Fig. 18), has recently been brought into operation in the Soudan Mine (Minnesota) [31]. A proposal for a 1000 ton tracking calorimeter (MAO) in the Soudan mine is also available [31].

The Mont Blanc and Soudan detectors, in some sense, are both still first-generation because they have small (or null) fiducial volumes according to our previous definition. However, the granularity and compactness are much improved.

2. The More Remote Future

1. The Fréjus experiment [32] This French-German experiment, recently approved and funded, will consist of a huge fine-grain

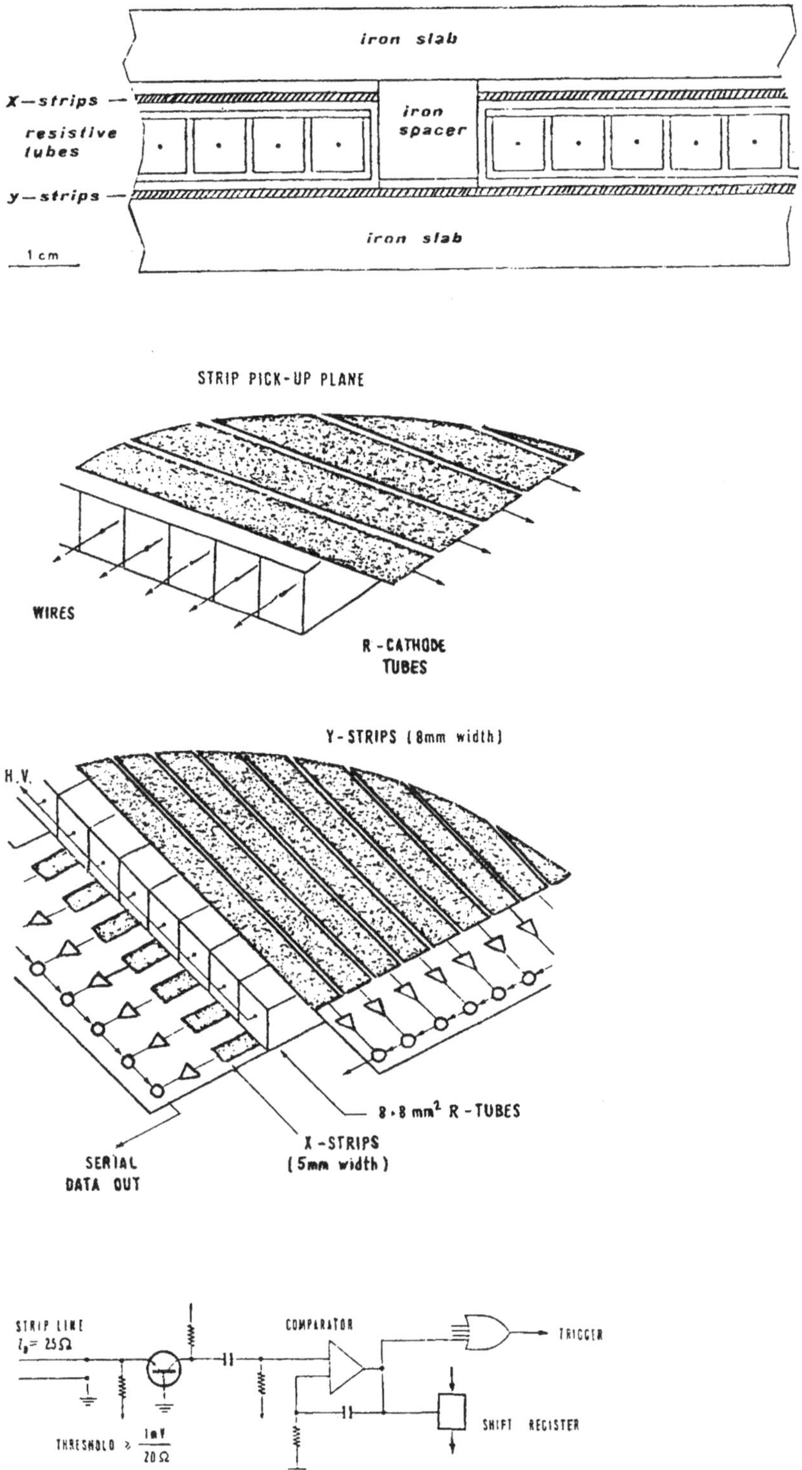
iron slab
X—strips
resistive tubes
iron spacer
y—strips
iron slab
1 cm
STRIP PICK-UP PLANE
WIRES
R-CATHODE TUBES
Y-STRIPS (8mm width)
H.V.
8·8 mm² R-TUBES
X-STRIPS (5mm width)
SERIAL DATA OUT
STRIP LINE
Z_0= 25Ω
COMPARATOR
TRIGGER
SHIFT REGISTER
THRESHOLD ≥ 1mV/20Ω

Fig. 16

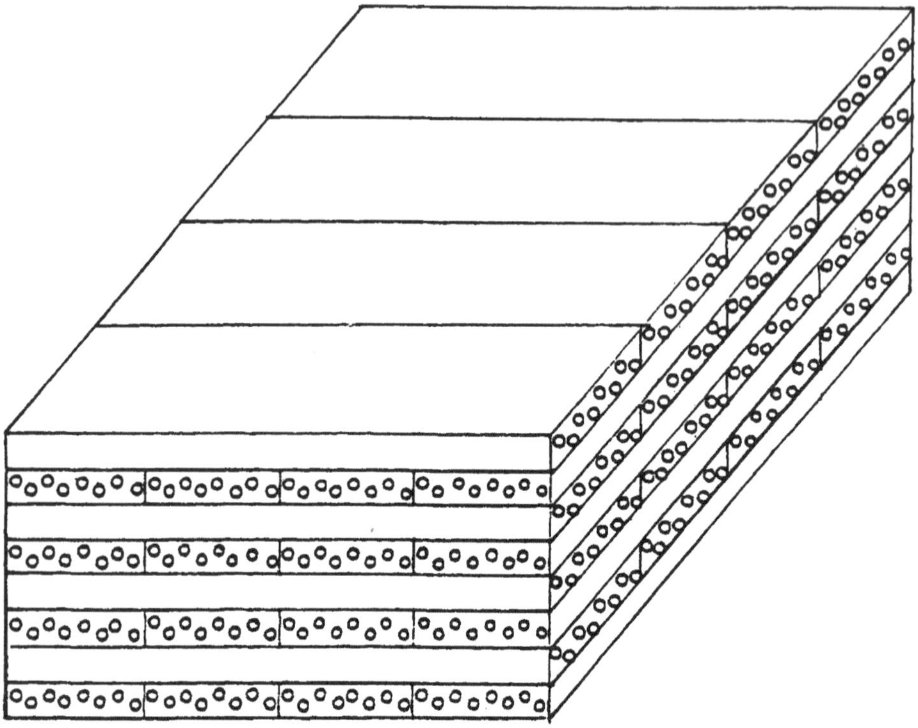

Fig. 18. Schematic of Detector Concept

calorimeter (Fig. 19). The size will be 6 x 6 x 17 m^3. The mass will come from iron plates, 3 mm thick. The trigger will be provided by 200 planes of Geiger tubes [$(15\ mm)^2$]. The detection will be ensured by 800 horizontal and 800 vertical planes of flash tubes [$(5\ mm^2)$]. The number of flash tubes hit is a linear estimator of energy. One should get for electrons a resolution of $15/\sqrt{E}$ and an e-π separation better than 80 %. The total mass will be 1500 tons, with 1000 tons in the fiducial volume. The hole for the calorimeter should be ready in autumn 1982, and some modules could already be installed at that time.

About the different types of tubes I referred to [33], let me distinguish:

- self-triggered ones, such as proportional tubes, limited streamer, and Geiger tubes,

- triggered ones, such as flash tubes, and recall the two important parameters of such devices;

- τ_s, the memory time of the passage of the particle, typically 1 μs;

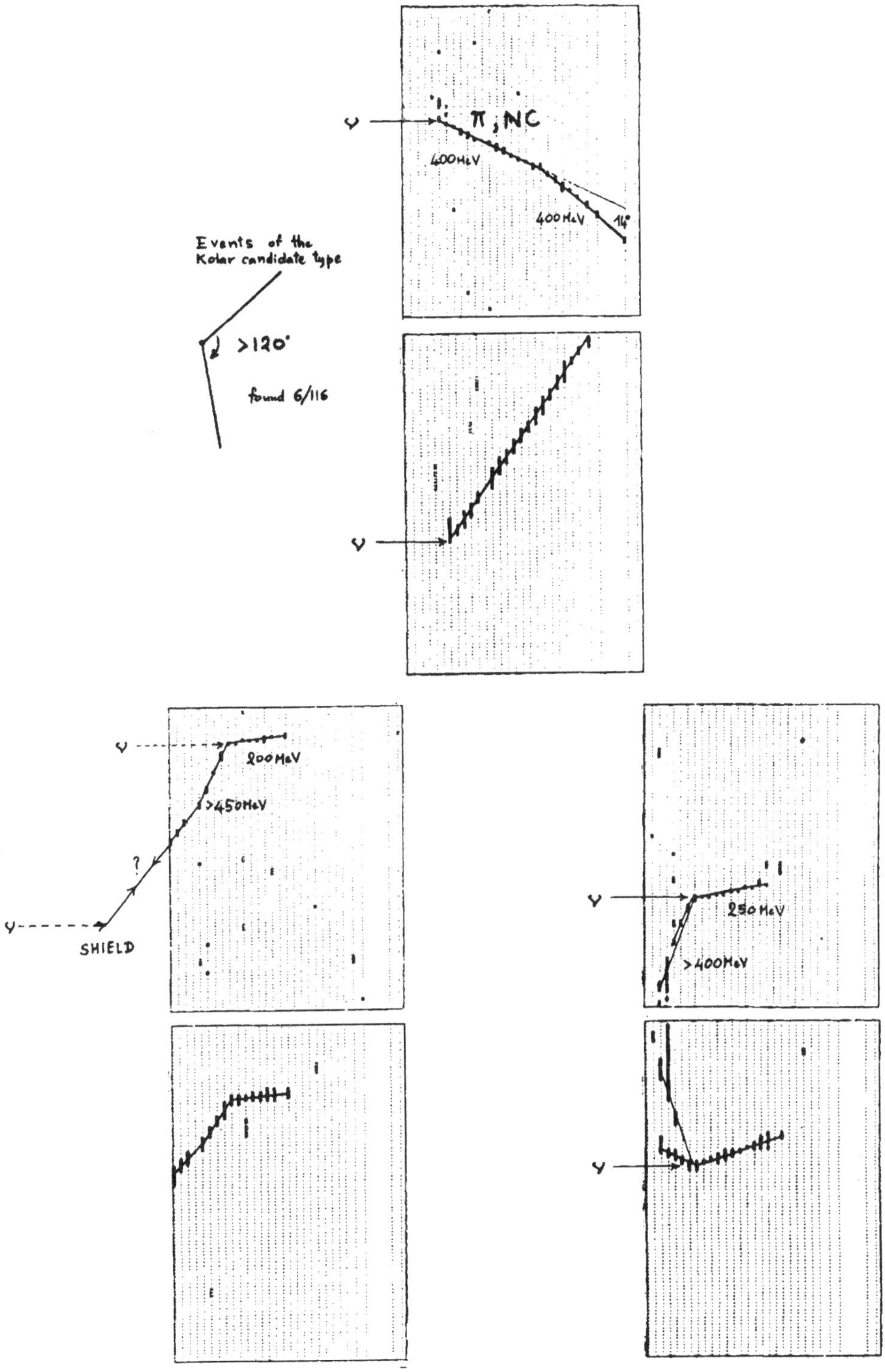
π ; NC
400MeV
400MeV
14°
Events of the Kolar candidate type
>120°
found 6/116
200MeV
>450MeV
?
SHIELD
250MeV
>400MeV

Fig. 17

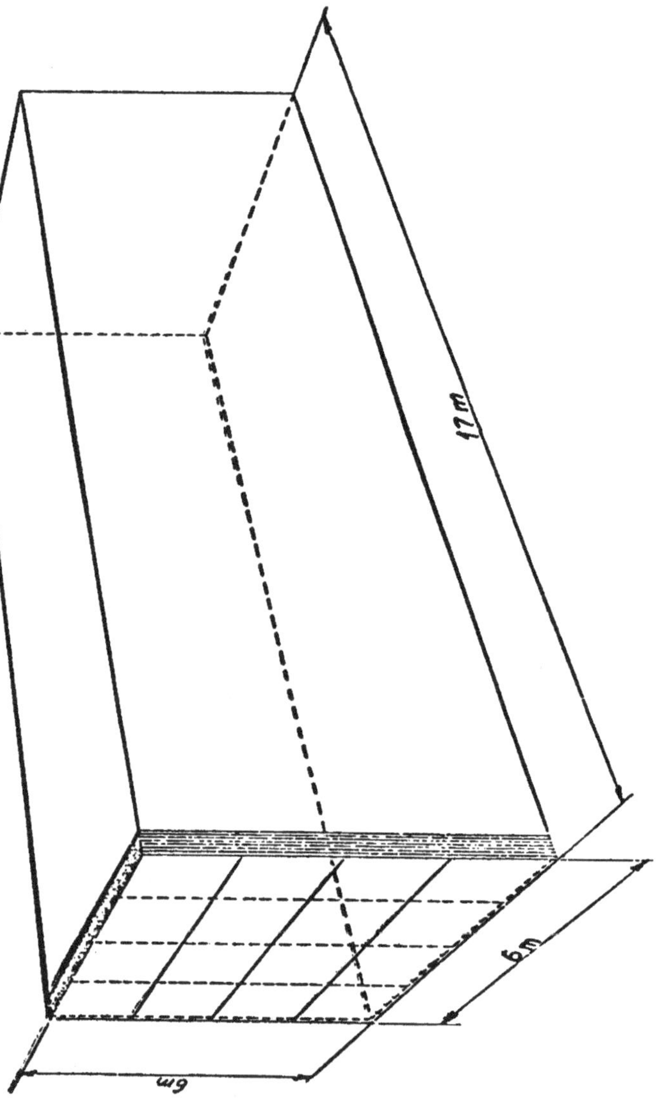

Fig. 19

- τ_R, the memory time of the discharge (if any), which can reach 100 ms.

For flash tubes, another important fact is that there is some probability of reignition during several seconds after a discharge. This, paradoxically enough, can cause a problem because of the high trigger rate due to ambient radioactivity.

2. The Čerenkov projects First a few basic facts about Čerenkov emission. The light is emitted along a cone of opening angle $\cos\theta = 1/\beta n$. The number of photons emitted is

$$n_\gamma \sim 500 \sin^2\theta \; \ell(\mathrm{cm}).$$

An electromagnetic shower is composed of $e^\pm$ which emit Čerenkov light over a total track length ℓ, which for a given initial energy is a constant in units of radiation length. So materials with a long radiation length in centimeters will give more light. Water (36 cm) is better than glass (11 cm) and lead glass (2.5 cm).

Figure 20 [34] shows how the measurement of Čerenkov light (position and time of arrival) on the edge of a detector -- a tank of water, for example -- can give information on the position, the length, and the direction of the emitting track.

Suppose the detector is a cube of $\sim (20\ \mathrm{m})^3$ and suppose a proton decays inside it to an $e^+\pi^0$. The total length over which $e^\pm$ from the showers emit Čerenkov radiation is a few meters, and this gives typically 6×10^4 γ's. Now suppose we neglect attenuation and collect 1 % of these γ's on the faces of the cube with 15 % of photoelectric conversion efficiency. We get $\sim$ 100 photoelectrons, i.e. a resolution in energy of $\sim$ 10 % just because of statistical fluctuation. Covering 1 % of the 2400 m^2 of outer surface implies that we use more than 2000 5-inch diameter phototubes, which is a major investment.

The above parameters are basically the ones of the Irvine-Michigan-Brookhaven (IMB) detector, which has already been well described [34]. Table 5 summarizes its expected properties. Water should be put in the tank quite soon.

I would now like to illustrate a project which is less known: the Japanese 1000 t Čerenkov detector[35]. The phototubes (Fig. 21) are huge, with nevertheless a quite respectable transit time spread (7.0 ns). The expected transparency of the water is shown in Fig. 22. The claim of this team is that because of the large photocathode coverage, many features are much improved compared to those of the IMB detector, e.g. topology, lepton identification (e-μ separation),

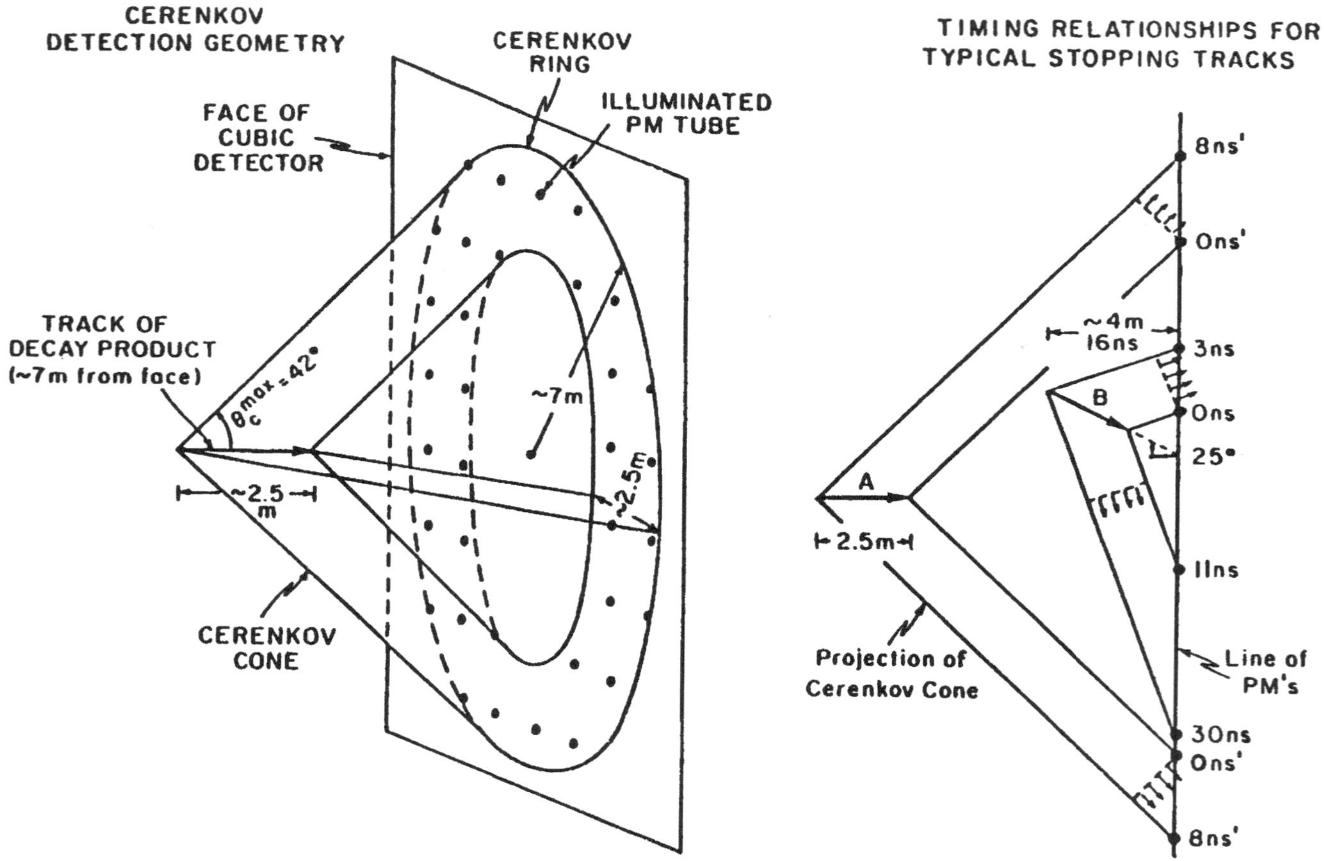

Illustration of how the arrival time of various parts of a Cerenkov cone can determine track direction in the IMB detector. Similar techniques can be used in the volume arrayed HPW detector.

Fig. 20

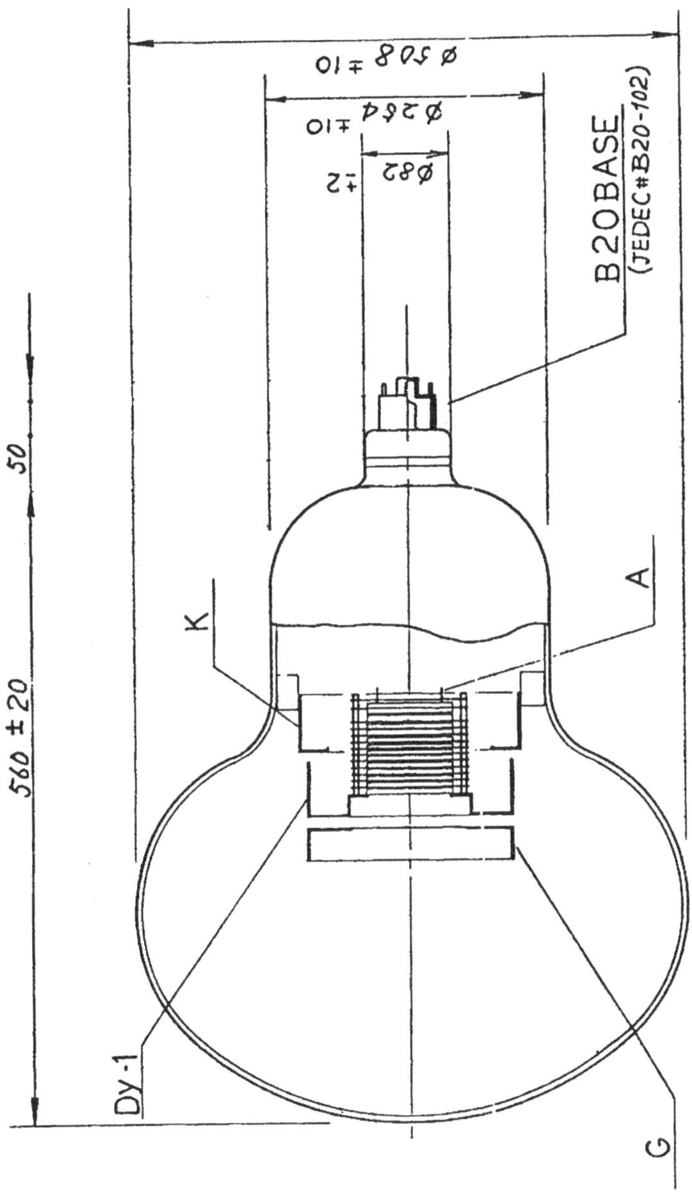

PIN CONNECTONS

pin #	1	2	3	4	5	6	7	8	9	10	11	12	13	14	15	16	17	18	19	20	socket #
	D 1	D 3	D 5	D 7	IC	D 9	D 11	D 13	A	D 12	IC	D 10	D 8	D 6	IC	D 4	D 2	K	G	IC	E678-20A

Fig. 21

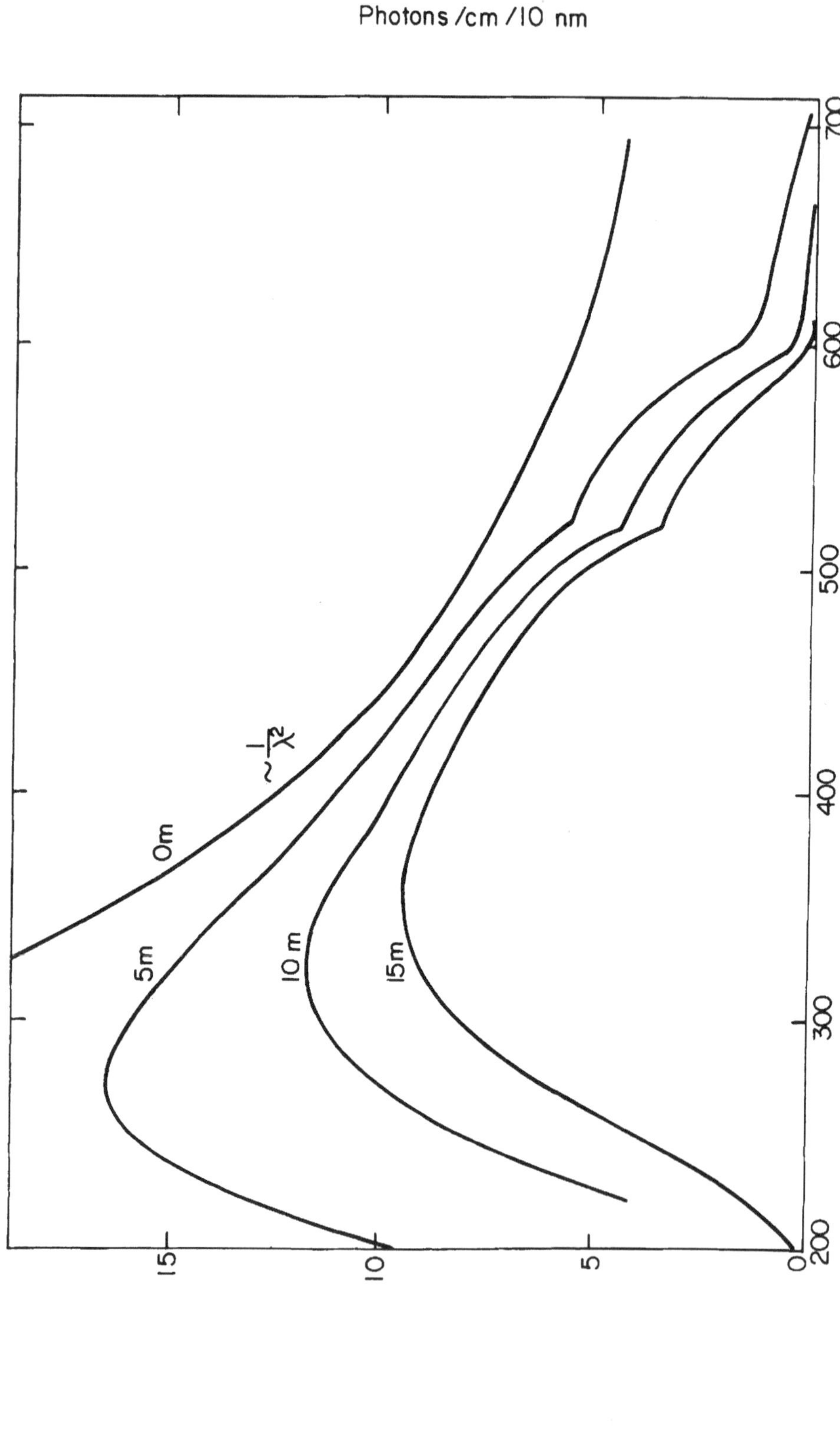
Photons /cm /10 nm
15
10
5
0
200
300
400
500
600
700
λ (nm)
0m
5m
10 m
15m
~ 1/λ²

Fig. 22.

Table 5. Summary of the IMB Detector Properties

Size:

Total volume = 6.5×10^{33} nucleons
Fiducial volume = $2.5 \times 10^{33} = (16\ m)^3$

Energy solution:

500 mev shower $\sigma = 11\ \%$
300-450 mev/c $\pi^{\pm}, \mu^{\pm}$ $\sigma = 10\text{-}15\ \%$

Vertex localization:

$\sigma = 0.5$ m

Track directionality:

No ambiguity

Angular resolution:

Showers $(e^{\pm}, \pi^{0})$ $\sigma = 10\text{-}20°$
Charged tracks $\sigma \lesssim 5°$

$\mu \to e$ detection:

$\gtrsim$ 80% after trigger (event thres. $\simeq$ 75 MeV)
Several $\mu \to e$: OK
Position correlation : 1.5 m
e direction : $\sim 25°$

hadron identification (π, η, ρ, ω separation), energy resolution. This is illustrated by Fig. 23.

Another huge Čerenkov project [36)] is the Harvard-Pennsylvania-Wisconsin (HPW) tank, installed at 1800 mwe, with 780 t of water, and veto systems on top of it. The photomultipliers here are regularly distributed throughout the whole volume of water. It is claimed that data will be recorded before autumn 1981. This team has also developed interesting ideas and prototypes for future large-area light intensifiers [36)].

We see that in a couple of years from now, several of these big detectors, with fiducial volumes of 1000 tons or so, will be in operation. Furthermore, two widely different techniques will be competing.

On the other hand, we should remember that all this gigantic enterprise will bring a gain of at most two orders of magnitude in sensitivity. GUTs, as they stand now, give to believers the promise that this small step will bring about a major revolution. It would

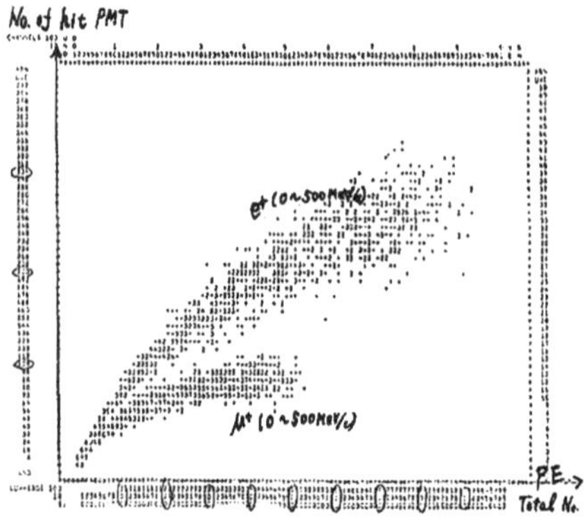

e-μ separation

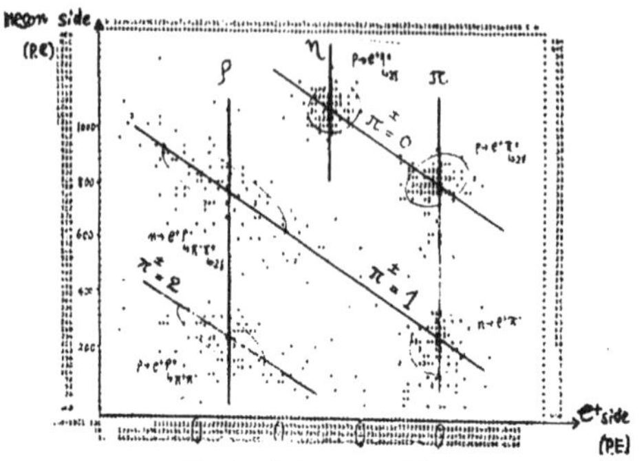

final state separation

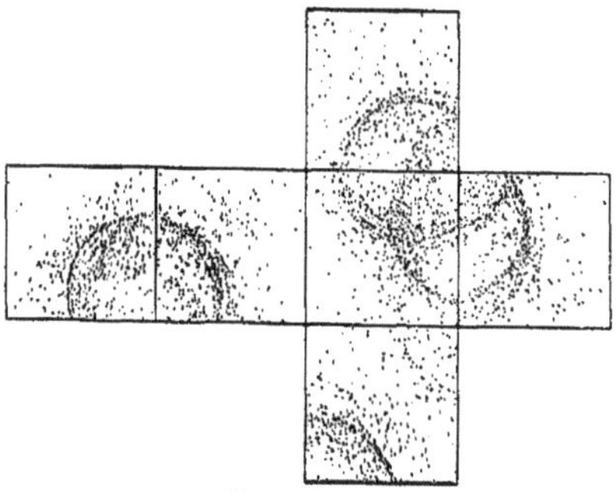

eπ° Monte Carlo

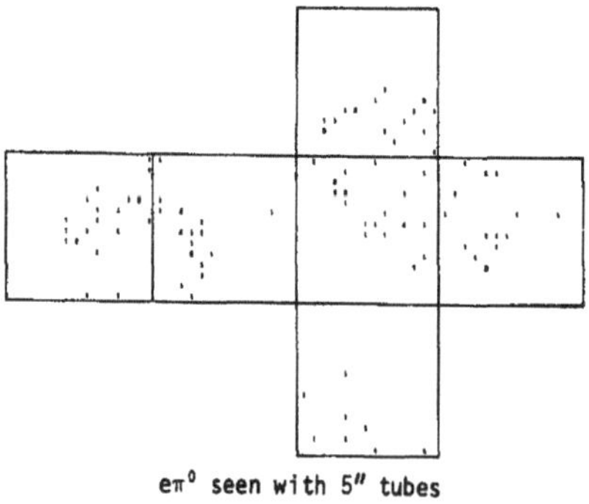

eπ° seen with 5" tubes

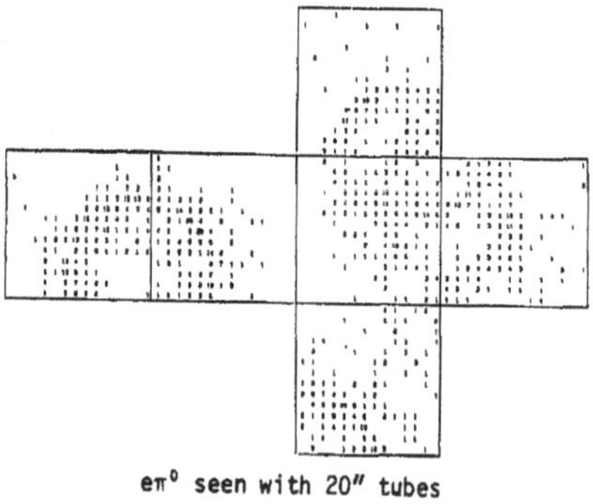

eπ° seen with 20" tubes

Fig. 23

certainly be the right attitude to enthusiastically share that faith.

V. OTHER POSSIBILITIES OFFERED BY THESE SET-UPS

Here I would like to mention three of them.

1. The Detection of GUTs Monopoles

The phenomenology of such monopoles is well known [37]. Their mass is $\sim 10^{16}$ GeV or so. Estimates of the abundance of their production in the early Universe give generally unacceptably large numbers but with large uncertainties. During the cosmological evolution, they took part in galaxy formation. They are accelerated in the magnetic fields of the galaxies (typically 5 μG on 5×10^{22} cm) and ejected from a galaxy when they reach $\beta \simeq 3 \times 10^{-2}\ (M/10^{16}\ \mathrm{GeV})^{-1/2}$. Limits on their abundance can be obtained in various ways. The most stringent come from the energetics of ejection (they should not have carried away more energy than the energy stored in the galactic field) and considerations on the steadiness of the galactic field[38]. For monopoles of $\beta \simeq 3 \times 10^{-2}$, with an anisotropic flux peaked in our galactic plane, the limit on the flux is

$$f \lesssim 10^{-7}\ \mathrm{cm}^{-2}\ \mathrm{y}^{-1}.$$

If one believes in such a limit, which is not a mandatory attitude, the first condition for a meaningful search is to cover an area larger than 1000 m^2 (10^7 cm^2): none of the set-ups considered approaches that.

How could one detect them ? To be ionizing they must have a velocity β larger than that of electrons in atoms, β_e. If $\beta \gg \beta_e$ the energy loss varies as $\sim \ln \beta^2$; if $\beta \sim \beta_e$ it varies as $\beta^2 \ln \beta^2$ [39]. For the velocity considered, the loss is substantial: 5 GeV/cm in H_2O for $\beta = 0.03$. A combination of time of flight and pulse height, at great depth since they are very penetrating (in spite of ionization, Zeeman, and ohmic losses, they can traverse the earth), should provide a fantastically clean signature. Direct detection of Zeeman effect looks desperate. What about sound waves in deep undersea experiments ?

Search for such monopoles trapped in samples of matter [40] may be one method, but we must try hard to convince ourselves that the monopoles did not have any good reasons for getting out of that sample at some time. For instance, a monopole trapped in a ferromagnetic meteorite is bound by an image force of 10 eV/Å. If the meteorite is accelerated or decelerated above g $\sim$ 10, the monopole will leave!

2. Relics

These are hypothetical heavy objects R of mass M_R and abundance n_R made in the early universe [41]. We will consider long-lived ones which could still be around nowadays. They must, however, be of pretty low abundance. If we define $f = n_R/n_{nucl}$, considerations on the mass of the universe imposes $f \lesssim 10\ (m_{nucl}/M_R)$. This rules out an asymmetric production, such as that of nucleons: they must have been produced in a rather symmetric way.

Proton decay set-ups are relevant because they can detect products originating from such relics decaying in the neighbourhood, for instance muons. The evaluation of the limit that can be given on their abundance for a given mass and lifetime is rather straightforward [42]. Figure 24 [42], [43] shows what the future "big experiments" will bring as an improvement. The striking fact is that a huge region is already forbidden. However, since here no theory is available to give a hint of the masses to expect, this huge region and the gain to be expected look desperately small compared to the decades of orders of magnitude where things could happen.

3. Neutrino Oscillations

It has been claimed [44] that large underground detectors provide a good way of observing ν oscillations. The source is still atmospheric ν's. The length over which oscillations can occur goes from a few to 10 thousand kilometers (downward- and upwardgoing ν, respectively). Neutrino events to be measured are essentially charged-current elastic and Δ production: their threshold on the one hand and the drop of the ν spectrum with energy on the other, naturally selects a rather narrow band of ν energies with $< p_\nu > \sim 0.5$ GeV. The mass region scanned is 10^{-1}-10^{-3} eV. The direction of the charged leptons indicates that of the incoming ν reasonably well. Figure 25 gives an idea of the sensitivity of such a measurement.

VI. n-$\bar{n}$ OSCILLATIONS

Oscillation of a neutron into an antineutron is a $\Delta B = 2$ process, quite reminiscent of well-known $\Delta S = 2$ processes [45]. It is relevant to the subject treated here because it is mediated by intermediate mass objects of the type discussed in Section I.

To oscillate, n and $\bar{n}$ should be quasi-degenerate (except for B). There should be no magnetic energy term, and one has to decrease the terrestrial magnetic field more than one hundred times with an appropriate shield. There should be no nuclear field, and one has to use free neutrons.

Starting from a beam of neutrons, the probability of observing

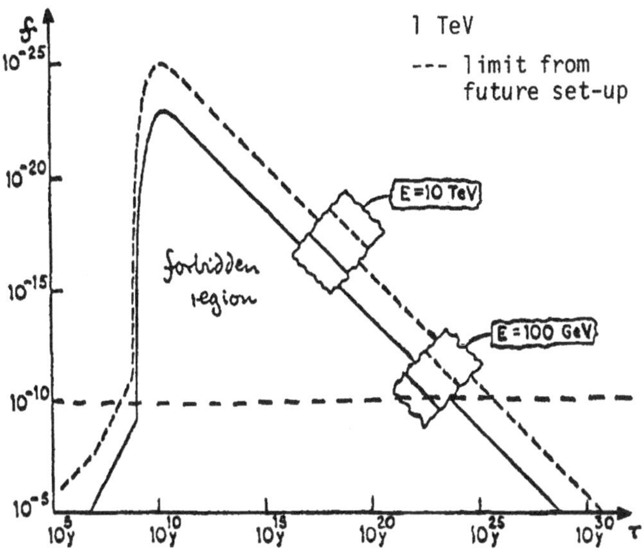
f
1 TeV
--- limit from future set-up
E=10 TeV
E=100 GeV
forbidden region
10^-25
10^-20
10^-15
10^-10
10^-5
10^5 y
10^10 y
10^15 y
10^20 y
10^25 y
10^30 y
τ

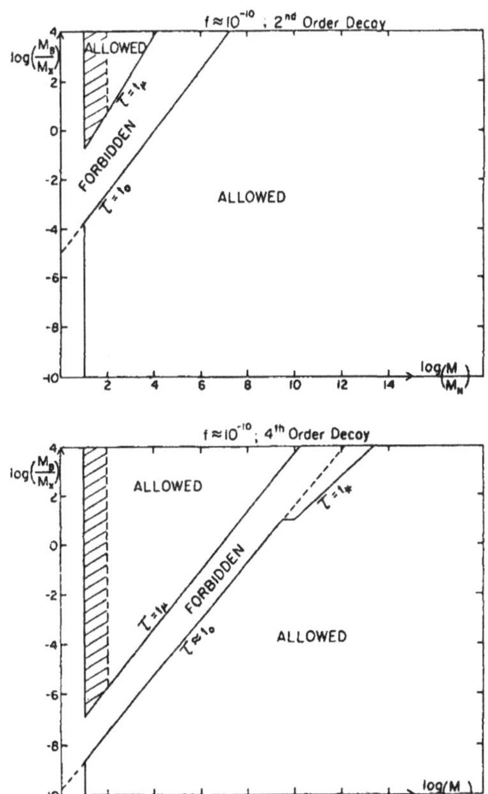
f ≈ 10^-10 ; 2nd Order Decay
log(M_B/M_X)
ALLOWED
τ = 1_p
FORBIDDEN
τ = 1_0
ALLOWED
log(M/M_N)
f ≈ 10^-10 ; 4th Order Decay
log(M_B/M_X)
ALLOWED
τ = 1_p
FORBIDDEN
τ ≈ 1_0
τ = 1*
ALLOWED
log(M/M_N)

Fig. 24

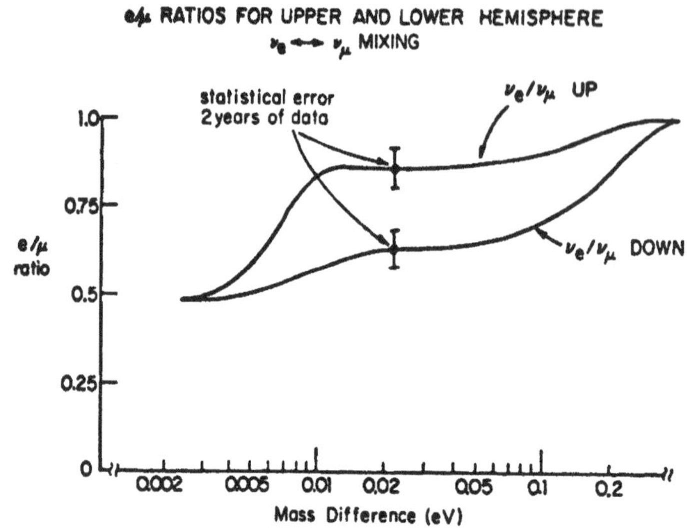

Fig. 25

an antineutron is

$$P_{\bar{n}}(t) \simeq \frac{t^2}{\tau^2_{n\bar{n}}} ,$$

where τ is a measure of the $\Delta B = 2$ coupling .

One has to maximize: $\Phi \equiv N \times t^2$, where N is the number of neutrons per second and t is the time of observation. For planned experiments [45)], using very slow neutrons from reactors, Φ can reach up to 3×10^{10} s; per day, one count corresponds to $\tau_{n\bar{n}} = 5 \times 10^7$ s. Preliminary results from the Laue-Langevin Institute [25)] give already

$$\tau_{n\bar{n}} > 10^5 \text{ s}.$$

Connecting this result to nuclear instability, i.e. two-nucleon annihilation in decay experiments, seems to be strongly model-dependent, as shown in Fig. 26.

I would like to thank R. Barloutaud, B. Degrange, J. Ellis, B. Pattison and especially M. Koshiba for very useful discussions.

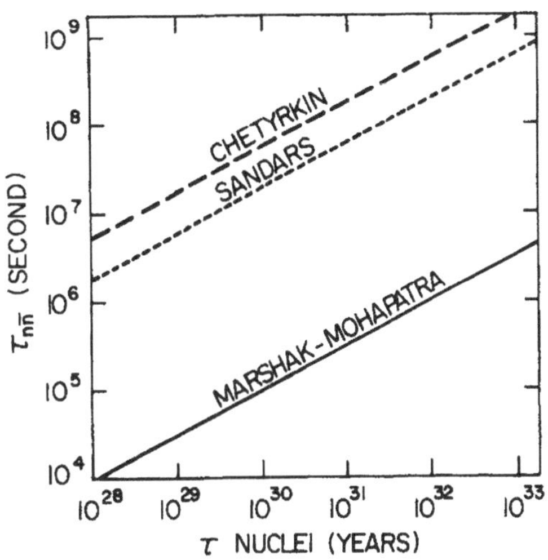

Fig. 26

REFERENCES

1. J. Ellis, Grand unified theories, CERN preprint TH-2942 (1980), in Scottish Universities Summer School in Physics, eds. K.C. Bowler and D.G. Sutherland (Univ. Edinburgh, 1981), p. 201

2. D.A. Ross, Phenomenology of unified gauge theories, CERN preprint TH-3174 (1981) and The calculation of the decay rate of the proton, CalTech report Calt 68779/C1 (1980)

3. M.B. Gavela et al., Phys. Rev. D 23, 1580 (1981)
   G. Kane and G. Karl, Branching ratios in baryon decay, Michigan University internal report (1980)
   J.F. Donoghue, Phys. Lett. 92B, 99 (1980)

4. M. Machacek, Nucl Phys. B159, 37 (1979)
   A. Din et al., Phys. Lett 91B, 77 (1980)

9. H. Georgi et al., μ^+ polarization in proton decay: a probe of flavour mixing in unified models, Harvard preprint HUTP 80/A064 (1980)

10. J.C. Pati and A. Salam, Quark lepton unification and proton decay, in Proc. First Workshop on Grand Unification, Durham, New Hampshire, (1980), eds. P.H. Frampton, S.L. Glashow and A. Yildiz (Math Sci Press, Brookline, 1980), p. 115

11. S. Weinberg, Expectations for baryon and lepton non-conservation, same Proc. as Ref. 10, p. 347

12. R.N. Mohapatra, Baryon and lepton non-conservation, Majorana neutrinos and neutron oscillations, same Proc. as Ref. 10, p. 69

13. See for instance: W.B. Rolnick, Scalar operators in nucleon decay, Wayne State Univ. preprint (Detroit) (1981)

14. L.W. Jones, Proton lifetime, quark abundance and age of the universe: a numerical coincidence ? Univ. Michigan preprint UMHE 81-27 (1981)

15. W. Fairbank, Talk given at the EPS Conf., Lisbon, 1981

16. H.B. Nielsen and M. Ninomiya, Nucl. Phys. B141, 153 (1978) See also J. Ellis et al., Nucl. Phys. B176, 61 (1980)

17. D.H. Perkins, Feasibility test for experiment on nucleon stability, Referee's report, CERN/EP/DP/mk (August 1979)

18. D.A. Sparrow, Phys. Rev. Lett. 44, 625 (1980)

19. B. Rossi, High-energy particles (Prentice Hall Inc., Englewood Cliffs, NJ, 1952), p. 163

20. D.H. Perkins, μ induced backgrounds in the Mont Blanc and Fréjus tunnels, CERN/EP/DP/mk (Sept. 1979)

21. A.L. Grant, Isolated neutron background in the Mont Blanc and Fréjus tunnels, CERN/EF/ALG/ed (Dec. 1979)

22. E. Belotti, Investigations of nucleon stability in Europe, in Unification of the fundamental particle interactions, eds. S. Ferrara, J. Ellis and P. Van Nieuwenhuizen (Plenum, New York, 1981), p. 673

23. M. Goldhaber et al., Is the proton stable ? Los Alamos report LA-UR/80-356 (1980)

24. L. Bergamasco and G. Cini, Nuovo Cimento 1C, 293 (1978)

25. R. Barloutaud, Talk given at the EPS Conf., Lisbon, 1981

26. M. Deakyne et al., the Homestake mine nucleon decay search, paper 716 submitted to the Madison Conf., 1980

27. J. Learned et al., Phys. Rev. Lett. 43, 907 (1979)

28. B. Sreekantan, Communication to the EPS Conf., Lisbon, (1981)
For the so-called KGF events, see: S. Miyake, Proc. 19th Int. Conf. on High-Energy Physics, Tokyo, (1978) (Phys. Soc. Japan, Tokyo, 1979), P. 433

29. G. Battistoni et al., Frascati-Milan-Turin Collab., NUSEX Proposal. See also Ref. 22

30. E. Iarocci, Communication to the EPS Conf., Lisbon, (1981)

31. H. Courant et al., A dense detector for baryon decay, Paper 0350 submitted to the Madison Conf. (1980)
E.A. Peterson, Communication to the EPS Conf., Lisbon, (1981)
Minnesota-Argonne-Oxford Collab., Soudan 2, a 1000 ton tracking calorimeter for nucleon decay, Proposal OUNP/42/81, COO-1764-410, ANL-HEP-PR 81-12 (1981)

32. P. Bareyre et al., Orsay-Palaiseau-Paris-Saclay Collab. Proposal Proposition d'une expérience pour l'étude de l'instabilité du nucléon au moyen d'un détecteur calorimétrique.

33. For flash tubes, see:
M. Conversi and L. Federici, Nucl. Instrum. Methods 151, 93 (1978)
For streamer tubes, see:
G. Battistoni et al., Nucl. Instrum. Methods, 164, 57 (1979)

34. L.R. Sulak, The Irving-Michigan-Brookhaven nucleon decay search in same book as Ref. 22, p. 639
J.C. Van der Velde, A review of proton decay experiments, Talk given at the SLAC Summer Institute on Particle Physics, July-Aug. 1980

35. M. Koshiba, Private communication

36. D.R. Winn, A search for baryon decay: plans for the HPW water Čerenkov detector, in same Proc. as Ref. 10, p. 189

37. G. Lazarides et al., Phys. Lett. 100B, 21 (1981)
M. Longo, Massive magnetic monopoles, Univ. Michigan preprint UMHE 81-33 (1981)

38. E.N. Parker, The origin of magnetic field, Astrophys. J. 160, 383 (1970)

39. K. Hayashi, Stopping power formula for a cosmological monopole, Kinki Univ. preprint
S.P. Ahlen, Phys. Rev. D 17, 229 (1978)

40. P. Eberhard et al., Phys. Rev. D 4, 3260 (1971)

41. C.B. Dover et al., Phys. Rev. Lett. 42, 1117 (1979)

42. P.H. Frampton and S.L. Glashow, Phys. Rev. Lett. 44, 1481 (1980)

43. J. Ellis et al., Exotic relics from the big bang, CERN preprint TH-2893 (1981)

44. B. Cortez and L. Sulak, A long baseline ν oscillation experiment sensitive to mass difference of hundredths of an eV, Univ. Michigan preprint UMPD 80-1 (1980)

45. R. Wilson, Possibilities of experiments to measure neutron-anti-neutron mixing, same Proc. as Ref. 10, p. 215

QUANTUM FIELD THEORY AND COSMOLOGY

F. Englert

Faculté des Sciences, Université Libre de Bruxelles
and
C.E.R.N. Geneva

I. INTRODUCTION

Experimental evidence points with ever increasing confidence towards an evolutionary cosmology. The standard framework used to confront this circumstance is the "hot big bang" theory : an initial classical singularity, possibly smeared by unknown quantum effects, produces a thermal distribution of the quanta of the matter field at a temperature close to the Planck temperature ($T_p \simeq 10^{19}$ Gev). This state then leads to an approximatively adiabatic free expansion. The model explains the background 2.7°K radiation observed in the presently expanding universe. It predicts, when combined with the known weak interaction physics, a primordial helium abundance in remarkable agreement with experimental data. More recently, the embedding of grand unified gauge theories in the big-bang cosmology has led to a plausible qualitative understanding of the small baryon to photon ratio in our universe.

The hot big bang theory is however incomplete in that it provides no natural explanation of the required "hot" initial condition. In fact we shall see that the stability of the large scale gravitational field requires the setting up of correlations which do not appear to be consistent with the existence of a thermal state at $T \simeq 10^{19}$ Gev. Hence the hot big-bang model is expected to break down when extrapolated back to the Planck temperature.

A theory of primordial quantum era, due to A. Casher and myself, which avoids these difficulties, will be presented [1]. A universe is conceived as a gravitational quantum fluctuation stabilized by the time dilation resulting from coherent black hole creation out of expansion energy. Subsequent black hole decay signals the onset of a more conventional adiabatic regime. The theory determines the initial conditions for the classical cosmological evolution [$T_{in} = 4.57 \times 10^{16}$ Gev; H_{in} (≡ initial expansion rate) = $4.36 \times 10^{41} s^{-1}$] explains in causal terms the adiabatic expansion, and predicts the present value of the deceleration parameter. An unexpected consequence of the theory will be a remarkable view of the universe, its beginning and its fate, which connects it to the structure of quantum gravity.

In these lectures, I shall first review the standard hot big-bang theory, its most successful predictions and its deadly theoretical disease. I shall the study the conjectured primordial creation process. Finally, I shall discuss the far reaching implications for quantum gravity which arise from the proposed cosmogenesis.

Whatever is new in these notes is the result of a collaboration between A. Casher and myself. We are very grateful to P. Spindel for explaining to us the de Sitter geometry. We thank R. Brout, W. Deans S. Rudaz, L. Van Hove, J. Weyers and J.A. Wheeler for helpful discussions. I thank Mrs. Loutz for her patience in deciphering and typing the manuscript.

Absolute units ($G = \hbar = c = k = 1$) will often be used; to convert to ordinary units, the following numbers are useful :

$$L_p \equiv \text{Planck length} \equiv (G\hbar/c^3)^{1/2} = 1.62 \times 10^{-33} \text{ cm}$$

$$t_p \equiv \text{Planck time} \equiv (G\hbar/c^5)^{1/2} = 5.39 \times 10^{-44} \text{ s}$$

$$m_p \equiv \text{Planck mass} \equiv (G/\hbar c)^{-1/2} = 2.17 \times 10^{-5} \text{ g}$$

$$E_p \equiv \text{Planck energy} \equiv (G/\hbar c^5)^{-1/2} = 1.95 \times 10^{16} \text{ ergs} = 1.22 \times 10^{19} \text{Gev}$$

$$T_p \equiv \text{Planck temperature} \equiv (Gk^2/\hbar c^5)^{-1/2} = 1.42 \times 10^{32} \,{}^\circ\text{K}.$$

II. THE HOT BIG BANG THEORY

1. The Cosmological Field Equations

On sufficiently large scales, our universe appears to us as isotropic and homogeneous. Before translating this information in the mathematical framework of general relativity it may be useful to describe such a situation in the Newtonian approximation. Consider therefore a set $\{i\}$ of material points of mass m placed on the sites of a large regular lattice. If the set has to represent for all times an approximate homogeneous medium (except for boundary effects), the configuration at two different times must be related by a similarity transformation. Hence

$$v_i(t) = H(t)\, r_i(t) \tag{2.1}$$

where $r_i(t)$ is the radial coordinate of the ith material point referred to an arbitrarily chosen point j : the Hubble constant H(t) must be independent of i and j. Defining, up to an arbitrary multiplicative constant, the scale factor a(t) by

$$H(t) = a^{-1}(t)\, \frac{da(t)}{dt} \tag{2.2}$$

we may find a coordinate ξ_i by

$$r_i(t) = a(t)\xi_i \tag{2.3}$$

where it follows from (2.1) that ξ_i is time independent. Equations (2.1), (2.2), (2.3) are purely kinematical. However if $\ddot{a}$ [$\equiv d^2a(t)/dt^2$] is small enough we may choose any point j as the origin of an inertial frame, for sufficiently small velocities, we may then use Newton's law to obtain a dynamical equation for the scale factor a(t). Thus

$$\frac{dv_i(t)}{dt} = -G\frac{M}{r_i^2(t)} \quad ; \qquad M \simeq \sigma(t)\,\frac{4\pi}{3}\, r_i^3(t) \tag{2.4}$$

As the total mass M is conserved we get

$$\frac{d}{dt}\left[\frac{v_i^2(t)}{2} - \frac{GM}{r_i(t)}\right] = 0 \tag{2.5}$$

or

$$\left[\dot{a}^2(t) - \frac{8\pi G}{3}\,\sigma(t)a^2(t)\right]\xi_i^2 = c_i \tag{2.6}$$

where c_i is an integration constant. Defining

$$c_i/\xi_i^2 = -k \tag{2.7}$$

we may write

$$\left(\frac{\dot{a}}{a}\right)^2 + \frac{k}{a^2} = \frac{8\pi G}{3}\,\sigma \tag{2.8}$$

With a little strech of imagination, we express the conservation of M as a particular case ($p = 0$) of the more general "thermodynamical" law.

$$(\sigma a^3)^{\cdot} = -p(a^3)^{\cdot} \tag{2.9}$$

The reason for introducing a non zero pressure in (2.9) is that (2.8) and (2.9) will turn out to be exact in the context of general relativity when σ and p will be suitably redefined.

The Newtonian "derivation" of these equations is of interest not only because of its simplicity but also because it yields a simple interpretation for the "curvature term" k/a^2. In fact, from (2.6) and (2.7) we see that

$$\frac{k/a^2}{\dot{a}/a^2} = -\frac{E_i}{K_i} \tag{2.10}$$

where K_i is the kinetic energy of the ith particle and E_i its total mechanical energy. Clearly, if $k > 0$ the world is "closed" in the sense that it eventually returns to its original state after having reached a maximal expansion; for $k < 0$, the world is "open" and expands indefinitely. The case $k = 0$ corresponds to the escape velocity condition in the Kepler problem and must therefore be considered only as a limiting case, a fact which should be kept in mind in the general relativistic generalization. This issue is sometimes obscured because one can always rescale a(t) by an arbitrary constant to reduce the admissible values of k to be + 1, - 1, or 0. We shall indeed do so in what follows but the relevant physical combination is k/a^2 which can take any value between $-\infty$ and $+\infty$.

In general relativity, the scale factor is introduced via the Robertson-Walker metrics which describe homogeneous isotropic gravitational backgrounds. For "closed" $k = +1$ or "open" $k = -1$ trajectories, the spatial sections of these metrics at a given proper time t can be obtained by adding a fictitious time dimension x_4 :

$$x_4^2 \mp (x_1^2 + x_2^2 + x_3^2) = a^2(t) \tag{2.11}$$

The spatial length element $d\sigma^2$ takes thus the form

$$\begin{aligned} k = -1;\ d\sigma^2 &= a^2(t)d\Sigma^2 \\ &= a^2(t)\,\{d\xi^2 + \sinh^2\xi\,(\sin^2\theta d\phi^2 + d\theta^2)\,\} \end{aligned} \tag{2.12}$$

$$k = +1;\ d\sigma^2 = a^2(t)d\Sigma^2$$

$$= a^2(t)\,\{d\xi^2 + \sin^2\xi\,(\sin^2\theta\, d\phi^2 + d\theta^2)\} \quad (2.13)$$

or equivalently, posing $\chi = \sinh\xi(\sin\xi)$ for $k = -1$ (+1)

$$d\sigma^2 = a^2(t)d\Sigma^2 = a^2(t)\{\frac{d\chi^2}{1 - k\chi^2} + \chi^2 d\Omega^2\} \quad (2.14)$$

The limit $k/a^2(t) \to 0$ is obtained by taking $\chi \ll 1$ in (2.14), (or equivalently $\xi \ll 1$ in (2.12) and (2.13)) with a(t) arbitrarily large.

The four dimensional length element is

$$ds^2 = dt^2 - a^2(t)d\Sigma^2 \quad (2.15)$$

Comoving particles would then follow the geodesics ξ = constant so that (2.1), (2.2) and (2.3) remain valid for material points when t is interpreted as proper time.

To get equations for a(t), one solves the Einstein's equations

$$R_{\mu\nu} - \frac{1}{2} g_{\mu\nu}R = 8\pi G < T_{\mu\nu} > \quad (2.16)$$

in the metric (2.15). For the expectation value of the matter energy momentum tensor one assumes, in accordance with the hypothesis of large scale homogeneity, the perfect fluid form

$$< T_{\mu\nu} > = (\sigma + p)u_\mu u_\nu - pg_{\mu\nu} \quad (2.17)$$

with $u_\mu = dx_\mu/ds$. It is easily verified that (2.15), (2.16) and (2.17) lead back to (2.8) and (2.9). The new results obtained from general relativity are thus the geometrical interpretation (2.15) and the correct generalized definition of σ and p : σ is the rest energy density of matter and p is defined by

$$< T > \equiv < T^\mu{}_\mu > \equiv \sigma - 3p \quad (2.18)$$

The cosmological evolution of the universe is determined from equations (2.8), (2.9) and from the (still uncompletely known) physical laws of interacting matter which relate p and σ. However "initial" conditions have to be specified. Quite generally if $p > 0$ when $a \to 0$, equation (2.9) and the positive definiteness of σ imply $\sigma \to \infty$ in this limit : this is the classical big bang singularity. As we shall now see, the existence of an isotropic microwave photon radiation at T = 2.7°K which is decoupled from all other form of matter[2] suggests, when the present expansion is extrapolated backwards in time, that the classical singularity is in fact realized as a thermal state, namely $T \to \infty$ as $a \to 0$. Such "initial" hot conditions cannot however be imposed for $T > T_P$, when quantum metric fluctuations are taken in account, because the classical equations (2.16) become meaningless in that region. The "hot big bang theory" is then defined by equations (2.8) and (2.9) and by the value of the scale foctor a_P at an "initial" time $t = 0$ when the temperature was everywhere of the order T_P.

2. The Adiabatic Expansion

Present observations confirm the expansion predicted by (2.8) and (2.9) with $p \simeq 0$, corresponding to the predominance in σ of ordinary massive matter at negligible temperature [3]. The present value of the Hubble constant, as measured from the Doppler effect of receding galaxies is estimated to be

$$H_0 \simeq 2 \times 10^{-18} s^{-1} \tag{2.19}$$

The observed matter density due to nucleons is thought to lay in the range

$$\sigma_0 \simeq 10^{-30} \text{ gr/cm}^3 \tag{2.20}$$

where uncertainties are mostly due to the possible presence of invisible cold matter in intergalactic space. The results (2.19) and (2.20) seem to favour the $k = -1$ cosmology, as is easily seen from (2.8); however the uncertainties in (2.19) and (2.20) could invalidate such a conclusion. A direct determination of k in (2.8) is in principle possible from a measurement of the "deceleration" parameter q defined by

$$q \equiv (\dot{H^{-1}}) - 1 \tag{2.21}$$

but present techniques for measuring q are not sufficiently precise to settle the issue. We thus leave k unspecified and as previously we shall not consider separately the limiting case $k/a^2 = 0$.

The very fact that we live in a state so close to spatial "flatness" that we are still unable to decide what is the sign of k has an interesting consequence. From (2.8), the present value of the scale factor a_0 is greater than H_0^{-1} if $k = -1$ and cannot be significantly smaller than H_0^{-1} if $k = +1$; thus we obtain the estimate

$$a_0 \gtrsim 10^{61} \tag{2.22}$$

This result will now be used in conjunction with the presently observed background photon radiation to characterize the scale factor in the early universe.

The black body formulae for a set of massless particles in thermal equilibrium at a temperature T are

$$\sigma = \nu(T)\,\frac{\pi^2}{30}\,T^4 \quad ; \quad p = \frac{\sigma}{3} \tag{2.23}$$

$$s \equiv \frac{\sigma + p}{T} = \nu(T)\,\frac{2\pi^2}{45}\,T^3 \tag{2.24}$$

Here $\nu(T)$ is the effective number of helicity states at temperature T

$$\nu(T) = \nu_{boson}(T) + \frac{7}{8}\,\nu_{fermion}(T) \tag{2.25}$$

and s the entropy density which is also an approximate measure of the number density

$$n = \frac{1.202}{\pi^2}\,T^3\,\{\nu_{boson}(T) + \frac{3}{4}\,\nu_{fermion}(T)\}$$

The photon temperature T_o = 2.7 °K shows then, when (2.24) (with ν = 2) is compared to (2.20), that

$$\frac{n_o^B}{n_o^\gamma} \simeq 10^{-9} \tag{2.26}$$

where n_o^γ and n_o^B are the presently observed photon and baryon number densities. The total number of baryons N_o^B and photons N_o^γ in a volume $4\pi/3\ H_o^{-3}$ are then from (2.19), (2.20) and (2.26)

$$N_o^B \simeq 10^{78} \tag{2.27}$$

$$N_o^\gamma \simeq 10^{87} \tag{2.28}$$

If we move away backwards in time from the present situation, the photon temperature T will increase according to

$$aT = \text{constant} \tag{2.29}$$

This is an immediate consequence of the conformal invariance of the free electromagnetic action and of the fact that (2.15) is conformally equivalent to a static metric

$$ds^2 = a^2(\bar{t})\,\{d\bar{t}^2 - d\Sigma^2\} \;;\quad a(\bar{t})d\bar{t} = dt$$

In the $\bar{t}$ coordinate system the expansion is not felt by the photon and hence in the proper time t, photon momenta must scale according to (2.29). A direct check of equation (2.29) is provided by (2.9) and (2.23) as applied to the decoupled photons.

According to (2.29) and (2.23), when a decreases,

$$\sigma_{photon}/\sigma_{baryon} \simeq T^4/n^B m^B$$

where m^B is a nucleonic mass, increases. From (2.26) σ_{photon} will dominate σ_{matter} at a temperature

$$T \simeq 10^{-9}\, m^B \simeq 10.000°K \tag{2.30}$$

where the uncertainties in (2.26) stemming from (2.20) fixes the temperature (2.30) only up to one order of magnitude. Above this temperature the main contribution to σ in (2.8) comes from radiation and p in (2.9) changes from zero to $p \simeq \sigma/3$. The universe has quitted the present matter dominated era to enter the radiation dominated era.

If we move further away in the radiation dominated era, the temperature rises and particle-antiparticle pairs in equilibrium with the photons are generated thermally. These particles become relativistic and the baryonic excess (2.26) becomes negligible. Similarly, presently unobserved neutrinos couple in. In the context of asymptotically free theories all these relativistic particles at high temperature should interact weakly so that at each stage of the expansion the dominant contribution to σ is that of a free gas of massless particles in thermal equilibrium. Thus σ is given by (2.23) where ν(T) depends on the detailed physics of the matter fields and is given by the number of "massless" degrees of freedom at the photon temperature T. (Other massless particles not in equilibrium with photons (i.e. gravitons) may also contribute significantly to σ and be characterized by a different temperature but we omit this complication in our qualitative discussion). This simple picture assumes that changes in free energy due to possible symmetry restoring phase transitions at high temperatures are small effects compared to the huge radiation background. For a discussion of this assumption and a review of alternatives the reader may consult reference 4.

The variation of ν(T) with T amounts to a departure from conformal invariance which invalidates the simple result (2.29) in the radiation dominated era. However the picture just discussed shows that the entropy S in a comoving volume $V = (4\pi/3)a^3$ is in very good approximation given by the equilibrium law

$$TdS = dE + pdV - \sum_i \mu_i dN_i$$

$$= dE + pdV = 0 \qquad (2.31)$$

where μ_i is the negligible chemical potential of the ith relativistic constituent present at temperature T and the last equality follows from (2.9). Hence the expansion is adiabatic and using (2.24) we simply generalize (2.29) to

$$\nu^{1/3}(T)aT = \text{constant} \qquad (2.32)$$

From (2.22) and from the fact that the present photon temperature is $\simeq 10^{-32}$ we get

$$\nu^{1/3}aT \gtrsim 10^{29} \simeq (N_o^\gamma)^{1/3} \qquad (2.33)$$

It may seem reasonable to extrapolate smoothly the increase of the temperature back to $T \simeq T_p$ where quantum metric fluctuations set in. One would then use such a thermal state as an initial condition for equations (2.8) and (2.9) : this is the hot big bang theory. We then expect (2.33) to hold for $T \lesssim T_p$; provided $\nu(T)$ remains small compared to 10^{29}, the initial scale factor a_p at $T \simeq T_p$ will then satisfy the inequality

$$a_p \gtrsim 10^{29} \simeq (N^\gamma)^{1/3} \qquad (2.34)$$

Before discussing the implications of such a large value for a_p, we briefly review the facts which militate in favour of the hot big bang theory.

3. Successful Predictions of the Hot Big Bang Theory

In the hot big bang theory, the adiabatic expansion and hence the existence of the presently observed background photon radiation appears as a consequence of the (unexplained) hot initial conditions. To test more directly the validity of the adiabatic expansion one must analyze the radiation dominated era where thermal effects are operative.

First note that the relation (2.33) implies that during this era space is effectively flat. Indeed, using (2.23) and (2.30) we get

$$\frac{8\pi\sigma}{3} a^2 = \frac{8\pi}{3} \nu(T) \frac{\pi^2}{30} T^4 a^2$$

$$= \frac{8\pi^3}{90} \nu^{1/3}(T)\, T^2 [\nu^{1/3}(T)\, aT]^2$$

$$\gtrsim \frac{8\pi^3}{90} \nu^{1/3}\, 10^{-56} . 10^{58} \gg 1. \tag{2.35}$$

Hence k/a^2 can be neglected in (2.8). This flatness condition yields a very simple time-temperature relation in the radiation dominated era. We have

$$H(T) = \frac{2\pi}{3} \sqrt{\frac{\pi\nu(T)}{5}}\, T^2 \tag{2.36}$$

and hence, using (2.32) in a domain where $\nu(T)$ is approximately constant

$$t_2 - t_1 = \frac{3}{4\pi\sqrt{\frac{\pi\nu(T)}{5}}} \left(\frac{1}{T_2^2} - \frac{1}{T_1^2}\right) \tag{2.37}$$

The term T_1^{-2} can in general be neglected in (2.37) so that if we choose the origin of the time at $T \simeq T_p$ we get for $T \ll T_p$

$$t = \frac{3}{4\pi\sqrt{\frac{\pi\nu(T)}{5}}} \frac{1}{T^2} \tag{2.38}$$

One has thus to explore a range $1 \gg T > 10^{-28}$ where t is less than 10^{56} ($\simeq$ 100.000 years). In this range two important observable phenomena appear to corroborate the adiabatic expansion.

A. Nucleosynthesis The most impressive evidence for adiabatic expansion is the prediction of primordial nucleosynthesis [5]. At $T \lesssim 10$ Mev ($\simeq 10^{11}$°K), the only particle present in thermal equilibrium are e^+, e^-, ν_e, ν_μ, (ν_τ.. ?) and γ together with a small left over neutron (n^N) and proton (n^P) densities which survived thermal annihilation because of baryon number conservation. The nucleons are maintained in thermal equilibrium by the weak interactions reactions

$$n + \nu_e \rightleftarrows p + e^- \; ; \; n + e^+ \rightleftarrows p + \bar{\nu}_e \; ; \; n \rightleftarrows p + e^- + \bar{\nu}_e \tag{2.39}$$

The time elapsed since t = 0 is from (2.38) $\simeq 10^{-2}$ s.

The small nucleonic density prevents nucleosynthesis to occur except via two-body processes and therefore the first step must be deuterium (^2H) formation. The small binding energy B of ^2H (B = 2.9 Mev) delays the onset of nucleosynthesis until the temperature reaches a critical temperature $T_{(^2H)}$ where the free energy of the deuterium state is equal to that of the free constituents. This can easily be estimated to be

$$T_{(^2H)} \simeq 0.9 \times 10^9 \,°K \tag{2.40}$$

Once this temperature is reached, nucleosynthesis proceeds very rapidly and nearly all existing neutrons are caught in the highly stable ^4H_e. The reason why the resulting helium abundance is much less than 100 % of the total massive matter is twofold : a) the mass difference between m^N and m^P ($m^N - m^P$ = 1.3 Mev) reduces rapidly the neutron to proton ratio when T drops in the 10^{10}°K range, b) the two-body weak processes in (2.39) can during this drop no more maintain the neutron and proton equilibrium densities because of the rapid expansion, and the left-over neutrons slowly decay. The detailed computation yields [5) 6)] at $T = T_{(^2H)}$

$$X^{(N)} \equiv \frac{n^N}{n^N + n^P} \simeq 12 \,\% \tag{2.41}$$

and hence the fraction of primordial ^4H_e (in weight) is in very good approximation

$$2X^{(N)} \simeq 24 \,\% \tag{2.42}$$

The remaining mass is, except for very small fractions, in the form of free protons which will later form neutral hydrogen atoms at the temperature $T_{(H)} \simeq$ 4 000 °K (this is also called the recombination temperature) below which the universe becomes transparent to photons and large scale structures such as galaxies can develop.

The exact value of $X^{(N)}$ depends slightly on n^B through $T_{(^2H)}$ and is influenced by the number of distinct neutrino species : more massless or low-mass neutrinos would speed up the expansion (2.36), resulting in a smaller neutron to proton conversion and hence higher value of $X^{(N)}$. The result (2.42) is in remarkable agreement with observable datas and with the theories of star formation [6)]. One may even use the data to put an upper bound of $\simeq$ 26 % of primordial ^4H_e which would imply that the number of (approximatively) massless distinct neutrinos should probably not exceed 3 or 4.

B. Baryon Imbalance An indication that the adiabatic expansion could hold up to temperature closer to T_p is suggested from the study of baryon imbalance. Where can the apparent baryon excess ($n^B \gg n^{\bar{B}}$) come from if particles originated from thermal excitations, and how can one explain the value of the baryon to photon ratio (2.26) ? The simplest answer to these questions would be that there is no baryon excess but that $n^B = n^{\bar{B}} \neq 0$. Complete baryon-antibaryon annihilation would have been prevented to occur because when $T < m^B$ the annihilation rate became too slow to compete with the expansion rate and baryon antibaryon froze out of equilibrium with $n^B = n^{\bar{B}}$. This mechanism is however ruled out because a direct computation yields only $n^B/n^\gamma \simeq 10^{-20}$. [7] Thus we shall seek for baryon number violating interactions which could explain the baryon imbalance from the desintegration of primordial particles X which were created symmetrically namely $n^X = n^{\bar{X}}$. When baryons-antibaryons decay products would then annihilate at lower temperature only the baryon excess would survive.

A natural framework in which such X particles would appear is provided by the Grand Unified phenomenologies which arise from tentative unification of the low energy gauge groups SU(2) x U(1) x SU(3) where SU(2) x U(1) is the electroweak group and SU(3) the colour group [8]. In these theories one tries to construct a representation of a simple gauge group which contains the fifteen 2-components (say left-handed) fermions forming a known flavour family (i.e. e_L^-, e_L^+, ν_e, u_{1L}, u_{2L}, u_{3L}, d_{1L}, d_{2L}, d_{3L}, $\bar{u}_{1L}$, $\bar{u}_{2L}$, $\bar{u}_{3L}$, $\bar{d}_{1L}$, $\bar{d}_{2L}$, $\bar{d}_{3L}$) and the gauge group is then spontaneously broken [9]. Because abelian and non abelian charges have opposite behaviour under the renormalization group, the electric and colour charges may then indeed merge at the high energy where the full symmetry will be restored, thereby also determining at that energy the Weinberg angle θ from group theory. A simple calculation yields for the mass scale of symmetry restoration $m^S \simeq 10^{15}$ Gev and the low energy Weinberg angle is determined to be $\sin^2\theta \simeq 0.2$, in remarkable agreement with experiment. The high mass m^S mediates only very weak quark lepton transitions (e.g. via "lepto-quark" gauge fields of mass $\simeq m^S$) yielding a proton lifetime of the order of 10^{31} years, an acceptable result which is moreover at the limit where experimental verification is conceivable.

Grand Unified theories contain "lepto-quark" gauge fields and the (effective?) scalar fields responsible for the spontaneous breaking at $\simeq 10^{15}$ Gev; all these fields mediate baryon number violating interactions. If these X-particles were created thermally at such high temperatures, can one expect the decay products to produce baryon imbalance? The answer is yes[10] owing to CP violating effects which are natural in the context of gauge field theory. However the computation of n^B/n^γ is model dependent and requires details of presently unknown physics. No reliable estimate exists, but the order of magnitude of the baryon excess (2.26) appears compatible with the proposed mechanism.

To conclude we may state that nucleosynthesis provides good evidence for an adiabatic expansion up to $T \simeq 10$ Mev ($t \gtrsim 10^{-2}$ s) and that there are indications from grand unified phenomenologies and baryon imbalance that this adiabatic expansion may be valid up to $T \simeq 10^{15}$ Gev ($t \gtrsim 10^{-36}$ s). The hot big bang theory apparently accounts for such a situation but are the (unexplained) initial conditions (2.34) required to define this theory acceptable ? This is the question we shall now discuss.

4. The Disease

The classical big bang singularity is in fact a limiting space-like surface, only part of which is seen by a comoving observer in the expanding universe. This feature is general but is easily visualized in the case of an open ($k = -1$) universe because the latter can be conformally mapped into the forward light cone of an event in ordinary Minkowski space; we therefore discuss for simplicity this case. Performing in (2.15) the coordinate transformation

$$a^{-1}(t)dt = \bar{\tau}^{-1}d\bar{\tau} \qquad (2.43)$$

where $\bar{\tau}$ is related to cartesian coordinates $(\vec{x}, \bar{t})$ by $\bar{\tau}^2 = \bar{t}^2 - \vec{x}^2 > 0$, $\bar{t} > 0$, we get

$$ds^2 = \frac{a^2(\bar{\tau})}{\bar{\tau}^2}\,[d\bar{\tau}^2 - \bar{\tau}^2 d\Sigma^2] = e^{\lambda(\bar{\tau})}ds_o^2 \qquad (2.44)$$

Here ds_o^2 is the Minkowskian metric and the conformal factor $\exp\lambda(\bar{\tau})$ is given by

$$e^{\lambda(\bar{\tau})/2} = \frac{a(\bar{\tau})}{\bar{\tau}} \qquad (2.45)$$

The expanding universe is depicted in Fig. 1 by the portion of the forward light cone bounded by a "lower hyperboloïd" corresponding to the classical singularity $a = 0$. The matter geodesics are straight lines originating from O. The acausal character of the big bang is now clear : different non overlapping region of the initial hyperboloïd $a = 0$ are causally disconnected. The same feature persists in the hot big bang theory where the surface $T = \infty$ (or $a = 0$) is simply "smeared" for a time t of order 1 and the "initial" classical surface is taken to be $T \simeq 1$, $a = a_p \gtrsim 10^{29}$. The absence of causal connections in the classical initial conditions may appear as an unpleasant feature of the theory but does not a priori constitute an unconsistency : it would simply translate our lack of understanding of these initial conditions.

A deeper difficulty appears when the "hot" nature of the initial conditions is taken into account. From (2.34), we see that the lower bound $a_p \gtrsim 10^{29}$ means that an initial volume $\simeq [10^{29}]^3$ roughly contains the total number of quanta $N \simeq 10^{87}$ presently observed. Fluctuation involving these quanta must remain small enough to allow for a description of our visible universe in terms of a mean cosmological field satisfying (2.8).

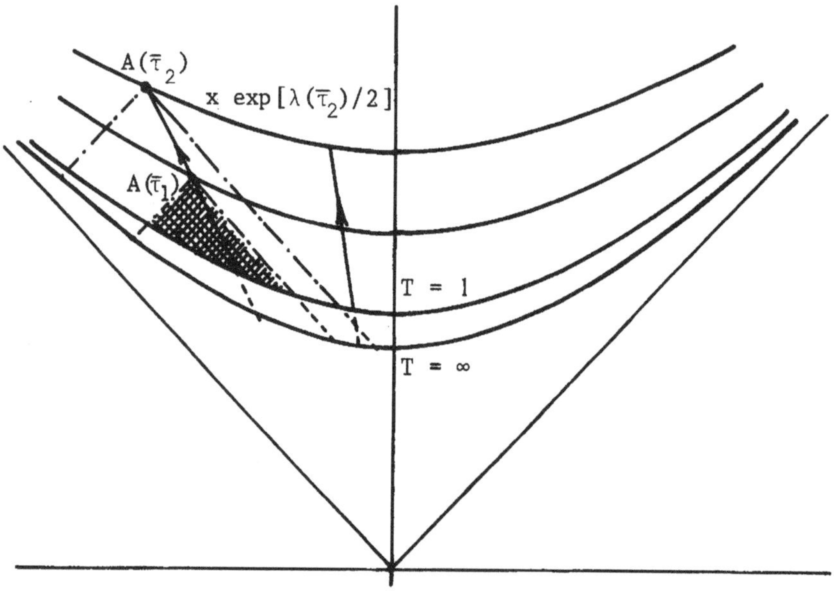

Fig. 1. Expanding open universe

: hyperboloïd of constant proper time

: geodesics of a comoving observer; the length rating two observers at a given time $\bar{\tau}$ is the segment of the hyperbolic joining them multiplied by the conformal factor $\exp\lambda(\bar{\tau})/2$

: past lights cones of the observer A

: visible portion of the universe seen by the observer A at time $\bar{\tau}_1$.

However an uncorrelated thermal state defined only by a canonical ensemble involving $N \simeq 10^{87}$ quanta will cause fluctuations in the thermodynamical variables. In fact such a state would yield, neglecting factors of order one,

$$\Delta\sigma \simeq \frac{\sigma}{\sqrt{N}} \simeq \nu(T)T^4N^{-1/2} \tag{2.46}$$

In order to keep the evolution equation (2.8) meaningful, we must have $\Delta\sigma a^2 < 1$; one thus gets from (2.46) and (2.33) the constraint

$$1 > \nu(T)T^4N^{-1/2}a^2 \gtrsim \nu^{1/3}(T)T^2N^{1/6} \tag{2.47}$$

or

$$T \lesssim \nu^{-1/6}(T)\ 10^{-7} \tag{2.48}$$

and this condition breaks down already way below T_p.

Hence the hot big bang theory makes sense only if the homogeneity of the initial thermal state is the result of strong correlations between the causally disconnected regions. This appears fatal for the theory because the terms $(\dot{a}/a)^2$ and $8\pi\sigma/3$ in the Einstein equations (2.8) become of order 1 at $T \simeq T_p$ while their difference k/a^2 as estimated from (2.34) remains of the order of

$$\frac{|k|}{a_p^2} \lesssim \nu^{2/3}(T_p)\ 10^{-58} \tag{2.49}$$

The setting up of an initial correlated state satisfying (2.49) seems absurd since uncontrolable quantum gravitational fluctuations of order 1 would have to be tuned everywhere to an accuracy of 10^{-58} in order to produce this initial state.

We thus conclude that the acausal character of the hot big bang theory is not only unpleasant but it is in fact dynamically unacceptable in view of large value of a_p. At some temperature close to 1, the dynamics of quantum matter fields coupled to gravity should provide a natural mechanism for building up the long range correlations which must extend at least over our visible universe in order to stabilize the classical cosmological field. The remaining part of these lectures will be devoted to an investigation of this dynamics and of its possible relevance for our quantum field theoretic framework.

III. ENTROPY GENERATION

1. The Exponential Expansion[11)]

The fundamental action functional of the coupled matter and

gravitational fields which leads to Einstein's equations (2.16) (which in the cosmological context reduce to (2.8) and (2.9)) is

$$S = -\frac{1}{16\pi}\int d^4x\,\sqrt{-g}\,R(g_{\mu\nu}) + \int d^4x\,\sqrt{-g}\{\mathcal{L}(\psi_i, g_{\mu\nu}) + \Lambda\} \tag{3.1}$$

where R is the Riemann curvature and $\mathcal{L} + \Lambda$ (Λ = constant) is the Lagrangian of the "matter" fields ψ_i in the background $g_{\mu\nu}$. We assume that the subtraction of the effects of vacuum fluctuations determine Λ from the requirement that when the average background is flat ($\langle g_{\mu\nu} \rangle = \eta_{\mu\nu}$) the vacuum expectation value of the effective matter energy momentum tensor is zero. In this way "empty" Minkowski space is a classical solution of the field equation, as suggested by the equivalence principle and we shall discuss the consistency of this requirement later on. In the conventional big bang framework such a procedure leads to the vanishing of the presently observed cosmological constant and thus corresponds to empirical facts. At higher teperatures some vacuum condensate of "mass" m may evaporate and an effective cosmological constant $\Lambda_{eff}(T)$ of order m^4 for $T > m$ would be generated. Thus, in principle, $\Lambda_{eff}(T)$ is determined from the physics of elementary particles and the above procedure, if fully justified, would dispose of the issue of "tuning" the possible large effective cosmological constant $\Lambda_{eff}(T)$ at early stages of the universe. Note that $\Lambda_{eff}(T)$ does anyway not affect significantly the description of the adiabatic expansion given in part II since $\sigma \gg m^4$ for $T > m$, as seen from (2.23).

In what follows we shall not require to specify further the "fundamental"matter Lagrangian in (3.1) so that we are working in the framework of conventional general relativity; we do not exclude of course that the full action (3.1) may contain additional symmetries, as is the case for instance in supergravity theories.

A first step towards solving the problem posed by the unacceptably large initial value of a in the hot big bang theory was made in reference 11. A theory was proposed in which the scale factor naturally grows from a size of order one to an exponentially large magnitude by generating self consistently a constant density of matter out of the expanding cosmological background. We shall indeed show that such a solution follows from the action (3.1) when the gravitational action is treated classically and all quantum effects of this classical background on a suitably chosen matter action are taken into account. We now study this self consistent mechanism which provides the necessary correlations missing in the hot big bang theory, and differ to section III.2 the discussion of the model matter action used here.

We first remark that if k/a^2 is neglected in the Einstein's equations (2.8) and (2.9), a constant σ implies

$$a(t) = e^{t/\tau} \quad ; \quad \sigma = \frac{3}{8\pi\tau^2} \tag{3.2}$$

$$\sigma = - p \tag{3.3}$$

Thus, the constant σ generates a de Sitter space-time characterized by the metric

$$ds^2 = dt^2 - (e^{t/\tau})^2 d\vec{x}^2 \tag{3.4}$$

We shall show that in a simple model where the matter Lagrangian describes a free massive neutral scalar field, the background (3.4) indeed creates matter [12] with a constant energy density and the pressure resulting from this quantum creation process is negative and equal to $- \sigma$. Thus for such a matter action the exponential expansion (3.2) is a non-trivial semi-classical solution to (3.1). The choice of massive matter is dictated by the fact that for spin 1/2 and spin 1 free massless fields are not produced by the cosmological expansion due to conformal invariance of the matter action (see the discussion following equation (2.29)). We use a scalar neutral field for simplicity but ensure conformal invariance in the $M = 0$ limit by using the non minimally coupled model matter action

$$S_{\text{matter}} = \int d^4x \sqrt{-g} \, \{ [\tfrac{1}{2}\partial_\mu\psi\partial_\nu\psi g^{\mu\nu} - \tfrac{1}{2}(M - \tfrac{R}{6})\psi^2] + \Lambda \} \tag{3.5}$$

Before computing the matter creation from (3.5) in the background (3.4) we should explain why a non trivial semi-classical solution can possibly exist. To this effect change coordinate in (3.4) to get the explicitely conformally flat form.

$$ds^2 = a^2(\bar{t}) \, [d\bar{t}^2 - d\vec{x}^2] \tag{3.6}$$

with

$$a \equiv e^{t/\tau} = (1 - \bar{t}/\tau)^{-1} \tag{3.7}$$

Perform now the conformal rescaling

$$g^{\mu\nu} = \eta^{\mu\nu}/a^2 \quad ; \quad \psi = \mathring{\psi}/a \tag{3.8}$$

to rewrite (3.1), using (3.5), as

$$S = -\frac{3}{8\pi}\int\sqrt{-\eta}\,(\eta^{\mu\nu}\partial_\mu a\,\partial_\nu a\,)\,d^4\bar{x}$$

$$+\frac{1}{2}\int\sqrt{-\eta}\,\{\,\eta^{\mu\nu}\partial_\mu\mathring{\psi}\partial_\nu\mathring{\psi} - M^2(\bar{t})\mathring{\psi}^2 + \Lambda a^4\}\,d^4\bar{x} \qquad (3.9)$$

where

$$M(\bar{t}) = a(\bar{t})M \qquad (3.10)$$

We now see clearly why matter can be created even if the total average Minkowskian energy is kept equal to zero as is in fact required by Einstein's equation: while the created matter energy is positive, <u>the Minkowskian expansion energy is negative definite</u>. It is this quite exceptional property of the gravitational action which leads to the solution which we now derive.

We see from (3.9) that in ordinary 3-momentum space $\vec{p}$, the matter system is equivalent to a set of free harmonic oscillators with time dependent frequencies

$$\omega_p(\bar{t}) = [\,\vec{p}^2 + M^2(\bar{t})]^{1/2} \qquad (3.11)$$

and the cosmological counterterm cancels the time dependent zero-point energies of the oscillators once Λ is chosen to cancel exactly the vacuum effects for a constant mass [11]. The latter requirement follows from our general anzats for Λ.

The varying frequencies will populate the modes, that is create particles. The resulting rest energy density σ_{matter} of the matter is computed as follows. Expand $\mathring{\psi}(\vec{x},\bar{t})$ as a sum of momentum space creation and annihilation operators,

$$\mathring{\psi}(x,\bar{t}) = \frac{1}{\sqrt{2}}\frac{1}{(2\pi)^{2/3}}\int d^3p\,[\,\xi_{\vec{p}}(\bar{t})e^{-i\vec{p}.\vec{x}}a^+(\vec{p}) + h.c.] \qquad (3.12)$$

The equation of motion and the canonical commutation relations yield

$$\frac{d^2\xi_{\vec{p}}(\bar{t})}{d\bar{t}^2} + \omega_p^2(\bar{t})\xi_{\vec{p}}(\bar{t}) = 0 \qquad (3.13)$$

and

$$\xi^*_{\vec{p}}(\bar{t})\,\frac{d\xi_{\vec{p}}(\bar{t})}{d\bar{t}} - \frac{d\xi^*_{\vec{p}}(\bar{t})}{d\bar{t}}\,\xi_{\vec{p}}(\bar{t}) = 2i \qquad (3.14)$$

The identification of $a^{+}(p)$ as the creation operator at an initial time $\bar{t}_{in}$ corresponding to $a(\bar{t}_{in})$ of order 1 is ensured by the initial condition

$$\left[i \frac{d}{d\bar{t}} + \omega_p(\bar{t}) \right] \xi_{\vec{p}}(\bar{t}) \Big|_{\bar{t} = \bar{t}_{in}} = 0 \tag{3.15}$$

The Heisenberg state of the universe is then defined by

$$a(\vec{p}) | \Omega > = 0 \tag{3.16}$$

and for convenience we may, with negligible error, take in (3.15) $\bar{t}_{in} = -\infty$. The solution of (3.13) satisfying (3.14) and (3.15) is then

$$\xi_{\vec{p}}(\bar{t}) = y^{1/2} \sqrt{\frac{\pi}{2i}} \frac{e^{\eta\pi/2} J_{-i\eta}(py) - e^{-\eta\pi/2} J_{i\eta}(py)}{\sinh \pi\eta} \tag{3.17}$$

with

$$y = \tau - \bar{t} \tag{3.18}$$

$$\eta = \frac{1}{2}[4M^2\tau^2 - 1]^{1/2} \tag{3.19}$$

and $J_{i\nu}$ is the Bessel function of order $i\nu$. From (3.12), (3.17) and (3.16) one can compute the rest energy density

$$\sigma_{matter} = < \Omega | T^t_t | \Omega > = < \Omega | \mathring{T}^{\bar{t}}_{\bar{t}} | \Omega > a^{-4}$$

Where $T^t{}_t$ is the energy density operator computed from (3.5) and $\mathring{T}^{\bar{t}}{}_{\bar{t}}$ the scaled minkowskian energy density as computed from (3.9). One gets

$$\sigma_{matter} = \left(\frac{\tau}{y}\right)^{-4} \int_0^\infty \frac{p^2 dp}{2\pi^2} n_p(y) \omega_p(y) \tag{3.20}$$

where

$$n_p(y) = \left| \left[-i \frac{d}{dy} + \omega_p(y) \right] \xi_{\vec{p}}(y) \right|^2 \frac{1}{4\omega_p(y)} \tag{3.21}$$

Equation (3.20) has a simple interpretation : $n_p(y)$ is the number of quanta created in the mode $\omega_p(y)$ and the prefactor in the conformal rescaling from $<\Omega|\overset{\circ}{T}{}^{\bar t}{}_{\bar t}|\Omega>$ to $<\Omega|T^{t}{}_{t}|\Omega>$. As already mentioned the zero point energy has been cancelled by the cosmological counterterm and is absent in (3.20). From (3.11) and (3.18) we see that $\omega_p(y) \equiv f(py)/y$ and $n_p(y) \equiv g(py)$; thus by changing variable in (3.20) to py we see that σ_{matter} is independent of y and is thus constant in time. Moreover one can compute $<\Omega|T|\Omega> \equiv \sigma_{matter} - 3p_{matter}$ and one finds $<\Omega|T|\Omega> = 4\sigma_{matter}$; hence

$$\sigma_{matter} = - p_{matter} \tag{3.22}$$

Comparing with (3.2) and (3.3) we see that we have indeed found a non trivial solution to the coupled Einstein-matter action provided.

$$\sigma = \sigma_{matter} \tag{3.23}$$

This equation determines in principle τ as a function of the only free parameter in the action, namely M

If the expansion rate is small compared to the mass, that is if

$$M\tau \gg 1 \tag{3.24}$$

the integral (3.20) is easily performed using the asymptotic form of the Bessel functions (3.17); one gets

$$\sigma_{matter}(M,\tau) \underset{\nu\to\infty}{\rightarrow} \frac{M^4}{32\pi^2(M\tau)^2} \int_0^\infty \frac{q^2dq}{(1+q^2)^{5/2}} = \frac{M^2}{96\pi^2\tau^2} \tag{3.25}$$

Comparing (3.25) to (3.2) we obtain

$$M = 6\sqrt{\pi}\ (=1.30 \times 10^{20}\text{Gev})\ ;\ M\tau \gg 1 \tag{3.26}$$

Hence for $M\tau \gg 1$ the mass turns out to be independent of the expansion rate! The existence of this asymptotic eigenvalue condition on the created mass can be inferred from a simple perturbative argument. When the expansion rate is slow the rate of creation should be proportional to the square of the effective coupling constant of the matter gravity system and the latter is proportional to $1/\tau$. Since σ is from Einstein's equation (2.8) also proportional to $1/\tau^2$, the factor τ^2 cancels in the self-consistency condition (3.23), thereby fixing the mass M.

Actually (3.26) constitutes a lower bound. The integral (3.20) can be performed analytically in general to yield [13])

$$\sigma_{matter} = \frac{M^4}{64\pi^2}[\psi(\frac{3}{2} - \eta) + \psi(\frac{3}{2} + \eta) - \ln M^2\tau^2] \tag{3.27}$$

where ψ is the digamma function. Equating (3.27) to (3.2) we get τ as a function of M. The resulting curve is plotted in Fig. 2. Note that, as expected from the above discussion, M is quite insensitive to τ for $\tau > 1$; note also that (3.26) constitutes a lower bound on M below which no self-consistent solution exists.

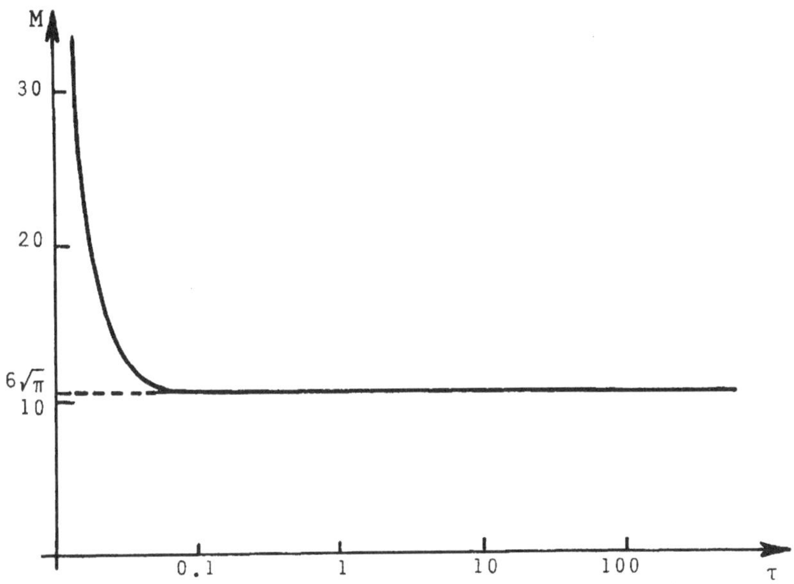

Fig. 2. M as a function of τ (reference 13)

We have thus completed the proof that a non trivial solution of the coupled matter-gravitational action exists in the semi-classical limit where the metric is treated classically, provided the quanta of the matter field have a sufficiently high mass. The resulting exponential expansion would provide a nautral mechanism for interpolating the scale factor from an initial value.

$$a_{in} \sim \tau \tag{3.28}$$

below which the curvation term k/a^2 must be taken into account (see equation (2.8)) and the huge value $a \gtrsim 10^{29}$ required to start the adiabatic expansion. If τ is indeed comparable to 1, the time required to build correlations over the size $\simeq N^{1/3}L_p$ of our visible universe is only, from (3.7), equal to a few hundred Plancks times and no unnaturally high numbers have to be invoqued. One could thus envisage decay of the massive particles into "ordinary matter" after this period and the universe would then enter an era of free expansion. The latter would thus be causally related to the natural initial condition (3.28). For times shorter than τ, one might then at least pose in scientific terms the problem raised by this initial condition itself.

As it stands however this theory is confronted with four major difficulties

1°) the result $M > 1$ renders the theory vulnerable to quantum gravitation fluctuations and the use of a classical cosmological background appears unjustified

2°) the stability of the result (3.26) for all possible classical expansion rate ($\tau > 1$), as examplified in Fig. 2, amounts to a determination of M. As M is however a free parameter in the action (3.9) this determination requires that the mass be explained in dynamical terms.

3°) because the "free" parameter M is in fact determined, the "dependent" variable τ can take any value between 1 and ∞ without changing the self consistency of the solution. Therefore we have no justification to assume, as we did, that τ is in fact comparable to 1.

4°) the points 2°) and 3°) clearly show that the theory is at best incomplete. In fact we do not even gain any understanding about the origin of the high temperature at the onset of the adiabatic expansion.

We shall now show that a solution to <u>all</u> these problems can be obtained from an analysis of strong gravitational effects in the early radiation dominated era; this analysis will recover (3.9) but only as an <u>effective</u> action describing the matter-gravitational system.

2. Quantum Creation of Black Holes

We argued in section II.4 that a description of a radiation dominated universe at $T \simeq 1$ in terms of a thermal distribution of quanta in a classical cosmological background field is inconsistent due to the appearance of strong gravitational quantum fluctuations. However, we shall see that a new phenomenon has to be taken into account below T_p, namely the existence of black holes configurations formed from the (nearly) massless quanta present at high temperatures. To understand this we shall first discuss the nature of a classical black hole and of its thermal properties.

For our purpose it will be sufficient to define a black hole as a classical material point of mass μ surrounded by its gravitational field. In this way it is the gravitational analog to a point electric charge surrounded by its Coulomb field in electromagnetic theory. The classical gravitational field can be described by the metric which solves Einstein's equation (2.16) for this problem. For spatial distance $\rho > 2\mu$ the result is the static Schwartzchild solution.

$$ds^2 = (1 - 2\mu/\rho)d\theta^2 - (1 - 2\mu/\rho)^{-1}d\rho^2 - \rho^2 d\Omega^2 \qquad (3.29)$$

where θ is a time variable. This metric is of course singular at $\rho = 0$ but the surface $\rho = 2\mu$ is not a metric singularity; rather it generates "horizons" across which signals can be sent at $\theta = +\infty$ (and from which signals may come out at $\theta = -\infty$); this is easily seen by changing variables

$$(1 - 2\mu/\rho)^{-1/2}d\rho = d\varepsilon \qquad (3.30)$$

or in the neighbourhood $\rho = 2\mu$ where we take $\varepsilon = 0$.

$$\varepsilon = 4\mu(1 - 2\mu/\rho)^{1/2} \qquad (3.31)$$

Thus in the vicinity of the horizons the metric takes the form (with $d\Omega = 0$)

$$ds^2 = \varepsilon^2 d(\frac{\theta}{4\mu})^2 - d\varepsilon^2 \qquad (3.32)$$

or equivalently

$$ds^2 = dv^2 - du^2 \; ; \qquad v = \varepsilon \sinh \frac{\theta}{4\mu}$$

$$u = \varepsilon \cosh \frac{\theta}{4\mu} \qquad (3.33)$$

The horizon properties now follow from the fact that $\varepsilon = 0$ $(\rho = 2\mu)$ $\theta = \pm\infty$ are just light cones in u, v - space as depicted in Fig. 3.

Thus, horizons are not singular metric surfaces and the propagator of a matter field in the background (3.35) can cross the horizons and come back. As shown by Hawking [14] this will in general appear to the external observer as emission or absorption of a thermal radiation. The radiation stemming from a horizon can be illustrated in ordinary Minkowski space by considering the portion of space-time available to all observers moving along the u ($\equiv$ x) direction with constant positive acceleration a [15].

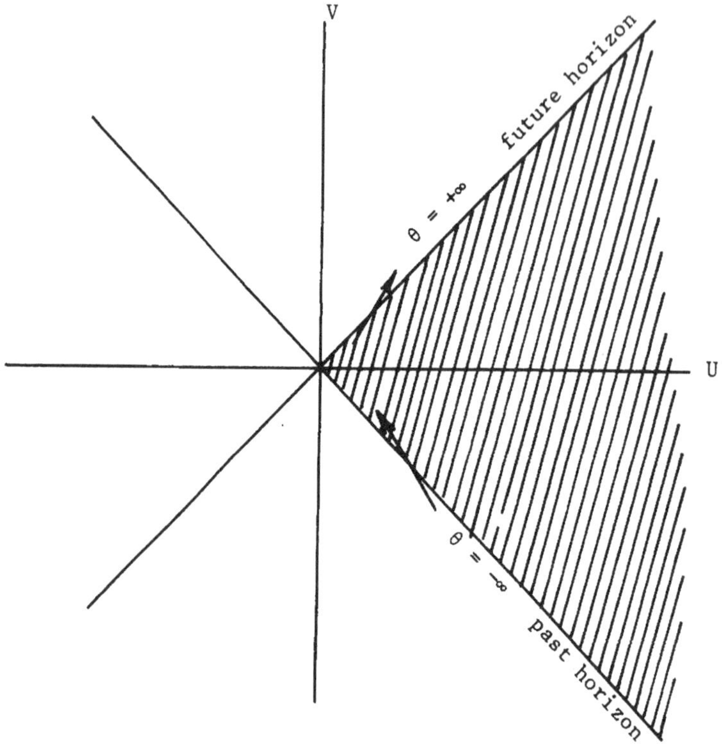

Fig. 3. Black hole horizons.

 : region of validity of the Schwartzchild metric or region available to the "external observer"

 : signals sent through horizons (for black holes formed by collapse the past horizon does not exist).

If θ is a coordinate time measured by the accelerating observers and v ($\equiv t$) the ordinary Minkowskian time, the trajectories can be parametrized by

$$v = \varepsilon \sinh(\lambda\theta)$$

$$u = \varepsilon \cosh(\lambda\theta) \tag{3.34}$$

Indeed we have

$$(\varepsilon\lambda)a \equiv \frac{d}{dt}\left[\frac{dx}{d\theta}\right] \equiv \frac{d}{dv}\frac{du}{d\theta} = \lambda \tag{3.35}$$

The Minkowskian metric can be written as

$$ds_0^2 = dv^2 - du^2 - dy^2 - dz^2 = \varepsilon^2 d(\lambda\theta)^2 - d\varepsilon^2 - dy^2 - dz^2 \tag{3.36}$$

where the ε, θ parametrization is valid in the domain $0 \leqslant \varepsilon < +\infty$. It is seen in Fig. 4 that the region of Minkowskian space time available to the accelerated observer is bounded as in Fig. 3 by future and past horizons.

As a consequence, the global stationary Minkowskian vacuum of the matter action will appear to the accelerated observer as a thermal state. This is because the propagator between (say) point 1 and 2 in Fig. 4 of any matter field described by a v - independent action is analytically continuable to Euclidian time iv. This analytically continued propagator in the variables ε, θ is then regular at the horizon $\varepsilon = 0$ because $i\theta$ becomes an euclidian angular variable in the metric (3.36) with periodicity $2\pi/\lambda$. The propagator in the coordinate system ε, θ therefore describes correlations in a heat bath with temperature T_{acc}

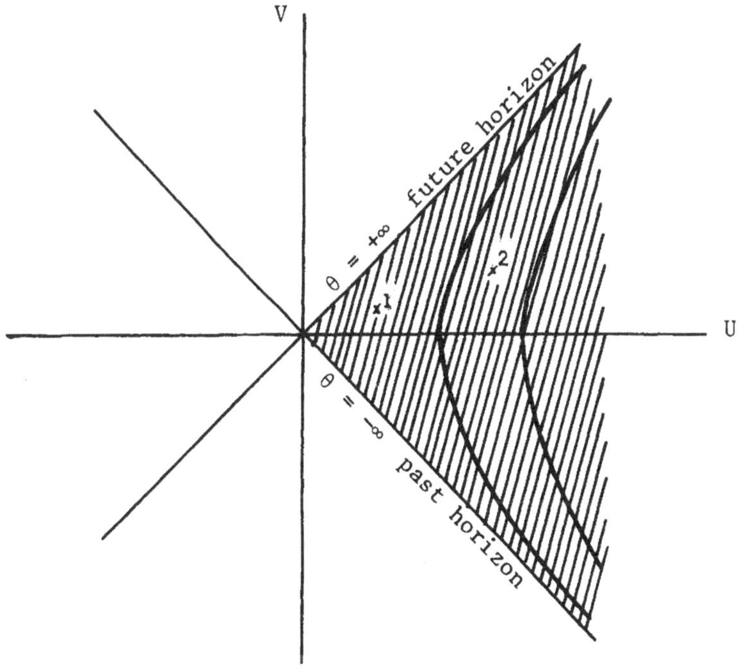

Fig. 4. Horizons for accelerating observers in Minkowski space-time (we take $y = z = 0$).

: region accessible to accelerated observers

: trajectories of observers.

$$T_{acc} = \frac{\lambda}{2\pi} \tag{3.37}$$

The same reasoning would formally immediately yields from (3.32) the black hole temperature T_{BH} felt by a static observer using the Schwartzchild metric (3.29), namely [14), 19)]

$$T_{BH} = \frac{1}{8\pi\mu} \tag{3.38}$$

However, in the black hole problem the situation is different : because of the singularity at $\rho = 0$ there is no a priori reason to believe that a global stationary ground state is available. In fact in the physical case in which the black hole singularity at $\rho = 0$ is formed by (say) a collapse, the past horizon does not exist and only the future horizon formed by the collapse can generate radiation. Nevertheless, we can describe the effect of this horizon by performing the thought experiment of putting an eternal black hole in a heat bath at the temperature (3.38) and assume that the system could then be described, as far as matter fields are concerned, by an (approximate?) global ground state. The radiation emitted or absorbed by the horizons would then be in equilibrium with this external world at the temperature (3.38). In the physical case where the external bath does not exist, an external observer can then describe the matter state by replacing the black hole by a black body source at the temperature (3.38).

This thermal emission causes black hole decay. Neglecting general relativistic geometry effects and taking only into account the emitted degrees of freedom with negligible mass (at the temperature (3.38)) we get, by using (2.23)

$$\frac{d\mu}{dt} = -\, 4\pi(2\mu)^2 \frac{\pi^2}{30} \nu(T)T^4 = -\, \frac{8}{15} \nu(T) \frac{1}{84\pi\mu^2} \tag{3.39}$$

A more reliable approximation yields 9/10 [16)] instead of 8/15, so we shall replace in (3.39) 8/15 by ζ_1 a parameter of order 1. We then write for the black hole life time

$$t_{BH} = (8\pi\mu)^3 \frac{8}{3\zeta\pi^2} \frac{1}{\nu(T)} \tag{3.40}$$

Black holes can be formed in many ways and their entropy, as measured by an external observer, is therefore very large [17)]; in fact the entropy S_{BH} of an isolated black hole is given by (neglecting translational entropy)

$$S_{BH} = 1/16\pi T^2 = 1/2T\ E_{BH} \tag{3.41}$$

as is immediately verified from

$$TdS_{BH} = -\frac{1}{8\pi T^2}\,dT = dE_{BH} \tag{3.42}$$

This large entropy may favour collapse of black body radiation below the Planck temperature. Consider indeed an isolated system of massless quanta (including gravitons) of total energy E and volume V. From (3.41) if the volume is sufficiently small for

$$\frac{1}{2}E_{BH} > \frac{4}{3}E_R \quad ; \; E_R = V\sigma \tag{3.43}$$

where σ is given by (2.23), the entropy of a single black hole would exceed the entropy of pure radiation at the same temperature. Taking into account the black body radiation in thermal equilibrium with the black hole, we find that a stable configuration involving a black hole is thermodynamically favoured if (3.43) is corrected to [17]

$$E_{BH} > 4E_R \qquad \text{or } V \leqslant \frac{5}{6}\,\frac{1}{\pi^3\nu(T)T^5} \tag{3.44}$$

This is easily understood by considering a fluctuation around the equilibrium state of energy

$$E = E_{BH} + E_R \tag{3.45}$$

We get indeed, if dT_{BH} and dT_R are temperature variations around the equilibrium temperature $T = T_{BH} = T_R$

$$-\frac{E_{BH}}{T}\,dT_{BH} + 4\,\frac{E_R}{T}\,dT_R = 0 \tag{3.46}$$

Thus if (3.44) is satisfied, a decrease in the black hole energy which yields from (3.38) an increase in T_{BH}, implies $dT_R > dT_{BH}$ so that the radiation temperature increases more than T_{BH} and equilibrium is reestablished by the resulting radiation flow into the black hole. Of course (3.44) is valid only if

$$V > \frac{4\pi}{3}(2E_{BH})^3 = \frac{1}{48\pi^2T^3} \tag{3.47}$$

so that in the isolated system, a collapse of black body radiation into a black hole is indeed thermodynamically favoured at sufficiently low temperature, namely if

$$T \leqslant \sqrt{40/\pi\nu(T)} \tag{3.48}$$

This result does not mean that the collapse actually occurs at low temperature because there is classically no mechanism available to collapse the black body radiation into isolated large black holes. In fact, according to (3.38) for a black hole to form from localized fluctuation its local temperature must exceed the average temperature of the medium; therefore a fluctuation can hardly grow and the forming black hole decays. The results (3.44) and (3.48) do however imply that "bound state" black hole configurations are available in the quantum vacuum of a system containing radiation and the gravitational field.

Thus when, extrapolating back in the radiation dominated era towards T_P we cross the temperature $1/8\pi M$ with M given by (3.26), we expect quantum creation of black holes out of expansion energy in a time comparable to M^{-1}. The created black holes behave as (quasi-) stable particles if

$$t_{BH}(M) \gg M^{-1} \tag{3.49}$$

and we see from (3.40) (with $\mu \equiv M$) that (3.49) is always satisfied provided $\nu(T)$ is not exceedingly large. We shall now study how the massive black holes can indeed be generated from the quantum vacuum.

To describe black holes interactions with gravity and other quantum fields we note that "short distance" quantum fluctuations will hide beyond the horizon and will not contribute to the interactions of the black holes with the external world; these will occur for low momentum fluctuations ($p \lesssim (8\pi\mu)^{-1}$) and hence a semi-classical description of these interactions, including gravity, is available. Moreover, as black holes interact with the gravitational and matter fields only through their conserved parameters (mass, angular momentum and gauge charges) we may represent black holes by massive fields with those parameters. In what follows we shall concentrate only on spinless neutral black holes interacting with gravity; asymptotic freedom justifies the neglect of all other interactions and one may explicitly verify that the results which follow are insensitive to spin. Thus black holes of a given mass M will be represented by an effective action (3.5) where $g^{\mu\nu}$ represents the long wavelength part of the external gravitational field. Note that the term R/12 in (3.5), which previously was justified only by consistency requirements is now implied because the black hole mass is here the only relevant departure from conformal symmetry.

We now show that the full effective action (3.9) describes a system of black holes including dynamical interactions through the large scale gravitational field. To this effect, we first prove that the non trivial de Sitter solution which is deduced from (3.9) in the semi-classical limit is indeed interpretable in terms of a large scale average over black hole metrics. We perform in (3.4) the coordinate transformation

$$e^{t/\tau} = (\tau^2 - \rho^2)^{1/2} \, e^{(\theta-\theta_o)/\tau} \quad ; \; \theta_o \text{ arbitrary} \tag{3.50}$$

$$|\vec{x} - \vec{x}_o| \, e^{t/\tau} = \rho \quad ; \; \vec{x}_o \text{ arbitrary} \tag{3.51}$$

to obtain the de Sitter metric in the form

$$ds^2 = (1 - \rho^2/\tau^2)d\theta^2 - (1 - \rho^2/\tau^2)^{-1} \, d\rho^2 - \rho^2 d\Omega^2 \tag{3.52}$$

valid in the patch

$$0 \leqslant \rho < \tau \tag{3.53}$$

In this way we can at any time t cover completely the de Sitter space by mutually overlapping patches of static metrics. From (3.51) we see that a comoving observer sitting on a given patch will describe the evolution in the $(\theta, \rho; \vec{x}_o, \theta_o)$ systems by a recession of the center of all patches with a Hubble constant $1/\tau$ and similarly will describe the flow of the cosmological fluid inside his patch by a velocity field ρ/τ (see Fig. 5); the surface $\rho = \tau$ $(\theta = +\infty)$ corresponds to the locus of points which escape with the velocity of light and is a future horizon in a $\rho - \theta$ coordinate system. The "reason" why the system is static in a $\rho - \theta$ coordinate system is that the outgoing cosmological fluid is constantly rebuilt by the creation process in order to maintain σ = constant; one in fact verifies immediately that the time dilation and space contraction factors for the static observers in (3.52) are just $\sqrt{1-v^2}$ where v is the velocity of a comoving observer with respect to a static observer located at $\rho = v\tau$.

For σ to be interpretable as an average black hole mass density, we should be allowed to replace the de Sitter metric (3.52) by the black hole metric (3.29) in the vicinity of a black hole of mass μ located at the origin of a given patch. Substituting in (3.52) $1 - 2\mu/\rho$ for $1 - \rho^2/\tau^2$ up to a radius ρ_c such that

$$\frac{4\pi}{3} \rho_c^3 \sigma = \mu \qquad (\rho_c > 2\mu) \tag{3.54}$$

we get using the de Sitter value for σ (3.2)

$$\frac{\rho_c^3}{\tau^2} = \frac{2\mu}{\rho_c} \tag{3.55}$$

This last relation precisely expresses the continuity of the two metrics at $\rho = \rho_c$ (Fig. 6). This is a manifestation of the general relativistic generalization [18] of Gauss'law which is known as Birkhoff's theorem. Thus we may indeed interpret the de Sitter space-time as an average over black-hole geometries for scales $\gtrsim \rho_c$; note that the constraint $\rho_c < \tau$ limits this description to masses

$$\mu < \frac{\tau}{2} \tag{3.56}$$

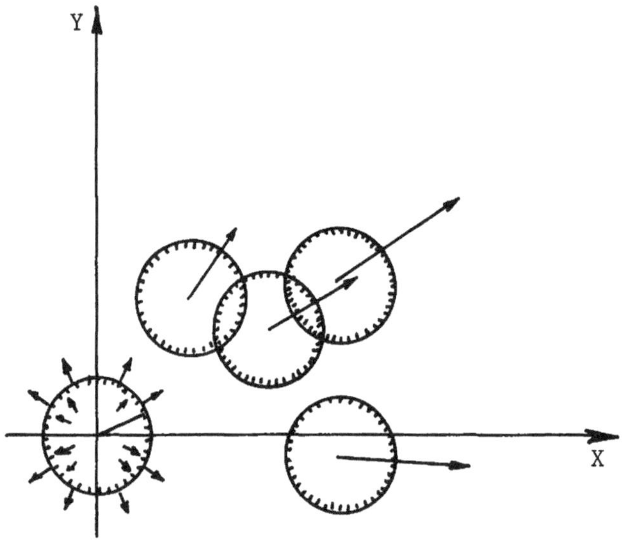

Fig. 5. Patches of static metrics in the plane z = 0 (t = constant)

: horizons for static observers

: velocity field of the comoving cosmological fluid.

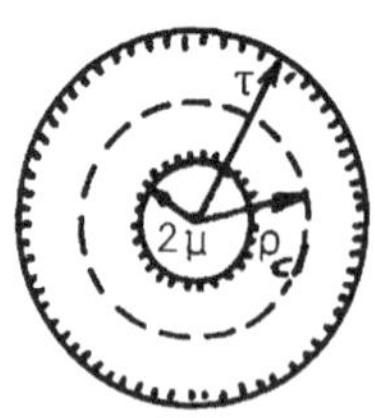

Fig. 6. Schwartzchild-de Sitter connexion

: Black hole horizon

: de Sitter horizon

However, consistency of the average de Sitter description with the local black hole structure for the low massless quanta produces another condition.

The effect of these massless quanta were not taken into account in the effective action (3.9) and in fact massless quanta are not produced in the comoving coordinate system of a de Sitter space-time because of the conformal flatness of the metric (3.6). Nevertheless a static observer would see a thermal radiation stemming from the horizon $\rho = \tau$. In the vicinity of $\rho = \tau$ we perform the change of variables analogous to (3.30)

$$(1 - \rho^2/\tau^2)^{-1/2}\, d\rho = d\varepsilon \qquad \varepsilon = 0 \text{ at } \rho = \tau \tag{3.57}$$

or

$$\varepsilon = - \tau(1 - \rho^2/\tau^2)^{1/2} \tag{3.58}$$

thus leading to ($d\Omega = 0$)

$$ds^2 = \varepsilon^2 d(\frac{\theta}{\tau})^2 - d\varepsilon^2 \tag{3.59}$$

The Euclidian analytically continued action is periodic with period $2\pi\tau$ so that repeating the reasoning made for black hole horizons leads to a de Sitter temperature T_{dS} [19),]

$$T_{dS} = \frac{1}{2\pi\tau} \tag{3.60}$$

Note that this radiation is in the pure de Sitter case, as for the accelerated observer, only an effect of the coordinate system but is in fact strictly correct only for the massless particles. For massive particles indeed the global hamiltonian defined in the comoving coordinate system is not stationary and therefore the analytic continuation is not justified.

In the present case, the massless quanta produced by (3.60) cannot be gauged away by a change of coordinate because of the local disturbances due to the black hole metrics. In particular, the black hole horizons at $\rho = 2\mu$ in Fig. 6 cannot expand and this fact introduces an explicit departure from conformal flatness. Thus, real massless quanta are present in the system as a consequence of the black hole horizons and the exchange of these massless quanta between the black holes can stabilize the system for times $\gtrsim \tau$. If an (approximate) ground state for the massless quanta exists, the black holes and the average de Sitter metrics must then be continuable to Euclidian space at least within a patch and the matching condition (3.55) must hold in Euclidian time as well. This is possible if and only if the periodicities stemming from the de Sitter and of the black hole horizons coincide or equivalently if both temperatures are the same. From (3.60) and (3.38) we get

$$\tau = 4\mu \tag{3.61}$$

a result consistent with the inequality (3.56)

To establish the validity of the semi-classical solution, we must now consider the continuous spectrum of available black hole masses μ which can be formed out of massless quanta. However we see from (3.61) that since τ is an average classical quantity the choice of an effective field with a single mass parameter M $(= \tau/4)$ is justified a posteriori. One may interpret the mass selected by (3.61) as follows : a "typical" black hole like configuration will in general have a non zero overlap with the state obtained by the self consistent solution and within an expansion time τ the only surviving mass will be $\tau/4$. The equation (3.26) now acquires the interpretation of an eigenvalue condition which selects out of the available mass spectrum μ the particular value M which dominates the path integral.

The relations (3.26) and (3.61) with $\mu = M$ determine entirely the initial cosmological state of the universe by providing the initial temperature T_{in} and Hubble constant H_{in} ((3.61) implies $M\tau \gg 1$)

$$T_{in} = \frac{1}{8\pi M} = [48\pi^{3/2}]^{-1} = 4.57 \times 10^{16} \text{ Gev} \tag{3.62}$$

$$H_{in} = \frac{1}{\tau} = [24\pi^{1/2}]^{-1} = 4.26 \times 10^{41} \text{ s}^{-1} \tag{3.63}$$

These results are clearly consistent with the existence of a classical background. The back reaction of the real massless quanta present, at the temperature (3.62), has however not been taken into account in determining the initial cosmological state. The justification of their neglect follows from the value of the matter (radiation) energy density which from (2.23) and (3.62) is

$$\sigma_{rad} = \nu(T_{in})\pi^2 T_{in}^4 /30 = \nu(T_{in})/30.48^4.\pi^4 \tag{3.64}$$

This will be negligible compared to the total de Sitter energy density σ, given by (3.7) and (3.63)

$$\sigma = \frac{3\pi T_{in}^2}{2} = 3/2.48^2\pi^2 \tag{3.65}$$

provided

$$\nu(T_{in}) << 45.48^2.\pi^2 \simeq 10^6 \tag{3.66}$$

That such a limitation on the total number of available degrees of freedom at $\simeq 5 \times 10^{16}$ Gev must be imposed anyhow follows from the constraint (3.49). Indeed, from (3.49), using (3.38), (3.40)

$$\nu(T_{in}) << T_{in}^{-4}/3\zeta\pi^3 = 48^4\pi^3/3 \simeq 10^7 \tag{3.67}$$

The inequality (3.67) puts a (very mild) constraint on the elementary particle physics which should be expected to hold in any reasonable model satisfying the asymptotic freedom requirement.

We know have solved all the four problems posed at the end of the section III.2. It remains however to understand the mechanism responsible for the transition to the adiabatic regime.

3. <u>The Transition to the Adiabatic Expansion</u>

As already mentioned the created black holes are thermally unstable because $d\mu/dT < 0$. The fraction η of space available for decay is rather large; indeed

$$\eta' = 1 - \frac{4\pi/3\ (2M)^3}{M/\sigma} = \frac{3}{4} \tag{3.68}$$

where we have used (3.2) and (3.61). Thus we expect radiation to be lost and a crude estimate of the black hole lifetime is provided by (3.40) with μ replaced by M. After a time of order t_{BH}, the slowing down of the expansion rate due to the decrease of σ resulting from black hole decay causes τ to depart from (3.61); fewer black holes are produced and incoherence sets in. The massless quanta (on the temperature scale $(8\pi M)^{-1}$) produced, hitherto neglected, become the dominant source of expansion and the universe enters the adiabatic era at a temperature close to (3.62). We can roughly estimate the relevant amount of entropy generated in the creation era by computing the value taken by $(aT_{in})^3$ after an exponential expansion lasting for a time t_{BH},

$$(aT_{in})^3 = a_{in}^3\ 48^{-3}\pi^{-9/2}e^{3t_{BH}/\tau} \simeq e^{4 \times 10^5/\nu(T_{in})} \tag{3.69}$$

Here we took $a_{in} \simeq \tau$ according to (3.28). The result (3.69) is consistent with present observation (2.33) if

$$e^{4 \times 10^5/\nu(T_{in})} \gtrsim 10^{87}/\nu(T_{in}) \text{ or } \nu(T_{in}) \lesssim 2000 \tag{3.70}$$

This agrees with the theoretical bound (3.66) and we then see, that provided $\nu(T_{in})$ is not too large we have a natural explanation for the large number aT. This in turn explains in causal terms the isotropy of the presently observed background radiation and the apparent flatness of the universe over the whole radiation dominated era.

The estimate (3.69) is perhaps to crude to take (3.70) better than an order of magnitude estimate of the constraint it imposes on model building for elementary particle physics. Note however that (3.70) is indeed verified in the simplest currently fashionable models (say SU(5)). A more precise evaluation of the created entropy requires a detailed study of the transition period between the creation era characterized by a constant $H = 1/\tau$ and the radiation dominated era where H varies according to (2.36). If the initial temperature of the adiabatic period can still be correctly estimated by (3.62) one would need a sliding down of the "instantaneous" de Sitter value τ to a final value τ_F (see Fig.7) such that, using (3.64) and (3.65)

$$\frac{\tau_F}{\tau} = \left[\frac{\sigma}{\sigma_{rad}}\right]^{1/2} = 144\pi \sqrt{\frac{5}{\nu(T_{in})}} \simeq \frac{10^3}{\sqrt{\nu(T_{in})}} \qquad (3.71)$$

This would lead to a ratio

$$\frac{\tau_F}{t_{BH}} \simeq \frac{10^3}{\nu^{1/2}(T_{in})} \; \frac{\nu(T_{in})}{10^5} \simeq 10^{-2}\nu^{1/2}(T_{in}) < 1 \qquad (3.72)$$

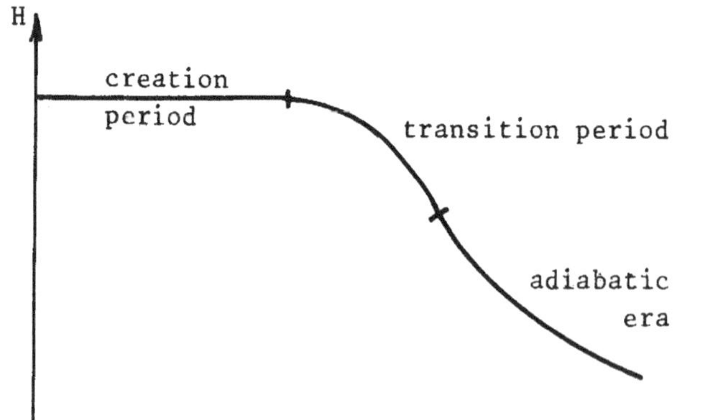

Fig. 7. The transition to the adiabatic era

Thus, slowing down of the expansion due to the black hole decay is indeed expected to occur smoothly on time scales comparable to the black hole lifetime. Therefore one expects some black holes to be still present at the onset of the adiabatic era. A more detailed analysis of the transition period, which is in principle determined from the theory, would provide an initial fluctuation spectrum which might lead to observational consequences.

To complete our study of cosmogenesis, it is now necessary to understand the nature of the original fluctuation which led to the initial classical condition $a_{in} \simeq \tau$ (3.28). This will drive us unavoidably to the realms of quantum gravity.

IV. QUANTUM GRAVITY

1. The Rise and Fall of a Universe

When a gets smaller than τ, spatial curvature effects in the Einstein equations are felt. The non trivial solution of the de Sitter space-time geometry cannot be sustained, as can be explicitely checked, and it does not seem possible to sustain a non trivial semi-classical solution at all [11]. This means that the initial condition $a_{in} \simeq \tau$ must find its origin in a quantum fluctuation [20]; from the structure of the creation period it is clear that metric fluctuations are needed so that the universe has to originate from a fluctuation of quantum gravity. Note that no initial classical singularity is required [11] because of the negative pressure $\rho = -\sigma$ in equation (2.9)

We shall make the following assumptions:

(a) at scales large compared to the Planck length, quantum metric fluctuations develop on an average background which is Minkowskian; this seems natural once the cosmological constant is taken as in (3.1) to ensure that Minkowskian space-time is always a classical solution to the field equation.

(b) no "unnaturally" tuned quantum fluctuations is admissible.

We just saw that for $a \lesssim \tau$ the curvature term k/a^2 is expected to fluctuate. However any Planckian fluctuation of the scale factor for $a \simeq \tau$ cannot survive in the exponential expansion without destroying the homogeneity of the solution itself. Therefore the homogeneous universe cannot be generated unless it is spatially flat! Note that for small enough spatial distances ($\xi \ll 1$ in (2.12) and (2.13)), any Robertson-Walker universe appears spatially flat. Hence a "typical" fluctuation k/a^2 (a of order one) can only be seized by the self consistent solution over a finite spatial extension, if space has to flatten in a time of order (Fig. 8). The universe is thus flat and finite in size with an initial classcial length L_{in} of order $a_{in} \simeq \tau$ [21]. Note that a_{in} can be defined only by reference to the scale of the external fluctuation [22] because an observer inside an Euclidian flat space can always redefine its scale factor by multiplying it by an arbitrary constant. From the above discussion however we see that $a_{in} \simeq L_{in}$ so that we may interpret the scale a(t) in our previous estimates as the actual size of the universe. The fact that no result hitherto obtained is affected by the finiteness of the universe is an immediate consequence of Birkhoff's theorem [18]. It is interesting to note that while the big bang theory precisely failed to account for the "apparent" flatness of the classical homogeneous universe, the present theory predicts that it is really flat as a consequence of the exponential expansion originated from a "typical" metric quantum fluctuation.

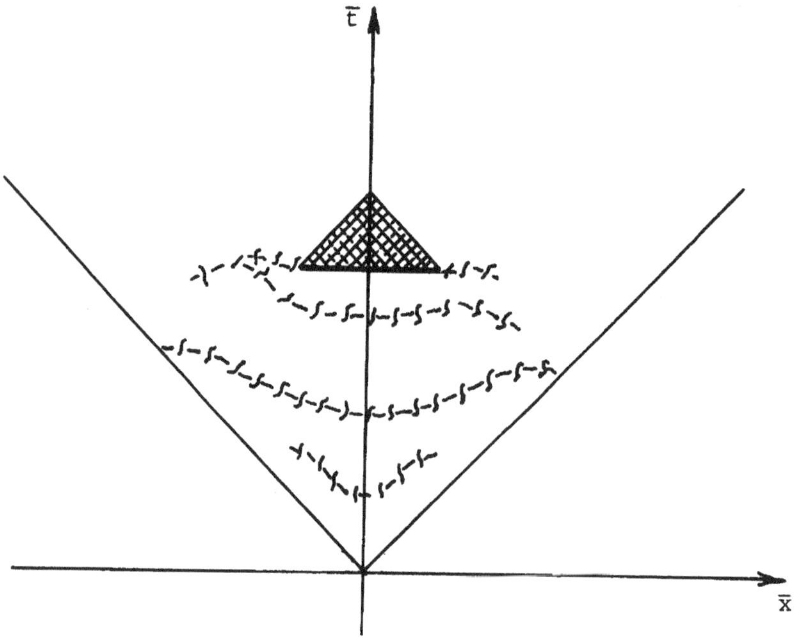

Fig. 8. Finite universe generated from (say) a negative local spatial curvature fluctuation. This should be compared to Fig. 1 with a classical solution starting in the vicinity of a hyperbola $a_{in} \simeq \tau$.

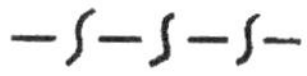 : fluctuating hyperbolae

 : classical universe in the creation era

 : region protected from external fluctuations in the subsequent classical era.

Thus an internal observer can contemplate an homogeneous flat cosmological background as long as its visible horizon does not exceed (at most) L_{in}. To understand the nature of the subsequent evolution we have to know the fate of the full initial fluctuation. To this effect we parametrize the external time by the time parameter $\bar{t}$ in (3.6) on scales comparable to τ ($\simeq$ 40 Planck lengths). Indeed $a(\bar{t})$ is comparable to one in the interval $0 < \bar{t} < \tau$ where the classical solution sets in. When $\bar{t}$ approaches τ, the scale factor inside the universe raises exponentially and, according to (3.7), $t \to \infty$; thus the whole creation era reduces to a quasi-instantaneous event in the external time $\bar{t}$. A huge metric gradient will then develop at the edge of the universe. This sharp drop in scale will produce violent fluctuations in the gravitational field which will hit the matter in the interior when the horizon of any internal observer exceeds the initial length. The protected region of space-time is depicted in Fig. 8; note that the conformal flatness of the inside Robertson-Walker metric permits to draw the protecting light cones as straight lines in the Minkowskian variables $\bar{t}$, $\vec{x}$. Outside this space-time region violent anisotropies develop which signal the "end of the universe" and its return to the quantum vacuum from which it originally emerged.

In the time $\bar{t}$, we see from Fig. 8 that the whole history subsequent to the creation period last for a time of order τ ($\simeq$ 40 Planck length $\simeq 10^{-43}$ s). Thus an external observer would describe the rise and fall of our universe as a "typical" unstable fluctuation of quantum gravity lasting for about 10^{-43} s. The reason "we" can exist during this seemingly short time, span is the enormous contraction of the proper time, as expressed in equation (3.7), generated by the self-consistent exponential expansion.

The theory thus leads to the prediction that the universe stays spatially flat as long as no significant anisotropy are detected. As this seems to be the case we conclude that "we" are still living in the protected flat region. As our universe is presently in the matter dominated era, the deceleration parameter q (equation (2.21)) is immediately checked from (2.8) and (2.9) (with $p = 0$) to be

$$q = \frac{1}{2} \tag{4.1}$$

up to small corrections due to the background radiation. Experiments cannot yet check directly (4.1) with sufficient accuracy but the spatial flatness expressed by (4.1) shows, as follows from the discussion of section II.1, that there is a serious missing mass problem. To satisfy (4.1) either a systematic error has to be discovered in astrophysical measurements of nuclear mass densities [3)] or surviving neutrinos must contribute a mass which would have to be about ten times larger than the nucleonic mass density. As low mass neutrinos should be present with a particle density of the order of n_γ this would then imply from (2.26) that surviving neutrinos have a mass in the 10 ev range.

2. The Structure of Quantum Gravity

The fact that a whole universe appears as a fluctuation of quantum gravity implies that micro-universe configurations which emerge and die within $\simeq$ 40 of (say) "our" Planck times do contribute to the path integral over metrics. These universes of course are small only when viewed by a "macroscopic external observer" ("us") but have a size comparable to "our" universe when contemplated by an inside observer. Their contributions to our vacuum energy is small (and may even be zero) because the are built out of a semi-classical solution to (3.9) where advantage has been taken from the negative energy due to the scale factor to compensate for the positive energy of matter. Thus they are not expected to affect ordinary field theoretical considerations at scale large compared to "our" Planck size.

Universe-like configurations lead to an interesting corrolary when the TCP theorem is applied to these quantum fluctuations. When "we" perform our path integral over such universe configurations, we must include also TCP reversed configurations. The latter would seem to us to live backwards in time. However the internal arrow of time is fixed by the structure of the internal correlations and would follow the direction in which entropy is generated. Thus there is no global arrow of time as the latter fluctuates on the scale of many universes [23]. Note that the classical "closed" universe cannot be stable against gravitational quantum fluctuation because of the flatness condition discussed in IV.1 : a finite spatial extension of the contracting phase of such a universe, if it persists at all, is in fact an expanding anti-universe. Because of the violent death suffered by any universe when its horizon covers the entire creation era, no paradoxe is generated and two separate sections of such a classical solution can meet only when they are being destroyed by the external vacuum. (Fig. 9).

We have thus been forced to contemplate implications of universes within universes within... in the path integral over metrics. This picture appears as a necessary consequence of the genesis of our universe from a Planckian fluctuation of gravity. It however leads to an intriguing speculation when the scope of the present discussion is enlarged : how can the path integral of quantum gravity be defined at all ?

It is well known that the gravitational action, analytically

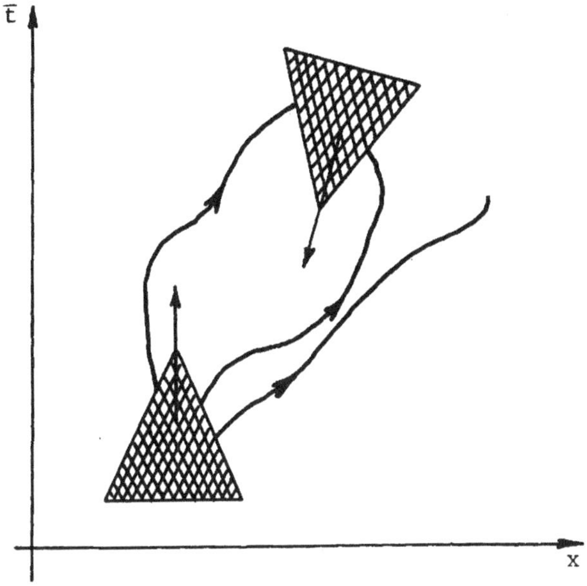

Fig. 9. Universes with opposite arrow of time.

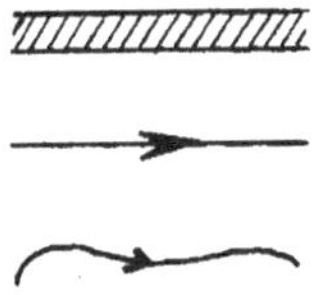

: classical universe (or anti-universe)

: "local" arrow of time

: quantum mechanical path.

continued to Euclidian space-time has an indefinite sign. This translates the fact that the theory has no ground state and indeed we have used the negative energy stored in the expansion to construct our theory. Thus a priori, violent fluctuations can occur at any scale and the path integral would appear unstable against fluctuations of infinite negative energy. However we have seen that the universe-like semi-classical configurations generate a huge entropy which may eventually be "absorbed" by anti-universes. In fact, when universes within universes are taken into account they may even generate an infinite entropy, only part of which is felt within a given universe. This large (infinite ?) entropy will tend to stabilize the paths in the vicinity of the paths of stationary phase built from these semi-classical solutions. We therefore conjecture that stabilization indeed occurs, namely we conjecture that the path integral over metrics is saturated [24] by a "foam of universes" of all

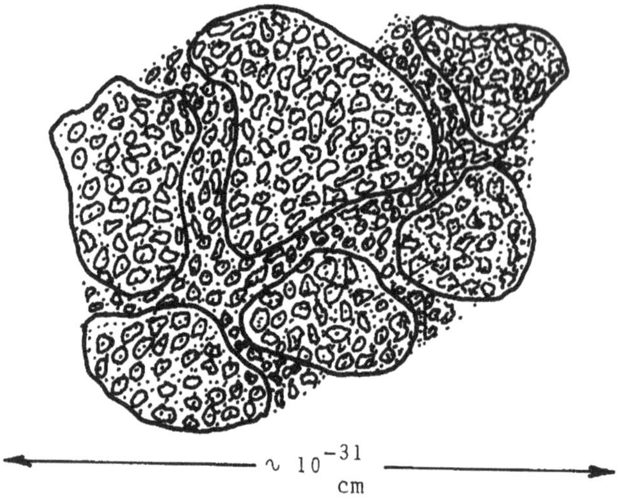

Fig. 10. A naive view of a foam configuration as seen by an "external" observer near its Planck length.

: (anti-) universe boundaries (the relative sizes of universes are grossly amplified).

sizes with respect to a given observer, except for infrared corrections. These are expected to be consistently cut-off at the Planck size of the universe considered because the fluctuating local metric would induce fluctuating dilations of the propagator from ultra-violct values to infrared ones.

If such a foam picture turns out to be correct the non-renormalizability of gravity would constitute the bonus that renders interactions between universes small and renormalizable field theory might be viewed as the phenomenological manifestation of long range interactions within a universe. However, in order to turn the above picture into a true theory of quantum gravity an operational device should be constructed which permits a systematic approach for uncovering successive levels of universes. We have not found such a procedure so the conjecture made here can only be viewed at an heuristic level. Note that even so, it seems hard to envisage a path integral which could connect all subuniverses if a measurable cosmological constant at low temperature could induce a space time curvature on a scale smaller than a whole universe. So the choice of Λ made in (3.1) might well be imposed by consistency requirements. More generally however, the foam conjecture does have an important corrolary.

In the foam, a given observer can define a background $g_{\mu\nu}$ only at the expense of projecting out correlations and infinite number of degrees of freedom, and this projection introduces an irreductible element of indeterminacy in the space-time description of quantum field theory.

This indeterminacy may be viewed as a generalization of the breakdown of predictability [25] which arises when a horizon hides a classical singularity, as in a black hole formed from a collapse : the singularity then implies that the thermal radiation cannot be gauged out by a change in coordinate and a pure quantum state does not appear available, at least in a space-time description. According to our conjecture, classical singularities are washed out by the foam but the indeterminacy due to the foam itself persists. An explicit manifestation of this basic indeterminacy would then be found in the appearance of a temperature in a universe which arose as a necessary concommitant of our dynamical contruction of the classical metric out of black holes.

Thus if the foam hypothesis is correct, the lack of a space-time description in terms of a pure global quantum state is fundamental and would suggest that four-dimensional space-time should be explained in terms of new concepts.

REFERENCES

1. A. Casher and F. Englert, Phys. Lett. 104B (1981), 117
   A. Casher and F. Englert, "Quantum cosmology and quantum gravity" in preparation.
2. R.H. Dicke, P.J.E. Peebles, P.G. Roll and D.T. Wilkinson, Ap. J. 142 (1965), 414.
   A.A. Penzias and R. W. Wilson, Ap. J. 142 (1965), 419.
3. For a review of observational data, see G.A. Tamman, A. Sandage and A. Yahil, Physical Cosmology Les Houches Session XXXII (North-Holland Publishing Co, Amsterdam) (1980) and references therein.
4. J. Ellis, Lectures presented at the 21st Scottish Summer School, ref. TH 2942-CERN (1980) and references therein.
5. P.J.E. Peebles, Ap. J. 146 (1966), 542.
6. J. Yang, D.N. Schramm, G. Steigman and R.T. Rood, Ap. J. 227, (1979), 697.

7. Ya. B. Zeldovich, Zh. Eksp. Theor. Fiz. 48 (1965), 986.

8. J.C. Pati and A. Salam, Phys. Rev. Lett. 31 (1973) 661; Phys. Rev. D8 (1973) 1240; D10 (1974), 275.
H. Georgi and S.L. Glashow, Phys. Rev. Lett. 32 (1974), 438.
H. Georgi, H.R. Quinn and S. Weinberg, Phys. Rev. Lett. 33, (1974), 451.
For more recent developments, see reference 4.

9. F. Englert and R. Brout, Phys. Rev. Lett. 13 (1964), 321.
P.W. Higgs, Phys. Rev. Lett. 12 (1964), 132; Phys. Rev. Lett. 13 (1964), 508.

10. M. Yoshimura, Phys. Rev. Lett. 41 (1978), 381; E 42 (1976); 746
S. Dimopoulos and L. Susskind, Phys. Rev. D18 (1978), 4500
D. Toussaint, S.B. Treiman, F. Wilczek and A. Zee, Phys. Rev. D19 (1979), 1036
J. Ellis, M.K. Gaillard and D.V. Nanopoulos, Phys. Lett. 80B (1979), 360; E 82B (1979), 464
S. Weinberg, Phys. Rev. Lett. 42 (1979), 850
A.D. Sakharow, Zh. Eksp. Theor. Fiz. 76 (1979), 1172
S. Dimopoulos and L. Susskind, Phys. Lett. 81B (1979), 416
M. Yoshimura, Phys. Lett. 88B (1979), 294.

11. R. Brout, F. Englert and E. Gunzig, G.R.G. 10 (1979), 1.
The quantitative solution discussed in section III.1 has been achieved by R. Brout, F. Englert and P. Spindel, Phys. Rev. Lett. 43 (1979), 417; E 43 (1979), 890
For a detailed analysis of the quantum effects in the model, see R. Brout, F. Englert, J.-M. Frère, E. Gunzig, P. Nardone, C. Truffin and P. Spindel, Nucl. Phys. B170 (F.S.1) (1980), 228
A summary of this work can be found in F. Englert, Physical Cosmology, Les Houches Session XXXII (North Holland Publishing Co. Amsterdam) (1980).

12. Study of particle creation from cosmological expansion was initiated by L. Parker, Phys. Rev. 183 (1969), 1057.
For further developments, see L. Parker "The production of elementary particles by strong gravitational fields" Proceedings of the Symposium on Asymptotic Properties of Space-Time (Plenum New York) (1977) and references therein.

13. P. Spindel, "Mass formula in a cosmological model", Mons University preprint (1981); to be published in Phus. Lett. B

14. S.W. Hawking, Nature 248 (1974), 30.
S.W. Hawking, Comm. Math. Phys. 43 (1975), 199.

15. W. Unruh, Phys. Rev. D14 (1976), 870.

16. B.S. De Witt, Phys. Reports 19 (1975), 295.

17. J.D. Bekenstein, Phys. Rev. D12 (1975), 3077.
S.W. Hawking, Phys. Rev. D13 (1976), 191.

18. G. Birkhoff, Relativity and Modern Physics, (Harvard University Press, Cambridge, Mass. (1923)).

19. G.W. Gibbons and S.W. Hawking, Phys. Rev. D15 (1977), 2738.

20. The idea that the universe could originate in a quantum fluctuation was proposed by E.P. Tryon, Nature 246 (1973), 396, who envisaged a global quantum effect. The possibility of generating a universe from a local quantum fluctuation which is used here was the basic motivation of reference 11.

21. In reference 11, the compatibility of an initial local fluctuation with the existence of an unbounded universe required the assumption of an unexplained phase transition. Tentative ways of avoiding this phase transition in the context of an unbounded universe have been recently proposed by
J.R. Gott III, Nature (1981)
D. Atkatz and H. Pagels "The origin of the universe as a quantum tunneling event", Rockefeller University preprint RU 81/B/2 (1981).

22. It is interesting to note that curvatures k/a^2 yield thermal distributions for massless particles with a temperature $T = 1/2\pi a$. This comes about if $k = -1$ from a coordinate transformation effect [11) 12)] and for $k = +1$ from a Casimir effect which is in fact equivalent to a thermal distribution [12)]. As for $a \simeq \tau$ this temperature is precisely what is required to seize black hole formation by the self-consistent process, the formation of a universe from the metric fluctuation at the scale $a_{in} \simeq \tau$ appears indeed "natural".

23. To our knowledge the idea that the arrow of time may fluctuate is due to Y. Aharonov (private communication).

24. This structure is reminiscent of "foams" discussed by
C.W. Misner, K.S. Thorne, J.A. Wheeler, Gravitation (W.H. Freeman and Co, San Francisco) (1973);
S.W. Hawking "The path integral approach to quantum gravity" published in General Relativity (Cambridge University Press, ed. S.W. Hawking and W. Israel) (1979)

25. S.W. Hawking, Phys. Rev. D14 (1976), 2460.

THE CONFINEMENT PHENOMENON IN QUANTUM FIELD THEORY

Gerard 't Hooft

Instituut voor Theoretische Fysica der Rijksuniversiteit
De Uithof
Utrecht, The Netherlands

0. ABSTRACT

In these written notes of four lectures it is explained how the phenomenon of permanent confinement of certain types of particles inside bound structures can be understood as a consequence of local gauge invariance and the topological properties of gauge field theories.

I. INTRODUCTION

The lagrangian of a quantum field theory describes the evolution of a certain number of degrees of freedom of the system, called *fields* as a function of space and time. In many cases this evolution is only straightforward in a small region of space-time and therefore these fields should be interpreted only as the *microscopic* variables of the theory. Only in the simplest cases these microscopic variables also correspond to actual physical particles (the *macroscopic* objects) but very often the connection is less straightforward for two reasons. One reason is that there may be a local gauge invariance. This is a class of transformations that transform a set of fields into another set of fields with the postulate that the new set describes the same physical situation as the old one. Therefore the physical fields only constitute some orthogonal subset of the original set of fields and one of the problems that we will study in these lectures is how to associate this subset to observable objects.

However even these observable objects ("transient particles") may in some cases not yet be the macroscopic physical particles. That is one second point : various kinds of Bose-condensation may take place after which the spectrum of physical particles may look entirely different once again. We will study these condensation phenomena as we go along.

The most challenging application of our theoretical considerations is "quantum chromodynamics" or generalizations thereof. The "microscopic" lagrangian there contains only vector particles ("gluons") and spinors ("quarks") but neither of these are really physical. The spectrum of physical particles always consists of bound states of certain numbers of quarks and/or antiquarks with unspecified numbers of gluons. We will obtain a qualitative understanding of this transition from the microscopic to the macroscopic dynamical variables.

The first models that we will consider may seem to be a far way off from that desired goal but studying them will turn out to be crucial for obtaining a suitable frame and language in order to put the more advanced systems in a proper perspective.

II. SCALAR FIELD THEORY

Let us consider the lagrangian of a complex scalar field theory :

$$\mathcal{L}(\phi,\phi^*) = -\partial_\mu\phi^*\partial_\mu\phi - m^2\phi^*\phi - \frac{\lambda}{2}(\phi^*\phi)^2 \qquad (2.1)$$

Equivalently we can use real variables :

$$\phi = \frac{1}{\sqrt{2}}(\phi_1 + i\phi_2)\ ; \qquad \phi^* = \frac{1}{\sqrt{2}}(\phi_1 - i\phi_2). \qquad (2.2)$$

We write ϕ^* rather than $\phi^\dagger$ because here the fields are c-numbers, not operators, and the lagrangian must be seen in the context of a functional integral as described by B. de Wit in his lectures at this School.

The lagrangian (2.1) describes a system with two species of Bose-particles both with the same mass m but distinguishable in the quantum mechanical sense. One can either consider ϕ_1 and ϕ_2 as the two species, both equal to their own antiparticles, or consider ϕ as a particle and ϕ^* as its distinct antiparticle. At this level these descriptions are equivalent.

Obviously there is a *symmetry* under exchange of ϕ_1 and ϕ_2, or more generally

$$\phi \to e^{i\Lambda}\phi$$
$$\phi^* \to e^{-i\Lambda}\phi^* \qquad (2.3)$$

This is a group of rotations in the complex plane called U(1) and is only global, that is, Λ must be independent of space-time.

By Noether's theorem the model contains a conserved current,

$$j_\mu = \phi^* \partial_\mu \phi - (\partial_\mu \phi^*)\phi \tag{2.4}$$

with $$\partial_\mu j_\mu = 0 \tag{2.5}$$

Classically (2.5) is only true if ϕ, ϕ^* are required to obey the Euler-Lagrange equations generated by (2.1). Quantum mechanically (2.5) follows if we substitute for ϕ and ϕ^* the corresponding operators ϕ and $\phi^\dagger$.

Our model is trivial in the sense that, if $m^2 > 0$, the microscopic fields ϕ_1 and ϕ_2 directly correspond to the expected macroscopic physical particles (apart from possible stable bound states) and the symmetry (2.3) is also a symmetry between these particles.

III. BOSE CONDENSATION

Bose condensation is a well-known phenomenon in quantum statistical physics. Just in order to make the connection with our case of interest let us first consider an ideal non-relativistic Bose gas. The states are

$$|n(\vec{k}_1), n(\vec{k}_2), \ldots > \tag{3.1}$$

where

$$\vec{k}_i = \frac{\pi}{L} (l_1, l_2, l_3) \tag{3.2}$$

and l_1, l_2, l_3 are positive integers. L is the length of a side of the box in which the particles are contained. The energy of the states is

$$E = \sum_i n(\vec{k}_i) k_i^2/2M \tag{3.3}$$

where M is the mass of the (non-relativistic) particles. We take for the thermodynamic free energy F,

$$e^{-\beta F} = \sum_{\{n(\vec{k})\}} e^{-\beta [E(\{n(\vec{k})\}) - \mu N]} \tag{3.4}$$

where $N = \sum_i n(\vec{k}_i)$, $\beta = 1/T$

T is the temperature in natural units, and μ is the chemical potential.

We can easily solve (3.4) :

$$e^{-\beta F} = \prod_k \sum_{n=0}^{\infty} e^{\beta(\mu - k^2/2M)n} = \prod_k \frac{1}{1 - \exp \beta(\mu - k^2/2M)} \qquad (3.5)$$

and in the limit of infinite volume :

$$F = (2\pi)^{-3} TV \int d^3k \log(1 - \exp\beta(\mu - k^2/2M)) \qquad (3.6)$$

$$V = L^3$$

The particle number density is

$$\frac{N}{V} = -\frac{1}{V}\frac{\partial F}{\partial \mu} = \int d^3k \left(\exp \beta(-\mu + k^2/2M) - 1\right)^{-1} \qquad (3.7)$$

One easily notes that the formulas (3.5) - (3.7) explode if the chemical potential μ becomes larger than or equal to zero, a typical property of Bose gases.

Since this is a non-relativistic system it is convenient to introduce a field $\phi(\vec{x})$ in the following way :

$$\phi(\vec{x}) = 2\sqrt{\frac{2}{V}} \sum_{\vec{k}} a(\vec{k}) \prod_{i=1,2,3} \sin \frac{\pi(k_i x_i)}{L} \qquad (3.8)$$

where $a(\vec{k})$ is an operator that annihilates one particle with momentum $\vec{k}$ in the usual way. The hamiltonian is then

$$H = \sum_{\vec{k}} \left(\frac{k^2}{2M} - \mu\right) a^\dagger(\vec{k})\, a(\vec{k}) \qquad (3.9)$$

$$= \int d^3\vec{x} \left(\frac{1}{2M} \vec{\partial}\phi^\dagger(\vec{x})\vec{\partial}\phi(\vec{x}) - \mu\phi^\dagger(\vec{x})\phi(\vec{x})\right) \qquad (3.10)$$

where for convenience we included the chemical potential term so that this hamiltonian describes the complete system :

$$e^{-\beta F} = \mathrm{Tr}\, e^{-\beta H}$$

It is obvious here that μ should not be allowed to be positive.

Now however we can take into account the *repulsive* forces between the particles. When many particles are close together we expect an extra, positive contribution to H. A simple model for that is

$$H = \int d^3x \left(\frac{1}{2M}\, \partial\phi^\dagger \partial\phi - \mu\phi^\dagger\phi + \frac{\lambda}{2}\, \phi^{\dagger 2}\phi^2\right) \tag{3.11}$$

As long as μ is negative the λ term is just a small perturbation. But if μ is positive then the λ term is only one that can stabilize the system.

Of course it is difficult to find the free energy of this revised system exactly, but an easy approximation, valid for λ not too large, is to substitute ϕ by a c-number (as defined in (3.8) it was an operator), and subsequently minimize H. We then find approximately the energy of the lowest eigenstate :

$$\phi^\dagger\phi \equiv \rho \tag{3.12}$$

$$\frac{\partial}{\partial\rho}\left(-\mu\rho + \frac{\lambda}{2}\rho^2\right) \simeq 0 \tag{3.13}$$

$$\rho \simeq \frac{\mu}{\lambda} \tag{3.14}$$

$$E_o \simeq -\frac{\mu^2}{2\lambda} V \tag{3.15}$$

$$e^{-\beta F} = \mathrm{Tr}\, e^{-\beta H} \simeq e^{-\beta E_o} \tag{3.16}$$

$$F \simeq E_o \tag{3.17}$$

$$\frac{N}{V} = \frac{1}{V}\int \phi^\dagger\phi\, d^3x = \rho \simeq \mu/\lambda \ , \quad (\mu > 0) \tag{3.18}$$

to be contrasted with (3.7), still approximately valid for $\mu < 0$. As μ turns from negative to positive a phase transition is said to take place; we suddenly get large values for the fields ϕ already in the lowest eigenstate of H. This is called Bose condensation. It takes place whenever the pressure is so high that the chemical potential becomes negative. Also the λ term must be sufficiently small.

The model described by our hamiltonian (3.11) resembles somewhat the model of the previous section. Note however that the mass

M comes in the derivative term and never vanishes. In the relativistic field theory M is replaced by 1/2, and $-\mu$ by the mass-squared, m^2. In that theory Bose-condensation can take place also : we must extend the allowed values for m^2 to negative values, clearly a more profound change in the particle properties. But what we get in return is that in the relativistic model, the vacuum itself, without any external pressure, can become a Bose-condensate.

IV. GOLDSTONE PARTICLES

In the previous section we discussed Bose-condensation in a statistical system only in a sketchy way because we will not need the details (for instance, ϕ and $\phi^\dagger$ were not canonical variables in the usual sense). We will be more precise for the case that is more relevant to us : the relativistic complex scalar case. We return to the lagrangian (2.1) and now assume m^2 to be negative. It is convenient to rewrite it as

$$\mathcal{L}(\phi,\phi^*) = -\partial_\mu\phi^*\partial_\mu\phi - \frac{\lambda}{2}(\phi^*\phi - F^2)^2 \tag{4.1}$$

where $m^2 = -\lambda F^2 < 0$ automatically, and an irrelevant constant, $-\lambda/2\ F^4$, has been added to the lagrangian.

The hamiltonian density of the system is

$$\mathcal{H} = \pi^\dagger\pi + \partial_i\phi^\dagger\partial_i\phi + \frac{\lambda}{2}(\phi^\dagger\phi - F^2)^2 \tag{4.2}$$

where $\pi,\pi^\dagger$ are the canonical momenta associated with $\phi,\phi^\dagger$ which are now operators rather than fields. The $\phi,\phi^\dagger$ dependent part has an extremum for

$$\phi^\dagger\phi = F^2 \tag{4.3}$$

from which

$$\phi = Fe^{i\omega}$$
$$\phi^\dagger = Fe^{-i\omega} \tag{4.4}$$

where ω is arbitrary but fixed to the same value everywhere in space.

Now if there were no $\pi^\dagger\pi$ term in $\mathcal{H}$ then (4.4) would be the exact solution to the Schrödinger equation for the lowest energy state. Every ω between 0 and 2π would describe a lowest-energy eigenstate of $H = \int \mathcal{H} d^3x$, each with eigenvalue $E_o = 0$. The question is whether the $\pi^\dagger\pi$ term causes sufficient fluctuations to lift this degeneracy. The answer is not so simple : in 1 space - 1 time dimension, yes; in 2 or more space dimensions, no. So let us limit ourselves to 3 space + 1 time dimensions. Then the vacuum (= lowest energy state) is degenerate and characterized by a phase angle ω. But (4.4) is not exactly valid due to the $\pi^\dagger\pi$ term. We replace it by

$$< 0|\phi|0 > = F' e^{i\omega} \tag{4.5}$$

where $|0>$ is the vacuum state and ω is the vacuum angle. From now on we will consider only the world surrounded by a vacuum with $\omega = 0$. Further, F' is close to F. In fact, because of ultraviolet divergences, subtractions must be made in F and ϕ, and we could choose these such that F' = F.

Actually our theory only makes sense if either λ is chosen to be rather small and F of order $1/\sqrt{\lambda}$, or if an adequate ultraviolet cutoff has been introduced. The reasons for this are deeper field theoretic arguments connected with the renormalization group that I will not go into. Let us assume that λ is rather small. Then the fluctuations of ϕ around F are also relatively small and it makes sense to split

$$\phi = F + \eta \tag{4.6}$$

and the lagrangian becomes

$$\mathcal{L} = - \partial_\mu \eta^* \partial_\mu \eta - \frac{\lambda}{2}(F(\eta + \eta^*) + \eta^* \eta)^2 \tag{4.7}$$

Writing

$$\eta = \frac{1}{\sqrt{2}}(\eta_1 + i\eta_2)$$

This becomes

$$\mathcal{L} = - \frac{1}{2}(\partial_\mu \eta_1^2 + \partial_\mu \eta_2^2) - \lambda F^2 \eta_1^2 + \text{int} \tag{4.8}$$

where "int" stands for higher order terms in η_1, η_2. Notice now that one particle, η_1, obtained a mass

$$M_\eta = F\sqrt{2\lambda} \tag{4.9}$$

But its companion η_2 became massless.

The occurrence of a massless particle as soon as the vacuum expectation value of a field ϕ is not invariant under the continuous symmetry (2.3) has been first observed by J. Goldstone[1], and it is an exact property of the system, not related to our perturbative approximation (no higher order mass corrections). We conclude that after the phase-transition caused by Bose-condensation, the symmetry (2.3) is *spontaneously broken* (the degeneracy of ϕ_1 and ϕ_2 is not reproduced in η_1, η_2) and at the same time a massless particle appears : the Goldstone particle. Here we see the first example where the microscopic fields ϕ,ϕ^* in the lagrangian do not reflect accurately the physical spectrum, but the transition towards the η fields was still very simple. It is correct to characterize the vacuum by

$$< 0|\phi(\vec{x})|0 > = F \neq 0$$

and the vacuum is infinitely degenerate. Characterization of the *Higgs* mode, next section (V) will be very different!

V. THE HIGGS MECHANISM

We now switch on electromagnetic interactions*) simply by adding the Maxwell term to the lagrangian and replacing derivatives by covariant derivatives :

$$\mathcal{L}(\phi,\phi^*,A_\mu) = -\frac{1}{4}F_{\mu\nu}F_{\mu\nu} - (D_\mu\phi)^* D_\mu\phi - \frac{\lambda}{2}(\phi^*\phi - F^2)^2 \tag{5.1}$$

where

$$F_{\mu\nu} = \partial_\mu A_\nu - \partial_\nu A_\mu$$

$$D_\mu\phi = (\partial_\mu + iqA_\mu)\phi \tag{5.2}$$

q is the electric charge of the particle ϕ.

Indeed the previously introduced current j_μ, eq (2.4), is now the conserved electromagnetic current. But the invariance (2.3) can now be replaced by

$$\phi \to e^{i\Lambda(x)}\phi\ , \quad \phi^* \to e^{-i\Lambda(x)}\ \phi^*$$

$$A_\mu \to A_\mu - 1/q\ \partial_\mu\Lambda(x) \tag{5.3}$$

---

*) It may seen to a superficial reader that these notes are just repeating the story[2] of the early 70's. However we are now not primarily interested in pertubative quantization but rather *non-perturbative characterization* of what happens.

which is a *local* invariance

As already mentioned in the Introduction, the consequence of this local invariance is that only a subspace of all (ϕ, ϕ^*, A_μ), namely the gauge-nonequivalent values, correspond to physical observables.

Traditionally, one now proceeds by choosing a gauge fixing procedure so that most, if not all, degeneracy is removed. One then performs perturbation expansions in λ and q^2. At zero λ and q^2 one can again ask whether or not

$$< \phi >_o = F \neq 0 \ ? \qquad (5.4)$$

and since the higher order corrections to $< \phi >_o$ are of higher order in λ and q^2 the qualitative distinction whether or not $< \phi_o > = 0$ remains valid at every order. And so we get the local variant on the Goldstone mechanism : after Bose-condensation of charged particles we get the Higgs mechanism [3]. However, there is a difficulty with (5.4), because ϕ is not gauge invariant. Smearing (5.4) over all of space-time may yield zero or not, depending on the gauge chosen. In a trivial gauge

$$\begin{aligned} Re(\phi) &> 0 \\ Im(\phi) &= 0 \end{aligned} \qquad (5.5)$$

we have

$$< \phi >_o > 0$$

always. We therefore propose to use criterion (5.4) *only* in perturbative considerations, where it is correct (as good as it can be) but *not* as an absolute non-perturbative criterion for the Higgs mode. Another criterion that cannot be used is whether or not the vacuum is degenerate. The problem there is that transformation (5.3) yields physically equivalent states, contrary to its global equivalent (2.3)!! Therefore all those vacuum states corresponding to different ω angles are now one and the same state. The vacuum is never degenerate if the symmetry is local. Local symmetries are *never* "spontaneously broken". Then why is this phrase so often used in connection with gauge theories ? Because, as I will show now, there certainly is such a thing as a Higgs mode and it usually can be described in some or other reasonable perturbation expansion around a Goldstone (= global) field theory.

Let us return to perturbation theory momentarily. We then write as usual

$$\phi(x) = F + \eta(x) \qquad (5.6)$$

$$\mathcal{L} = -\frac{1}{4}F_{\mu\nu}F_{\mu\nu} - q^2F^2A_\mu^2 - \sqrt{2}\,qFA_\mu D_\mu \eta_2 - (D_\mu\eta)^* D_\mu\eta - \lambda F^2\eta_1^2 + \text{int} \tag{5.7}$$

A convenient renormalizable gauge is obtained by adding the gauge fixing term

$$\mathcal{L}^c = -\frac{1}{2}(\partial_\mu A_\mu + \sqrt{2}\,qF\eta_2)^2 \tag{5.8}$$

so that

$$\mathcal{L} + \mathcal{L}^c = -\frac{1}{2}(\partial_\mu A_\nu)^2 - \frac{1}{2}M_A^2A^2 - D_\mu\eta^* D_\mu\eta - \frac{1}{2}M_\eta^2\eta_1^2 - \frac{1}{2}M_A^2\eta_2^2 + \text{int} \tag{5.9}$$

with

$$M_A = \sqrt{2}\,qF \qquad M_\eta = \sqrt{2\lambda}\,F \tag{5.10}$$

It is easily read off from this lagrangian that the vector particle A has a mass M_A and η_1 a mass M_η. The longitudinal component of the vector field A_μ and η_2 are ghosts, which both cancel against the Faddeev-Popov-De Wit ghost [4,5] all having the same mass M_A.

So perturbation theory suggests that the Higgs theory behaves in a way very different from the symmetric or Coulomb theory : one of the two scalar fields ϕ disappears and the vector field obtains a mass so that the photon field is short-range only. This *is* a distinction that should survive beyond perturbation theory. Thus the criterion that electromagnetic forces become short-range is much more fundamental than either the vacuum value of the scalar field (5.4) or the "degeneracy of the vacuum". But, there is yet another new phenomenon in the Higgs mode contrary to the "unbroken" or Coulomb mode. This is important because the above does not yet distinguish a Higgs theory from just any non-gauge theory with massive vector particles.

VI. VORTEX TUBES

The non-relativistic version of the theory of the previous section is the superconductor : if electrically charged bosons (Cooper's bound state of an electron pair) Bose-condense then there the electric fields become short-range. Also magnetic fields are repelled completely (Meissner effect). Except when they become too strong. Then, because of magnetic flux conservation, they have to be allowed in. What happens is that a penetrating magnetic field forms narrow flux tubes. These flux tubes carry a multiple of a precisely defined quantum of magnetic flux. A little amendement to the perturbative Higgs theory can explain this.

By gauge transformations the "vacuum value" of the Higgs field ϕ can be changed into

$$< \phi(x) >_{o} = Fe^{i\omega(x)} \tag{6.1}$$

if $\omega(x)$ is a continuous, differentiable function of space and time. However we can also consider perturbation theory around field configurations

$$\phi(x) = \rho(x)\, e^{i\omega(x)} \tag{6.2}$$

where $\rho(x) \simeq F$ nearly everywhere, but at some points ρ may be zero. At such points ω needs not be well defined and therefore in all the rest of space ω could be multivalued. For instance, if we take a closed contour C around a zero of $\rho(x)$ then following ω around C could give values that run from 0 to 2π instead of back to zero. The energy of such a field configuration is only finite provided that $D_i\phi(x)$ goes to zero sufficiently rapidly at ∞. Since $\partial_i\phi$ does not go to zero fast enough there must be a supplementary vector potential $A_i(x)$.

The easiest way to find that is by taking a gauge transformation that is regular at ∞ but singular at the origin.

$$e^{-i\Lambda} = \phi(x)/\rho(x) \tag{6.3}$$

In the new gauge $\partial_i\phi$ may vanish rapidly at ∞ and therefore $A_i(x)$ also. So in the old gauge

$$\oint A_i dx^i = \oint \frac{d\lambda}{q} = \frac{2\pi}{q} \tag{6.4}$$

which is the magnetic flux. One can compute the energy of the flux tube by assuming cylindrical symmetry and substituting (6.2) into (5.1). One then varies A_μ and ρ with the boundary condition (6.4) fixed, and minimizes the hamiltonian derived from (5.1). One typically finds that the energy of a flux with length ℓ is [6])

$$E = \alpha\ell \tag{6.4}$$

$$\text{with } \alpha = O(F^2) = O(M_A^2/q^2) \tag{6.5}$$

$$\text{if } q^2 = O(\lambda) \tag{6.6}$$

If finally a magnetic field is admitted inside a superconductor it can only come in some multiple of these vortices, never spread out because of the Meissner effect. Classically one may consider this as an aspect of the infinite conductivity of the material that is only broken down in sufficiently strong magnetic fields. The stable vortex configurations that we discussed here were first derived in the relativistic theory by Nielsen, Olesen and Zumino[6]. The existence of these macroscopic stable objects can be used as another characterization of the Higgs mechanism. They should also survive beyond perturbation expansion.

VII. DIRAC'S MAGNETIC MONOPOLES

At this stage it is useful to introduce the notion of a single magnetic charge à la Dirac[7]. It is not (yet) a dynamic particle but just a source or sink of magnetic flux, a spectator particle not dynamically involved in the lagrangian of the theory. A Dirac monopole can be visualized as the end point of an infinitely thin coil carrying a large electric current. The vector potential $\vec{A}$ is very large close to the coil, because of this electric current :

$$\oint \vec{A} d\vec{x} = \Phi \qquad (7.1)$$

where Φ is the magnetic flux of the coil and the integral is over any contour going closely around it. Close to the coil dx is small, therefore $\vec{A}$ becomes large.

Nevertheless the effect of the coil on its surroundings comes only through the end points, if a gauge transformation exists that removes this large vector potential :

$$\oint \frac{d\Lambda}{q} = \Phi \qquad (7.2)$$

Such gauge transformations Λ would be multivalued, but we require that $e^{i\Lambda}$ in (5.3) remains single-valued. So the jumps that Λ is allowed to make are multiples of 2π. Therefore the gauge transformation (7.2) turns single-valued field configurations into single-valued field configurations if

$$\Phi = 2\pi n/q \qquad (7.3)$$

This is how Dirac found that the total amount of magnetic flux carried by a magnetic monopole must be quantized in units $2\pi/q$ where q is the smallest possible electric charge in the universe. This condition must be satisfied whenever we want a rotationally invariant quantized theory with magnetic monopoles and single valued fields. It is illustrative now to see what would happen with such a spectator particle inside a Higgs theory (or superconductor).

It is not accidental that the monopole quantum $2\pi/q$ coincides with the Nielsen-Olesen-Zumino vortex quantum. This implies that the monopole will be sitting at the end of an integer number of such vortices. Antimonopoles may be sitting at the other ends. Now the energy of such configurations is approximated by eq. (6.4). It is proportional to their separation distance. And so we notice that the monopoles inside a superconductor are kept together by an infinite potential well, the potential being simply linearly proportional to their separation. This is the first observation of a confinement feature in quantum field theory, although the confined objects were as yet spectators, not any of the participants of the field equations. That will come later (sect. XIII).

VIII. THE UNITARY GAUGE

So far we limited ourselves strictly to Abelian gauge theories. We knew what the microscopic field variables are, and now we know what particles and vortices survive at macroscopic distance scales. The latter depend critically on what kind, if any, of Bose condensation took place. Now in our introduction we also mentioned the *microscopic physical* variables. Formally the space H of these variables is given by

$$H = R/G \tag{8.1}$$

where R is the space of field variables and G is the (local) gauge group. How do we enumerate the variables in H ?

Traditionally one imposes a gauge condition on the fields in R, thus obtaining a subspace in R which could be representative for H. In section V we used the gauge fixing term (5.8). This is good enough if one intends to do perturbation expansion[2,5]. The ghost particles one obtains cancel each other and can be dealt with. However we claim that if a non-perturbative characterization of the physical variables is required then this is not good enough. Imagine that one tries to solve the Dyson-Schwinger equations of the theory in some nonperturbative way. Whenever a computed S-matrix element shows a pole one can never be sure whether or not this is due to a ghost or whether it is physical. Furthermore as we will see the ghosts will produce their own topological features called "phantom solitons" which are entirely non-physical. Therefore if we want some understanding of the physical variables we must go to a "unitary gauge" (a gauge with no ghosts)

Often the axial gauge

$$A_o = 0 \tag{8.2}$$

is used in order to understand the physical Hilbert space. However, this leaves invariance with respect to time-independent gauge transformations :

$$\Lambda(\vec{x},t) = \Lambda(\vec{x}) \qquad (8.3)$$

and so there is still a redundancy in our set of variables. It is not suitable for our purposes.

A completely ghost-free gauge can be formulated if we have a charged scalar field ϕ (if no such field is present one may consider building such a field by composing, say, two fermion fields). We do *not* require the Higgs phenomenon to take place. Regardless what condensation takes place at large distance scales one can look at the gauge

$$\begin{aligned} \mathrm{Re}(\phi) &= \rho > 0 \\ \mathrm{Im}(\phi) &= 0 \end{aligned} \qquad (8.4)$$

This fixes the gauge function Λ locally, point by point in space-time, contrary to gauges such as eq. (5.8) where the condition on Λ requires solving a second order partial differential equation (the cause of the ghosts).

Within the unitary gauge (8.4) all components of the vector field A_μ are entirely observable. The complex scalar field ϕ is reduced to a real field ρ that can only take positive values. This would be a convenient description of the space of microscopic physical field variables were it not for one deficiency in the condition (8.4) : the original space of variables R certainly allows the scalar field ϕ to vanish at certain points in space-time. These points, defined by (for any $\phi \in R$)

$$\begin{aligned} \mathrm{Re}(\phi) &= 0 \\ \mathrm{Im}(\phi) &= 0 \end{aligned} \qquad (8.5)$$

have the topological structure of a set of closed curves in 3-space, or closed surfaces in 3 + 1 dimensional space-time. At these points the condition (8.4) becomes singular : if $\phi = \rho e^{i\theta}$ then we must choose

$$\Lambda = -\theta \qquad (8.6)$$

but the gradient of θ is easily seen to explode close to a zero of ϕ and therefore the vector potential $\vec{A}$, transforming as the gradient of Λ, will grow as the inverse power of the distance to this zero. Thus we find that the string-like structures, defined by (8.5) are separate degrees of freedom, giving a boundary condition on ρ ($\rho = 0$) and a prescribed singular boundary behavior of A_μ. This completes our discussion of the microscopic physical degrees of freedom for any Abelian gauge theory : we have observable vector fields A_μ, a truncated scalar field ρ ($\rho > 0$) and all possible closed strings *), on which there is a boundary condition for both ρ and A_μ

Note that only in the Higgs theory these physical variables are in the same time the *macroscopic* physical variables, although of course the macroscopic variables will be "dressed with a bound of virtual particles" (the string becomes a vortex with finite thickness). In the "unbroken" Abelian theory of electromagnetism the macroscopic variables are harder to discuss. It appears that our vortices "Bose-condense" to form long range, non-energetic magnetic field lines : the ordinary magnetic field $\vec{B}$.

IX. PHANTOM SOLITONS

The gauge (8.5) is called "unitary gauge" because in that gauge all surviving fields will be physically observable. Their quanta will all contribute in the unitarity relation

$$\sum_n < a|S^\dagger|n >< n|S|b > = < a|b > \tag{9.1}$$

However as soon as practical calculations are considered smoother gauge conditions are required. (8.5) is hard to implement if ϕ oscillates wildly at small distances. We will now argue that after a transition towards "smoother" gauge conditions not only ghost particles arise but also what we call "phantom solitons" : extended structures which are stable for topological reasons but nevertheless unphysical gauge artefacts [8].

Intermediate between the "renormalizable gauge" (5.8) and the unitary gauge we could choose

$$\arg(\phi) + \kappa\partial_\mu A_\mu = 0 \tag{9.2}$$

where $\arg(\phi) = \mathrm{Im}(\log\phi)$ and κ is an arbitrary gauge parameter. The gauge condition (9.2)is smoother than the unitary gauge because at small distances, by power counting, the second term dominates and we come close to the renormalizable Lorentz gauge. One finds ghosts in this gauge which propagate with a mass

$$m_{gh} = (q/\kappa)^{1/2} \tag{9.3}$$

so for small κ they become unimportant.

---

*) That is, strings without ends; they could run from ∞ to ∞.

Now imagine a Nielsen-Olesen-Zumino vortex tube in the form of a closed curve. What do the field configurations in the gauge (9.2) look like ? The gauge(9.2) is that particular gauge for which

$$W = \int d^4x(q^{-1}(\arg(\phi))^2 + \kappa A_\mu^2) \tag{9.4}$$

has an extremum. Let us assume this is a minimum. The system then likes to arrange $\arg(\phi)$ to be zero as much as possible but not with too large vector potentials A_μ.

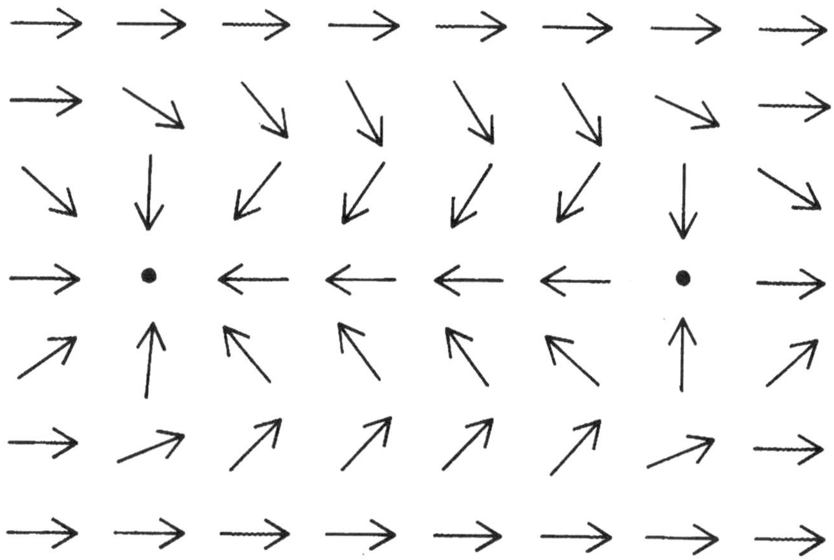

Fig 1

In fig 1 we pictured a cross section of the vortex. The plane is intersected twice, in opposite directions. At the intersection points the scalar field ϕ makes one complete rotation, again in opposite directions. The configuration in the figure is close to the optimal gauge (9.2), keeping ϕ as much as possible oriented towards the positive real axis. We see that the topology of the complete "twist" from top to bottom had to be preserved. This twist will cover an entire sheet spanned by the vortex. Certainly (9.2) will be obtained (i.e. (9.4)will be minimal) if that sheet has minimal surface. The equations for the field configurations inside the sheet are easy to solve if the sheet is considered to be locally sufficiently flat. Clearly the sheet is a gauge artifact. We think that structures of this sort will further obscure the physical interpretation of whatever solutions will be found to the Dyson-Schwinger equations in the gauge (9.2) or completely renormalizable gauges. For instance, bubbles made out of these sheets will form a whole Regge-like family of phantom particles.

X. NON-ABELIAN GAUGE THEORY

We now intend to perform the same procedure in non-Abelian gauge theories. For simplicity we restrict ourselves to the case that the gauge group is SU(N), with arbitrary N. The *microscopic* field variables are : a matrix vector field $A_\mu^{ij}(x)$ where μ is a Lorentz index and i, j run from 1 to N; and there may be Dirac spinor fields ψ_{iF}, $\bar{\psi}_{iF}$ where F is a flavor index. Scalar fields are not assumed to play a significant role but may be present too.

An element of the gauge group G (here SU(N)) is a space-time dependent unitary matrix $\Omega(x)$. Any transformation of the form

$$A'_\mu = \Omega(x)(A_\mu + \frac{i}{g}\partial_\mu)\Omega^{-1}(x)$$

$$\psi'_F = \Omega(x)\psi_F$$

$$\bar{\psi}'_F = \bar{\psi}_F\Omega^{-1}(x) \tag{10.1}$$

is postulated to describe the same physical situation as before, but of course gives a different set of values to the microscopic variables. If R is the space of microscopic variables, then R/G is the space of microscopic *physical* variables. Again we ask the question how to categorize or enumerate these physical variables.

In perturbation theory it is customary to impose a gauge condition, which implies that we find a subspace of R (the set of fields in R that satisfy the gauge condition) representative for R/G. A renormalizable gauge condition is

$$\partial_\mu A_\mu = 0 \tag{10.2}$$

but it is easy to see that this subspace of R does not accurately describe the physical degrees of freedom in R/G (even though it is accurate enough for perturbation theory). To see this, consider an infinitesimal perturbation in R

$$A_\mu \to A_\mu + \delta A_\mu \tag{10.3}$$

where δA_μ is entirely localized in a particular region δV of space-time. We may however have

$$\partial_\mu \delta A_\mu = f(x) \neq 0 \tag{10.4}$$

In order to impose the gauge condition (10.2) we now find an infinitesimal gauge transformation $\Omega = e^{ig\Lambda}$ that restores (10.2). It must satisfy

$$\partial_\mu D_\mu \Lambda = f(x) \tag{10.5}$$

but the inverse of the operator $\partial_\mu D_\mu$ is non-local. So in our subspace of R satisfying the gauge condition we find a perturbation which is spread out all over space-time, well outside δV. Clearly this perturbation outside δV is unphysical and in perturbation theory we learnt how to deal with this : the theory has "ghosts".

We know claim that beyond perturbation theory these ghosts obscure the physical contents of our theory. Therefore we shall look for a "unitary gauge". Our strategy is to determine this gauge in two steps. Let L be one of the largest Abelian subgroups of G, in our case $L = U(1)^{N-1}$.

$$L = U(1)^{N-1} \tag{10.6}$$

We call G/L the "non-Abelian part" of the gauge group and our first step will be to fix the "non-Abelian part" of the gauge redundancy. We choose a gauge condition C that reduces the space R into a subspace that could be called

$$H_1 = R/(G/L) \tag{10.7}$$

If the gauge group G has N^2-1 generators and L has N-1 generators, then we choose N^2-N real components for the gauge condition C, all invariant under $L \subset G$ but not G itself. The second step is the choice of an N-1 component gauge condition that fixes the remaining invariance under L :

$$H = H_1/L = R/G \tag{10.8}$$

But this second step is precisely the same as fixing the gauge in an ordinary Abelian gauge theory such as electromagnetism and is therefore much more trivial. To understand the physical contents of the theory one could just as well stop after obtaining (10.7) which is expected to describe Abelian charged particles and photons.

A variant on the Lorentz condition that reduces R to R/(G/L) is easy to find :

$$D^o_\mu A^{ch}_\mu = 0 \tag{10.9}$$

there D^o_μ is the L-covariant derivative, containing the diagonal part of the vector field A_μ only. A^{ch}_μ is the set of off-diagonal elements of A_μ only. Eq. (10.9) has indeed N^2-N components.

This gauge suffers from the ghost problem as much as the ordinary Lorentz gauge, and is therefore not suitable for understanding all physical degrees of freedom.

XI. UNITARY GAUGE

A unitary gauge must be picked in a way similar to the Abelian case. We need a field that transforms without derivatives under gauge transformations. We will limit ourselves to the case that this field, call it X, transforms as the adjoint representation under G.

$$X \to \Omega X \Omega^{-1} \tag{11.1}$$

Such a field namely can always be found. The simplest choice would be

$$X^{ij} = G^{ij}_{12} \tag{11.2}$$

which is one of the components of the covariant curl $G_{\mu\nu}$. This choice has the disadvantage of not being Lorentz-invariant. One may choose a composite field :

$$X^{ij} = G^{ik}_{\mu\nu} G^{kj}_{\mu\nu} \tag{11.3}$$

This however does not work if G = SU(2) because then X would be proportional to the identity matrix. We need a non-vanishing isovector part. We could choose

$$X^{ij} = G^{ik}_{\mu\nu} D^2 G^{jk}_{\mu\nu} \tag{11.4}$$

but this choice looks rather complicated. Perhaps the most practical choice would be to take ones refuge to an extra scalar field in the theory, giving it a sufficiently high mass value so that the theory is not changed perceptibly at low energies.

Our gauge condition will be that X is diagonal :

$$X = \begin{pmatrix} \lambda_1 & & & & 0 \\ & \lambda_2 & & & \\ & & \cdot & & \\ & & & \cdot & \\ & & & & \cdot \\ 0 & & & & \lambda_N \end{pmatrix} \tag{11.5}$$

where the eigenvalues λ_i may be ordered :

$$\lambda_1 > \lambda_2 > \ldots . > \lambda_N \tag{11.6}$$

What is the subgroup of the gauge transformations Ω under which (11.5) is invariant ? If we require

$$X' = \Omega X \Omega^{-1} = X$$

then

$$[X, \Omega] = 0 \tag{11.7}$$

therefore, Ω is also diagonal :

$$\Omega = \begin{pmatrix} e^{i\omega_1} & & & 0 \\ & \cdot & & \\ & & \cdot & \\ & & & \cdot \\ 0 & & & e^{i\omega_N} \end{pmatrix} \tag{11.8}$$

and since $\det\Omega = 1$, we have

$$\sum_{i=1}^{N} \omega_i = 0 \tag{11.9}$$

Indeed, this is the largest Abelian subgroup L of G. If we write

$$\Omega = e^{i\omega} \tag{11.10}$$

then the diagonal part A^o_μ of A_μ transforms as

$$A^o_\mu \to A^o_\mu - \frac{1}{g}\partial_\mu \omega \tag{11.11}$$

and the off-diagonal part A^{ch}_μ as

$$A^{ch\ ij}_\mu \to e^{i(\omega_i - \omega_j)} A^{ch\ ij}_\mu \tag{11.12}$$

Apart from the gauge transformations (11.11) and (11.12) all our fields are physically observable. So our physical degrees of freedom are

- N-1 "massless" photons
- 1/2 N(N-1) "massive" charged vector fields
- N scalar fields λ_i with the restriction : $\lambda_1 > \lambda_2 > \ldots > \lambda_N$.

There is of course another constraint : depending on our choice for

X, we have that

$$\begin{pmatrix} \lambda_1 & & & \\ & \cdot & & \\ & & \cdot & \\ & & & \cdot \\ & & & & \lambda_N \end{pmatrix}$$

satisfies (11.2), (11.3), or (11.4). Of course we still have the local $U(1)^{N-1}$ symmetry to be removed by either conventional gauge fixing, or the procedure described in the previous chapters.

XII. A TOPOLOGICAL OBJECT

So far we assumed that the eigenvalues $\lambda_1, \ldots, \lambda_N$ coincide nowhere. What if they do, at some set of points in space-time ? At those points, the invariance group is larger and a problem emerges with our enumeration procedure. We now argue that these exceptionnal points : a) are pointlike in 3-space, describing particle-like trajectories in space-time, and b) correspond to singularities in the fields A_μ and ψ. The argument for the singularity is similar to the Abelian case, but the fact that these things are particle-like differs from the Abelian case, where the singular points were string-like.

The reason why the singularities in the generic case are pointlike is that the dimensionality of the space of field variables with two coinciding eigenvalues λ is three less than the space with non-coinciding eigenvalues. Therefore the dimensionality of the points in space or space-time where two eigenvalues coincide is three less than that of space-time itself. This statement holds for any generic N x N hermitian matrix field X(x).

What is the physical nature of such a particle-like singularity ? Since the eigenvalues λ_i were ordered, we only need to consider the case that two successive λ's coincide :

$$\lambda_j = \lambda_{j+1} \tag{12.1}$$

for certain j. Let us consider a close neighbourhood of such a point. Prior to the gauge-fixing we may take X to be (12.2), where D_1 and D_2 may safely be considered to be diagonalized because the other eigenvalues did not coincide. The three fields $\varepsilon_a(x)$ are small because we are close to the point where they vanish. With respect to that SU(2) subgroup of SU(N) that corresponds to rotations among the jth and $j + 1^{st}$ components, the fields ε_a form an isovector.

$$X = \begin{array}{r} \\ \\ j \\ \\ j+1 \\ \\ \\ \end{array} \begin{array}{c} \left[\begin{array}{c|cc|c} D_1 & & 0 & 0 \\ \hline & \lambda+\varepsilon_3 & \varepsilon_1 - i\varepsilon_2 & \\ 0 & & & 0 \\ & \varepsilon_1 + i\varepsilon_2 & \lambda - \varepsilon_3 & \\ \hline 0 & & 0 & D_2 \end{array}\right] \\ \qquad j \qquad\qquad j+1 \qquad\qquad \end{array} \tag{12.2}$$

One may write the center block as

$$X = \lambda I + \vec{\varepsilon}.\vec{\sigma} \tag{12.3}$$

where σ_a are the Pauli spin matrices. Close to a zero point of this $\vec{\varepsilon}$ field, the field $\vec{\varepsilon}$ has a hedgehog configuration. But gauge fixing i.e. diagonalization of X, corresponds to rotating ε_1 and ε_2 away such that ε_3 is positive ($\lambda_j > \lambda_{j+1}$). Thus is our unitary gauge,

$$\vec{\varepsilon} = \begin{pmatrix} 0 \\ 0 \\ +|\varepsilon_3| \end{pmatrix} \tag{12.4}$$

By now the reader may recognize this field configuration as the one for a magnetic monopole[9]. Indeed, fixing the L-gauge as well cannot be done without accepting a string-like singularity connecting zeros of opposite signature : the Dirac string.

The magnetic charges of the monopole can most easily be characterized with respect to the $U(1)^N$ subgroup of the extended gauge group U(N) :

$$\vec{m} = (\, 0,\, \ldots,\, 0,\, \frac{2\pi}{g},\, -\frac{2\pi}{g},\, 0,\, \ldots,\, 0) \tag{12.5}$$

where the $\pm\, 2\pi/g$ are at the jth and $j{+}1^{st}$ position. g is here the fundamental electric charge of the elementary representation. We then see that $\vec{m}$ actually only acts in the subgroup $U(1)^N/U(1)$ of SU(N) because the sum of all its charges vanishes. It is constructive to notice a subtile difference between this magnetic charge spectrum and the spectrum of the electrically charged gauge particles A_μ^{ij} : from (11.12) we read off that these have electric charges

$$\vec{Q} = (0, \ldots, 0, +g, 0, \ldots, 0, -g, 0, \ldots, 0) \qquad (12.6)$$

where $\pm$ g occur at arbitrary, not necessarily adjacent, positions i and j. Again however the sum of the charges vanishes, so that we are really working in the Cartan group $U(1)^{N-1}$, not $U(1)^N$. Magnetic monopoles with

$$\vec{m} = (0, \ldots, 0, \frac{2\pi}{g}, 0, \ldots, 0, -\frac{2\pi}{g}, 0, \ldots, 0) \qquad (12.7)$$

are possible only if three or more eigenvalues coincide. The dimensionality of such points is at least 8 less than space-time so in general they do not occur at single points. Rather, they should be considered as bound states of "elementary" monopoles (12.5).

We conclude that we arrive at a picture where an Abelian gauge theory is enriched with magnetic monopoles, but because of the slightly different charge spectrum this picture is in general not "self dual". Contrary to the case discussed in section VII, the monopoles we have here are "dynamical", that is, they will inevitably take part in the dynamics of the system.

XIII. THE MACROSCOPIC VARIABLES

We have now arrived at a point where we could sketch a possible strategy for precise calculations for the dynamics of the system :

1) Consider the physical degrees of freedom in the space $H_1 = R/(G/L)$. We find N-1 sets of Maxwell fields, electrically charged fields (among which vector fields), and magnetically charged particles. The particular case of interest is now the possibility that magnetically charged particles "Bose-condense". If we ever are to understand such a mechanism in detail, the following step is probably necessary :

2) Eliminate the electric charges. With as much precision as possible we must compute all light-by-light scattering amplitudes and express them in term of an effective interaction lagrangian for the photon fields :

$\Gamma(A_\mu) = \sum$ 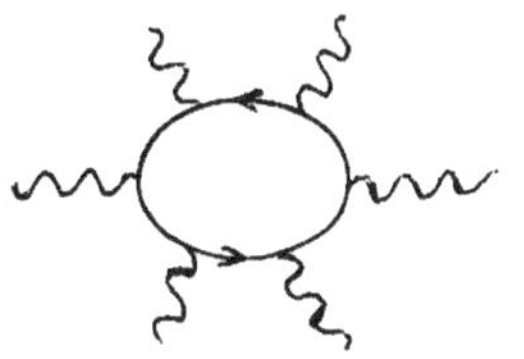 + higher orders (13.1)

3) Now perform the "dual transformation". Since we have only Maxwell fields and magnetic charges interacting with them, we could replace $\vec{B}$ by $\vec{E}$ and $\vec{E}$ by $-\vec{B}$, then introduce operator fields in the usual way for the monopole particles, which now look like ordinary electrically charged objects.

4) Work out the self-interactions among these magnetic monopoles. Set up a perturbation theory now in terms of $2\pi/g$. Then the question is :

5) Does, in terms of this perturbation theory, Bose condensation occur among these monopoles ? Is it reasonable to start with

$$< 0|\phi_{mon}|0 > \neq 0 \quad ? \tag{13.2}$$

If so, then the vacuum is a magnetic superconductor. The monopoles formally have a negative mass-squared. In this magnetic superconductor electric charges are confined. The descriptions of section VII apply qualitatively, after the interchange electric $\leftrightarrow$ magnetic.

XIV. THE DIRAC CONDITION IN THE ELECTRIC-MAGNETIC CHARGE SPECTRUM

Fig 2 represents the spectrum of possible charges in the case that the gauge group G is SU(2).

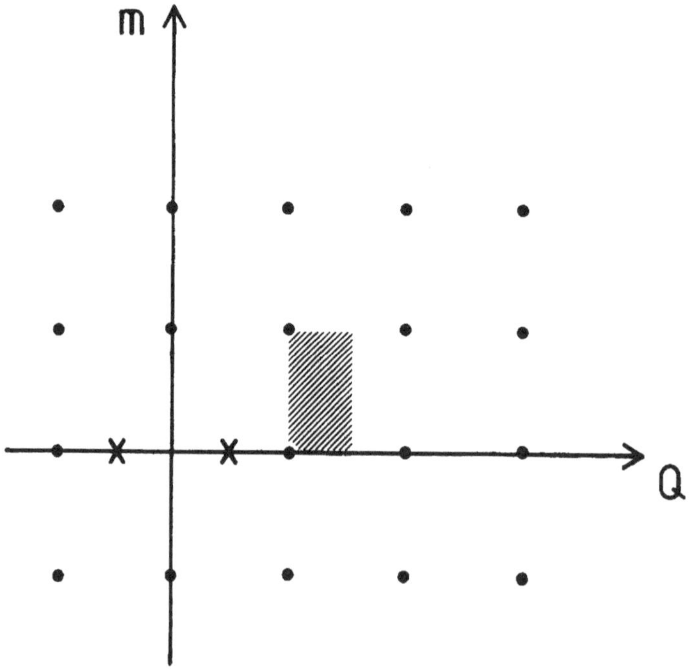

Fig 2

Horizontally is plotted the electric charge Q, vertically the magnetic charge $\vec{m}$. Elementary and bound state charges are indicated. The crosses represent a fundamental SU(2) doublet which may or may not have been added to the theory. The pure Maxwell field equations are now invariant under rotations of the figure about the origin. This is why one is always free to postulate that the fundamental fields carry no magnetic monopole charge. But does the lattice obtained in fig 2 have to be rectangular ? We will now argue that Dirac's quantization condition allows more kinds of lattices and "oblique" lattices indeed may result in a non-Abelian gauge theory.

The Dirac condition for a magnetic charge quantum m and the electric charge quantum q was

$$qm = 2\pi n \tag{14.1}$$

where n is integer. To be precise this corresponds to a quantization condition for the Lorentz force that the magnetically charged particle exerts on the electrically charged object. But now consider two particles 1 and 2 both with various kinds of magnetic and electric charges. Then the Lorentz force quantization corresponds to

$$\sum_i (q_i^{(1)} m_i^{(2)} - q_i^{(2)} m_i^{(1)}) = 2\pi n_{12} \tag{14.2}$$

where the index i refers to the label of the species of photons and n_{12} is an integer relevant for particles 1 and 2, to be referred to as the Dirac quantum of particle (1) with respect to particle (2).

In our specific case of SU(N) broken down to $U(1)^{N-1}$ we usually require

$$\sum_{i=1}^{N} q_i = 0 \qquad\qquad \sum_{i=1}^{N} m_i = 0 \tag{14.3}$$

Our charge lattice in this case will be spanned by 2(N-1) basic charges, to be labelled by an index A = 1, ..., 2N-2. Because of the invariance

$$\begin{pmatrix} m_i \\ q_i \end{pmatrix} \to \begin{pmatrix} \cos\phi_i & \sin\phi_i \\ -\sin\phi_i & \cos\phi_i \end{pmatrix} \begin{pmatrix} m_i \\ q_i \end{pmatrix} \tag{14.4}$$

we may always take

$$m_i^{(A)} = 0 \qquad \text{for } A = 1, \ldots, N-1 \tag{14.5}$$

The gluons will provide us with a basis of electric charges:

$$q_i^{(A)} = g\delta_i^A - g\delta_i^{A+1} \quad \text{for } A = 1, \ldots, N-1 \tag{14.6}$$

(The fundamental representation, if it occurs, could have

$$q_i^{(k)} = g\delta_{ik} - g/N \text{)}. \tag{14.7}$$

The magnetic monopoles give the remaining basic charges:

$$m_i^{(A)} = \frac{2\pi}{g}\delta_i^{A+1-N} - \frac{2\pi}{g}\delta_i^{A+2-N} \quad \text{for } A = N, \ldots, 2N-2 \tag{14.8}$$

It was Witten[10] who observed that monopoles may also carry electric charges. He found

$$q_i^{(A)} = \frac{\theta g^2}{4\pi^2} m_i^{(A)}, \quad \text{for } A = N, \ldots, 2N-2 \tag{14.9}$$

where θ is the instanton angle of the theory, $0 \leqslant \theta < 2\pi$. Notice that for any value of θ the Dirac condition (14.2) is fulfilled.

It can be seen that this phenomenon, eq. (14.9), follows from the lagrangian

$$\mathcal{L} = -\frac{1}{4}G^a_{\mu\nu}G^a_{\mu\nu} + \frac{\theta i g^2}{32\pi^2}G^a_{\mu\nu}\tilde{G}^a_{\mu\nu} + A^a_\mu J^a_\mu \tag{14.10}$$

there

$$\tilde{G}^a_{\mu\nu} = \frac{1}{2}\varepsilon_{\mu\nu\alpha\beta}G^a_{\alpha\beta} \tag{14.11}$$

and $G^a_{\mu\nu}\tilde{G}^a_{\mu\nu}$ corresponds to

$$4\sum_a \vec{E}^a.\vec{B}^a \tag{14.12}$$

The canonical argument can be found in refs 10), 11), 8).

In the case of SU(2) the charge lattice indeed becomes tilted now (Fig 3). It is remarkable that if θ runs from 0 to 2π then the charge lattice indeed turns back into itself, but the "elementary" monopole labelled by (2) in fig 3 is replaced by the "monopole gluon bound state" labelled (3). It seems that no fundamental distinction

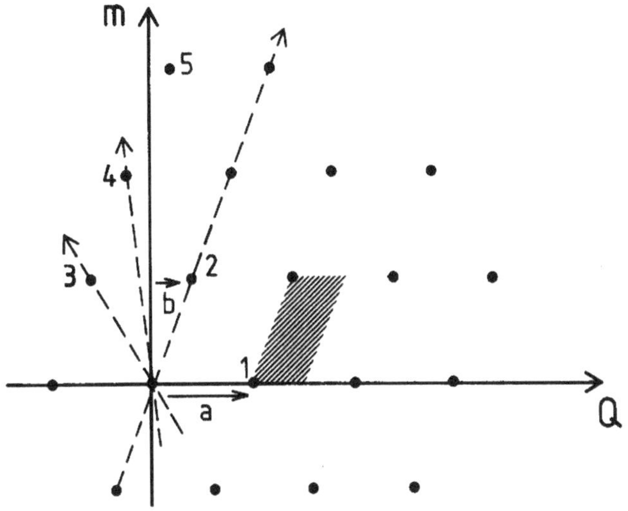

Fig 3

will be possible between monopoles and monopole-gluon bound states.

XV. OBLIQUE CONFINEMENT

We can now ask which of the objects in the lattice of fig 3 will form a Bose-condensate. If it is a purely electrically charged object, (1) in fig 3, then we get the familiar Higgs theory or electric superconductor. All charges that are not on the horizontal axis will then be confined by linear potentials, because of the arguments presented in section VII. By duality we now expect that also monopoles may conceivably undergo Bose condensation, for instance charge #(2) in fig 3. One cannot however have both electric and magnetic charges Bose-condense because if the electric ones condense then the magnetic ones have $m^2 \to +\infty$, not negative, and vice versa. In the case of larger N, Bose condensation can only take place among charges forming any linear sublattice of the original charge lattice, as long as its members all have vanishing Dirac quanta with respect to each other.

Now if θ is switched on, then point (3) gradually takes the place of (2). (2) and (3) cannot both Bose condense because they have a non-vanishing relative Dirac quantum. So it is either (2) or (3). It is likely that Bose condensation in the (2)-direction is replaced by condensation in the (3)-direction at $\pi < \theta < 3\pi$. This would then be a *phase transition* in θ, possibly of first order, just like the transition between Higgs and confinement.

Various attempts at dynamical calculations however indicate that at $\theta \simeq \pi$ the confinement mechanism is not strong. An explanation could simply be that the monopoles then carry large electric charges and therefore may have larger self-energies contributing positively to their mass-squared. Suggestions have been made that at $\theta \simeq \pi$ the Higgs mode reappears [12] or a Coulomb mode (no Bose condensation at all).

I suggest yet a different condensation mode that could possibly occur in theories with θ close to π. If neither (2) nor (3) condense because they carry large (but opposite) electric charges, then perhaps (4) which is a bound state of these two with much smaller electric charge condenses. This would only be possible if the lattice is oblique ($\theta \neq 0$) so this mode is referred to as "oblique confinement". A theory with oblique confinement shows some peculiar features. We stress that these will not occur in ordinary QCD because there we know that $\theta \simeq 0$. Our observations may be relevant for certain models with "technicolor" as we will show shortly.

Returning to the case that our gauge group was SU(2), we first argue that the "quarks" (or "preons") in this oblique confinement mode are not confined in the usual sense. That is, if we attach a flavor quantum number to every type of preon, then physical particles transforming as the fundamental representation of the flavor group do occur. The preons are the crosses in fig 4.

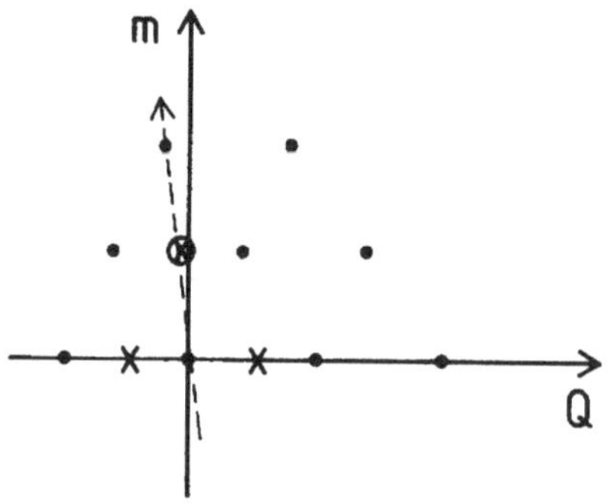

Fig 4.

Since they are not on the line connecting the origin to the condensed object (arrow) they are confined. However, a bound state of a monopole (without flavor quantum numbers) and one preon does occur on that line (⊗), and hence is physical, and has the same flavor as the preon itself. But a price had to be paid : spin and statistics-properties of the physical object are opposite to that of the original preon! If the preon is a fermion, the liberated object is a boson and vice-versa. This is a consequence of a general rule : if two particles have an odd relative Dirac quantum, then the *orbital* angular momentum in any bound state of the two is half-odd-integer, a well-known property of the Schrödinger equation of an electrically charged particle in the field of a magnetic point-source[13]. That also the *statistics* of the bound state gets an extra fermionic contribution has been shown by Goldhaber[14]. Let me give an outline of the argument.

If we wish to consider the statistics of two identical composite states A and B, both composed of a magnetic monopole M and an electric charge Q, (fig 5) then we would like to separate the center-of-mass motions of A and B from the orbital motions of M and Q inside A and B.

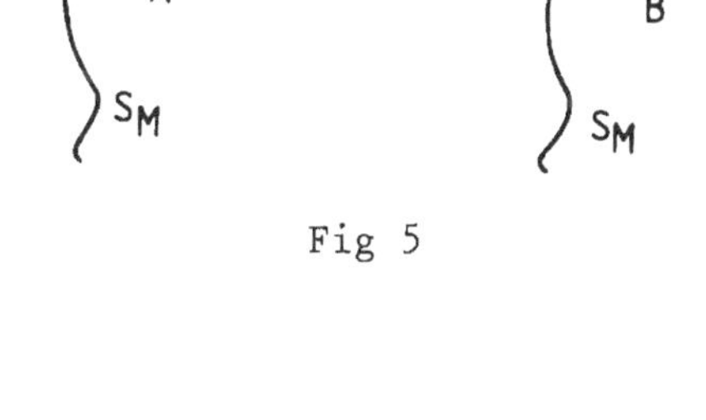

Fig 5

The particles Q see two Dirac strings, ending at both M's. The M's see two Dirac strings ending at both Q's. Now the center-of-mass motion of A can only be split from the orbital motion of M and Q if the magnetic strings run in a direction opposite to that of the electric strings (see fig 5), so that if M hits an electric string, in the same time Q hits a magnetic string. Now let us ignore the motion of M and Q inside A and B, but only consider A and B as a whole. Then A feels both strings at B (and vice versa), in fact, these two are connected in such a way that one string results, running from infinity to infinity. This could be expected because A and B have the same electromagnetic charge combination (they were identical) so *their* relative Dirac quantum vanishes. Their relative motion (A against B) is as if they only were electrically charged. Therefore we obtain the familiar Coulomb Schrödinger equation by removing this complete string by a single gauge transformation :

$$\psi_{AB} \rightarrow e^{i\phi_{AB}} \psi_{AB} \tag{15.1}$$

where ϕ_{AB} is the angle by which A rotates around the B-string.

Now notice that if we interchange A and B this angle is 180°, so that

$$e^{i\phi_{AB}} = -1$$

This is how one can see an extra minus sign appearing in the commutation properties of the particles A and B.

XVI. FERMIONS OUT OF BOSONS AND VICE-VERSA

Some exotic models can be constructed if we use oblique confinement as a starting point. Consider for instance an SU(3) gauge theory without fermions but with a scalar field ϕ in the fundamental representation. Let θ be close to π, and let us assume that the familiar Higgs mechanism takes place, described by

$$< \phi > = \begin{pmatrix} 0 \\ 0 \\ F \end{pmatrix} \tag{16.1}$$

(through formally incorrect, as explained in section V, the notation is useful and adequate for our purpose). SU(3) then "breaks spontaneously" into SU(2). One heavy neutral Higgs particle arises. The original octet of gauge fields splits into one neutral heavy vector particle and one SU(2) complex doublet of heavy vector particles.

The SU(2) triplet remains massless, and now we will assume that it produces oblique confinement in the SU(2) sector. The SU(2) doublet of heavy vector bosons will appear as physical particles but disguised as fermions! Here we have an example of fermionic gauge "bosons" without invoking anything resembling supersymmetry.

Another model worth considering is a simplistic weak interaction theory, based upon $SU(2)^{tc} \times (SU(2) \times U(1))^{ew}$. There "tc" stands for "technicolor" and ew for "electro-weak". Quarks and fermions are all in the usual representation of $(SU(2) \times U(1))^{ew}$ and singlets under $SU(2)^{tc}$. Now add one fermion multiplet transforming just like all other fermions under $(SU(2) \times U(1))^{ew}$ but also as a 2 under tc. Assume oblique confinement. Then this fermion will be liberated, but disguised as a boson. Probably it will have a spin-zero component. Since it is in line with the condensing monopole bound states it may Bose-condense itself. So we obtain a scalar field transforming just as the fermions under $(SU(2) \times U(1))^{ew}$ and a non-vanishing vacuum expectation value : a model for the Higgs particle. Indeed, its Bose condensation could well be responsible for the oblique confinement mode in the first place, so here it was not even necessary to consider the monopole-dyon bound state.

Unfortunately, elegant as it may be, this model seems to suffer from the same shortcomings as the more conventional technicolor ideas : it is hard to reproduce the required Yukawa couplings between this scalar field and the other fermions.

XVII. OTHER CONDENSATION MODES

It will be clear from the previous sections, by looking at the electric-magnetic charge lattice, that even more exotic forms of oblique confinement can be imagined. Just assume condensation of bound states with three or more monopoles. Such a condensation mode would be required for instance if we would wish to liberate the fundamental triplets in an SU(3) gauge theory. A fundamental exercise tells us then that these triplets do not switch their spin-statistics properties [8]. In any case, all these different confinement modes will be separated from each other by sharp phase transition boundaries, which should show up in the solutions of the theory when the parameter θ is varied.

In principle each point on the electric-magnetic charge lattice may correspond to a possible phase of the system. There may however be features which cannot easily be understood in terms of our intermediate physical degrees of freedom with Abelian electric and magnetic charges. We have in mind a condensation mode studied in more detail by Bais[15]. The simplest example is an SU(2) gauge theory with an isospin-two Higgs field, ϕ^{ab} (a, b = 1, 2, 3).

Let us assume

$$< \phi^{ab} > = F^{(a)} \delta^{ab} \tag{17.1}$$

If we were just dealing with a global symmetry, we would say that SU(2) is spontaneously broken into a subgroup D (the invariance group of eq. (17.1)). Now D is the discrete subgroup of SU(2) corresponding to rotations of spinors over 90° :

$$D = (\pm I, \pm i\sigma^1, \pm i\sigma^2, \pm i\sigma^3) \tag{17.2}$$

If the symmetry is local then D is not really a global invariance of the vacuum. What we do see is that magnetic vortex tubes can be constructed which are characterized by the following boundary condition at infinity :

$$A_\mu \rightarrow \frac{i}{g} \Omega \partial_\mu \Omega^{-1} \tag{17.3}$$

where Ω is multivalued. If we go around the vortex once, then Ω turns into itself multiplied with an element D_1 of D. These vortices are non-commuting. Physically this means the following (fig 6)

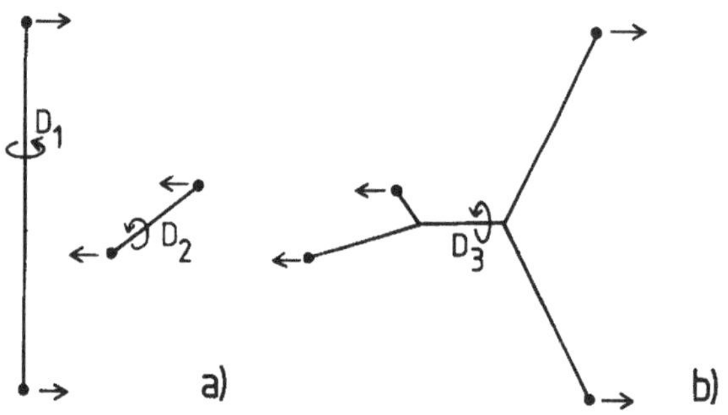

Fig 6

If two strings, characterized by different, non commuting elements D_1 and D_2, approach each other at right angles (fig 6a), then they cannot pass each other without leaving a connecting string D_3 (fig 6b). The element D_3 is given by

$$D_3 = D_1 D_2 D_1^{-1} D_2^{-1} \tag{17.4}$$

Clearly the non-commuting properties of the original gauge group were crucial for understanding this phenomenon, so that our Abelian physical variables are not useful here. Indeed, we could ask the question whether the dual of this "Bais mode" exists, with electric strings having similar properties ? As yet, the answer to that question is unknown.

REFERENCES

1. J. Goldstone, Nuovo Cim. 19 154 (1961)
2. F.S. Abers and B.W. Lee, Phys Reports 9C, 1 (1973) and references therein.
3. P.W. Higgs, Phys. Lett. 12, 132 (1969), Phys. Lett. 13, 508 (1964), Phys. Rev. 145, 1156 (1966)
   F. Englert, R. Brout, Phys. Rev. Lett. 13, 321 (1964)
   G.S. Guralnik, C.R. Hagen and T.W.B. Kibble, Phys. Rev. Lett. 13 585 (1964)
   T.W.B. Kibble, Phys. Rev. 155, 627 (1967)
4. B.S. De Wit , Phys. Rev. 162, (1967), 1195, 1239
5. G. 't Hooft and M. Veltman, "DIAGRAMMAR", CERN report 73/9 (1973)
6. H.B. Nielsen, P. Olesen, Nucl. Phys. B61, 45 (1973)
   B. Zumino, in "Renormalization and Invariance in Quantum Field Theory ed. E.R. Caianiello Plenum Press New York (1974) p. 367
7. P.A.M. Dirac, Proc. Roy. Soc. A133, 60 (1931)
   Phys. Rev. 74, 817 (1948)
8. G. 't Hooft, Nucl. Phys. B, to be published
9. A.M. Polyakov JETP Lett. 20, 194 (1974)
   G. 't Hooft, Nucl. Phys. B79, 276 (1974)
10. E. Witten, Phys. Lett. 86B, 283 (1979)
11. A. Salam and J. Strathdee, Lett. in Mathematical Physics 4, 505 (1980)
12. C. Callan, private communication (1979)
13. R. Jackiw and C. Rebbi, Phys. Rev. Lett. 36, 1116 (1976)
    G. 't Hooft and P. Hasenfratz, Phys. Rev. Lett 36, 1119 (1976)
14. A.F. Goldhaber, Phys. Rev. Lett. 36, 1122 (1976)
15. F.A. Bais, private communication.

DEVELOPMENTS IN PARTICLE PHYSICS

Sheldon Lee Glashow

Lyman Laboratory of Physics
Harvard University
Cambridge, MA 02138 U.S.A.

I. THE FERMION FAMILY

Leptons and quarks (the only known fermions) come in families, each consisting of fifteen chiral fields. Today, there are two known families which are complete, and an incomplete third one. Only the top quark remains to be discovered, and we will have three complete families. Toponium has not yet been discovered, so that the top quark must weigh more than 18 GeV. I have predicted that it weighs 20 GeV, and this prediction will be put to the test in the near future at PETRA. My prediction follows from a single simple and plausible but not certain hypothesis: that the matrices describing the masses of up quarks, down quarks, and charged leptons are linearly dependent. Are there more than three families in Nature ? Each known family contains a massless or nearly massless neutrino. There cannot be many of these, we are assured by the astrophysicists. They affect the expansion rate of the universe at the time of helium synthesis. We cannot exclude the possible existence of a fourth family, but there surely cannot be ten families with ten very light neutrinos. When LEP is built, we can measure directly the number of neutrinos. Then, perhaps, we shall know how many fermion families there are. Why do all the fermion families transform as equivalent 15-dimensional representations of SU(3) x SU(2) x U(1), the gauge group of strong and electroweak interactions ? We do not know. Other representations are conceivable: quixes and quaits have been discussed by theorists. Here are more questions we cannot answer: why are most (if not all) of the fermions so light ? There is a natural mass scale, M_F = 250 GeV, which characterizes the electroweak interaction and which provides the source of fermion masses through the intervention of Yukawa couplings of Higgs bosons. For the electron, the coupling constant is only 2×10^{-6}. Why is it so small ? Should not the W and Z have fermionic cousins with comparable masses ? To find out, we shall need a 10 TeV high-luminosity hadron collider.

Each family is a reducible representation of SU(3) x SU(2) x U(1) with five irreducible components. In particular, here is the first family:

(u,d)	[3,2;1]	A
$(\bar{u})$	$[\bar{3},1;-4]$	B
$(\bar{d})$	$[\bar{3},1;2]$	C
(ν,e^-)	[1,2;-3]	D
(e^+)	[1,1;6]	E.

All fifteen fields are left -handed chiral fields. In the bracket are shown the SU(3) x SU(2) x U(1) transformation properties, the last integer denoting six-times the conventional value of the U(1) quantum number. The representation is evidently complex, being inequivalent to its conjugate. However, as a representation of the unbroken electro-chromo SU(3) x U(1) subgroup, it is evidently real. Moreover, the representation is anomaly-free, a necessary condition for the renormalizability of the underlying gauge theory. It is easily verified that $Tr(T_i\{T_j,T_k\})$ vanishes for any choice of the generator T of SU(3) x SU(2) x U(1). More elegantly, this condition is equivalent to $TrT^3 = 0$ where T is any 15-dimensional matrix in the associated Lie algebra. Not only does $TrT^3 = 0$, but $TrT = 0$, and in particular, the sum of the electric charges vanishes. This is a necessary condition for unification in terms of a simple group without the real need for extra unobserved fermions.

Tracelessness of T and T^3 and electro-chromo reality are not sufficient to pin down the U(1) assignments. At this pre-unification level, the curious values of quark and lepton charges remain mysterious.

We shall now establish the existence of a unique current S_μ which commutes with SU(3) x SU(2) x U(1), of which each fermion family is a representation, which is linearly independent of the U(1) current, and which is free of gauge-gauge triangle anomaly. The S_μ charges of each irreducible component are indicated by capital letters, thus by construction satisfying the first two conditions. We satisfy the third condition by requiring that S and U(1) are orthogonal:

$$A - 2B + C - D + E = 0.$$

To avoid gauge-gauge anomalies, we must attend to each of the simple components of the gauge group:

$[SU(3)]^2 \quad 2A + B + C = 0$

$[SU(2)]^2 \quad 3A + D = 0$

$[U(1)]^2 \quad A + 8B + 2C + 3D + 6E = 0.$

These are four conditions among five parameters. The assignments of S_μ are thus determined up to an overall scale. We may choose

$A = B = E = 1, \; C = D = -3.$

This unique anomaly-free commuting current has a surprising property. It commutes with all the generators of SU(5). Consequently, it has a second surprising property. There are no S^2-gauge anomalies, since U(1) is traceless on the $\bar{5}$ and 10 of SU(5). On the other hand, there is an S^3 anomaly since $TrS^3 = -125$. Thus, S_μ cannot be a gauge symmetry unless we introduce more fermions. But, S_μ can be a global symmetry of the Lagrangian, and in simple realizations of SU(5) it is automatically so. There are simply no renormalizable interactions which are not S invariant, unless one introduces further complexity, such as a deciment of scalar mesons, or a richer unified theory than SU(5).

A word about unification. What are the possibilities for extending SU(3) x SU(2) x U(1) to a simple group G of which the family is a representation ? If the gauge theory is to be renormalizable, the condition $TrT^3 = 0$ must extend to the Lie algebra of G. This would not be the case, for example, for the obvious choice G = SU(15). Indeed, there is only one possible choice of a unifying group, G = SU(5). Of course, the 15-dimensional family may not be complete. Other fermions may be needed to constitute an anomaly-free representation of G. With one extra fermion (an SU(5) singlet), we obtain O(10) as a possible unifying group. In this case, the extra neutrino state is assigned $S = -5$ and S_μ becomes completely free of anomalies. It is a generator of SO(10), of which SU(5) x U(1) is a regular subgroup. With twelve extra fermions per family, we may choose E_6 or its subgroup $[SU(3)]^3$ as unifying groups. And so on.

A word about fermion masses in the context of SU(3) x SU(2) x U(1). No member of the 15 dimensional family may have gauge-invariant masses. Quark and charged lepton masses, which are non-zero, transform as weak doublets. Neutrino Majorana masses, which are very small or zero, transofrm as weak triplets. It has been conjectured that fermions whose masses are gauge invariant should have masses of order $M_G \sim 10^{15}$ Gev , the unification mass. Such fermions, unless they are stable, will never be detected.

What is the meaning of S_μ, the automatic symmetry of naive SU(5) ? A linear combination of S_μ and the U(1) current is baryon number minus lepton number:

$$B - L = \frac{1}{5} S + \frac{2}{15} U(1).$$

The reader may verify that this quantum number assigns $(B - L) = 1/3$ to quarks and $(B - L) = -1$ to leptons. Moreover, the vev of the Higgs boson responsible for electroweak symmetry breaking is annihilated by B - L. This symmetry remains an exact global symmetry even after spontaneous symmetry breakdown, and it ensures the vanishing of neutrino masses in naive SU(5).

Of course, naive SU(5) may be wrong. From a general point of view, there are five possibilities:

(1) B - L is an exact local symmetry. This is not a realistic possibility, for there does not exist a second species of massless photon.

(2) B - L is an exact global symmetry. Perhaps naive SU(5) is correct. Perhaps Majorana neutrino masses are identically zero. Finite Dirac neutrino masses can be introduced, but it then becomes a mystery as to why they are so small.

(3) B - L is a spontaneously-broken local symmetry. This is the case for SO(10), wherein the B - L gauge bosons must be very heavy, presumably of order M_G. Majorana neutrino masses are expected with typical magnitude $M_F^2/M_G \sim 10^{-2\pm4}$ eV.

(4) B - L is a spontaneously broken global symmetry. In this case, there must exist a true Goldstone boson, the Majoron. This case has recently been studied by several workers at Monaco di Baviera. We shall consider it in detail in section IV.

(5) The last possibility is that B - L is not a symmetry of the Lagrangian at all. This is a viable but unappealing alternative.

II. ON SYMMETRY BREAKING

To get from the unifying group to the cold and crystalline low-energy world, we must proceed through at least two stages of symmetry breaking. At $M_G \sim 10^{15}$ GeV, the unifying group breaks down to SU(3) x SU(2) x U(1) (or, perhaps, a larger group), while at $M_F \sim 10^2$ GeV, another stage of symmetry breakdown leaves only SU(3) x U(1) (or, with technicolor, a larger group). We shall presume that the electro-chromo group is sacrosanct: that there is no low-temperature breakdown of charge conservation, and that there are neither integer charged quarks, nor fractionally charged leptons, nor isolable colored states. The existence of two widely disparate mass scales is a problem of large numbers, much like many of the problems Paul Dirac considers. It is called the gauge hierarchy problem, and it remains unsolved.

There are two popular approaches to the problem, each representing a radical theoretical departure: Supersymmetry and Technicolor. There is even such a thing as supersymmetric technicolor. Supersymmetry transformations do not commute with the Lorentz group and mix fermions with mesons. Supersymmetric multiplets contain fermions and bosons in matched pairs. While there is excellent empirical data for the existence of multiplets corresponding to isospin, the eightfold way, the electroweak group, and even SU(5), there is none at all for the existence of supersymmetric pairing. None of the 13 known fundamental "particles" has an observed supersymmetric mate. Given, say, 48 "particles" all told, with random masses, the probability of picking out 13 particles and no mated pairs is 10 %. The probability of picking out 11 fermions with no associated boson is 0.05 %. Either there is some conspiracy separating the supersymmetric pairs, or we must conclude that supersymmetry is silly. Technicolor asserts the existence of a stronger confining interaction borne by particles not yet seen. Most of them lie at masses near M_F and, reasonably, have not yet been seen. Some "pseudo-Goldstone" states lie at lower energies and can be looked for now. Technicolor offers a neat and natural explanation for the origin of the electroweak mass scale which is perfectly analogous to the appearance of the QCD mass scale. The problem lies in the origin of hard quark and lepton masses. Extended technicolor was invented to address just this question, but, so far, there has been little success in constructing realistic models. We have nothing to add to this subject. We simply adopt the conventional (and unsatisfactory) view that there is a hierarchy of symmetry breaking induced by appropriate systems of Higgs bosons. Outhouses were ugly but necessary, and were used until Thomas Crapper had a better idea.

How do we break SU(5) down to SU(3) x SU(2) x U(1) ? Perhaps SU(5) is exact at no level, but is manifestly broken by terms of low dimension. Simply introduce a quintet of Higgs bosons with manifest symmetry breaking in its mass term: let the triplet have mass M_G and the doublet have mass M_F. This is ugly and unnatural, and perhaps unrenormalizable. However, we may not care so much about renormalizability so close to the Planck mass where we know so little.

More conventionally, we look for a convenient Higgs multiplet whose vev has SU(3)xSU(2)xU(1) as a little group. Let us prove that such a multiplet of SU(5) must have fivality zero. Every irreducible representation is contained in the direct product of N fundamental representations, 5^N, and it will have fivality f = N (mod 5). The irreducible representations of SU(3) x SU(2) x U(1) contained in 5^N will have SU(3) triality p (mod 3) and SU(2) biality q (mod 2) where p and q are positive integers satisfying p + q = N. We are searching for a term which is SU(3) x SU(2) invariant, which therefore must satisfy p = 3a, q = 2b with a and b integers. U(1) invariance implies that $2p - 3q = 0$ and hence that $a = b$. Thus, we obtain $N = 5a$, which was our assertion.

The smallest representation of SU(5) which has vanishing fivality is the 24 dimensional adjoint R, which indeed contains an SU(3) x SU(2) x U(1) singlet.

The most general $R \to -R$ invariant Hamiltonian is

$$a(R^4) + b(R^2)^2 - \mu^2(R^2)$$

with parentheses denoting traces. Positivity requires $30a > 7b$ and $20b > 13a$. In the domain for which $a > 0$, the energy is minimized for a vev whose little group is SU(3) x SU(2) x U(1). For $a < 0$, the little group is SU(4) x U(1). The existence of two distinct symmetry breaking phases may have cosmological significance. Guth's inflationary universe proceeds from a hot SU(5) symmetric phase, to an SU(4) x U(1) phase, to and SU(3) x SU(2) x U(1) phase.

At this stage, quarks and leptons are still massless. The X boson mediating proton decay has mass M_G. So do any hypothetical fermions with SU(3) x SU(2) x U(1) invariant masses. What Higgs bosons can be used to give quark and lepton their masses ? Since these fermions tranform as $\overline{5} + 10$ under SU(5), we must examine the following products:

$$\overline{5} \times 10 = 45 + 5$$

$$10 \times 10 = (45)_{AS} + (50 + \overline{5})_{SYM}$$

$$\overline{5} \times \overline{5} = (\overline{10})_{AS} + (15)_{SYM}$$

Neither the $\overline{10}$ nor the 50 contain states which are electro-chromo neutral. The fifteen contains an SU(2) triplet whose electrically neutral member can give neutrinos small Majorana masses. We are left with a 45 and a 5. The vev of the 45 cannot be sole source of mass for the up quarks because the up-quark masses would then be degenerate in pairs. So, there must be a 5. Simplicity suggests that there should only be a 5, and the resulting naive mass formulae were discussed in 1974 by Georgi and me:

$$m_e = m_d; \; m_\mu = m_s; \; m_\tau = m_b.$$

The equalities pertain at M_G and not at low energies. If there are 3 fermion families, they must be corrected by a factor of about 3, giving

$$m_d \sim 3m_e; \; m_s \sim 3m_\mu; \; m_b \sim 3m_\tau.$$

Of these three relations, only the third is evidently correct. From the others, we obtain $m_s/m_d \sim m_\mu/m_e \sim 200$, which is in violent contradiction with current algebra. We examine three possible resolutions to this problem.

(1) Add the 45. It is not unreasonable to have both a 45 and a 5. In the O(10) context, after all, a 126 is obligatory to give the dead neutrino a large Majorana mass, and the O(10) - 126 contains an SU(5) - 45. Moreover, with a complex O(10) - 10 and a 126, and with only two sets of Yukawa couplings, we may obtain the result $m_t m_\mu \simeq m_c m_\tau$, which, when renormalization-group corrected, predicts toponium at 40 GeV This is one way to solve the puzzle.

(2) We may appeal to our lack of understanding of physics at the Planck mass. We envisage the existence of SU(5) gauge invariant but non-renormalizable effective interactions. Their scale is controlled by the Planck mass $M_P \sim 10^{19}$ GeV. Nothing we can write down produces evident problems: we can have anomalous electric or magnetic moments of fermions of the form

$$M_P^{-1}\,\bar{\psi}\sigma_{\mu\nu}\psi(F^{\mu\nu} \text{ or } \varepsilon^{\mu\nu\lambda\sigma}F_{\lambda\sigma}),$$

which will contribute 10^{-20} to g - 2 for the muon, and a neutron electric dipole moment of 10^{-33} cm. No problem here.

Neutrino masses may arise through couplings of the form

$$M_P^{-1}\psi(\bar{5})\phi(5)\phi(5)\psi(\bar{5}),$$

and will give neutrino masses of order $M_F^2/M_P \sim 10^{-5}$ eV. No problem here.

Ellis has suggested that such renormalizable couplings can contribute an effective 45 to quark and lepton masses, from the term

$$M_P^{-1}\,\psi(\bar{5})\phi(\bar{5})\phi(24)\psi(10).$$

This contribution is of order $M_F M_G/M_P \sim 10$ MeV. This is just the right size to contribute significantly to the masses of the first family. It is just what the doctor ordered to repair the false prediction of naive SU(5).

(3) The third potential solution to the problem is neither a surrender as in (1), nor a confession of ignorance as in (2), but a consequence of a rich fermionic structure at M_G. Suppose that there are N fermion families and N $\psi_f(10)$ with f = 1, ... N. Suppose that there are M > N $\psi_\beta(\bar{5})$ with β = 1, ..., M and M - N $\psi_\gamma(5)$ with

$\gamma = 1, \ldots, M - N$. This structure is anomaly free and all of the fermions save N families develop masses of order M_G. The large mass matrix may transform as an SU(5) singlet or as the neutral member of a 24. This M x (M - N) matrix can be quite different for color triplet fermions than for weak doublet fermions. Even for M - N = 1, this is enough to entirely disrupt the naive mass relations. Heavy split multiplets can do the trick, and there is no way to exclude their existence, nor any way to see them.

III. NEUTRINOS

In recent months, interest has grown in the possibility that neutrinos have mass. Two fundamental principles have evolved: that the only good symmetry is a gauge symmetry, and that everything which is not forbidden is compulsory. From these principles, it is easily shown that neutrinos must have masses. In several plausible, unified models, these masses come out of order $M_F^2/M_G \sim 10^{-2}$ eV. Given the observed four decade spread between m_e and m_b, we have "theoretically expected" neutrino masses in the domain 10^{-4} eV to 1 eV. How convenient that this lies within the domain of "experimentally observable" neutrino masses between 10^{-6} eV (below which not even solar neutrinos may oscillate) and 10^2 eV (above which we run into severe trouble from what is known of the age of the universe).

A second explanation for the interest in neutrino masses today has to do with the possibility, with modern technologies, to vastly increase the sensitivity of experiments to search for masses or oscillations. In fact, some of these experiments seem to indicate a positive effect. There are five distinct lines of endeavor:

1. Solar Neutrino Oscillations

The Davis data on Chlorine is well known. There are too few high energy solar electron-neutrinos now incident on Earth. All underlined modifiers are of interest. Perhaps the central solar temperature is inconstant and cannot be judged by the mean solar luminosity. A million years ago, the flux of neutrinos may have been significantly higher. A rock experiment has been proposed to test this hypothesis, by Cowan and Haxton. The ^{8}B neutrinos can be expected to transmute Molybdenum to Technetium, which is unstable but lives for several million years. In principle (and, let us hope, in practice) the time-averaged flux of high-energy electron neutrinos can be measured in this way. If there is no discrepancy between this experiment and the Chlorine experiment, another explanation than time-dependence must be invoked. Experiments have been suggested (using Gallium or Indium) which are sensitive to the lower-energy solar neutrinos. These must be carried out with dispatch. Perhaps the deficit of ^{8}B neutrinos reflects an error in our solar model. It is less likely that this error applies to lower energy neutrinos.

Except for the possibility of a fluctuating central temperature, the low energy neutrinos must be there. But are they electron neutrinos ? The Gallium experiment will tell. The solar experiments are the most sensitive conceivable searches for neutrino oscillations with large amplitude (large mixing angles) and long wavelength (small neutrino masses). They simply must be done.

2. Oscillations of Terrestrial Neutrinos

Here, there are many possible types of experiments. Accelerators can produce intense beams of muon neutrinos (or antineutrinos) (from charged meson decays) or 50/50 muon neutrino-electron neutrino (from K° decays). Energies from a hundred MeV to a hundred GeV are available at the various operative facilities. Some experiments have been done which do not show evident oscillations. Much more sensitive experiments can and will be done. Reactors are a copious source of MeV electron antineutrinos. Reines and Sobel have published an indication of an effect, one which has not at this time been confirmed. Again, the experiments continue. Artificial radioactive sources of megacurie magnitude can serve as intense sources of monochromatic electron neutrinos. Not only can such a source be used to calibrate solar neutrino detectors, but it can also be used effectively in the search for neutrino oscillations. Lastly, we mention the search for oscillations of atmospheric neutrinos, those produced in the atmosphere by primary cosmic rays. From the comparison of upward bound neutrinos to downward ones, we have an effective oscillation length of an Earth diameter. A dedicated experiment for the study of atmospheric neutrinos deep underground should have high priority.

3. Endpoint Experiments

A decade ago, Bergkvist put a limit of 65 eV on the mass of the electron antineutrino from a careful study of the endpoint of the electron spectrum in Tritium beta decay. More recently, a Soviet experiment on the same process tentatively claims a determination of the antineutrino mass: $14 \text{ eV} < m < 40 \text{ eV}$. This provocative result demands confirmation, hopefully with a tritium source in a less chemically complex environment than valine. However, there is more to be done that a simple repeat of the Tritium experiment. One can study the photon endpoint in radiative electron capture. De Rujula has persuasively argued that this is a promising approach for the study of the neutrino mass. Suffice it to say that experiments are now under way to study the photon spectrum in the beta capture process in $^{163}H_0$ and ^{193}Pt. A tough assignment at the low-energy frontier!

4. Double Beta Decay

Although interesting in its own right as an observable

second-order weak decay, it is even more interesting in the light of possible neutrino Majorana masses. Such masses make possible a no-neutrino transition, wherein all of the available energy is transferred to the electrons. Searches for this process in ^{76}Ge and ^{82}Se have been done with no positive result. They must be done with far greater sensitivity. The present limits correspond to an upper limit of $\sim$ 15 eV on the relevant component of the electron-neutrino mass. It should be noted that the present sensitivity of endpoint measurements and double beta determinations are comparable so that we have a horse race to enjoy. There is indirect evidence for the double beta decay of ^{128}Xe and ^{130}Te coming from the Xe abundances in antique Tellurium ores. It seems to be that a confirming experiment can and should be done in real time. A ton of Tellurium deep underground should produce 10^3 atoms of ^{128}Xe and 10^6 atoms of ^{130}Xe per year if the rock experiments are correct. All we need is an efficient (laser ?) technology for the detection of small numbers of Xenon isotopes. Here is why the issue is so important: Doi et al., in analyzing the rock Tellurium data conclude that a neutrino mass of $\sim$ 30 eV is required. A prerequisite to a critical analysis of their study would be confirmation of the Tellurium double beta-decay lifetimes.

5. Cosmological Missing Mass

At several levels in the universe, there is a missing mass problem. Some would like the universe to be neither open nor closed, but parabolic. This would require considerably more mass than can be accounted for. This mass can be in the form of relic neutrinos, of which we know the present mean number density to be of order 100 per cubic centimeter per neutrino species. If the sum of the neutrino masses were 50-150 eV, this could be accomplished. There is also a missing mass problem at smaller scales. Schramm and Steigman call for neutrino masses of 6 ± 3 eV which gather at the scale of galactic clusters and explain their apparent missing mass. There is also a well documented mass deficit in galactic haloes. Rotation curves are roughly constant and extend well outside the luminous galaxy. There all invisible and extended galactic haloes containing the majority of the galactic mass. Perhaps these are made up of neutrinos with masses of order 30 eV. Further data on the rotation curves are needed to see how far out the haloes extend. It must be determined whether the haloes are made up of neutrinos or of non-luminous stars. Perhaps the resolving power of gravitational lenses can be measured. Perhaps galactic neutrinos suffer radiative decay, and produce an observable flux of monochromatic UV photons. Perhaps such radiative decays affect the ionization state of the intergalactic medium. Perhaps a theory of galactic evolution will demand a neutrino dominated universe, or one in which the neutrino contribution to the universal mass density is negligible. Perhaps.

IV. MAJORONS

There are the Goldstone bosons of a spontaneously-broken local B - L symmetry. Two different models involving Majorons have been invented in the past year. The first was due to Chicasige, Mohapatra and Peccei. They add to each fermion family a "dead" neutrino state which is electroweak neutral but carries B - L = - 1. They also introduce a "dead" complex Higgs singlet ϕ, whose only admissable Yukawa coupling is bilinear in the added neutrinos. Thus ϕ carries B - L = 2. Two things happen. The dead neutrinos play the role of right-handed partners of ordinary neutrinos. Neutrinos thus develop Dirac (SU(2)-doublet) masses of conventional magnitudes (MeV to GeV). However, B - L is spontaneously broken at high energy so that $< \phi > = F \neq 0$. This leads to the existence of a true Goldstone boson with Yukawa couplings suppressed by F^{-1}. It also leads to Majorana mass terms for the dead neutrinos which are of order F. In the usual "seesaw" fashion, this induces Majorana masses of ordinary neutrinos of order m^2/F where m is the magnitude of the Dirac mass. In the limit $F \to \infty$, ordinary neutrinos become massless and the Goldstone boson decouples from the low-energy world. For F large but finite, say $F \sim M_G$, we obtain interesting neutrino masses, but the Goldstone boson remains hopelessly unobservable and irrelevant: the Majoron is invisible. For this reason, I do not much like the GMP model.

Gelmini and Roncadelli have devised a more amusing variant of the Majoron which has richer physical consequences. They need not to introduce added neutrino states. Instead, they introduce a complex Higgs triplet T with Yukawa couplings $gT\ell\ell$, e.g.,

$$T^{++}e^-e^-,\ T^+e^-\nu,\ T^0\nu\nu.$$

Once again, T carries B - L = 2, and it develops a vev $< T^0 > = v$ which leads to neutrino masses of size gv. Since B - L has no $(\text{gauge})^2$ anomalies, a true Goldstone boson appears. But, T is non trivial under SU(2) x U(1). This means that $v \ll M_F$, lest the success of the empirical relation $M_W/M_Z = \cos\theta$ is disturbed. We shall see, for other reasons, that this inequality must obtain. In the GR model, it is a small vev that breaks B - L symmetry, not a large vev as in the CMP model. Notice that the conventional Higgs doublet D carries B - L = 0. The trilinear meson self-coupling DTD would manifestly violate B - L, and would preclude the appearance of a Goldstone boson. Such a term must be excluded by fiat. B - L symmetry is not "automatic" in the GR model. Nor is it automatic in an SU(5) extension wherein T belongs to a SU(5) - 15. The $\bar{5}.15.5$ coupling must be excluded. This is an aesthetic, not a substantive, objection to the GR model.

Let us briefly describe the properties of the mesons of the GR model. Firstly, there is a doubly-charged meson T^{++} whose mass

M is no greater than M_F = 250 GeV. It may be considerably lighter. It couples to pairs of charged leptons, may be reasonably produced in electron-electron collisions, and may mediate $\mu \to 3e$.

There are two singly-charged fields D^+ and T^+. One linear combination, principally D^+ with a $v\, M_F^{-1}$ admixture of T^+ becomes a part of W^+. The orthogonal state, principally T^+ with a $v\, M_F^{-1}$ admixture of D^+, survives as an observable particle. Its mass, up to small corrections, is $2^{-1/2}$ M. Here is a remarkable (and testable) prediction. Its couplings are to pairs of leptons, and via the D^+ admixture, to quark-antiquark.

There are four neutral fields. They are, schematically,

$\mathrm{Im}[D^0 + v\, M_F^{-1} T^0]$	Part of Z^0,
$\mathrm{Im}[T^0 + v\, M_F^{-1} D^0]$	The Majoron,
$\mathrm{Re}[D^0 + v\, M_F^{-1} T^0]$	The ordinary Higgs ($m \lesssim M_F$),
$\mathrm{Re}[T^0 + v\, M_F^{-1} D^0]$	The light Higgs ($m \lesssim v$).

It is, of course the Majoron and the light Higgs which are of interest to us. They have primary couplings to neutrinos, and couplings to quarks suppressed by $v\, M_F^{-1}$.

One of the ways these new particles may be discovered is through their contribution to the decays of the weak interaction intermediaries. The following table gives the estimated partial widths for these decay modes, and the condition on M for them to be kinematically allowed:

Decay Mode	Partial Width(MeV)	Constraint on M(GeV)
$Z^0 \to T^{++}T^{--}$	110	M < 47
$W^+ \to T^{++}T^-$	240	M < 49
$Z^0 \to T^+T^-$	20	M < 66
$W^+ \to T^+T^0$	240	M < 117
$Z^0 \to T^0\bar{T}^0$	340	Any M.

Notice that the TT decay modes may be quite copious for the case of $W^{\pm}$ decays. Such events could be detected in hadron-hadron colliders because of the presence of as many as six high $p_{\perp}$ quark jets. Even more importantly, notice the decay mode $Z^o \rightarrow T^o\bar{T}^o$. This is, in fact, a decay into a Majoron plus a light (and essentially stable) Higgs meson. It is an invisible decay mode, just like each neutrino mode. The partial decay width for this decay mode is exactly twice the decay width into each neutrino species. LEP, by comparing the actual Z^0 width with the width as deduced from the leptonic branching ratio and partial width, will measure the partial decay width into unobservable modes. It will count "neutrino species". If the GR theory is correct, this count must be no less than five.

Let us turn to the question of the magnitude of v. The GR Majoron has a Yukawa coupling to electrons which is that of the ordinary Higgs boson, reduced by $v\, M_F^{-1}$,

$$\phi\bar{e}\gamma_5 e m_e v M_F^{-1} .$$

It is therefore like a weakly-coupled light axion. In the next section, we demonstrate that the coupling of such a particle, for astrophysical reasons, must be less than 10^{-12}. We may conclude that $v < 100$ KeV. We have a small gauge hierarchy separating the vevs of D and T.

Since Yukawa couplings must be less than unity, we have learned from the stars that neutrino masses cannot exceed 100 KeV. Moreover, we have learned that most elementary particle decays into states with Majorons are too negligible to be measured, decays such as $\mu \rightarrow e\phi$ or $K \rightarrow \pi\phi$.

Majoron emission, however, may be relevant in double beta decay. There are now three possible modes: two-neutrino, no-neutrino, and no-neutrino plus a Majoron. Doi, et al., from a comparative study of ^{128}Te and ^{130}Te decays deduce that $m(\nu_e) = gv = 30$ eV. Taking into account Majoron emission, we obtain a modified estimate

$$g^2[v^2 + (50 \text{ KeV})^2] = (30 \text{ eV})^2,$$

where the first term in brackets is the no-neutrino term and the second term corresponds to Majoron emission for the particular case of ^{128}Te decay. This result, as well as others in this section, may be found in a recent paper by Georgi, Nussinov, and myself. If v approaches its maximal value of 100 KeV, then the no-neutrino process dominates the Majoron process. In the more likely case that v is significantly smaller, it is the Majoron process which dominates, and we may deduce that $g \sim 10^{-3}$.

Incidentally, the "Majoron process" yields a Majoron half the time and a light Higgs boson the other half. The two processes are indistinguishable. Note that there is no potential conflict between the modified Doi et al. estimate above and the Haxton et al. result that $m(\nu_e) \lesssim 15$ eV. The latter result is deduced from the failure to observe calorimetrically monoenergetic double beta decays in Ge and Se. These are characteristic of the no-neutrino process, but not of the Majoron process. In the GR model, things are more complicated.

A word about the double β^+ process. There are three conventional modes with two antineutrinos produced: double capture, single capture and two-positron. There are also three no-neutrino processes: single capture, two positron, and the strange case of double-capture resonant atomic mixing. The latter process, rather analogous to $K_0\bar{K}_0$ mixing or $n\bar{n}$ mixing, is characterized by a decay rate of ε^2/Δ, where ε is the magnitude of the atomic mixing (and is proportional to m_ν) and Δ is the mass difference between an atom with atomic number Z and an excited atom with Z - 2. Finally, there are three possible modes involving Majoron emission. Surely, there are instances wherein some of these nine different double β^+ modes may be studied.

With the estimate $g = 10^{-3}$ for the coupling of T to electron neutrinos, we deduce the partial width for $e^-e^- \to T^{--}$ to be of order 10^{-7} M. For M $\sim$ 50 GeV, this gives M = 5 KeV. If the T mesons are discovered in W or Z decay, or via pair production in e^+e^- collisions, it is not inconceivable to design an e^-e^- collider to produce them resonantly.

It has not escaped our notice that T^{++} can mediate the decay of muons into three electrons. If g' is the coupling of T^{++} to the μe system, it must satisfy $g' < (20)^{-1}(M/M_F)^2$ in order that there be no conflict with the current experimental limit. In this connection, a stronger limit on the process $\mu \to 3e$ would be most useful.

We turn to neutrino cosmology in the GR model. Three new neutrino interactions come into play. A heavier neutrino may decay into a lighter neutrino and a Majoron, a process which proceeds through a W loop correction. The result, in terms of a generic Yukawa coupling g is

$$\tau = \tau_0 g^{-3}(100/v)^5 > \tau_0$$

where v is in KeV and τ_0 is the age of the universe. It follows that the GR neutrinos are stable under this decay scheme over any cosmologically relevant time scale.

Several things can happen in neutrino-neutrino collisions

$$\nu_H + \nu_H \to \nu_L + \nu_L \qquad \nu + \nu \to 2\phi .$$

The second process plays a dominant role in GR neutrino cosmology: it makes a neutrino-dominated universe impossible. Its cross section is given approximately by $\sigma = g^4\beta(24\pi s)^{-1}$ where s is the center-of-mass energy variable and $\beta = v/c$. The equation for the number N of neutrinos in a comoving volume is simply $\dot{N} = - N^2V^{-1}\sigma v$ where V is the volume of the comoving volume. The initial condition is known at a temperature large compared to the neutrino mass. It is straightforward to compute the necessary conditions on m_ν and v such that the present neutrino mass density is at least 1 % of the critical (closure) mass density of the universe. They are:

$$m_\nu > 1\ \text{eV}$$

$$(m_\nu /1\ \text{eV})^5 < 10^{-7}(v/100\ \text{KeV})^8.$$

Since v < 100 KeV, it is clear that these conditions are incompatible. GR neutrinos with sufficient mass to have dominated (or significantly contributed to) the universal mass density will have annihilated soon after they cooled to a temperature comparable with their mass. They will have annihilated into massless, unobservable Majorons.

The physics of the Gelmini-Roncadelli model is bizarre, but very interesting. Only time, and future experiments, will show whether or not the model is correct.

V. AXIONS

Once upon a time, there was a strong CP problem, connected with the appearance of a term $\Theta F\tilde{F}^D$ in the QCD Lagrangian, and associated with the phase of the determinant of the quark mass matrix. One way to solve the problem is to make this determinant vanish. So it is that several authors have proposed the unsavory hypothesis that the up quark has zero hard mass. Another way, suggested by Peccei and Quinn, is to introduce a symmetry under which the phase of the quark matrix changes. The strong CP problem is thereby solved. But a pseudo-Goldstone boson appears because the PQ symmetry is spontaneously broken by the Higgs vevs. It is not a true Goldstone boson because the PQ symmetry, by construction, has anomalies. This new particle, which has been looked for but not unambiguously found, is known as the Axion.

The simplest realization of the PQ symmetry involves the use of separate Higgs bosons to give masses to up quarks and to down quarks. In an SU(5) context, we introduce two Higgs quintets ϕ and η with Yukawa couplings

$$\psi(10)\psi(10)\phi(5);\quad \psi(10)\psi(\bar{5})\bar{\eta}(\bar{5})$$

with a PQ symmetry under which

$$\delta\psi(10) = i\theta\psi(10), \quad \delta\psi(\overline{5}) = i\theta\psi(\overline{5})$$

$$\delta\phi(5) = -2i\theta\phi(5), \quad \delta\eta(5) = +\, 2i\theta\eta(5).$$

This PQ symmetry is broken by the vevs of ϕ and η at a mass scale M_F. The axion mass is of order $f_\pi m_\pi / M_F \sim 10$ KeV and its couplings are of order $f_\pi / M_F \sim 10^{-4}$. The existence of such a particle is probably ruled out by reactor searches, accelerator searches and astrophysical data. On the other hand, Faissner and his coworkers report positive evidence.

The astrophysical argument pertains to relatively light axions, and it is relevant to our preceding discussion of Majorons. Let us examine this argument with the neglect of axion masses compared to central stellar temperatures, and in the weak-coupling limit where axion rescattering is negligible. Consider a red supergiant with 100 solar masses, a central temperature of 10 KeV, a core radius R of 3×10^9 cm and a core density of 10^4 g/cm^3. Let us calculate the probability P that a photon-electron collision produces an axion

$$P = \frac{\sigma(\gamma e \to \phi e)}{\sigma(\gamma e \to \gamma e)} .$$

It is simple to deduce that $P = g^2/50$ where g is the Yukawa coupling constant of the axion ϕ to electrons. Now, the photon may be regarded as executing a random walk on its way out of the star. The length of each jog is the collisional mean free path $L = (\sigma n)^{-1}$ where σ is the Compton cross section and n the electron number density. In the core, we find $L = 4 \times 10^{-4}$ cm. The number of jogs in a photon's outward journey is of order $N = (R/L)^2 = 1/2\ 10^{26}$. The net ratio of axion emission to photon emission by the star is simply $NP = 10^{24}\, g^2$. If the energy emitted by the star is not to be axion dominated, we must conclude that $g < 10^{-12}$. Confidence in this limit depends, to a certain extent, on confidence in our theory of red supergiant evolution. Surely however, it cannot be grossly violated. This is one argument against the existence of reasonably light axions, and it is our basis for constraining $v < 100$ KeV in the preceding section.

Now, let us see how easy it is to modify the PQ axion so that it becomes an invisible artifact of the world at M_G. It cannot be seen. It exists only to solve the strong CP puzzle. The trick was suggested by Dine, Fischler and Srednicki. Why not let the PQ symmetry be broken at very large mass scale ? The mass of the axion, and more importantly, its coupling strength, depend reciprocally on the scale of symmetry breaking. The axion can thereby be made harmless.

If the scale of PQ breaking is not merely high, but is simply M_G, then the axion is not only harmless but it is quite invisible. Wise, Georgi and I, in a simple model, show how the SU(5)-breaking scale can naturally coincide with the PQ breaking scale. In addition to the $\phi(5)$ and $\eta(5)$ defined previously, we introduce a complex SU(5) adjoint representation, Σ, whose PQ transformation behaviour is defined to be

$$\delta\Sigma = 2i\theta\Sigma.$$

Allowed quartic couplings include such terms as $(\bar{\eta}\ \Sigma\bar{\Sigma}\eta)$ and $(\bar{\eta}\Sigma^2\phi)$.

Now, the Higgs structure which breaks SU(5) down to SU(3) x SU(2) x U(1) also leads to a breakdown of PQ symmetry at M_G. The mass of our modified axion is $f_\pi m_\pi/M_G \sim 10^{-8}$ eV and its Yukawa couplings are $f_\pi/M_G \sim 10^{-16}$. Such an axion plays a negligible role even in red supergiants. It is quite invisible.

Notice that our invisible axion solves the (unadvertised) Guth-Pi problem. They consider the thermally induced sequence of phase transitions SU(5) $\to$ SU(4) x U(1) $\to$ SU(3) x SU(2) x U(1) in an SU(5) model with $\Sigma \to -\Sigma$ symmetry. Their doubly-degenerate vacuum leads to a domain structure with too much boundary energy. In our model, the $\Sigma \to -\Sigma$ symmetry is embedded in a continuous group, and this problem may not arise.

We are not wedded to our particular realization of the PQ symmetry. We would prefer that the PQ symmetry is automatic, like B - L symmetry in naive SU(5). We want to construct a model in which the PQ symmetry is forced upon us in the presence of any gauge invariant renormalizable coupling of the fields in the model. A first stab at a model was suggested by Georgi, Hall and Wise. Their unifying group is SU(9) with each fermion family transforming as $5\psi_\alpha \pm \psi^{\alpha\beta}$ where ψ_α is a 9 and $\psi^{\alpha\beta}$ is the antisymmetric product of two $\bar{9}$s. It is anomaly free and reduces to one conventional family plus 66 chiral fields which develop masses of order M_G. The Higgs structure involves a real adjoint, five distinct ψ_α and a four index antisymmetric field $\phi_{\alpha\beta\gamma\delta}$. The Yukawa Higgs coupling,

$$\psi^{\alpha\beta}\psi^{\gamma\delta}\phi_{\alpha\beta\gamma\delta}$$

$$\psi^{\alpha\beta}\psi_\alpha\phi_\beta$$

suffice to give large masses to the heavy fermions and small masses to the light fermions. Define a potential PQ symmetry such that $S(\psi^{\alpha\beta}) = 1$. It follows that $S(\phi_{\alpha\beta\gamma\delta}) = -2$ and that $S(\psi_\alpha)+1+S(\phi_\beta) = 0$. The existence of the meson self-coupling $\varepsilon\phi_{\alpha\beta\gamma\delta}\phi_{abcd}\phi_e$ compels

$S(\phi_\alpha) = 4$ and hence $S(\psi_\alpha) = -5$. This automatic symmetry does have an anomaly and can serve as a PQ symmetry. It is broken at M_G and leads to an invisible axion. It is contrived. It does not address the flavour question. It is merely an example of an automatic PQ symmetry with an invisible axion.

Here is another example, no less appealing, for which I hear responsibility. The gauge group is SU(5) x SU(5) with a discrete interchange. The first factor is conventional SU(5), the second is completely broken down.

The following simple example based on U(1) x U(1) plus interchange shows that this is not so unnatural as it seems. The most general Hamiltonian is

$$\lambda_1(\phi^2 + \chi^2 - a^2)^2 + \lambda_2\phi^2\chi^2.$$

For $\lambda_2 < 0$, $\langle\phi\rangle = \langle\chi\rangle$ at the minimum and the symmetry is completely broken. On the other hand, for $\lambda_2 > 0$, we have $\langle\phi\rangle = a$ and $\langle\chi\rangle = 0$ and one U(1) factor survives but not the other.

The fermion representation $R = (10, \bar{5}) + (\bar{5}, 10) + (5, 5)$ is readily verified to be anomaly free. The requisite Yukawa couplings are

$$\psi(10,\bar{5})\psi(10,\bar{5})\phi(5,15)$$
$$\psi(10,\bar{5})\psi(\bar{5},10)\phi(\bar{5},\bar{5})$$
$$\psi(10,\bar{5})\psi(5,5)\phi(10,1)$$

plus those required by interchange symmetry. Assign the PS quantum number to the mesons in accordance to their overall fivality, i.e.

$$S[\phi(\bar{5},\overline{15})] = S[\phi(5,5)] = S[\phi(10,1)] = 2.$$

This guarantees that all of the many meson self-couplings, which, of course, must have fivality zero, are also PQ-invariant. The Yukawa couplings are also PQ-invariant with the assignments

$$S[\psi(10,\bar{5})] = 1$$
$$S[\psi(5,5)] = -3.$$

Thus, we have an automatic PQ symmetry which has anomalies and which is broken at the unification mass.

The details are amusing. The representation is 125 dimensional, of which there are five heavy $5 + \bar{5}$ multiplets and five conventional families. The masses of the superheavies are related: two at A, two at B, and one at $(A^2 + B^2)^{1/2}$. The five light up quarks can receive arbitrary masses in tree approximation. Of the five light down quarks, four remain degenerate in tree approximation, and one remains massless. Not yet physics, but fun.

Georgi and I went so far as to invent a group with 621 fermions transforming under SU(6) x SU(6) as $(15,15) + (\bar{6},\bar{6}) + (\overline{15},6) + (6,\overline{15}) + (\bar{6},\overline{15})$. One SU(6) factor contains ordinary SU(5), the other is completely broken down. Our anomaly-free representation contains just three light families, and a lot of heavy stuff. The reader can fill in the details, and find an automatic PQ symmetry which leads to an invisible axion. The author will summarize with some Morals:

(1) Invisible axions, like proton decay and small neutrino masses, should be regarded as relics of grand unification.

(2) Who wants automatic B - L symmetry ?

(3) We need an anomaly-ridden PQ symmetry which is broken by the same mechanism which breaks the grand unification symmetry.

(4) We want an automatic PQ symmetry.

(5) Automatic PQ can be used as a guideline for the construction of a proper GUT which addresses (solves ?) the flavour problem.

(6) The models we have at hand are mere toys, not realistic theories.

VI. THE FUTURE

This is a good time to be an experimental physicist, for the theorists are (for the moment) ahead. There is so much to be done: at high energy, at higher energy, and at low energy. Surprises are sure to come:

(1) Am I right, and will PETRA find the top quark ? Or, will its discovery await KEK, or even LEP ?

(2) Will there be lots of new particles on the way to Z^0 in e^+e^- collisions ? The desert from 10 GeV to 36 GeV is a poor augury.

(3) Technicolor and/or supersymmetry predict a wealth of new particles at 1 TeV. To test this prediction, a high-luminosity 10 TeV hadron collider is necessary. Whether it is proton-proton or proton-antiproton makes very little difference. Who will build it, and when. Mere on TeV colliders (like the last three "big" machines, ISR, PEP, and PETRA) will make no great discoveries, say I.

(4) Proton decay, if it exists, will be a grand adventure as important as any in high-energy physics. First, we must find it. We must test the predicted $\Delta B = \Delta L$ selection rule. We must measure 24 two-body partial decay widths. Measure muon polarization. Do a mode-independent measurement of the lifetime.

(5) Ed Fireman suggests an approach to the latter problem. He considers the transmutation of stable Potassium to Argon by nucleon decay. A very deep underground experiment is called for, using a great deal of Potassium, and employing techniques that can detect a very few Argon atoms. Good luck, Ed.

(6) So much needs to be done on the solar neutrino problem: Gallium, Indium, and the Technetium search. The latter experiment can also search for heavy technifermions which invade nuclei to form superheavy isotopes. See my work with Bob Cahn in Science.

(7) Double beta decay is not a virgin field, but there is much that remains to be done. Seeing double beta decay, for example. Because of the recent reanalyses of Haxton, Doi, Rosen and their coworkers, the community has been alerted to the importance of a new generation of more sensitive experiments.

(8) Will there be $m\bar{n}$ oscillations ? I doubt it, but we must look if we can. Proposals aim towards a transition time of 10^9 seconds. This is a far more sensitive limit than we get, indirectly, from proton decay experiments.

(9) Monopoles and free quarks. We have got to look for these. As well as $\mu \to e\gamma$, $\mu \to 3e$, the neutron electric dipole moment, axions, departures from the superweak model of CP violation, neutrinos from astrophysical calamity, neutrino masses, neutrino oscillations, fundamental scalar mesons, a fourth family, pseudo-Goldstone Techni-particles, rishons or maons, more than one Z^0. The theorists are ahead, and we are thereby clay pigeons for the enterprising experimenter.

INDEX

Kolibri 360 GmbH
Ferdinand-Jühlke-Straße 7
99095 Erfurt, Deutschland
produktsicherheit@kolibri360.de